T0265409

# ADDITIVE MANUFACTURING

Handbooks in Advanced Manufacturing

# ADDITIVE MANUFACTURING

Series Editors

J. PAULO DAVIM

KAPIL GUPTA

Edited by

JUAN POU

ANTONIO RIVEIRO

J. PAULO DAVIM

Elsevier
Radarweg 29, PO Box 211, 1000 AE Amsterdam, Netherlands
The Boulevard, Langford Lane, Kidlington, Oxford OX5 1GB, United Kingdom
50 Hampshire Street, 5th Floor, Cambridge, MA 02139, United States



**Library of Congress Cataloging-in-Publication Data**
A catalog record for this book is available from the Library of Congress

**British Library Cataloguing-in-Publication Data**
A catalogue record for this book is available from the British Library

ISBN: 978-0-12-818411-0

For information on all Elsevier publications visit our website at
https://www.elsevier.com/books-and-journals

*Publisher:* Matthew Deans
*Acquisitions Editor:* Brian Guerin
*Editorial Project Manager:* Megan Healy
*Production Project Manager:* Sojan P. Pazhayattil
*Cover Designer:* Victoria Pearson

Typeset by TNQ Technologies

# Contents


*Contributors* *xiii*
*Foreword* *xix*
*Preface* *xxi*

**1. Introduction to additive manufacturing** **1**

Xinchang Zhang *and* Frank Liou

**1.1** Basic concepts of additive manufacturing 1
**1.2** Basic procedure of additive manufacturing 2
**1.3** Categories of additive manufacturing 5
**1.4** Applications of additive manufacturing 13
**1.5** Comparison of additive manufacturing and subtractive manufacturing 23
**1.6** Hybrid manufacturing 25
**1.7** Challenges and limitations of current additive manufacturing 26
References 28

**2. Introduction to powder bed fusion of polymers** **33**

Andreas Wegner

**2.1** Introduction 33
**2.2** Processes, machines, technologies 33
**2.3** Postprocessing and surface treatment 38
**2.4** Materials and powder production techniques for powder bed fusion 39
**2.5** Parameter settings and influences 44
**2.6** Process monitoring 60
Symbols 63
Abbreviations 63
References 64

**3. Selective laser melting: principles and surface quality** **77**

Evren Yasa

**3.1** Introduction to selective laser melting 77
**3.2** Surface integrity in the selective laser melting process 83
**3.3** Treatments applied during the selective laser melting process 89
**3.4** Treatments applied after the selective laser melting process 96
References 112

**4. Laser-directed energy deposition: principles and applications** **121**

F. Arias-González, O. Barro, J. del Val, F. Lusquiños, M. Fernández-Arias, R. Comesaña, A. Riveiro *and* J. Pou

4.1 Introduction 121
4.2 Principles of laser-directed energy deposition 123
4.3 Industrial applications of laser-directed energy deposition 128
4.4 Biomedical applications of laser-directed energy deposition 137
4.5 Summary 149
Acknowledgments 151
References 151

**5. Vat photopolymerization methods in additive manufacturing** **159**

Ali Davoudinejad

5.1 Introduction 159
5.2 Vat photopolymerization process 161
5.3 Postprocessing 166
5.4 Direct fabrication of parts by vat photopolymerization 168
5.5 Additive manufacturing technologies for tooling 174
References 179

**6. Polymer and composites additive manufacturing: material extrusion processes** **183**

Vidya Kishore *and* Ahmed Arabi Hassen

6.1 Introduction 183
6.2 Extrusion additive manufacturing processes 185
6.3 Hybrid systems 193
6.4 In-line monitoring and automation for smart manufacturing 194
6.5 Materials development 196
6.6 Current applications and path forward 210
Acknowledgments 212
References 212

**7. Introduction to fused deposition modeling** **217**

Przemysław Siemiński

7.1 Historical outline and used labels 217
7.2 The RepRap project—history and models of 3D printers 222
7.3 Model and support materials 228
7.4 Extrusion head structure 248

7.5 Selected details about heads in open systems 256
7.6 An example of head construction in a Stratasys device 260
7.7 Fiber deposition strategy and finishing process 265
7.8 Conclusions 271
References 273

**8. Electron beam melting process: a general overview** **277**

Manuela Galati

8.1 Introduction 277
8.2 Electron beam melting physical mechanisms 280
8.3 Process control and process parameters 282
8.4 Part features 285
8.5 Process monitoring 292
8.6 Numerical simulation 293
8.7 Summary and scientific and technological challenges 295
References 296

**9. Introduction to 4D printing: methodologies and materials** **303**

Xiao Kuang

9.1 Introduction 303
9.2 Fundamentals of 4D printing 305
9.3 Material extrusion-based 4D printing 315
9.4 4D printing by polyjet printing 324
9.5 Vat photopolymerization-based 4D printing 327
9.6 Summary 331
References 333

**10. Laser polishing of additive-manufactured Ti alloys and Ni alloys** **343**

Yingchun Guan, Yuhang Li *and* Huaming Wang

10.1 Introduction 343
10.2 Laser polishing LMD TC11 345
10.3 Laser polishing SLM TC4 350
10.4 Laser polishing SLM inconel 718 superalloy 355
10.5 Conclusions 363
Acknowledgments 364
References 364

**11. On surface quality of engineered parts manufactured by additive manufacturing and postfinishing by machining** **369**

M. Pérez, A. García-Collado, D. Carou, G. Medina-Sánchez *and* R. Dorado-Vicente

11.1 Introduction 369
11.2 Surface roughness 375
11.3 Experimental studies on additive manufacturing and machining 383
11.4 Challenges and opportunities 388
11.5 Conclusions 389
References 389

**12. Standards for additive manufacturing technologies: structure and impact** **395**

Asunción Martínez-García, Mario Monzón *and* Rubén Paz

12.1 Introduction 395
12.2 Structure of additive manufacturing standardization working groups 397
12.3 Published AM standards in ISO/ASTM 403
12.4 Impact of standards for additive manufacturing 405
12.5 Conclusions 408
References 408

**13. Metal matrix composites processed by laser additive manufacturing: microstructure and properties** **409**

Anne I. Mertens

13.1 Introduction 409
13.2 In situ synthesis of metal matrix composites by laser additive manufacturing 411
13.3 Laser additive manufacturing for the production of "pure" ex situ metal matrix composites 413
13.4 Laser additive manufacturing of hierarchical metal matrix composites 414
13.5 Microstructural characterization of metal matrix composites produced by laser additive manufacturing 416
13.6 Properties and applications of metal matrix composites produced by laser additive manufacturing 419
13.7 Concluding remarks 420
References 421

**14. Laser aided metal additive manufacturing and postprocessing: a comprehensive review** **427**

Rajkumar Velu, Arun V. kumar, A.S.S. Balan *and* Jyoti Mazumder

14.1 Introduction 427
14.2 Laser additive metal manufacturing 429
14.3 Postprocessing techniques for additive manufactured components 440
14.4 Future scope of laser-based additive manufacturing 450
Acknowledgment 450
References 451

**15. Nanofunctionalized 3D printing** **457**

Maria P. Nikolova, K. Karthik *and* Murthy S. Chavali

15.1 Introduction 457
15.2 Nanoenhancement of structural materials 465
15.3 3D-printed electronics, optics, and energy conversion devices 471
15.4 3D-printed nanocomposites used in medicine 478
15.5 Additive-manufactured nanofunctionalized surfaces 489
15.6 Conclusions 493
Abbreviations 495
References 496

**16. Additive manufacturing for the automotive industry** **505**

Joel C. Vasco

16.1 Introduction 505
16.2 Complementary methods and techniques 506
16.3 Overview of additive manufacturing applications on automotive production tools 511
16.4 Hard tools for mass-scale replication of automotive components 512
16.5 Emerging additive manufacturing applications on automotive components 513
16.6 Economic impact of additive manufacturing applications on the automotive industry 522
16.7 Conclusions 524
References 525

**17. Additive manufacturing of large parts** **531**

G.A. Turichin, O.G. Klimova-Korsmik, K.D. Babkin *and* S. Yu. Ivanov

17.1 Direct laser deposition process simulation 531
17.2 Residual stresses and distortion in direct laser deposition of large parts 542
17.3 Technological equipment in direct laser deposition of large parts 548
17.4 Structure and properties of products obtained by direct laser deposition technology 555
17.5 Conclusions 565
References 566

**18. 3D printing of pharmaceutical products** **569**

Iria Seoane-Viaño, Francisco J. Otero-Espinar *and* Álvaro Goyanes

18.1 Introduction 569
18.2 Pharmaceutical 3D printing technologies 570
18.3 3D printing: a new era of personalized medicine 580
References 591

**19. 3D bioprinting: a step forward in creating engineered human tissues and organs** **599**

O. Alheib, L.P. da Silva, Yun Hee Youn, Il Keun Kwon, R.L. Reis *and* V.M. Correlo

19.1 Introduction 599
19.2 Bioprinting technologies 600
19.3 Bioinks 604
19.4 Tissue and organ bioprinting 614
19.5 Limitation and technical challenges 624
19.6 Conclusions and future directions 627
References 628

**20. Additive processing of biopolymers for medical applications** **635**

Rajkumar Velu, Dhileep Kumar Jayashankar *and* Karupppasamy Subburaj

20.1 Introduction 635
20.2 Biopolymers and their biomedical applications 637
20.3 Summary and need for 3D printing techniques 644
20.4 Additive manufacturing/3D printing strategies and challenges 644
20.5 Future and prospects 653
References 655

**21. Additive manufacturing using space resources** **661**

Athanasios Goulas, Daniel S. Engstrøm *and* Ross J. Friel

21.1 Additive manufacturing: a tool to support future activities on another planet 661
21.2 Indigenous space material resources 661
21.3 Additive manufacturing using indigenous space resources 668
21.4 The potential of laser-based powder bed fusion additive manufacturing 673
21.5 Conclusions 679
References 680

**22. Modeling and simulation of additive manufacturing processes with metallic powders—potentials and limitations demonstrated on application examples** **685**

Loucas Papadakis

22.1 Introduction 685
22.2 Selective laser melting and laser metal deposition process cases 696
22.3 Heat input modeling in case of selective laser melting 698
22.4 Geometrical accuracy calculation on industrial relevant selective laser melting and laser metal deposition examples 702
22.5 Potentials and limitation of modeling approaches for additive manufacturing 714
Acknowledgment 718
References 718

*Index* *723*

# Contributors

**O. Alheib**
3B's Research Group, I3Bs — Research Institute on Biomaterials, Biodegradables and Biomimetics, University of Minho, Headquarters of the European Institute of Excellence on Tissue Engineering and Regenerative Medicine, AvePark, Parque de Ciência e Tecnologia, Zona Industrial da Gandra, Barco, Portugal; ICVS/3B's—PT Government Associate Laboratory, Braga, Portugal

**F. Arias-González**
Universitat Internacional de Catalunya (UIC), Sant Cugat del Vallès, Spain

**K.D. Babkin**
Saint-Petersburg State Marine Technical University, Institute of Laser and Welding Technologies, Saint-Petersburg, Russian Federation

**A.S.S. Balan**
Department of Mechanical Engineering, National Institute of Technology Karnataka, Mangaluru, Karnataka, India

**O. Barro**
CINTECX, Universidade de Vigo, LaserON Research Group, School of Engineering, Vigo, Spain

**D. Carou**
Department of Mechanical and Mining Engineering, EPS de Jaén, University of Jaén, Campus Las Lagunillas, Jaén, Spain; Centre for Mechanical Technology and Automation (TEMA), University of Aveiro, Aveiro, Portugal; Departamento de Deseño na Enxeñaría, Universidade de Vigo, Campus As Lagoas, Ourense, Spain

**Murthy S. Chavali**
Shree Velagapudi Rama Krishna Memorial College (PG Studies; Autonomous), Guntur, Andhra Pradesh, India

**R. Comesaña**
CINTECX, Universidade de Vigo, LaserON Research Group, School of Engineering, Vigo, Spain; Materials Engineering, Applied Mechanics and Construction Department, School of Engineering, Universidade de Vigo, Vigo, Spain

**V.M. Correlo**
3B's Research Group, I3Bs — Research Institute on Biomaterials, Biodegradables and Biomimetics, University of Minho, Headquarters of the European Institute of Excellence on Tissue Engineering and Regenerative Medicine, AvePark, Parque de Ciência e Tecnologia, Zona Industrial da Gandra, Barco, Portugal; ICVS/3B's—PT Government Associate Laboratory, Braga, Portugal

**L.P. da Silva**
3B's Research Group, I3Bs — Research Institute on Biomaterials, Biodegradables and Biomimetics, University of Minho, Headquarters of the European Institute of Excellence

on Tissue Engineering and Regenerative Medicine, AvePark, Parque de Ciência e Tecnologia, Zona Industrial da Gandra, Barco, Portugal; ICVS/3B's—PT Government Associate Laboratory, Braga, Portugal

**Ali Davoudinejad**
Technical University of Denmark (DTU), Mechanical Department, Manufacturing Section, Copenhagen, Denmark; Mechanical Engineering, Manufacturing and Production, Politecnico di Milano, Milan, Italy

**J. del Val**
CINTECX, Universidade de Vigo, LaserON Research Group, School of Engineering, Vigo, Spain

**R. Dorado-Vicente**
Department of Mechanical and Mining Engineering, EPS de Jaén, University of Jaén, Campus Las Lagunillas, Jaén, Spain

**Daniel S. Engstrøm**
Loughborough University, Wolfson School of Mechanical, Electrical & Manufacturing Engineering, Loughborough, United Kingdom

**M. Fernández-Arias**
CINTECX, Universidade de Vigo, LaserON Research Group, School of Engineering, Vigo, Spain

**Ross J. Friel**
ITE, Halmstad University, Halmstad, Sweden

**Manuela Galati**
Department of Management and Production Engineering, Torino, Italy

**A. García-Collado**
Department of Mechanical and Mining Engineering, EPS de Jaén, University of Jaén, Campus Las Lagunillas, Jaén, Spain

**Athanasios Goulas**
Loughborough University, Wolfson School of Mechanical, Electrical & Manufacturing Engineering, Loughborough, United Kingdom

**Álvaro Goyanes**
FabRx Ltd., Ashford, Kent, United Kingdom; Departamento de Farmacología, Farmacia y Tecnología Farmacéutica, I+D Farma Group (GI-1645), Universidade de Santiago de Compostela, Santiago de Compostela, Spain

**Yingchun Guan**
School of Mechanical Engineering and Automation, Beihang University, Beijing, China; National Engineering Laboratory of Additive Manufacturing for Large Metallic Components, Beihang University, Beijing, China; International Research Institute for Multidisciplinary Science, Beihang University, Beijing, China

**Ahmed Arabi Hassen**
Oak Ridge National Laboratory (ORNL), Oak Ridge, TN, United States

**S. Yu. Ivanov**
Saint-Petersburg State Marine Technical University, Institute of Laser and Welding Technologies, Saint-Petersburg, Russian Federation

**Dhileep Kumar Jayashankar**
Digital Manufacturing and Design Centre, Singapore University of Technology and Design, Singapore

**K. Karthik**
Department of Physics, Bharathidasan University, Tiruchirappalli, Tamil Nadu, India

**Vidya Kishore**
Oak Ridge National Laboratory (ORNL), Oak Ridge, TN, United States

**O.G. Klimova-Korsmik**
Saint-Petersburg State Marine Technical University, Institute of Laser and Welding Technologies, Saint-Petersburg, Russian Federation

**Xiao Kuang**
The George W. Woodruff School of Mechanical Engineering, Georgia Institute of Technology, Atlanta, GA, United States

**Arun V. kumar**
Department of Mechanical Engineering, VIT University, Vellore, Tamil Nadu, India

**Il Keun Kwon**
Department of Dental Materials, School of Dentistry, Kyung Hee University, Dongdaemun-gu, Seoul, Republic of Korea

**Yuhang Li**
School of Mechanical Engineering and Automation, Beihang University, Beijing, China; National Engineering Laboratory of Additive Manufacturing for Large Metallic Components, Beihang University, Beijing, China; International Research Institute for Multidisciplinary Science, Beihang University, Beijing, China

**Frank Liou**
Department of Mechanical & Aerospace Engineering, Missouri University of Science and Technology, Rolla, MO, United States

**F. Lusquiños**
CINTECX, Universidade de Vigo, LaserON Research Group, School of Engineering, Vigo, Spain; Galicia Sur Health Research Institute (IIS Galicia Sur). SERGAS-UVIGO, Vigo, Spain

**Asunción Martínez-García**
Innovative Materials and Manufacturing Area, AIJU, Ibi (Alicante), Spain

**Jyoti Mazumder**
Centre for Laser Aided Intelligent Manufacturing, University of Michigan, Ann Arbor, MI, United States

**G. Medina-Sánchez**
Department of Mechanical and Mining Engineering, EPS de Jaén, University of Jaén, Campus Las Lagunillas, Jaén, Spain

**Anne I. Mertens**
University of Liège, Faculty of Applied Science, Aerospace and Mechanical Engineering Department, Metallic Materials Science (MMS), Liège, Belgium

**Mario Monzón**
Departamento de Ingeniería Mecánica, ULPGC, Las Palmas de Gran Canaria, Spain

**Maria P. Nikolova**
Department of Material Science and Technology, University of Ruse "A. Kanchev", Ruse, Bulgaria

**Francisco J. Otero-Espinar**
Departamento de Farmacología, Farmacia y Tecnología Farmacéutica, Facultade de Farmacia, Universidade de Santiago de Compostela, Santiago de Compostela, Spain; Paraquasil Group. Health Research Institute of Santiago de Compostela (IDIS), Santiago de Compostela, Spain

**Loucas Papadakis**
Department of Mechanical Engineering, Frederick University, Nicosia, Cyprus

**Rubén Paz**
Departamento de Ingeniería Mecánica, ULPGC, Las Palmas de Gran Canaria, Spain

**J. Pou**
CINTECX, Universidade de Vigo, LaserON Research Group, School of Engineering, Vigo, Spain; Galicia Sur Health Research Institute (IIS Galicia Sur). SERGAS-UVIGO, Vigo, Spain

**M. Pérez**
Department of Mechanical and Mining Engineering, EPS de Jaén, University of Jaén, Campus Las Lagunillas, Jaén, Spain; Department of Manufacturing Engineering, Universidad Nacional de Educación a Distancia (UNED), Madrid, Spain

**R.L. Reis**
3B's Research Group, I3Bs — Research Institute on Biomaterials, Biodegradables and Biomimetics, University of Minho, Headquarters of the European Institute of Excellence on Tissue Engineering and Regenerative Medicine, AvePark, Parque de Ciência e Tecnologia, Zona Industrial da Gandra, Barco, Portugal; ICVS/3B's—PT Government Associate Laboratory, Braga, Portugal; Department of Dental Materials, School of Dentistry, Kyung Hee University, Dongdaemun-gu, Seoul, Republic of Korea

**A. Riveiro**
CINTECX, Universidade de Vigo, LaserON Research Group, School of Engineering, Vigo, Spain; Materials Engineering, Applied Mechanics and Construction Department, School of Engineering, Universidade de Vigo, Vigo, Spain

**Iria Seoane-Viaño**
Departamento de Farmacología, Farmacia y Tecnología Farmacéutica, Facultade de Farmacia, Universidade de Santiago de Compostela, Santiago de Compostela, Spain; Paraquasil Group. Health Research Institute of Santiago de Compostela (IDIS), Santiago de Compostela, Spain

**Przemysław Siemiński**
Academy of Fine Arts in Warsaw, Faculty of Design and Warsaw University of Technology, The Faculty of Automotive and Construction Machinery Engineering, Warsaw, Poland

**Karupppasamy Subburaj**
Engineering Product Development Pillar, Singapore University of Technology and Design, Singapore

**G.A. Turichin**
Saint-Petersburg State Marine Technical University, Institute of Laser and Welding Technologies, Saint-Petersburg, Russian Federation

**Joel C. Vasco**
School of Technology and Management, Polytechnic of Leiria, Leiria, Portugal; Institute for Polymers and Composites, University of Minho, Guimarães, Portugal

**Rajkumar Velu**
Centre for Laser Aided Intelligent Manufacturing, University of Michigan, Ann Arbor, MI, United States; Department of Mechanical Engineering, Indian Institute of Technology Jammu, Jammu & Kashmir, India

**Huaming Wang**
School of Mechanical Engineering and Automation, Beihang University, Beijing, China; National Engineering Laboratory of Additive Manufacturing for Large Metallic Components, Beihang University, Beijing, China; International Research Institute for Multidisciplinary Science, Beihang University, Beijing, China

**Andreas Wegner**
AM POLYMERS GmbH, Willich; Chair for Manufacturing Technology, University of Duisburg-Essen, Germany

**Evren Yasa**
Department of Mechanical Engineering, Eskisehir Osmangazi University, Eskisehir, Turkey

**Yun Hee Youn**
3B's Research Group, I3Bs — Research Institute on Biomaterials, Biodegradables and Biomimetics, University of Minho, Headquarters of the European Institute of Excellence on Tissue Engineering and Regenerative Medicine, AvePark, Parque de Ciência e Tecnologia, Zona Industrial da Gandra, Barco, Portugal; ICVS/3B's—PT Government Associate Laboratory, Braga, Portugal; Department of Dental Materials, School of Dentistry, Kyung Hee University, Dongdaemun-gu, Seoul, Republic of Korea

**Xinchang Zhang**
Department of Mechanical & Aerospace Engineering, Missouri University of Science and Technology, Rolla, MO, United States

# Foreword

Dear Readers,

This series of handbooks on advanced manufacturing covers four major areas, namely, advanced machining and finishing, advanced welding and deforming, additive manufacturing, and sustainable manufacturing. The series aims to not only present the advancements in various manufacturing technologies but also provide a fundamental and detailed understanding about them. It encompasses a wide range of manufacturing technologies with their mechanisms, working principles, salient features, applications, and research, development, and innovations in there. Fundamental research, latest developments, and case studies conducted by international experienced researchers, engineers, managers, and professors are mainly presented. *Handbook 1* on advanced machining and finishing majorly covers advanced machining of difficult-to-machine materials; hybrid, high-speed, and micromachining; and burnishing, laser surface texturing, and advanced thermal energy—based finishing processes. *Handbook 2* on advanced welding and deforming covers ultrasonic welding, laser welding, and hybrid welding-type advanced joining processes and also describes advanced forming techniques such as microwave processing, equal channel angular pressing, and energy-assisted forming. *Handbook 3* additive manufacturing sheds light on 3D and 4D printing, rapid prototyping, laser-based additive manufacturing, advanced materials, and postprocessing in additive manufacturing. *Handbook 4* on sustainable manufacturing presents advancements, results of experimental research, and case studies on sustainability interventions in production and industrial technologies.

We hope that this series of handbooks would be a good source of knowledge and encourage researchers and scientists to conduct research, developments, and innovations to establish these fields further.

J. Paulo Davim and Kapil Gupta

# Preface

Additive manufacturing (AM) has become the new paradigm of manufacturing and promises to revolutionize industry due to its endless range of applications. Rather than a new manufacturing process, this is a collection of technologies where parts are built up by adding material in a layer-wise fashion according to a predefined geometry extracted from a 3D computerized model. Unlike part production using conventional manufacturing processes, AM technologies allow the production of customized parts on demand, with an unprecedent flexibility in geometries and materials, not possible to reach by any other conventional manufacturing process. However, many challenges must be addressed to exploit the full potential of these technologies, bringing new applications and opportunities.

This handbook covers the working principles of the different technologies involved in AM, as well as the finishing operations commonly involved after the fabrication of parts with these technologies. The production of new feedstock materials and the study of the particularities involved in their processing, the development of standards to guarantee the level of quality, safety, and reliability of these technological processes, and the technical details involved in selected applications (automotive, pharmaceutical, biomedical, etc.) are also subject matter of this handbook. The latest research in this area and possible avenues of future research are also highlighted to encourage the researchers.

This handbook consists of 22 chapters grouped in three sections dealing with the main aspects related to AM: (1) Additive Manufacturing Technologies, (2) Postprocessing and Standardization, and (3) Materials and Applications. It starts with Section 1 (*Additive Manufacturing Technologies*) that sheds light on the working principles, processing parameters, feedstock materials, quality, and applications of the main AM technologies. Chapter 1 summarizes the philosophy under AM, and the main technologies are briefly discussed. Chapter 2 describes the powder bed fusion of polymers; it describes the basics of these processes with a focus on laser sintering. Chapter 3 provides some insights on powder bed fusion of metallic materials; it summarizes the basics of these processes with a focus on selective laser melting. Chapter 4 covers directed energy deposition processes based on laser; it describes the basics and its application into the industry and biomedical fields. Chapter 5 focuses on AM by vat photopolymerization technologies where methods, applications, and technologies are reviewed. Chapter 6 focuses on material extrusion AM processes; the different extrusion AM platforms, their operating mechanisms, processing capabilities, and limitations are reviewed. One particular material extrusion process, fused deposition modeling, widely recognized by the industry, academia, and costumers, is the main topic of Chapter 7. Chapter 8 deals with electron beam melting; the description of the processing technique, main processing parameters, features of the manufactured parts, and monitoring and simulation tools for this technique are reviewed. Chapter 9 deals with the new paradigm of AM: 4D printing. This refers to AM of parts whose shape, properties, or functionalities change with time in response to external stimuli.

Section 2 (*Postprocessing and Standardization*) deals with the postprocessing techniques required after part production by any of the precedent AM technologies. Chapter 10 discusses the application of laser polishing to tailor the surface roughness of metallic parts produced by AM. Chapter 11 deals with the surface quality of parts produced by AM technologies and introduces conventional machining (turning, milling, and grinding) as a postprocessing technique. Chapter 12 highlights the necessity and impact of AM standards and presents an update of the existing structure of the ISO/ASTM AM standardization working groups and their liaisons with other relevant committees.

Section 3 (*Materials and Applications*) deals with the main topics involved in AM of different materials (metals, polymers, composites, etc.) and with their potential applications in different fields. Chapter 13 focuses on laser AM of metal matrix composites, focusing on the characterization of their complex microstructures and properties. Chapter 14 reviews the laser AM of metals. Chapter 15 deals with AM of functional nanocomposite materials and presents how different nanoparticles could yield functional materials for medical, bionic, electronic, sensing, energy storage, or structural devices. Chapter 16 highlights the AM for the automotive industry and focuses on its technical and economic impact. Laser AM of large-size products is reviewed in Chapter 17, pointing out the difficulties and particularities found when dealing with parts of large size. Chapter 18 focuses on the production of personalized medicines by AM of pharmaceutical products; manufacturing methods, possibilities, and challenges are reviewed. Chapter 19 sheds light on AM for the biomedical field; the most common 3D bioprinting techniques used in tissue engineering, applications, and associated challenges are presented. Chapter 20 discusses the AM of biopolymers and composites for medical applications. Chapter 21 provides an insight into AM within a future planetary manufacturing scenario; prospective material space resources, their simulants, and potential processing methods are described. Chapter 22 presents the state-of-the-art of different modeling approaches for replicating physical effects in AM.

We believe that this handbook provides a comprehensive review on AM and will be a reference source not only to researchers, engineers, and technical experts working in the field but also to students and young scientists who plan to work in this research area. We sincerely acknowledge Elsevier Inc. for this opportunity and their professional support. This handbook has been only possible due to the invaluable efforts from renowned researchers working in AM from around the globe. The editors would like to express their appreciation for the cooperation and contribution of all the authors. Their time and efforts have resulted in this comprehensive handbook on the many topics that comprise the field of AM. We hope that this will contribute to the rapid advancement of this exciting field.

Vigo, Aveiro
January 2021

Antonio Riveiro
Juan Pou
J. Paulo Davim

# CHAPTER 1

# Introduction to additive manufacturing

**Xinchang Zhang, Frank Liou**
Department of Mechanical & Aerospace Engineering, Missouri University of Science and Technology, Rolla, MO, United States

## 1.1 Basic concepts of additive manufacturing

Additive manufacturing (AM), also known as 3D printing, is a process in which a three-dimensional object is built from a computer-aided design (CAD) model, usually by successively adding materials in a layer-by-layer fashion. AM was originated in the 1980s used to print a solid model demonstrated by Hideo Kodama of Nagoya Municipal Industrial Research Institute [1,2]. Ciraud was the first to introduce the use of powders to manufacture 3D objects like the way of modern 3D sintering machines [3]. With decades of development, AM grows rapidly and has been used in a wide range of applications. AM is different from conventional manufacturing processes such as machining, casting, and forging, where materials are removed from a block or injected into a mold to form the product. For traditional subtractive machining, detailed process planning must be made to determine machining steps to obtain the physical geometries. For example, in computer numerical control (CNC) machining, proper tools must be selected for specific materials, and the tool path must be carefully designed to prevent tool crash. In contrast, AM is a tool-free process and therefore can reduce both wear and machine setup times. AM also provides more flexibility in product design. In general, there are no limits in terms of design since AM is a building-up process with materials added layer by layer. Components that are difficult to fabricate by conventional processing such as parts with hollow features are easy to build by AM. The complexity of components no longer complicates the AM process. Besides, an assembly that consists of several components is now able to be constructed by AM. AM also provides more customized and personalized solutions because the geometry of a component can be easily adjusted. It makes the creation of personalized components such as implant and dental bridges easier and faster.

*Additive Manufacturing*
ISBN 978-0-12-818411-0
https://doi.org/10.1016/B978-0-12-818411-0.00009-4

The key of AM is that components are fabricated by adding materials layer by layer. AM uses CAD models or scanned models as the input to guide the hardware to deposit materials to obtain the final shape in a layer-upon-layer fashion. Instead of using cutting tools in machining processes, AM uses a heat source (laser or electron beam), ultraviolet (UV) light, or binding agent to achieve the deposition of materials. After one layer is built, the AM stage moves by a layer thickness so that another layer can be deposited. This process continues until the final part is fabricated. Since AM is a layer-by-layer fabrication process, the finally obtained physical object possesses step features as shown in Fig. 1.1. A thinner layer thickness improves surface finish but increases building time. Besides layer thickness, many other processing parameters should be considered, including heat power, traveling speed, material feed rate, hatch spacing (spacing between two adjacent tracks, etc.). The following of this chapter discusses the basic procedures of the AM process, detailed classification of AM processes and their principles, applications of AM, the distinction between AM and CNC machining, hybrid manufacturing, and the challenges and limitations of AM.

## 1.2 Basic procedure of additive manufacturing

Although there are various terminologies based on AM technique, the fabricating principle is similar in which physical object is built layer by layer. The general process obeys the following steps as shown in Fig. 1.2.

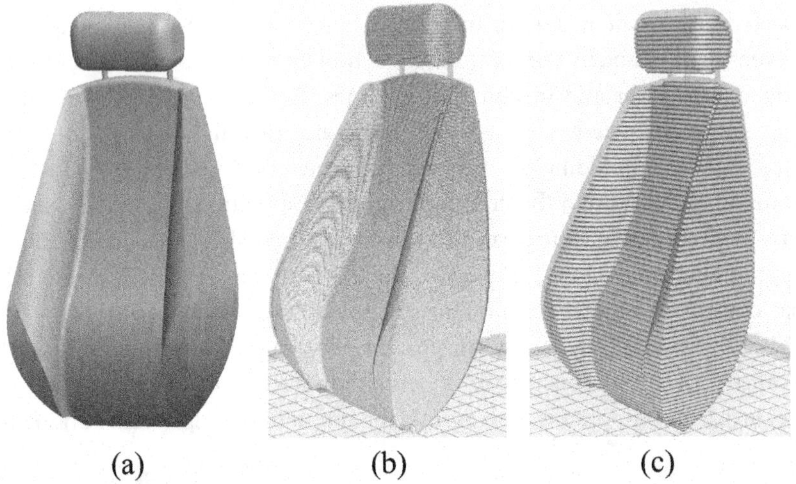


**Figure 1.1** (A) CAD model of a car seat. The sliced model with (B) small layer thickness and (C) large layer thickness.

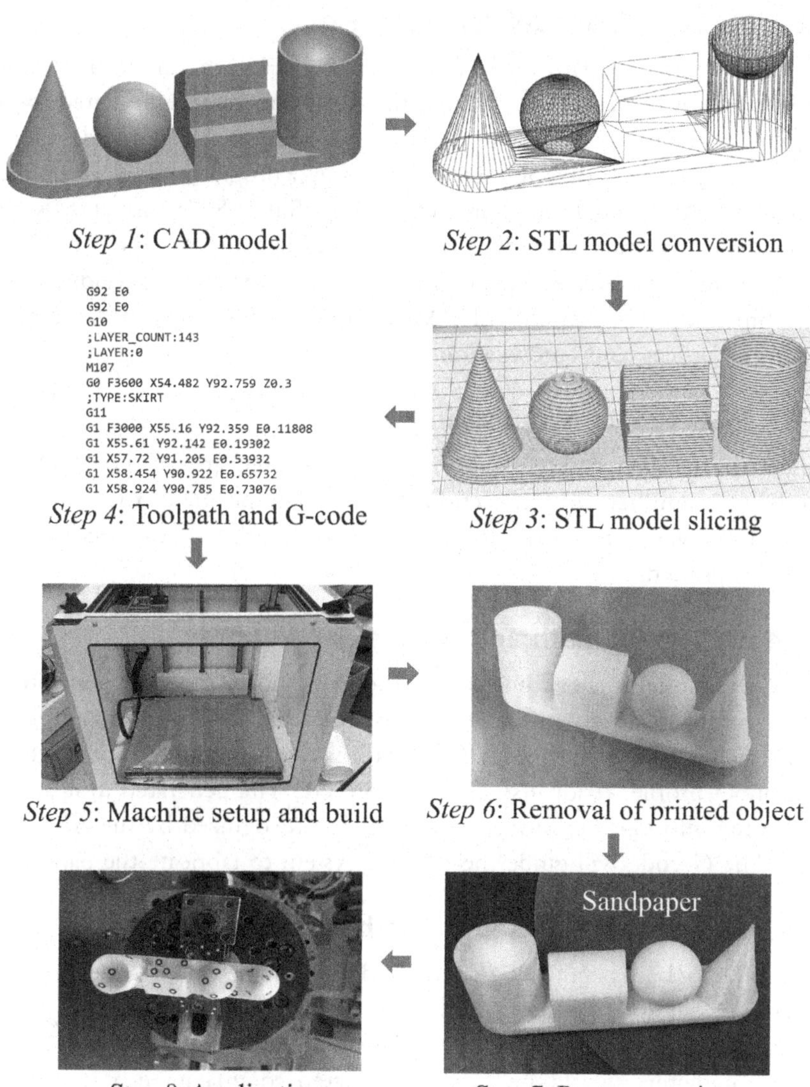


**Figure 1.2** General procedure for additive manufacturing.

## 1.2.1 Step 1: CAD model

The first step in the AM process is to produce a three-dimensional model. In general, the most widely used approach for creating a 3D digital solid model is using CAD software. Reverse engineering tools such as 3D scanner can also be used to regenerate the model of an existing object if the original model is missing. The goal of this step is to generate the geometry of the part that will be additively manufactured.

### 1.2.2 Step 2: STL model conversion

Currently, STL (standard tessellation language) is still the file format most widely used in AM. In STL format, the CAD model is represented by a number of triangular facets with normals and vertices. STL only describes the surface geometry of a 3D object. Most CAD programs can convert the designed CAD model into an STL model. Since STL format fails to represent many other properties of a CAD model such as color and material, additive manufacturing file format (AMF) is recently developed for AM processes. The benefits of AMF format include support for color, materials, lattices, and constellations.

### 1.2.3 Step 3: STL model slicing

Since AM is a layer-based fabrication process, it is required to slice the STL model to a number of cross-sections (layers). The printing process is performed in layers one on the top of a previously printed layer until a full model is printed.

### 1.2.4 Step 4: toolpath and G-code generation

Once the model is sliced into layers, toolpath and G-code are generated. There are many algorithms in generating toolpath. The algorithms consider many factors such as the selection of infill patterns (contour, zigzag, etc.), heat-input profile, residual stress evolution, etc. The generated toolpath is converted into G-code format, which can be recognized by the AM system. The G-code will guide the printing system to fabricate the part.

### 1.2.5 Step 5: machine setup and build

The AM machine needs to be configured properly before printing. Several parameters should be considered such as material type, power, layer thickness, traveling speed, building plate temperature, support type, etc. Such parameters should be carefully selected to guarantee a successful printing. Once such parameters are determined, printing can be started. Most AM machines do not need to be monitored once printing has begun, and the printing is largely automated. Attention is only needed when materials run out or printing is failed.

### 1.2.6 Step 6: removal of printed objects

For some AM technologies, the removal of the printed object is simply separating it from the build plate. For some industrial AM machines, the removal of the printed part is a technical process involving many complicated removal procedures. For example, in the Binder Jetting process, the printed parts are encapsulated in the powder and are left to cure and gain strength before removal. After that, the part is removed from the powder bin and excess powder is cleaned via pressurized air.

### 1.2.7 Step 7: post-processing

Post-processing procedures vary by printing techniques and materials. Some AM processes may not require post-processing, but some processes require careful and long-time post-processing. For example, photosensitive resins printed by stereolithography (SLA) require the component to cure under the UV environment before the final application. Support structures are often used when parts have steep overhangs or unsupported areas. For example, if an arch is printed, the center of the arch may need support structure because when the printer tries to print that top layer, there is no material supporting it from below. Printing without support could cause drop of layers. If support structures are used during printing, removal of such support is required. The support can be removed mechanically by cutting it from the printed object, or chemically by dissolving in a solvent that will not damage the target part.

### 1.2.8 Step 8: application

The post-processed part may be ready for real application. However, for some applications, additional processing may be required. For example, some metallic parts may need heat treatment to improve mechanical properties. Certain critical surfaces may need machining or polishing to meet surface roughness requirements. Such further treatment is based on applications and can be performed according to requirements.

## 1.3 Categories of additive manufacturing

According to the standard terminology for AM technologies by American Society for Testing and Materials (ASTM) F2792-12a, AM can be classified into the following seven categories: (1) vat photopolymerization; (2) material jetting; (3) material extrusion; (4) binder jetting; (5) powder bed fusion (PBF); (6) sheet lamination; and (7) directed energy deposition (DED). Table 1.1 summarizes the classification of AM processes and characteristics of each process. Each process was introduced in the following of this chapter.

**Table 1.1** Classification of additive manufacturing processes by American Society for Testing and Materials.

| Categories | Technologies | Power source | Materials | Advantages | Disadvantages |
|---|---|---|---|---|---|
| Vat photopolymerization | Stereolithography, digital light processing, continuous liquid interface production, daylight polymer printing | Ultraviolet light | Photosensitive resin, ceramics | High accuracy, good resolution, fully automation | Overcuring lengthy post-processing, single composition, high cost of materials |
| Material jetting | Drop on demand, PolyJet, nanoparticle jetting | Thermal energy | Photopolymer resins, metals, ceramics | High accuracy, smooth surface finish, multimaterial | Low mechanical strength |
| Material extrusion | Fused deposition modeling, fused filament fabrication | Thermal energy | Thermoplastics (ABS, PLA, PC, nylon) | Inexpensive (both machine and feedstock), multimaterial, easy to operate | Poor resolution and surface finish, poor bonding |
| Binder jetting | Binder jetting | Binder/thermal energy | Polymer/ceramic/metal powder | Wide material selection, relatively fast printing | Lengthy post-processing, porosities within parts |
| Powder bed fusion | Direct metal laser sintering, electron beam melting, selective laser melting, selective laser sintering | Laser, electron beam | Polymer/ceramic/metal powder | High accuracy, high resolution, fully dense parts, high strength | Powder recycling, support structures, single material, residual stress |
| Sheet lamination | Laminated object manufacturing | Laser | Plastic/metal/ceramic foil | High surface finish | Material limitation |
| Directed energy deposition | Laser engineered net shaping, direct metal deposition, laser metal deposition, laser cladding, laser consolidation | Laser | Metal/ceramic powder | Repair of worn components, multi-material (functionally graded materials) | Low accuracy, low surface finish, residual stress, require post-machining |

### 1.3.1 Vat photopolymerization

Vat photopolymerization encompasses several different processes including stereolithography (SLA), digital light processing (DLP), continuous liquid interface production (CLIP), and daylight polymer printing (DPP) [4,5]. Such different processes share the same basic principle that the parts are built by selectively curing layers of photosensitive polymer resin with a source such as UV light. Fig. 1.3 shows a schematic diagram of the vat photopolymerization process. The photosensitive resin that is stored in the container is cured by a UV light source to form a layer of the printed object, after which the stage moves and another layer of material is cured. In this layer-by-layer manner, the 3D physical object is built until completion. Based on the use of different light source and printing setup, vat photopolymerization can be classified into SLA, DLP, CLIP, and DPP. During SLA, a UV light is used to selectively cure the liquid photopolymer resin. The UV light scans on the top of the resin according to the tool path. For DLP, instead of using UV light or laser, a digital projector screen is used to flash an image of each layer according to the geometry of the designed model. Because each image is composed of a number of pixels, DLP can achieve faster printing since each layer can be cured at once in comparison with SLA where a light source needs to draw a number of lines for a single layer. Rather than using a laser or a projector to solidify the resin, DPP uses an LCD (liquid crystal display) to cure the polymer. For CLIP, part of the resin container bottom is transparent to UV light, making the light be able to cure a layer of resin. After that, the object rises layer thickness and the resin flow under the part which is cured afterward.

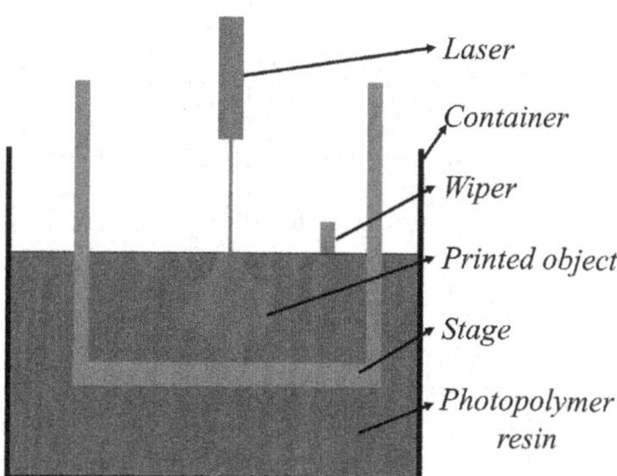


**Figure 1.3** Schematic diagram of the vat photopolymerization process.

Vat photopolymerization can print a part with high accuracy and good surface finish, but since the cured part needs to be further cured in a UV chamber, it requires lengthy post-processing.

### 1.3.2 Material jetting

Fig. 1.4 illustrates the principle of the material jetting process. Similar to the two-dimensional inkjet printing process, in material jetting, liquid materials are delivered into an extruder and sprayed to form numerous tiny droplets that are solidified when exposing to light [6]. This process requires support that can be printed at the same time during printing modeling material. The support material can be dissolved in a solvent making the removal of support convenient. Since modeling materials can be injected into the building area through multiple extruders, the material jetting process allows for different materials to be printed within the same object [7]. The material jetting process contains several techniques including drop on demand (DOD), PolyJet (PJ), and nanoparticle jetting (NPJ) [8]. DOD process injects materials at points where needed to build the layers of a component. The principle of PJ is comparable to inkjet 2D printing, in which materials are sprayed on a stage that is cured by light source immediately. NPJ uses nanoparticles as feedstock and jets them on a stage in very thin layers of droplets. The materials for material jetting are mostly photopolymer resins, but the process can also print metals and ceramics. The fabricated part can be featured with high accuracy and smooth surface but relatively poor mechanical properties.

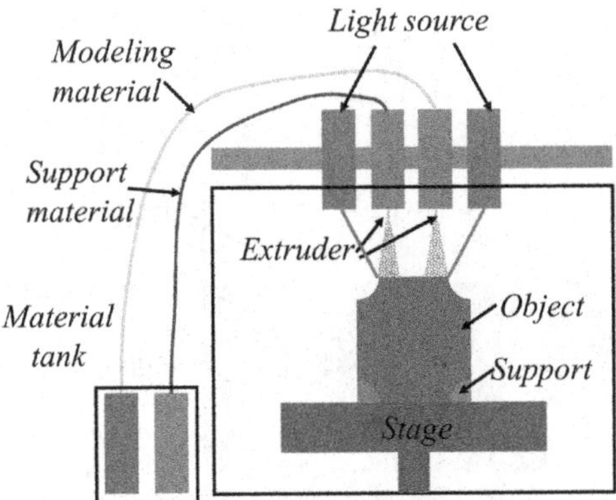


**Figure 1.4** Schematic diagram of the material jetting process.

### 1.3.3 Material extrusion

Material extrusion is an AM technique in which a polymer filament is continuously fed into the printing area through extruding nozzles where it will be heated (usually in the range of 150–250°C) and then deposited onto the building platform layer by layer. A widely used printing process based on the material extrusion technique is fused deposition modeling (FDM), also known as fused filament fabrication. Fig. 1.5 shows an illustration of the FDM process [9]. The filament from the material spool is fed along guiding tubes into a liquefier head and is then heated and deposited on the printed bed or previously deposited materials. The liquefier head and extrusion nozzles move along X–Y directions to realize printing a layer of the object. After finishing the deposition of one layer, the print bed moves in the Z direction in a layer thickness for the printing of the next layer. This process continues so that the physical object can be obtained based on the CAD model. FDM requires support structure to build over-hang features, but the support material can be easily removed either mechanically (removed by stripping them from the printed part) or chemically (dissolved in a solvent solution). FDM can be used to print a wide variety of materials including acrylonitrile butadiene styrene, polylactic acid, polycarbonate, thermoplastic polyurethane, and aliphatic polyamides (nylon). Other advantages of FDM include low initial and running costs, easily understandable printing techniques, automated printing without supervision, and

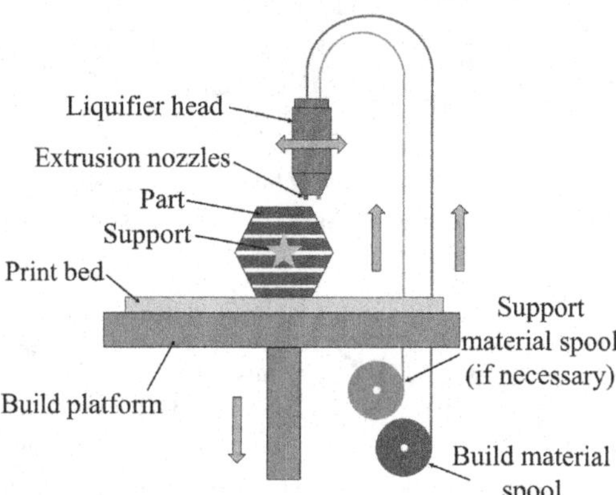


**Figure 1.5** Illustration of the fused deposition modeling process. *(Reproduced with kind permission of Elsevier from F. Ning, W. Cong, J. Qiu, J. Wei, S. Wang, Additive manufacturing of carbon fiber reinforced thermoplastic composites using fused deposition modeling, Compos. B Eng. 80 (2015) 369–378.)*

office-friendly process. FDM also has many drawbacks. For example, since the filament is relatively large in diameter (typically, 1.75 or 2.85 mm) and the layer thickness is large, the layers are clearly visible and thus resulting in poor surface finish and accuracy [10]. Finer resolution, however, will increase printing time significantly. Another disadvantage is since the material filament is heated and deposited on previously deposited materials, the bonding strength along building direction (Z-direction) is relatively poor.

### 1.3.4 Binder jetting

Fig. 1.6 shows general procedures in the binder jetting AM process. The binder jetting system shown in Fig. 1.6 consists of two beds: powder bed and build bed [11]. During the binder jetting process, powder stored in the powder bed is sprayed onto a build table in the build bed by a roller. After that, liquid binder is injected from the print head and selectively deposited on the area of powder and consolidate powder together. When this layer is finished, the build stage moves down a layer thickness and another layer of powder is sprayed on the previously deposited layer by the roller. Such powder is again solidified selectively by adhesive binder. The process is

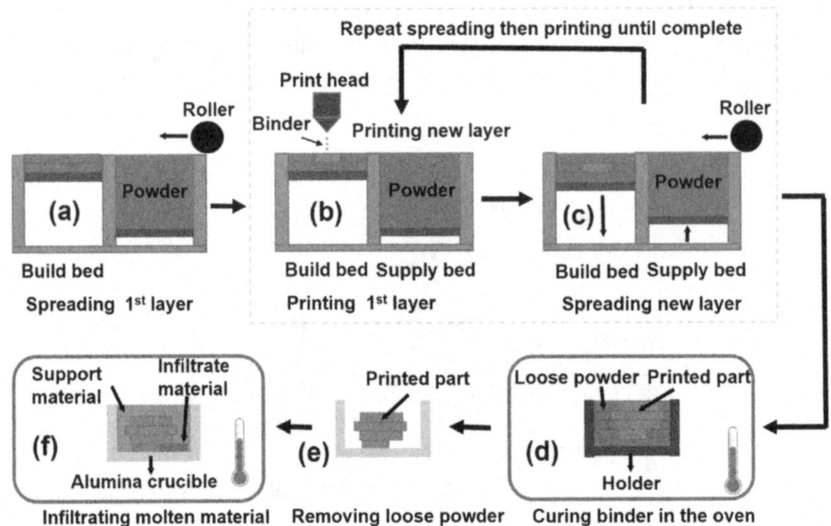


**Figure 1.6** Processing steps of a typical binder jetting process. *(Reproduced with kind permission of Elsevier from T. Do, P. Kwon, C.S. Shin, Process development toward full-density stainless steel parts with binder jetting printing, Int. J. Mach. Tools Manuf. 121 (2017) 50–60.)*

repeated layer by layer until the designed geometry is printed. After this process, the printed part (also known as green part) is removed from unbonded loose powder. The part is further sintered in a furnace so that the binder material can be burned out. The debinding temperature is typically in the 200−600°C range. The selection of binder material is determined by which material is printed. Typical binder materials include furan binder (for sand), phenolic binder (for sand molds and cores), and aqueous-based binder (for metals). Many materials can be processed using binder jetting technique including metals, sands, and ceramics. Parts made of sands usually contain porosities, making the parts not suitable for functional application. However, metal-based parts usually have good mechanical properties that benefited from the infiltration process, making them ready to use as functional components. Compared with other AM processes such as DED, in the binder jetting process, no heat is employed in making the green parts, and therefore, residual stresses in the parts can be minimized.

### 1.3.5 Powder bed fusion

PBF includes the following processes: direct metal laser sintering (DMLS), electron beam melting (EBM), selective laser melting (SLM), and selective laser sintering (SLS). Instead of using binder in the binder jetting process, PBF uses a laser source (SLS, SLM, DMLS) or electron beam (EBM) to directly and selectively melt or sinter layers of materials together to form a solid part. If a laser source is adopted, the deposition process is carried out in an inert atmosphere such as argon or helium chamber to prevent material from oxidation at elevated temperature. The usage of an electron beam requires a vacuum chamber. Fig. 1.7 shows the principle of the SLM process [12]. At first, metal particles are sprayed on the top of a base plate (substrate). These powders are then melted by a laser and subsequently solidified to form a cross-section. After that, the base plate moves down a layer thickness, and another layer of powders is sprayed on the top of the printed part by a powder recoating mechanism. The material is again selectively melted and solidified to form a cross-section. The continuous melting and solidifying multiple layers of powder results in the fabrication of the final part. PBF can be used to process multiple powder-based materials, but commonly used materials are metals and polymers. Powders are usually spherical and the particle size is usually in the range of 15−40 μm for SLM [13], 20−80 μm for SLS [14], and 40−100 μm for EBM [15].

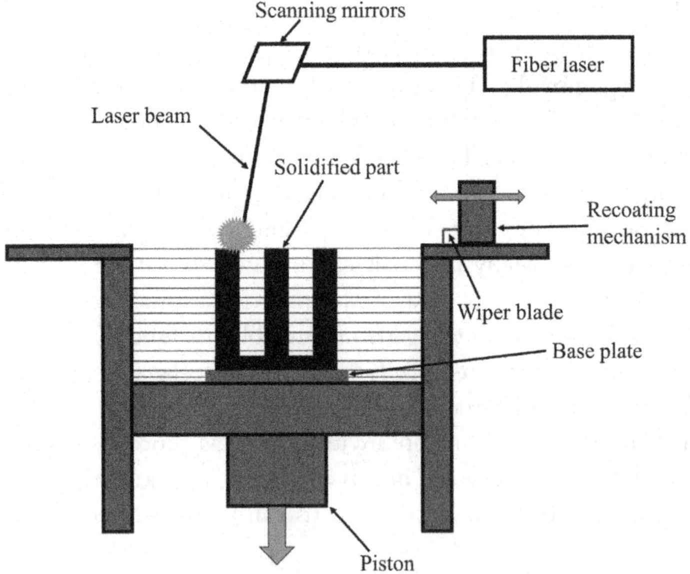


**Figure 1.7** Illustration of the selective laser melting process. *(Reproduced with kind permission of Elsevier from E. Louvis, P. Fox, C.J. Sutcliffe, Selective laser melting of aluminium components, J. Mater. Process. Technol. 211 (2011) 275–284.)*

## 1.3.6 Sheet lamination

Sheet lamination, also known as laminated object manufacturing, shares similar building principles with other AM processes, but instead of using powder or wire as feedstock, sheet lamination uses foil to make an object [16]. This AM technique was first developed by Helisys Inc. There are three types of processes using sheet lamination technique including paper-based lamination, composite-based lamination, and selective lamination. In the paper-based lamination, papers/foils are glued together layer by layer and precisely cut to the designed geometry to make the final object. Potentially any sheet material that can be precisely cut using cutting tools (laser or mechanical cutter) and that can be bonded can be used for part construction. For example, metal sheets can be used to make AM products. These products are laser cut from flat sheets of metal and then piled up and weld to form a metallic 3D part. The paper thicknesses usually range from 0.07 to 0.2 mm [17]. Composite lamination has a similar principle, but some reinforcement can be added to the buck material to improve the strength of the fabricated part. Rather than placing glue to add layers together, in the selective lamination process, a binder can be selectively applied in certain locations to fabricate the part.

### 1.3.7 Directed energy deposition

DED belongs to the AM family that is used to create parts by directly melting materials and depositing them on a substrate layer by layer. Many processes are based on the DED principle, including laser engineered net shaping (LENS), direct metal deposition, laser metal deposition, laser deposition welding, 3D laser cladding, and laser consolidation. Materials typically used in the DED process are metal particles. A schematic diagram of a typical DED system is illustrated in Fig. 1.8A. It includes a laser system, gas feeding system, powder feeding system, motion control system, and an enclosure to realize the continuous material printing. A diagram of the material deposition mechanism is shown in Fig. 1.8B, where a laser is used to create a molten pool on the substrate, and metallic powder is delivered from a powder feeder into the molten pool passing a powder feeding nozzle. The particles are melted by the laser beam and then solidified to form a layer. The powder used for DED is usually spherical and in the range of 50−150 μm [18]. Once a single layer has been built, the laser head moves on to the next layers for the deposition of successive layers until the designed geometry is obtained. To avoid material oxidation under high temperature, the deposition is usually performed in an inert atmosphere such as nitrogen, argon, and helium. Comparing with the PBF process where powder is sprayed on the platform at the beginning of printing, in the DED process, metal powders are injected into the laser-generated molten pool through nozzles simultaneously with printing [19]. This feature makes the DED process suitable for repairing damaged components.

## 1.4 Applications of additive manufacturing

### 1.4.1 Fabrication of functionally graded materials

Functionally graded material (FGM) refers to a kind of composites that has different materials over different positions of a single component, which leads to the variation of properties including chemical property (corrosion, oxidation, etc.), physical property (density, conductivity), and mechanical property (tensile strength, toughness, stiffness, ductility, hardness, etc.) [20]. The rapid growth of technologies increases the requirement of varied properties in a single component and thus making it hard or impossible for a single material to meet all required properties. Joining of two or more materials together is an ideal approach to take advantage of each individual material. Therefore, interest in fabricating FGM is recently escalated in

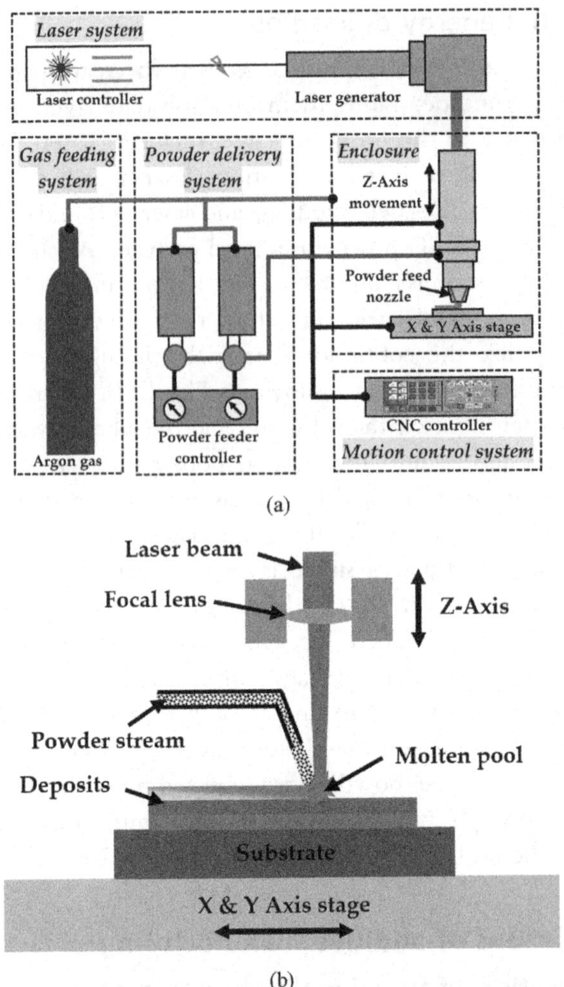


**Figure 1.8** Schematic diagram of (A) the major components in a directed energy deposition system, and (B) laser-aided material deposition process.

advanced manufacturing. Several manufacturing processes have been proposed to fabricate FGM. For example, Kim et al. coated SiC on carbon fiber-reinforced carbon (C/C) through C/SiC intermediate layers using chemical vapor deposition method to relieve residual thermal stress and enhance oxidation resistance [21]. Rezapoor et al. fabricated graded Fe–TiC layers on carbon steel using plasma spray by increasing weight fraction of TiC from 25% to 100% and reducing Fe in the same amount [22]. Comparing with direct coating TiC on carbon steel, the graded sample yields improved bonding strength and wear resistance. Thermal spray is extremely popular for thermal barrier coatings to achieve better resistance to wear, erosion, and oxidation [23]. Diffusion bonding was used to build graded WC-Co/Ni composite on 410 stainless steel (SS), and it showed Ni contributes to the release of residual stress across the joint [24]. Askari et al. successfully fabricated $Al_2O_3/SiC/ZrO_2$ graded materials using electrophoretic deposition [25]. Spark plasma sintering was used to produce a four-layer TiB—Ti system with well-bonded and defect-free interfaces [26]. In addition to the aforementioned approaches, AM is an ideal method to fabricate FGM due to its layer-by-layer fabrication [27]. Among AM family, DED process in which particle feedstock is delivered to a laser-generated molten pool is particularly suitable for making FGM [28]. Comparing with powder-bed based techniques such as SLM [29] where a powder layer is placed on a substrate before laser sintering, in the DED process, metal powders are fed from powder feeders into the deposition area simultaneously with the operation of printing [30]. The incoming materials can be easily changed using a deposition system equipped with multiple powder feeders [31].

Fig. 1.9A shows the basic procedures in fabricating FGM. Four major tasks should be considered in making FGM: design, build strategy, manufacturing, and characterization [32]. Fig. 1.9B shows different types of strategies in joining material *A* with *B*. In general, there are three types of approaches to join material *A* with *B*. Fig. 1.9B(i) shows the direct joining of material *A* with *B* without any gradient regions or intermediate layers. Fig. 1.9B(ii) shows a gradual change of material composition along the graded direction in making FGM while Fig. 1.9B(iii) joins material *A* with *B* through *C* intermediate layers.

For the design of FGM, the appropriate selection of materials should be taken into consideration. The selection of materials for fabricating FGM is mainly determined by the requirement of the component, such as mechanical properties, thermal conductivity, weight, and cost. For instance, Cu possesses excellent thermal conductivity, but the strength is low, while SS has good mechanical properties but low thermal conductivity. The requirement of both thermal conductivity and strength requires the target component has both Cu and SS. Once materials are selected, build strategy should be considered to join dissimilar materials. One way is to directly deposit material *A* on material *B* without gradient layers. However, due to the mismatch of lattice parameters, susceptibility to intermetallic phase—induced cracking, and coefficients of thermal expansion of two dissimilar

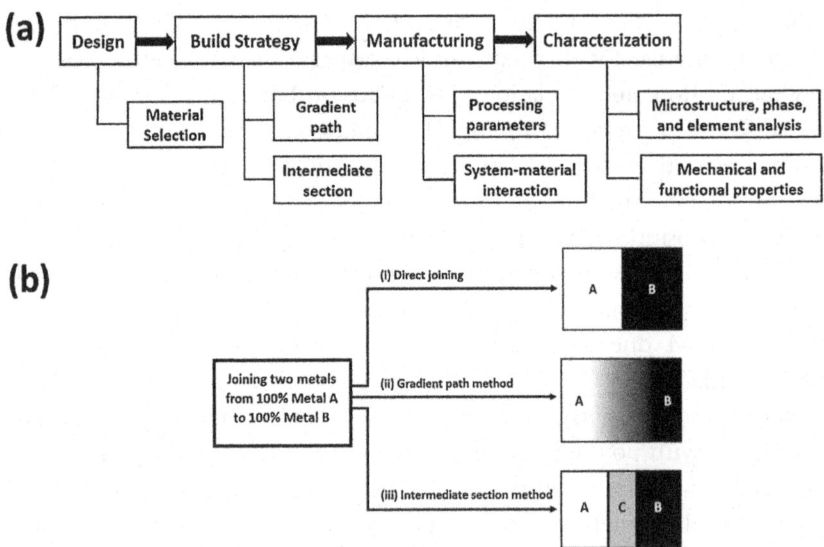


**Figure 1.9** (A) Overview of fabricating metallic functionally graded materials by additive manufacturing. (B) Build strategies of joining two pure metals: (i) Joining 100% metal A and 100% metal B directly, (ii) Joining 100% metal A and 100% metal B with a gradient path, (iii) Joining 100% metal A and 100% metal B with a third type of material C as an intermediate section. *(Reproduced with kind permission of Elsevier from L. Yan, Y. Chen, F. Liou, Additive manufacturing of functionally graded metallic materials using laser metal deposition. Addit. Manuf. 31 (2020) 100901.)*

materials, direct joining of two different materials may encounter potential issues. For example, it is reported that it is unfeasible to direct deposit Ti−6Al−4V on 316L stainless steel (SS316L) due to the formation of a large number of Ti-based intermetallic phases, such as TiFe, TiFe2, and TiCr [33]. Therefore, it is of significance to design an appropriate material transition route to join Ti−6Al−4V with SS316L. There are two available solutions in making such FGM, one is to change material composition gradually (gradient path), and another is to insert intermediate layers of other materials (intermediate path). In Ref. [33], a Ti−6Al−4V → V → Cr → Fe → SS316 path was designed to successfully join Ti−6Al−4V with SS316. The goal is to find fitting materials that can alloy with both materials without forming cracks. Inserting intermediate layers between two distinct materials may face other issues. Due to the variation in thermal expansion, a sharp change of material may give rise to a significant level of residual stress that, if not properly controlled, will result in severe distortion of the FGM. Fortunately, AM process parameters can be modified to control the evolution of residual stress. For example, Denlinger et al. studied the effect of interlayer dwell time on residual stress in additively manufactured Ti6Al4V and Inconel 625 [34]. Deposition pattern also plays an important role in residual stress and deflection as investigated in Ref. [35]. Residual stress in as-deposited and heat-treated AISI 4340 steel was evaluated in Ref. [36]. Once the build strategy is determined, the AM processing parameters can be optimized to obtain FGM with optimized properties. The final step is to characterize the fabricated FGM to validate the design and test the FGM samples for revealing mechanical and functional properties.

### 1.4.2 Repair and remanufacturing of damaged components

AM is promising in component repair and remanufacturing, in which the damaged region of a component is refilled/repaired with filler materials to recover the missing geometry [37]. The critical surfaces of the repaired parts can undergo final machining to acquire an acceptable surface finish. Repair of existing damaged components is critical to maximize their service life and reduce the costs of replacements.

A typical remanufacturing process generally obeys the procedure illustrated in Fig. 1.10, where the critical steps are (1) pre-repair process, (2) worn model acquisition, (3) repair volume reconstruction, (4) repair experiments (both AM and subtractive machining), and (5) repair quality inspection. In the beginning, a complete inspection of worn parts is performed to assess the feasibility of repair. Considering the excessive variety of locations and geometries of defects, the inspection is highly a

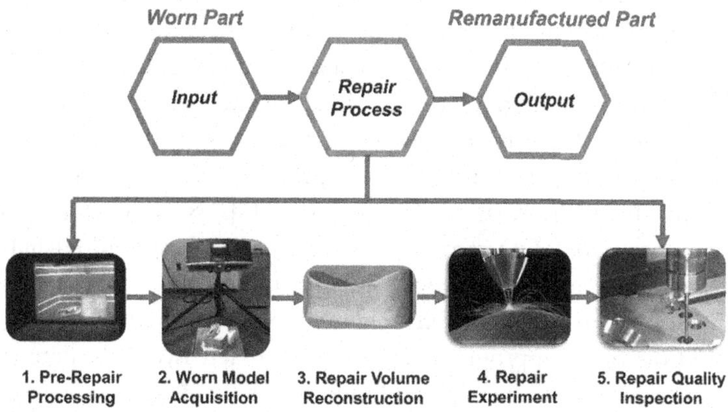


**Figure 1.10** The general procedure of a typical remanufacturing process.

case to case basis but following a general rule. The objective of pre-repair processing is to guarantee the worn parts are ready to repair. Reconstruction of the repair volume is essential because the tool path is generated based on the repair volume. The repair volume must be precisely obtained to guarantee a successful repair. Repair experiments usually involve depositing materials on the damaged zone by AM process such as laser-aided DED, as well as post-repair machining using CNC machining. Finally, the repaired part is inspected to assess the repair quality. This step usually includes geometric inspection as well as material and mechanical testing.

In general, damaged components cannot be repaired without pre-repair processing. There are several reasons that pre-repair processing is required. At first, the worn area cannot be directly accessed by the AM system such as laser beam and powder feed nozzle; thus, melt pool cannot be generated [38]. Besides, worn components usually have heat-checks or damaged layers due to corrosion, erosion, and wear. Such contaminated layers need to be removed before AM process because direct depositing materials on such regions could introduce contamination and thus the solid bimaterial interfacial bonding cannot be guaranteed. Contaminated inclusion between filler material and substrates decimates the mechanical properties of parts and therefore leaves a great threat in service. Pre-repair heat treatment is also important for component remanufacturing. That is because many parts such as die/mold undergo heating and cooling cycles in service and can result in thermal fatigue failure. It is reported that with the increasing number of thermal cycles, the hardness and microstructure of tool steel change, which eventually results in loss of mechanical strength and plastic deformation [39].

It is necessary to acquire a damaged model and compare it with the nominal model to define the missing geometry [40]. Therefore, reconstructing the worn model with high accuracy is critical. Reverse engineering tools are usually used to regenerate the model of existing parts [41]. Reverse engineering digitizers can be classified into two categories: contact and non-contact. Coordinate measuring machines such as touch-trigger probes can be integrated into CNC machines easily and yields precise results, but the measuring process is slow [42]. Non-contact scanning equipment includes CCD cameras, laser scanners, and structured-light 3D scanner, etc. The capturing process using a non-contact approach is fast, but the result can be easily affected by color and surface finish [43]. The selection of an appropriate scanning approach is mainly determined by the measurement speed, resolution, cost, and output data type.

Nowadays, there is a trend of transition of scanning method from touch probe to more efficient approaches such as laser scanning system or topometric digitizing system. For example, Gao et al. used GOM ATOS II optical scanner to reconstruct the model of damaged aircraft engine blades [44]. The scanner uses CCD cameras to capture the profile of damaged parts with high accuracy (15—30 μm). The blade is fixed on a rotary table and the scanner is positioned in front of the blade. Recognizing targets are put on the fixture to realize multiple scans in different orientations. This setup can also be used to analyze repaired part quality by capturing the repaired part and then comparing it with the nominal model. Yilmaz also used a similar scanner (GOM ATOS II-400) to obtain data from the blade surface [45]. Wang et al. collected 3D point cloud data of aero-engine blades by a Leica laser tracker in association with a T-Scan [46]. In Ref. [47], blade with damaged edge was scanned using CCD cameras.

Reconstructing the worn geometry is essential in the remanufacturing process. That is because, similar to desktop 3D printer where STL model of a part is needed to generate a tool path, for component repair, the geometry of damage is also crucial to generate a tool path so that materials can be precisely built up at the desired region. It is possible to manually generate tool paths for very sample cases. However, for complex geometries, it is important to acquire the worn geometry automatically. Many automated approaches have been demonstrated to generate the worn volume as discussed in Refs. [42,44,48,49].

Once the worn geometry is defined, repair experiments can be conducted to deposit materials in the damaged zone. Many factors should be considered in planning experiments. For example, the tool path should be carefully designed so that low residual stress will be induced to prevent the base part from distortion. Because the part after AM may have a rough surface, subtractive machining may be needed to smooth the critical surface. The finally repaired part then undergoes geometry inspection and mechanical testing before reuse.

### 1.4.3 Fabrication of advanced materials using additive manufacturing

One objective of AM is to produce parts with exceptional properties and thus results in high performance. Therefore, the development of new and advanced materials is particularly interested in the evolution of AM. In general, advanced materials refer to all materials that have superior performance in applications. Such materials include metals that can work in extreme conditions (high temperature, high pressure, corrosion environment, radiation atmosphere, etc.), composites, high-performance polymers, and ceramics. Unlike traditional subtractive machining techniques where the machinability of the newly developed materials must be considered, in AM, since materials are joined together layer by layer, no tools are necessarily needed. Therefore, AM gives much more freedom for the design of new materials.

Most AM processes such as DED and PBF use powders as the feedstock. Hence, the properties of the printed part are determined by the elemental composition of these powder materials. In general, there are two available options in the current powder processing techniques, pre-alloyed powder and pre-mixed powder. Pre-alloyed powder indicates the powder itself is already alloyed with different materials, and therefore, can provide precise and reliable chemical composition, which is desired for AM. However, there are issues associated with pre-alloyed powder. For example, pre-alloyed powder has fixed elemental compositions; therefore, it might be difficult to find available powder with desired chemical compositions. As a result, the design of new materials requires customized powders from powder supplier, which can be time-consuming and expensive. Owing to such issues, pre-mixed powder can offer more flexible ways in the design of new advanced materials, although separation occurred due to the weight difference of powders [50]. In making pre-mixed powder, several kinds of elemental powders (e.g., Fe, Cr, Ni) are blended with a specific composition ratio. The preparation of pre-mixed powder is quite easy and in a fast manner, so it is used in many literatures. For example, a new Fe–Cr–Ni alloy was designed with different elemental mixing ratios (Fe–16Cr–8Ni, Fe–14Cr–16Ni, Fe–12Cr–23Ni, Fe–9Cr–28Ni) in Ref. [51] using pre-mixed powder. The change of chemical composition results in a variation of microstructures as shown in Figs. 1.11 and 1.12. The microstructure of deposits with Ni content of 8 and 16 wt% contains lathy morphology with minor skeletal morphology. By increasing Ni content to 23 and 28 wt%, the microstructure is dominated by cellular and dendritic structures. The fabricated materials also show the transition of phases by change compositions. It shows a phase transition from ferrite to austenite by adding Cr and decreasing Ni. Because of variation in microstructure and phases, the fabricated material can be tuned by adjusting chemical composition to meet the required performance.

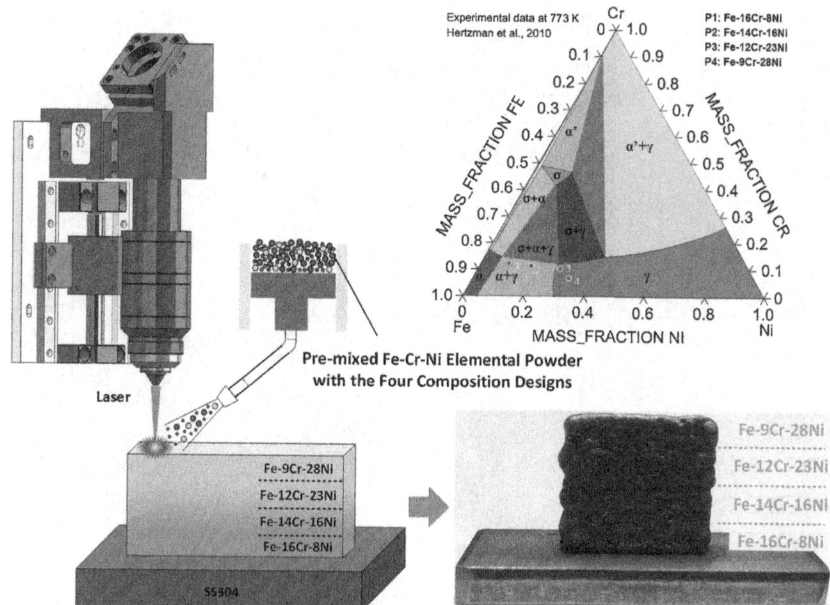


**Figure 1.11** Four Fe–Cr–Ni composition designs and schematic illustration of the fabricating sample. *(Reproduced with kind permission of Elsevier from W. Li, L. Yan, X. Chen, J. Zhang, X. Zhang, F. Liou, Directed energy depositing a new Fe-Cr-Ni alloy with gradually changing composition with elemental powder mixes and particle size' effect in fabrication process, J. Mater. Process. Technol. 255 (2018) 96–104.)*

The easily mixing of elemental powders also contributes to the fabrication of high-entropy alloys (HEAs). HEA indicates the mixing of several principal elements in the alloy system that leads to a substantial change in entropy. Some HEAs such as AlCoCrFeNi have high hardness and excellent wear resistance and therefore show good application where such properties are required [52]. AM is an ideal process to fabricate HEAs, and the powder feedstock is easily accessible and the process requires no tooling. Besides, AM can realize the fabrication of bulk metallic glass [53] and transparent glass [53]. In the nuclear field, AM has been used to fabricate nuclear fuels [54]. AM even enables the fabrication of organs and tissues using cells or other biomaterials as the feedstock [55]. Therefore, it can be seen that AM has already significantly changed and will continue moving the current concepts to practical application.

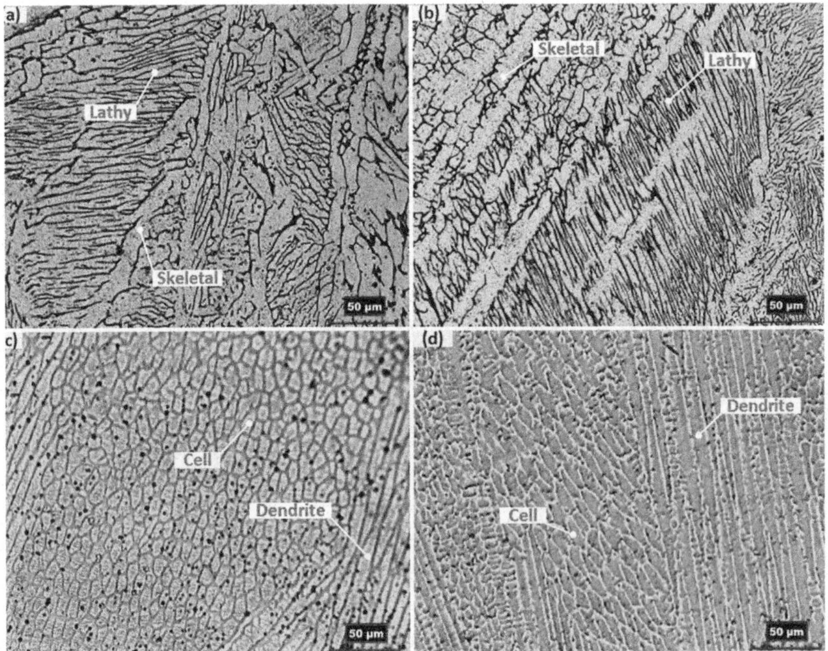


**Figure 1.12** Optical metallography of four composition regions: (A) Fe—16Cr—8Ni; (B) Fe—14Cr—16Ni; (C) Fe—12Cr—23Ni; (D) Fe—9Cr—28Ni. *(Reproduced with kind permission of Elsevier from W. Li, L. Yan, X. Chen, J. Zhang, X. Zhang, F. Liou, Directed energy depositing a new Fe-Cr-Ni alloy with gradually changing composition with elemental powder mixes and particle size' effect in fabrication process, J. Mater. Process. Technol. 255 (2018) 96—104.)*

## 1.4.4 Fabrication of smart structures with embedded sensors

AM process can be adopted to embed sensors in components to fabricate smart structures without compromising the part's complex geometries. An embedded sensor system contains sensing elements (such as thermocouple) and data processing capabilities embedded in a component. Sensors embedded can be used for in-situ monitoring of critical parameters such as temperature, pressure, and strain/stress and providing real-time feedback to secure critical infrastructure. AM is expanding the applicability of embedded sensors into areas that are difficult to measure by traditional measuring equipment. Moreover, wireless sensors embedded in multiple components can send signals to the central processor and thus providing data collection and control capabilities, advancing the industrial 4.0. Research has demonstrated embedding sensors for sensing capabilities using AM.

For example, researchers demonstrated embedding piezoelectric sensors in Ti–6Al–4V metallic component using EBM AM process [56]. The embedded sensor could be used to measure force/pressure and hence provide in-situ monitoring of components under harsh conditions. Optical fiber can also be embedded in the AM-fabricated component. For example, optical fiber sensors were inserted into ceramic components by extrusion-based AM process as shown in Fig. 1.13. The optical fiber can be placed in the deposits during printing and offer good adhesion to the part without damaging the mechanical properties of the parts [57].

## 1.5 Comparison of additive manufacturing and subtractive manufacturing

The key difference between AM and subtractive manufacturing (CNC machining) is that AM is a material building-up process while subtractive manufacturing is a material removal process. This means in AM process, materials are usually in the form of powder or wire and such materials are joined together to get the part. In CNC machining, a block of material is machined to cut away certain material to create the final object. Such manufacturing principle distinguishes AM and CNC machining and results in different aspects in terms of material and tool, speed, design freedom, accuracy, and cost.

### 1.5.1 Material and tool

The materials that can be machined by CNC machining is mainly determined by the selection of tools. CNC machining is usually used to fabricate

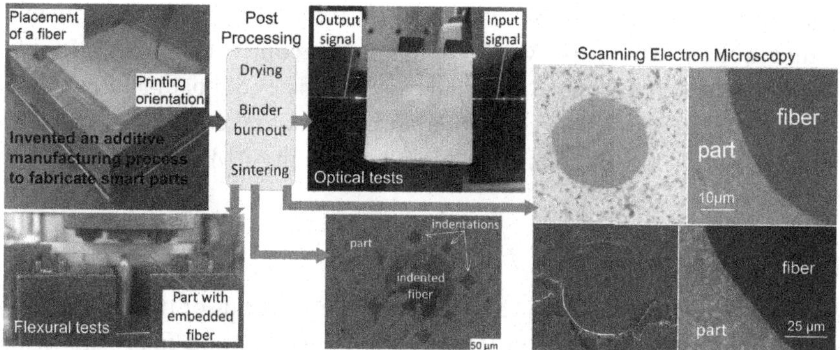


**Figure 1.13** Embedding optical fiber in ceramic components using additive manufacturing. *(Reproduced with kind permission of Elsevier from A. Ghazanfari, W. Li, M.C. Leu, Y. Zhuang, J. Huang, Advanced ceramic components with embedded sapphire optical fiber sensors for high temperature applications. Mater. Des. 112 (2016) 197–206.)*

metallic parts with superior surface finish and geometric accuracy. Common materials such as steel, aluminum, copper, and titanium can be machined by selecting proper machining tools. However, many advanced materials with very high hardness and wear resistance such as nickel-based alloys, cobalt-based alloys, and HEAs are difficult for CNC machining. These materials are often adopted in aerospace, automotive, and nuclear applications; therefore, the feasibility of processing such materials is critical. In contrast, no cutting tool is used in AM; therefore, AM is capable of processing materials that are difficult or impossible to machine by CNC machining. Desktop 3D printers are usually restricted to a few materials such as thermoplastics or resins. Industrial AM machines, however, can print a variety of metals, alloys, and composites. Consideration of tool selection in CNC machining is now transformed into fabrication feasibility by considering thermal properties in AM. Although AM can process many advanced materials, due to the local rapid heating and cooling cycles during printing, the fabricated component may have a high level of residual stress that distorts the part. Besides, the microstructure of materials processed by AM may have anisotropy and therefore results in anisotropy in mechanical properties (higher strength along with specific orientation).

### 1.5.2 Speed

High fabricating speed is always desired for various applications. As discussed earlier, there are many processes based on AM technique, and their manufacturing speed is different. SLA can print a part in hours due to very fine layer thickness. Binder jetting, on the other hand, can fabricate multiple parts at the same time and therefore, the time for each part is less. Since AM is a layer-based fabricating process and material has to be added together layer by layer, CNC machining usually takes less time to manufacture a part. For parts with complicated geometries, a considerable amount of time is necessary to generate appropriate tool path and machine setup for CNC machining, and therefore, it may take a longer time. In contrast, complexity in geometry does not affect the AM process a lot since the part is sliced into layers.

### 1.5.3 Design freedom

A significant advantage of AM is it unlocks new worlds of design freedom and allows engineers to create geometries, structures, and final parts with complexity impossible to create through CNC machining. For example, AM enables the optimization of topology with complex lattice structures or cellular structures to reduce the weight of a part. These structures are beneficial for achieving high strength at low weights. Such complex structures are very difficult and time-consuming for CNC machining. AM can also realize the fabrication of components with multimaterial to take advantage of each material. CNC machining is more suitable to deal with homogeneous material. Besides, with AM, many subcomponents can be integrated into a single component without assembly, reducing weight and manufacturing cost and benefiting system reliability.

### 1.5.4 Accuracy

In general, CNC machining gives higher accuracy than AM. AM is a layer-upon-layer manufacturing process. Therefore, the layers can be viewed and this results in a "step" feature. Because of such appearance, the printed object has rough surface roughness and compromised accuracy. One solution to improve the accuracy and surface finish is using thinner layer thickness. However, this will result in a significant increase in printing time. CNC machining is a subtractive manufacturing process that cut off materials from a substrate. The rough and finishing cutting processes can provide excellent accuracy. This is the reason that in repairing expensive components, materials deposited by AM will undergo machining to obtain exact dimensions and surface finish.

### 1.5.5 Cost

In terms of material waste, there is less waste in AM as this technology only requires the material needed for creating the object. In many AM-based processes such as PBF and DED, waste powder can be easily recycled. In contrast, in CNC machining, a large amount of material is removed, and such waste is often cannot be recycled. Such difference makes AM more efficient to produce components. However, CNC machining usually is capable of removing materials much faster than AM (additively manufactured components may also require post-processing to clean and smooth surfaces and edges); therefore, CNC is desired if a large number of products are wanted.

## 1.6 Hybrid manufacturing

Hybrid manufacturing is a process that combines several processes in a single machine. These processes include joining, dividing, subtractive, transformative, additive, etc. Most frequently, hybrid manufacturing refers to the combination of AM and CNC machining in a single machining system typically for processing metal. This integration enables each process to work together on the same machine and the same component. Among many of the AM-based processes, DED is most often hybridized because its powder-fed feature makes it easier to integrate with CNC machining. In a hybrid manufacturing process, a near-net-shape component can be printed by AM, and its subsequent machining can be realized by a subtractive process. Besides, it can work alternately in the way that CNC machining can machine a blank, and AM can be used to add needed features onto the part, and then CNC machining is used to machine such 3D printed features for surface improvement. Hybrid manufacturing is widely used for repairing damaged components to extend their service time. A multi-axis laser deposition process has the capability of depositing materials onto the damaged parts to restore them to the near net-shaped parts. The repaired parts after the AM process tend to have a poor surface finish and low dimensional accuracy. The multi-axis hybrid manufacturing process can restore a damaged part to the machined quality without post-processing. This process could also be a robotic deposition process, which can repair damaged structures or machines in the field.

Many studies have been conducted to further develop hybrid manufacturing process. For example, researchers developed a 6-DOF (degree of freedom) hybrid additive-subtractive manufacturing process by combining a robot with multiple changeable heads [58]. For the objective of repair existing damaged parts, a hybrid process that integrated reverse engineering, pre-repair machining, DED, and post-repair machining was proposed in Ref. [59]. The configuration enables the repair of metallic parts in a single setup. In addition to CNC machining, studies have been done to integrate wire and arc additive manufacturing (WAAM) with micro-rolling [60]. The proposed hybrid process can be used to eliminate material anisotropy in AM materials and also achieve outstanding mechanical properties.

## 1.7 Challenges and limitations of current additive manufacturing

With the rapid growth of AM, this technology has allowed expedited component manufacturing with revolutionized design and application-tailoring capabilities. As mentioned in this chapter, AM-based techniques have been used to achieve products with increased customization, improved design, multifunctionality, and performance. However, despite the significant changes AM has already affected in many manufacturing industries, there are challenges and limitations to its wide application.

AM is currently used to fabricate components and products with small quantities and relatively high cost compared with traditional manufacturing processes. Because component design has higher levels of freedom for AM, and the design can be easily modified, it is more suitable for manufacturing customized components, for example, personalized medical devices. It also shows great potential in rapid prototyping and remanufacturing. However, most AM processes are not suitable and cost-effective for mass production [61]. For building plastic components, the AM techniques such as SLA and FDM are slower compared with injection molding. In terms of metallic parts, AM-based techniques like DED and SLM are also slower than traditional machining. Although AM eliminates tooling which is required in traditional manufacturing, it is still time-consuming and inefficient for mass production. Besides, such AM processes require post-processing of printed products, which takes a lengthy time. For example, critical surfaces of SLM-fabricated components require post-machining to achieve high tolerance. Besides, AM usually requires specified raw material (e.g., metal powder for DED, SLM and Binder Jetting, photosensitive resin for SLA). These materials are usually expensive which makes the process not cost-effective. The future generation of AM lies in its capacity to achieve high-speed manufacturing capability and lower cost for expanding its industrial application.

Defects, anisotropy in properties, and excess residual stress are common issues within AM-fabricated metallic components due to inappropriate selection of processing parameters and its layer-upon-layer fabrication nature. For example, lack of fusion and gas porosities may exist in DED parts that damage the performance. Most AM parts have structure and mechanical anisotropies resulted from the layer-by-layer laying-down process [62]. Since DED is a fusion-based process in which materials are locally melted and solidified with the aid of heat source, the heat source can develop a large thermal gradient near the beam spot. Due to its layer-by-layer deposition fashion, materials will also experience repeated rapid heating and cooling. These facts can give rise to a significant level of residual stress. These issues need to be addressed to deliver satisfactory and consistent products. The good news is that these drawbacks are being studied and surmounted by R&D projects. With the aid of strategies such as in-situ monitoring and feedback control, future AM is believed to achieve parts with repeatable and enhanced performance.

In addition to the abovementioned limitations, there are many other challenges AM faces. For example, AM is usually less repeatable, which means it can still lead to differences in producing the same part although the settings are the same. This can impact the quality of the final product. Besides, the availability of suitable materials remains one of the barriers for AM. Although a broad range of materials is available for AM, including metals, ceramics, polymers, and composites, material diversity is still limited. AM also needs a new set of skills such as design for AM, heat-transfer, topology optimization, powder handling, post-processing knowledge. Education and training are significant for the widespread adoption of AM. Looking ahead, AM has seen rapid development in recent years, there are still challenges and limitations AM faces. As AM mature and these limitation addresses, AM will exploit its unique capabilities for product design and manufacture.

## References


[1] H. Kodama, Automatic method for fabricating a three-dimensional plastic model with photo-hardening polymer, Rev. Sci. Instrum. 52 (1981) 1770–1773.

[2] H. Kodama, A scheme for three-dimensional display by automatic fabrication of three-dimensional model, IEICE Trans. Electron. (Jpn. Ed) (1981). J64-C:237–41.

[3] A. Ciraud, Process and Device for the Manufacture of Any Objects Desired from Any Meltable Material, FRG Discl Publ, 1972, p. 2263777.

[4] A. Medellin, W. Du, G. Miao, J. Zou, Z. Pei, C. Ma, Vat photopolymerization 3D printing of nanocomposites: a literature review, J. Micro Nano-Manuf. 7 (2019).

[5] N.A. Chartrain, C.B. Williams, A.R. Whittington, A review on fabricating tissue scaffolds using vat photopolymerization, Acta Biomater. 74 (2018) 90–111.

[6] Y.L. Yap, C. Wang, S.L. Sing, V. Dikshit, W.Y. Yeong, J. Wei, Material jetting additive manufacturing: an experimental study using designed metrological benchmarks, Precis. Eng. 50 (2017) 275–285.

[7] D.V. Kaweesa, D.R. Spillane, N.A. Meisel, Investigating the impact of functionally graded materials on fatigue life of material jetted specimens, in: Proc 28th Annu Int Solid Free Fabr Symp Austin, Texas, USA, 2017, pp. 578–592.

[8] H. Bikas, A.K. Lianos, P. Stavropoulos, A design framework for additive manufacturing, Int. J. Adv. Manuf. Technol. 103 (2019) 3769–3783.

[9] F. Ning, W. Cong, J. Qiu, J. Wei, S. Wang, Additive manufacturing of carbon fiber reinforced thermoplastic composites using fused deposition modeling, Compos. B Eng. 80 (2015) 369–378.

[10] O.S. Carneiro, A.F. Silva, R. Gomes, Fused deposition modeling with polypropylene, Mater. Des. 83 (2015) 768–776.

[11] T. Do, P. Kwon, C.S. Shin, Process development toward full-density stainless steel parts with binder jetting printing, Int. J. Mach. Tool Manufact. 121 (2017) 50–60.

[12] E. Louvis, P. Fox, C.J. Sutcliffe, Selective laser melting of aluminium components, J. Mater. Process. Technol. 211 (2011) 275–284.
[13] C.Y. Yap, C.K. Chua, Z.L. Dong, Z.H. Liu, D.Q. Zhang, L.E. Loh, et al., Review of selective laser melting: materials and applications, Appl. Phys. Rev. 2 (2015) 41101.
[14] M. Schmid, A. Amado, K. Wegener, Polymer powders for selective laser sintering (SLS), AIP Conf. Proc. 1664 (2015) 160009.
[15] C. Körner, Additive manufacturing of metallic components by selective electron beam melting — a review, Int. Mater. Rev. 61 (2016) 361–377.
[16] S. Yi, F. Liu, J. Zhang, S. Xiong, Study of the key technologies of LOM for functional metal parts, J. Mater. Process. Technol. 150 (2004) 175–181.
[17] I. Gibson, D.W. Rosen, B. Stucker, Sheet lamination processes, in: Addit. Manuf. Technol. Rapid Prototyp. to Direct Digit. Manuf, Springer US, Boston, MA, 2010, pp. 223–252.
[18] D.D. Gu, W. Meiners, K. Wissenbach, R. Poprawe, Laser additive manufacturing of metallic components: materials, processes and mechanisms, Int. Mater. Rev. 57 (2012) 133–164.
[19] X. Zhang, W. Cui, W. Li, F. Liou, Effects of tool path in remanufacturing cylindrical components by laser metal deposition, Int. J. Adv. Manuf. Technol. 100 (2019) 1607–1617.
[20] X. Zhang, Y. Chen, F. Liou, Fabrication of SS316L-IN625 functionally graded materials by powder-fed directed energy deposition, Sci. Technol. Weld. Join. 00 (2019).
[21] J.I. Kim, W.-J. Kim, D.J. Choi, J.Y. Park, W.-S. Ryu, Design of a C/SiC functionally graded coating for the oxidation protection of C/C composites, Carbon N Y 43 (2005) 1749–1757.
[22] M. Rezapoor, M. Razavi, M. Zakeri, M.R. Rahimipour, L. Nikzad, Fabrication of functionally graded Fe-TiC wear resistant coating on CK45 steel substrate by plasma spray and evaluation of mechanical properties, Ceram. Int. 44 (2018) 22378–22386.
[23] M. Ivosevic, R. Knight, S.R. Kalidindi, G.R. Palmese, J.K. Sutter, Solid particle erosion resistance of thermally sprayed functionally graded coatings for polymer matrix composites, Surf. Coating. Technol. 200 (2006) 5145–5151.
[24] K. Feng, H. Chen, J. Xiong, Z. Guo, Investigation on diffusion bonding of functionally graded WC–Co/Ni composite and stainless steel, Mater. Des. 46 (2013) 622–626.
[25] E. Askari, M. Mehrali, I.H.S.C. Metselaar, N.A. Kadri, M.M. Rahman, Fabrication and mechanical properties of $Al_2O_3$/SiC/$ZrO_2$ functionally graded material by electrophoretic deposition, J. Mech. Behav. Biomed. Mater. 12 (2012) 144–150.
[26] Z. Zhang, X. Shen, C. Zhang, S. Wei, S. Lee, F. Wang, A new rapid route to in-situ synthesize TiB–Ti system functionally graded materials using spark plasma sintering method, Mater. Sci. Eng. A 565 (2013) 326–332.
[27] V.A. Popovich, E.V. Borisov, A.A. Popovich, V.S. Sufiiarov, D.V. Masaylo, L. Alzina, Functionally graded Inconel 718 processed by additive manufacturing: crystallographic texture, anisotropy of microstructure and mechanical properties, Mater. Des. 114 (2017) 441–449.
[28] W. Liu, J.N. DuPont, Fabrication of functionally graded TiC/Ti composites by laser engineered net shaping, Scripta Mater. 48 (2003) 1337–1342.
[29] L. Thijs, F. Verhaeghe, T. Craeghs, J Van Humbeeck, J.-P. Kruth, A study of the microstructural evolution during selective laser melting of Ti–6Al–4V, Acta Mater. 58 (2010) 3303–3312.
[30] M. Zietala, T. Durejko, M. Polański, I. Kunce, T. Płociński, W. Zieliński, et al., The microstructure, mechanical properties and corrosion resistance of 316L stainless steel fabricated using laser engineered net shaping, Mater. Sci. Eng. A 677 (2016) 1–10.

[31] B.E. Carroll, R.A. Otis, J. Paul, J. Suh, R.P. Dillon, A.A. Shapiro, et al., Functionally graded material of 304L stainless steel and inconel 625 fabricated by directed energy deposition: characterization and thermodynamic modeling, Acta Mater. 108 (2016) 46–54.

[32] L. Yan, Y. Chen, F. Liou, Additive manufacturing of functionally graded metallic materials using laser metal deposition, Addit. Manuf. 31 (2020) 100901.

[33] W. Li, S. Karnati, C. Kriewall, F. Liou, J. Newkirk, K.M.B. Taminger, et al., Fabrication and characterization of a functionally graded material from Ti-6Al-4V to SS316 by laser metal deposition, Addit. Manuf. 14 (2017) 95–104.

[34] E.R. Denlinger, J.C. Heigel, P. Michaleris, T.A. Palmer, Effect of inter-layer dwell time on distortion and residual stress in additive manufacturing of titanium and nickel alloys, J. Mater. Process. Technol. 215 (2015) 123–131.

[35] A.H. Nickel, D.M. Barnett, F.B. Prinz, Thermal stresses and deposition patterns in layered manufacturing, Mater. Sci. Eng. A 317 (2001) 59–64.

[36] G. Sun, R. Zhou, J. Lu, J. Mazumder, Evaluation of defect density, microstructure, residual stress, elastic modulus, hardness and strength of laser-deposited AISI 4340 steel, Acta Mater. 84 (2015) 172–189.

[37] X. Zhang, W. Li, K.M. Adkison, F. Liou, Damage reconstruction from tri-dexel data for laser-aided repairing of metallic components, Int. J. Adv. Manuf. Technol. 96 (2018) 3377–3390.

[38] X. Zhang, T. Pan, W. Li, F. Liou, Experimental characterization of a direct metal deposited cobalt-based alloy on tool steel for component repair, JOM 71 (2019) 946–955.

[39] Z. Jia, Y. Liu, J. Li, L.-J. Liu, H. Li, Crack growth behavior at thermal fatigue of H13 tool steel processed by laser surface melting, Int. J. Fatig. 78 (2015) 61–71.

[40] X. Zhang, W. Li, W. Cui, F. Liou, Modeling of worn surface geometry for engine blade repair using Laser-aided Direct Metal Deposition process, Manuf. Lett. 15 (2018) 1–4.

[41] X. Zhang, W. Li, F. Liou, Damage detection and reconstruction algorithm in repairing compressor blade by direct metal deposition, Int. J. Adv. Manuf. Technol. 95 (2018) 2393–2404.

[42] J. Zheng, Z. Li, X. Chen, Worn area modeling for automating the repair of turbine blades, Int. J. Adv. Manuf. Technol. 29 (2006) 1062–1067.

[43] X. Zhang, W. Li, X. Chen, W. Cui, F. Liou, Evaluation of component repair using direct metal deposition from scanned data, Int. J. Adv. Manuf. Technol. 95 (2018) 3335–3348.

[44] J. Gao, X. Chen, O. Yilmaz, N. Gindy, An integrated adaptive repair solution for complex aerospace components through geometry reconstruction, Int. J. Adv. Manuf. Technol. 36 (2008) 1170–1179.

[45] O. Yilmaz, N. Gindy, J. Gao, A repair and overhaul methodology for aeroengine components, Robot. Comput. Integrated Manuf. 26 (2010) 190–201.

[46] T. Wang, Y.L. Liu, L.W. Wang, H. Wang, J. Tang, Digitally reverse modeling for the repair of blades in aero-engines, in: Funct. Manuf. Mech. Dyn. II, Trans Tech Publications, 2012, pp. 258–263.

[47] J. Li, F. Yao, Y. Liu, Y. Wu, Reconstruction of broken blade geometry model based on reverse engineering, in: 2010 Third Int. Conf. Intell. Networks Intell. Syst, 2010, pp. 680–682.

[48] J.M. Wilson, C. Piya, Y.C. Shin, F. Zhao, K. Ramani, Remanufacturing of turbine blades by laser direct deposition with its energy and environmental impact analysis, J. Clean. Prod. 80 (2014) 170–178.

[49] J. He, L. Li, J. Li, Research of key-technique on automatic repair system of plane blade welding, in: 2011 Int. Conf. Control. Autom. Syst. Eng, 2011, pp. 1–4.

[50] W. Li, J. Zhang, X. Zhang, F. Liou, Effect of optimizing particle size on directed energy deposition of Functionally Graded Material with blown Pre-Mixed Multi-Powder, Manuf. Lett. 13 (2017) 39–43.

[51] W. Li, L. Yan, X. Chen, J. Zhang, X. Zhang, F. Liou, Directed energy depositing a new Fe-Cr-Ni alloy with gradually changing composition with elemental powder mixes and particle size' effect in fabrication process, J. Mater. Process. Technol. 255 (2018) 96–104.

[52] W. Cui, S. Karnati, X. Zhang, E. Burns, F. Liou, Fabrication of AlCoCrFeNi high-entropy alloy coating on an AISI 304 substrate via a $CoFe_2Ni$ intermediate layer, Entropy 21 (2018).

[53] H.Y. Jung, S.J. Choi, K.G. Prashanth, M. Stoica, S. Scudino, S. Yi, et al., Fabrication of Fe-based bulk metallic glass by selective laser melting: a parameter study, Mater. Des. 86 (2015) 703–708.

[54] A. Bergeron, J.B. Crigger, Early progress on additive manufacturing of nuclear fuel materials, J. Nucl. Mater. 508 (2018) 344–347.

[55] F.P.W. Melchels, M.A.N. Domingos, T.J. Klein, J. Malda, P.J. Bartolo, D.W. Hutmacher, Additive manufacturing of tissues and organs, Prog. Polym. Sci. 37 (2012) 1079–1104.

[56] M.S. Hossain, J.A. Gonzalez, R.M. Hernandez, M.A.I. Shuvo, J. Mireles, A. Choudhuri, et al., Fabrication of smart parts using powder bed fusion additive manufacturing technology, Addit. Manuf. 10 (2016) 58–66.

[57] A. Ghazanfari, W. Li, M.C. Leu, Y. Zhuang, J. Huang, Advanced ceramic components with embedded sapphire optical fiber sensors for high temperature applications, Mater. Des. 112 (2016) 197–206.

[58] L. Li, A. Haghighi, Y. Yang, A novel 6-axis hybrid additive-subtractive manufacturing process: design and case studies, J. Manuf. Process. 33 (2018) 150–160.

[59] X. Zhang, W. Cui, W. Li, F. Liou, A hybrid process integrating reverse engineering, pre-repair processing, additive manufacturing, and material testing for component remanufacturing, Materials 12 (2019).

[60] Y. Fu, H. Zhang, G. Wang, H. Wang, Investigation of mechanical properties for hybrid deposition and micro-rolling of bainite steel, J. Mater. Process. Technol. 250 (2017) 220–227.

[61] P. Reeves, C. Tuck, R. Hague, Additive manufacturing for mass customization, in: F.S. Fogliatto, G.J.C. da Silveira (Eds.), Mass Cust. Eng. Manag. Glob. Oper., Springer London, London, 2011, pp. 275–289.

[62] S. Guessasma, W. Zhang, J. Zhu, S. Belhabib, H. Nouri, Challenges of additive manufacturing technologies from an optimisation perspective, Int. J. Simul. Multidiscip. Des. Optim. 6 (2015). A9.

CHAPTER 2

# Introduction to powder bed fusion of polymers

**Andreas Wegner**
AM POLYMERS GmbH, Willich; Chair for Manufacturing Technology, University of Duisburg-Essen, Germany

## 2.1 Introduction

Additive manufacturing processes are becoming increasingly important also for the manufacture of series components in small series. In the field of polymer processing, this potential is particularly attributed to powder bed fusion processes, as demonstrated by numerous standardization activities for these processes [1–7]. These processes standardized according to DIN EN ISO/ASTM 52900 include in particular laser sintering (LS) as the first process developed in this category, but also newer processes such as mask sintering, high-speed sintering (HSS), or Multi Jet Fusion. As examples for the successful application of these methods or LS as series production methods, some examples can be given at present: applications such as the serial production of aircraft components at Boeing [8–10] or Northrop Grumman [11]; the production of components for the aerospace industry [12]; the hundredfold production of hearing aid shells [13,14]; the production of complex gripper systems with integrated bellows constructions [15–17] as well as the production of small series in automation technology [18] or in special machine construction [17]. Studies on the economic efficiency and energy efficiency of LS compared with injection molding prove the special suitability of the process especially for the production of small series up to a certain limit quantity, which can be in the upper four-figure or low five-figure ranges for small parts and in the range of several hundred parts for larger components [19–22]. But even larger series are possible today, such as tens of thousands of eyeglass frames [23] or even the production of millions of mascara brushes [24].

## 2.2 Processes, machines, technologies

The first approaches to the development of LS were taken by Ciraud in 1972 and Housholder in 1979 [25,26]. However, the development and

*Additive Manufacturing*
ISBN 978-0-12-818411-0
https://doi.org/10.1016/B978-0-12-818411-0.00011-2

implementation of LS of plastics, which is known and widespread today, is based on the work of Carl Deckard at the University of Texas in Austin. In his master's thesis and subsequently deepened in his doctoral thesis, Deckard developed the first LS system for the layer-by-layer sintering of meltable polymer powders by laser radiation and carried out fundamental investigations for black-colored acrylonitrile butadiene styrene (ABS) powder using Nd:YAG laser radiation [27,28]. 1989/90 DTM's first LS system was developed on this basis in the form of the Sinterstation 125. In 1992, the Sinterstation 2000 was commercialized as the world's first LS system. In 1994, the first system of the German manufacturer EOS followed. Since then, the systems have been further developed several times and supplemented by further variants and sizes, including special systems for high-temperature thermoplastics. In 2001, DTM's business was taken over by the company 3D Systems, which since then has been optimizing it [29–31].

The usual LS process is divided into three process phases: preheating phase, build phase, and cooling down phase [32,33]. For optimized process conditions, the powder material is preheated in the first process phase close to the melting point or up to the glass transition temperature of the polymer material [29]. Heating systems installed above the building envelope or attached to the building envelope are used to regulate the heat supply or heat dissipation during the process [34]. At the same time, the process chamber is heated to process temperature and usually inerted by flushing with nitrogen. This process serves to prevent powder explosion or material oxidation and to reduce process influences due to inhomogeneous preheating conditions [31,34,35]. In contrast to conventional sintering, the process works without pressure increase at standard ambient pressure or a slight overpressure [29].

The subsequent construction phase is subdivided into four single steps (Fig. 2.1).

In the first step, the building platform is lowered by a layer thickness, usually between 60 and 180 μm, by a lowering mechanism. Powder is then conveyed from a storage container and, depending on the manufacturer, spread via a roller (3D systems) or recoating system (EOS) to a flat powder layer on the building platform. Depending on the system, the storage containers are located either below or above the construction level. Accordingly, the powder is provided in front of the spreading system by lifting or pouring. After application, the powder layer is preheated to process temperature by an infrared radiant heater located above it until a

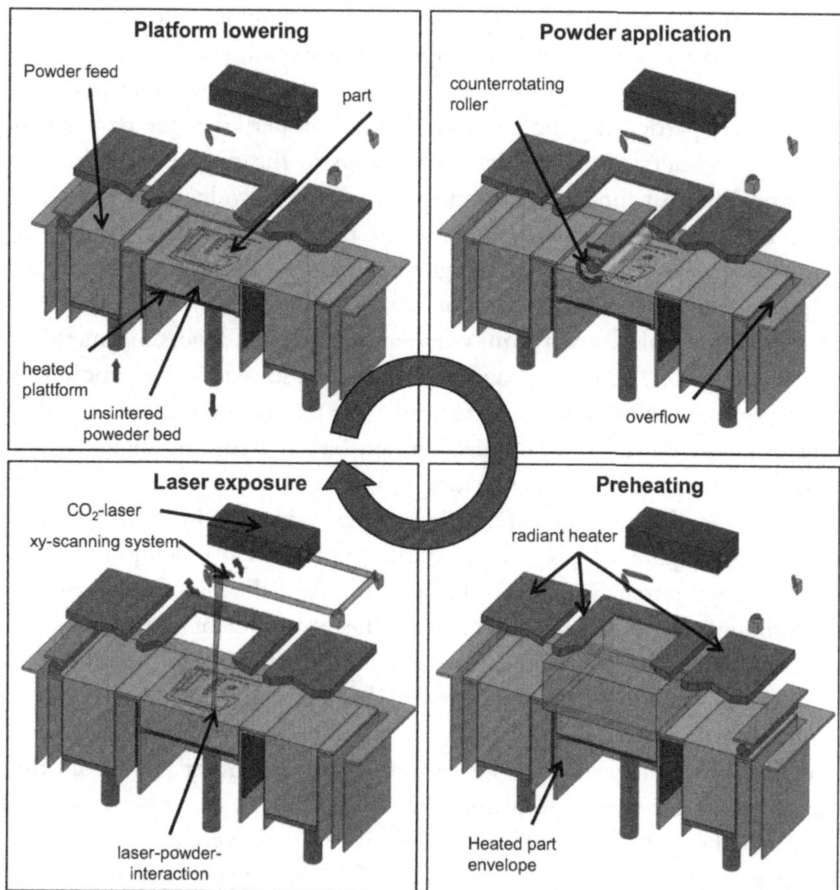


**Figure 2.1** Repetitive individual steps of the process cycle in LS of plastics.

stable temperature level with a preferably homogeneous temperature distribution is achieved. After reaching this temperature, the different part contours in the current layer are usually traced line by line by a focused $CO_2$ laser which is directed by an XY scanning system. Depending on the energy input and the material properties, the polymer is sintered or melted by the resulting rise in temperature, and the individual layers are joined together. These individual steps are repeated until all layers containing the parts have been generated. The preheated but unsintered powder bed compacted by the coating mechanism has a sufficient supporting function for fixing the parts, so that there is no need for further supporting structures as in other additive manufacturing processes [16]. After completion of the build phase and completion of all parts, a few millimeters of additional, empty powder layers are applied as thermal insulation [16,29–31,34].

The advantages of LS compared with alternative additive manufacturing processes are particularly the good mechanical properties of the parts when processing semicrystalline plastics. This results in the special suitability for the production of functional components and end products. At the same time, the process offers increased design freedom with maximum part complexity, fine structures, and integrated functions as well as the possibility of producing large-volume components. Compared with other additive processes, materials known from conventional plastics processing methods are processed. The absence of support structures also reduces the amount of postprocessing required, so that a ready-to-use part is immediately available after cooling. At the same time, this enables a dense arrangement of components in the building envelope [16,29–31].

However, the necessary preheating of the systems and the use of lasers for melting require a high energy consumption, which in addition to the cost-intensive purchase of a suitable machine and the significant material consumption has a negative impact on the economy of the process. In addition, laser-sintered components have a high surface roughness in contrast to components manufactured with other additive or conventional manufacturing processes. This results on the one hand from the high layer thicknesses and the associated step effect and on the other hand from the use of comparatively coarse plastic powder [16,30,31].

While the companies EOS and DTM/3D Systems were the only players on the market for a long time, the expiry of the basic patents has resulted in a significant increase in the number of suppliers on the market. The market was divided into industrial systems and desktop systems.

Further companies have joined EOS and DTM/3D Systems as classic providers without being able to list all the machine manufacturers here:

- 3D Systems (former DTM)
- Aspect
- EOS
- Farsoon
- Integra
- LSS
- Nxtfactory
- Prodways
- RICOH
- Shining

- Weirather
- XYZprinting

In addition, a number of manufacturers have developed desktop systems:

- Formlabs
- Natural Robotics
- Red Rock
- Sharebot
- Sinterit
- Sintratec

In addition to LS, alternative methods using a powder bed have been developed over the years. The first to be developed was the HSS, developed by Neil Hopkinson at Loughborough University. Although a heated powder bed with corresponding units is used for powder spreading as in LS, melting is carried out by an alternative process. The powder is fused using a high-energy radiant heater instead of a laser [36,37]. The selectivity in the layers or the contouring is achieved by the application of a solid or liquid mostly black absorbent in the areas of the parts cross section in the layer. This selectively increases the absorption in the powder bed. The areas with absorbents have a sufficient absorption to be melted with the high-energy radiant heater. On the other hand, the surrounding powder does not absorb sufficient energy for melting, so that it remains in the solid state. In contrast to LS, a cross section is not run line by line. A layer is exposed and melted during a translative movement of the radiant heater. The translation speed and the power of the radiant heater regulate the energy input. In contrast to LS, the layer time does not depend on the area of the scanned parts but on the translation speed and the corresponding size of the build envelope. Up to now, the process has been running without inerting. The companies Xaar3D and Voxeljet are each working on the development and commercialization of corresponding systems.

HP announced the Multi Jet Fusion process in 2015. The process procedure is essentially the same as HSS, but there are minor differences. While the direction of HSS powder spreading and the direction of movement of the radiant heater are the same, the directions of movement cross in Multi Jet Fusion. At the same time, a so-called detailing agent is used in addition to an absorber. This is intended to increase the edge sharpness of the parts.

Both processes have the advantage compared with LS of a constant layer time over the overall height. The materials used here correspond to the material portfolio for LS, although not all polymers available there have yet been qualified for the processes. The main influences here are in particular the power of the lamp used, the speed of traverse, and the amount of absorber introduced. In some cases, better mechanical characteristics and elongation at break can be achieved compared with LS. However, some studies also show a poorer performance of the parts [38–45].

Mask sintering was developed as a further variant. In contrast to the other powder bed processes, the light is exposed using a light mask, which is generated by applying a toner to a glass panel or other variants [46]. The advantage here is the exposure of the entire layer in one step, which has a high potential for increasing productivity. This technology has unfortunately disappeared from the market because commercialization was not possible.

## 2.3 Postprocessing and surface treatment

Since the powder bed fusion processes work close to the material melting temperature, the completed building cycle must first cool to at least the unpacking temperature. Usually the parts can only be removed after reaching the glass transition temperature. The cooling time is almost identical to the process time, whereby with increasing acceleration of the building process the cooling time partly becomes predominant. Cooling can take place within the machine or by removing the build envelope from the machine. This has the advantage of higher machine availability. For some materials, especially materials sensitive to oxidation such as polyamide 6, the cooling must always take place in an inert gas atmosphere.

After cooling, the parts can be removed from the powder cake during unpacking. Usually, this is done manually by wiping off or brushing off the surrounding powder. Powder that adheres directly to parts is usually separated and disposed of. The remaining used powder can usually be returned to the process and is refreshed with virgin powder [16]. In the context of process automation, more and more systems for automatic or semiautomatic powder removal are being developed, as demonstrated by systems from Pulvermeister or CNC-Speedpart. After the separation of powder and parts, the parts are usually cleaned by blasting with glass beads, and the remaining adhering powder is removed. This process can be carried out manually in a classic blasting cubicle or automated for several parts at the same time. In the second case, the blasting cubicles are equipped with a rotating drum. The components are then blasted via several blast nozzles.

There are now numerous suppliers of such systems. Among others, CNC-Speedpart and Dye-Mansion offer such systems. After blasting, parts made by powder bed fusion are usually ready for use.

However, various other processes can follow to finish the parts. Parts are often dyed by dip coloring in dye baths. There are numerous in-house developments by individual users as well as commercial solutions by companies such as Dye Mansion or Cipres/Thies. Furthermore, the components are often smoothened to reduce the high roughness of the component. For years, vibratory finishing has been used to reduce the roughness $R_Z$ to values in the range of 35–50 μm [47–50]. However, this surface quality is often still not sufficient for application. For this reason, chemical processes are increasingly being used to smooth the surface of parts. Here, considerably low roughness $R_Z$ in the range of 10–20 μm can be achieved [51]. In the meantime, various commercial systems such as those from Additive Manufacturing Technologies, LuxYours or Dye-Mansion, are also available for this purpose. The surfaces of the parts are chemically etched using a liquid or a vapor medium. This smoothens out peaks as well as the step effect. However, solvents and thus often hazardous chemicals are used for this purpose. In addition, further after-treatments such as painting steps, coating, or metallization may be carried out [16,29,30,34].

## 2.4 Materials and powder production techniques for powder bed fusion

In general, all materials with thermoplastic behavior are suitable for LS [29]. The starting point for material development was amorphous thermoplastics such as acrylonitrile butadiene styrene (ABS), polycarbonate (PC), or polystyrene (PS) [27,35,52]. These can be processed in principle and can be used to produce illustrative models or as pattern models for casting processes. However, dense parts and good mechanical properties are not achieved, or only by exceeding the degradation limit, which has been confirmed by numerous studies [52–56].

Therefore, most commercial amorphous LS materials have disappeared from the market. Only PS has remained for a long time in the product ranges of EOS and 3D Systems as a material for the production of patterns for investment casting. The materials have been replaced by semicrystalline thermoplastics, especially polyamide 11 and 12 (PA11 and PA12). Their use led to components with good mechanical properties and high density for the first time [56–59]. Due to their outstanding suitability for the LS process, they still dominate the market with a share of about 90% [60–62], as shown by the summary of commercial materials in Table 2.1. Thus, PA12 is sold by most suppliers both as unfilled and as material filled with glass, carbon fiber, or aluminum. Besides PA12, PA11 has the highest availability. In addition, various other unfilled standard thermoplastics as well as special materials for the LS process have been commercialized, but without achieving a large market share so far. In recent years, there have been significant additions to the material portfolio (Table 2.1). In the area of standard plastics and commodity plastics, various polypropylene (PP) materials in particular have been commercialized, while polyethylene (PE) is hardly available. In the area of engineering plastics, various new thermoplastic polyurethanes (TPU) materials and especially polyamide 6 (PA6) materials have been developed and commercialized. Other plastics such as polybutylene terephthalate (PBT) or polycarbonate (PC) are about to be launched on the market. Various materials have also been developed in the field of high-performance plastics. While initially only polyetherketone (PEK) was available, polyetherketoneketone (PEKK), polyetheretherketone (PEEK) polyphenylene sulfide (PPS), and polyetherimide (PEI) are now available or are about to be introduced to the market. However, their pure availability has not yet resulted in any significant change in the materials market. The materials market is still dominated by PA11 and especially PA12. This is partly due to worse processing of the materials in terms of powder application and part production, insufficient properties compared with injection molding and poor surfaces, excessively high prices above PA12, limited availability, or insufficient batch consistency. The differences between the materials offered are sometimes very large. However, general statements that alternative materials for powder bed fusion are not comparable with the processability of polyamide 12 are incorrect. In this case, the properties of the individual materials on the market have to be examined.

The reasons for this market development and the slow establishment of alternative materials can be found in the special requirements which materials for powder bed fusion have to meet and which PA12 fulfills particularly well [63,64]. Alscher and Schmachtenberg summarize these requirements on the basis of investigations on injection molding granulates as follows [63,65]:

1. Narrow melting range over a small temperature range for complete melting when the melting temperature is reached

**Table 2.1** Commercial laser sintering materials.

| | Machine manufacturers | | | | | | | | | | Material manufacturers | |
|---|---|---|---|---|---|---|---|---|---|---|---|---|
| | 3D-systems | EOS | Farsoon | HP | Prodways | Ricoh | Sintratec | Sinterit | Voxeljet | Weirather | AM POLYMERS | ALM |
| PA12 | X | X | X | X | X | X | X | X | X | X | | X |
| PA12 + Fillers | X | X | X | X | X | X | | | | X | | X |
| PA1212 | | | X | | | | | | | | | |
| PA11 | X | X | X | X | X | | | X | | | | X |
| PA11 + Fillers | | | | | | | | | | | | X |
| PA6 | | | X | | X | X | | | | | A | |
| PA6 + Fillers | | | | | | | | | | | | |
| PA612 | | | | | X | | | | | | | |
| PA613 | | | | | | | | | | | | |
| PBT | | | | | | A | | | | | X | |
| PP | | | | A | X | X | A | | A | | X | |
| PE | | | | | | | | | | | X | |
| PS | X | X | | | | | | | | | | |
| PC | | | | | | | | | | | | |
| PEI | | | | | | | | | | | | |
| PEK | | X | | | | | | | | | | |
| PPS | | | X | | | | | | | | | |
| PEKK | | A | | | | | | | | | | |
| PEKK + Fillers | | | | | | | | | | | | X |
| PEEK | | | | | | | | | | | | |
| TPE | X | X | | | | | | X | | | | X |

| | | | | | | | | | | | | |
|---|---|---|---|---|---|---|---|---|---|---|---|---|
| TPU | X | X | | X | X | | X | X | A | | X | X |

*A*, announced; *X*, available.

| Material manufacturers | | | | | | | | | | | | |
|---|---|---|---|---|---|---|---|---|---|---|---|---|
| Arkema | BASF | Covestro | Diamond plastic | DSM | Evonik | Lehmann & Voss | Lubrizol | Mistubishi chemicals | Sabic | Solvay | Victrex | Windform |
| X | | | | | X | X | | | | | | |
| | | | | | A | | | | | | | X |
| X | | | | | | | | | | | | |
| | X | | | | | | | | | X | | |
| | X | | | | | | | | | X | | |
| | | | | | X | | | | | | | |
| | | | | A | | | | X | | | | |
| | X | | X | | | X | | | | | | |
| | | | X | | | | | | | | | |
| | | | | | | | | | A | | | |
| | | | | | | | | | A | | | |
| X | | | | | | | | | | | | |
| | | | | | | | | | | | X | |
| | | | | | X | | | | | | | |
| | X | X | | | | X | X | | | | | |

2. High melting enthalpy to achieve a high edge sharpness through distinct separation of melt and powder
3. Large temperature window between melting peak and crystallization peak or between the melting and crystallization onset points
4. Low crystallization rates or amorphous solidification
5. Minimum surface tension value required to fuse particles
6. Preferably a large difference between the surface energy of the powder and the surface tension of the melt to achieve a high edge sharpness
7. Melt viscosity determines the speed of fusion
8. Narrow particle size distribution for homogeneous energy absorption

The poor suitability of amorphous thermoplastics can be deduced from these requirements. These have a softening range above the glass transition temperature. During preheating, it is therefore not possible to exceed this temperature without simultaneously sintering the entire powder bed. Accordingly, the material solidifies again immediately after exposure, so that, unlike with semicrystalline materials, complete densification cannot be achieved due to the lack of isothermal process control [53,55,66]. New developments of classical amorphous materials such as polycarbonate or PEI avoid such problems by modifying the material to achieve an almost semicrystalline behavior [67].

Based on the requirement profile described earlier, numerous investigations and attempts were carried out to qualify new materials for the LS process. Different processes are used to produce suitable powder materials. Commercial PA12 powder systems are usually produced from solution by a precipitation process or by bead polymerization and therefore have approximately rounded particle shapes [68–70]. This process is also occasionally used to generate test powders [71]. Alternative powder production processes are cryogenic grinding or wet grinding of plastics without [53,64,72–75] or with subsequent rounding [76–78]. This is used for material such as TPU, PP, or PA11. Another process for the production of commercial powders is the copolymerization of the aim product with a soluble second polymer, which is then put into solution to obtain the powder [70]. This very expensive process is used, for example, for some PP powders. Other experimental processes for powder production are stretching of extruded polymer melts until melt breakage due to Rayleigh turbulence [79,80], atomization with supercritical $CO_2$ [81,82], or cutting conversion of spun human-made fibers [80,83]. [70].

The aim of many scientific studies was to qualify standard plastics known from injection molding, such as PE [64,84,85], PP [64,86–91], or technical plastics, such as UHMWPE [92,93], POM [64,72,85,94], PBT [64,82,95,96], TPU [97], or PA6 [98–102], for the LS process. In contrast to the various unfilled materials, studies in the field of filled materials usually focus on the modification of PA12 to achieve better component properties [103–106] or the adjustment of certain properties such as the achievement of electrical conductivity [107–109]. However, most of the materials have not yet been commercialized, as a thorough understanding of the process for processing the materials is still lacking. The thermal and rheological properties of the material as well as the powder rheological properties are decisive for processing in the LS process. The influence of the individual properties on the process is discussed in the following on the basis of the known state of the art, taking into account the material requirements.

## 2.5 Parameter settings and influences

The LS process or the modified processes are determined by a large number of influencing variables which, individually or in interaction, influence the process result in terms of part quality and economic efficiency. In various studies on LS and the adjacent beam melting, numerous influencing variables on the process are mentioned [4,32,33,35,56,57,75,110–113]. A total of 276 influences affecting part quality and process economy have been identified [114]. The most important influences are described in more detail in the following.

### 2.5.1 Material properties

#### *2.5.1.1 Melting and crystallization behavior*

For the evaluation of the basic suitability of materials for the LS process, the melting or crystallization behavior, which is determined by differential scanning calorimetry (DSC), is usually used. First theoretical investigations to evaluate different thermoplastic materials with respect to their melting and crystallization behavior as well as their crystallization speed were carried out by Alscher in Ref. [63] on various injection molding granulates. The results show that the characteristic values determined depend very strongly on the test conditions, such as heating and cooling rate. Based on these results, Rietzel carried out DSC investigations with POM, HD-PE, PP, and PA 12 in Refs. [64,73,115], in which the test parameters were adapted to real conditions in the LS process with higher heating rates and low cooling rates of up to 1 K/min. Here, the start of crystallization is shifted toward higher temperatures due to the low cooling rates in LS, which reduces the overall process window and makes processing more difficult. Due to the inhomogeneity of the temperature distribution in the construction field, the processability of materials with small theoretical process windows, such as POM or HD-PE, is negatively affected. In addition, the tendency of materials to isothermal crystallization at different holding temperatures was investigated by means of isothermal DSC measurements [64]. The results show a particular tendency to crystallize for POM, while the lowest tendency was determined for PP. Overall, the process is significantly influenced by the isothermal crystallization.

DSC measurements for polyamide 12 were carried out in numerous studies. Typical melting temperatures and crystallization temperatures for unaged material are between 181.8 and 190.7°C and between 142.6 and 151.5°C [64,116–122], with an accumulation in each case at approximately 185 and 146°C. The deviations of the measured values are caused by diverging heating and cooling rates or different measuring equipment.

Storage at high temperatures during the process also causes a change in the material, which has an increased effect on the material quality with increasing storage time. In numerous studies, an increase in the melting temperature was observed with increasing aging time [116,121–126], whereby the effect is even stronger at higher storage temperatures [125]. The temperature rise is explained by an increase in the degree of crystallization in the material due to postcrystallization [124,126] and by thermally induced postpolymerization [116,126], which is superimposed with chain degradation during long storage periods [126]. This increases the processing window for aged material. For alternative materials, studies have so far only been carried out for polypropylene, in which a very strong increase in the melting temperature due to storage at higher process temperatures [88] and a reduction in the width of the melting range due to temperature effects [64,127] can be observed. As a result, the isothermal process window widens, and PP can be processed at higher preheating temperatures. This increases process reliability.

#### *2.5.1.2 Viscosity*

In addition to the melting and crystallization behavior, the LS process is largely determined by the viscosity of the melt due to its influence on the ability to form a melt film. This is illustrated, for example, by investigations with amorphous polystyrene of different viscosity [128]. A lower viscosity results in better sintering and a higher part density.

Numerous analyses therefore consider the relationship between material quality and melt viscosity of PA12, but the focus is primarily on the analysis and evaluation of material aging effects and the identification of suitable measurement methods. Investigations by Seul using rotational viscometry illustrate this problem in the form of strong changes in melt viscosity during the measurement [122]. For new material in particular, very strong changes and large viscosity differences between new and old material are already apparent in the first few minutes of the measurement, which were also observed by Haworth in Ref. [129]. In the meantime, melt flow index measurement has established itself as a suitable measuring method, which is also recommended in VDI 3405 Part 1.1 or ISO/ASTM 52925 as a suitable measuring method for characterizing the material quality of PA12 powders under strict measuring conditions [1]. The suitability of the method has been demonstrated in numerous investigations by simulated furnace storage tests [116,125,126,130,131] or aging tests during the process [121,123, 130–133]. The results of the individual tests show a strong increase in viscosity with increasing aging time, whereby the effect is intensified with increasing storage temperature [125,130]. The change in material is particularly significant with short aging periods and decreases steadily with increasing duration, finally approaching a constant viscosity level [121,123, 125,126,130,132]. At the same time, powder from the center of a building envelope shows a stronger increase in viscosity than powder from the edges [130,134].

The strong changes in viscosity are attributed to a postpolymerization or cross-linking of the material. This thesis has been proven by gel permeation chromatography or indirect methods for characterizing the molecular weight such as solution viscosity measurements of material of different aging states in the form of an increase in molecular weight [116,122,126,129,135,136]. The increase in the molecular weight of the material causes a deterioration of the part quality when reused in the process [136,137]. Distinct signs of inadequate material quality and strong aging effects are in particular the deterioration of the surface quality in connection with the occurrence of the so-called "orange peel" effect [125,130–132] as well as the deterioration of the mechanical component properties [112,121,123,131,132].

The aim of various studies was therefore to determine limit values for material quality or an evaluation system to determine at which point in quality one of the two described effects occurs [125,130–132,138]. The occurrence and predominance of the individual effects depends on the machine and parameters [131,132]. To avoid these effects, material suppliers of PA 12 recommend the use of material mixtures of new powder and recycled waste material with refresh rates of 30%–50% [130], although in the meantime, also materials requiring a lower refresh rate are available. To save resources and to achieve constant reproducible material qualities, the melt flow index measurement is currently being increasingly qualified [135]. The aging of alternative materials was only considered by Keller in Ref. [57] using PA 11, by Schmid in Ref. [127] for a PP/PE blend and by Wegner in Refs. [90,139] for PP, whereby the viscosity increase for PA 11 was significantly lower than for PA12 [57]. PP shows no significant change on viscosity caused by aging. For PP, changes in powder flowability were found.

#### 2.5.1.3 Powder properties

In addition to the thermal material properties, the LS process—and here above all the powder application or the packing density in the powder bed—is determined by the powder rheological properties. The densest possible powder packing can be achieved by high particle sphericity and smooth particle surfaces [140]. For continuous particle size distributions, such as those usually exhibited by LS powders, a theoretical packing density of 66.3% is possible [140]. However, real bulk densities of commercial LS powders are usually much lower at 41%–45.3% for PA12 [64,131,132,141–143]. Other commercial materials such as PE or PP lie in the same range [91]. For many experimental powders but also some commercial powders, even lower values between 27.5% and 39.2% are achieved in some cases due to the angular particle shape of ground powders or small average particle sizes [64,91].

As a further powder parameter, the flowability of the powder has a decisive influence on the process characteristics in LS. This has a significant effect on the homogeneous layer formation during powder application. In Ref. [64], Rietzel investigated the powder flow behavior of various experimental LS powders in comparison with PA12. Influences on the flow behavior result from the particle shape and thus the manufacturing process, the particle size, the material itself as well as from possible additives, such as aerosils, to improve the flow properties. In Ref. [60], Amado qualified a new test method for LS powders using a revolution powder analyzer from Mercury Scientific Inc. which allows a simulation of the powder application during LS and measurements at elevated temperatures [144]. The investigations performed compared different commercial and experimental powders with respect to their flowability and fluidization behavior. The best results were obtained with powders of high sphericity and a small average size [60]. In addition, Amado et al. [144] showed that the flow behavior of LS powders is influenced by the preheating temperature. Material aging also influences the flowability of LS powders, as demonstrated by Hausner value measurements in Refs. [131,132] for PA 12 powders of different ages. With some alternative LS materials, such as the PP/PE blend (PP-R201) from the Japanese company Trial, material aging in the process has no influence on the flow behavior of the powder [127].

### 2.5.2 Energy input

The LS process is essentially determined by the interaction of the laser energy introduced with the preheated powder material. To describe the energy introduced by the laser, Nelson defined the area energy density $E_A$ as the ratio of the laser power P, the scanning speed v, and the hatch distance h (Eq. 2.1) [35]. This was extended in later works by Meiners [145], Starr [146], and Kaddar [50], among others, to the volume energy density $E_V$, Eq. (2.2), including the layer thickness s. Today, these equations are often used to describe the energy input and to correlate process interrelations [50,124,146–148].

$$E_A = \frac{P}{v \cdot h} \tag{2.1}$$

$$E_V = \frac{P}{v \cdot h \cdot s} \tag{2.2}$$

When laser radiation interacts with a material, some of the radiation introduced is reflected at the surface. A further part enters the material, is absorbed there by the material, or exits on the back of the irradiated sample. The relationship between reflectance R, absorbance A, and transmittance T is described by Eq. (2.3) [149]. The proportion is determined by the properties of the incident radiation and of the material. Thus, the interaction is influenced on the one hand by the laser wavelength and on the other hand by the angle of incidence and its polarization. In addition, surface morphology and composition of the material as well as material thickness and temperature affect the proportions [57,149–152]. Through interaction with the material, the intensity that has entered the material is weakened. In the case of plastics, this can be attributed to the interaction of the radiation with the molecules and in particular individual molecular groups, the spherulitic structure and fillers contained in the material, such as glass or carbon fibers, or additives, such as absorbers or dyes [151–154]. For example, the individual groups of molecules are stimulated by medium- and long-wave infrared radiation to form framework or group oscillations, which explains the high basic absorption of plastics in these wavelength ranges [153,154].

$$R + A + T = 1 \tag{2.3}$$

LS of polymer materials powders with particle sizes typically between 0 and 150 μm results in additional interaction possibilities. Due to the porosity in the powder layer, laser radiation can either be transmitted through the layer by multiple reflections on the particle surfaces and thus contribute to heating the underlying layers or be reflected back in areas close to the surface. In addition, the powder bed can form a so-called radiation trap within which the radiation is completely absorbed [145,155, 156].

Numerous studies have examined the optical properties of LS materials. To measure the absorption behavior, a measuring setup with an integrating sphere is often used to measure the diffuse reflection on a powder surface. In most studies, the absorption coefficient for ABS, PC, SAN, and PEEK is more than 94%. PE, PMMA, or even partially PC only show values of 75%–91% [35,55,156–159]. With PA 12, the effect of $CO_2$ laser radiation results in an absorption coefficient of 95%–98% with an optical penetration depth of 80–103 μm [75,159,160]. However, a good interlayer bonding in the laser-sintered part can only be achieved with a sufficient penetration depth, which is specified at Wilkening [56] as about 150%–180% of the layer thickness used.

The author's own studies on absorption were carried out on the basis of emissivity measurements on preheated powder at 170°C and in the molten state at 220°C using a measuring system at DLR Berlin for white and black PA12 [114]. Compared with other measurements, characteristic values can thus be determined directly under process conditions for a wavelength range from 2.5 to 16 μm. A comparison of the complete emissivity curves of PA2200 in powder form and in the molten state (Fig. 2.2A) shows that in the long wavelength range from 8.75 to 15 μm, the melt has an emissivity that is up to 0.1 higher than that of the powder. Thus, more heat can be absorbed by the melt in this wavelength range than by the powder. A similar effect can be seen in medium-wave infrared. The emissivity and the associated absorption coefficient of the melt are thus higher than that of the

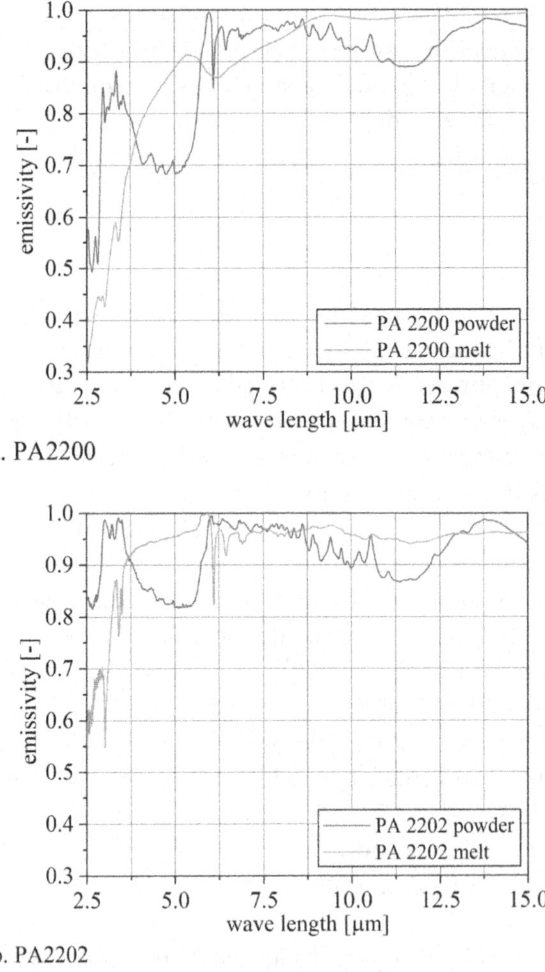


**Figure 2.2** Wavelength-dependent emissivity of PA2200 (a.) and PA2202 (b.) in powder and molten form.

powder over wide wavelength ranges. Since the infrared heating emitters used in LS emit medium- and long-wave infrared radiation over a very wide wavelength range, the melt and powder can absorb different amounts of heat radiation depending on the wavelength. In the wavelength range under consideration, the melt seems to be able to absorb more radiation than the powder due to a higher emission or absorption. PA2202 black (Fig. 2.2B) also shows a similar emission behavior, although the size of the areas in the medium-wave and long-wave infrared ranges varies somewhat.

In the range of the $CO_2$ laser wavelength of 10.6 μm, emissivities of the powder are 0.94 for PA2200 and 0.93 for PA2202, which corresponds to the literature measurements for $CO_2$ laser radiation by Laumer and Rechtenwald for PA12 [75,159].

After coupling into the powder layer, the heat introduced is distributed within the layer, in the molten part, and in the surrounding powder bed. At the same time, heat is released into the processing atmosphere [64,160,161]. The main mechanisms are based on heat conduction, heat radiation, and convection [35,52,64,160,162–164].

By heating the building platform and the building cylinder, excessive heat dissipation from the powder bed is prevented. In addition, the low thermal conductivity of the powder prevents rapid heat dissipation from the molten or hot areas close to the surface as well as homogenization of the temperatures in the overall system. Within the powder bed, a thermodynamic equilibrium between heat dissipation and heat supply is established by the formation of temperature gradients, whereby the temperature level decreases with increasing distance from the powder bed surface due to the increasing heat dissipation to the container [160,165].

Based on this theory of heat balance and the material properties of LS materials, Alscher and Schmachtenberg have developed the theory of isothermal LS as an idealized process model [63,65]. This theory states that at the same temperature, the powder bed temperature, viscous melt, and solid powder are simultaneously present. This is possible due to the special melting and crystallization behavior of semicrystalline polymers (Fig. 2.3). In theory, the laser energy applied is only used to transfer the powder, which has been preheated to below the melting temperature, into the molten state by applying the melting enthalpy without increasing the temperature. The melt afterward remains in the molten state without crystallizing. Only after completion of all layers is the temperature lowered and the part begins to solidify. With this idealized process control, hardly any residual stresses are formed between the layers, and the generation of low-distortion parts becomes possible [63,65].

The validity of this theory has so far only been confirmed on the basis of the microstructure of laser-sintered parts. In Ref. [72], Rietzel demonstrated for POM on the basis of spherulites with a size of several layer thicknesses that an isothermal process above the crystallization temperature must be present at least over a certain period of time. Investigations by Shen [166], Steinberger [160], Wegner [165], and Josupeit [167] on the temperature profile in the powder bed indicate that the crystallization

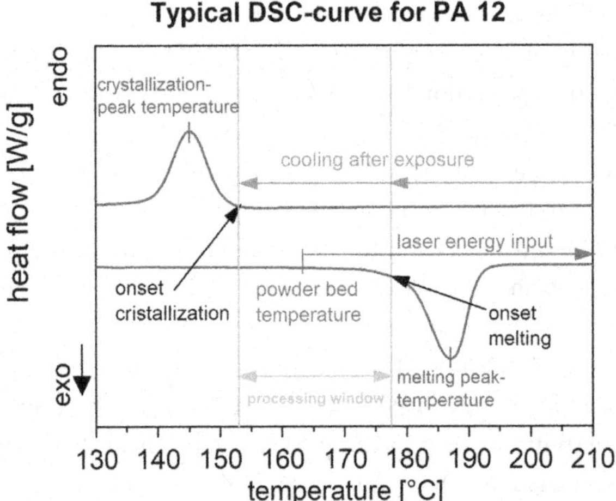


**Figure 2.3** Typical DSC curve for PA12 describing the possibility of isothermal processing.

temperature limit is already exceeded during the process, at least for higher building heights. During the process, the build envelope slowly cools down from the outside to the inside and from the bottom to the top [160,165,166], which causes a strongly position-dependent cooling and thus also solidification behavior. This is also evident in the results on the position-dependent shrinkage behavior in the form of significant nonlinear relationships [160,165,166,168], which can cause distortion due to the existing temperature gradients, especially in large components (Fig. 2.4A) [169]. The results of Wegner in Ref. [165] prove that the theory of isothermal LS can be applied at least to smaller components. The temperature curves measured in the building envelope show a decrease of the temperature in the powder bed from the preheating temperature to the crystallization temperature only about 150 mm below the powder surface, whereby locally very different cooling gradients can occur.

In addition, shrinkage effects can occur directly below the powder bed surface, resulting in component distortion, known as curling (Fig. 2.4B). If successive single layers shrink due to strong supercooling of the melt, stresses can build up between the solidifying layers (Fig. 2.4B). These can lead to curling of the layers [52,57,64,163]. Possible causes for this are, for example, too low preheating temperatures or materials with a small processing window or a high tendency to crystallize immediately after laser input. If curling is particularly strong, the generated layers can be pulled out

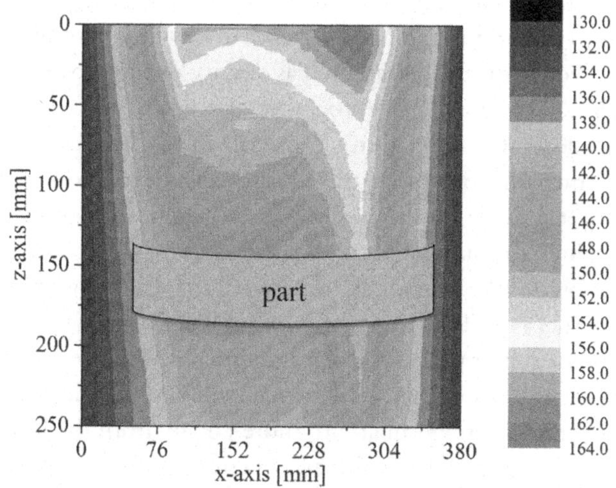


a. Distortion of large parts in the powder bed due to inhomogeneous cooling conditions

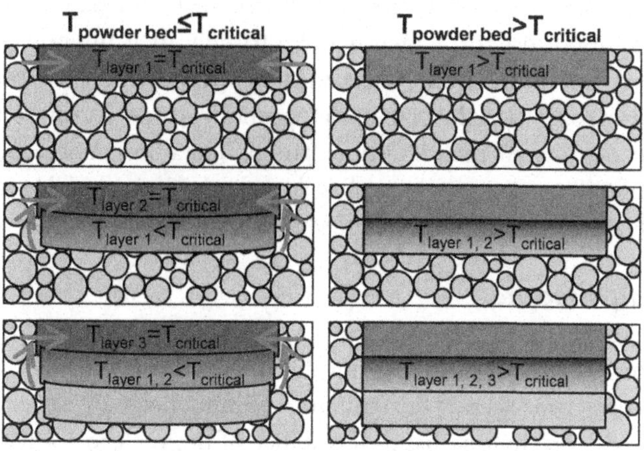


b. Curl development close to the surface due to isothermal crystallization processes caused by melt supercooling

**Figure 2.4** Mechanisms for part distortion during laser sintering due to temperature gradients (a.) and shrinkage (b.).

of the powder bed by the recoater during the following powder application.

To increase the understanding of the processes involved in LS of polymers, the time- and temperature-dependent processes have been simulated as simulation models in numerous publications. In most cases, the aim of the studies was to determine the time-dependent development of the sintering process to predict part properties such as density or the process cycle on this basis [35,66,141,160,170,171]. Altogether, the described models for process simulation allow the simulation of partial aspects of the LS process in sufficient approximation. Nevertheless, an exact prediction of the process cycle is not yet possible, so that in some cases, significant deviations between measured values and simulation results still occur.

In addition to the theory of isothermal LS, the author was able to establish the theory of the continuation of melting processes in LS as a basic requirement for robust processing: An additional material requirement in this context is that the absorption behavior of melt and powder has to be such that the melt after melting can absorb more energy from the heater than the surrounding powder bed. Only this allows constant melt temperatures above the material melting point to be achieved in the period between exposure and powder application, Fig. 2.5A. This allows melting processes to continue. The postmelting processes taking place at these temperatures homogenize the temperature distribution in the melt film (Fig. 2.5B). However, energy input has to be high enough to fully melt the powder. Then, the standard deviations of the mean melt temperature drop to below 1 K, which characterizes the achievement of optimum layer-to-layer bonding and a high part density. At the same time, the temperatures above the material melting point allow the newly applied layer to be heated and melted from below, which further improves the layer-to-layer bonding [114].

#### 2.5.2.1 Part structure

Laser-sintered parts are generated layer by layer by fusing powder layers. The structure of the polymer parts produced in this way differs from known manufacturing processes. The structure is influenced by several factors. These include, on the one hand, the ability of the melt to form a flat melt film due to the temperature gradient and its viscosity and, on the other hand, the melting depth of the laser. The different behavior of amorphous and semicrystalline thermoplastics and the associated different processing conditions lead to the formation of different sintering mechanisms. Liquid-phase sintering occurs mainly during the processing of amorphous materials, as the particles are only melted on the surface during the interaction with the laser. As a result, sinter necks are formed between the particles. Higher energy input leads to a partial melting of the particles and the formation of sintered surfaces, whereby the cores remain unmelted. A similar behavior

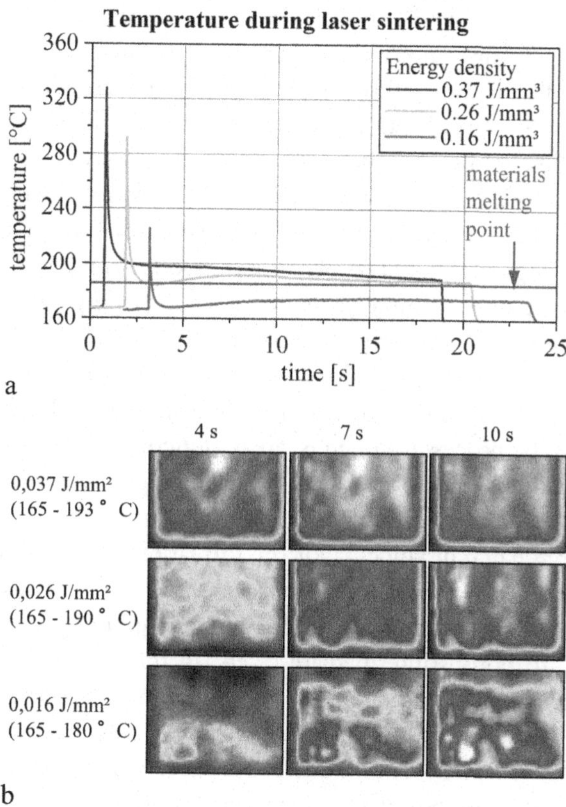


**Figure 2.5** Temperature development in the part during laser energy input and after melting (a) as well as development of the temperature distribution in the cross-section (b).

also results for semicrystalline plastics. As the energy input increases, the proportion of unmelted areas decreases, as has been demonstrated in investigations conducted by Zarringhalam at the University of Loughborough [120,172–174]. The results show that the mechanical behavior and crystallinity of the component correlate with the proportion of unmelted core areas. An increasing proportion of melted material results in a decrease of the crystallinity, because especially unmelted particles have a high degree of crystallinity [120,172]. With sufficient energy input by the laser, the powder of one layer is completely melted, and the melt overlaps with the underlying layer.

To better understand and be able to describe the component structure, depending on the process conditions, various studies were performed on the development of suitable models. Ho in Ref. [55] and Childs et al. in Ref. [66] provide first approaches for a structural model of laser-sintered parts for PC. Both studies show an increase of the sintering degree with increasing energy input. First, melt films are formed within the layers. The porosity in these films decreases with increasing energy density. Only a further increase of the energy input reduces the porosity between layers. In case of a too high energy input, an increased formation of round pores occurs due to degradation effects. The mechanical component properties as well as the part density increase with increasing energy input until the degradation limit is reached [55,66]. Rechtenwald et al. define in Refs. [75,175,176] the intermediate layer as a potential weak point in the component. The investigations for the semicrystalline PEEK show that, depending on the process parameters, areas of increased porosity are formed particularly at the intermediate layer. Especially high layer thicknesses, low preheating temperatures as well as low energy densities due to low laser powers or large hatch distances favor the formation of porous intermediate layers. Rechtenwald distinguishes between three different structure categories. The first is characterized by incomplete sintering with a lamellar structure due to porous intermediate layers. In the second and third categories, this structure is eliminated by the formation of a dense, completely sintered structure with low residual porosity. The third category is characterized by shape deviations due to excessive energy input. The author was able to demonstrate the correlations described in Ref. [114] for polyamide 12 by means of SEM images and microscopy images on thin sections.

Based on the studies, the development of the part structure in LS can be summarized in six structural stages, which are set up depending on the various process parameters and especially the energy input (Fig. 2.6). After exceeding the sintering threshold in stage 2 (Fig. 2.6A and B), the degree of sintering increases with increasing energy input up to stage 4 (Fig. 2.6C and D), to form a dense and homogeneous structure with low residual porosity and optimum mechanical properties (stage 5, Fig. 2.6E) when the optimum energy density is reached. If the energy density is further increased, the degradation limit is exceeded during exposure, which causes round pores with enclosed decomposition gas in the part, and the porosity rises again, stage 6 (Fig. 2.6F) [55,64,66,75,114,175,177–180].

The mechanics of laser-sintered parts depends strongly on the achieved part structure. Depending on the structure, different stress–strain curves are formed. The structure has a significant influence on the ductility of the

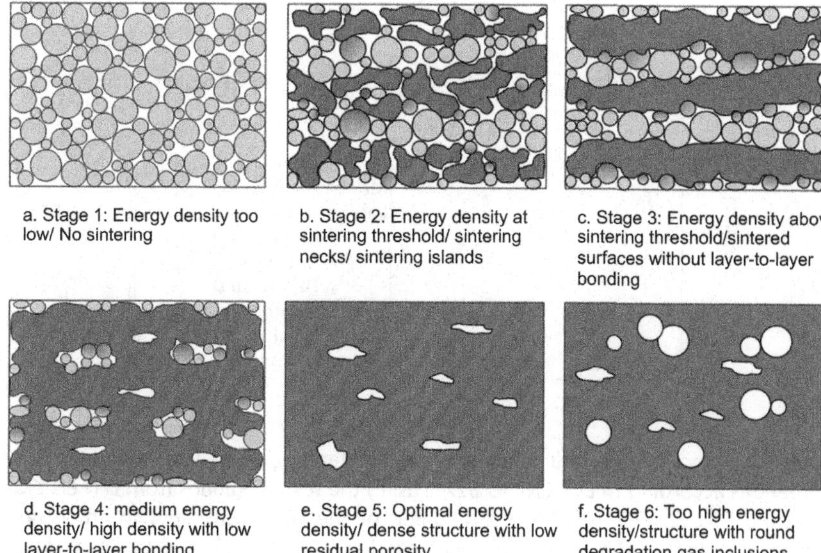


**Figure 2.6** Schematic part structure of laser-sintered parts depending on the energy input from stage 1 (a.) to 6 (f.).

components, especially in the building direction (Fig. 2.7). Curve 1 represents the optimum here, as there is simultaneously a good layer formation in the building plane and a good layer-to-layer bonding in the building direction, both characterized by high breaking elongation values. Curves 2 and 3 result in good mechanical properties within the build plane. In the building direction, the anisotropy typical for LS is shown with poorer characteristic values. Curve 3 in particular clearly illustrates the problem of insufficient layer-to-layer bonding. The parts structure models described explain the causes for the occurrence of a significant anisotropy during LS.

A comparison of various studies on the characteristic values in different directions shows in Refs. [114,181,182] that anisotropy of the tensile properties occurs in particular for the tensile strength and the elongation at break depending on the respective part structure. The deviations between building plane and building direction for the Young's modulus are always less than 5%, and thus, an almost isotropic behavior is present here. In contrast, the average deviation for strength is 14.1% and for elongation at break even 29.1% [114,182,183]. If the deviations and characteristic values are evaluated as a function of the volume energy density, it can be seen that for strength from an energy input of 0.25 $J/mm^3$ in most studies, almost identical values are achieved for both orientations. Compared with the

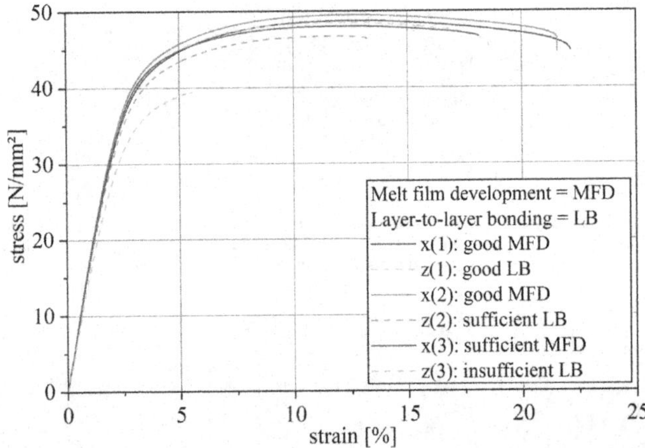


**Figure 2.7** Typical stress—strain curves of laser-sintered specimens for different part orientations according to DIN EN ISO 527-1 using the test conditions from DIN EN ISO 527-2 and DIN EN ISO 10350-1, conditioning of the specimens until the moisture balance according to DIN EN ISO 291 is reached at 23°C and 50% relative humidity.

strength, elongation at break shows significantly higher deviations between the orientations. A significant reduction of the anisotropy to values between 2.6% and 25% only occurs from an energy density of over 0.33 J/mm$^3$. Nevertheless, even in the range of high energy densities, deviations of up to 50% occur in single cases. Due to the dependence on the quality of the layer-to-layer bonding, the elongation at break in the building direction in particular is a value that reacts particularly strongly to process influences and thus represents an important quality parameter for the LS process [58,110,112,132,146,148,165,183—187].

Studies by Grießbach [124,188] or Usher [147] show that anisotropy can be reduced by adjusting the process parameters and especially by double exposure. However, the additional or mostly double scanning time significantly reduces the economic efficiency. The lowest deviations determined by Wegner in Ref. [183] using double exposure strategies and the results of [114] show that almost isotropic mechanical properties also in regard on elongation at break can be achieved for laser-sintered parts. High elongations at break in z-direction can also be achieved for Multi Jet Fusion parts [40].

#### 2.5.2.2 Parameter influences

The high number of potential influences illustrates the complexity of the influences on the LS process. Some of the factors have a particularly strong effect on the process, while others have only minor effects or act as quality-reducing disturbance variables. There are numerous studies on the influence of individual parameters on various component characteristics. Most studies deal with polyamide 12, while alternative LS materials have seldom been considered [52,55,66,75,189−193]. In addition, there is a clear focus on the factors investigated, which concentrate in particular on the values of the energy density formula (Eq. 2.2) [50,58,110,147]. In most cases, therefore, the laser power, the hatch distance, the scanning speed, and the layer thickness or the energy density value itself were varied. For the mechanical properties, different orientations and powder qualities were often considered [112,131,132,146]. The evaluations show that these parameters often have a very large influence on the process result. However, significant process influences also result for other factors, such as the number of exposures or the powder bed temperature [50,124,183,194].

From the studies listed earlier, it can be concluded for the mechanical properties as well as the part density that in particular an increase in energy density through variation of one of the included parameters leads to an increase in the characteristic values, whereby for very high energy inputs, a decrease in the characteristic values can be observed [50,64,124,148]. In addition, high powder bed temperatures [75,112,124], horizontally oriented components [112,124,146], multiple exposures [50, 120], and good material qualities [112,132,195] lead to high characteristic values for part density and especially for mechanical properties. The evaluation of the individual optimizations shows that for PA12, highest mechanical properties can be expected in an energy density range between 0.25 and 0.45 $J/mm^3$ and maximum densities above 0.2 $J/mm^3$.

Own results of the author from Ref. [114] prove the correlations from the literature. In particular, it was shown that for each target value, there is a system-specific threshold energy density above which an approximately constant level of characteristic values can be achieved. Among the variables considered, elongation at break requires the highest minimum energy input to achieve optimum characteristic values. For PA12, these values are in the range of 0.325−0.35 $J/mm^3$ depending on the system. However, the energy input is also limited at the upper end. Too high energy input causes degradation effects in the material. These can be demonstrated in particular by the formation of large round pores in the component but can also cause a drop in mechanical properties if the energy limit determined is significantly exceeded. The investigation results show that the limit for the occurrence of such effects lies at an area energy density of about

0.038–0.040 J/mm$^2$. Based on the tests, the main parameters influencing the process control could also be identified. The main influences are the hatch distance, the scanning speed, and the layer thickness with their respective linear effects. Furthermore, the complexity of the process interrelationships, which are characterized by numerous parameter interactions and nonlinearities, was demonstrated.

Typical parameter ranges of industrial LS machines are for the hatch distance between 0.15 and 0.3 mm, for the layer thickness between 0.06 and 0.18 mm, and for the scan speed between 3000 and 20,000 mm/s. The laser power of conventional $CO_2$ lasers is usually between 30 and 150 W. Recently, 500 W are also used with a fiber laser. Experimentally, a $CO_2$ laser with 600 W was used in Ref. [196] to work with focus diameters of up to 2 mm.

In addition, there are influences from the layer time. In the case of large layer time jumps within a part, different shrinkage behavior can result in the formation of steps along the buildup direction. The layer time therefore has a significant influence on the temperature balance in the part and thus also affects the part properties, especially dimensional stability [114,177]. However, elongation at break and its anisotropy are also strongly influenced by the layer time. At the same time, due to the variation of the degree of crystallinity, these two properties behave in the opposite direction to the other mechanical properties, their anisotropy, and the part density. Short layer times lead to a high elongation at break due to a low degree of crystallinity and to good layer-to-layer bonding due to the higher melt temperatures prior to powder application.

## 2.6 Process monitoring

With increasing use as a series production technique, there is an increasing demand for proof of part quality in additive processes [18,197,198] to ensure zero-defect generation of contours. For example, VDI Guideline 3405 Part 1 mentions numerous process variables that should be recorded online for quality verification [4]. The focus here is on monitoring the function of the laser scanner system as well as the process temperatures and temperature distributions resulting from the various heating elements. In addition, the environmental conditions and the times of individual process steps are to be documented. Further process steps to be monitored are the quality of the layer application, the accuracy of the platform lowering as well as the optical or thermal detection of the melt image. The guideline thus suggests that online monitoring of these parameters should be made possible in future machine systems.

Based on these trends, various process monitoring systems and the first control systems for LS process have been developed in recent years. At the same time, measuring systems introduced in numerous publications have been used as inline process analysis for recording process data.

To monitor the LS process, Hausotte is setting up a system based on stripe light projection and white light interferometry to measure the melted areas [199,200]. EOS is developing an optical system for monitoring the powder application and the molten contour areas, which allows the detection of contours on the powder bed surface [201]. Alternatively, patented monitoring systems include the monitoring of the temperature by an IR camera instead of the usual pyrometer [202,203] or the inspection of the coating condition and the solidified areas by IR radiation images [204]. The Fraunhofer IPA has developed a camera-based solution that uses image processing algorithms to automatically identify errors and defects in the LS process [205]. Current machine systems are increasingly equipped with optical or thermal camera systems. For example, the Prodways ProMaker P1000X or EOS P500 is equipped with an optical camera as standard. In the Integra P400, even an infrared camera is built in, which allows monitoring of the process.

Noncontact temperature measurement systems are used not only to monitor the process but also to characterize the temperature differences on the surface of the building area. Typical temperature variations on the powder bed surface are 8 K [61,206] for an EOSINT P390, 7 K [168,169] for an EOSINT P700, 9 K for a Formiga P100 (Fig. 2.8A) and 10–15 K [207,208] or 11 K for a DTM Sinterstation 2500 (Fig. 2.8B). Various investigations have shown the effect of such large temperature variations on the LS process, as studies by Goodridge [92], Wegner [165], Drummer [142,143], Stein [11], or Grießbach [124] illustrate. The temperature differences result in varying mechanical properties, part densities, and part dimensions, which partly depend on the temperature profile on the powder bed surface [165]. In addition, for alternative LS materials such as PA11 or UHMWPE, these temperature differences can lead to particularly strong variations in the characteristic values or to general problems in part production due to the occurrence of strong curling [92]. To reduce such effects, optimized multizone heating systems with lower temperature variations on the powder bed surface are becoming increasingly common, such as those developed by Integra [10,61,203], LSS, or EOS for the

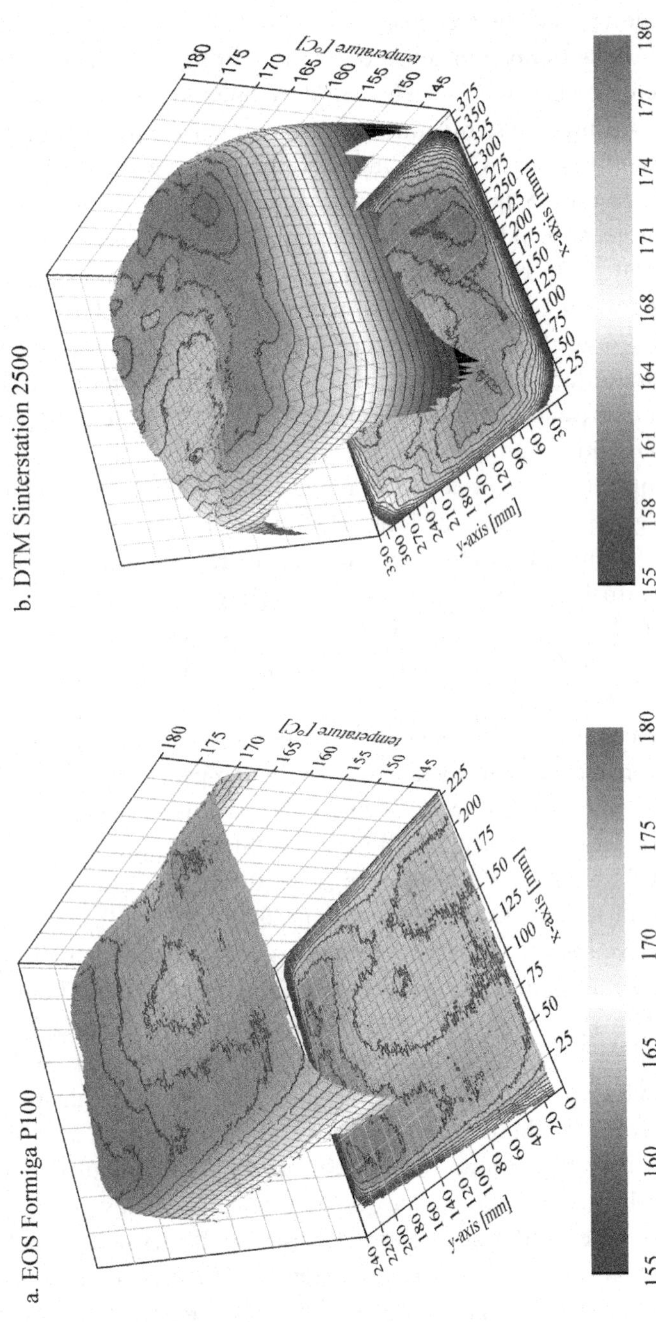


**Figure 2.8** Measured temperature distributions on the powder bed surface for EOS Formiga P100 (a.) and DTM Sinterstation 2500 (b.).

Formiga P110 but also by other companies such as Farsoon, Prodways, or 3D Mectronic.

## Symbols

| | | |
|---|---|---|
| A | [-] oder [%] | Absorption degree |
| $E_A$ | [J/mm$^2$] | Area energy density |
| $E_V$ | [J/mm$^3$] | Volume energy density |
| h | [mm] | Hatch distance |
| P | [W] | Laser power |
| R | [-] oder [%] | Degree of reflection |
| $R_Z$ | [μm] | Average roughness depth |
| s | [mm] | Layer thickness |
| T | [°C] | Temperature |
| T | [-] oder [%] | Degree of transmission |
| $T_M$ | [°C] | Material melting temperature |
| v | [mm/s] | Scan speed |

## Abbreviations

| | |
|---|---|
| **ABS** | Acrylonitrile butadiene styrene |
| **ALMs** | Advanced laser materials |
| **ASTM** | American Society for Testing and Materials |
| **$CO_2$** | Carbon dioxide |
| **DIN** | Deutsches Institut für Normung |
| **DLR** | Deutsches Zentrum für Luft-und Raumfahrt |
| **DSC** | Differential scanning calorimetry |
| **EOS** | Electrooptical systems |
| **HDPE** | High-density polyethylene |
| **HSS** | High-speed sintering |
| **IR** | Infrared |
| **LS** | Laser sintering |
| **LSS** | Laser sinter service |
| **MVR** | Melt volume flow rate |
| **Nd:YAG** | Neodymium-doped yttrium aluminum garnet |
| **$O_2$** | Sauerstoff |
| **PA6** | Polyamide 6 |
| **PA613** | Polyamide 613 |
| **PA11** | Polyamide 11 |
| **PA12** | Polyamide 12 |
| **PA1212** | Polyamide 1212 |
| **PBT** | Polybutylene terephthalate |

| | |
|---|---|
| **PC** | Polycarbonate |
| **PE** | Polyethylene |
| **PEI** | Polyetherimide |
| **PEEK** | Polyetheretherketone |
| **PEK** | Polyetherketone |
| **PEKK** | Polyetherketoneketone |
| **PMMA** | Polymethyl methacrylate |
| **PP** | Polypropylene |
| **PPS** | Polyphenylene sulfide |
| **POM** | Polyoxymethylene |
| **PS** | Polystyrene |
| **SEM** | Scanning electron microscopy |
| **SAN** | Styrene-acrylonitrile |
| **TPE** | Thermoplastic elastomers |
| **TPU** | Thermoplastic polyurethane |
| **UHMWPE** | Ultrahigh-molecular-weight polyethylene |
| **VDI** | Verein deutscher Ingenieure |

## References

[1] N. N.,VDI 3405 Blatt 1.1, Additive Manufacturing Processes - Laser Sintering of Polymer Parts - Qualification of Materials, 2018.
[2] N. N., VDI 3405 Blatt 7, Additive Fertigungsverfahren - Güteklassen für additiv gefertigte Kunststoffbauteile, 2019.
[3] N. N., VDI 3405 Blatt 6.2, Additive Fertigungsverfahren - Anwendersicherheit beim Betrieb der Fertigungsanlagen - Laser-Sintern von Kunststoffen, 2019.
[4] N. N., VDI 3405 Blatt 1, Additive Fertigungsverfahren, Rapid Manufacturing – Laser-Sintern von Kunststoffbauteilen: Güteüberwachung, 2012.
[5] N. N., ISO/ASTM CD 52924, Additive Manufacturing — Qualification Principles — Classification of Part Properties for Additive Manufacturing of Polymer Parts, 2019.
[6] N. N., ISO/ASTM WD 52925, Additive Manufacturing — Qualification Principles — Qualification of Polymer Materials for Powder Bed Fusion Using a Laser, 2019.
[7] N. N., DIN EN ISO 27547-1, Plastics - Preparation of Test Specimens of Thermoplastic Materials Using Mouldless Technologies, 2010.
[8] J. Wooten, Aeronautical Case Studies Using Rapid Manufacture, Rapid Manufacturing - an Industrial Revolution for the Digital Age, Wiley Publishing, Weinheim, 2006, pp. 233–239.
[9] G.N. Levy, R. Schindel, J.P. Kruth, Rapid manufacturing and rapid tooling with layer manufacturing (LM) Technologies, state of the art and future perspectives, CIRP Ann. - Manuf. Technol. 52 (2) (2003) 589–609.
[10] B. Lyons, Additive manufacturing in aerospace: examples and research outlook, Bridge 42 (1) (2012) 13–19.
[11] G.N. Stein, Development of high-density nylon using direct manufacturing-selective laser sintering, Technol. Rev. J. 14 (2) (2006) 67–79.
[12] R. Spielman, Space Applications, Rapid Manufacturing - an Industrial Revolution for the Digital Age, Wiley Publishing, Weinheim, 2006, pp. 241–246.

[13] M. Masters, T. Velde, F. McBagonluri, Rapid Manufacturing in the Hearing Industry, Rapid Manufacturing - an Industrial Revolution for the Digital Age, Wiley Publishing, Weinheim, 2006, pp. 195–208.
[14] U. Lindemann, R. Reichwald, M.F. Zäh, Individualisierte Produkte – Komplexität beherrschen in Entwicklung und Produktion, Springer-Publishing, Berlin - Heidelberg, 2006.
[15] A. Grzesiak, R. Becker, Light Weight Design for Additive Manufacturing of Plastic Components in the Automation - Design Rules and Industrial Test Cases, Additive Fertigung - vom Prototyp zur Serie, University Erlangen-Nürnberg, Institute for Polymer Technology, 2009, pp. 83–96.
[16] J. Breuninger, R. Becker, A. Wolf, S. Rommel, A. Verl, Generative Fertigung mit Kunststoffen, Springer-Publishing, Berlin - Heidelberg, 2013.
[17] H. Kuhn, Funktionsintegration mittels additiver Fertigung für den Sondermaschinenbau, Tagungsband 17. Augsburger Seminar für additive Fertigung - Funktionsintegration und Leichtbau, Augsburg, iwb Anwenderzentrum Augsburg, 2013.
[18] K. Müller-Lohmeier, M.-M. Speckle, Licht und Schatten beim industriellen Einsatz des Kunststoff-Lasersinterns, Tagungsband 17. Augsburger Seminar für additive Fertigung - Funktionsintegration und Leichtbau, Augsburg, iwb Anwenderzentrum Augsburg, 2013.
[19] N. Hopkinson, Production Economics of Rapid Manufacturing, Rapid Manufacturing - an Industrial Revolution for the Digital Age, Wiley Verlag, Weinheim, 2006, pp. 147–156.
[20] C. Telenko, C.C. Seepersad, A comparison of energy efficiency of selective laser sintering and injection molding of nylon parts, Rapid Prototyp. J. 18 (6) (2012) 472–481.
[21] M. Ruffo, C. Tuck, R. Hague, Cost estimation for rapid manufacturing - laser sintering production for low to medium volumes, Proc. IME B J. Eng. Manufact. 220 (2006) 1417–1427.
[22] N. Hopkinson, P. Dickens, Analysis of rapid manufacturing – using layer manufacturing processes for production, Proc. IME C J. Mech. Eng. Sci. 217 (2003) 31–39.
[23] J. Bräunlein, 3D-Druck: augenoptik im Umbruch? Eyebizz 6 (2016).
[24] E. Lai, Chanel Announces Plan to Mass-Produce a 3D Printed Mascara Brush, 3d Printing Industry, March 20, 2018.
[25] P.A.L. Ciraud, Verfahren und Vorrichtung zur Herstellung beliebiger Gegenstände aus beliebigem schmelzbaren Material, 1972. Patent DE2263777.
[26] R.F. Housholder, Molding Process, 1981. Patent US4247508A.
[27] C. Deckard, Selective Laser Sintering, Dissertation, The University of Texas at Austin, 1988.
[28] C.R. Deckard, Method and Apparatus for Producing Parts by Selective Sintering, 1986. Patent US4863538A.
[29] A. Gebhardt, Generative Fertigungsverfahren – Rapid Prototyping – Rapid Tooling – Rapid Manufacturing, 3. edition, Carl Hanser Publishing, Munich, 2007.
[30] R. Noorani, Rapid Prototyping – Principles and Applications, John Wiley & Sons, Hoboken, 2006.
[31] C.K. Chua, K.F. Leong, C.S. Lim, Rapid Prototyping: Principles and Applications, 2, edition, World Scientific Publishing, Singapore, 2003.
[32] N. N., Machine Manual Formiga P100, EOS GmbH, 2011.
[33] N. N., The Sinterstation System 2000, 2500, 2500$^{plus}$ – User's Guide, DTM Corporation, 2000.
[34] I. Gibson, D.W. Rosen, B. Stucker, Additive Manufacturing Technologies – Rapid Prototyping to Direct Digital Manufacturing, Springer-Publishing, New York - Heidelberg - Dordrecht - London, 2010.

[35] J.C. Nelson, Selective Laser Sintering: A Definition of the Process and an Empirical Sintering Model, Dissertation, University of Texas at Austin, 1993.
[36] N. Hopkinson, P. Dickens, P. Erasenthiran, High Speed Sintering Process – Final Report, Loughborough University, 2004.
[37] N. Hopkinson, P. Erasenthiran, Method and Apparatus for Combining Particulate Material, 2003. Patent US7879282B2.
[38] R. Brown, C.T. Morgan, C. E: Majewski, Not Just Nylon Improving the Range of Materials for High Speed Sintering, Proceedings of the 29th International Solid Freeform Fabrication Symposium (SSF 2018), The University of Texas at Austin, 2018.
[39] A. Ellis, C.J. Noble, L. Hartley, C. Lestrange, N. Hopkinson, C. Majewski, Materials for high speed sintering, J. Mater. Res. 29 (17) (2014) 2080–2085.
[40] H.J. O' Connor, D.P. Dowling, Comparison between the properties of polyamide 12 and glass bead filled polyamide 12 using the multi jet fusion printing process, Addit. Manufact. 31 (2020) 1–8.
[41] S. Morales-Planas, J. Minguella-Canela, J. Lluma-Fuentes, J.A. Travieso-Rodriguez, A.-A. García-Granada, Multi jet fusion PA12 manufacturing parameters for watertightness, strength and tolerances, Materials 11 (1472) (2018) 1–11.
[42] F. Sillani, R.G. Kleijnen, M. Vetterli, M. Schmid, K. Wegener, Selective Laser Sintering and Multi Jet Fusion: process-induced modification of the raw materials and analyses of parts performance, Addit. Manufact. 27 (2019) 32–41.
[43] D. Rouholamin, N. Hopkinson, Understanding the efficacy of micro-CT toanalyse high speed sintering parts, Rapid Prototyp. J. 22 (1) (2016) 152–161.
[44] P. Erasenthiran, N. Hopkinson, High speed sintering – early research into a new rapid manufacturing process, in: Proceedings of the 15th International Solid Freeform Fabrication Symposium (SFF 2004), The University of Texas at Austin, 2004, pp. 312–320.
[45] H.J. O' Connor, A.N. Dickson, D.P. Dowling, Evaluation of the mechanical performance of polymer parts fabricated using a production scale multi jet fusion printing process, Addit. Manufact. 22 (2018) 381–387.
[46] R. Larsson, Method and Device for Manufacturing Three-Dimensional Bodies, 1998. Patent EP1015214B1.
[47] L. Wiedau, L. Meyer, A. Wegner, G. Witt, Einflussuntersuchung von verschiedenen Nachbehandlungsmethoden auf die Oberflächentopologie von laser-gesinterten Polyamid 12 Proben, Rapid.Tech + FabCon 3.D International Hub for Additive Manufacturing: Exhibition + Conference + Networking: Proceedings of the 16th Rapid.Tech Conference Erfurt, München, Carl Hanser Verlag, 2019.
[48] T. Reinhardt, Entwicklung einer ganzheitlichen Verfahrenssystematik bei der Qualifizierung neuer Werkstoffe für das Laser-Sintern am Beispiel Polypropylen, University of Duisburg-Essen, Dissertation, 2016.
[49] M. Schmid, C. Simon, G.N. Levy, Finishing of SLS-Parts for Rapid Manufacturing (RM) - A Comprehensive Approach, Proceedings of the 20th International Solid Freeform Fabrication Symposium (SSF 2009), The University of Texas at Austin, 2009, pp. 1–10.
[50] W. Kaddar, Die generative Fertigung mittels Laser-Sintern: Scanstrategien, Einflüsse verschiedener Prozessparameter auf die mechanischen und optischen Eigenschaften beim LS von Thermoplasten und deren Nachbearbeitungsmöglichkeiten, University of Duisburg-Essen, Dissertation, 2010.
[51] L. Wiedau, L. Meyer, A. Wegner, G. Witt, Chemisches Nachbehandeln von Laser-Sinter-Proben – Einflussuntersuchung von verschiedenen Säuren auf die Oberflächentopologie, Rapid.Tech – International Trade Show & Conference for

Additive Manufacturing - Proceedings of the 14th Rapid.Tech Conference Erfurt, München, Carl Hanser Publishing, 2018.
[52] S. Nöken, Technologie des Selektiven Lasersinterns von Thermoplasten, Dissertation, RWTH Aachen, 1997.
[53] C. Yan, Y. Shi, L. Hao, Investigation into the differences in the selective laser sintering between amorphous and semi-crystalline polymers, Int. Polym. Process. 26 (4) (2011) 416–423.
[54] H.C.H. Ho, I. Gibson, W.L. Cheung, Effect of energy density on morphology and properties of selective laser sintered polycarbonate, J. Mater. Process. Technol. 89–90 (1999) 204–210.
[55] H.C.H. Ho, Properties and Morphological Development of Laser Sintered Polycarbonate and its Composites, Dissertation, University of Hong Kong, 2001.
[56] C. Wilkening, Lasersintern als Rapid Prototyping Verfahren – Möglichkeiten und Grenzen, Dissertation, TU München, 1997.
[57] B. Keller, Rapid Prototyping: Grundlagen zum selektiven Lasersintern von Polymerpulver, Dissertation, University, Stuttgart, 1998.
[58] I. Gibson, D. Shi, Material properties and fabrication parameters in selective laser sintering process, in: Rapid Prototyping Journal, vol. 3, 1997, pp. 129–136, 4.
[59] E.D. Dickens, B. Lin Lee, G.A. Taylor, J. Magistro, H. Ng, Sinterable Semi-crystalline Powder and Near-Fully Dense Article Formed Therewith, 1994. Patent US5342929A.
[60] A. Amado, M. Schmid, G. Levy, K. Wegener, Advances in SLS Powder Characterization, Proceedings of the 22nd International Solid Freeform Fabrication Symposium (SFF 2011), The University of Texas at Austin, 2011, pp. 438–452.
[61] R.D. Goodridge, C.:J. Tuck, R.J.M. Hague, Laser sintering of polyamides and other polymers, Prog. Mater. Sci. 57 (2) (2012) 229–267.
[62] M. Schmid, Laser Sintering with Plastics: Technology, Processes, and Materials, München, Carl Hanser Publishing, 2018.
[63] G. Alscher, Das Verhalten Teilkristalliner Thermoplaste Beim Lasersintern, Dissertation, University of Essen, 2000.
[64] D. Rietzel, Werkstoffverhalten und Prozessanalyse beim Laser-Sintern von Thermoplasten, Friedrich-Alexander-University Erlangen-Nuremberg, Dissertation, 2011.
[65] E. Schmachtenberg, T. Seul, Model of Isothermic Laser-Sintering, Proceedings of the Antec Conference, Society of Plastics Engineers, San Francisco Kalifornien, 2002, 2002.
[66] T.H.C. Childs, M. Berzins, G.R. Ryder, A. Tontowi, Selective laser sintering of an amorphous polymer - simulations and experiments, Proc. IME B J. Eng. Manufact. 213 (1999) 333–349.
[67] C.A. Leenders, Method of Producing Crystalline Polycarbonate Powders, 2017. Patent WO2017033146A1.
[68] F.E. Baumann, G. Dreske, N. Wilczok, Verfahren zur Herstellung von Polyamid-Feinstpulvern, 1994. Patent DE4421454A1.
[69] W. Christoph, H. Scholten, Verwendung eines Polyamids 12 für selektives Laser-Sintern, 1997. Patent DE19747309A1.
[70] M. Schmid, Selektives Lasersintern (SLS) mit Kunststoffen – Technologie, Prozesse und Werkstoffe, München, Carl Hanser Verlag, 2015.
[71] C.Z. Yan, Y.S. Shi, J.S. Yang, J.H. Liu, An origanically modified montmorillonite nylon-12 composite powder for selective laser sintering, Rapid Prototyp. J. 17 (1) (2011) 28–36.
[72] D. Rietzel, B. Wendel, R.:W. Feulner, E. Schmachtenberg, Neue Kunststoffpulver für das Selektive Lasersintern, in: Kunststoffe, vol. 2, 2008, pp. 65–68.

[73] D. Rietzel, F. Kühnlein, D. Drummer, D., Selektives Lasersintern von teilkristallinen Thermoplasten, Additive Fertigung - vom Prototyp zur Serie, Universität Erlangen-Nürnberg, Lehrstuhl für Kunststofftechnik, 2009, pp. 57–71.

[74] C. Yan, Y. Shi, J. Yang, J.H. Liu, Investigation into the selective laser sintering of styrene-acrylonitrile copolymer and postprocessing, Int. J. Adv. Manuf. Technol. 51 (2010) 973–982.

[75] T. Rechtenwald, Quasi-isothermes Laserstrahlsintern von Hochtemperatur-Thermoplasten – Eine Betrachtung werkstoff- und prozessspezifischer Aspekte am Beispiel PEEK, Dissertation, Friedrich-Alexander-Universität Erlangen-Nürnberg, 2011.

[76] A. Pfister, Neue Materialsysteme für das Dreidimensionale Drucken und das Selektive Lasersintern, Dissertation, Univertität Freiburg, 2005.

[77] J. Schmidt, M. Sachs, C. Blümel, S. Fanselow, B. Winzer, K.-E. Wirth, A Novel Process Route for the Production of Polymer Powders of Small Size and Good Flowability for Selective Laser Sintering of Polymers, Proceedings of the Fraunhofer Direct Digital Manufacturing Conference, 2014. Berlin, 2014.

[78] J. Schmidt, W. Peukert, K.-E. Wirth, M. Sachs, C. Blümel, Strahlschmelzverfahren – Einstellung der Dispersität und Funktionalisierung von Partikeln, Industriekolloquium des Sonderforschungsbereichs 814 - Additive Fertigung, Erlangen, Sonderforschungsbereich 814 - Additive Fertigung, 2012, pp. 11–25.

[79] W. Aquite, M. Launhardt, T. Osswald, T., Manufacturing of Micropellets Using Rayleigh Disturbances, Proceedings of the 44th Conference on Manufacturing Systems, University of Wisconsin-Madison, 2011, pp. 1–4.

[80] D. Rietzel, W. Aquite, D. Drummer, T. Osswald, T., Polymer powders for selective laser sintering - production and characterization. Proceedings of the 44th Conference on Manufacturing Systems, University of Wisconsin-Madison, 2011, pp. 1–6.

[81] C. Eloo, M. Rechberger, Neue Technologien zur Herstellung thermoplastischer Pulver, 5. Symposium Partikeltechnologie, Pfiztal, Fraunhofer Verlag, 2011, pp. 353–366.

[82] M. Rechberger, C. Eloo, M. Renner, Neue thermoplastische Werkstoffe für das SLS – vom Polyamid zum Funktionswerkstoff, Tagungsband 14. Anwenderforum Rapid Product Development 2009, Stuttgart, Fraunhofer IPA, 2009.

[83] C.J. Fruth, R. Feulner, E. Schmachtenberg, D. Rietzel, Fasern zur Verwendung bei der Herstellung eines schichtweise aufgebauten Formkörpers, 2008. Patent EP2282884A2.

[84] A. Wegner, T. Ünlü, Polypropylen und Polyethylen für das Laser-Sintern: Eigenschaften und Anwendungen, Tagungsband 20. Augsburger Seminar für additive Fertigung - Meilensteine der Forschung und Zukunftsthemen für die Industrie, Augsburg, iwb Anwenderzentrum Augsburg und Fraunhofer IWU, 2016.

[85] A. Wegner, New Materials for Laser Sintering: Processing Conditions of Polyethylene and Polyoxymethylene, Proceedings of the Fraunhofer Direct Digital Manufacturing Conference 2016, Berlin, 2016.

[86] L. Fiedler, A. Hähndel, A. Wutzler, J. Gerken, H.-J. Radusch, H.-J., Development of New Polypropylene Based Blends for Laser Sintering, Proceedings of the Polymer Processing Society 24th Annual Meeting PPS-24, Salerno (Italien), Polymer Processing Society, 2008, pp. 203–207.

[87] H.-J. Radusch, L. Fiedler, A. Hähndel, J. Gerken, Polypropylene Based Blends for Laser Sintering Technology, Tagungsband: Fortschritte in der Kunststofftechnik - Theorie und Praxis, Fachhochschule Osnabrück, 2011, pp. 1–9.

[88] L. Fiedler, R. Androsch, D. Mileva, H.-J. Radusch, A. Wutzler, J. Gerken, Experimentelle simulation der Physikalischen alterung von Lasersinterpulvern, Zeitschrift Kunststofftechnik, Bd. 6 (1) (2010) 20–32.

[89] L. Fiedler, L.O. Garcia Correa, H.-J. Radusch, Evaluation of polyproplene powder grades in consideration of the laser sintering processibility, Zeitschrift Kunststofftechnik 3 (4) (2007) 1–14.
[90] A.: Wegner, T. Ünlü, Powder Life Cycle Analyses for a New Polypropylene Laser Sintering Material, Proceedings of the 27th International Solid Freeform Fabrication Symposium (SSF 2016), The University of Texas at Austin, 2016, pp. 834–846.
[91] A. Wegner, New polymer materials for the laser sintering process: polypropylene and others, Phy. Proc. 83 (2016) 1003–1012.
[92] R.D. Goodridge, R.J.M. Hague, C.J. Tuck, An empirical study into laser sintering of ultra-high molecular weight polyethylene (UHMWPE), J. Mater. Process. Technol. 210 (2010) 72–80.
[93] J.T. Rimell, P.M. Marquis, Selective laser sintering of ultra high molecular weight polyethylene for clinical applications, J. Biomed. Mater. Res. 53 (4) (2000) 414–420.
[94] C. Dallner, S. Funkhauser, F. Müller, J. Demeter, M. Völkel, Lasersinterpulver auf Polyoxymethylen, Verfahren zu dessen Herstellung und Formkörper, hergestellt aus diesem Lasersinterpulver, 2011. Patent WO 2011/051250 A1.
[95] J. Blömer, Material und Prozessentwicklung für das Lasersintern: neue Materialien und Mehrkomponenten-Werkstoffe, Tagungsband Fachtagung Rapid Prototyping 15.11, 2012. Lemgo, 2012.
[96] A. Wegner, M. Oehler, T. Ünlü, Development of a new polybutylene terephthalate material for laser sintering process, Proc. CIRP 74 (2018) 254–258.
[97] A. Wegner, T. Ünlü, Adjustment of Part Properties for an elastomeric laser sintering material, J. Occup. Med. 70 (3) (2017) 419–424.
[98] M.S. Wahab, K.W. Dalgarno, R.F. Cochrane, Selective laser sintering of polymer nanocomposites. Proceedings of the 18th International Solid Freeform Fabrication Symposium (SFF 2007), The University of Texas at Austin, 2007, pp. 358–366.
[99] M.S. Wahab, K.W. Dalgarno, R.F. Cochrane, S. Hassan, Development of Polymer Nanocomposites for Rapid Prototyping Process, Proceedings of the World Congress on Engineering, vol. II, WCE, London UK, 2009, 2009, 2009.
[100] G.V. Salmoria, J.L. Leite, L.F. Vieira, A.T.N. Pires, C.R.M. Roesler, Mechanical properties of PA6/PA12 blend specimens prepared by selective laser sintering, Polym. Test. 31 (2012) 411–416.
[101] A.: Wegner, M. Oehler, T. Ünlü, Entwicklung alternativer Polyamidwerkstoffe für das Laser-Sintern, Rapid.Tech – International Trade Show & Conference for Additive Manufacturing - Proceedings of the 14th Rapid.Tech Conference Erfurt, München, Carl Hanser Verlag, 2018.
[102] A. Wegner, Development of a New Polyamide 6 Material for Laser Sintering Process, Proceedings of the Polymer Processing Society 33rd Annual Meeting PPS-33, Cancun, Mexico, Polymer Processing Society, 2017, pp. S19–S176.
[103] S. Kenzari, D. Bonina, J.M. Dubois, V. Fournee, Quasicrystal-polymer composites for selective laser sintering technology, Mater. Des. 35 (2012) 691–695.
[104] C.K. Kim, T. Saotome, H.T. Hahn, Y.G. Bang, W.B. Sung, Development of Nanocomposite Powders for the SLS Process to Enhance Mechanical Properties, Proceedings of the 18th International Solid Freeform Fabrication Symposium (SFF 2007), The University of Texas at Austin, 2007, pp. 367–376.
[105] J. Bai, R.D. Goodridge, R.J. Hague, M. Song, Carbon Nanotube Reinforced Polyamide 12 Nanocomposites for Laser-Sintering, Proceedings of the 23rd International Solid Freeform Fabrication Symposium (SFF 2012), The University of Texas at Austin, 2012, pp. 98–107.
[106] R.D. Goodridge, M.L. Shofner, R.J.M. Hague, M. McClelland, M.R. Schlea, R.B. Johnson, Processing of a polyamide-12/carbon nanofibre composite by laser sintering, Polym. Test. 30 (2011) 94–100.

[107] S.R. Athreya, Processing and Characterization of Carbon Black-Filled Electricaly Conductive Nylon-12 Nanocomposites Produced by Selective Laser Sintering, Georgia Institute of Technology, Dissertation, 2010.

[108] S.R. Athreya, K. Kalaitzidou, S. Das, Processing and Characterization of a Carbon Black-Filled Electrically Conductive Nylon-12 Nanocomposite Produced by Selective Laser Sintering, Proceedings of the 20th International Solid Freeform Fabrication Symposium (SFF 2009), The University of Texas at Austin, 2009, pp. 538–546.

[109] K. Kalaitzidou, S. Athreya, C. Chun, S. Das, Laser Sintering vs Melt Compounding: A New Approach for Functionally Graded Polymer Nanocomposites, Proceedings of the ICCM17, Edinburgh UK, 2009.

[110] A. Sauer, G. Witt, Optimierung der Eigenschaften von thermoplastischen Lasersinter-Bauteilen, RTejournal 2 (2005) 1–11.

[111] J.T. Sehrt, Möglichkeiten und Grenzen bei der generativen Herstellung metallischer Bauteile durch das Strahlschmelzverfahren, Disseration, University of Duisburg-Essen, 2010.

[112] A. Sauer, Optimierung der Bauteileigenschaften beim Selektiven Lasersintern von Thermoplasten, Dissertation, University of Duisburg-Essen, 2005.

[113] O. Rehme, Cellular Design for Laser Freeform Fabrication, Dissertation, Technical University Hamburg, 2010.

[114] A. Wegner, A.: Theorie über die Fortführung von Aufschmelzvorgängen als Grundvoraussetzung für eine robuste Prozessführung beim Laser-Sintern von Thermoplasten, University of Duisburg-Essen, Dissertation, 2015.

[115] D. Drummer, D. Rietzel, F. Kühnlein, Development of a characterization approach for the sintering behavior of new thermoplastics for selective laser sintering, Phys. Proc. 5 (2010) 533–542.

[116] D.: Drummer, F. Kühnlein, D. Rietzel, G. Hülder, Untersuchung der Materialalterung bei Pulverbasierten Schichtbauverfahren, RTejournal 7 (2010).

[117] M. Vasquez, B. Haworth, N. Hopkinson, Methods for quantifying the stable sintering region in laser sintered polyamide-12, Polym. Eng. Sci. 53 (3) (2012) 1–11.

[118] M. Vasquez, N. Hopkinson, B. Haworth, Laser Sintering Processes: Practical Verification of Particle Coalescence for Polyamides and Thermoplastic Elastomers, Proceedings of the Antec Conference 2011, Society of Plastics Engineers, 2011, pp. 1–5.

[119] M. Vasquez, B. Haworth, N. Hopkinson, Optimum sintering region for laser sintered nylon-12, Proc. IME B J. Eng. Manufact. 225 (12) (2011) 2240–2248.

[120] H. Zarringhalam, Investigation into Crystallinity and Degree of Particle Melt in Selective Laser Sintering, Loughborough University, Disseration, 2007.

[121] T. Gornet, Materials and Process Control for Rapid Manufacturing, Rapid Manufacturing - an Industrial Revolution for the Digital Age, Wiley Publishing, Weinheim, 2006, pp. 125–146.

[122] T. Seul, Ansätze zur Werkstoffoptimierung beim Lasersintern durch Charakterisierung und Modifizierung grenzflächenenergetischer Phänomene, Dissertation, RWTH Aachen, 2004.

[123] T.J. Gornet, K.R. Davis, T.L. Starr, K.M. Mulloy, Characterization of Selective Laser Sintering Materials to Determine Process Stability, Proceedings of the 13th International Solid Freeform Fabrication Symposium (SFF 2002), The University of Texas at Austin, 2002, pp. 546–553.

[124] S. Grießbach, Korrelation zwischen Materialzusammensetzung, Herstellungsbedingungen und Eigenschaftsprofil von lasergesinterten Polyamid-Werkstoffen, Dissertation, Martin-Luther-University Halle-Wittenberg, 2012.

[125] D.T. Pham, K.D. Dotchev, W.A.Y. Yusoff, Detoriation of polyamide powder properties in the laser sintering process, Proc. IME C J. Mech. Eng. Sci. 222 (11) (2008) 2163–2176.

[126] F. Kühnlein, D. Drummer, K. Wudy, M. Drexler, Alterungsmechanismen von Kunststoffpulvern bei der Verarbeitung und deren Einfluss auf prozessrelevante Materialeigenschaften, Industriekolloquium des Sonderforschungsbereichs 814 - Additive Fertigung, Erlangen, Sonderforschungsbereich 814 - Additive Fertigung, 2012, pp. 49–66.
[127] M. Schmid, F. Amado, G. Levy, iCoPP – A New Polyolefin for Additive Manufacturing (SLS), Proceedings of the International Conference on Additive Manufacturing, Loughborough University, 2011.
[128] Y. Shi, Z. Li, H. Sun, S. Huang, F. Zeng, Effect of the properties of the polymer materials on the quality of selective laser sintering parts, Proc. IME J. Mater. Des. Appl. 218 (2004) 247–252.
[129] B. Haworth, N. Hopkinson, D. Hitt, X. Zhong, Shear viscosity measurements on polyamide-12 polymers for laser sintering, Rapid Prototyp. J. 19 (1) (2013) 28–36.
[130] K. Dotchev, W. Yusoff, Recycling of polyamide 12 based powders in the laser sintering process, Rapid Prototyp. J. 15 (3) (2009) 192–203.
[131] A. Wegner, C. Mielicki, T. Grimm, B. Gronhoff, G. Witt, J. Wortberg, Determination of robust material qualities and processing conditions for laser sintering of polyamide 12, Polym. Eng. Sci. 54 (7) (2014) 1540–1554.
[132] A. Wegner, G. Witt, Betrachtung zur Pulvernutzungsdauer beim Laser-Sintern und Einfluss der Prozessführung auf die Entstehung von Ausschussbauteilen, RTejournal 9 (2012).
[133] D. Drummer, K. Wudy, M. Drexler, Influence of Energy Input on Degradation Behavior of Polyamide 12 during Selective Laser Melting Process, Proceedings of the Fraunhofer Direct Digital Manufacturing Conference 2014, Berlin, 2014.
[134] S. Rüsenberg, H.-J. Schmid, Advanced Characterization Method of Nylon 12 Materials for Application in Laser Sinter Processing, Proceedings of the Polymer Processing Society 29th Annual Meeting PPS-29, Nürnberg (Deutschland), Polymer Processing Society, 2013.
[135] S. Rüsenberg, R. Weiffen, F. Knoop, H.-J. Schmid, M. Gessler, H. Pfisterer, Controlling the Quality of Laser-Sintered Parts along the Process Chain, Proceedings of the 23rd International Solid Freeform Fabrication Symposium (SFF 2012), The University of Texas at Austin, 2012, pp. 1024–1044.
[136] B. Haworth, D.J. Hitt, N. Hopkinson, M. Vasquez, Laser Sintering Process for Polymers: Influence of Molecular Weight and Definition of a Stable Sintering Region, Proceedings of the Polymer Processing Society 29th Annual Meeting PPS-29, Nürnberg (Deutschland), Polymer Processing Society, 2013.
[137] G.W. Ehrenstein, Polymer Werkstoffe, 3. edition, Munich Carl Hanser Publishing, 2011.
[138] D.T. Pham, K.D. Dotchev, W.A.Y. Yusoff, Improvement of Part Surface Finishing in Laser Sintering by Experimental Design Optimization (DOE), Proceedings of Third Virtual International Conference on Innovative Production Machines and Systems (IPROMS 2007), CRC Press, 2007.
[139] A. Wegner, Processing Conditions and Aging Effects of a New Polypropylene Material for the Laser Sintering Process, Proceedings of the 6th International Conference on Additive Technologies (iCAT 2016), Interesansa – zavod, Ljubljana, 2016, pp. 161–169.
[140] R.M. German, Particle Packing Characteristics, Metal Powder Industries Federation, 1989.
[141] A.E. Tontowi, Selective Laser Sintering of Crystalline Polymers, Dissertation, University of Leeds, 2000.
[142] D. Drummer, M. Drexler, F. Kühnlein, Influence of Powder Coating on Part Density as a Function of Coating Parameters, Proceedings of the 4th International Conference on Additive Technologies (ICAT 2012), DAAAM International Organisation, 2012.

[143] D. Drummer, M. Drexler, F. Kühnlein, Effects on the density distribution of SLS-parts, Phys. Proc. 39 (2012) 500–508.
[144] A. Amado, M. Schmid, K. Wegener, Flowability of SLS powders at elevated temperature, Tagungsband Rapid.Tech. (2013). Erfurt, Desotron-Verlags-Gesellschaft, 2013.
[145] W. Meiners, Direktes Selektives Laser Sintern Einkomponentiger Metallischer Werkstoffe, Dissertation, RWTH Aachen, 1999.
[146] T.L. Starr, T.J. Gornet, J.S. Usher, The effect of process conditions on mechanical properties of laser-sintered nylon, Rapid Prototyp. J. 17 (6) (2011) 418–423.
[147] J.S. Usher, T.J. Gornet, T.L. Starr, Weibull growth modeling of laser-sintered nylon 12, Rapid Prototyp. J. 19 (4) (2013) 300–306.
[148] B. Caulfield, P.E. McHugh, S. Lohfeld, Dependence of mechanical properties of polyamide components on build parameters in the SLS process, J. Mater. Process. Technol. 182 (2007) 477–488.
[149] R. Poprawe, Lasertechnik für die Fertigung, 1. edition, Springer-Publishung, Berlin - Heidelberg, 2005.
[150] C. Wagner, Untersuchungen zum Selektiven Lasersintern von Metallen, Dissertation, RWTH Aachen, 2002.
[151] D. Hänsch, Die optischen Eigenschaften von Polymeren und ihre Bedeutung für das Durchstrahlschweißen mit Diodenlaser, Dissertation, RWTH Aachen, 2001.
[152] T. Frick, M. Hopfner, Laserstrahlschweißen von Kunststoffen – Band 1, 2. edition, Erlangen, Bayerisches Laserzentrum, 2005.
[153] C. Bonten, C. Tüchert, Welding of plastics - introduction into heating by radiation, J. Reinforc. Plast. Compos. 21 (8) (2002) 699–709.
[154] R.M. Klein, Bearbeitung von Polymerwerkstoffen mit Infraroter Laserstrahlung, Dissertation, RWTH Aachen, 1990.
[155] J.P. Kruth, X. Wang, T. Laoui, L. Froyen, Lasers and materials in selective laser sintering, Assemb. Autom. 23 (4) (2003) 357–371.
[156] M.-S.M. Sun, Physical Modeling of the Selective Laser Sintering Process, Dissertation, University of Texas at Austin, 1991.
[157] K.M. Fan, K.W. Wong, W.L. Cheung, Reflectance and transmittance of True-Form powder and its composites to $CO_2$-laser, Rapid Prototyp. J. 13 (3) (2007) 175–181.
[158] N.K. Tolochko, T. Laoui, Y.V. Khlopkov, et al., Absorptance of powder materials suitable for laser sintering, Rapid Prototyp. J. 6 (3) (2000) 155–160.
[159] T. Laumer, T. Stichel, P. Appel, P. Amend, Untersuchungen zum Absorptionsverhalten von Pulverschüttungen für das Laserstrahlschmelzen von Kunststoffen, Tagungsband Rapid.Tech 2013, Desotron-Verlags-Gesellschaft, Erfurt, 2013.
[160] J. Steinberger, Optimierung des Selektiven-Laser-Sinterns zur Herstellung von Feingußteilen für die Luftfahrtindustrie, Dissertation, TU München, 2001.
[161] Y.A. Song, Selektives Lasersintern Metallischer Prototypen, Dissertation, RWTH Aachen, 1995.
[162] S. Baehr, K. Stephan, Wärme- und Stoffübertragung, 5 edition, Springer Publishing, Berlin - Heidelberg, 2006.
[163] N.M. Jamal, Finite Element Analysis of Curl Development in the Selective Laser Sintering Process, Dissertation, University of Leeds, 2001.
[164] H. Herwig, A.-Z. Wärmeübertragung, Berlin - Heidelberg, Springer-Publishing, 2000.
[165] A. Wegner, G. Witt, Ursachen für eine mangelnde Reproduzierbarkeit beim Laser-Sintern von Kunststoffbauteilen, RTejournal 10 (2013).
[166] J. Shen, Inhomogeneous Shrinkage of Polymer Materials in Selective Laser Sintering, Proceedings of the 11th International Solid Freeform Fabrication Symposium (SFF 2000), The University of Texas at Austin, 2000, pp. 298–305.

[167] S. Josupeit, H.-J. Schmid, Temperature history within laser sintered partcakes and its influence on process quality, Rapid Prototyp. J. 22 (5) (2016) 788–793.
[168] S.P. Soe, D.R. Eyers, R. Setchi, Assessment of non-uniform shrinkage in the laser sintering of polymer materials, Int. J. Adv. Manuf. Technol. 68 (1–4) (2013) 111–125.
[169] S.P. Soe, Quantitative analysis on SLS Part Curling using EOS P700 maschine, J. Mater. Process. Technol. 212 (2012) 2433–2442.
[170] J.P. Schultz, Modeling Heat Transfer and Densification during Laser Sintering of Viscoelastic Polymers, Dissertation, Virginia Polytechnic Institute and State University, 2003.
[171] G. Bugeda, M. Cervera, G. Lombera, Numerical prediction of temperature and density distributions in selective laser sintering processes, Rapid Prototyp. J. 5 (1) (1999) 21–26.
[172] C.E. Majewski, H. Zarringhalam, N. Hopkinson, Effects of Degree of Particle Melt and Crystallinity in SLS Nylon-12 Parts, Proceedings of the 19th International Solid Freeform Fabrication Symposium (SFF 2008), The University of Texas at Austin, 2008, pp. 45–54.
[173] N. Hopkinson, C. Majewski, H. Zarringhalam, Quanitfying the degree of particle melt in selective laser sintering, CIRP Ann. - Manuf. Technol. 58 (2009) 197–200.
[174] H. Zarringhalam, C. Majewski, N. Hopkinson, Degree of particle melt in nylon-12 selective laser-sintered parts, Rapid Prototyp. J. 15 (2) (2009) 126–132.
[175] M. Schmidt, D. Pohle, T. Rechtenwald, Selective laser sintering of PEEK, CIRP Ann. - Manuf. Technol. 56 (1) (2007) 205–208.
[176] T. Rechtenwald, S. Roth, D. Pohle, Fuktionsprototypen aus PEEK, Kunststoffe 96 (11) (2006) 62–68.
[177] M. Blattmeier, Strukturanalyse von lasergesinterten Schichtverbunden mit werkstoffmechanischen Methoden, University of Duisburg-Essen, Dissertation, 2012.
[178] D.K. Leigh, A Comparison of Polyamide 11 Mechanical Properties between Laser Sintering and Traditional Molding, Proceedings of the 23rd International Solid Freeform Fabrication Symposium (SFF 2012), The University of Texas at Austin, 2012, pp. 574–605.
[179] E. Schmachtenberg, R.W. Feulner, D. Rietzel, B. Wendel, Wechselwirkungen kunststoff - laserprozess, Kunststofftechnik 4 (3) (2008) 1–19.
[180] S. Dupin, Etude Fondamentale de la Transformation du Polyamide 12 par Frittage Laser: Mécanismes Physico-Chimiques et Relations Microstructures/Propriétés, University Lyon, Dissertation, 2012.
[181] A. Wegner, G. Witt, Laser Sintered Parts with Isotropic Mechanical Properties, RAPID 2012 and 3D Imaging Conferences & Exposition: May 22 - 25, 2012, Society of Manufacturing Engineers, Atlanta, GA, 2012, pp. 1–16.
[182] A. Wegner, G. Witt, Adjustment of isotropic Part Properties in laser sintering based on adapted double laser exposure strategies, Optic Laser. Technol. 109 (2019) 381–388.
[183] A. Wegner, G. Witt, Laser Sintered Parts with Isotropic Mechanical Properties, Technical Paper - Society of Manufacturing Engineers TP12PUB43, 2012, pp. 1–16.
[184] T.J. Silverman, A. Hall, B. South, W. Yong, J.H. Koo, Comparison of Material Properties and Microstructure of Specimens Built Using the 3D Systems Vanguard HS and Vanguard HiQ+HS SLS Systems, Proceedings of the 18th International Solid Freeform Fabrication Symposium (SFF 2007), The University of Texas at Austin, 2007, pp. 392–401.
[185] U. Ajoku, N. Saleh, N. Hopkinson, R. Hague, P. Erasenthiran, Investigating mechanical anisotropy and end-of-vector effect in laser-sintered nylon parts, Proc. IME B J. Eng. Manufact. 220 (2006) 1077–1086.

[186] S. Wartzack, D. Drummer, S. Wittmann, J. Stuppy, D. Rietzel, S. Tremmel, F. Kühnlein, Besonderheiten bei der Auslegung und Gestaltung lasergesinterter Bauteile, RTejournal 7 (2010).

[187] C. Majewski, N. Hopkinson, Effect of section thickness and build orientation on tensile properties and material charakteristics of laser sintered nylon-12 parts, Rapid Prototyp. J. 17 (3) (2011) 176–180.

[188] S. Grießbach, R. Lach, Hochbelastbare Lasersinterteile mit homogenen Materialeigenschaften, RTejournal 5 (2008).

[189] T. Gill, B. Hon, Selective Laser Sintering of SiC/Polyamide Matrix Composites, Proceedings of the 13th International Solid Freeform Fabrication Symposium (SFF 2002), The University of Texas at Austin, 2002, pp. 538–545.

[190] P.J. Hadro, A Design of Experiment Approach to Determine the Optimal Process Parameters for Rapid Prototyping Machines, Proceedings of the 15th International Solid Freeform Fabrication Symposium (SFF 2004), The University of Texas at Austin, 2004.

[191] H. Chung, S. Das, Processing and properties of glass bead particulate-filled functionally graded nylon-11 composites produced by selective laser sintering, Mater. Sci. Eng., A 437 (2006) 226–234.

[192] A.K. Singh, R.S. Prakash, Response surface-based simulation modeling for selective laser sintering process, Rapid Prototyp. J. 16 (6) (2010) 441–449.

[193] M.M. Savalani, L. Hao, P.M. Dickens, Y. Zhang, K.E. Tanner, R.A. Harris, The effects and interactions of fabrication parameters on the properties of selective laser sintered hydroxyapatite polyamide composite biomaterials, Rapid Prototyp. J. 18 (1) (2012) 16–27.

[194] W. Kaddar, G. Witt, Die festigkeit in abhängigkeit von Scanstrategien & -optionen beim lasersintern vom kunststoff, RTejournal 7 (2010).

[195] P.K. Jain, M. Pulak, P.V. Pandey, M. Rao, Experimental investigation for improving Part Strength in selective laser sintering, Virt. Phys. Prototyp. 3 (3) (2008) 177–188.

[196] N.N., Ergebnissbericht zum BMBF-Verbundprojekt HiPer-LS - Ressourceneffizientes und reproduzierbares Hochleistungs-Laser-Sintern zur Herstellung von Kunststoffbauteilen, 2018.

[197] E. Abele, G. Reinhart, Zukunft der Produktion - Herausforderungen, Carl Hanser Publishing, Forschungsfelder, Chancen, Munich, 2011.

[198] T. Uihlein, Potenziale additiver Fertigung im Triebwerksbau und notwendige Voraussetzungen, Tagungsband 17. Augsburger Seminar für additive Fertigung - Funktionsintegration und Leichtbau, Augsburg, iwb Anwenderzentrum Augsburg, 2013.

[199] T. Hausotte, W. Hartmann, M. Timmermann, B. Galovskyi, Optische Messsysteme zur In-Line-Prüfung im additiven Fertigungsprozess, Industriekolloquium des Sonderforschungsbereichs 814 - Additive Fertigung, Erlangen, Sonderforschungsbereich 814 - Additive Fertigung, 2012, pp. 67–85.

[200] W. Hartmann, T. Hausotte, F. Kühnlein, D. Drummer, Incremental In-Line Measurement Technique for Additive Manufacturing, Proceedings of the Fraunhofer Direct Digital Manufacturing Conference 2012, Berlin, 2012.

[201] F. Pfefferkorn, Anlagentechnik zur Herstellung funktionsintegrierter Kunststoffbauteile mittels Lasersintern, Seminarbericht (2013) 108: 17. Augsburger Seminar für additive Fertigung - Funktionsintegration und Leichtbau, iwb, TU München, 2013.

[202] M. Chung, A.L. Allanic, Sintern unter Verwendung von Thermobild-Rückkopplung, 2004. Patent DE102004017769B4.

[203] C.S. Huskamp, Methods and Systems for Controlling and Adjusting Heat Distribution over a Part Bed, 2007. Patent US 7515986 B2.

[204] J. Philippi, Verfahren zum Herstellen eines dreidimensionalen Objekts mittels Lasersintern, 2007. Patent DE102007056984A1.

[205] I. Effenberger, Optisches Inline-Inspektionssystem IQ4AP für das Selektive Laser-sintern - Automatisierung ist der nächste Schritt, Quality Engineering, 2018.
[206] U. Ajoku, Investigating the Compression Properties of Selective Laser Sintered Nylon-12, Loughborough University, Dissertation, 2008.
[207] T. Diller, R- Sreenivasan, J. Beaman, D. Bourell, J. LaRocco, Thermal Model of the Build Environment for Polyamide Powder Selective Laser Sintering, Proceedings of the 21st International Solid Freeform Fabrication Symposium (SSF 2010), The University of Texas at Austin, 2010, pp. 539–548.
[208] M. Yuan, D. Bourell, Efforts to Reduce Part Bed Thermal Gradients during Laser Sintering Processing, Proceedings of the 23rd International Solid Freeform Fabrication Symposium (SFF 2012), The University of Texas at Austin, 2012, pp. 962–974.

# CHAPTER 3

# Selective laser melting: principles and surface quality

**Evren Yasa**
Department of Mechanical Engineering, Eskisehir Osmangazi University, Eskisehir, Turkey

## 3.1 Introduction to selective laser melting

Selective laser melting (SLM) is a powder-bed-fusion AM process whereby a high-density-focused laser beam selectively scans a powder bed and those scanned and solidified layers are stacked upon each other to build a fully functional three-dimensional part, tool, or prototype [1]. As shown in Fig. 3.1, SLM is very similar to the selective laser sintering (SLS) process, which uses sintering or partial melting mechanism for binding powder particles rather than fully melting [2]. Among SLS, selective heat sintering, and electron beam melting (EBM), SLM belongs to powder-bed fusion processes in AM groups according to ASTM F42 categorization [3]. Powder-bed fusion technologies, in which either laser, heat or electron beam is used, as the energy source, to melt and fuse the powder particles together to form a three-dimensional object, can be used for a diverse range of applications. A comparison of most widely used powder-bed fusion technologies is presented in Table 3.1. In the powder-bed fusion processes, generally a similar process chain is followed as shown in Fig. 3.1 although some significant differences are present such as the energy source, processing temperature, or binding mechanisms. Unlike conventional manufacturing processes, pre-processing starts directly with the CAD model converted to a special file format (.stl) to represent the geometry with triangles and special algorithms are used to slice the 3D model into very thin layers, typically of about 20−40 μm in the SLM process. After setting process parameters such as scan strategy, laser power, hatch distance, and scan speed depending on the material, required specifications, and geometry, the SLM machine's software gets this slicing information and scans every layer on top of each other until the part geometry is completely built [4].

Most research on SLM of metallic materials focuses on ferrous-, nickel-, and titanium-based alloys [5−7]. Typically, spherical powders with 10−45 μm in diameter are used in the process. The SLM of aluminum,

*Additive Manufacturing*
ISBN 978-0-12-818411-0
https://doi.org/10.1016/B978-0-12-818411-0.00017-3

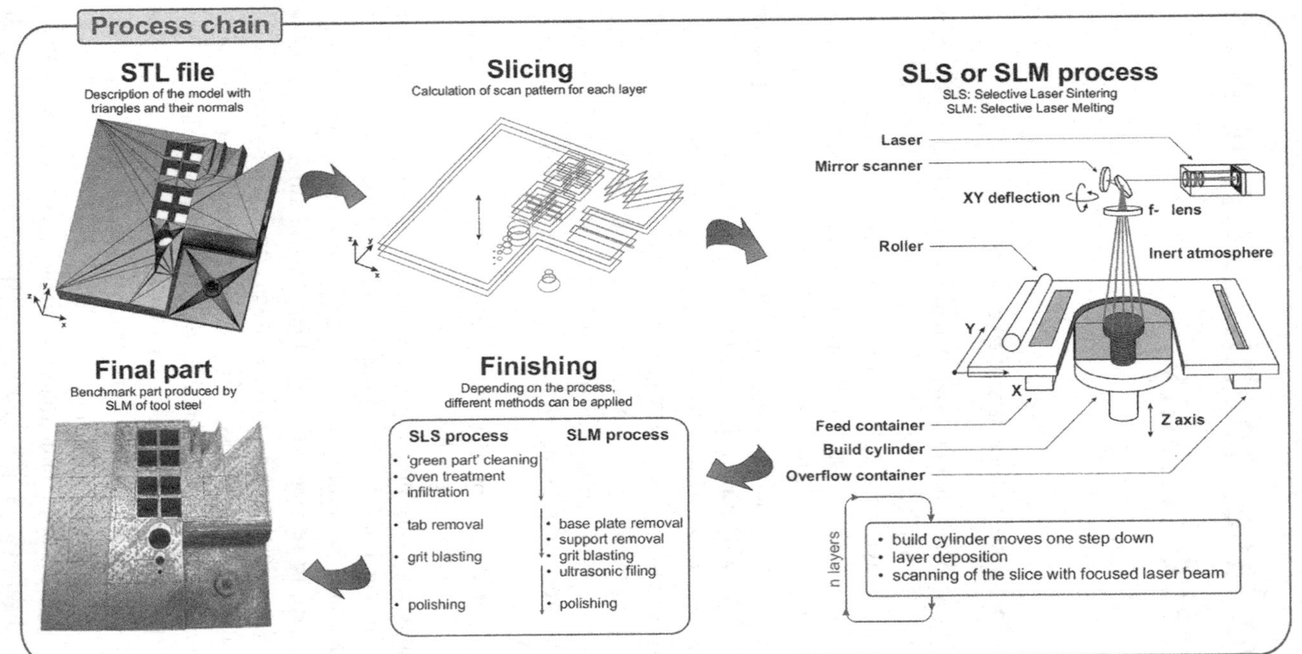


**Figure 3.1** Schema of the selective laser melting process. *(Reproduced with kind permission of Catholic University of Leuven from J.P. Kruth, B. Vandenbroucke, J. Van Vaerenbergh, P. Mercelis, Benchmarking of different SLS/SLM processes as rapid manufacturing techniques, in: Proceedings of the International Conference Polymers & Moulds Innovations PMI 2005 (April 20-23, 2005, Gent, Belgium). 2005.)*

**Table 3.1** Comparison of most widely used powder-bed fusion technologies.

| | **Selective laser sintering** | **Selective laser melting** | **Electron beam melting** |
|---|---|---|---|
| Energy source | Laser | Laser | Electron |
| Materials | Thermoplastics, ceramics, and metals | Tool steels, stainless steels, Al alloys, Ni superalloys, Ti and its alloys, CoCr, precious metals | Ti and Ti alloys, titanium aluminides, Ni superalloys, CoCr |
| Powder particle size | 10–250 μm | 10–45 μm | 45–105 μm |
| Atmosphere | Protective | Protective | Vacuum |
| Process temperatures | Preheating up to a temperature in between crystallisation and melting | Preheating below 250°C or room temperature | High temperature preheating |
| Prone to residual stresses | Low | High | Low |
| Geometrical complexity | High | High | Medium |
| Support need | None | High | Almost none |
| Typical layer thickness | 100–120 μm | 20–50 μm | 50–100 μm |
| Build rate | Low | Medium | High |

copper-, and magnesium-based materials have become one of the focus areas with the development of technology [8–11]. Widening the range of possible materials in SLM has always been an interesting topic in research because each group has its own advantages and limitations. For example, the poor fluidity of Al alloys leading to the formation of agglomeration when spreading the powder or the high laser reflectivity reaching 91% combined with high thermal conductivity, high oxidation and moisture absorption capacity make Al alloys difficult to process. Thus, for the time being, AlSi10Mg and Al–Si12 are the most widely studied Al alloys because castability and weldability are relatively good [12].

With the developments of AM processes, especially in the last two - decades, SLM as well as other AM techniques are broadly applied in the manufacturing of functional end-products in diverse industries from aerospace, automobile, to biomedical applications in addition to prototyping and rapid tooling. For example, in biomedical applications, using the advantages of the SLM process of high customization and geometrical complexity, different case studies have successfully been accomplished such as drug-delivering implants [13], biodegradable implants [14], metallic scaffolds [15], and bone replacement parts [16]. In aerospace applications, combined with advanced design techniques such as generative design, SLM and other AM processes offer weight reduction, manufacturability of highly complex geometries, reduction of waste material in the form of chips in conventional manufacturing, part consolidation as well as reduced lead time from design to testing, which accelerates the product development stages significantly [17,18]. An example geometry obtained by using topology optimization and suitable for the SLM process is shown in Fig. 3.2 [19].

The powder-bed fusion AM technologies differ from conventional manufacturing techniques in many ways. One of the most important differences is the generation of the microstructure together with the geometry simultaneously in contrast to machining to the final geometry following forging or casting processes whereby the microstructure is mainly determined. Thus, the process parameters mainly controlling the energy input and cooling rates are very critical to ensure a high density and good mechanical properties as well as minimum deformations resulted due to residual stresses. As shown in Fig. 3.3, the obtained density highly depends on the selected parameters, especially at low energy inputs. However, for very large energy inputs, it is possible that the mode of melting is changed to keyhole-mode melting rather than conduction-mode known with semicircular melt pool geometry. In the keyhole-mode melting, the energy density of the laser beam is sufficiently high to evaporate the material and to form plasma leading to the result that the melt pool is much deeper than its

**Figure 3.2** Topologically optimized aerospace bracket. *(Reproduced with kind permission of Elsevier from M. Seabra, J. Azevedo, A. Araújo, L. Reis, E. Pinto, N. Alves, R. Santos, J.P. Mortágua, Selective laser melting (SLM) and topology optimization for lighter aerospace componentes, Proced. Struct. Integr. 1 (2016) 289–296.)*

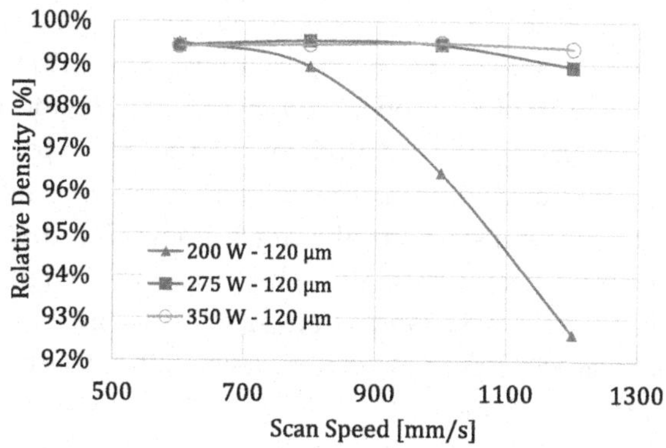


**Figure 3.3** Obtained density at different process parameters: scan speed set to 600–1200 mm/s, laser power set to 200–350 W at a hatch distance of 120 μm [24].

width and the depth-to-width ratio exceeds 0.5 [20–23]. Conduction-mode melting is more stable and does not cause formation of large voids in the material unlike keyhole-mode melting [20].

In the SLM process, both continuous and pulsed lasers are used although continuous lasers currently dominate the market. Typically, ytterbium fiber lasers ($\lambda \approx 1.03-1.1$ μm) with an average laser power <1 kW, and producing laser spots ≤100 μm are used in SLM machines; however, other laser sources, such as $CO_2$ ($\lambda = 10.6$ μm) and Nd:YAG ($\lambda \approx 1.06$ μm) lasers, have been used in the past. The utilization of fiber laser is a consequence of the higher absorptance of metallic powders to this infrared laser radiation (in contrast to $CO_2$ lasers).

Depending on the laser type, different process parameters need to be studied and optimized for the required material properties. In SLM, the common processing parameters to consider are laser power, scan speed, hatch spacing, layer thickness (LT), and scan strategy.

In both laser types (continuous and pulsed lasers), the basic mechanism to create a melt pool is similar: the high energy input causes large melt pool dimensions. Yet, pulsed mode lasers present a higher number of process parameters such as point distance, frequency, peak power density, or pulse duration in comparison with continuous mode lasers. For both pulsed and continuous mode lasers, volumetric energy density (VED), which combines laser power, scan speed, LT, and beam diameter as a single parameter, is commonly used in literature although it cannot be always used as an accurate criterion for melt pool dimensions [25,26]. For example, VED is not the right metric to express melt pool depth and to define the threshold between different modes of melting. King et al. [20] proposed the use of normalized melt pool depth as a function of normalized enthalpy, and this presents a more appropriate model in describing this change from conduction to key-hole melting mode. Moreover, VED is not able to fully capture the complex physical phenomena in the SLM process such as Marangoni flow, hydrodynamic instabilities, and recoil pressure that mainly drive heat and mass transport in the melt pool [27]. Before using a combined single parameter such as VED, the relations of single factors, that is, process variables, and their interactions need to be investigated to fully understand the change of process outputs such as the density, mechanical properties, or microstructure.

In the SLM process, the scan strategy is one of the most important parameters affecting the process outputs such as residual stresses. Some scan strategies, used in the SLM for different purposes, are shown in Fig. 3.4 [28]. The general scan strategy calls for regular rotation of scanning patterns between layers to minimize periodic material porosity as well as to reduce residual stresses (see Fig. 3.5). The meander strategy is based on scanning the area with neighboring vectors having opposite directions with a constant

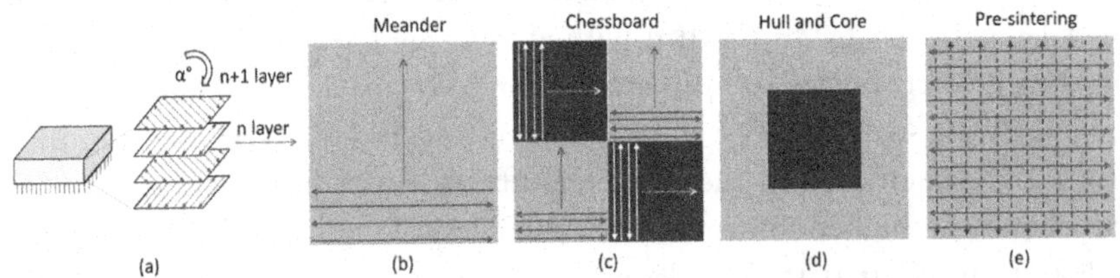


**Figure 3.4** Various scan strategies used in the selective laser melting process. *(Reproduced with kind permission of MDPI from D. Koutny, D. Paloušek, P. Libor, H. Christian, P. Rudolf, T. Lukas, K. Jozef, Influence of scanning strategies on processing of aluminum alloy EN AW 2618 using selective laser melting, Materials 11 (2018) 298.)*

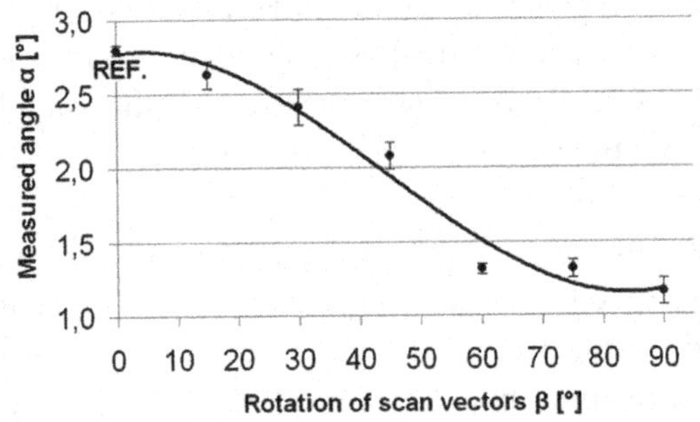


**Figure 3.5** Influence of the orientation of the scan vectors on the curling of the test part [29].

hatch distance, or scan spacing. To avoid long scan vectors leading to high residual stresses, the area to be scanned can be divided into bands or islands so that the whole area is scanned with the same and short vector length. Chessboard scanning, also known as island scanning, is proved to reduce the residual stresses compared to other scan strategies by employing a random sequence of islands [29]. When the core of the part does not necessitate high density or fine details, hull and core strategy can be usedd. In this strategy, the part is divided into two main parts where different parameters can be allocated. Sometimes, re-melting or pre-melting can be used for increasing density, reducing residual stresses, etc. This strategy calls for scanning the same layer twice with possibly different parameters without powder deposition although this may increase the production time. It should be noted that many parameters, such as residual stresses, production time, cracking, microstructure, and edge effects need to be simultaneously considered to determine the right scan strategy depending on the application requirements.

Moreover, surface quality issues of the SLM-fabricated parts still possess a major challenge for a wider adoption of this process since surface quality has a massive effect on the service life and reliability of the manufactured parts. Moreover, it is not an independent parameter to be optimized on its own, but accomplishing the best surface is rather an interconnected process with many other process outputs since it can affect other properties such as mechanical properties and obtained density [30]. The following section details the main issues in terms of surface quality and integrity in the SLM process proceeded with the surface enhancement treatments applied either during or after the SLM process.

## 3.2 Surface integrity in the selective laser melting process

One of the several barriers currently hindering the potential of SLM is the surface-quality-related issues. Due to the layered manufacturing nature of the process, there are some outcomes related to the surface integrity, which can be defined as effects such as edge effect, staircase (stepping) effect, peel effect, and balling. Moreover, down-facing surfaces inherently lead to deteriorated surface quality. Thus, the underlying mechanisms and the results will be addressed in the following paragraphs.

The overhanging surfaces possess the hardest design limitation of a part to be produced in the SLM process. Moreover, it is an important factor to determine the final product's dimensional accuracy and surface quality as shown in Fig. 3.6 [31]. In comparison with other surfaces, overhangs are

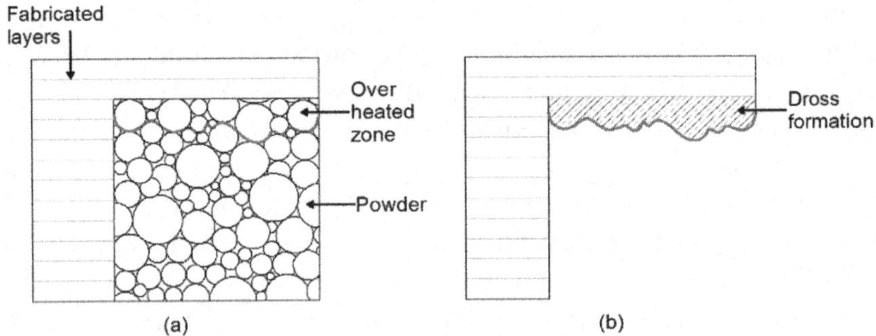


**Figure 3.6** A depiction of overheated zone above loose powder and the resulting dross formation. *(Reproduced with kind permission of MDPI from A. Charles, A. Elkaseer, L. Thijs, V. Hagenmeyer, S. Scholtz, Effect of process parameters on the generated surface roughness of down-facing surfaces in selective laser melting, Appl. Sci. 9 (6) (2019) 1256.)*

characterized by the highest surface roughness due to the combined effect of several underlying mechanisms such as stair-stepping effect, reduced conduction, warping, and dross formation [32,33]. Stair-stepping, in other words staircase, effect is an unavoidable phenomenon occurring during the fabrication of overhang surfaces. Before the SLM process starts, as a pre-processing step, the CAD file of the part to be produced should be sliced into several thin layers. The overhanging length between consecutive two layers depends on the LT and the angle of inclination [34]. A decrease of the inclination angle leads to a larger overhanging length and thus a dominant staircase effect. Besides, the utilization of a thinner powder layer, meaning a lower LT, would reduce the staircase effect.

When the parts to be produced by the SLM process have overhang surfaces, then the process parameters used for down-skin surfaces become critical. Fig. 3.7 shows two parts orientated at different positions with

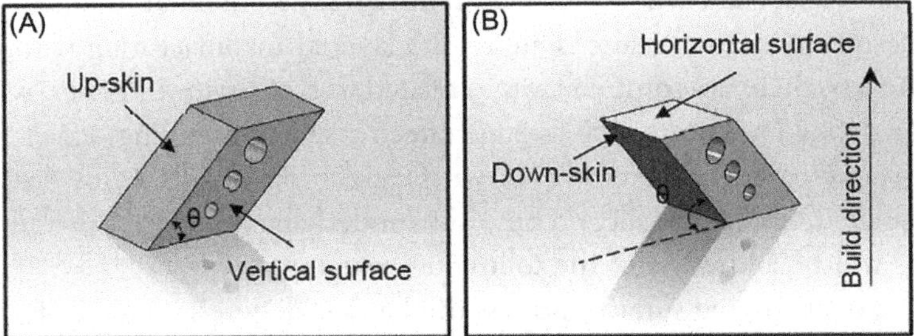


**Figure 3.7** Schematic drawings of parts with different orientations showing down-skin and up-skin surfaces. *(Reproduced with kind permission of Elsevier from Z. Chen, S. Cao, X. Wu, C.H.J. Davies, Surface roughness and fatigue properties of selective laser melted Ti–6Al–4V alloy, in: F. Fores, R. Boyer (Eds.), Additive Manufacturing for the Aerospace Industry, Elsevier, 2019, pp. 283–299, ISBN 9780128140628.)*

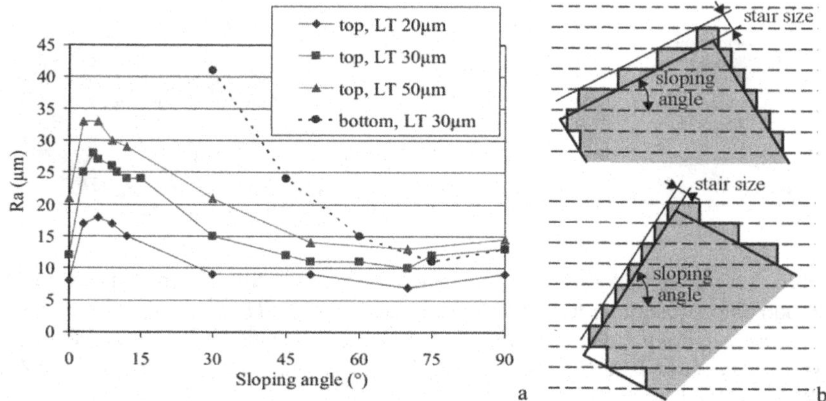


**Figure 3.8** Dependence of the stair effect on the layer thickness and sloping angle. *(Reproduced with kind permission from B. Vandenbroucke, J.P. Kruth, Selective Laser Melting of Biocompatible metals for rapid manufacturing of medical parts, in: Proceedings of Solid Freeform Fabrication Symposium, Austin, TX, USA. 2006.)*

respect to the base plate showing up-skin and down-skin regions [35]. Both up-skin and down-skin surfaces are affected by the stair-effect, which is inherent to additive manufacturing (AM) processes. This effect is well addressed in the literature in terms of modeling its influence on the surface quality [36–38] or studying the effect of LT or inclination angle [39,40]. As shown in Fig. 3.8, as the sloping angle is increased or LT is reduced, the surface quality is improved. However, these parameters, the LT and sloping angle, are generally constrained by many other factors to be considered such as part orientation, residual stresses, part design, and production rate. Thus, it is not easy to alter these variables to lessen the stair effect. As shown in the graph, the stair effect is much more pronounced on down-skin surfaces (see bottom, LT-30 μm in Fig. 3.8).

Besides the staircase effect, warping is another non-negligible fabricating defect encountered during the SLM process. The temperature gradient mechanism (TGM) addressed by Kruth et al. [41] can be applied to the SLM process to explain warping. In this process, TGM acts on previously fabricated layers lying beneath the processed powder layer. A critically steep temperature gradient develops owing to the rapid heating of the upper surface scanned by the high-energy-focused laser beam. The thermal stresses form during the process of the rapid solidification of melt pool, and a bending angle toward the laser beam develops as shown in Fig. 3.9 [41]. For reducing the thermal deformations during processing in the SLM process, effective methods are using preheating the base plate to a preset

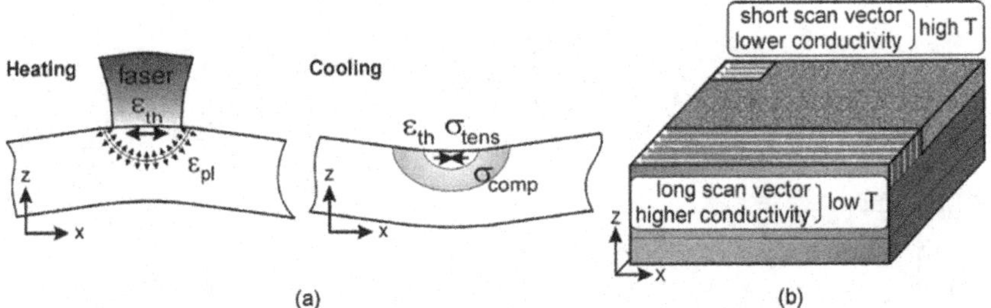


**Figure 3.9** (A) Temperature gradient mechanism and (B) different temperatures through different scan vector lengths. *(Reproduced with kind permission of Elsevier from J.-P. Kruth, L. Froyen, J. Van Vaerenbergh, P. Mercelis, M. Rombouts, B. Lauwers, Selective laser melting of iron-based powder, J. Mater. Process. Technol. 149 (1–3) (2004) 616–622, ISSN 0924-0136.)*

value or use of laser re-melting as a stress relief in-process heat treatment [42–45]. Another solution to avoid the undesired results of residual stresses such as cracks and warping is to use support structures [46,47]. In addition, these support structures used to dissipate the heat from the overhang surfaces are used to fix the part to the building base plate to resist the forces of the metallic recoating blade. It is a compromise between having a sufficient volume of supports to avoid distortions and cracking and having minimum volume of supports to be able to easily remove them after the process. One of the approaches to minimize the supports in the SLM process is to optimize the build orientation [48,49]. Despite the fact that changing the part orientation is one of the measures to overcome the adverse effect of overhang surfaces, proposed orientations should generally be optimized for more than one objective, such as to reduce residual stresses. Reducing staircase effect and minimizing the build time may be other optimization criteria [50]. Needless to say, in a part with a complex geometry, there are usually more overhang surfaces than only one that may be conflicting. Therefore, it is generally not very effective.

The design of support structures is an important topic unless they are completely eliminated. Although there are various types of support types used in the SLM process such as point, line, spider, and contour, block-type supports are the most widely used ones. In the study by Poyraz et al. [46], the effects of different design parameters of the block support structures were investigated in terms of warping and microstructure formation. As shown in Fig. 3.10, the influence of hatch distance, fragmentation, top tooth length, and Z-offset on the thermally induced deformations was studied in SLM of Inconel 625. The obtained results indicate that hatching

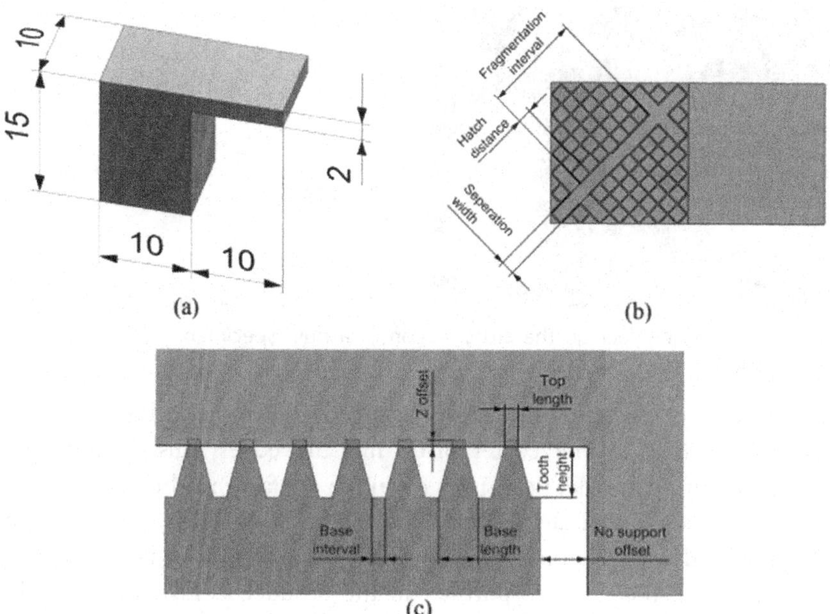


**Figure 3.10** Block support design parameters. *(Reproduced with kind permission from [47].)*

parameters have greater influence on the support structures compared with tooth parameters. Increased hatch distance increased the warping on the part, and even more it caused separation after a certain level. On the other hand, in terms of the easiness of removing the supports, higher hatch distance and fragmentation interval shall be preferred to detach supports from the part [46].

The study by Hussein et al. described an application of lightweight support structures where they showed that the type of support, the volume fraction, and the size of the cells are the main factors that influence the manufacturability, the amount of support, and the total build time [51]. In another study carried out with Hastelloy X, it was concluded that it is critical to have a right combination of support structure type and process parameters to overcome residual stress-induced part distortion in the areas of interaction between support and part, and to optimally conduct excess heat away from the part [52].

Another serious effect that deteriorates the surface topology and dimensional accuracy is the edge-effect, which is defined as the formation of elevated edges in SLM, as shown in Fig. 3.11. Additionally, the existence of the elevated edges may also worsen the staircase effect. Elevated edges on

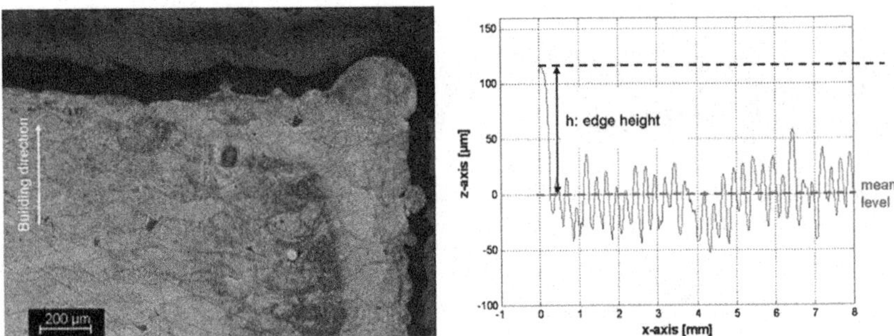


**Figure 3.11** Edge-effect on the cross-section of a built specimen. *(Reproduced with kind permission from [54].)*

the contours of the parts also leads to the collision of these edges with the hard coater blades. The coater blade hits the edges causing undesired vibrations during powder deposition owing to the fact that the height of the produced edges is typically higher than one-layer thickness. This leads to the undesired waviness on the surface of deposited powder layer, thereby possibility to cause periodic porosity in the produced SLM parts [53,54]. The studies show that laser erosion applied with the same laser of SLM or depositing the powder (see Fig. 3.12) or scanning the last layer twice by removing the contour scanning can minimize the edge-effect if not eliminated completely.

Balling effect, which is defined as the shrinking tendency to reduce the surface energy under the action of surface tension during the SLM is detrimental to part quality in terms of dimensional accuracy and surface quality [55]. This leads to necessity of post-process finishing operations and probability of occurrence of porosities between many discontinuous metallic balls. Moreover, if the balling phenomenon is severe, the bellied metal balls tend to hinder the movement of coater blades resulting in an undesired stop. There are many studies in the literature studying the effect of various process variables on the balling effect [55—61]. Li et al. investigated the influence of the oxygen content, line energy density, and scan intervals on the severity of the balling. The results show that balling is enhanced by the increase of the oxygen content, whereas a high linear energy density results in a nicely continuous track due to good wetting. As shown in Fig. 3.13, the LT deteriorates the balling effect. They also pointed out that the scan interval (or scan spacing) does not change the balling [55].

Humping, also seen in welding, is another phenomenon encountered in the SLM process depending on the process parameters and leads to swelling deteriorating the surface quality [62]. Swelling can be defined as the rise of solid material above the plane of powder level and melting, and it occurs

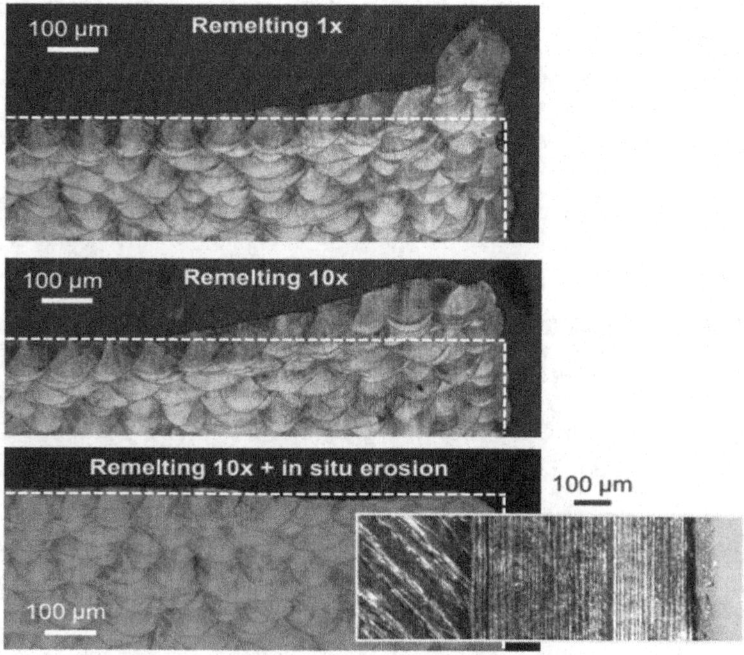


**Figure 3.12** The effect of re-melting and erosion on the edge-effect. *(Reproduced with kind permission of Catholic University of Leuven from J. Metelkova, Y. Kinds, de C. Formanoir, A. Witrouw, B. Van Hooreweder, In situ combination of selective laser melting and selective laser erosion for edge effect removal, in: Presented at the Solid Freeform Fabrication (SFF), Austin, Texas, USA, 12 Aug 2019–14 Aug 2019. 2019. Available: lirias.kuleuven.be/2843118?limo=0.)*

due to surface tension effects related to the melt pool geometry. Similar to balling, wetting, and capillary characteristics of a melt pool have been identified as possible contributors to humping (or swelling) [63].

### 3.2.1 Surface enhancement treatments

There are many diverse treatments applied after or during the SLM process to improve the surface quality and integrity. Therefore, the applied treatments are investigated under two different groups.

## 3.3 Treatments applied during the selective laser melting process

One of the most influential remedies to improve the surface quality is laser re-melting. This treatment has been addressed by many researchers in the

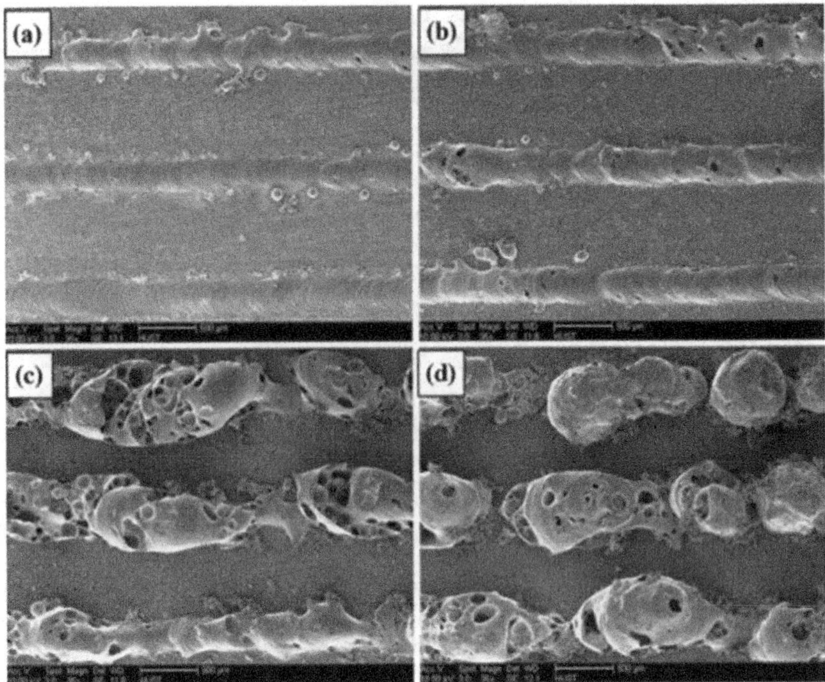


**Figure 3.13** As the layer thickness is increased from A to D, the balling effect becomes more dominant. *(Reproduced with kind permission of Springer Nature from R. Li, J. Liu, Y. Shi, L. Wang, W. Jiang, Balling behavior of stainless steel and nickel powder during selective laser melting process, Int. J. Adv. Manuf. Technol. 59 (2012) 1025. https://doi.org/10.1007/s00170-011-3566-1.)*

literature with different materials and is now used by most of the machine vendors for improving the uppermost and horizontal layers of produced specimens [30,64–76]. During the SLM process, after depositing a new layer of powder, scanning the geometry, and selectively melting the powder, a second scanning takes place before the deposition of the powder for the next layer starts. This treatment, called as laser re-melting, aims to reduce the surface pores and increase the smoothness of the final layers. This is not limited to only top layers, at the cost of increased production time, laser re-melting can be applied after every layer or after every (n) layers to almost totally eliminate the pores. Some researchers focused on the effect of the laser re-melting parameters on the improvements of the surface quality and obtained density. Yasa and Kruth [68] applied laser re-melting while processing AISI 316L stainless steel to understand the effect of scan speed, scan spacing, laser power, and number of re-melting layers on the porosity

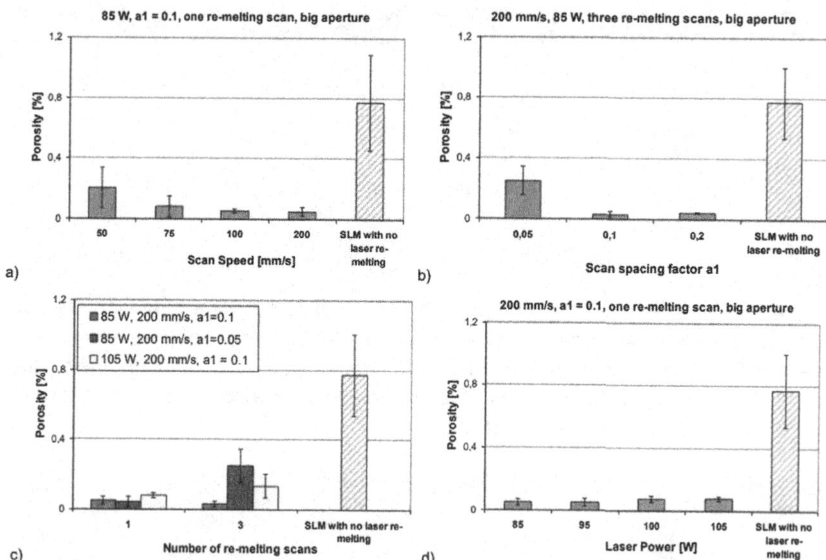


**Figure 3.14** Effect of process parameters on the remaining porosity (A) scan speed, (B) scan spacing, (C) number of re-melting scans, and (D) laser power. *(Reproduced with kind permission from E. Yasa, J.-P. Kruth, Application of laser re-melting on selective laser melting parts. Adv. Prod. Eng. Manag. 6 (2011) 259–270.)*

as shown in Fig. 3.14. The average porosity of SLM parts without any laser re-melting is about 0.77%, whereas the densest re-molten part obtained with a scan speed of 200 mm/s, a laser power 85 W, and a scan spacing factor of a1 = 0.1 has a porosity of 0.032%. The graphs show that the obtained density is significantly improved once the laser re-melting is applied. Moreover, the higher scan speed together with higher scan spacing and lower laser power leads to lower density. To further decrease the porosity, it may not always be critical to increase the number of re-melting layers (see Fig. 3.14C). The effect of laser re-melting can clearly be observed from Fig. 3.15. The cross-sections of two specimens showing the top layers of produced parts, either with or without laser re-melting are displayed. This figure shows the enhancement of the surface quality after applying laser re-melting in comparison with a part produced only by SLM. Moreover, as shown, a re-melted zone, clearly distinguishable at the top of the part is observed. This fully dense zone without any porosity or other defects is desired for enhanced surface properties while its thickness depends on the re-melting parameters and is increased with higher energy densities. Additionally, as evident from Fig. 3.16, the cellular/dendritic

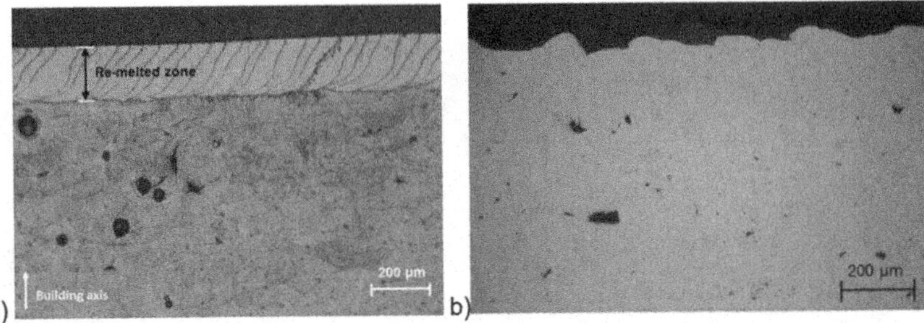


**Figure 3.15** A cross-sectional optical microscopy image of (A) a part with laser re-melting (50 mm/s scan speed, 85 W laser power, a1 = 0.1 scan spacing) and (B) only selective laser melting part with no laser re-melting. *(Reproduced with kind permission from E. Yasa, J.-P. Kruth, Application of laser re-melting on selective laser melting parts. Adv. Prod. Eng. Manag. 6 (2011) 259–270.)*

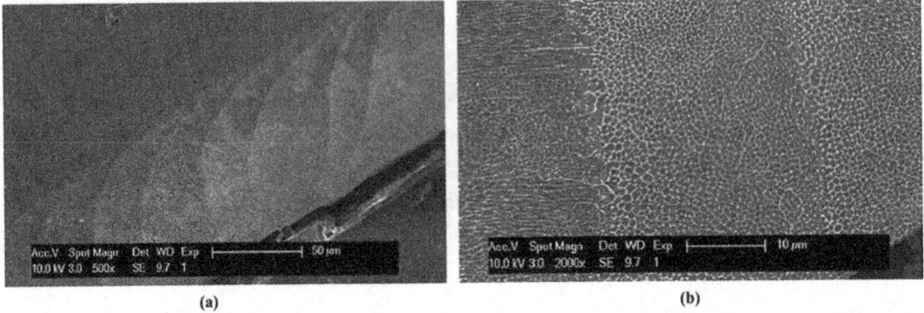


**Figure 3.16** Scanning electron microscopy pictures of the selective laser melting part with laser re-melting of a scan speed of 200 mm/s, laser power of 105 W, scan spacing factor of 0.1 and 10 re-melting layers. *(Reproduced with kind permission of Emerald from E. Yasa, J. Deckers, J.-P. Kruth, The investigation of the influence of laser re-melting on density, surface quality and microstructure of selective laser melting parts, Rapid Prototyp. J. (17) 5 (2011) 312–327.)*

microstructure is refined as the laser re-melting is applied. The cell size was finer in parts with laser re-melting than the only SLM parts [69].

The surface improvement is not obtained only with AISI 316L stainless steel but also with SLM of Ti6Al4V as shown in Fig. 3.17 [69]. During these tests, the LT was set to 30 μm. Depending on the laser power and scan speed, the average roughness ($R_a$) value of only SLM parts, which was about 15 μm could significantly be reduced to around 2.9 μm with a scan speed of 640 mm/s and 256 W as shown in Fig. 3.17B. In a study by Demir and Previtali, the effect of laser re-melting is demonstrated on the SLM of 18Ni300 maraging steel as well [74]. In their study, it is concluded that the re-melting strategies are effective in improving the part density and surface quality, which are closely linked to each other. Their results prove that

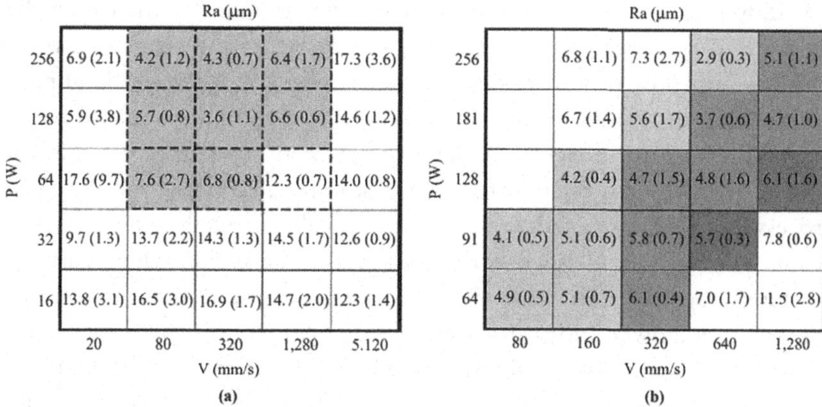


**Figure 3.17** The average roughness results, $R_a$, of the Ti6Al4V specimen standard deviations expressed in micrometers in parentheses for the (A) first set of tests; (B) second set of tests. *(Reproduced with kind permission of Emerald from E. Yasa, J. Deckers, J.-P. Kruth, The investigation of the influence of laser re-melting on density, surface quality and microstructure of selective laser melting parts, Rapid Prototyp. J. (17) 5 (2011) 312–327.)*

surface smoothing effect provided by the laser re-melting pass is also beneficial for avoiding pore formation and its further propagation through the layers.

Laser re-melting is beneficial in terms of surface quality and density improvement as pointed out by many researchers studying different materials. However, there are also some weaknesses accompanied by the laser re-melting such as dimensional errors due to edge effect [69,74] or the increase of the production time [30,64–69].

Laser re-melting is not the only treatment that can be applied during the process to improve the surface quality of SLM parts. The surface roughness of an SLMed part is determined by a combination of process parameters, powder characteristics, and part geometry. The most influential SLM process parameters on the surface roughness are laser power, spot size, scanning speed, scan strategy, hatch distance, and LT [64–66]. These parameters can be changed within each build, even for each part to be produced with some limitations. Majeed et al. addressed the change of the surface roughness with respect to the laser power in the range of 320–400 W for AlSi10Mg. It is concluded that with increasing laser power, the surface roughness increased while the best results are obtained with a laser power of 320 W [30]. Another important process variable that cannot be easily changed during SLM processing is the laser beam size. As shown in the study by Koutiri et al. the roughness is deteriorated by increasing the beam size [77]. Actually, the reason of using different process parameters for contours, up-skin or down-skin vectors is also to improve the surface quality and to eliminate the surface defects during the SLM process [78–80]. Fig. 3.18 shows the difference between up-skin, down-skin, or in-skin, which is also referred to as core, scanning. As shown, in the SLM process, the diameter of the laser beam is smaller than the effective laser beam, which can be defined as the diameter of the region where particles are fused together. This effective laser beam diameter depends on the selected laser power and scan speed. To compensate for the dimensional accuracy, a beam compensation needs to be used as is the case for the tool's diameter in machining. This is known as beam offset and defined with respect to the nominal part boundary. During the SLM process, first of all, all contours are scanned with contour parameters, and this is followed by the hatching of the inner areas. During hatching, the laser beam moves line after line several times to fill the area to be scanned. The distance between successive laser tracks are known as the scan spacing or hatch distance. After the whole area is scanned and solidified, a second exposure on the contours may be exposed due to two reasons. First, sharper part contours can be generated with higher temperature gradients due to the higher thermal conductivity of the already solidified material in the area of the first contour in comparison with powder. This leads to reduced roughness on the vertical planes. Secondly, the building accuracy in the xy plane is increased [80,81]. Two samples with and without contour scans are shown in Fig. 3.19 where it is shown that the irregularities at the edges of samples reduced when a contour scan was applied around hatches [82]. To obtain a good surface quality, a better understanding and knowledge of the process parameters for contours and hatch lines is critical. In the study by Calignano et al. [80], it was concluded that scan speed was found to have the greatest influence on the surface roughness for aluminum samples. In another study by Ref. [77], it is shown that globally, the nonfiltered surface finish value $S_a$ was shown to increase with lower scan speeds, to decrease with higher powers, and to increase severely on down-skin sides for large building angles for Inconel 625 specimens.

One of the process variables to enhance the surface quality on vertical planes is the skywriting option to be employed during the SLM process. A certain amount of time is needed to accelerate the scanning head to the intended speed during laser scanning due to its inertia. The laser travels some distance with non-constant speed during this time, sacrificing surface

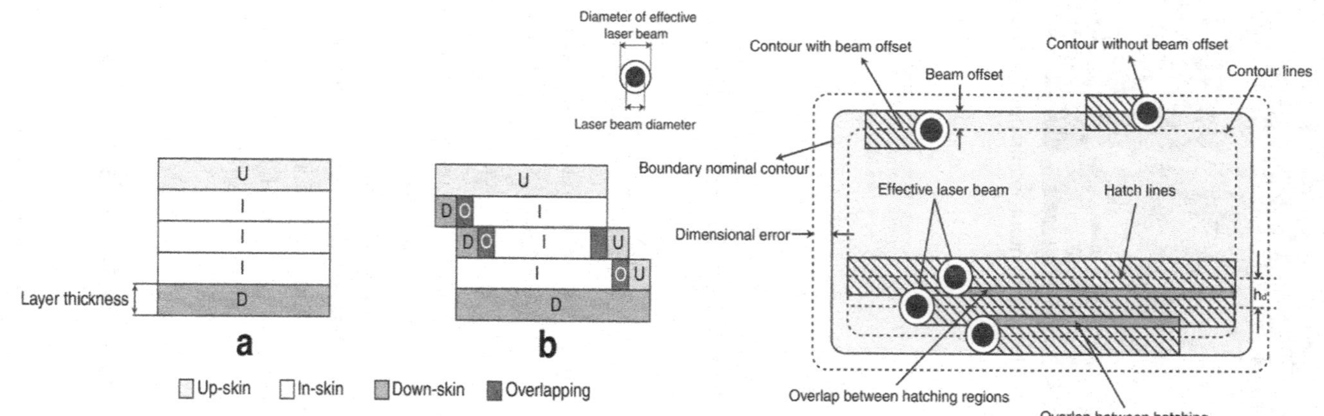


**Figure 3.18** The difference between up-skin and down-skin scanning with overlap regions (cross-sectional view); contour and hatch scanning. *(Reproduced with kind permission of Springer Nature from F. Calignano, D. Manfredi, E.P. Ambrosio, L. Iuliano, P. Fino, Influence of process parameters on surface roughness of aluminum parts produced by DMLS, Int. J. Adv. Manuf. Technol. 67 (2013) 2743–2751.)*

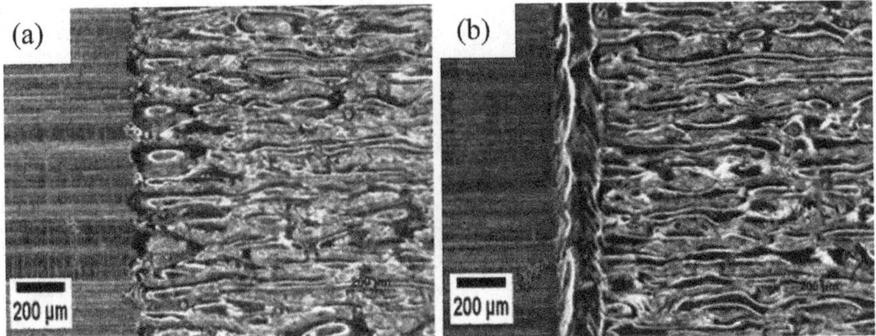


**Figure 3.19** Without (A) and with (B) contour scanning. *(Reproduced with kind permission of Elsevier from [83].)*

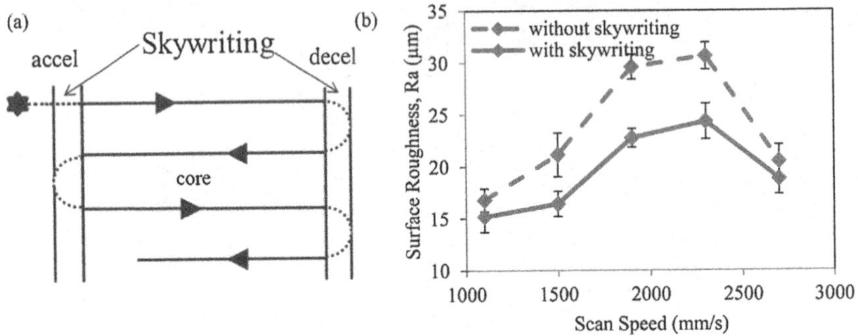


**Figure 3.20** Effect of skywriting on the surface roughness. *(Reproduced with kind permission of Elsevier from [83].)*

quality and dimensional accuracy. Provided that the skywriting option is selected, an approach as shown in Fig. 3.20 is used. During acceleration and deceleration phases when the scanning direction is about to be altered, the laser is switched off. Thus, no powder material is melted during positioning in between two neighboring laser tracks. The surface roughness results obtained with and without skywriting option are depicted at a laser power of 370 W, hatch distance of 90 μm, and LT of 40 μm. The positive influence of the skywriting is obvious across all the scan speeds investigated as shown in Fig. 3.20B.

## 3.4 Treatments applied after the selective laser melting process

Although there are many measures taken during the SLM process to improve the surface quality and eliminate the surface defects, the need for further processing of the surfaces is not totally ruled out. Depending on the application and technical requirements, after the process is completed and the part is removed from the machine together with the base plate it is built on, some further finishing steps may be necessary for surface enhancement. There are various types of finishing operations applied after the SLM process as shown in Fig. 3.21, and these may be broadly categorized into mechanical, thermal, and electrochemical/chemical processes adapting the classification of deburring operations by Gillespie [83,84]. Under mechanical processes, machining, blasting, grinding, or water jetting processes may be considered while thermal processes include laser polishing and electron beam irradiation. Electrochemical or chemical processes are mainly etching and polishing methods using chemicals [84].

The effect of machining processes on the part performance has been evaluated by different researchers. Dumas et al. studied the effect of the finish cutting operations on the fatigue performance of Ti–6Al–4V parts produced by the SLM process [85].This study shows that the fatigue strength of additively manufactured samples was half the fatigue strength of conventional ones mainly due to the porosity levels and brittle phases. None of the finishing techniques, turning, milling, or polishing were able to lead to a fatigue resistance comparable to wrought titanium part. They also concluded that the most widely used roughness parameters such as $S_a$ (areal arithmetic mean height) and $S_{vk}$ (areal reduced valley depth) do not well correlate with fatigue. They recommend that the roughness parameter $S_{sk}$ (areal skewness), which is a good representation of probability of stress concentrations sites, is a better parameter to show the link between the roughness and fatigue performance. However, as expected, the applied machining operations have increased the surface quality significantly as shown in Fig. 3.22. The high improvement of the surface quality after machining, as expected, is consistent with other studies as shown in Fig. 3.23 [86,87]. However, the improvement rate highly depends on the initial surface quality. Thus, it is important to develop right SLM process parameters resulting in the least roughness values.

In another work, Milton et al. [87] studied the effect of finish milling on the roughness, hardness, and obtained cutting forces of SLMed parts. They concluded that SLM samples resulted in 22% greater axial cutting force than the conventional sample, which would lead to a higher wear rate and lower tool life [88]. Moreover, depending on the build direction, it is concluded that the obtained surface quality may vary even after finishing operations.

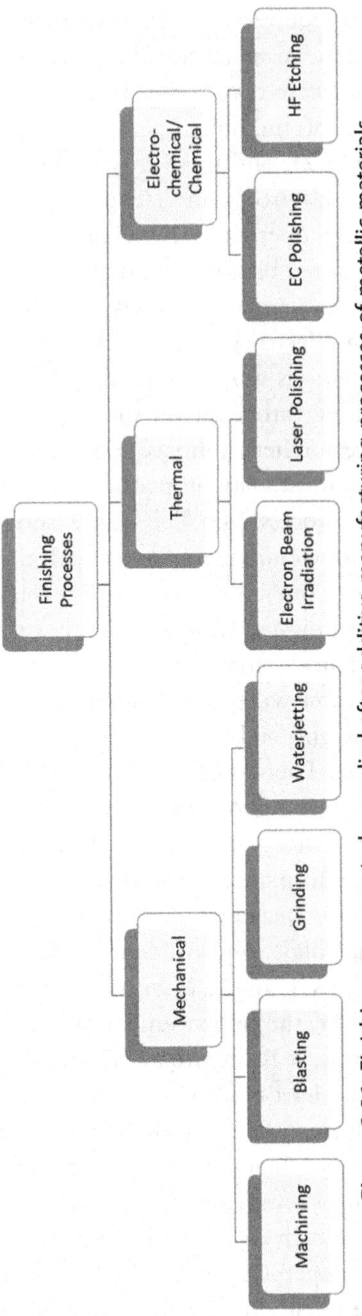


**Figure 3.21** Finishing processes to be applied after additive manufacturing processes of metallic materials.

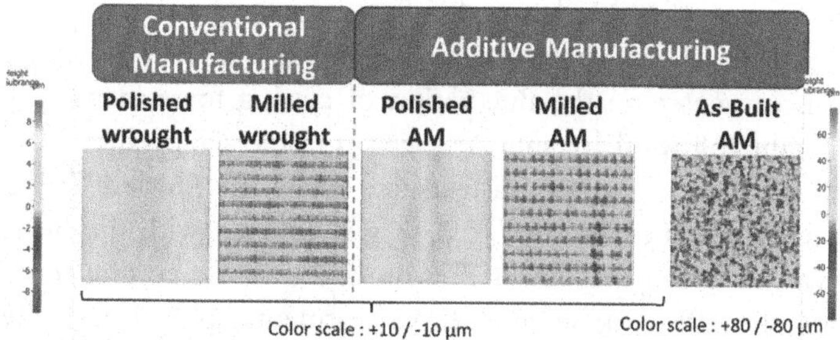


**Figure 3.22** 3D topographies for the different processes. *(Reproduced with kind permission of Elsevier from M. Dumas, F. Cabanettes, R. Kaminski, F. Valiorgue, E. Picot, F. Lefebvre, C. Grosjean, J. Rech, Influence of the finish cutting operations on the fatigue performance of Ti-6Al-4V parts produced by Selective Laser Melting, Proced. CIRP 71 (2018) 429–434, ISSN 2212-8271.)*

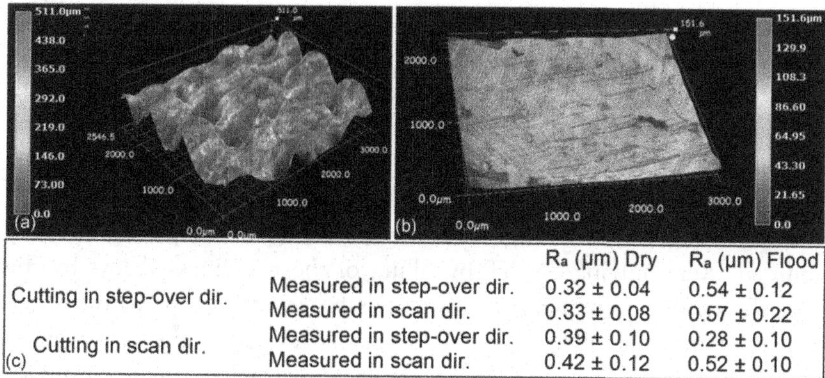


(c)

| | | $R_a$ (µm) Dry | $R_a$ (µm) Flood |
|---|---|---|---|
| Cutting in step-over dir. | Measured in step-over dir. | 0.32 ± 0.04 | 0.54 ± 0.12 |
| | Measured in scan dir. | 0.33 ± 0.08 | 0.57 ± 0.22 |
| Cutting in scan dir. | Measured in step-over dir. | 0.39 ± 0.10 | 0.28 ± 0.10 |
| | Measured in scan dir. | 0.42 ± 0.12 | 0.52 ± 0.10 |

**Figure 3.23** (A) Surface topography of sample 1 prior to milling; (B) after milling; (C) surface roughness $R_a$ values of the machined samples. *(Reproduced with kind permission of Taylor & Francis Online from D. Brown, C. Li, Z.Y. Liu, X.Y. Fang, Y.B. Guo, Surface integrity of Inconel 718 by hybrid selective laser melting and milling, Virtual Phys. Prototyp. 13 (1) (2018) 26–31. https://doi.org/10.1080/17452759.2017.1392681.)*

Various studies confirm that the effect of the build orientation on the machinability is significant. In a study by Shunmugavel et al. [89], it is observed that cutting forces are higher during orthogonal cutting of samples perpendicular to the build direction compared with other directions or wrought samples. Not only the cutting forces were different for different build orientation but also different chip curling was observed due to sticking and friction occurred between the tool-chip interfaces. There are other studies in the literature whereby the chip formation is studied in addition to effect of feed rate, cutting speed, and depth of cut on cutting forces in turning of SLMed parts [87,90]. In milling, similar studies have been carried out. For example, Khorosani et al. concluded that the most influential parameter on the fluctuation of cutting force in milling is the scallop height followed by spindle speed, finishing allowance, tool path, heat treatment temperature, and feed rate when the SLMed Ti−6Al−4V parts are milled. When linear tool path is compared with the helical tool path, it was observed that linear one produced lower cutting forces [91]. Moreover, the effect of different heat treatments applied after the SLM process is studied on the cutting forces. Gomez et al. found that the most influential factor was annealing [92].

Other than machining, there are various mechanical surface treatments used for reducing surface roughness such as shot peening (SP), laser shock peening (LSP), cavitation peening (CP), surface mechanical attrition treatment (SMAT), ultrasonic impact treatment (UIT), and barrel finishing (BF). Depending on the application requirements, these mechanical surface treatments can be used for the small-sized parts (SP, SMAT, LSP, ultrasonic shot peening [USP], and BF), large parts (SP, LSP, and UIT), and parts with a complex geometry (SP, SMAT, LSP, USP, BF, and CP) [93]. Being one of them, surface mechanical attrition treatment is defined as hitting the surface with steel balls at a defined vibration frequency for a defined period of time [94]. As shown in Fig. 3.24, the surface topography is significantly modified after applying SMAT by filling of the roughness valleys by the plastically deformed material of roughness peaks and the partially melted particles. Sun et al. [94] concluded that SMAT is an effective way to enhance surface quality. The required surface microrelief, increased surface hardness, and compressive residual macrostress are provided by the modification of the uppermost layers led by plastic deformation [94]. Even after 10 min of application, the surface roughness is reduced by 88%. By increasing the treatment time, the surface roughness can be further reduced to below 0.5 μm matching the values achieved by surface grinding. However, when the method is applied to highly reactive materials, caution shall be taken against contamination of the SLMed part.

Moreover, some studies focus on comparing various post-processes in terms of enhancing the surface quality on real applications. Lesyk et al. comprehensively compared BF, SP, USP, and UIT on turbine blades geometries in terms of surface topography, residual porosity, microhardness, and residual macrostress (see Fig. 3.25) [93]. The results showed that the surface average roughness ($S_a$) of the specimens were reduced by 20.6%, 26.2%, and 57.4% after the BF, USP, and UIT processes, respectively in

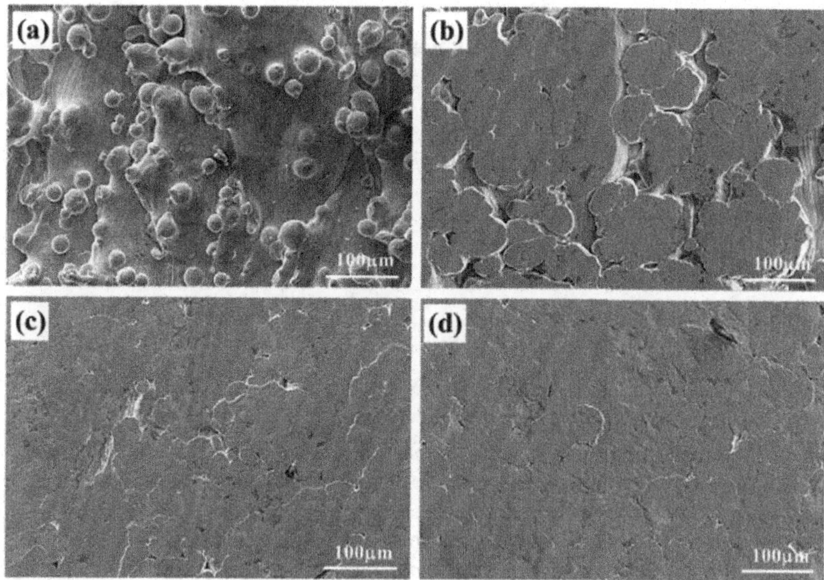


**Figure 3.24** SEM images showing the as-selective laser melting (SLM) surface (A) and the SLM surfaces after surface mechanical attrition treatment for (B) 10 min, (C) 30 min, and (D) 80 min. *(Reproduced with kind permission of Elsevier from Y. Sun, R. Bailey, A. Moroz, Surface finish and properties enhancement of selective laser melted 316Lstainless steel by surface mechanical attrition treatment, Surf. Coating. Technol. 378 (2019) 124993.)*

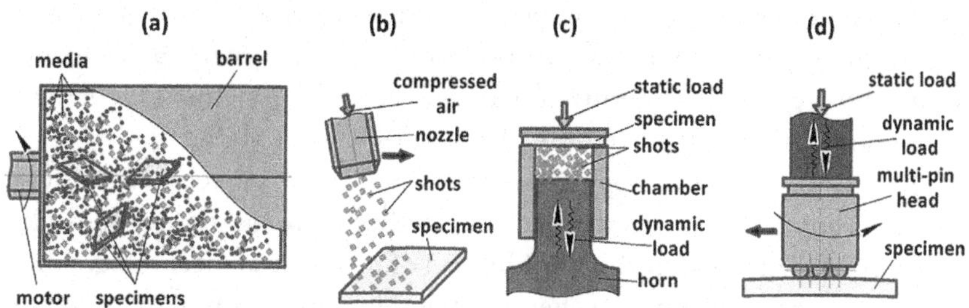


**Figure 3.25** Demonstrations of the barrel finishing (A), shot peening (B), ultrasonic shot peening (C), and ultrasonic impact treatment (D) processes. *(Reproduced with kind permission of Elsevier from D.A. Lesyk, S. Martinez, B.N. Mordyuk, V.V. Dzhemelinskyi, A. Lamikiz, G.I. Prokopenko, Post-processing of the Inconel 718 alloy parts fabricated by selective laser melting: effects of mechanical surface treatments on surface topography, porosity, hardness and residual stress, Surf. Coating. Technol. 381 (2020) 125136, ISSN 0257-8972.)*

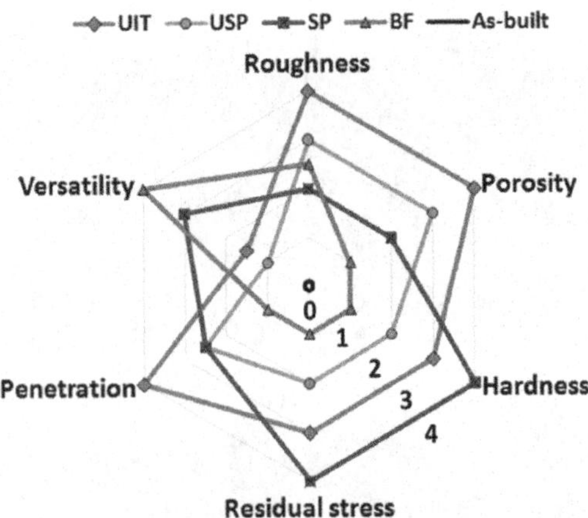


**Figure 3.26** The effectiveness of various post-processes in terms of given aspects. *(Reproduced with kind permission of Elsevier from D.A. Lesyk, S. Martinez, B.N. Mordyuk, V.V. Dzhemelinskyi, A. Lamikiz, G.I. Prokopenko, Post-processing of the Inconel 718 alloy parts fabricated by selective laser melting: effects of mechanical surface treatments on surface topography, porosity, hardness and residual stress, Surf. Coating. Technol. 381 (2020) 125136, ISSN 0257-8972.)*

comparison to the SLM-only parts. These processes also improved the relative density of the parts. BF reduced the porosity level by 23% whereas SP decreased it by 41%. The greatest improvement in porosity reduction was achieved by the UIT by 84%. Moreover, all processes transformed the tensile surface stresses to compressive stresses which is a significant improvement, especially for fatigue performance. The general performance of these post-processes in terms of hardness, porosity, roughness, versatility, penetration, and residual stresses is shown in Fig. 3.26 [93].

Due to geometric complexity, thin walls and low removable-stock reserves, chemical and electrochemical methods can be used for surface quality enhancement after the SLM instead of mechanical machining. Some studies focus only on applying chemical polishing. In the study by Balaykin, the chemical etching of Ti6Al4V samples by HF solutions is addressed [95]. Chemical etching of titanium alloys with HF has some disadvantages due to the fact that hydrogen gas is produced by the reaction. It is flammable and explosive, thus needs to be carefully handled. Moreover, chemical polishing is quite a slow process. The results investigated chemical solutions with different amounts of HF and $HNO_3$ leading to different improvement rates in the surface quality. It was seen that with the best combination, the $R_a$ value has decreased from 3.99 to 1.69 μm [95].

They may be used as a single operation [96–102] or a combination may be used. It is observed that electrochemical polishing, a non-conventional surface treatment, which is based on the localized anodic dissolution of any conductive material, is quite fast to significantly improve the surface quality and reduce the variability of the roughness values. Baicheng et al. studied the electrochemical polishing of Inconel 718 specimens and improved the surface roughness more than 50% in less than 5 min. Even, after 1 min of the treatment, the partially melted powder particles were fully removed. However, microstructural changes and residual stress relief on the surface result in a relatively lower hardness and elastic modulus [97]. The results obtained with Ti–6Al–4V inside a 1:9 vol. ratio of perchloric acid (60%) and glacial acetic acid show that the roughness can be decreased by 92% in some cases with this method with a uniform distribution of peaks and valleys both on internal and external surfaces. The starting surface roughness is important for electrochemical polishing, also known as electropolishing, anodic polishing, or electrolytic polishing, and it has been found that surfaces built at orientations between 22.5 and 112.5 degrees possess a similar surface finish leading to the fact that surface finish and thickness reduction evolve similarly throughout the treatment [100,101]. Although having some disadvantages such as the necessity of special tools and difficulty in determining the net shape due to mass losses related to the electropolishing processes, studies show their efficiency in the surface quality enhancement. Sarkar et al. also studied electropolishing on the SLMed parts in order to improve components' fatigue life [103]. In this study, electropolishing of as-built SLM specimens drastically reduced (~97.4%) the $R_a$ value. The reduction in the surface roughness leads to a significant improvement in the fatigue life of about 100%. Moreover, the fractography analysis shows surface roughness, subsurface defects, and micronotches are primary crack initiating factors for SLM specimens [103]. As a combined process, Pyka et al. studied the surface modification of open porous structures produced by the SLM process with a novel protocol as a combination of chemical and electrochemical etching using HF-based solutions [104]. The starting roughness values and surfaces obtained with the scaffolds are shown in Fig. 3.27. The combination of chemical and electrochemical polishing led to a controlled and homogenous roughness distribution decreasing the average roughness more than 30%. The developed method was also efficient in removing the partially melted powder particles. However, it should be noted that the reduction of the strut size

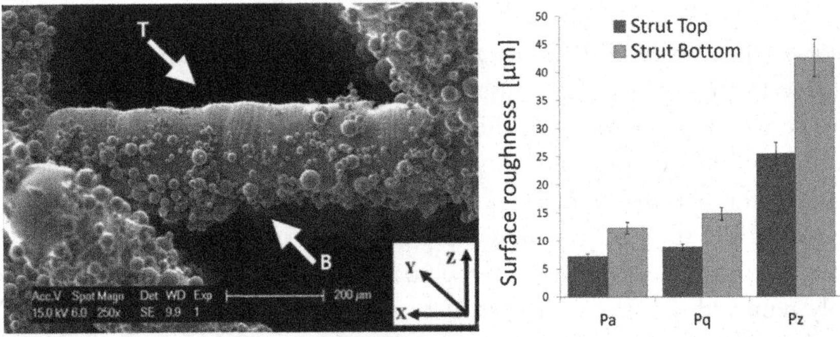


**Figure 3.27** Left: SEM micrograph of a typical strut where non-melted powder grains are present on the top (T) and bottom (B) strut surfaces, Y-direction: build direction. Right: Roughness results on top and bottom of the struts. *(Reproduced with kind permission of John Wiley and Sons from G. Pyka, A. Burakowski, G. Kerckhofs, M. Moesen, S. Van Bael, J. Schrooten, Surface modification of Ti6Al4V open porous structures produced by additive manufacturing, Adv. Eng. Mater. 14 (6) (2012) 363–370.)*

and the increase in the internal porosity after surface modification led to the decrease in the mechanical properties [104].

Abrasive flow machining (AFM) is an interior surface finishing process characterized by flowing a liquid consisting of abrasive particles through a workpiece. This makes AFM a good choice to improve the surface quality of SLMed parts with internal features and cavities. The main parameters to be investigated with AFM are the abrasive concentration and medium viscosity. The study by Duval-Chaneac et al. shows that with the increase of the abrasive concentration and media viscosity, the improvement of the areal roughness values was higher [105]. Other studies in the literature confirm that AFM is an effective method to be applied especially on internal features [106–108]. It is shown that turbine blades are one of the applications where AFM is being demonstrated due to their complex internal and external freeform geometries [109]. AFM, being a flow driven process, presents some challenges with uniform polishing of complex passages, that is, passages that contain internal chambers and expanding cross-sections (see Fig. 3.28). Moreover, sharp or angular features experience rounding and significant variability in final finish exists for internal features of changing cross-sectional area [108].

Another modification of AFM is chemical-AFM for highly complex components with inaccessible areas with hard polishing tools. In the work of Mohammadian et al. the use of chemical and abrasive flows for surface polishing of SLMed IN625 specimens was studied [110,111]. The results on

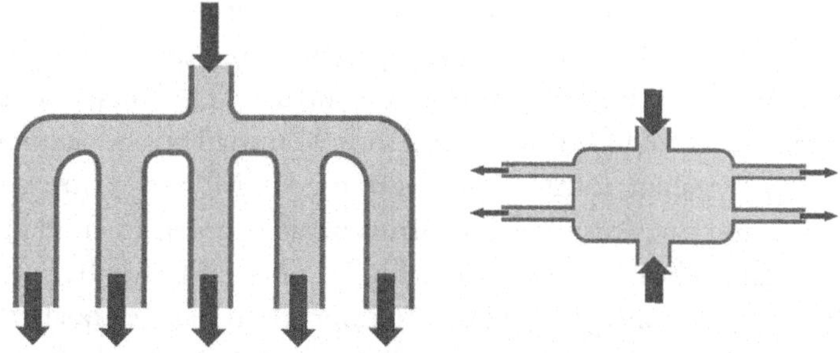

**Figure 3.28** Left: An selective laser melting produced manifold like this represents an ideal geometry for abrasive flow machining polishing. Right: This combustion chamber design would achieve successful polishing of inlets and outlets, but the chamber itself may be difficult to polish entirely. *(Reproduced with kind permission from R.W. Gilmore, An Evaluation of Ultrasonic Shot Peening and Abrasive Flow Machining as Surface Finishing Processes for Selective Laser Melted 316L, Master Thesis of Science in Industrial Engineering, California Polytechnic State University, 2018. Available at: https://digitalcommons.calpoly.edu/cgi/viewcontent.cgi?article=3236&context=theses)*

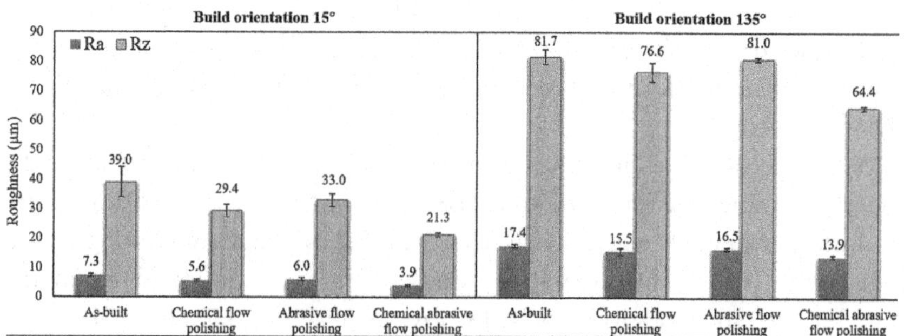


**Figure 3.29** Roughness improvement for static chemical polishing and chemical flow polishing processes for different part orientations of 15 and 135 degrees. *(Reproduced with kind permission of Elsevier from [111].)*

two differently oriented parts are shown in Fig. 3.29 where the effect of chemical abrasive flow polishing proves to be effective in comparison with individual chemical or abrasive flow polishing treatments. Moreover, the polishing time is reduced from 3 to 1 h for a given $R_a$ improvement when the combine chemical-abrasive flow treatment is used.

Abrasive blasting, widely known as sandblasting, is carried out for cleaning the surfaces of parts from loosely connected powder particles. Various media, such as sand, abrasives, metallic materials, and nut shells are used. The sandblasting as one of the post-processes is generally applied after SLM for surface quality enhancement. De Wild et al. investigated the use of sandblasting to finish porous orthopedic Ti implants as a step before acid etching [112]. They have shown that applying sandblasting removed the powder particles and smoothened the strut surfaces. The surface roughness of the implant was reduced from 3.33 to 0.94 μm after sandblasting with corundum (a crystalline form of aluminum oxide). Moreover, the strut size was reduced, and cavities in between struts became larger. In another study for biomedical applications, the tests with Ti13Zr13Nb alloy have focused on sandblasting, grinding, and etching to improve the surface quality [113]. As shown in Fig. 3.30B, during sandblasting treatment, the stream of accelerated particles removed most of the spherical particles and significantly expanded the surface with many irregular microcracks in comparison to only SLMed part shown in Fig. 3.30A. After etching, the formation of a more uniform fine-textured titanium alloy surface with most of the spherical grains being dissolved is observed, whereas the surface is relatively smoother and irregularities with evenly distributed depression areas. The ground SLM specimens, as shown in Fig. 3.30D, presents a flat surface with clearly visible shallow grooves generated by grinding in comparison to a solid bar used as a reference in the study [113]. Sandblasting was also used to improve the aesthetic appearance of maraging steel specimens produced by SLM [114].

Magnetic abrasive finishing, or alternatively named as magnetic-field-assisted finishing, is another method used after SLM for the modification of the surface roughness and residual stresses [115,116]. As shown in Fig. 3.31, a magnetic particle brush, which consists either a mixture of magnetic particles and abrasive slurry or balls, presses against a target surface in a magnetic field. The results show that the roughness $R_z$ of SLMed parts was modified by controlling the contacts of the magnetic brush against the target surface from over 100–0.1 μm. Moreover, magnetic-assisted finishing (MAF) led to compressive stresses at the SLM parts' surfaces [115]. Although the studies in the literature are focusing on roughness reduction on flat surfaces by MAF, the process is flexible to allow enhancement of complex geometries with an adaptation to multi-axis platforms. In addition, as a result of this study, it is concluded that the initial surface roughness for complex geometries built by the SLM process shall be taken into account to adjust the finishing parameters [116].

Another modified version of abrasive machining treatment is the ultrasonic abrasive polishing for fabricating complex shapes, deep holes, and high-aspect-ratio microstructures on hard and brittle materials. The ultrasonic abrasive polishing depends on the ultrasonic machining process,

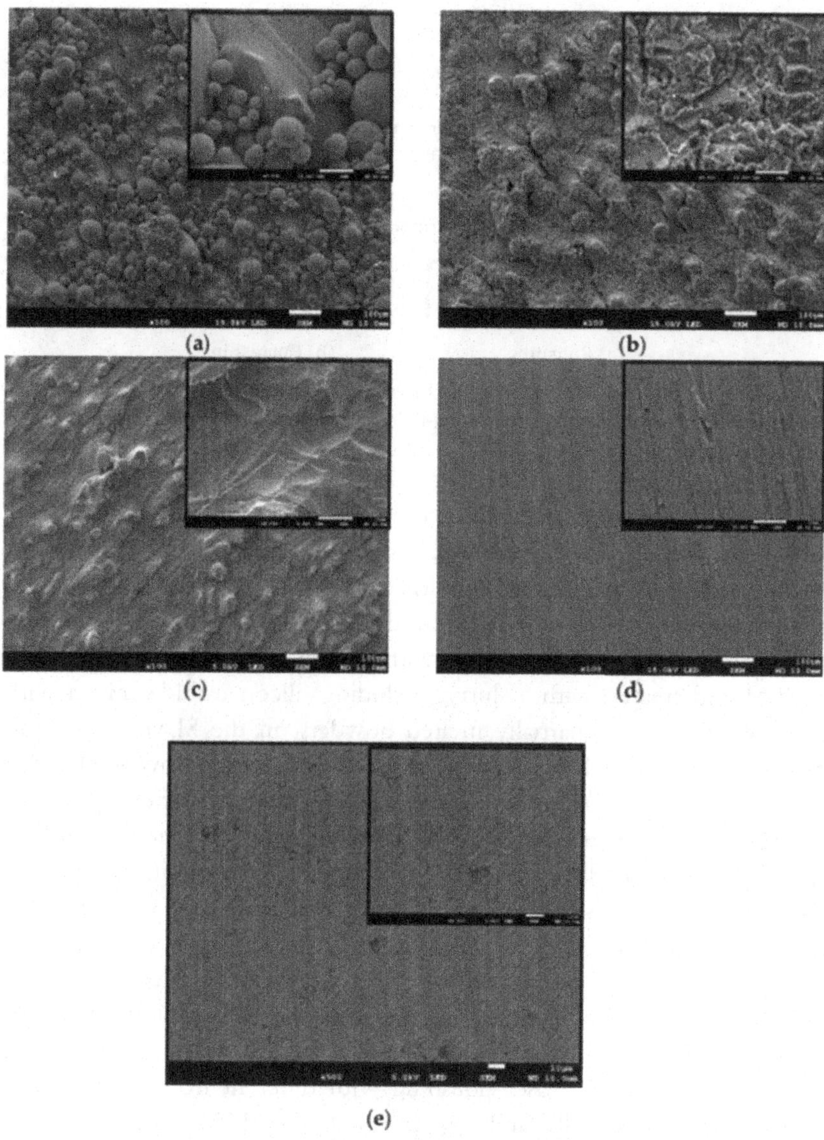

(a) (b) (c) (d) (e)

**Figure 3.30** SEM images of selective laser melting (SLM) part's surfaces after different surface treatments and at two different magnifications (A) untreated SLM; (B) sandblasted SLM; (C) etched SLM; (D) ground SLM; and (E) solid bar. *(Reproduced with kind permission of PubMed from M. Dziaduszewska, M. Wekwejt, M. Bartmański, A. Pałubicka, G. Gajowiec, T. Seramak, A.M. Osyczka, A. Zieliński, The effect of surface modification of Ti13Zr13Nb alloy on adhesion of antibiotic and nanosilver-loaded bone cement coatings dedicated for application as spacers. Materials 12 (2019) 2964.)*

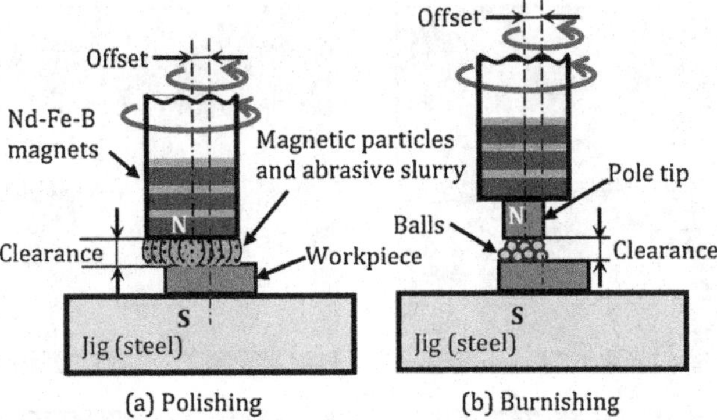


**Figure 3.31** Schematics of magnetic-field-assisted finishing processes (A) magnetic-field-assisted polishing and (B) magnetic-field-assisted burnishing. *(Reproduced with kind permission of Elsevier from [116].)*

whereby an ultrasonically vibrated tool is used in conjunction with a liquid slurry for material removal. By doing so, a severe number of microsized abrasive particles suspended in the slurry flow through the working area and hammer the workpiece repeatedly by the propulsion of the vibrated tool [117,118]. Experiments carried out with Inconel 625 specimens produced by SLM and treated with a slurry, including silicon carbide mixed with water, show that the partially melted powders on the SLMed parts can efficiently be removed, with a further surface quality improvement of $R_a$ reduced to 2.93 μm from the initial value of 9.48 μm after 30 min of polishing [118]. Other studies in the literature confirm that the partially melted powder particles are efficiently removed, and a significant increase of the surface quality is achieved [117]. Moreover, it has been presented that an increase in the ultrasonic output power and abrasive concentration within a certain range lead to a more desirable polishing effect.

In addition to other techniques, laser polishing is a technique to reduce surface roughness of metals, alloys, and ceramics with no significant distortion to the parts. Laser polishing with different materials has been studied both numerically and experimentally in the literature. Due to various effects of other processes such as environmental impact of chemical and electrochemical processes, laser polishing offers an ecologically friendly method of achieving good finishes being suitable for automation. Even for laser-based AM techniques, the same laser can be employed as explained in the previous section. Yet, it can be applied as a post-process, too. Laser polishing depends on re-melting a thin surface layer and smoothing of surface due to surface tension with nearly no material removal. The laser can be operated either in pulsed or continuous mode depending on the initial surface quality. Depending on the material and the initial surface roughness, the average roughness can be decreased down to 0.05 μm [119]. The surface topography, solidification microstructure, and mechanical performance of the SLMed Ti alloy after laser polishing have been investigated by Li et al. [120]. In this study, surface roughness ($R_a$) of SLMed parts was reduced to 0.32 μm from 6.53 μm, whereas both microhardness and wear resistance were enhanced by 25% and 39% in comparison with only SLM parts, respectively. However, microstructural changes, for example, the formation of the martensitic layer with poor ductility, were observed leading to the reduction of fatigue life [120]. In another study by Ma et al. [121] with titanium alloys, laser polishing hardens the surface from 345 HV (TC4 alloy) and 400 HV (TC11 alloy) by 32% and 42% due to the formation of martensite phase on the surface, leading to better wear characteristics. Moreover, the roughness decreases from over 5 μm to less than 0.8 μm [121]. This is clearly evident in Fig. 3.32. Marimuthu et al. [122] addressed laser polishing of Ti−6Al−4V components with a continuous wave fiber laser without an important change to the mechanical or metallurgical characteristics. In their study, it is mentioned that the

**Figure 3.32** SEM micrograph of the boundary between laser-polished region and as-received region. *(Reproduced with kind permission of Elsevier from C.P. Ma, Y.C. Guan, W. Zhou, Laser polishing of additive manufactured Ti alloys, Optic Laser. Eng. 93 (2017) 171−177, ISSN 0143-8166.)*

precise control of the laser energy input as well as productivity are key requirements for achieving the surface quality improvement. Moreover, it was observed that surface quality of the laser polished area relies on the melt pool velocity, which itself is influenced by the laser power and speed [122]. Not only titanium alloys were studied with laser polishing but also the surface quality of AISI 316L stainless steel parts produced by SLM was enhanced by laser polishing whereby a reduction of the average roughness $R_a$ from 10.4 to 2.7 μm was achieved [123]. Rosa et al. also addressed the laser polishing of AISI 316L stainless steel parts. They concluded that the increase of the number of passes improves the smoothening as well as elimination of the cracks. Moreover, the laser parameters in laser polishing must be chosen in correlation to the initial topography and material [124]. Other materials were also addressed such as cobalt-chrome parts where the surface quality improvement was improved by a minimum of 85% depending on the initial surface finish [125] and aluminum alloys with pulsed or continuous wave lasers [126]. The study by Zhihao et al. [127] shows that laser polishing is an effective way to reduce the roughness ($R_a$) for Inconel 718, a nickel superalloy, down to 0.1 μm. Moreover, based on the microstructure and XRD results, it is found that the laser polishing reduces the grain size leading to a finer microstructure. Additionally, gamma double prime phase precipitate forms in the polished layer due to high cooling rates. The microhardness is also enhanced to 440 HV from 345 HV after laser polishing, which results in a harder surface, and thus better friction and wear properties [127].

Mass finishing technologies like BF and vibratory grinding are widely applied to SLMed parts since they are easy to operate and the investment costs are generally lower compared with other finishing methods. Batch processing is possible with no tooling requirements although no localized finishing can be applied [128]. In a study carried out by Bagehorn et al. [129], it is concluded that the main disadvantage of the vibratory grinding process, reducing the roughness down to less than 1 μm, is the long treatment time necessary to reduce the high roughness ranges. Similar to machining, the accessibility is also very limited regarding complex geometries [129]. Another form of mass finishing, BF, studied in Ref. [130], reduces the roughness by peak cutting. Rounding sharp and angular features is considered as a limitation in BF.

The post-processes that are generally applied after the SLM process are summarized in Table 3.2, which is significantly modified based on the study given in Ref. [64]. Various processes with their capabilities, advantages, and

**Table 3.2** Comparison of various processes for surface enhancement after selective laser melting.

| Process | Capabilities | Advantages | Limitations |
|---|---|---|---|
| CNC machining | $R_a < 1$ μm [84]<br>$R_a < 0.5$ μm [86,87]<br>$R_a \approx 0.4$ μm [126] | Effective for simple geometries, a mirror finish, availability of equipment | Difficult to machine complex shapes and internal features |
| Chemical etching/ Electrochemical polishing (ECP) | $R_a \approx 0.5$ μm [104,131]<br>$R_a \approx 1.7$ μm [95]<br>$R_a < 0.5$ μm [102] | Easy to process internal channels and difficult-to-access areas<br>Very effective in removing partially melted or sintered powder particles | Dimensional inaccuracies due to material erosion<br>Slow process (chemical etching)<br>Necessity of special tools (ECP) |
| Laser polishing | $R_a \approx 0.35$ μm [120]<br>$R_a \approx 0.4$ μm [121]<br>$R_a \approx 1.5$ μm [75]<br>$R_a \approx 2.4$ μm [121] | Same lasers can be used<br>Less footprint<br>Improved wear properties due to modified microstructure | Difficult for internal features and inclined surfaces<br>Residual stresses and changes in surface chemistry |
| Abrasive blasting | $R_a \approx 0.9$ μm [112]<br>$R_z \approx 0.1$ μm (magnetic) [113,116]<br>$R_a \approx 2.93$ μm [117,118] (ultrasonic) | Simple and flexible | Poor process repeatability |
| Abrasive flow machining | $R_a < 1$ μm [106] | Suitable for internal channels and cavities | Difficult to uniformly polish of complex passages<br>Rounding sharp and angular features<br>Limited finish due to abrasive flow direction |

*Continued*

**Table 3.2** Comparison of various processes for surface enhancement after selective laser melting.—cont'd

| Process | Capabilities | Advantages | Limitations |
|---|---|---|---|
| Mass finishing | $R_a < 1$ μm [129,130] | No tooling requirements<br>Batch processing | No localized finishing<br>Rounding sharp and angular features<br>High process cycle time |

limitations are presented. The selection of the suitable finishing process shall be done depending on many factors such as the application requirements, part size and shape, material, and initial surface roughness. Moreover, these post-processes may be applied in series. Not only the surface roughness reduction is considered but also side effects such as elimination of manufacturing defects and porosity, hardening, change of microstructure and surface properties as well as the thickness of modified/affected layers are of great importance. Moreover, flexibility, controllability, and practicality are other important considerations [84]. For example, one process' being suitable for only specific materials or its capability to process freeform surfaces or internal features can be used to determine how flexible it is.

## References

[1] J.-P. Kruth, B. Vandenbroucke, J. Van Vaerenbergh, P. Mercelis, Benchmarking of different SLS/SLM processes as rapid manufacturing techniques, in: Proceedings of the International Conference Polymers & Moulds Innovations PMI 2005 (April 20-23, 2005, Gent, Belgium), 2005.

[2] J.-P. Kruth, P. Mercelis, L. Froyen, M. Rombouts, Binding mechanisms in selective laser sintering and selective laser melting, Rapid Prototyp. J. 11 (1) (2004) 26–36.

[3] ASTM F2792, 12a Standard Terminology for Additive Manufacturing Technologies, 2013.

[4] E. Yasa, T. Craeghs, J.-P. Kruth, Selective laser sintering/melting and selective laser erosion with Nd:YAG lasers, in: W. Arkin (Ed.), Advances in Laser and Optics Research, vol. 7, Nova Publishers, 2012. Chapter 1.

[5] H. Shipley, D. McDonnell, M. Culleton, R. Lupoi, G. O'Donnell, D. Trimble, Optimisation of process parameters to address fundamental challenges during selective laser melting of Ti-6Al-4V: a review, Int. J. Mach. Tool Manufact. 128 (2018) 1–20.

[6] C.Y. Yap, C. Chua, Z. Dong, Z. Liu, D. Zhang, L.E. Loh, S.L. Sing, Review of selective laser melting: materials and applications, Appl. Phys. Rev. 2 (4) (2015) 041101.

[7] M. Rombouts, J.P. Kruth, L. Froyen, P. Mercelis, Fundamentals of selective laser melting of alloyed steel powders, CIRP Ann. 55 (1) (2016) 87–192.

[8] Z. Mao, D. Zhang, P. Wei, K. Zhang, Manufacturing feasibility and forming properties of Cu-4Sn in selective laser melting, Materials 10 (4) (2017) 333.
[9] I. Yadroitsev, A. Gusarov, I. Yadroitsava, I. Smurov, Single track formation in selective laser melting of metal powders, J. Mater. Process. Technol. 210 (2010) 1624–1631.
[10] C.C. Ng, M.M. Savalani, M.L. Lau, H.C. Man, Microstructure and mechanical properties of selective laser melted magnesium, Appl. Surf. Sci. 257 (2011) 7447–7454.
[11] E.O. Olakanmi, Selective laser sintering/melting (SLS/SLM) of pure Al, Al-Mg, and Al-Si powders: effect of processing conditions and powder properties, J. Mater. Process. Technol. 213 (8) (2013) 1387–1405.
[12] J. Zhang, B. Song, Q. Wei, D. Bourell, Y. Shi, A review of selective laser melting of aluminum alloys: processing, microstructure, property and developing trends, J. Mater. Sci. Technol. 35 (2) (2019) 270–284.
[13] H. Hassanin, L. Finet, S.C. Cox, P. Jamshidi, L.M. Grover, D.E.T. Shepherd, O. Addison, M.M. Attallah, Tailoring selective laser melting process for titanium drug-delivering implants with releasing micro-channels, Addit. Manuf. 20 (2018) 144–155.
[14] A.G. Demir, L. Monguzzi, B. Previtali, Selective laser melting of pure Zn with high density for biodegradable implant manufacturing, Addit. Manuf. 15 (2017) 20–28.
[15] K.S. Munir, Y. Li, C. Wen, 1 - Metallic scaffolds manufactured by selective laser melting for biomedical applications, in: C. Wen (Ed.), Metallic Foam Bone, Woodhead Publishing, 2017, pp. 1–23.
[16] J. Vaithilingam, E. Prina, R.D. Goodridge, R.J.M. Hague, S. Edmondson, F.R.A.J. Rose, S.D.R. Christie, Surface chemistry of Ti6Al4V components fabricated using selective laser melting for biomedical applications, Mater. Sci. Eng. C 67 (2016) 294–303.
[17] J.C. Najmon, S. Raeisi, A. Tovar, Review of Additive Manufacturing Technologies and Applications in the Aerospace Industry, Additive Manufacturing for the Aerospace Industry, 2019, pp. 7–31.
[18] S. Torres-Carrillo, H.R. Siller, C. Vila, C. López, C.A. Rodríguez, Environmental analysis of selective laser melting in the manufacturing of aeronautical turbine blades, J. Clean. Prod. 246 (2020) 119068.
[19] M. Seabra, J. Azevedo, A. Araújo, L. Reis, E. Pinto, N. Alves, R. Santos, J.P. Mortágua, Selective laser melting (SLM) and topology optimization for lighter aerospace componentes, Proced. Struct. Integr. 1 (2016) 289–296.
[20] W.E. King, H.D. Barth, V.M. Castillo, G.F. Gallegos, J.W. Gibbsa, D.E. Hahn, C. Kamath, A.M. Rubenchik, Observation of keyhole-mode laser melting in laser powder-bed fusion additive manufacturing, J. Mater. Process. Technol. 214 (2014) 2915–2925.
[21] C. Tuck, N.T. Aboulkhair, I. Maskery, C. Tuck, I. Ashcroft, N.M. Everitt, On the formation of AlSi10Mg single tracks and layers in selective laser melting: microstructure and nano-mechanical properties, J. Mater. Process. Technol. 230 (2016) 88–98.
[22] T. Qi, H. Zhu, H. Zhang, J. Yin, L. Ke, X. Zeng, Selective laser melting of Al7050 powder: melting mode transition and comparison of the characteristics between the keyhole and conduction mode, Mater. Des. 135 (2017) 257–266.
[23] K.Q. Le, C. Tang, C.H. Wong, On the study of keyhole-mode melting in selective laser melting process, Int. J. Therm. Sci. 145 (2019) 105992.
[24] E. Yasa, İ. Atik, İ. Bayraktar, Selective laser melting process development for enhanced productivity with Ph 17-4 stainless steel, in: AMCTURKEY Additive Manufacturing Conference, İstanbul, Turkey, 17–18 October 2019, 2019.

[25] J. Kim, S. Ji, Y.S. Yun, J.S. Yeo, A review: melt pool analysis for selective laser melting with continuous wave and pulse width modulated lasers, Appl. Sci. Converg. Technol. 27 (6) (2018) 113–119.
[26] U.S. Bertoli, A.J. Wolfer, M.J. Matthews, J.P.R. Delplanque, J.M. Schoenung, On the limitations of volumetric energy density as a design parameter for selective laser melting, Mater. Des. 113 (2017) 331–340.
[27] K.G. Prashanth, S. Scudino, T. Maity, J. Das, J. Eckert, Is the energy density a reliable parameter for materials synthesis by selective laser melting? Mater. Res. Lett. 5 (6) (2017) 386–390.
[28] D. Koutny, D. Paloušek, P. Libor, H. Christian, P. Rudolf, T. Lukas, K. Jozef, Influence of scanning strategies on processing of aluminum alloy EN AW 2618 using selective laser melting, Materials 11 (2018) 298.
[29] J.P. Kruth, J. Deckers, E. Yasa, R. Wauthle, Assessing and comparing influencing factors of residual stresses in selective laser melting using a novel analysis method, in: 16th International Symposium on Electromachining (ISEM XVI), Shanghai, China, 19–23 April 2010, 2010. ISBN: 978-1-63266-648-2.
[30] A. Majeed, A. Ahmed, A. Salam, M.Z. Sheikh, Surface quality improvement by parameters analysis, optimization and heat treatment of AlSi10Mg parts manufactured by SLM additive manufacturing, Int. J. Lightweight Mater. Manuf. 2 (4) (2019) 288–295. ISSN 2588-8404.
[31] A. Charles, A. Elkaseer, L. Thijs, V. Hagenmeyer, S. Scholtz, Effect of process parameters on the generated surface roughness of down-facing surfaces in selective laser melting, Appl. Sci. 9 (6) (2019) 1256.
[32] I. Yadroitsev, I. Shishkovsky, P. Bertrand, I. Smurov, Manufacturing of fine-structured 3D porous filter elements by selective laser melting, Appl. Surf. Sci. 255 (10) (2009) 5523–5527. ISSN 0169-4332.
[33] B. Vandenbroucke, J.-P. Kruth, Selective laser melting of biocompatible metals for rapid manufacturing of medical parts, Rapid Prototyp. J. 13 (4) (2007) 196–203.
[34] H. Chen, D. Gu, J. Xiong, M. Xia, Improving additive manufacturing processability of hard-to-process overhanging structure by selective laser melting, J. Mater. Process. Technol. 250 (2017) 99–108. ISSN 0924-0136.
[35] Z. Chen, S. Cao, X. Wu, C.H.J. Davies, Surface roughness and fatigue properties of selective laser melted Ti–6Al–4V alloy, in: F. Froes, R. Boyer (Eds.), Additive Manufacturing for the Aerospace Industry, Elsevier, 2019, pp. 283–299. ISBN 9780128140628.
[36] E. Yasa, O. Poyraz, E. Ugur Solakoglu, G. Akbulut, S. Oren, A study on the stair stepping effect in direct metal laser sintering of a nickel-based superalloy, Proced. CIRP 45 (2016) 175–178, https://doi.org/10.1016/j.procir.2016.02.068. ISSN 2212-8271.
[37] A. Boschetto, L. Bottini, F. Veniali, Roughness modeling of AlSi10Mg parts fabricated by selective laser melting, J. Mater. Process. Technol. 241 (2017) 154–163. ISSN 0924-0136.
[38] G. Strano, L. Hao, R.M. Everson, K.E. Evans, Surface roughness analysis, modelling and prediction in selective laser melting, J. Mater. Process. Technol. 213 (4) (2013) 589–597. ISSN 0924-0136.
[39] B. Vandenbroucke, J.P. Kruth, Selective Laser Melting of Biocompatible metals for rapid manufacturing of medical parts, in: Proceedings of Solid Freeform Fabrication Symposium, Austin, TX, USA, 2006.
[40] T. Schwanekamp, M. Bräuer, M. Reuber, Geometrical and topological potentialities and restrictions in selective laser sintering of customized carbide precision tools, in: Proceedings of Lasers in Manufacturing Conference, Munich, Germany, 2017.

[41] J.-P. Kruth, L. Froyen, J. Van Vaerenbergh, P. Mercelis, M. Rombouts, B. Lauwers, Selective laser melting of iron-based powder, J. Mater. Process. Technol. 149 (1–3) (2004) 616–622. ISSN 0924-0136.

[42] B. Zhang, L. Dembinski, C. Coddet, The study of the laser parameters and environment variables effect on mechanical properties of high compact parts elaborated by selective laser melting 316L powder, Mater. Sci. Eng. A 584 (2013) 21–31. ISSN 0921-5093.

[43] H. Ali, H. Ghadbeigi, F. Hosseinzadeh, J. Oliveira, K. Mumtaz, Effect of pre-emptive in-situ parameter modification on residual stress distributions within selective laser-melted Ti6Al4V, Int. J. Adv. Manuf. Technol. 103 (2019) 4467.

[44] R. Mertens, S. Dadbakhsh, J. Van Humbeeck, J.-P. Kruth, Application of base plate preheating during selective laser melting, Proced. CIRP 74 (2018) 5–11. ISSN 2212-8271.

[45] R. Mertens, B. Vrancken, N. Holmstock, Y. Kinds, J.-P. Kruth, J. Van Humbeeck, Influence of powder bed preheating on microstructure and mechanical properties of H13 tool steel SLM parts, Phys. Proced. 83 (2016) 882–890. ISSN 1875-3892.

[46] Ö. Poyraz, E. Yasa, G. Akbulut, A. Orhangül, S. Pilatin, Investigation of support structures for direct metal laser sintering of IN625 parts, in: Proceedings of Solid Freeform Fabrication Symposium, Austin, TX, USA, 2015.

[47] Ö. Poyraz, M.C. Kushan, Residual stress-induced distortions in laser powder bed additive manufacturing of nickel-based superalloys, Strojniski Vestnik-J. Mech. Eng. 65 (6) (2019).

[48] L. Cheng, A. To, Part-scale build orientation optimization for minimizing residual stress and support volume for metal additive manufacturing: theory and experimental validation, Comput. Aided Des. 113 (2019) 1–23.

[49] P. Das, R. Chandran, R. Samant, S. Anand, Optimum part build orientation in additive manufacturing for minimizing part errors and support structures, Proced. Manuf. 1 (2015) 343–354.

[50] G. Moroni, W.P. Syam, S. Petrò, Functionality-based part orientation for additive manufacturing, in: Proceedings of 25th CIRP Design Conference, Haifa, Israel, 2015.

[51] A. Hussein, L. Hao, C. Yan, R. Everson, P. Young, Advanced lattice support structures for metal additive manufacturing, J. Mater. Process. Technol. 213 (2013) 1019–1026.

[52] F. Calignano, P. Minetola, Influence of process parameters on the porosity, accuracy, roughness and support structures of Hastelloy X produced by laser powder bed fusion, Materials 12 (2019) 3178.

[53] E. Yasa, J. Deckers, T. Craeghs, M. Badrossamay, J.-P. Kruth, Investigation on occurrence of elevated edges in SLM, in: Proceedings of SFF Symposium 2009, Austin, TX, USA, 2009.

[54] J. Metelkova, Y. Kinds, C. de Formanoir, A. Witrouw, B. Van Hooreweder, In situ combination of selective laser melting and selective laser erosion for edge effect removal, in: Presented at the Solid Freeform Fabrication (SFF), Austin, Texas, USA, 12 Aug 2019–14 Aug 2019, 2019. Available: lirias.kuleuven.be/2843118?limo=0.

[55] R. Li, J. Liu, Y. Shi, L. Wang, W. Jiang, Balling behavior of stainless steel and nickel powder during selective laser melting process, Int. J. Adv. Manuf. Technol. 59 (2012) 1025, https://doi.org/10.1007/s00170-011-3566-1.

[56] X. Zhou, X. Liu, D. Zhang, Z. Shen, W. Liu, Balling phenomena in selective laser melted tungsten, J. Mater. Process. Technol. 222 (2015) 33–42. ISSN 0924-0136.

[57] D. Gu, Y. Shen, Balling phenomena in direct laser sintering of stainless steel powder: metallurgical mechanisms and control methods, Mater. Des. 30 (8) (2009) 2903–2910. ISSN 0261-3069.

[58] V. Gunenthiram, P. Peyre, M. Schenider, M. Dal, F. Coste, R. Fabbro, Analysis of laser–melt pool–powder bed interaction during the selective laser melting of a stainless steel, J. Laser Appl. 29 (2) (2017), 022303.

[59] Y.F. Shen, D.D. Gu, Y.F. Pan, Balling process in selective laser sintering 316 stainless steel powder, Key Eng. Mater. 315–316 (2006) 357–360. ISSN: 1662-9795.

[60] N.A. Sheikh, M. Khan, K. Alam, H.I. Syed, A. Khan, L. Ali, Balling phenomena in selective laser melting (SLM) of pure Gold (Au), in: Proceedings of the Int. Matador Conference, 2013, pp. 369–372.

[61] P. Oyar, Laser sintering technology and balling phenomenon, Photomed. Laser Surg. 36 (2) (Febuary 2018) 72–77.

[62] D. Dai, D. Gu, Effect of metal vaporization behavior on keyhole-mode surface morphology of selective laser melted composites using different protective atmospheres, Appl. Surf. Sci. 355 (2015) 310–319. ISSN 0169-4332.

[63] W.J. Sames, F.A. List, S. Pannala, R.R. Dehoff, S.S. Babu, The metallurgy and processing science of metal additive manufacturing, Int. Mater. Rev. 61 (5) (2016) 315–360.

[64] I. Yadroitsev, P. Bertrand, I. Smurov, Parametric analysis of the selective laser melting process, Appl. Surf. Sci. 253 (19) (2007) 8064–8069.

[65] I. Yadroitsev, I. Smurov, Surface morphology in selective laser melting of metal powders, Phys. Proced. 12 (2011) 264–270.

[66] Y. Pupo, J. Delgado, L. sereno, J. Ciurana, Scanning space analysis in selective laser melting for CoCrMo powder, Proced. Eng. 63 (2013) 370–378.

[67] E. Yasa, Manufacturing by Combining Selective Laser Melting and Selective Laser Erosion/laser Re-melting, PhD thesis, KU Leuven, Leuven, Belgium, January 2011.

[68] E. Yasa, J.-P. Kruth, Application of laser re-melting on selective laser melting parts, Adv. Prod. Eng. Manag. 6 (2011) 259–270.

[69] E. Yasa, J. Deckers, J.-P. Kruth, The investigation of the influence of laser re-melting on density, surface quality and microstructure of selective laser melting parts, Rapid Prototyp. J. 17 (5) (2011) 312–327.

[70] R.H. Morgan, A.J. Papworth, C. Sutcliffe, P. Fox, W. O'neill, High density net shape components by direct laser re-melting of single-phase powders, J. Mater. Sci. 37 (15) (2002) 3093–3100.

[71] Q. Han, Y. Jiao, Effect of heat treatment and laser surface remelting on AlSi10Mg alloy fabricated by selective laser melting, Int. J. Adv. Manuf. Technol. 102 (2019) 3315–3324, https://doi.org/10.1007/s00170-018-03272-y.

[72] G. Yu, D. Gu, D. Dai, M. Xia, C. Ma, K. Chang, Influence of processing parameters on laser penetration depth and melting/re-melting densification during selective laser melting of aluminum alloy, Appl. Phys. A 122 (2016), 891, https://doi.org/10.1007/s00339-016-0428-6.

[73] B. Liu, B.-Q. Li, Z. Li, Selective laser remelting of an additive layer manufacturing process on AlSi10Mg, Results Phys. 12 (2019) 982–988.

[74] A.G. Demir, B. Previtali, Investigation of remelting and preheating in SLM of 18Ni300 maraging steel as corrective and preventive measures for porosity reduction, Int. J. Adv. Manuf. Technol. 93 (2017) 2697–2709.

[75] A. Lamikiz, J.A. Sanchez, L.N. Lopez de Lacalle, J.L. Arana, Laser polishing of parts built up by selective laser sintering, Mach. Tools Manuf. 47 (12–13) (2007) 2040–2050.

[76] W. Yu, S.L. Sing, C.K. Chua, X. Tian, Influence of re-melting on surface roughness and porosity of AlSi10Mg parts fabricated by selective laser melting, J. Alloys Compd. 792 (2019) 574–581.

[77] I. Koutiri, E. Pessard, P. Peyre, O. Amlou, T. de Terris, Influence of SLM process parameters on the surface finish, porosity rate and fatigue behavior of as-built Inconel 625 parts, J. Mater. Process. Technol. 255 (2018) 536–546.
[78] E. Ramirez-Cedillo, J.A. Sandoval-Robles, L. Ruiz-Huerta, A. Caballero-Ruiz, C.A. Rodriguez, H.R. Siller, Process planning guidelines in selective laser melting for the manufacturing of stainless steel parts, Proced. Manuf. 26 (2018) 973–982. ISSN 2351-9789.
[79] R. Vrána, D. Koutný, D. Paloušek, L. Pantělejev, J. Jaroš, T. Zikmund, J. Kaiser, Selective laser melting strategy for fabrication of thin struts useable in lattice structures, Materials 11 (9) (2018), 1763, https://doi.org/10.3390/ma11091763.
[80] F. Calignano, D. Manfredi, E.P. Ambrosio, L. Iuliano, P. Fino, Influence of process parameters on surface roughness of aluminum parts produced by DMLS, Int. J. Adv. Manuf. Technol. 67 (2013) 2743–2751.
[81] EOS, EOSint M 270 User Manual, 2009.
[82] Y. Tian, D. Tomus, P. Rometsch, X. Wu, Influences of processing parameters on surface roughness of Hastelloy X produced by selective laser melting, Addit. Manuf. 13 (2017) 103–112. ISSN 2214-8604.
[83] L.K. Gillespie, Deburring and Edge Finishing Handbook, Society of Manufacturing Engineers: American Society of Mechanical Engineers, Michigan; New York, 1999.
[84] E. Gordon, V. Dhokia, Experimental framework for testing the finishing of additive parts, in: Proceedings of 13th International Conference on Manufacturing Research, Bath, UK United Kingdom, 8/09/15–10/09/15, 2015, pp. 57–62.
[85] M. Dumas, F. Cabanettes, R. Kaminski, F. Valiorgue, E. Picot, F. Lefebvre, C. Grosjean, J. Rech, Influence of the finish cutting operations on the fatigue performance of Ti-6Al-4V parts produced by Selective Laser Melting, Proced. CIRP 71 (2018) 429–434. ISSN 2212-8271.
[86] D. Brown, C. Li, Z.Y. Liu, X.Y. Fang, Y.B. Guo, Surface integrity of Inconel 718 by hybrid selective laser melting and milling, Virtual Phys. Prototyp. 13 (1) (2018) 26–31, https://doi.org/10.1080/17452759.2017.1392681.
[87] Y. Kaynak, E. Tascioglu, Finish machining-induced surface roughness, microhardness and XRD analysis of selective laser melted Inconel 718 alloy, Proced. CIRP 71 (2018) 500–504.
[88] S. Milton, A. Morandeau, F. Chalon, R. Leroy, Influence of finish machining on the surface integrity of Ti6Al4V produced by selective laser melting, Proced. CIRP 45 (2016) 127–130. ISSN 2212-8271.
[89] M. Shunmugavel, A. Polishetty, J. Nomani, M. Goldberg, G. Littlefair, Metallurgical and machinability characteristics of wrought and elective laser melted Ti-6Al-4V, J. Metall. 2016 (2016), 7407918, https://doi.org/10.1155/2016/7407918.
[90] G. Struzikiewicz, W. Zebala, A. Matras, M. Machno, L. Slusarczyk, S. Hichert, F. Laufer, Turning research of additive laser molten stainless steel 316L obtained by 3D printing, Materials 12 (182) (2019), https://doi.org/10.3390/ma12010182.
[91] A.M. Khorasani, I. Gibson, M. Goldberg, G. Littlefair, The effect of machining parameters on cutting forces of SLM curved component, in: International Conference of Additive Manufacturing, at Maribor, Slovenia, 2018.
[92] M. Gomez, J. Heigel, T. Schmitz, Force modeling for hybrid manufacturing, Proced. Manuf. 26 (2018) 790–797.
[93] D.A. Lesyk, S. Martinez, B.N. Mordyuk, V.V. Dzhemelinskyi, A. Lamikiz, G.I. Prokopenko, Post-processing of the Inconel 718 alloy parts fabricated by selective laser melting: effects of mechanical surface treatments on surface topography, porosity, hardness and residual stress, Surf. Coating. Technol. 381 (2020) 125136. ISSN 0257-8972.
[94] Y. Sun, R. Bailey, A. Moroz, Surface finish and properties enhancement of selective laser melted 316Lstainless steel by surface mechanical attrition treatment, Surf. Coating. Technol. 378 (2019) 124993.

[95] A. Balaykin, Chemical polishing of samples obtained by selective laser melting from titanium alloy Ti6Al4V, MATEC Web Conf. 224 (2018) 01031.

[96] J. Lohser, Evaluation of Electrochemical and Laser Polishing of Selectively Laser Melted 316L Stainless Steel, Master of Science Thesis, Polytechnic State University, USA, 2018.

[97] Z. Baicheng, L. Xiaohua, B. Jiaming, G. Junfeng, W. Pan, S. Chen-nan, N. Muiling, Q. Guojun, W. Jun, Study of selective laser melting (SLM) Inconel 718 part surface improvement by electrochemical polishing, Mater. Des. 116 (2017) 531–537. ISSN 0264-1275.

[98] V. Urlea, V. Brailovski, Electropolishing and electropolishing-related allowances for powder bed selectively laser-melted Ti-6Al-4V alloy components, J. Mater. Process. Technol. 242 (2017) 1–11, https://doi.org/10.1016/j.jmatprotec.2016.11.014.

[99] V. Cruz, Q. Chao, N. Birbilis, D. Fabijanic, P.D. Hodgson, S. Thomas, Electrochemical Studies on the Effect of Residual Stress on the Corrosion of 316L Manufactured by Selective Laser Melting, Corrosion Science, 2019, p. 108314. ISSN 0010-938X.

[100] A. Lassell, The Electropolishing of Electron Beam Melting, Additively Manufactured TI6AL4V Titanium: Relevance, Process Parameters and Surface Finish, Master of Engineering Thesis, Industrial Engineering. University of Louisville, 2016, p. 76.

[101] L. Yang, A. Lassell, G.P.V. Paiva, Further study of the electropolishing of Ti6Al4 V parts made via electron beam melting, in: The Twenty-Sixth Annual International Solid Freeform Fabrication (SFF) Symposium —An Additive Manufacturing Conference, University of Texas, Austin, August, 2015, pp. 10–12.

[102] S. Jain, M. Corliss, B. Tai, W.N. Hung, Electrochemical polishing of selective laser melted Inconel 718, Proced. Manuf. 34 (2019) 239–246. ISSN 2351-9789.

[103] S. Sarkar, C.S. Kumar, A.K. Nath, Effects of different surface modifications on the fatigue life of selective laser melted 15–5 PH stainless steel, Mater. Sci. Eng. A 762 (138109) (2019). ISSN 0921-5093.

[104] G. Pyka, A. Burakowski, G. Kerckhofs, M. Moesen, S. Van Bael, J. Schrooten, Surface modification of Ti6Al4V open porous structures produced by additive manufacturing, Adv. Eng. Mater. 14 (6) (2012) 363–370.

[105] S. Duval-Chaneac, S. Han, C. Claudin, F. Salvatore, J. Bajolet, J. Rech, Experimental study on finishing of internal laser melting (SLM) surface with abrasive flow machining (AFM), Precis. Eng. 54 (2018) 1–6. ISSN 0141-6359.

[106] X. Wang, S. Li, Y. Fu, H. Gao, Finishing of additively manufactured metal parts by abrasive flow machining, solid freeform fabrication 2016, in: Proceedings of the 26th Annual International, Austin, TX, USA, 2016.

[107] C. Peng, Y. Fu, H. Wei, S. Li, X. Wang, H. Gao, Study on improvement of surface roughness and induced residual stress for additively manufactured metal parts by abrasive flow machining, Proced. CIRP 71 (2018) 386–389. ISSN 2212-8271.

[108] R.W. Gilmore, An Evaluation of Ultrasonic Shot Peening and Abrasive Flow Machining as Surface Finishing Processes for Selective Laser Melted 316L, Master Thesis of Science in Industrial Engineering, California Polytechnic State University, 2018. Available at: https://digitalcommons.calpoly.edu/cgi/viewcontent.cgi?article=3236&context=theses.

[109] E. Uhlmann, C. Schimedel, J. Wendler, CFD simulation of the abrasive flow machining process, 15th CIRP conference on modelling of machining operations, Proced. CIRP 31 (2015) 209–214.

[110] N. Mohammadian, S. Turenne, V. Brailovski, Surface finish control of additively-manufactured Inconel 625 components using combined chemical-abrasive flow polishing, J. Mater. Process. Technol. 252 (2018) 728–738. ISSN 0924-0136.

[111] N. Mohammadian, Surface Finish Control of Inconel 625 Components Produced by Additive Manufacturing Using Combined Chemical-Abrasive Flow Polishing, Master of Applied Sciences Thesis, École Polytechnique De Montréal, 2017.
[112] M. De Wild, R. Schumacher, K. Mayer, E. Schkommodau, D. Thoma, M. Bredell, A.K. Gujer, K.W. Graetz, F.E. Weber, Bone regeneration by the osteoconductivity of porous titanium implants manufactured by selective laser melting: a histological and micro computed tomography study in the rabbit, Tissue Eng. A 19 (23–24) (2013) 2645–2654, https://doi.org/10.1089/ten.tea.2012.0753.
[113] M. Dziaduszewska, M. Wekwejt, M. Bartmański, A. Pałubicka, G. Gajowiec, T. Seramak, A.M. Osyczka, A. Zieliński, The effect of surface modification of Ti13Zr13Nb alloy on adhesion of antibiotic and nanosilver-loaded bone cement coatings dedicated for application as spacers, Materials 12 (2019) 2964.
[114] G. Galimberti, E.L. Doubrovski, M. Guagliano, B. Previtali, J.C. Verlinden, Investigating the links between the process parameters and their influence on the aesthetic evaluation of selective laser melted parts, in: Solid Freeform Fabrication 2016: Proceedings of the 27th International Solid Freeform Fabrication Symposium, Austin, TX, USA, 2016.
[115] H. Yamaguchi, O. Fergani, P.-Y. Wu, Modification using magnetic field-assisted finishing of the surface roughness and residual stress of additively manufactured components, CIRP Ann. 66 (1) (2017) 05–308. ISSN 0007-8506.
[116] J. Zhang, A. Chaudhari, H. Wang, Surface quality and material removal in magnetic abrasive finishing of selective laser melted 316L stainless steel, J. Manuf. Process. 45 (2019) 710–719. ISSN 1526-6125.
[117] K.L. Tan, S.H. Yeo, Surface modification of additive manufactured components by ultrasonic cavitation abrasive finishing, Wear 378 (2017) 90–95. ISSN 0043-1648.
[118] J. Wang, J. Zhu, P.J. Liew, Material removal in ultrasonic abrasive polishing of additive manufactured components, Appl. Sci. 9 (24) (2019) 5359, https://doi.org/10.3390/app9245359.
[119] Laser polishing by Fraunhofer Institute of Laser Technology (ILT), Available at: http://doras.dcu.ie/23212/1/Laser%20Polishing%20of%20Additive%20Manufactured%20316L%20Stainless%20Steel%20Synthesized%20by%20Selective%20Laser%20Melting_DORAS.pdf.
[120] Y. Li, B. Wang, C. Ma, Z. Fang, L. Chen, Y. Guan, S. Yang, Material characterization, thermal analysis, and mechanical performance of a laser-polished Ti alloy prepared by selective laser melting, Metals 9 (2019) 12.
[121] C.P. Ma, Y.C. Guan, W. Zhou, Laser polishing of additive manufactured Ti alloys, Optic Laser. Eng. 93 (2017) 171–177. ISSN 0143-8166.
[122] S. Marimuthu, A. Triantaphyllou, M. Antar, D. Wimpenny, H. Morton, M. Beard, Laser polishing of selective laser melted components, Int. J. Mach. Tool Manufact. 95 (2015) 97–104. ISSN 0890-6955.
[123] M.A. Obeidi, E. McCarthy, B. O'Connell, I. Ul Ahad, D. Brabazon, Laser polishing of additive manufactured 316L stainless steel synthesized by selective laser melting, Materials 12 (6) (2019) 991, https://doi.org/10.3390/ma12060991.
[124] B. Rosa, P. Mognol, J-y. Hascoet, Laser polishing of additive laser manufacturing surfaces, J. Laser Appl. 27 (2015) S29102.
[125] W.S. Gora, Y. Tian, A.P. Cabo, M. Ardron, R.R.J. Maier, P. Prangnell, N.J. Weston, D.P. Hand, Enhancing surface finish of additively manufactured titanium and cobalt chrome elements using laser based finishing, in: 9th Int. Conf. on Photonic Technologies, Phys. Proced., 83, 2016, pp. 258–263.

[126] M. Hofele, J. Schanz, B. Burzic, S. Lutz, M. Merkel, H. Riegel, Laser based post processing of additive manufactured metal parts, in: Lasers in Manufacturing Conference 2017, Munich, Germany, June 26–29, 2017. https://www.wlt.de/lim/Call-for-papers-2017-v10_extended.pdf.

[127] F. Zhihao, L. Libin, C. Longfei, G. Yingchun, Laser polishing of additive manufactured superalloy, Proced. CIRP 71 (2018) 150–154. ISSN 2212-8271.

[128] B. Nagarajan, N. Hu, X. Song, W. Zhai, J. Wei, Development of micro selective laser melting: the state of the art and future perspectives, Engineering 5 (2019) 702–720.

[129] S. Bagehorn, J. Wehr, H.J. Maier, Application of mechanical surface finishing processes for roughness reduction and fatigue improvement of additively manufactured Ti-6Al-4V parts, Int. J. Fatig. 102 (2017) 135–142. ISSN 0142-1123.

[130] A. Boschetto, L. Bottini, F. Veniali, Surface roughness and radiusing of Ti6Al4V selective laser melting-manufactured parts conditioned by barrel finishing, Int. J. Adv. Manuf. Technol. 94 (2018) 2773–2790, https://doi.org/10.1007/s00170-017-1059-6.

[131] K. Alrbaey, D.I. Wimpenny, A.A. Al-Barzinjy, A. Moroz, Electropolishing of re-melted SLM stainless steel 316L parts using deep eutectic solvents: 3x3 full factorial design, J. Mater. Eng. Perform. 25 (7) (2016) 2836–2846.

CHAPTER 4

# Laser-directed energy deposition: principles and applications

F. Arias-González[1], O. Barro[2], J. del Val[2], F. Lusquiños[2,3], M. Fernández-Arias[2], R. Comesaña[2,4], A. Riveiro[2,4], J. Pou[2,3]
[1]Universitat Internacional de Catalunya (UIC), Sant Cugat del Vallès, Spain; [2]CINTECX, Universidade de Vigo, LaserON Research Group, School of Engineering, Vigo, Spain; [3]Galicia Sur Health Research Institute (IIS Galicia Sur). SERGAS-UVIGO, Vigo, Spain; [4]Materials Engineering, Applied Mechanics and Construction Department, School of Engineering, Universidade de Vigo, Vigo, Spain

## 4.1 Introduction

Over the past decades, additive manufacturing (AM) processes have been investigated and significant progress has been made in their development and commercialization [1,2]. ASTM F42 Technical Committee defines AM as the *"process of joining materials to make objects from three-dimensional (3D) model data, usually layer upon layer, as opposed to subtractive manufacturing technologies"* [3]. AM technologies can be used in the fabrication of prototypes (for design verification), tools, concept parts, and functional components. AM technology is also known as additive fabrication, additive processing, direct digital manufacturing, rapid prototyping, rapid manufacturing, additive layer manufacturing, and solid freeform fabrication.

AM technology was firstly applied to create polymer prototypes, and many processes have been developed to produce parts with this material. However, nowadays, it is not just restricted to rapid prototyping, and it is also capable of producing fully functional complex components made of metals, ceramics, or composites [1,2]. AM processes offer the ability to efficiently manufacture components with geometric or material complexity and the flexibility to manufacture low quantities of products, down to a single unit.

Directed energy deposition (DED) is an AM process to build a component by delivering energy and material simultaneously [4]. The material is supplied in the form of particles or wire, and it is melted by means of a heat source (laser beam, electron beam, or electric arc). The melted material is selectively deposited on a specified surface, where it solidifies (see Fig. 4.1). DED has the ability to produce relatively large parts requiring minimal tooling and relatively little secondary processing. It is a complex AM process that can be used to repair or add additional material to

*Additive Manufacturing*
ISBN 978-0-12-818411-0
https://doi.org/10.1016/B978-0-12-818411-0.00003-3

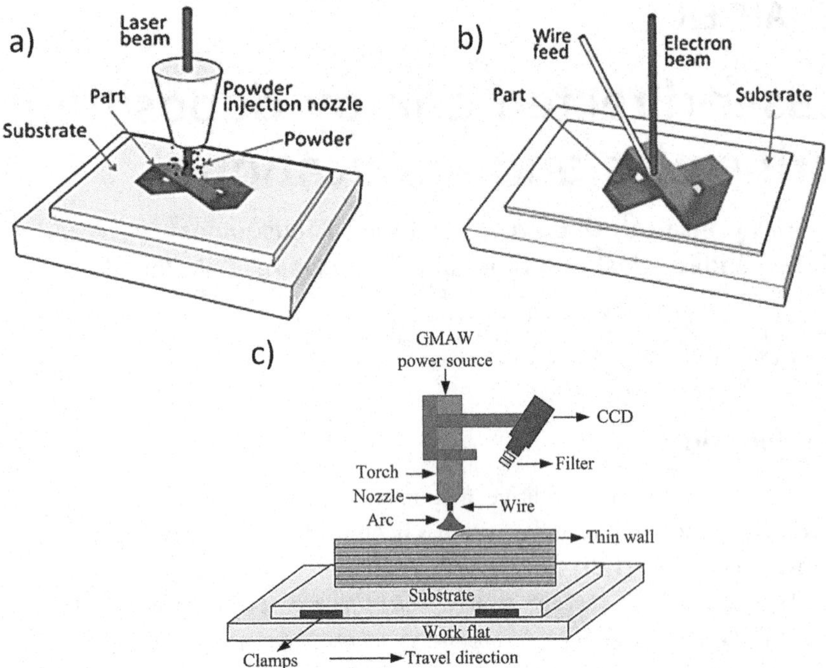


**Figure 4.1** Scheme of directed energy deposition processes [4,5]: (A) laser-directed energy deposition; (B) electron beam additive manufacturing; (C) shaped metal deposition or wire arc additive manufacturing. *(Modified from Reference T. DebRoy, H.L. Wei, J.S. Zuback, T. Mukherjee, J.W. Elmer, J.O. Milewski, et al., Additive manufacturing of metallic components - process, structure and properties, Prog. Mater. Sci. 92 (2018) 112–224. doi:10.1016/j.pmatsci.2017.10.001 with permission from Elsevier.)*

existing components. In addition, DED processes can be used to produce components with composition gradients, or hybrid structures consisting of multiple materials having different compositions and structures.

DED processes can be classified by the heat source used to melt the material (electron beam, electric arc, or laser beam). This technology is known as electron beam additive manufacturing (EBAM) when the heat source is an electron beam [6]. EBAM uses commercial filler wire as feedstock material, which is molten by an electron beam, so the process has to be done in a vacuum chamber [6]. Shaped metal deposition or wire arc additive manufacturing (WAAM) uses an electric arc as heat source to deposit material from filler wires in a similar way to fusion welding [5]. EBAM and WAAM are both restricted to be used with conductive materials (metals) in form of wire.

## 4.2 Principles of laser-directed energy deposition

Laser-directed energy deposition (LDED) is a DED method that uses a laser beam source to generate a molten pool in a substrate or workpiece (see Fig. 4.2) [4,8]. The laser beam is moved relatively to the workpiece melting

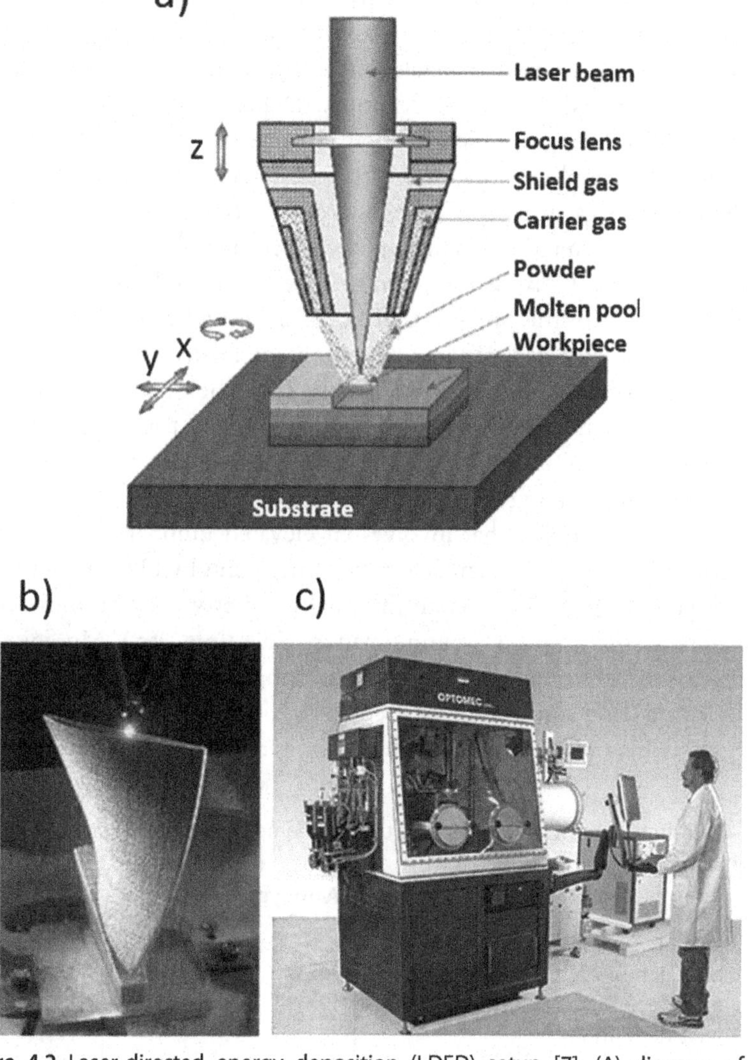


**Figure 4.2** Laser-directed energy deposition (LDED) setup [7]: (A) diagram of the process; (B) detail of turbine blade manufactured by LDED; and (C) LDED system (Optomec LENS 850). *(Modified from Reference K.H.Chang, Rapid Prototyp., e-Design, 14 (2015) 743–786. doi.org/10.1016/B978-0-12-382038-9.00014-4 with permission from Elsevier.)*

its surface. Feedstock material, in the form of particles or wire, is delivered into the molten pool. The laser moves leaving backward the molten material that immediately solidifies transmitting the heat mainly by conduction to the substrate. The relative movement between the laser head and the workpiece makes possible depositing material to generate single clad tracks with dimensions of millimeters or microns [9,10]. The overlapping of single clad tracks allows creating a bidimensional layer or coating. Finally, a component can be build-up by depositing material layer-by-layer. This process is typically used with metals, but it can be used with any material that can be melted including ceramics. Depending on the material, shielding gas like argon or even an inert chamber is required to avoid the oxidation in the molten pool; and, if the material is supplied in form of particles, an inert gas is used to carry and deliver them.

An LDED system requires a high-power laser source, a material feeder, a positioning system (CNC workstation or robot arm), and other systems to monitor and control the process. LDED has been widely explored during the last decades, and independent development teams have used a large variety of designations for very similar techniques. Their differences lie in its own unique system equipment, setup, and processing parameters. Therefore, a general definition of LDED includes technologies named under a wide range of terminology and acronyms showed in Table 4.1.

LDED is a technique that involves an elevated number of parameters summarized in Table 4.2. Manufacturing components by LDED requires to define a path strategy to generate the final geometry and the movement parameters (scanning speed, scanning pattern, idle time, etc.). Moreover, it is necessary to determine and control processing parameters related to the laser radiation (laser source, power, frequency, beam profile, spot size, etc.). Finally, to deliver material into the molten, it is crucial to define its preform

**Table 4.1** Different terms and acronyms used to name laser-directed energy deposition (LDED).

| Term | Acronym | References |
|---|---|---|
| Laser cladding | — | [11,12] |
| Direct laser deposition | DLD | [13] |
| Direct metal deposition | DMD | [14,15] |
| Laser metal deposition | LMD | [16,17] |
| Laser powder deposition | LPD | [18] |
| Laser engineered net shaping | LENS | [19,20] |
| Laser consolidation | LC | [21] |

**Table 4.2** Parameters that need to be defined in laser-directed energy deposition.

| Laser source | Power<br>Wavelength<br>Spot size<br>Beam profile<br>CW/pulsed<br>Pulse shape | Delivered material | Composition<br>Properties<br>Wire diameter<br>Powder size<br>Powder geometry<br>Powder flowability |
|---|---|---|---|
| Positioning system | Scanning speed<br>Acceleration<br>Accuracy | Substrate material | Geometry<br>Composition<br>Properties |
| | Scanning pattern<br>Idle time | Environment | Preheating<br>Shielding gas<br>Air/inert chamber |
| Material feeding system | Powder/wire<br>Mass flow rate<br>Carrying gas (powder) | Laser head setup | Lateral<br>Coaxial |

(powder size and geometry or wire diameter), the material feeding system and the laser head setup used to feed the material into the molten pool (lateral or coaxial, see Fig. 4.3).

Some processing parameters have to be defined in advance to manufacturing process because they are inherent to the equipment used (e.g., laser source, laser head) or to the final geometry of the component (e.g., path strategy). However, some parameters can be adjusted to control the process, even it is possible sometimes to modify them during the processing of the new piece. Typically, these processing parameters that are controlled use to be laser irradiation, scanning speed, and material flow rate. Processing parameters have a strong influence on the geometry of the final component and its microstructure/properties. Moreover, the repeatability in the properties/microstructure of deposited material requires mechanisms to stabilize or control the molten pool by adjusting processing parameters.

Despite the fact that in the last years a new industrial device to produce 3D parts using coaxial metal wire has become available: the COAXwire by Fraunhofer IWS (Dresden, Germany); the vast majority of the work on AM by LDED methods has been carried out using powder as precursor material. LDED is the only DED technology that can use powder as feedstock material and not only wire [4,8]. Wire-based LDED provides a high

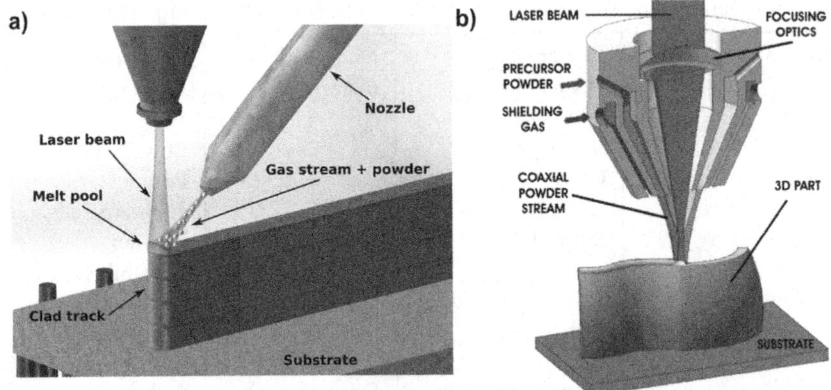


**Figure 4.3** Two types of laser head setup for laser-directed energy deposition: (A) lateral [12] and (B) coaxial [22]. *(Modified from Reference R. Comesaña, F. Lusquiños, J. del Val, T. Malot, M. López-Álvarez, A. Riveiro, et al., Calcium phosphate grafts produced by rapid prototyping based on laser cladding, J. Eur. Ceram. Soc. 31 (2011) 29–41. doi:10.1016/j.jeurceramsoc.2010.08.011 with permission from Elsevier and from Reference A. Riveiro, J. del Val, R. Comesaña, F. Lusquiños, F. Quintero, M. Boutinguiza, et al., Laser additive manufacturing processes for near net shape components. In: Gupta K. (eds) Near Net Shape Manufacturing Processes. Materials Forming, Machining and Tribology. Springer, Cham, 2019 with permission from Springer.)*

deposition efficiency of the material reaching almost 100%. The deposition efficiency that provides powder-based LDED is lower because the powder is typically blown through nozzles; so, some particles are not caught by the molten pool resulting in unused powder accumulation in the machine. These unused particles may be recovered and reused in future production; however, they can be affected someway by the process; therefore, the influence of their reutilization in the properties of the final component should be analyzed.

LDED can be used to process any material that can be melted, particularly metals and some ceramics. There are other AM methods capable of producing metallic and ceramic components, and they are classified in direct or indirect AM methods [23,24]. Direct methods are those that directly produce a fully dense part like LDED or selective laser melting (SLM). Meanwhile, indirect methods do not allow creating a fully dense component because particles are only partially melted or a binder is used to bond them. Examples of indirect methods to produce metal and ceramic parts are stereolithography, fused deposition modeling, 3D printing (3DP), and selective laser sintering. Some of these indirect methods are compatible with a postprocessing like the removal of the polymer binder, thermal sintering, or liquid-metal infiltration to achieve a fully dense part. However, direct methods are more promising because they can produce fully dense parts with nearly the same or even better mechanical properties compared with bulk material.

The AM processes are also categorized by the dimensional and geometrical accuracy of the manufactured product, their capacity to produce geometrically intricate parts, the absence of defects in the components, the production times, and the maximum size of the component that can be manufactured [4]. In general, SLM is more suitable for small and geometrical complex parts and LDED is more suitable for fabrication of bigger near net shape parts with a simpler geometry.

The ability to produce components with intricate geometries is considered the main strength of the SLM technique due to inherent support mechanism in the form of the remaining powder [4]. Moreover, the dimensional accuracy and surface finish can be relatively better than competing techniques because powder particles, layer thickness, and molten pool are smaller. However, the production time of the SLM processes is high because of the low powder feeding rate and small layer thickness.

LDED processes have the capacity for producing larger components faster than SLM because they allow relatively higher deposition rates [4,23]. However, the high deposition rates require large molten pools resulting in rough beaded surfaces. Machining is often required to achieve the final functional geometry. LDED does not rely on predeposited layer of metallic powder like SLM, and explicit support mechanisms are required to produce complex geometries.

Within the available processes for additively manufacture components, powder-based LDED is the primary method to manufacture functionally graded materials (FGMs) [18]. Powder-based LDED can easily feed different materials simultaneously to the processing area, and the composition of the delivered powder can be modified and controlled while the component is being built. However, SLM systems have also been explored to produce FGM components [4]. Finally, LDED systems combining coaxial powder delivering with lateral wire feeding have also been proposed to accomplish a system that can benefit from the utilization of both preforms [25,26].

## 4.3 Industrial applications of laser-directed energy deposition

LDED technology is particularly suitable for producing low volumes of products in comparison with subtractive manufacturing methods. Compared with other AM methods, it has the ability to produce near net shape large ceramic or metallic components. Moreover, it can be used to repair or add material to existing parts (including metallic or ceramic coatings). In addition, this technology can be used to generate FGMs. These capabilities are interesting to produce unique high-value components with applications in industry.

### 4.3.1 Additive manufacturing of metals and alloys

LDED can produce functional metallic components through a layer-upon-layer deposition without using molds or dies. It has been evaluated with a variety of metals and alloys, including nickel-based superalloys (IN 625, IN 718, IN 738, and Waspaloy), stainless steels (austenitic SS316L and martensitic SS420), and lightweight alloys (Ti6Al4V titanium alloy and 4047 aluminum alloy) [21,27]. Fully dense and crack-free parts with mechanical properties close to (or even superior) the conventionally processed material can be achieved. In Fig. 4.4, different geometries that were produced by means of LDED are shown.

LDED of Ti6Al4V has been extensively studied [27–29]. Clark et al. have produced a 3D, thin-walled structure of representative aerospace component geometry, and fabricated directly by LDED of Ti6Al4V powder. This fabrication technique is an economic, commercial process that can add features such as bosses or flanges to existing forms of gas turbine components [30]. Kobryn et al. determined the effect of microstructure and texture of Ti6Al4AV generated by LDED on mechanical properties [30]. Yield strength showed a noticeable anisotropy due to residual deposition porosity and crystallographic texture. Microstructures obtained in Ti6Al4V by means of LDED are highly influenced by directional heat extraction, and there is a tendency for this alloy to form long columnar grains shown in Fig. 4.5, as it was reported by Wu et al. [31].

Nickel-based superalloys are widely used in high-temperature applications, especially in land-based and aerospace gas turbine engines. It has been demonstrated the feasibility of LDED to produce near net shape turbine blade of a nickel-based superalloy (Inconel 738) showed in Fig. 4.6. Crack-free single-wall components of Inconel 738 can be successfully

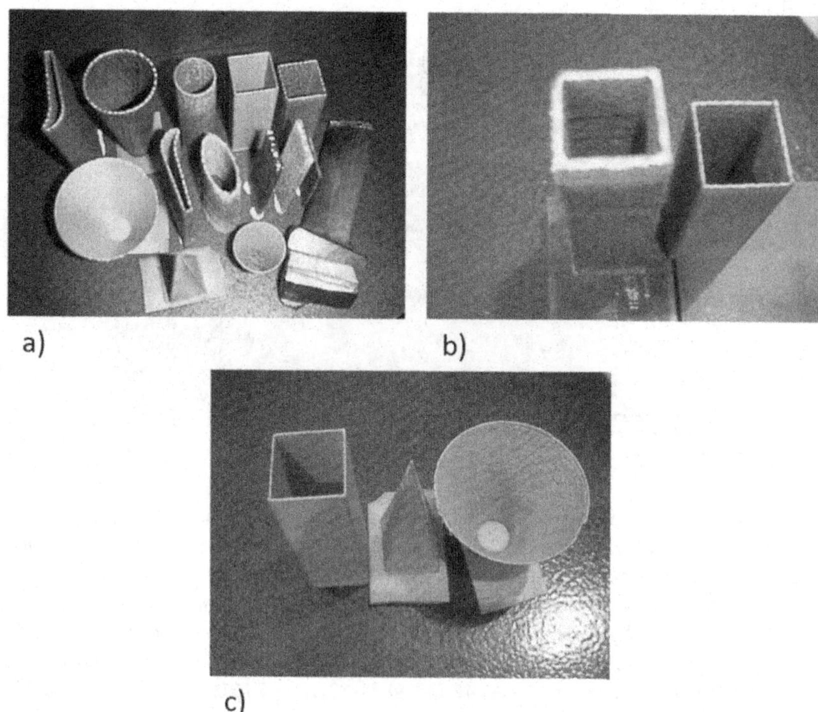


**Figure 4.4** Components manufactured by laser-directed energy deposition [27]: (A) collection of different geometries; (B) Ti–6Al–4V tubes of different thickness; and (C) geometries with straight or inclined thin walls. *(Reproduced with kind permission of Elsevier from Reference X. Wu, J. Mei, Near net shape manufacturing of components using direct laser fabrication technology, J. Mater. Process. Technol. 135 (2003) 266–270. doi.org/10.1016/S0924-0136(02)00906-8.)*

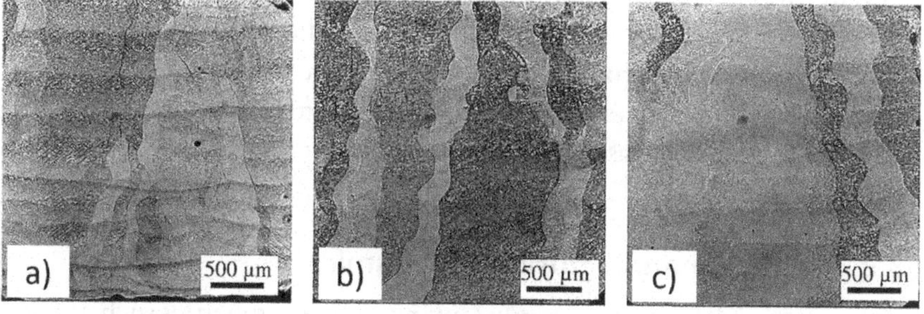


**Figure 4.5** Optical micrographs obtained at different locations of Ti6Al4V samples generated by laser-directed energy deposition showing columnar grains that dominate the entire sample: (A) bottom of the sample; (B) center of the sample; and (C) top of the sample [31]. *(Modified from Reference X. Wu, J. Liang, J. Mei, C. Mitchell, P.S. Goodwin, W. Voice, Microstructures of laser-deposited Ti-6Al-4V, Mater. Des. 25 (2004) 137–144. doi:10.1016/j.matdes.2003.09.009 with permission from Elsevier.)*

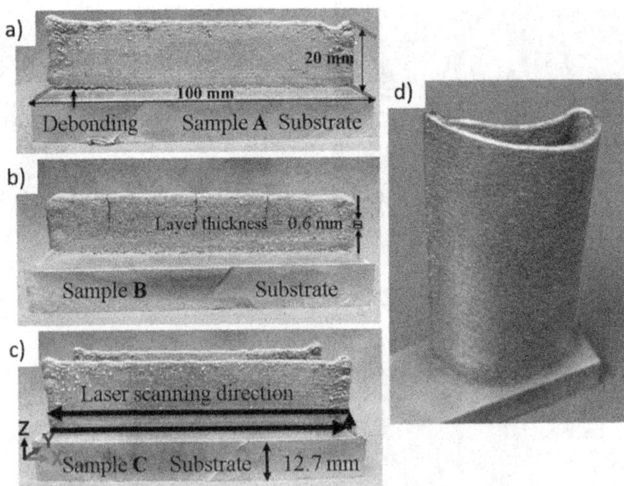


**Figure 4.6** (A) As-deposited IN 738 sample A showing debonding at the substrate/deposit interface; (B) as-deposited IN 738 sample B with cracks at almost equal distance; (C) crack free as-deposited IN 738 sample C showing the laser scanning direction of the single wall specimen; and (D) prototype of a turbine blade developed using IN 738 alloy [32]. *(Reproduced with kind permission of Elsevier from Reference A. Ramakrishnan, G.P. Dinda, Materials science & engineering A direct laser metal deposition of Inconel 738, Mater. Sci. Eng. 740–741 (2019) 1–13. doi:10.1016/j.msea.2018.10.020.)*

manufactured by maximizing the energy density and by reducing the scanning speed [32]. Due to heat sink effect caused by the substrate, the process is inherent with rapid directional solidification leading to unique microstructures that cannot be achieved by conventional processes such as casting [33].

LDED was also used to fast generate AISI 316L stainless steel parts with large size and excellent mechanical properties. The effect of building direction was found to produce anisotropy on microstructure, mechanical properties, and machinability of the stainless steel. Anisotropy in the machinability could be used to increase efficiency and reduce production cost [34].

## 4.3.2 Additive manufacturing of ceramics

Ceramics are widely used in industry owing to their properties such as high wear resistance, chemical inertness, and excellent mechanical properties at elevated temperatures. These materials are, however, not easy to process by conventional manufacturing methods. LDED using high-power-density laser beams has been developed to manufacture ceramic components [24]. In Fig. 4.7, ceramic parts with different geometries that were produced by means of LDED are shown.

Niu et al. successfully prepared dense, large-sized $Al_2O_3$ ceramic samples using LDED [35]. Microstructure of these ceramic samples consists of grains that grow along the deposition height direction as it is shown in Fig. 4.8. Property test results show that flexural strength and compressive strength of deposited $Al_2O_3$ reaches the level of traditional sintered $Al_2O_3$. Balla et al. fabricated dense and net-shaped structures (e.g., cylinder, cube, gear, etc.) of $Al_2O_3$ by LDED [36]. As processed structures showed that columnar grains formed along the build direction and they exhibited anisotropy in mechanical properties.

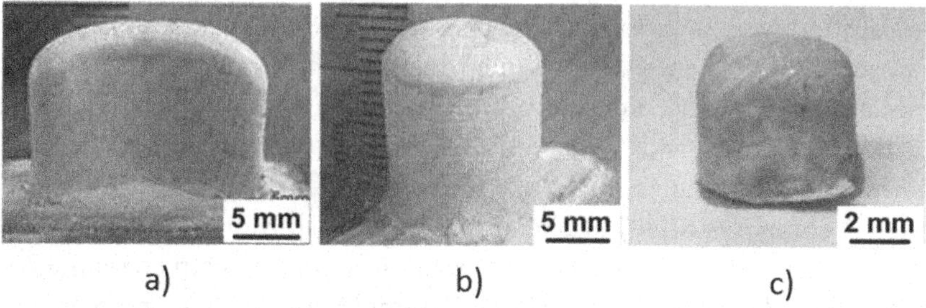


**Figure 4.7** $ZrO_2-Al_2O_3$ parts with different shapes manufactured by laser-directed energy deposition [24]: (A) arc wall; (B) cylinder; and (C) cube. *(Reproduced with kind permission of Elsevier from Reference Y. Hu, W. Cong, A review on laser deposition-additive manufacturing of ceramics and ceramic reinforced metal matrix composites, Ceram. Int. 44 (2018) 20599–20612. doi:10.1016/j.ceramint.2018.08.083.)*

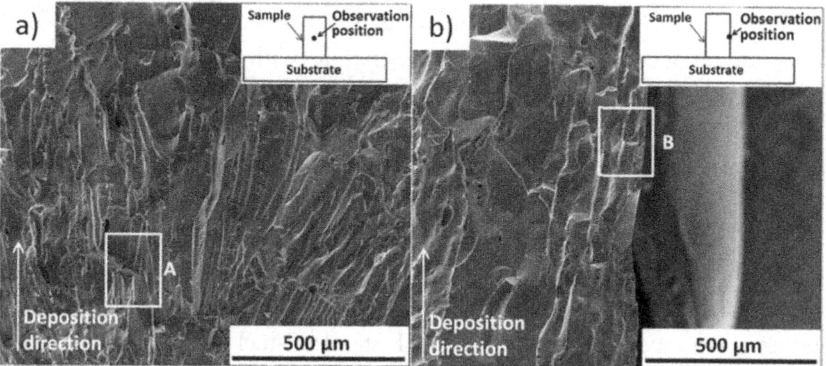


**Figure 4.8** Directionally grown microstructure of $Al_2O_3$ sample [35]: (A) middle of the sample and (B) edge of the sample. *(Reproduced with kind permission of Elsevier from Reference F. Niu, D. Wu, F. Lu, G. Liu, G. Ma, Z. Jia, Microstructure and macro properties of $Al_2O_3$ ceramics prepared by laser engineered net shaping, Ceram. Int. 44 (2018) 14303–14310. doi:10.1016/j.ceramint.2018.05.036.)*

Zirconia–alumina ($ZrO_2$–$Al_2O_3$) ceramics are gaining popularity among researchers due to their excellent mechanical properties at high temperature and chemical stability. Niu et al. used LDED to fabricate cylindrical and arc-shaped $Al_2O_3$–$ZrO_2$ eutectic ceramic structures directly from pure powders without any binders [37]. During the process, mixed ceramic powders are completely melted yielding fully dense ceramic with fine-grained microstructure. Li et al. studied walled structures of $Al_2O_3$–$ZrO_2$ ceramic were deposited using LDED with different $Al_2O_3/ZrO_2$ powder weight ratios [38]. The results of this research provide some basis for attaining high-quality $Al_2O_3$–$ZrO_2$ ceramics by LDED. To reduce cracking and inhomogeneous material dispersion, Hu et al. propose a novel ultrasonic vibration-assisted LDED process for fabrication of bulk ZrO2–$Al_2O_3$ parts [39]. Results showed that the initiation of cracks and the crack propagation were suppressed in the parts fabricated by ultrasonic vibration-assisted LDED. These parts presented higher microhardness, higher wear resistance, and better compressive properties.

Besides $Al_2O_3$ and $ZrO_2$–$Al_2O_3$, LDED process was studied to generate other ceramic materials such as $Al_2O_3$-YAG- $ZrO_2$ ternary eutectic components [40] or zirconate titanate [41]. Moreover, LDED has been explored to produce ceramic reinforced metal matrix composites (MMCs) with improved properties. Examples of MMCs are titanium matrix composites reinforced with TiC [42], TiN [43], or TiB [44] and Ni matrix composites reinforced with TiC [45].

### 4.3.3 Surface treatments and repairing components

LDED is not limited to AM of components. It can be used to modify the surface of materials by depositing a coating, which is also commonly known as laser cladding or laser alloying. Full-dense coatings with a strong bond between the substrate and the coating can be achieved by means of laser cladding. This technique is particularly interesting in industrial applications to improve the surface properties of components, by adding a layer of a different material, and to repair worn or damaged parts. In Fig. 4.9, it is shown the employment of LDED technique to add material to an existing component.

Stellite alloys are a group of cobalt-based superalloys with excellent wear resistance. Two modified Stellite 21, low-carbon high-molybdenum Stellite alloys, were deposited on 316 L stainless steel via laser cladding by Yao et al. [46]. The intermetallic compounds generated can increase hardness and sliding wear resistance of the hardfacings. Moreover, Stellite

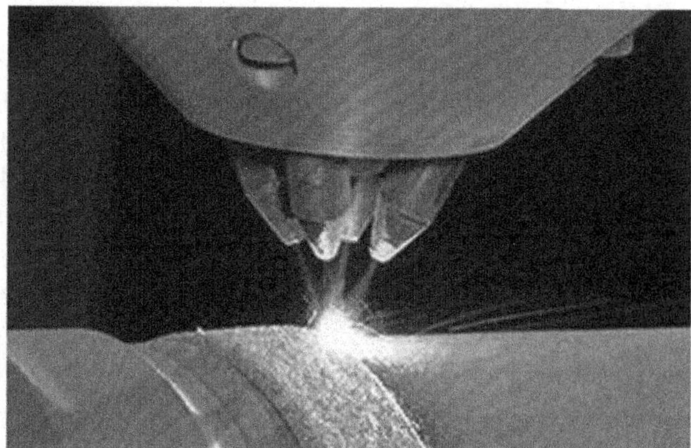

**Figure 4.9** Laser-directed energy deposition technique used to deposit material on the surface of a component [24]. *(Reproduced with kind permission of Elsevier from Reference Y. Hu, W. Cong, A review on laser deposition-additive manufacturing of ceramics and ceramic reinforced metal matrix composites, Ceram. Int. 44 (2018) 20599–20612. doi:10.1016/j.ceramint.2018.08.083.)*

can protect against corrosion of components because a protective oxide film is formed in this material in corrosive environments. A Stellite alloy mixture hardfacing was created via laser cladding for control valve seat sealing surfaces by Ding et al. [47]. The real industrial test on the butterfly valve sealing surface with this Stellite alloy mixture hardfacing exhibited superior sealing performance to the existing sealing surfaces. Ocelík et al. also explored the generation of thick cobalt-based coatings on cast iron by laser cladding [48]. Laser cladding of cobalt-based alloys represents a promising way for an improvement of local wear and corrosion resistance of cast iron industrial components.

Nickel-based superalloys (e.g., Inconel) are alloys that exhibit a high corrosion resistance. Laser cladding was used to create coatings of Inconel 625 on medium alloy steel by Verdi et al. [49]. The coatings presented a good adhesion to the substrate and high hardness. In other work developed by Xu et al. [50], it was analyzed the corrosion performance of Inconel 625 coatings obtained by laser cladding on the surface of 316 L Stainless Steel. The corrosion performance of coating area was superior to substrate in different solutions, indicating an excellent protecting effect of Inconel 625 coating. Fiber laser cladding of NiCrBSi alloy on cast iron was studied by Arias-González et al. [51]. NiCrBSi coatings present a similar elastic modulus compared with the cast iron substrate but a significantly higher hardness, so it can improve the surface properties of the cast iron.

Laser cladding has also been developed to generate coatings of other multiple materials like phosphor bronze [52], aluminum [53], or ceramic materials [24,54]. Phosphor bronze is a suitable bearing material because of its good fatigue strength and excellent wear properties under corrosive conditions, high temperatures, and high loads. Laser cladding was proposed by Arias-González et al. [52] as a method to create a bronze surface in an area of a shaft as a substitute of warm shrink fitting of machined bronze bushes (see Fig. 4.10). Riveiro et al. studied the production of aluminum-based

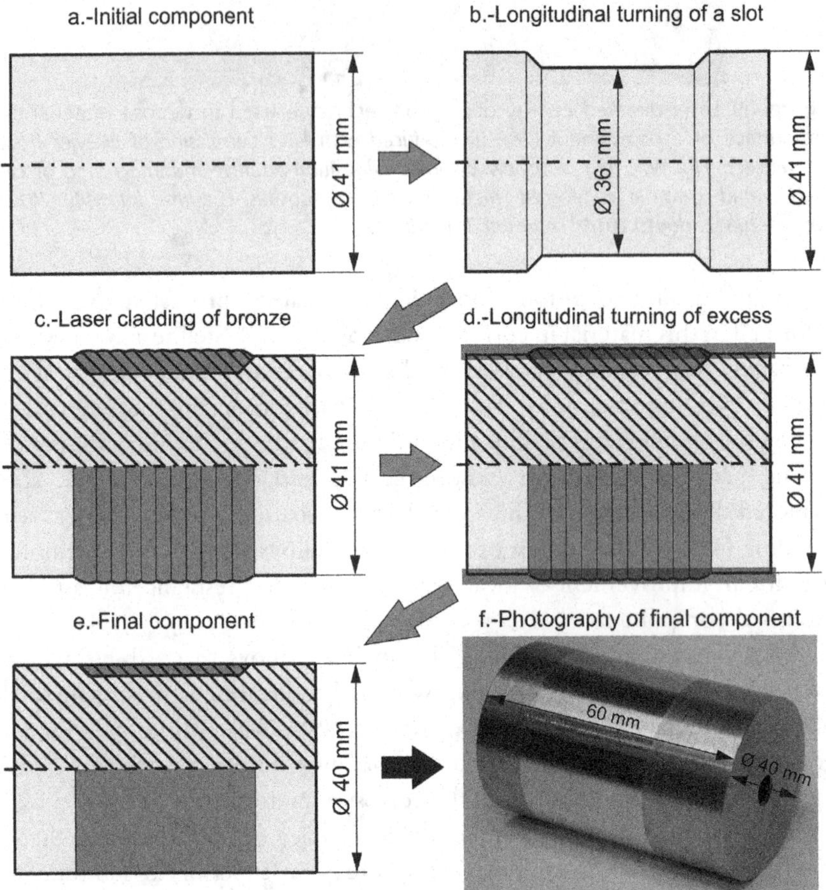


**Figure 4.10** Method proposed to create a bronze surface in an area of a shaft [52]: (A) initial alloy steel shaft; (B) generation of a slot by means of longitudinal turning; (C) deposition of phosphor bronze by means of laser cladding; (D) removal of the excess material by means of longitudinal turning; (E) sketch of the final component; and (F) photography of the final component generated following the previous steps. *(Reproduced with kind permission of Elsevier from Reference F. Arias-González, J. del Val, R. Comesaña, J. Penide, F. Lusquiños, F. Quintero, et al., Laser cladding of phosphor bronze, Surf. Coatings Technol. 313 (2017) 248–254. doi:10.1016/j.surfcoat.2017.01.097.)*

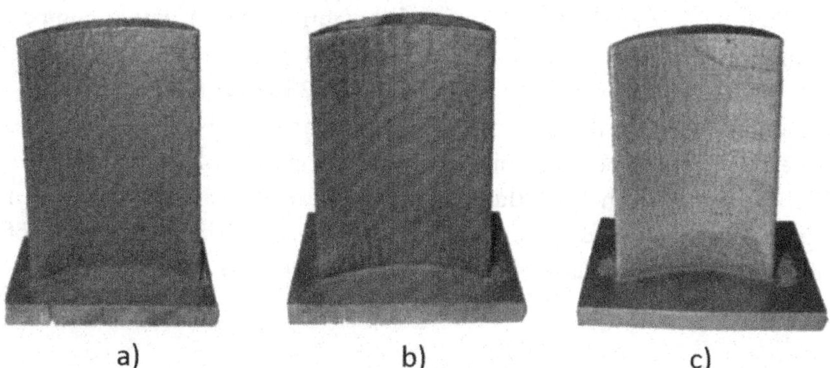


**Figure 4.11** Turbine blades built with the laser-directed energy deposition process [55]: (A) baseline undamaged blade; (B) first restored blade; and (C) second restored blade. *(Reproduced with kind permission of Elsevier from Reference J.M. Wilson, C. Piya, Y.C. Shin, F. Zhao, K. Ramani, Remanufacturing of turbine blades by laser direct deposition with its energy and environmental impact analysis, J. Clean. Prod. 80 (2014) 170–178. doi:10.1016/j.jclepro.2014.05.084.)*

coating on 304 Stainless Steel by laser cladding [53]. These coatings are very interesting for potential application such as fuel cells or catalytic converters.

Moreover, LDED is also an interesting and cost-effective method to refurbish worn or damaged high-value components [55]. Michael Wilson et al. used this technique to successfully repair defective voids in turbine airfoils (see Fig. 4.11. Their experimental results demonstrate the feasibility of LDED in remanufacturing and its potential to adapt to a wide range of part defects. In other work, thin sections of Ti6Al4V-based aeroengine structural material was refurbished using LDED technique by Raju et al. achieving desirable mechanical strength and metallurgical properties [56].

### 4.3.4 Functionally graded materials

FGM parts are heterogeneous objects with material composition and microstructure that change gradually into the component [57]. This concept has been proposed in the 1980s, and they give the possibility of selecting the distribution of properties to achieve the desired functions. These multimaterial parts offer great potential for industrial applications because it is possible to adapt their properties. LDED is a promising technique to manufacture FGM parts because it is possible to gradually change the supplied material [58].

Chen et al. used LDED to successfully fabricate FGM with a composition varying from 100% 316L stainless steel to 100% Inconel 625 alloy (see Figs. 4.12 and 4.13) [59]. The 316L/Inconel 625 FGM combines the significant properties of the two materials and can be widely used in aerospace, nuclear power, and petrochemical industries. Continuously graded Stainless Steel 316L and Inconel 718 thin wall structures made by LDED have been explored by Shah et al. [60]. They have demonstrated that generation of carbides can be controlled by processing parameters to tailor the hardness and wear resistance of the FGM.

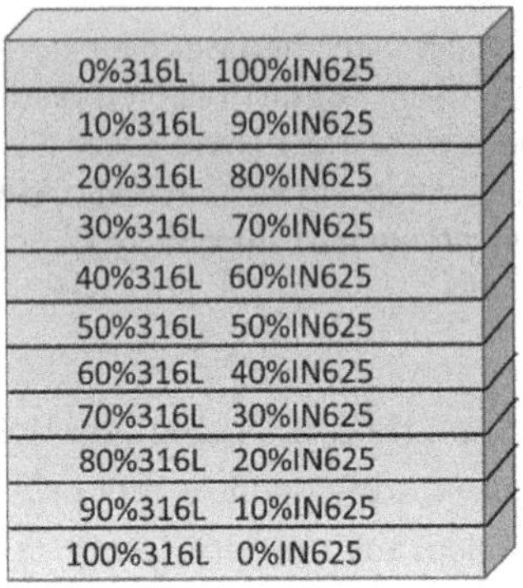

0%316L 100%IN625
10%316L 90%IN625
20%316L 80%IN625
30%316L 70%IN625
40%316L 60%IN625
50%316L 50%IN625
60%316L 40%IN625
70%316L 30%IN625
80%316L 20%IN625
90%316L 10%IN625
100%316L 0%IN625

**Figure 4.12** Schematic of a 316L/Inconel 625 functionally graded material [59]. *(Reproduced with kind permission of Elsevier from Reference B. Chen, Y. Su, Z. Xie, C. Tan, J. Feng, Development and characterization of 316L/Inconel625 functionally graded material fabricated by laser direct metal deposition, Opt. Laser Technol. 123 (2020) 105916. doi:10.1016/j.optlastec.2019.105916.)*

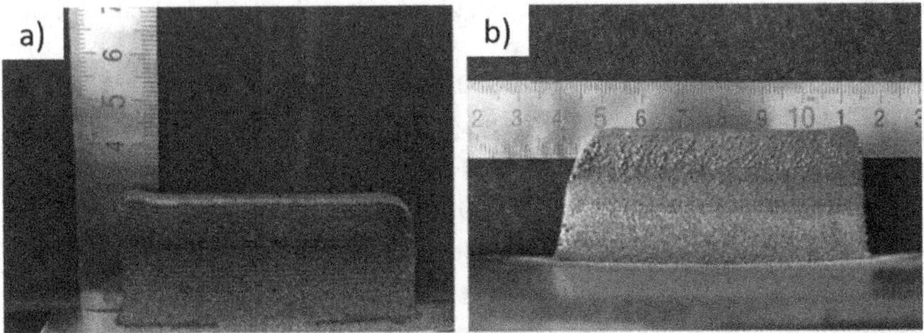


**Figure 4.13** Photograph of 316L/Inconel 625 functionally graded material manufactured by laser-directed energy deposition [59]: (A) the measurement of sample height and (B) the measurement of sample length. *(Reproduced with kind permission of Elsevier from Reference B. Chen, Y. Su, Z. Xie, C. Tan, J. Feng, Development and characterization of 316L/Inconel625 functionally graded material fabricated by laser direct metal deposition, Opt. Laser Technol. 123 (2020) 105916. doi:10.1016/j.optlastec.2019.105916.)*

The manufacturing of multiple other combinations of pure metals and alloys by LDED have been explored and are still been explored. Simple parts with a composition varying smoothly from 100% 316L Stainless Steel to 100% cobalt base superalloy have been produced using LDED by del Val et al. [61]. The base of the part has identical characteristics to the pure steel parts: similar values of microhardness (175–245 HV) and similar austenitic microstructure. On the other hand, the top of the part has the same values of hardness (600–700 HV) and dendrite microstructure than pure cobalt base superalloy parts. A novel Ti/Al lightweight graded material was fabricated successfully using LDED by Liu et al. [62]. The composition of the material continuously changes from 100% Ti6Al4V to 100% AlSi10Mg resulting in a gradual change of properties.

Nonetheless, as LDED is not limited to metallic materials, there are promising research for the production of metal/ceramic FGMs. FGM consisting of Inconel 625 (IN 625) and yttria-stabilized zirconia were deposited on an A516 steel substrate via LDED by Rao et al. [63]. These combinations of materials can have applications as thermal barrier coatings that are regularly used today to protect and extend the service life of several superalloys, which are extensively used in high-temperature applications. Wang et al. produced Ti6Al4V reinforced with TiC as compositionally graded material by LDED using TiC powder and Ti6Al4V wire, which were fed simultaneously [26]. Preliminary sliding wear test shows that the Ti6Al4V tribological properties have been improved by the reinforced TiC particles. Last example is the Ti/$Al_2O_3$ compositionally graded structure manufactured by Zhang et al. using an LDED process [64].

## 4.4 Biomedical applications of laser-directed energy deposition

With the development of modern society, life expectancy has increased, but elderly people have a higher risk of hard tissue failure. The number of aged people demanding failed tissue replacement is rapidly increasing. Advanced manufacturing of biomedical implants is seeing as a way to satisfy this demand and improve the performance of the biomaterials (materials that present biological compatibility with the human body in the long term).

In this context, LDED is being explored as an advanced technique for manufacturing orthopedic and dental implants. The research carried out during last decades in the biomedical applications of LDED has tackled the

AM of metallic and ceramic biomaterials. Furthermore, LDED has being used to produce specific coatings to modify the surface of implants enhancing their properties. Moreover, LDED is an especially interesting technique in this field to generate FGMs and heterogeneous multimaterial components that allow new implants design with special functionalities.

### 4.4.1 Additive manufacturing of metallic biomaterials

Metals and their alloys are widely used as biomedical materials, and it is estimated that 70%–80% of biomedical implants are made of metallic materials [65]. Metallic biomaterials like commercially pure titanium (cp-Ti), Ti6Al4V, and CoCrMo alloys are extensively used because of their excellent mechanical properties like mechanical strength and toughness. At present, ceramics or polymers cannot replace metals as biomaterials for load-bearing surgical implants. However, metallic materials sometimes show toxicity, and they fracture because of corrosion and mechanical damages.

Titanium and its alloys are generally considered as one of the most suitable materials for biomedical applications owing to their relatively low elastic modulus, high corrosion resistance, and excellent biocompatibility [65]. Particularly, Ti6Al4V is the most commonly used in biomedical industries because of its excellent strength and fracture toughness. Ti6Al4V scaffold has been fabricated by LDED technology for patient-specific bone tissue engineering by Dinda et al. [66]. The ductility of the as-deposited tensile sample was below the ASTM limit for Ti6Al4V implants. After a suitable heat treatment (sample annealed at 950°C followed by furnace cooling), mechanical properties become higher than ASTM limits for medical implants.

Although Ti6Al4V is the most commonly used titanium alloy, almost all commercially available dental implants are made of cp-Ti [67]. LDED has been used to generate thin walls of commercially pure Ti by Arias-González et al. [68]. The microstructure observed in the samples consists of large elongated columnar prior beta grains, which have grown epitaxially from the substrate to the top, in parallel to the building direction. The prior beta grains are formed by alpha Ti lamellae and lamellar colonies, which are structured in a basket-weave pattern. LDED has been explored to fabricate net shape porous Ti for load-bearing bone implants by Xue et al. [69]. In vitro tests showed that Ti porous structures improved cell adhesion and proliferation. Vamsi Krishna et al. demonstrated that porosity and mechanical properties of cp-Ti structures generated by LDED can be tailored by changing the process parameters as it is shown in Fig. 4.14. Young's modulus and 0.2% proof strength of porous Ti samples having porosity in the range 35–42 vol.% are close to those of human cortical bone [70].

Mechanical properties of cp-Ti cannot satisfy the requirements of biomaterials in some cases when high strength is necessary, such as hard tissue replacement or underintensive wear use [65]. Ti6Al4V possesses ideal mechanical properties, but the diffusion of cytotoxic V and Al ions may cause long-term health issues once released inside the human body [71]. Recently, significant thrust has been generated toward the development of new beta titanium alloys elaborated from a combination of noncytotoxic elements for applications in biomedical implants. They present superior biocompatibility, corrosion resistance, high strength, and lower elastic modulus, which can facilitate to reduce the stress shielding effect [71].

LDED is proposed as a promising technology for the near-net shape processing of load-bearing orthopedic implants from beta Ti alloys with a relatively low elastic modulus (55 GPa, as it is shown in Fig. 4.15) [72].

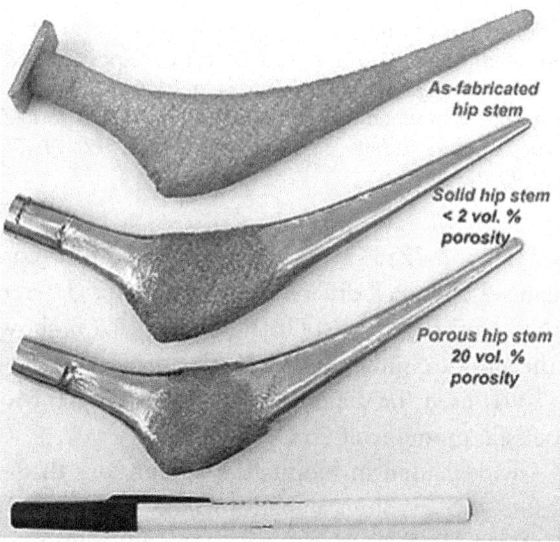


**Figure 4.14** Net shape functional hip stems with designed porosity fabricated using laser-directed energy deposition [70]. *(Reproduced with kind permission of Elsevier from Reference B.V. Krishna, S. Bose, A. Bandyopadhyay, Low stiffness porous Ti structures for load-bearing implants, Acta Biomater. 3 (2007) 997–1006. doi:10.1016/j.actbio.2007.03.008.)*

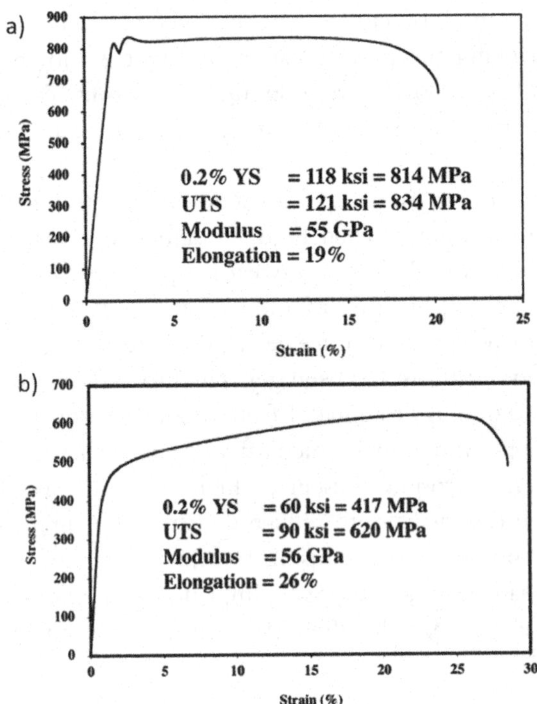


**Figure 4.15** A representative engineering stress versus strain graph from a tensile-tested laser deposited Ti35Nb7Zr5Ta sample in (A) as-deposited and (B) beta solutionized conditions [72]. *(Reproduced with kind permission of Elsevier from Reference S. Nagn, R. Banerjee, Laser deposition and deformation behavior of Ti-Nb-Zr-Ta alloys for orthopedic implan, J. Mech. Behav. Biomed. Mater. 16 (2012) 21–28. doi:10.1016/j.jmbbm.2012.08.014.)*

Laser-deposited Ti35Nb7Zr5Ta (wt.%) exhibits excellent corrosion resistance and enhanced bone cell differentiation [73]. In a different work, Ti-27.5 (at. %) Nb synthesized by LDED (see Fig. 4.16) showed that the properties of the alloy are similar or better after the deposition process [74]. Finally, LDED was used for deposition of beta type Ti15Mo biomedical alloy with an elastic modulus of 73 GPa [75].

CoCrMo is widely used in biomedical applications that require wear resistance because the tribological performance of Ti-based materials is rather poor. CoCrMo is the hardest known biocompatible metal alloy with good tensile and fatigue properties. Fully dense, metallurgically sound CoCrMo deposits have been produced by LDED [76]. Deposited CoCrMo exhibits very thin, long interconnected carbide particles and continuous

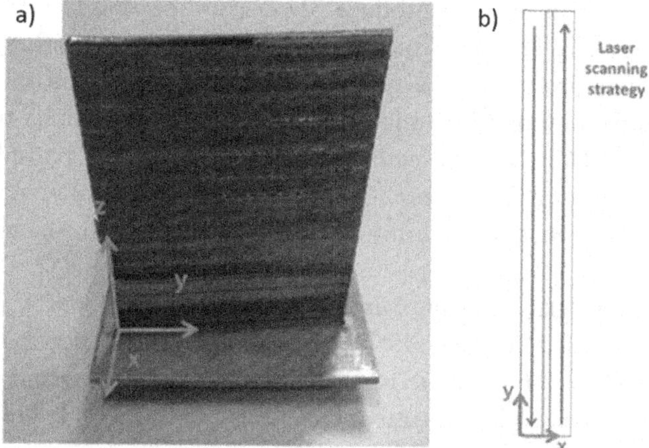


**Figure 4.16** (A) Laser-directed energy deposition manufactured wall and (B) its fabrication scanning strategy [74]. *(Reproduced with kind permission of Elsevier from Reference M. Fischer, P. Laheurte, P. Acquier, D. Joguet, L. Peltier, T. Petithory, et al., Synthesis and characterization of Ti-27.5Nb alloy made by CLAD® additive manufacturing process for biomedical applications, Mater. Sci. Eng. C. 75 (2017) 341–348. doi:10.1016/j.msec.2017.02.060.)*

carbide networks leading to inferior abrasive wear resistance compared with wrought CoCrMo material. LDED is potentially suitable for fabrication of CoCrMo biomedical implants, but further work is required to determine the effects of this microstructure on properties. Recently, O. Barro et al. performed a comparative study on the characteristics of CoCr pieces produced by LDED, SLM, milling and casting for the application in dental prosthesis [77]. The results show that LDED pieces have a higher toughness than the materials obtained by competing techniques (SLM and casting), while fulfilling the requirements posed by the ISO standard 22674 to be used in dental restorations and appliances.

### 4.4.2 Additive manufacturing of ceramic biomaterials

Almost 70% of bone is made of ceramic material [78]. Regenerative bone treatments require materials with a noncytotoxic behavior and the ability to boost the growth and recovery of lost bone. Resorbable and osteoconductive bioceramics, such as silicate bioactive glass and calcium phosphate, are used in applications where high mechanical resistance is not an essential requirement. To fulfill the complete function, the ceramic implant geometry must resemble the original bone, supporting and guiding the growth of the surrounding healthy bone [79].

LDED has been used to produce 3D calcium phosphate grafts, as it is shown in Fig. 4.17 [22]. Hydroxyapatite (HA) was selected as the precursor material. Laser processing leads to complete dihydroxylation of the precursor material resulting in a microstructure composed by alpha-tricalcium phosphate (α-TCP) matrix with nucleated tetracalcium phosphate and amorphous calcium phosphate. The produced bioceramic grafts were observed to be bioactive and promoted pre-osteoblastic cell attachment and proliferation.

3D bioactive glass implants were produced by LDED [80,81]. $CO_2$ laser radiation was used to melt 45S5 and S520 bioactive glass particles to generate crack-free fully dense implants (see Fig. 4.18). Implants are noncytotoxic in contact with pre-osteoblastic cells and the obtained results state that the bioactive properties of the starting material remain generally unaffected by the process. S520 bioactive glass implants keep the characteristic mechanical properties of bulk bioactive glasses, but mechanical properties of 45S5 implants are lower than those of the precursor glass because of extensive crystallization during laser processing [81].

LDED has demonstrated its capability to produce unique implants for low-load-bearing bone repair by combining calcium phosphates and bioglass particles [82]. Implants with a tailored distribution of low-resorbability

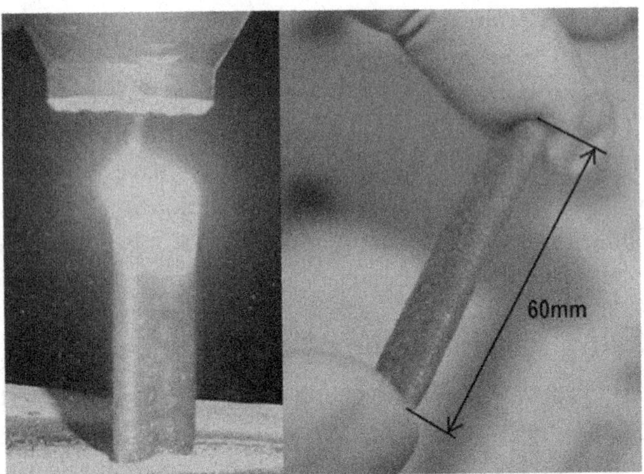


**Figure 4.17** Calcium phosphate sample produced by laser-directed energy deposition (complete sample height 60 mm, laser power 160 W, scanning speed 3.0 mm/s, mass flow 10 mg/s, 320 stacked layers) [22]. *(Reproduced with kind permission of Elsevier from Reference R. Comesaña, F. Lusquiños, J. del Val, T. Malot, M. López-Álvarez, A. Riveiro, et al., Calcium phosphate grafts produced by rapid prototyping based on laser cladding, J. Eur. Ceram. Soc. 31 (2011) 29–41. doi:10.1016/j.jeurceramsoc.2010.08.011.)*

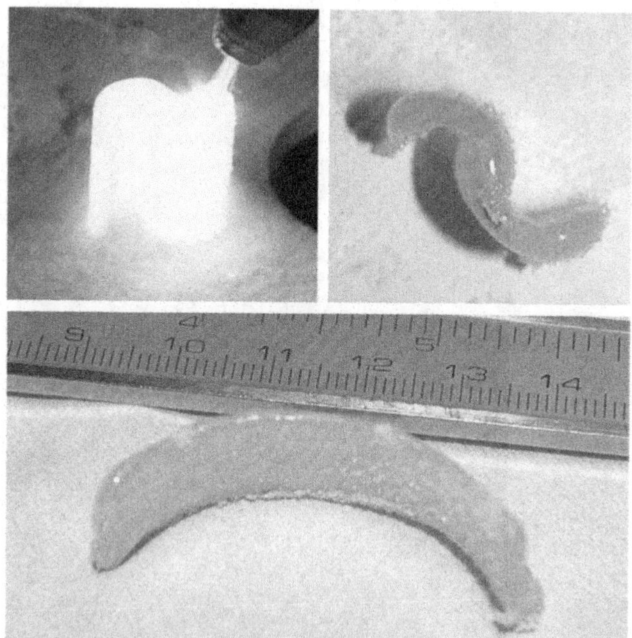

**Figure 4.18** Fabrication of S520 bioactive glass implants by laser-directed energy deposition [81]. *(Reproduced with kind permission of Elsevier from Reference J. del Val, R. López-Cancelos, A. Riveiro, A. Badaoui, F. Lusquiños, F. Quintero, et al., On the fabrication of bioactive glass implants for bone regeneration by laser assisted rapid prototyping based on laser cladding, Ceram. Int. 42 (2016) 2021–2035. doi:10.1016/j.ceramint.2015.10.009.)*

calcium phosphate at the core and highly reactive bioglass at the surface were produced. The bioactivity and degradation rates of the bioglass are preserved after deposition, and this novel type of implant can lead to new research to determine the optimum proportions of calcium phosphate and bioglass to match the new bone ingrowth for each particular case.

### 4.4.3 Surface treatments of biomaterials

Laser cladding is particularly interesting in biomedical applications to enhance the corrosion resistance and bioactivity behavior of titanium alloys [83]. Laser cladding of Ta coatings on Ti have been explored to enhance the osteointegration of implants (see Fig. 4.19) [84]. An in vitro biocompatibility study showed excellent cellular adherence and growth on the Ta coating surface compared with the Ti surface. The enhanced cell attachment and proliferation on the Ta surface were a direct consequence of its

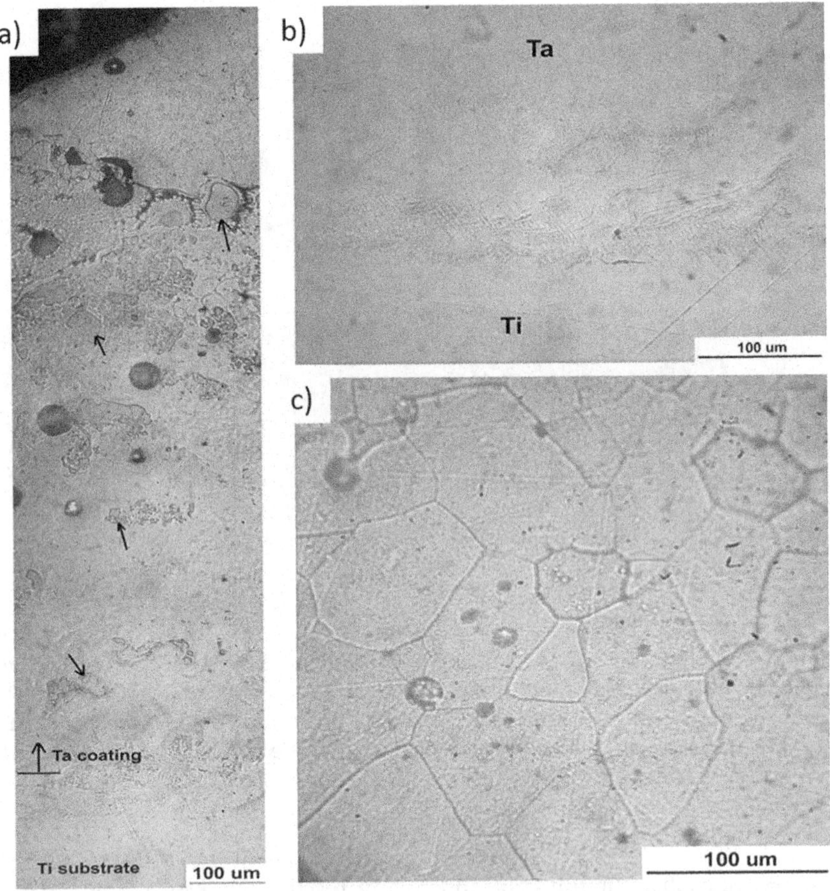


**Figure 4.19** Cross-sectional microstructures of Ta coating on Ti [84]: (A) full coating. Arrows indicate unmelted/partially melted Ta powder; (B) coating–substrate interface; and (C) high-magnification SEM image showing the grain size of the laser-processed Ta coating. *(Reproduced with kind permission of Elsevier from Reference V.K. Balla, S. Banerjee, S. Bose, A. Bandyopadhyay, Acta Biomaterialia Direct laser processing of a tantalum coating on titanium for bone replacement structures, Acta Biomater. 6 (2010) 2329–2334. doi:10.1016/j.actbio.2009.11.021.)*

high wettability and surface energy. Moreover, dense Ta coatings do not suffer from low fatigue resistance, which is a major concern for porous coatings used for enhanced/early biological fixation.

Since the invention of bioactive materials, which can generate chemical bond with bones, the researchers proposed combining the superior mechanical properties of metals and bioactivity of bioactive materials.

Successful coatings of calcium phosphate on a titanium alloy have been manufactured by means of laser cladding by Lusquiños et al. [85]. The method allows the generation of a strong fusion bonding between the calcium phosphate coating and the titanium substrate. The formation of calcium titanates and titanium phosphides at the coating—substrate interface is responsible for the sound join [86].

LDED has been applied to coat titanium substrate with uniformly distributed tricalcium phosphate ceramics to improve bone cell—material interactions, as it is shown in Fig. 4.20 [87]. The fine Ti grains along with tricalcium phosphate at the grain boundaries increased the coating hardness to 1049 ± 112 HV compared with a substrate hardness of 200 ± 15 HV. In vitro studies indicated that the TCP coatings had good biocompatibility and promoted biomineralization.

Laser cladding technology has been explored to produce bioactive glass coatings on Ti6Al4V substrates [88]. Coatings are formed by crystallized calcium silicate at the surface with uniform composition along the cross-section as it can be seen in Fig. 4.21. No dilution is observed and the coatings present similar bioactivity to that of the precursor material.

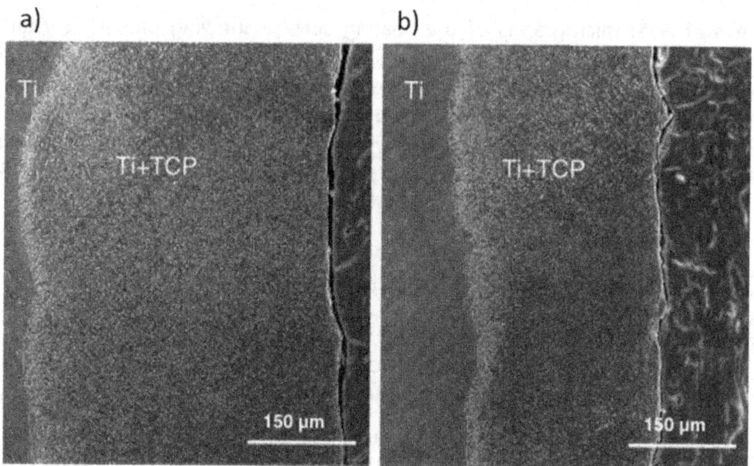


**Figure 4.20** SEM micrographs of TCP coatings on Ti fabricated by laser-directed energy deposition [87]. *(Reproduced with kind permission of Elsevier from Reference M. Roy, B. Vamsi Krishna, A. Bandyopadhyay, S. Bose, Laser processing of bioactive tricalcium phosphate coating on titanium for load-bearing implants, Acta Biomater. 4 (2008) 324—333. doi:10.1016/j.actbio.2007.09.008.)*

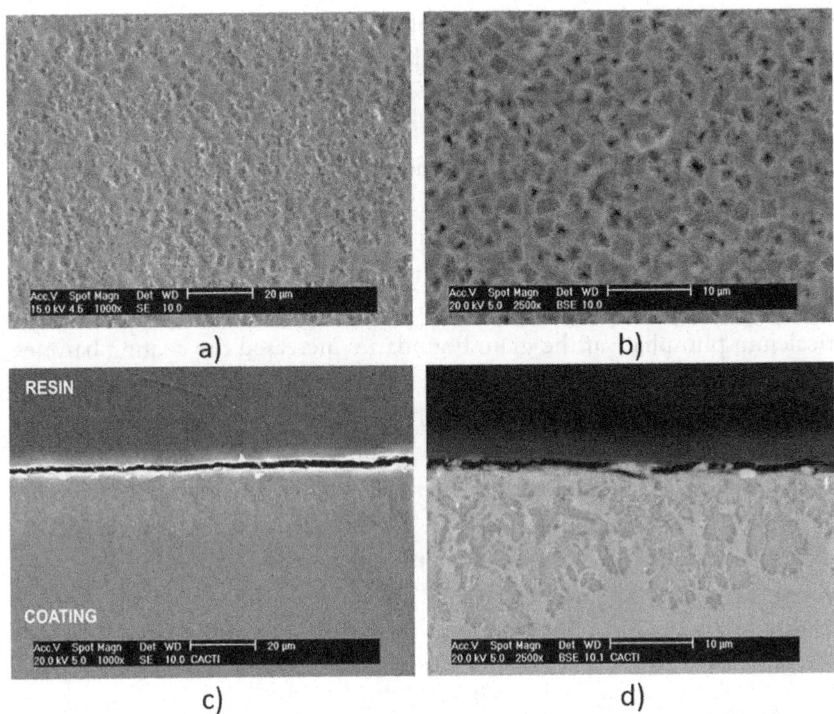


**Figure 4.21** SEM micrographs of the coating surface showing partially crystallized S520 bioactive glass coating [88]: (A) secondary electrons and (B) backscattered electrons micrograph detail. Coating cross-section showing surface crystallization. (C) Secondary electrons and (D) backscattered electrons micrograph detail. *(Reproduced with kind permission of Elsevier from Reference R. Comesaña, F. Quintero, F. Lusquiños, M.J. Pascual, M. Boutinguiza, A. Durán, et al., Laser cladding of bioactive glass coatings, Acta Biomater. 6 (2010) 953–961. doi:10.1016/j.actbio.2009.08.010.)*

## 4.4.4 Functionally graded biomaterials

In the biomedical field, there is still a challenge to develop an implant that has the optimal combination of biocompatibility, mechanical properties, and wear resistance. One approach to solve this issue are FGMs to combine the biocompatibility and mechanical properties of Ti alloys with the hardness and wear resistance of CoCrMo. Krishna et al. tested in vitro biocompatibility of functionally graded CoCrMo coating on Ti6Al4V generated by LDED (see Fig. 4.22) [89]. A combination of 50% CoCrMo and 50% Ti6Al4V showed to have the best combination of wear resistance and biocompatibility.

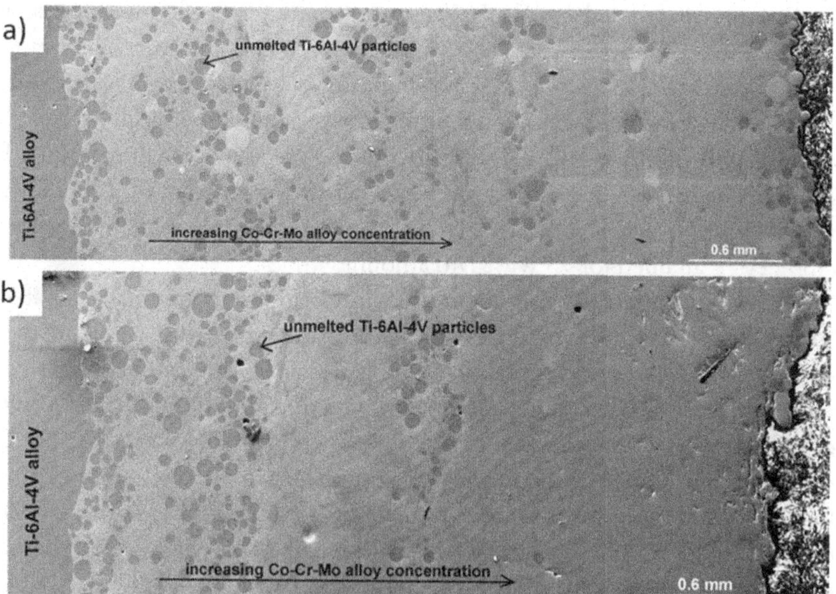


**Figure 4.22** Typical microstructures of laser-processed CoCrMo graded coatings on Ti6Al4V alloy [89]: (A) 50% CoCrMo alloy at the surface and (B) 86% CoCrMo alloy at the surface. *(Reproduced with kind permission of Elsevier from Reference B. Vamsi Krishna, W. Xue, S. Bose, A. Bandyopadhyay, Functionally graded Co-Cr-Mo coating on Ti-6Al-4V alloy structures, Acta Biomater. 4 (2008) 697–706. doi:10.1016/j.actbio.2007.10.005.)*

LDED of functionally gradient biocoatings of CoCrMo over Ti6Al4V have been built and analyzed by Wilson et al. [90]. CoCrMo material was deposited on a Ti6Al4V substrate transitioning from 0% to 100%. Control over the cooling rate is shown to be a key to reduce the effects of thermal expansion differences of the materials and to avoid the formation of a brittle intermetallic phases. The microhardness of the samples is enhanced, and their bonding strength meets the standard requirements.

Ti/Mo FGM with compositions varying from 0% to 19% wt. Mo were produced by means of LDED [91]. The objective of the work was to study the effect of Mo as beta-stabilizer of Ti for the different compositions with microcharacterization techniques. Results showed that alloys formed mainly by a beta phase presents lower values of Young's modulus and hardness than those containing $\alpha'$ or $\alpha''$ phases. Nevertheless, minimum values of Young's modulus (75 GPa) and hardness (240 HV) were attained for the Ti13Mo (wt.%) alloy.

Functionally graded Ti6Al4V—Mo manufactured by LDED were studied by Schneider-Maunoury et al. [92]. A sample with five gradients of composition, from 0 to 100 wt.% Mo, was manufactured using a coaxial nozzle (see Fig. 4.23). Satisfactory metallurgical bonding was achieved between deposited material and substrate and between deposited layers. Microhardness varied along the deposition due to the increase in molybdenum amount. Minimum value (265 HV) is reached for 25 wt.% Mo amount, while maximum value (450 HV) is attained for 75 wt.% Mo amount.

An application of FGM in the biomedical field is fracture fixation plates (also called osteosynthesis plates) with improved properties to avoid the stiffness mismatch between the bone and plate. This issue can be addressed with a stiffness-graded (compositionally graded) titanium alloys with low elastic modulus (or stiffness) at the ends, comparable with the bone modulus; and higher elastic modulus at the center, close to the site of the fracture [93]. LDED has been used to develop a stiffness-graded plate with a low-modulus Ti35Nb15Zr (wt.%) alloy and the higher modulus, commercially pure Ti near the center of the plate (see Fig. 4.24). The control of composition and phases on the material leads to control of mechanical properties in different areas of the implant.

The production of functionally graded Ti/Nb by means of LDED has been studied to generate custom-made parts with tailored properties [94]. In this work, titanium and niobium ratios were modified in different steps to create the variation in alloy composition (see Fig. 4.25). Mechanical test showed a decrease of the microhardness with the increase of Nb content. The elastic modulus was found to be the lowest for Ti40Nb (wt.%).

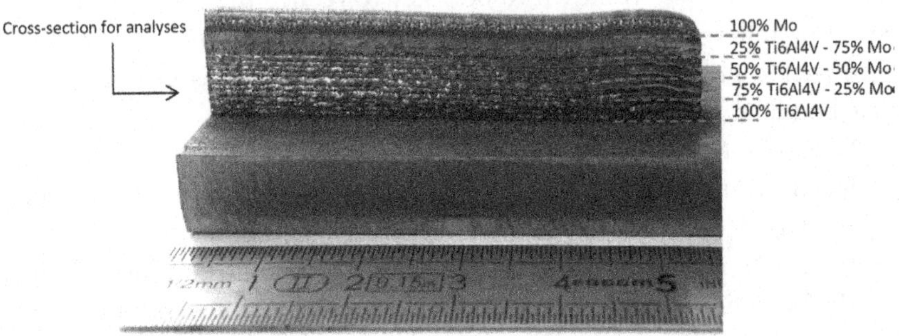


**Figure 4.23** Functionally graded Ti6Al4V/Mo alloy with 25 wt.% Mo increment [92]. *(Reproduced with kind permission of Elsevier from Reference C. Schneider-Maunoury, L. Weiss, P. Acquier, D. Boisselier, P. Laheurte, Functionally graded Ti6Al4V-Mo alloy manufactured with DED-CLAD® process, Addit. Manuf. 17 (2017) 55–66. doi:10.1016/j.addma.2017.07.008.)*

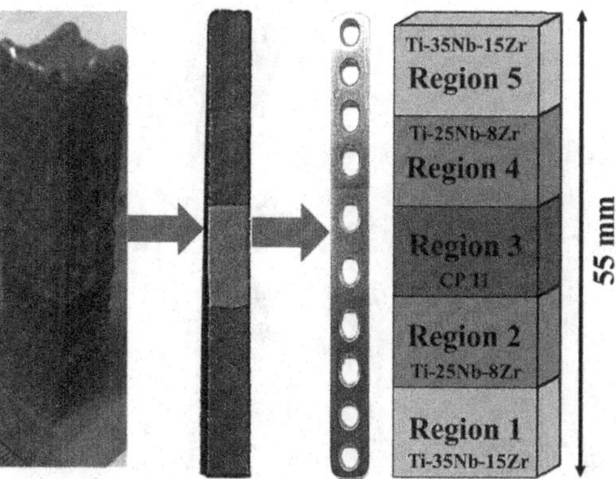


**Figure 4.24** (A) Functionally graded material component produced by laser-directed energy deposition in the as-deposited state; (B) sectioned component on which the characterization was done; (C) final component being shown as the representative bone plate; and (D) dimensions of the printed part with the expected composition profile along its length [93]. *(Reproduced with kind permission of Elsevier from Reference D.D. Lima, S.A. Mantri, C. V. Mikler, R. Contieri, C.J. Yannetta, K.N. Campo, et al., Laser additive processing of a functionally graded internal fracture fixation plate, Mater. Des. 130 (2017) 8–15. doi:10.1016/j.matdes.2017.05.034.)*

## 4.5 Summary

LDED is an AM process to build a component by delivering energy and material simultaneously. A laser beam is used to melt material that is selectively deposited on a specified surface, where it solidifies. LDED technology is particularly suitable for producing low volumes of products in comparison with subtractive manufacturing methods. Compared with other AM methods has the ability to produce near net shape large ceramic or metallic components. Moreover, it can be used to repair or add material to existing parts (including metallic or ceramic coatings). In addition, this technology can be used to generate FGMs. These capabilities are interesting to produce unique high-value components with applications in the industrial and the biomedical fields.

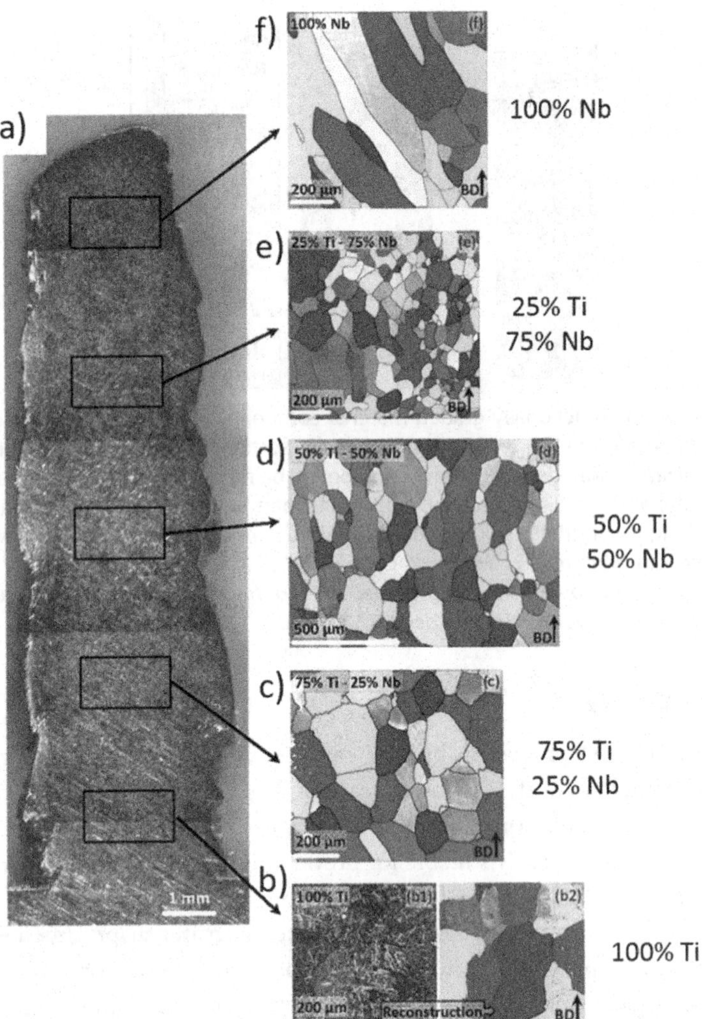


**Figure 4.25** (A) Optical macrograph of functionally graded material Ti/Nb sample (B)–(F) Crystallographic orientation of Ti/Nb alloys with the Nb content. (b1, b2) Measured before and after beta phase reconstruction [94]. *(Reproduced with kind permission of Elsevier from Reference C. Schneider-Maunoury, L. Weiss, O. Perroud, D. Joguet, An application of differential injection to fabricate functionally graded Ti-Nb alloys using DED-CLAD® process, J. Mater. Process. Tech. 268 (2019) 171–180. doi:10.1016/j.jmatprotec.2019.01.018.)*

## Acknowledgments

This work was partially supported by the EU research project Bluehuman (EAPA_151/2016 Interreg Atlantic Area), Government of Spain [RTI2018-095490-J-I00 (MCIU/AEI/FEDER, UE)], and by Xunta de Galicia (ED431B 2016/042, ED481D 2017/010, ED481B 2016/047-0).

## References

[1] N. Guo, M.C. Leu, Additive manufacturing: technology, applications and research needs, Front. Mech. Eng. 8 (2013) 215–243, https://doi.org/10.1007/s11465-013-0248-8.

[2] F. Calignano, D. Manfredi, E.P. Ambrosio, S. Biamino, M. Lombardi, E. Atzeni, et al., Overview on additive manufacturing technologies, Proc. IEEE 105 (2017) 593–612, https://doi.org/10.1109/JPROC.2016.2625098.

[3] ISO/ASTM 52900-15, Standard Terminology for Additive Manufacturing – General Principles – Terminology, ASTM International, West Conshohocken, PA, 2015.

[4] T. DebRoy, H.L. Wei, J.S. Zuback, T. Mukherjee, J.W. Elmer, J.O. Milewski, et al., Additive manufacturing of metallic components - process, structure and properties, Prog. Mater. Sci. 92 (2018) 112–224, https://doi.org/10.1016/j.pmatsci.2017.10.001.

[5] J. Xiong, Y. Lei, H. Chen, G. Zhang, Fabrication of inclined thin-walled parts in multi-layer single-pass GMAW-based additive manufacturing with flat position deposition, J. Mater. Process. Technol. 240 (2017) 397–403, https://doi.org/10.1016/j.jmatprotec.2016.10.019.

[6] M.S. Weglowski, S. Błacha, J. Pilarczyk, J. Dutkiewicz, L. Rogal, Electron beam additive manufacturing with wire - analysis of the process, AIP Conf. Proc. 1960 (2018) 140015, https://doi.org/10.1063/1.5035007.

[7] K.H. Chang, Rapid Prototyp., e-Design 14 (2015) 743–786, doi.org/10.1016/B978-0-12-382038-9.00014-4.

[8] S.M. Thompson, L. Bian, N. Shamsaei, A. Yadollahi, An overview of direct laser deposition for additive manufacturing; Part I: transport phenomena, modeling and diagnostics, Addit. Manuf. 8 (2015) 36–62, https://doi.org/10.1016/j.addma.2015.07.001.

[9] F. Lusquiños, R. Comesaña, A. Riveiro, F. Quintero, J. Pou, Fibre laser micro-cladding of Co-based alloys on stainless steel, Surf. Coating. Technol. 203 (2009) 1933–1940, https://doi.org/10.1016/j.surfcoat.2009.01.020.

[10] J. del Val, R. Comesaña, F. Lusquiños, M. Boutinguiza, A. Riveiro, F. Quintero, et al., Laser cladding of Co-based superalloy coatings: comparative study between Nd:YAG laser and fibre laser, Surf. Coating. Technol. 204 (2010) 1957–1961, https://doi.org/10.1016/j.surfcoat.2009.11.036.

[11] E. Toyserkani, A. Khajepour, S.F. Corbin, Laser Cladding, CRC Press, Boca Ratón, Florida, 2005.

[12] A. Riveiro, J. del Val, R. Comesaña, F. Lusquiños, F. Quintero, M. Boutinguiza, et al., Laser additive manufacturing processes for near net shape components, in: K. Gupta (Ed.), Near Net Shape Manufacturing Processes. Materials Forming, Machining and Tribology, Springer, Cham, 2019.

[13] D. Clark, M.T. Whittaker, M.R. Bache, Microstructural characterization of a prototype titanium alloy structure processed via direct laser deposition (DLD), Metall. Mater. Trans. B Process Metall. Mater. Process. Sci. 43 (2012) 388–396, https://doi.org/10.1007/s11663-011-9599-x.

[14] M. Gharbi, P. Peyre, C. Gorny, M. Carin, S. Morville, P. Le Masson, et al., Influence of various process conditions on surface finishes induced by the direct metal deposition laser technique on a Ti−6Al−4V alloy, J. Mater. Process. Technol. 213 (2013) 791−800, https://doi.org/10.1016/j.jmatprotec.2012.11.015.

[15] M. Gharbi, P. Peyre, C. Gorny, M. Carin, S. Morville, P. Le Masson, et al., Influence of a pulsed laser regime on surface finish induced by the direct metal deposition process on a Ti64 alloy, J. Mater. Process. Technol. 214 (2014) 485−495, https://doi.org/10.1016/j.jmatprotec.2013.10.004.

[16] M. Thomas, T. Malot, P. Aubry, Laser metal deposition of the intermetallic TiAl alloy, Metall. Mater. Trans. 48 (2017) 3143−3158, https://doi.org/10.1007/s11661-017-4042-9.

[17] J.I. Arrizubieta, S. Martínez, A. Lamikiz, E. Ukar, K. Arntz, F. Klocke, Instantaneous powder flux regulation system for Laser Metal Deposition, J. Manuf. Process. 29 (2017) 242−251, https://doi.org/10.1016/j.jmapro.2017.07.018.

[18] R. Vilar, Laser powder deposition, Compr. Mater. Process 10 (2014) 163−216, https://doi.org/10.1016/B978-0-08-096532-1.01005-0.

[19] L. Wang, S.D. Felicelli, J.E. Craig, Experimental and numerical study of the LENS rapid fabrication process, J. Manuf. Sci. Eng. 131 (2009) 41019, https://doi.org/10.1115/1.3173952.

[20] H. Attar, S. Ehtemam-Haghighi, D. Kent, X. Wu, M.S. Dargusch, Comparative study of commercially pure titanium produced by laser engineered net shaping, selective laser melting and casting processes, Mater. Sci. Eng. 705 (2017) 385−393, https://doi.org/10.1016/j.msea.2017.08.103.

[21] J. Chen, L. Xue, S.-H. Wang, Experimental studies on process-induced morphological characteristics of macro- and microstructures in laser consolidated alloys, J. Mater. Sci. 46 (2011) 5859−5875, https://doi.org/10.1007/s10853-011-5543-3.

[22] R. Comesaña, F. Lusquiños, J. del Val, T. Malot, M. López-Álvarez, A. Riveiro, et al., Calcium phosphate grafts produced by rapid prototyping based on laser cladding, J. Eur. Ceram. Soc. 31 (2011) 29−41, https://doi.org/10.1016/j.jeurceramsoc.2010.08.011.

[23] K.P. Karunakaran, A. Bernard, S. Suryakumar, L. Dembinski, G. Taillandier, Rapid manufacturing of metallic objects, Rapid Prototyp. J. 18 (2012) 264−280, https://doi.org/10.1108/13552541211231644.

[24] Y. Hu, W. Cong, A review on laser deposition-additive manufacturing of ceramics and ceramic reinforced metal matrix composites, Ceram. Int. 44 (2018) 20599−20612, https://doi.org/10.1016/j.ceramint.2018.08.083.

[25] W.U.H. Syed, A.J. Pinkerton, L. Li, Combining wire and coaxial powder feeding in laser direct metal deposition for rapid prototyping, Appl. Surf. Sci. 252 (2006) 4803−4808, https://doi.org/10.1016/j.apsusc.2005.08.118.

[26] F. Wang, J. Mei, X. Wu, Compositionally graded Ti6Al4V + TiC made by direct laser fabrication using powder and wire, Mater. Des. 28 (2007) 2040−2046, https://doi.org/10.1016/j.matdes.2006.06.010.

[27] X. Wu, J. Mei, Near net shape manufacturing of components using direct laser fabrication technology, J. Mater. Process. Technol. 135 (2003) 266−270, doi.org/10.1016/S0924-0136(02)00906-8.

[28] A. Saboori, D. Gallo, S. Biamino, P. Fino, M. Lombardi, An overview of additive manufacturing of titanium components by directed energy deposition: microstructure and mechanical properties, Appl. Sci. 7 (2017) 883, https://doi.org/10.3390/app7090883.

[29] A. Azarniya, X. Garmendia, M.J. Mirzaali, Additive manufacturing of Ti-6Al-4V parts through laser metal deposition (LMD): process, microstructure, and mechanical properties, J. Alloys Compd. 804 (2019) 163−191, https://doi.org/10.1016/j.jallcom.2019.04.255.

[30] P.A. Kobryn, S.L. Semiatin, Mechanical properties of laser-deposited Ti-6Al-4V, in: 2001 International Solid Freeform Fabrication Symposium, 2001, https://doi.org/10.26153/tsw/3261.

[31] X. Wu, J. Liang, J. Mei, C. Mitchell, P.S. Goodwin, W. Voice, Microstructures of laser-deposited Ti-6Al-4V, Mater. Des. 25 (2004) 137–144, https://doi.org/10.1016/j.matdes.2003.09.009.

[32] A. Ramakrishnan, G.P. Dinda, Materials science & engineering A direct laser metal deposition of Inconel 738, Mater. Sci. Eng. 740–741 (2019) 1–13, https://doi.org/10.1016/j.msea.2018.10.020.

[33] J. Chen, L. Xue, Process-induced microstructural characteristics of laser consolidated IN-738 superalloy, Mater. Sci. Eng. 527 (2010) 7318–7328, https://doi.org/10.1016/j.msea.2010.08.003.

[34] P. Guo, B. Zou, C. Huang, H. Gao, Journal of Materials Processing Technology Study on microstructure , mechanical properties and machinability of efficiently additive manufactured AISI 316L stainless steel by high-power direct laser deposition, J. Mater. Process. Technol. 240 (2017) 12–22, https://doi.org/10.1016/j.jmatprotec.2016.09.005.

[35] F. Niu, D. Wu, F. Lu, G. Liu, G. Ma, Z. Jia, Microstructure and macro properties of $Al_2O_3$ ceramics prepared by laser engineered net shaping, Ceram. Int. 44 (2018) 14303–14310, https://doi.org/10.1016/j.ceramint.2018.05.036.

[36] V.K. Balla, S. Bose, A. Bandyopadhyay, Processing of bulk alumina ceramics using laser engineered net shaping, Int. J. Appl. Ceram. Technol. 5 (2008) 234–242, https://doi.org/10.1111/j.1744-7402.2008.02202.x.

[37] F. Niu, D. Wu, G. Ma, J. Wang, B. Zhang, Nanosized microstructure of $Al_2O_3$–$ZrO_2$ ($Y_2O_3$) eutectics fabricated by laser engineered net shaping, Scripta Mater. 95 (2015) 39–41, https://doi.org/10.1016/j.scriptamat.2014.09.026.

[38] F. Li, X. Zhang, C. Sui, J. Wu, H. Wei, Y. Zhang, Microstructure and mechanical properties of $Al_2O_3$-$ZrO_2$ ceramic deposited by laser direct material deposition, Ceram. Int. 44 (2018) 18960–18968, https://doi.org/10.1016/j.ceramint.2018.07.135.

[39] Y. Hu, F. Ning, W. Cong, X. Wang, H. Wang, Ultrasonic vibration-assisted laser engineering net shaping of $ZrO_2$-$Al_2O_3$ bulk parts: effects on crack suppression, microstructure, and mechanical properties, Ceram. Int. 44 (2018) 2752–2760, https://doi.org/10.1016/j.ceramint.2017.11.013.

[40] Z. Fan, Y. Zhao, Q. Tan, N. Mo, M. Zhang, M. Lu, H. Huang, Nanostructured $Al_2O_3$-YAG-$ZrO_2$ ternary eutectic components prepared by laser engineered net shaping, Acta Mater. 170 (2019) 24–37. org/10.1016/j.actamat.2019.03.020.

[41] S.A. Bernard, V.K. Balla, S. Bose, A. Bandyopadhyay, Direct laser processing of bulk lead zirconate titanate ceramics, Mater. Sci. Eng. B 172 (2010) 85–88, https://doi.org/10.1016/j.mseb.2010.04.022.

[42] R.M. Mahamood, E.T. Akinlabi, M. Shukla, S. Pityana, Scanning velocity influence on microstructure , microhardness and wear resistance performance of laser deposited Ti6Al4V/TiC composite, Mater. Des. 50 (2013) 656–666, https://doi.org/10.1016/j.matdes.2013.03.049.

[43] T. Borkar, S. Gopagoni, In situ nitridation of titanium–molybdenum alloys during laser deposition, J. Mater. Sci. 47 (2012) 7157–7166, https://doi.org/10.1007/s10853-012-6656-z.

[44] Y. Hu, F. Ning, X. Wang, H. Wang, B. Zhao, W. Cong, Laser Deposition-Additive Manufacturing of in Situ TiB Reinforced Titanium Matrix Composites: TiB Growth and Part Performance, 2017, https://doi.org/10.1007/s00170-017-0769-0.

[45] C. Hong, D. Gu, D. Dai, M. Alkhayat, W. Urban, P. Yuan, et al., Laser additive manufacturing of ultrafine TiC particle reinforced Inconel 625 based composite parts: tailored microstructures and enhanced performance, Mater. Sci. Eng. 635 (2015) 118–128, doi.org/10.1016/j.msea.2015.03.043.

[46] J. Yao, Y. Ding, R. Liu, Q. Zhang, L. Wang, Wear and corrosion performance of laser-clad low-carbon high- molybdenum Stellite alloys, Optic Laser. Technol. 107 (2018) 32–45, https://doi.org/10.1016/j.optlastec.2018.05.021.
[47] Y. Ding, R. Liu, J. Yao, Q. Zhang, L. Wang, Stellite alloy mixture hardfacing via laser cladding for control valve seat sealing surfaces, Surf. Coating. Technol. 329 (2017) 97–108, https://doi.org/10.1016/j.surfcoat.2017.09.018.
[48] V. Ocelík, U. de Oliveira, M. de Boer, J.T.M. de Hosson, Thick Co-based coating on cast iron by side laser cladding: analysis of processing conditions and coating properties, Surf. Coating. Technol. 201 (2007) 5875–5883, https://doi.org/10.1016/j.surfcoat.2006.10.044.
[49] D. Verdi, M.A. Garrido, C.J. Múnez, P. Poza, Mechanical properties of Inconel 625 laser cladded coatings: depth sensing indentation analysis, Mater. Sci. Eng. 598 (2014) 15–21, https://doi.org/10.1016/j.msea.2014.01.026.
[50] X. Xu, G. Mi, L. Chen, L. Xiong, P. Jiang, X. Shao, et al., Research on microstructures and properties of Inconel 625 coatings obtained by laser cladding with wire, J. Alloys Compd. 715 (2017) 362–373, https://doi.org/10.1016/j.jallcom.2017.04.252.
[51] F. Arias-González, J. del Val, R. Comesaña, J. Penide, F. Lusquiños, F. Quintero, et al., Laser cladding of nickel-based alloy on cast iron, Appl. Surf. Sci. 374 (2016) 197–205, https://doi.org/10.1016/j.apsusc.2015.11.023.
[52] F. Arias-González, J. del Val, R. Comesaña, J. Penide, F. Lusquiños, F. Quintero, et al., Laser cladding of phosphor bronze, Surf. Coating. Technol. 313 (2017) 248–254, https://doi.org/10.1016/j.surfcoat.2017.01.097.
[53] A. Riveiro, A. Mejías, F. Lusquiños, J. del Val, R. Comesaña, J. Pardo, et al., Laser cladding of aluminium on AISI 304 stainless steel with high-power diode lasers, Surf. Coating. Technol. 253 (2014) 214–220, https://doi.org/10.1016/j.surfcoat.2014.05.039.
[54] F. Lusquiños, J. Pou, F. Quintero, M. Pérez-Amor, Laser cladding of SiC/Si composite coating on Si–SiC ceramic substrates, Surf. Coating. Technol. 202 (2008) 1588–1593, https://doi.org/10.1016/j.surfcoat.2007.07.011.
[55] J.M. Wilson, C. Piya, Y.C. Shin, F. Zhao, K. Ramani, Remanufacturing of turbine blades by laser direct deposition with its energy and environmental impact analysis, J. Clean. Prod. 80 (2014) 170–178, https://doi.org/10.1016/j.jclepro.2014.05.084.
[56] R. Raju, M. Duraiselvam, V. Petley, S. Verma, R. Rajendran, Microstructural and mechanical characterization of Ti6Al4V refurbished parts obtained by laser metal deposition, Mater. Sci. Eng. 643 (2015) 64–71, https://doi.org/10.1016/j.msea.2015.07.029.
[57] P. Muller, P. Mognol, J.Y. Hascoet, Modeling and control of a direct laser powder deposition process for Functionally Graded Materials (FGM) parts manufacturing, J. Mater. Process. Technol. 213 (2013) 685–692, https://doi.org/10.1016/j.jmatprotec.2012.11.020.
[58] L. Yan, Y. Chen, F. Liou, Additive manufacturing of functionally graded metallic materials using laser metal deposition, Addit. Manuf. 31 (2020) 100901, https://doi.org/10.1016/j.addma.2019.100901.
[59] B. Chen, Y. Su, Z. Xie, C. Tan, J. Feng, Development and characterization of 316L/Inconel625 functionally graded material fabricated by laser direct metal deposition, Optic Laser. Technol. 123 (2020) 105916, https://doi.org/10.1016/j.optlastec.2019.105916.
[60] K. Shah, A. Khan, S. Ali, M. Khan, A.J. Pinkerton, Parametric study of development of Inconel-steel functionally graded materials by laser direct metal deposition, Mater. Des. 54 (2014) 531–538, https://doi.org/10.1016/j.matdes.2013.08.079.

[61] J. del Val, F. Arias-González, O. Barro, A. Riveiro, R. Comesaña, J. Penide, F. Lusquiños, M. Bountinguiza, F. Quintero, J. Pou, Functionally graded 3D structures produced by laser cladding, Proc. Manuf. 13 (2017) 169–176, https://doi.org/10.1016/j.promfg.2017.09.029.

[62] Y. Liu, C. Liu, W. Liu, Y. Ma, C. Zhang, Q. Cai, B. Liu, Microstructure and properties of Ti/Al lightweight graded material by direct laser deposition, Mater. Sci. Technol. 34 (2017) 945–951, https://doi.org/10.1080/02670836.2017.1412042.

[63] H. Rao, R.P. Oleksak, K. Favara, A. Harooni, B. Dutta, D. Maurice, Behavior of yttria-stabilized zirconia (YSZ) during laser direct energy deposition of an Inconel 625-YSZ cermet, Addit. Manuf. 31 (2020) 100932, https://doi.org/10.1016/j.addma.2019.100932.

[64] Y. Zhang, A. Bandyopadhyay, Direct fabrication of compositionally graded Ti-$Al_2O_3$ multi-material structures using Laser Engineered Net Shaping, Addit. Manuf. 21 (2018) 104–111, https://doi.org/10.1016/j.addma.2018.03.001.

[65] Y. Li, C. Yang, H. Zhao, S. Qu, X. Li, Y. Li, New developments of Ti-based alloys for biomedical applications, Materials 7 (2014) 1709–1800, https://doi.org/10.3390/ma7031709.

[66] G.P. Dinda, L. Song, J. Mazumder, Fabrication of Ti-6Al-4V scaffolds by direct metal deposition, Metall. Mater. Trans. A Phys. Metall. Mater. Sci. 39 (2008) 2914–2922, https://doi.org/10.1007/s11661-008-9634-y.

[67] A. Sidambe, Biocompatibility of advanced manufactured titanium implants—a review, Materials 7 (2014) 8168–8188, https://doi.org/10.3390/ma7128168.

[68] F. Arias-González, J. del Val, R. Comesaña, J. Penide, F. Lusquiños, F. Quintero, et al., Microstructure and crystallographic texture of pure titanium parts generated by laser additive manufacturing, Met. Mater. Int. 24 (2018), https://doi.org/10.1007/s12540-017-7094-x.

[69] W. Xue, B.V. Krishna, A. Bandyopadhyay, S. Bose, Processing and biocompatibility evaluation of laser processed porous titanium, Acta Biomater. 3 (2007) 1007–1018, https://doi.org/10.1016/j.actbio.2007.05.009.

[70] B.V. Krishna, S. Bose, A. Bandyopadhyay, Low stiffness porous Ti structures for load-bearing implants, Acta Biomater. 3 (2007) 997–1006, https://doi.org/10.1016/j.actbio.2007.03.008.

[71] L.C. Zhang, L.Y. Chen, A review on biomedical titanium alloys: recent progress and prospect, Adv. Eng. Mater. 21 (2019) 1–29, https://doi.org/10.1002/adem.201801215.

[72] S. Nagn, R. Banerjee, Laser deposition and deformation behavior of Ti-Nb-Zr-Ta alloys for orthopedic implan, J. Mech. Behav. Biomed. Mater 16 (2012) 21–28, https://doi.org/10.1016/j.jmbbm.2012.08.014.

[73] S. Samuel, S. Nag, S. Nasrazadani, V. Ukirde, M. El Bouanani, A. Mohandas, et al., Corrosion resistance and in vitro response of laser-deposited Ti-Nb-Zr-Ta alloys for orthopedic implant applications, J. Biomed. Mater. Res. 94 (2010) 1251–1256, https://doi.org/10.1002/jbm.a.32782.

[74] M. Fischer, P. Laheurte, P. Acquier, D. Joguet, L. Peltier, T. Petithory, et al., Synthesis and characterization of Ti-27.5Nb alloy made by CLAD® additive manufacturing process for biomedical applications, Mater. Sci. Eng. C 75 (2017) 341–348, https://doi.org/10.1016/j.msec.2017.02.060.

[75] T. Bhardwaj, M. Shukla, C.P. Paul, K.S. Bindra, Direct energy deposition - laser additive manufacturing of titanium-molybdenum alloy: parametric studies, micro-structure and mechanical properties, J. Alloys Compd. 787 (2019) 1238–1248, https://doi.org/10.1016/j.jallcom.2019.02.121.

[76] G.D.J. Ram, C.K. Esplin, B.E. Stucker, Microstructure and wear properties of LENS deposited medical grade CoCrMo, J. Mater. Sci. Mater. Med. 19 (2008) 2105–2111, https://doi.org/10.1007/s10856-007-3078-6.

[77] O. Barro, F. Arias-González, F. Lusquiños, R. Comesaña, J. del Val, A. Riveiro, F. Gómez-Baño, J. Pou, Effect of manufacturing techniques (Casting, milling, selective laser melting and laser directed energy deposition) on microstructural, mechanical and electrochemical properties of Co-Cr dental alloys, J. Mater. Process. Technol. (2020) submitted.

[78] S. Bose, D. Ke, H. Sahasrabudhe, A. Bandyopadhyay, Additive manufacturing of biomaterials, Prog. Mater. Sci. 93 (2018) 45–111, https://doi.org/10.1016/j.pmatsci.2017.08.003.

[79] F. Lusquiños, J. del Val, F. Arias-González, R. Comesaña, F. Quintero, A. Riveiro, et al., Bioceramic 3D implants produced by laser assisted additive manufacturing, Phys. Proc. 56 (2014) 309–316, https://doi.org/10.1016/j.phpro.2014.08.176.

[80] R. Comesaña, F. Lusquiños, J. Del Val, M. López-Álvarez, F. Quintero, A. Riveiro, et al., Three-dimensional bioactive glass implants fabricated by rapid prototyping based on $CO_2$ laser cladding, Acta Biomater. 7 (2011) 3476–3487, https://doi.org/10.1016/j.actbio.2011.05.023.

[81] J. del Val, R. López-Cancelos, A. Riveiro, A. Badaoui, F. Lusquiños, F. Quintero, et al., On the fabrication of bioactive glass implants for bone regeneration by laser assisted rapid prototyping based on laser cladding, Ceram. Int. 42 (2016) 2021–2035, https://doi.org/10.1016/j.ceramint.2015.10.009.

[82] R. Comesaña, F. Lusquiños, J. Del Val, F. Quintero, A. Riveiro, M. Boutinguiza, et al., Toward smart implant synthesis: bonding bioceramics of different resorbability to match bone growth rates, Sci. Rep. 5 (2015) 10677, https://doi.org/10.1038/srep10677.

[83] F. Weng, C. Chen, H. Yu, Research status of laser cladding on titanium and its alloys: a review, Mater. Des. 58 (2014) 412–425, https://doi.org/10.1016/j.matdes.2014.01.077.

[84] V.K. Balla, S. Banerjee, S. Bose, A. Bandyopadhyay, Acta Biomaterialia Direct laser processing of a tantalum coating on titanium for bone replacement structures, Acta Biomater. 6 (2010) 2329–2334, https://doi.org/10.1016/j.actbio.2009.11.021.

[85] F. Lusquiños, J. Pou, J.L. Arias, M. Boutinguiza, B. Léon, M. Pérez-Amor, et al., Production of calcium phosphate coatings on Ti6Al4V obtained by Nd:Yttrium-aluminum-garnet laser cladding, J. Appl. Phys. 90 (2001) 4231–4236, https://doi.org/10.1063/1.1402975.

[86] F. Lusquiños, J. Pou, M. Boutinguiza, F. Quintero, R. Soto, B. León, et al., Main characteristics of calcium phosphate coatings obtained by laser cladding, Appl. Surf. Sci. 247 (2005) 486–492, https://doi.org/10.1016/j.apsusc.2005.01.134.

[87] M. Roy, B. Vamsi Krishna, A. Bandyopadhyay, S. Bose, Laser processing of bioactive tricalcium phosphate coating on titanium for load-bearing implants, Acta Biomater. 4 (2008) 324–333, https://doi.org/10.1016/j.actbio.2007.09.008.

[88] R. Comesaña, F. Quintero, F. Lusquiños, M.J. Pascual, M. Boutinguiza, A. Durán, et al., Laser cladding of bioactive glass coatings, Acta Biomater. 6 (2010) 953–961, https://doi.org/10.1016/j.actbio.2009.08.010.

[89] B. Vamsi Krishna, W. Xue, S. Bose, A. Bandyopadhyay, Functionally graded Co-Cr-Mo coating on Ti-6Al-4V alloy structures, Acta Biomater. 4 (2008) 697–706, https://doi.org/10.1016/j.actbio.2007.10.005.

[90] J.M. Wilson, N. Jones, L. Jin, Y.C. Shin, Laser deposited coatings of Co-Cr-Mo onto Ti-6Al-4V and SS316L substrates for biomedical applications, J. Biomed. Mater. Res. B Appl. Biomater. 101 (2013) 1–9, https://doi.org/10.1002/jbm.b.32921.

[91] A. Almeida, D. Gupta, C. Loable, R. Vilar, Laser-assisted synthesis of Ti-Mo alloys for biomedical applications, Mater. Sci. Eng. C 32 (2012) 1190–1195, https://doi.org/10.1016/j.msec.2012.03.007.

[92] C. Schneider-Maunoury, L. Weiss, P. Acquier, D. Boisselier, P. Laheurte, Functionally graded Ti6Al4V-Mo alloy manufactured with DED-CLAD® process, Addit. Manuf. 17 (2017) 55–66, https://doi.org/10.1016/j.addma.2017.07.008.

[93] D.D. Lima, S.A. Mantri, C.V. Mikler, R. Contieri, C.J. Yannetta, K.N. Campo, et al., Laser additive processing of a functionally graded internal fracture fixation plate, Mater. Des. 130 (2017) 8–15, https://doi.org/10.1016/j.matdes.2017.05.034.

[94] C. Schneider-Maunoury, L. Weiss, O. Perroud, D. Joguet, An application of differential injection to fabricate functionally graded Ti-Nb alloys using DED-CLAD® process, J. Mater. Process. Technol. 268 (2019) 171–180, https://doi.org/10.1016/j.jmatprotec.2019.01.018.

## CHAPTER 5

# Vat photopolymerization methods in additive manufacturing

**Ali Davoudinejad**
Technical University of Denmark (DTU), Mechanical Department, Manufacturing Section, Copenhagen, Denmark; Mechanical Engineering, Manufacturing and Production, Politecnico di Milano, Milan, Italy

## 5.1 Introduction

Additive manufacturing (AM) processes, unlike other manufacturing processes (subtractive, casting, forming, etc.) are fundamentally different for fabrication of structures. AM allows the entirely customized manufacturing of parts with a high complexity level without traditional manufacturing limitations with a decreased production time and cost. AM is also known as 3D printing due to the fabrication method of objects through the layer-wise fashion from scratch.

An additional potential of AM processes is the capability to fabricate parts with different type of materials that are challenging to process in conventional machining methods such as hard metals, ceramics, and composites. During the past few decades, remarkable progress has been achieved in developing a range of materials for AM processes, which are still developing [1,2].

Some of the main potential advantages [1–3] of AM can be presented as follows:

- Rapid prototyping.
- Ability of direct manufacturing of final parts or near final with minimal to no additional processing.
- Economic in medium to low volume production.
- Design freedom for manufacturing (e.g., topology optimization [TO], lightweighting).
- Ability to respond to specific customer needs (e.g., customization).
- A considerable decrease in overall product development and fabrication time leading to faster transfer to market.
- Tool for other processes such as injection molding (IM).

*Additive Manufacturing*
ISBN 978-0-12-818411-0
https://doi.org/10.1016/B978-0-12-818411-0.00007-0

- Fabrication of high aspect ratio features.
- Merge production and consumption.

Nowadays, one of the most important application areas of AM is rapid prototyping, although, other areas, such as product development, direct and indirect manufacturing are growing. This suggests a large use of AM in a wider range of applications [2]. Fig. 5.1 shows a classification of manufacturing processes in terms of the complexity and scale features. Moreover, 4D printing for multiscale engineering of high-complexity structures is also compared. 4D printing (4DP) technology combines tool-driven processing with the "precision character" of the material themselves that self-assembles by a code of geometrical instructions via adventitious energy sources. 3D printing is revolutionary in increasing the efficiency of manufacturing processes in relation to fabrication cost, mass production product, and its effect on addressing some of the most challenging issues in healthcare [4].

The selection of proper manufacturing processes is subjected to different factors, such as part dimensional accuracy and precision, quality of surface finish, design and fabrication limitation, and cost and speed of production. For instance, in subtractive manufacturing, high precision milling requires an accurate management of all the involved resources, like machine tool, tool, fixture, and workpiece [5]. Fig. 5.2 shows different dimensional accuracy for various manufacturing processes used with polymeric and

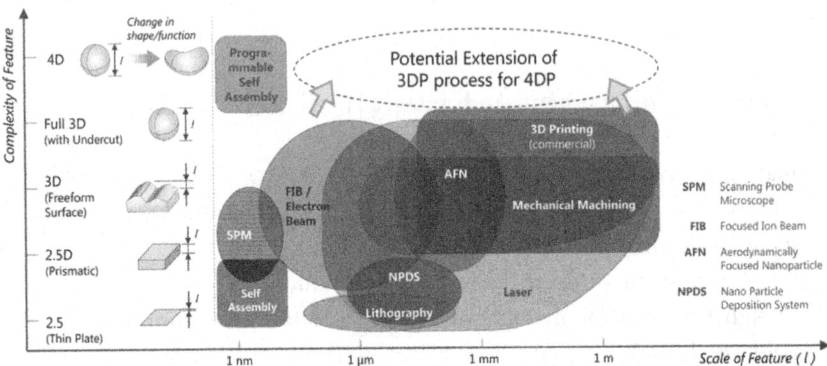


**Figure 5.1** Complexity versus scale of features showed for different manufacturing processes. *(Reproduced with kind permission of Springer Nature from Reference V. Khare, S. Sonkaria, G. Y. Lee, S. H. Ahn, and W. S. Chu, "From 3D to 4D printing – design, material and fabrication for multi-functional multi-materials," Int. J. Precis. Eng. Manuf. - Green Technol., 2017).*

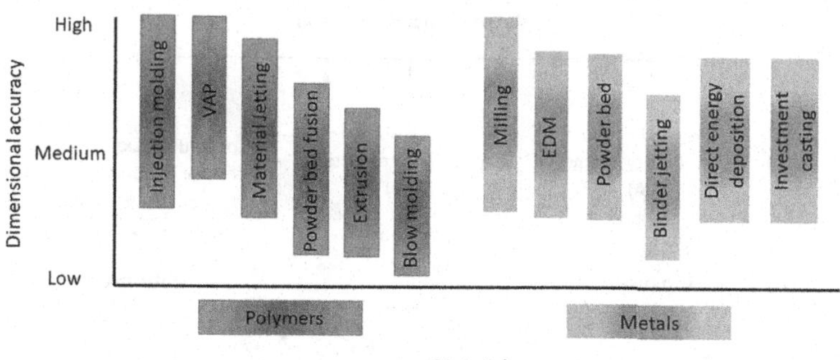


**Figure 5.2** Comparison of dimensional accuracy for different manufacturing processes using polymeric and metallic materials. (The dimensional accuracy is assumed to represent the best case of build parameters and working conditions of each process).

metallic materials [1]. The vat photopolymerization (VP) process shows a high dimensional accuracy; this is comparable with that of IM for fabrication of polymeric parts, or high precision milling for metallic components [6].

This chapter summarizes the VP AM process setup and primary configurations for different technologies (stereolithography [SLA], digital light processing [DLP], and continuous liquid interface production [CLIP]) to better understanding the process and expand the possible application of the AM process. Additionally, direct manufacturing of polymeric components, as well as fabrication of tools for other manufacturing processes is presented.

## 5.2 Vat photopolymerization process

In the Standard Terminology for Additive Manufacturing Technologies [7], VP is defined as an AM process in which a liquid photopolymer in a vat is selectively cured by light-activated polymerization. VP-based AM approach is categorized into three different technologies, namely, SLA, DLP, and CLIP.

There are two possible structures for a VP machine depending on the manufacturer design: bottom-up and top-down orientations, as shown in Fig. 5.3. All VP technologies use almost similar mechanisms for curing and fabrication of structures. Bottom-up printers have the main light engine positioned below the resin vat and the part print upside-down. The build plate moves up from the resin vat to print the part layer by layer. In contrast, in top-down printers, the light engine is positioned above the built platform. The printing starts from the top of the vat in the resin and moves

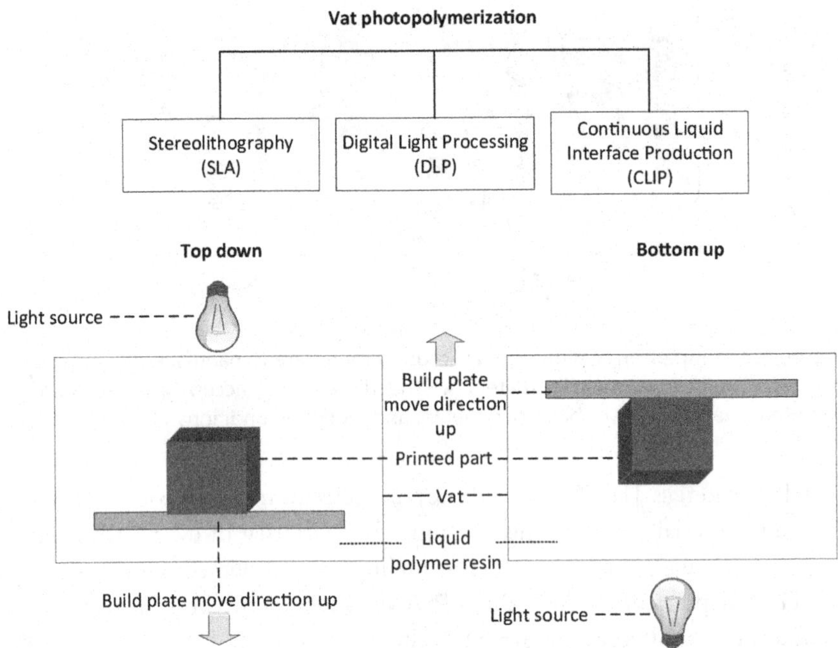


**Figure 5.3** Bottom-up and top-down orientations in vat photopolymerization printers.

down as the layers are cured to fabricate the features. The light source can be different, such as, UV lamps, laser radiation, gamma rays, LED screen, X-rays, or electron beams; however, UV light is commonly applied for curing the photopolymer resin. The bottom-up VP printers is the more common method to print microparts with finely detailed geometries. More examples will be presented in the following sections.

### 5.2.1 Photopolymer materials

A photopolymer is any type of material that experiences a direct or indirect interaction with light to change its physical or chemical properties [8]. Photopolymer materials were developed in the late 1960s and rapidly applied in several commercial areas, mainly in the coating, electrical, and printing industries. However, using UV-curable materials date back to the mid-1980s when Chuck Hull exposed resin to a scanning laser, similarly to in those systems recently used in the laser printers [9]. In contrast to the thermoplastic polymers, photopolymers have a cross-linked molecular structure that do not melt and exhibit much less creep and stress relaxation. The liquid light-activated resin alters its properties when exposed to light,

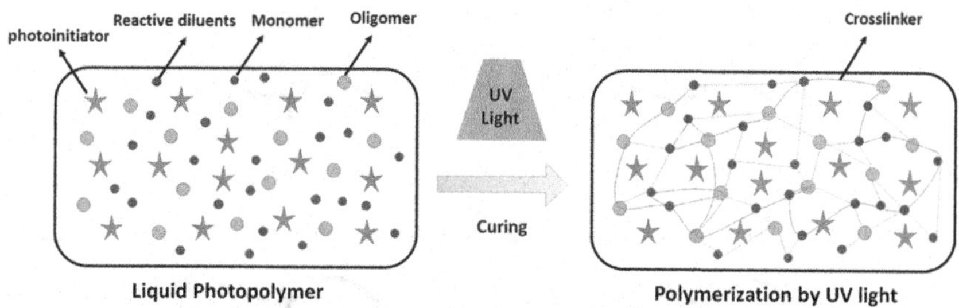


**Figure 5.4** Basic schematic of the photopolymerization process.

and it consists of different type of elements, such as photoinitiators, reactive diluents, flexibilizers, stabilizers, and liquid monomers [10]. Fig. 5.4 shows a general schematic of the modification of properties in the liquid photopolymer when exposed to the UV light, and the subsequent polymerization after irradiation.

Crivello and Reichmanis [8] categorized photopolymers into five different basic types, as shown in Fig. 5.5. Each type of the photopolymer consists of components that cause the chemical reaction. In type 1, photopolymers undergo a photoinitiated radical, cationic, or anionic chain polymerization on irradiation to yield cross-linked network polymers. Type 2 photopolymers undergo step growth reactions. Type 3 functional polymers undergo photo-induced cross-linking between X and Y. Type 4 photopolymers undergo a deep-seated photochemical reaction modification under irradiation. Type 5 photopolymers undergo photoinduced cleavage reaction.

### 5.2.2 Stereolithography

SLA is the one of the oldest VP technologies in printing as it was established by Hull in 1984 [9]. It uses a UV laser beam to harden the surface of a vat of photopolymer resin. In the commercial SLA machines, a laser is used for curing the resin point by point, and the part slices are generated from a CAD model. The optics system includes a laser, focusing and adjustment optics, and two galvanometric mirrors that move the laser beam over the vat. The build platform raises after each slice is printed, and the surface of the vat is repositioned for the next layer printing; then, the laser starts to trace the next slice of the CAD model [11]. Fig. 5.6 shows the schematic of a typical SLA machine and a commercial bottom-up SLA machine.

**Type 1. Photopolymers that undergo photoinitiated chain growth.**

**Type 2. Photopolymers that undergo step-growth polymerization**

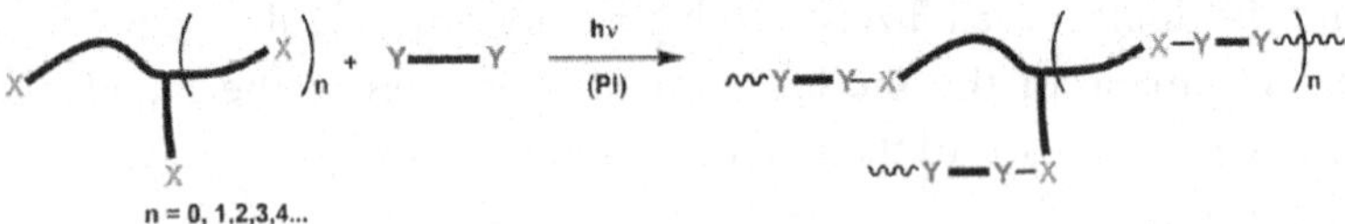


**Type 3. Functional polymers that undergo photoinduced crosslinking**

**Type 4. Polymers that undergo functional group modification under irradiation.**

**Type 5. Polymers that undergo photoinduced cleavage reactions.**

**Figure 5.5** Five different types of photopolymers. *(Reproduced with kind permission of ACS Publications from Reference J. V Crivello and E. Reichmanis, "Photopolymer materials and processes for advanced technologies," Chem. Mater., vol. 26, no. 1, pp. 533–548, January, 2014).*

## 5.2.3 Digital light processing

In DLP, a liquid photopolymer resin is contained in a shallow vat. The DLP method uses a video projector that contains a micro-opto-electro-mechanical mirror array, a digital micro-mirror device, to modulate a collimated UV light source, which is subsequently focused to an imaging plane placed on the bottom surface of a transparent vat. Fig. 5.7 shows the schematic of the DLP process. The desired geometry is built up layer by layer by modulating image masks corresponding to a sliced representation of the fabricated geometry as the vertical stage of the machine, and thus the workpiece is moved upward layer by layer. The image for each layer is projected over the whole surface with UV light based on the layer shape. This method of VP is a development of SLA, which allows for more control on the process [13].

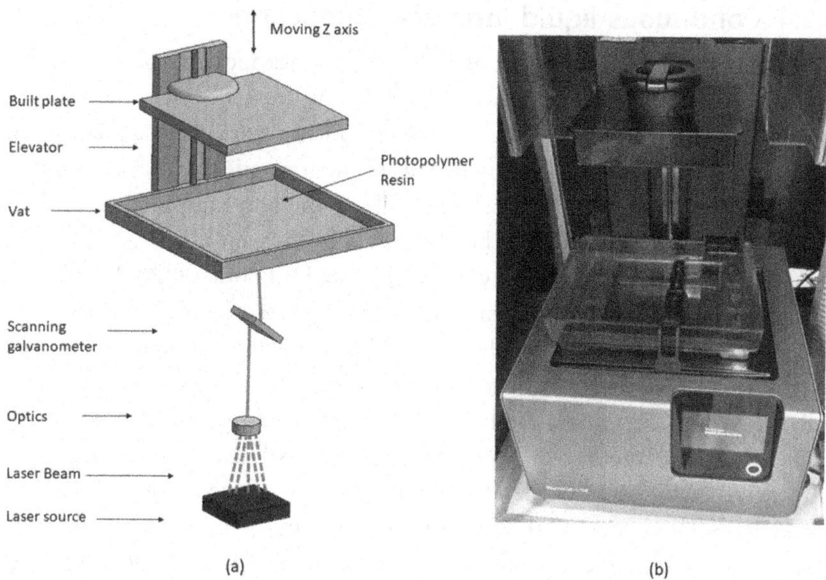


**Figure 5.6** (A) Schematic of bottom-up stereolithography (SLA) technology 3D printing process (B) a commercial SLA machine. *(Reproduced with kind permission of Elsevier from Reference A. Davoudinejad et al., "Geometrical and feature of size design effect on direct stereolithography micro additively manufactured components," Procedia Struct. Integr., vol. 13, pp. 1250–1255, January 2018).*

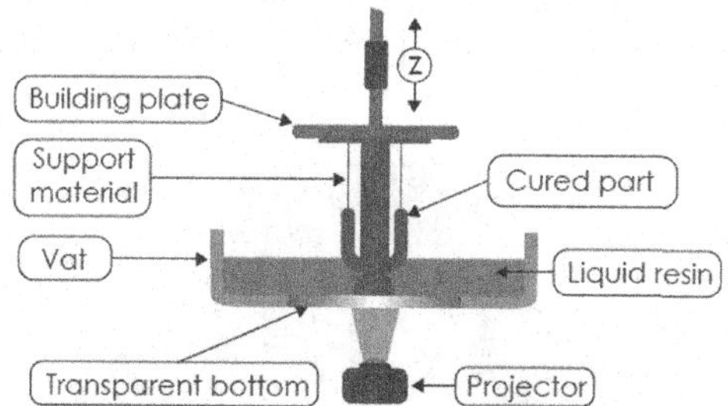


**Figure 5.7** Schematic representation of the digital light processing vat polymerization process, in a bottom-up configuration. *(Reproduced with kind permission of IOP Publishing from Reference A. Davoudinejad, M. M. Ribo, D. B. Pedersen, A. Islam, and G. Tosello, "Direct fabrication of bio-inspired gecko-like geometries with vat polymerization additive manufacturing method," J. Micromechanics Microengineering, vol. 28, no. 8, 2018, PP 1–10).*

## 5.2.4 Continuous liquid interface production

CLIP relies on the inhibition of free radical photopolymerization in the presence of atmospheric oxygen. The dead zone created above the window maintains a liquid interface below the advancing part. Consequently, an oxygen-permeable build window results in the formation of a dead zone, or a region of uncured liquid resin, which allows the continuous fabrication of features. Fig. 5.8 shows the schematic of a button-up CLIP 3D-printing process. The CLIP-process is very similar to DLP but differs by the architecture of the transparent window in the bottom of the resin container. By using an oxygen-permeable and UV-transparent window below the resin vat, oxygen can travel through the window and mix into the liquid polymer resin. Therefore, the build plate does not move up and down for each layer, and there is a continuous growth of the part. The part is considered layer less, and the traditional trade-off between speed and layer thickness is eliminated with this continuous growth of the features. The resolution in the build direction is not determined by the layer thickness but is limited to the slicing conditions of the part and the optical absorption-height of the resin [14].

# 5.3 Postprocessing

Manufacturing by VP is a process chain that involves three main stages. The additive manufacture of the part, the postprint cleaning of the part, and the postprint curing of the part to ensure that no residual uncured resin are left on the manufactured components [13].The next step after printing is the

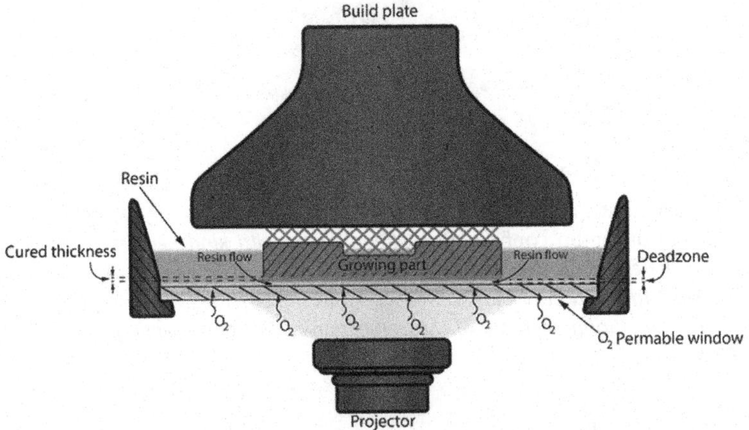


**Figure 5.8** Schematic of the continuous liquid interface production technology with bottom-up projection method.

postprocessing, which has a high impact on the final part shape. The postprocessing of the manufactured features consists of two main steps: cleaning and postcuring. Cleaning should be applied to the part that is covered with uncured polymer resin. Isopropanol is commonly used for cleaning to remove the excess resin from the sample. In this step, cleaning the surface is performed to be sure that any additional leftover resins are removed from the part. The next step is postcuring. Depending on the resin, it should be cured by UV light, or with in a high-temperature oven. This ensures that no reactive resin residue is left on the samples, and optimal mechanical properties are achieved [13–15]. Fig. 5.9 shows the post-processing order applied to prepare the final printed part.

### 5.3.1 Postprocessing challenges

The postprocessing depending on the complexity of the printed features has some challenges in both cleaning and postcuring steps, such as taking out the part from the build plate and removing the support structures, which may cause collision with the features. In addition, in the cleaning step for removing the excess resin in between the features, mainly, the high-aspect-ratio features are prone to be deformed. Thus, printing small-scale features shows more difficulties for cleaning and removing the leftover resin in between the features or thin wall structures. Fig. 5.10B shows the

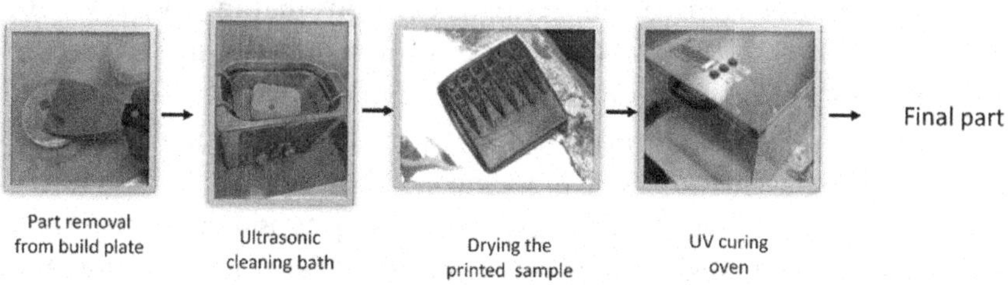


**Figure 5.9** Postprocessing process chain.

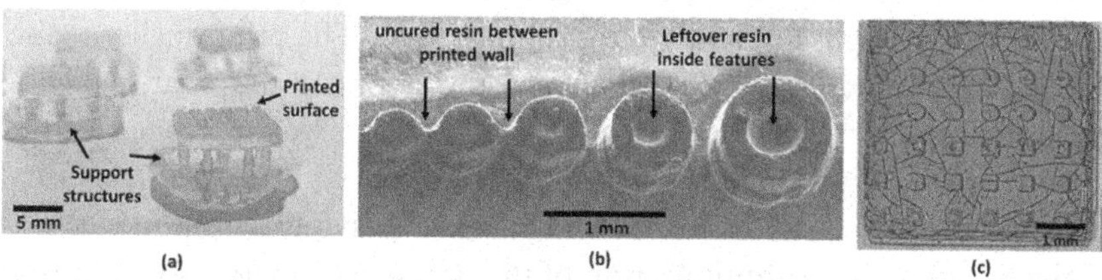


**Figure 5.10** (A) Printed samples with support structures. (B) Leftover resins between and inside the features due to incomplete cleaning process. (C) Overcuring the printed part in the flash chamber with highly intensive light radiation.

printed structures with leftover resins between and inside the features. Moreover, overcuring the parts may cause the cracks or breakage in the printed part. Fig. 5.10C shows an overcured part and the formation of cracks due to the application of a highly intensive light radiation. Thus, applying the flashlight continuously on the small parts produces high temperatures and the cracking of the samples.

## 5.4 Direct fabrication of parts by vat photopolymerization

The direct high-precision fabrication capability of components via VP methods was presented in a number of studies. A study by Charalambis et al. contributes to the field of AM by proposing a cost estimation method that can be applied on DLP technology. More specifically, the cost estimation model for DLP can be easily adapted to the fabrication of different parts in size, material, and level of complexity as long as those parts have been manufactured by DLP. Creating a method to estimate the cost of a part manufactured with DLP technology has a great potential in the continuously evolving field of AM. In an industrial context, many decisions are driven by the cost advantages that a technology or solution can offer. Therefore, estimating with a good level of confidence the cost of a part manufactured with an AM technology has the opportunity to offer a tool to use in management-based decisions that involve AM [16]. Davoudinejad et al. [17] investigated the accuracy of the printable features with DLP on a test part that designed with different sizes and aspect ratios to evaluate the additively manufactured parts. The printing parameters selected for the evaluation are considered as exposure time, light intensity, and layer thickness. In terms of printed features by considering the relation between layer thicknesses, which was a significant factor for the number of all produced features, it was observed that the thinner layer leads to the more details printed. The best results were obtained with a low layer thickness, high light intensity, and exposure time. The print quality is also affected by the transparent film between the resin and the light, which is directly in contact with the part [17]. Fig. 5.11 shows the process of the improvement of parameters for printing samples with small microstructures. The direct production of miniaturized polymer components was analyzed. The process optimization of the part quality features was carried out to highlight the potential and challenges of the micro-AM process. The geometry of the features affects the printing results, for instance, smaller features were printed with the cylindrical shape geometry than with a square-shape geometry [18].

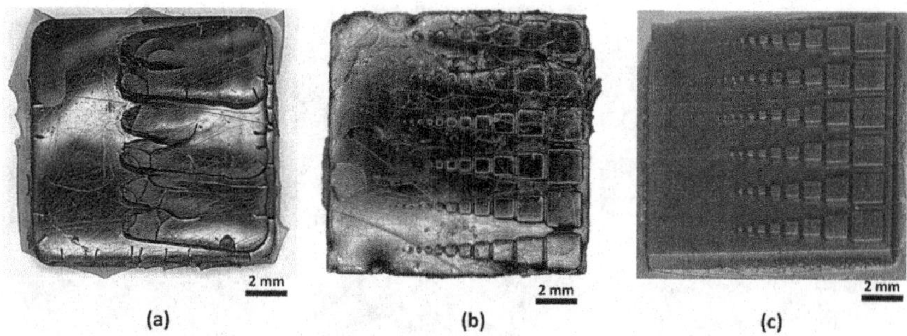


**Figure 5.11** Printed samples (A) initial printing parameters (B) improvement with some sensitivity analysis (C) optimized printing parameters.

Davoudinejad et al. [19] evaluated the printable features with DLP method on a test part that designed with multihierarchical surfaces to evaluate the additively manufactured parts for functional surfaces with microfeature size in different geometries [19]. It was revealed a simpler and less time-consuming process chain than previous methods used for manufacturing similar surfaces. Moreover, the functionality of the surface was determined with a wettability test. This revealed the high hydrophobicity of the obtained surfaces due to the higher density of the features. The results of this study prove the possibility of 3D-printed biosurfaces, which may become commercially viable in the near future. SEM images of different designs are shown in Fig. 5.12. Arrays of pillars, with straight and regular features, were successfully printed (see Fig. 5.12A); however, in other geometries with smaller structures the surface quality was diminished. This was traced due to number of reasons such as different sample designs (smaller nominal dimensions and higher feature density), optimization process, and cleaning process [13]. In the surfaces of samples, the leftover of the resin, which could not be properly cleaned in the postprocessing, is clearly visible; however, with smaller features and less distance between them, it is even a more challenging procedure. Measurements were conducted in a stratified sampling, to find out how uniform the features were printed all over the surface. After a process optimization, the samples printed were within a ±9 μm tolerance range from the nominal values of the CAD model; this is supported by the measurement of the circular hair features.

Direct fabrication of microstructure surfaces by DLP method with different feature sizes (50–150 μm) was investigated. Measurements were applied to evaluate the printed surface (micropillars and holes array) to

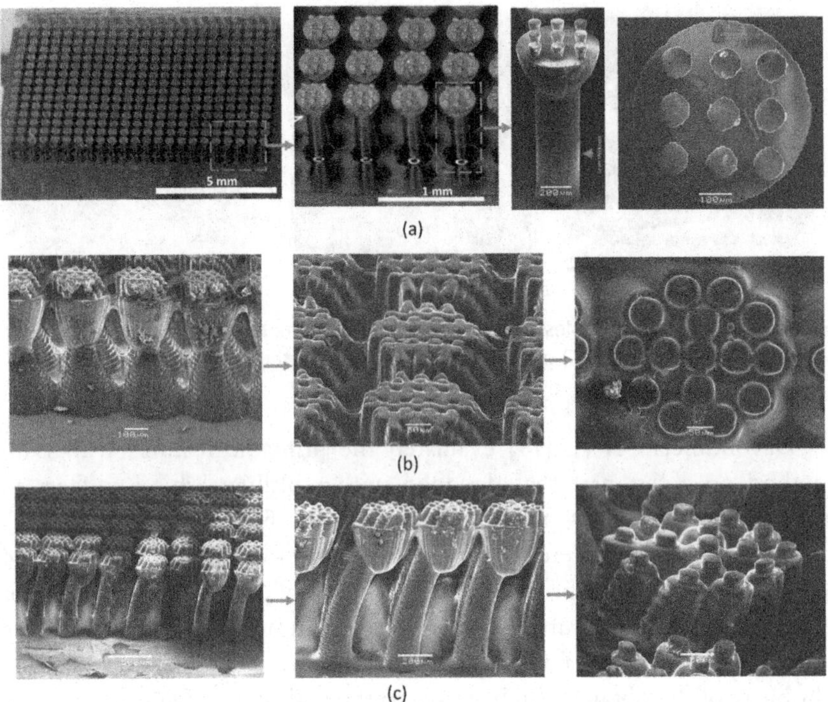


**Figure 5.12** SEM images of three different printed surfaces with digital light processing method. *(Reproduced with kind permission of IOP Publishing from Reference A. Davoudinejad, M. M. Ribo, D. B. Pedersen, A. Islam, and G. Tosello, "Direct fabrication of bio-inspired gecko-like geometries with vat polymerization additive manufacturing method," J. Micromechanics Microengineering, vol. 28, no. 8, 2018, PP 1–10).*

analyze the uniformity of the microfeatures. A precision AM machine, with projection mask pixel spacing 7.6 μm, was used in the image plane. In terms of postprocessing a bath of isopropanol in a vibration plate, ultrasonic cleaner was used to remove uncured resin. The samples were cleaned in different positions to find out the best position for removing the excess resin from the components with microstructures. The structures were fabricated in the designed shaped; however, the geometry of smaller features (with a pitch of 50 μm or 70 μm) was not accurately manufactured, and their circular shape was not properly printed [15–20]. Fig. 5.13 shows the printed samples (pillars and holes) and water contact angle measurements on the printed surface, and the SEM image of the microholes arrays in different pitches with different magnifications.

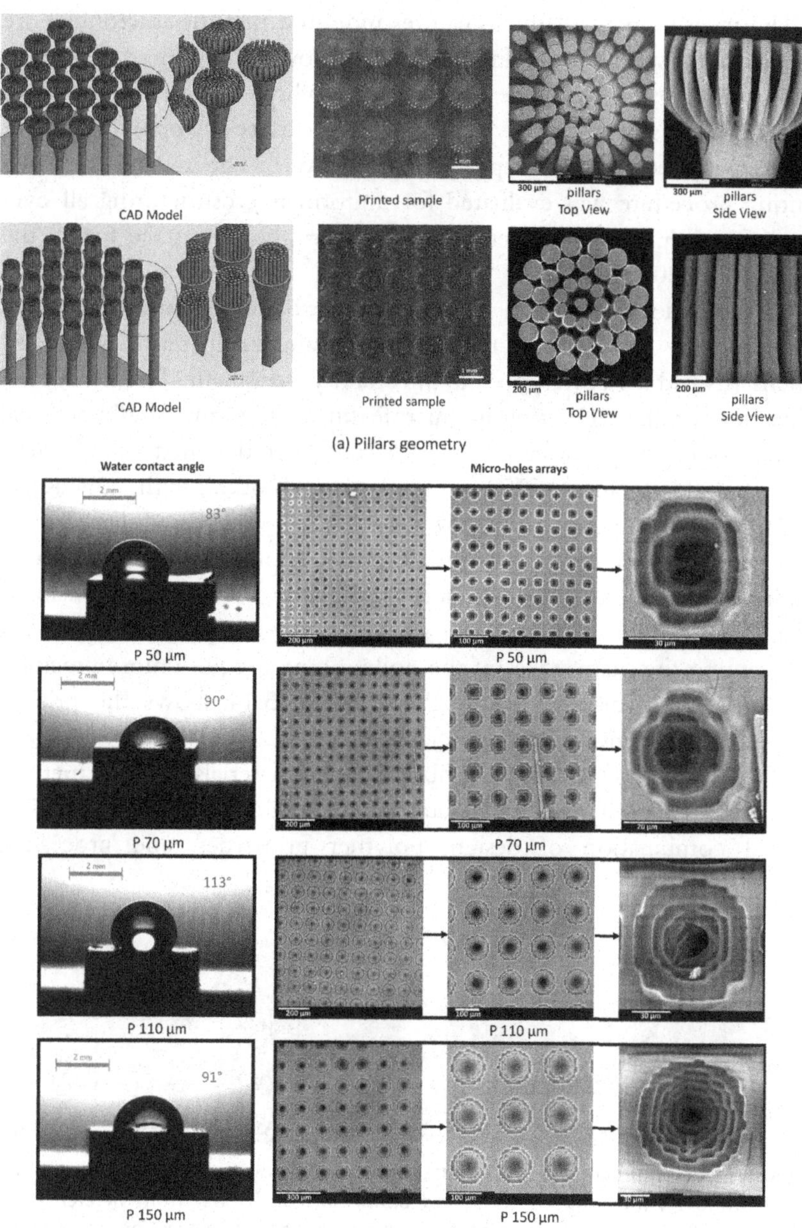


**Figure 5.13** (A) Printed pillars with different dimensions (diameter from 90 μm down to 60 μm) (B) Gaussian-shaped surface (holes array 50–150 μm). *(Reproduced with kind permission of Elsevier from References A. Davoudinejad, Y. Cai, D. B. Pedersen, X. Luo, and G. Tosello, "Fabrication of micro-structured surfaces by additive manufacturing, with simulation of dynamic contact angle," Mater. Des., vol. 176, p. 107839, August, 2019; A. Davoudinejad, D. B. Pedersen, and G. Tosello, "Direct fabrication of microstructured surfaces by additive manufacturing," in euspen's 19th International Conference & Exhibition, 2019).*

The measurement of microfeatures indicated that the microholes are almost evenly separated in different areas all over the printed surface. It was found that this method enables the possibility of direct fabrication of the components with micro-/nanostructured surfaces by applying proper process parameters and postprocessing methods. The reliability of the printing procedure was evaluated for uniform microstructuring all over the surfaces. A wettability test revealed hydrophobic surface for all the samples [15–20].

The precision fabrication of polymer microcomponents was investigated, and the capability of the machine evaluated in terms of printed dimensions and the corresponding uncertainty assessment. To find out the replication of the 3D printing microfeatures, in terms of dimensional accuracy of the parts, different test parts have been designed. The features are separated a distance of 250 μm between each other, with a 3:4 aspect ratio for the lateral size. In the rows, the features are ordered decreasingly from top (1.5 mm in diameter/width) to bottom (6 μm in diameter/width). The test part of the cylinders presents the smallest printed feature, having better results than the test part of the boxes. This is due to the fact that printing the sharp edges of the hollow boxes require more precision than printing a rounded shape [21–23] Fig. 5.14 shows the printed samples for different designs.

The VP process was used for fabricating biomaterials (lattice structure) with ability to print details in a custom-built micro-SLA machine. The optimal combination of design, polymer properties, and processing

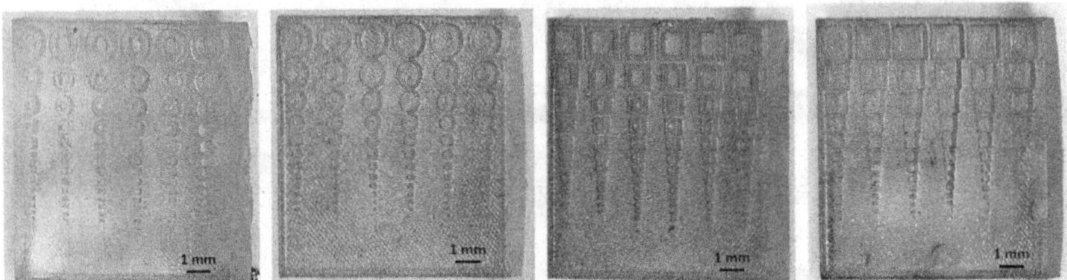


**Figure 5.14** Printed samples with digital light processing method in different size and geometries. *(Reproduced with kind permission of Elsevier from References A. Davoudinejad et al., "Additive manufacturing with vat polymerization method for precision polymer micro components production," Procedia CIRP, vol. 75, pp. 98–102, January, 2018; A. Davoudinejad et al., "Geometric and fature size design effect on vat photopolymerization micro additively manufactured surface deatures," in Special Interest Group Meeting: Additive Manufacturing: Advancing Precision in Additive Manufacturing, 2018).*

capabilities were investigated to fabricate the structures. 3D printing of biomaterials offers potential for drug delivery, tissue engineering scaffolds, and wound dressing applications [24].

The consideration of the manufacturing restrictions can clarify the feasibility of the designs to be produced by AM. In a study, the capability and limitations of feature size and geometry, of the VP method by producing various components were investigated. In this specific case, the technology used is SLA [12]. To analyze the performance of the SLA machine in terms of geometry and size, when printing two different kinds of microgeometry with different sizes, have been designed. Different batches of samples were printed to find out the limit for micropolymer components manufacturing with different geometries. The variability of the results in a single print and different batch was also evaluated. The smallest printed feature of size with hollow shape was 630 μm for both geometries, and the features smaller than 355 μm were completely solid [12]. Fig. 5.15A shows 15 different features fabricated in each row, and with a smallest diameter dimension of 26 μm for both the hollow box external side and the hollow cylinder external diameter. Fig. 5.15B shows cylindrical and pyramidal steps with 100 μm height of each step printed with SLA method.

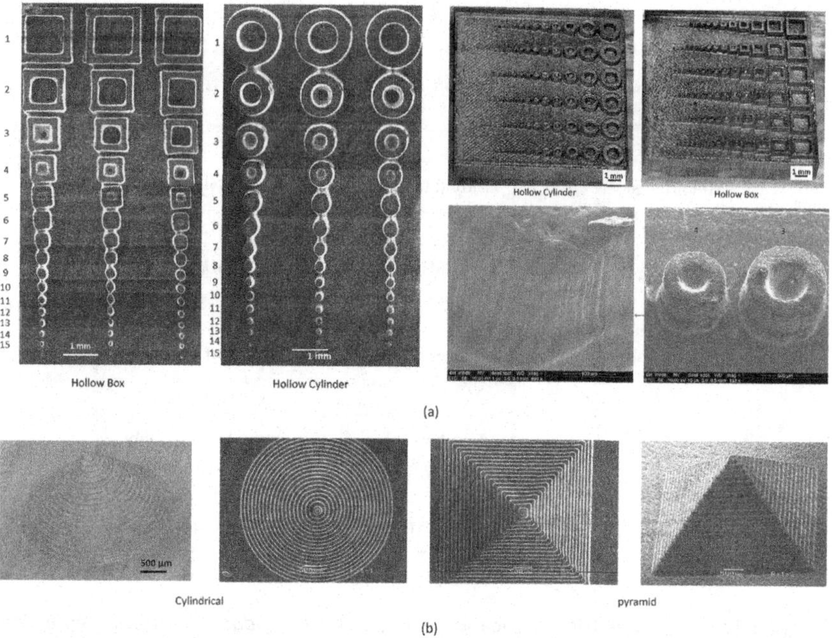


**Figure 5.15** The printed samples with hollow box and cylinder geometry.

Another study investigated quality assurance of reference specimens manufactured by CLIP using coordinate metrology. To evaluate the capabilities of this new process, a CLIP AM machine was used to manufacture a series of reference specimens developed following the design of existing metrological transfer standards. The reference specimens were measured by a tactile coordinate measuring machine to assess the manufacturing capability of the CLIP printer. Eventually, the data collected will enable adjustments for improving the manufacturing accuracy of the CLIP machine and for paving the way toward the quality assurance of production based on the CLIP technique [25].

## 5.5 Additive manufacturing technologies for tooling

AM technology transformation resulted in increasing the efficiency of AM in the production cycle and fabrication of tooling for medium-large volume production processes. This change intended at tools that maintain their consistency over a long term is a critical aspect for AM tooling processes [26]. In contrast to conventional tool fabrication with subtractive manufacturing methods, AM applies for both hard and soft tooling, as shown in Fig. 5.16. AM technique has been used to produce injection

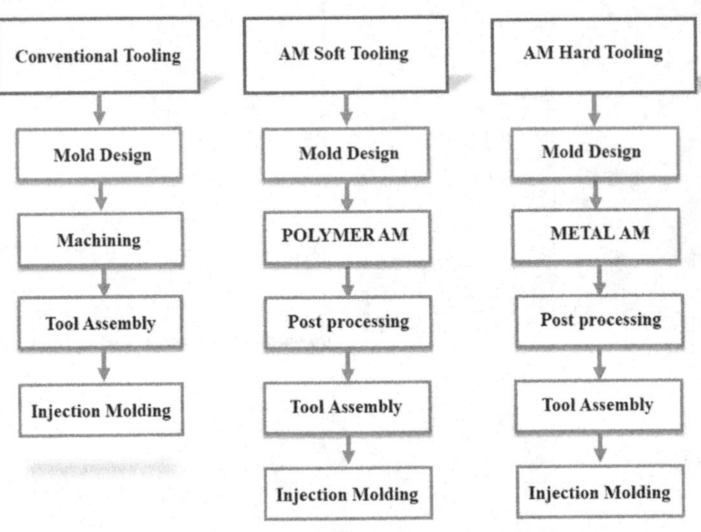


**Figure 5.16** Tooling method fabrication flow charts. *(Reproduced with kind permission of Elsevier from Reference A. Davoudinejad, M. R. Khosravani, D. B. Pedersen, and G. Tosello, "Influence of thermal ageing on the fracture and lifetime of additively manufactured mold inserts," Eng. Fail. Anal., vol. 115, p. 104694, 2020).*

molded parts for extensive functionality tests of the parts produced with the actual material of the final product. For this purpose, several design improvements and modifications can be applied in the mold whose production and overall lead-time are drastically reduced by using AM methods. It should be noted that the soft tooling method (i.e., the production of tool inserts by polymer AM rather than using metal AM, also known as AM hard tooling, according to the definition given in Ref. [27] could be used for several thousand IM cycles [28]. The mold inserts fabricated by AM are suitable for intermediate production volumes in IM processes, and a relative economic advantage in comparison with conventional tooling can be achieved in selected conditions [29].

Fabricated IM inserts by VP have become a serious option for significantly faster and more economical prototyping and pilot production due to technological progress and advancements in photopolymer materials in the recent years. Manufacturing of 10.000 parts of a geometry, including microfeatures, have been injection-molded in acrylonitrile butadiene styrene (ABS) with a single $20 \times 20 \times 2.5\ mm^3$ IM insert manufactured in a photopolymer composite material. This research investigates the dimensional accuracy of the injection-molded parts as a function of inserts wearing and deformation with increasing shot number. An SLA printer (3D Systems, SLA 3500, Solid State Nd:$YVO_4$ laser, 354.7-nm laser wavelength) was used to manufacture the IM insert in a ceramic composite photopolymer. Next, a heat treatment was performed after the printing to increase the photopolymer's heat deflection temperature [31].

Since printed molds experience various conditions during their service life, investigations of these working conditions are necessary to determine the behavior of the molds in production, and to predict their performance. A study evaluates IM inserts fabricated by the AM vat photopolymerization method. The inserts are directly manufactured with a photopolymer material, integrated on an IM tool, and subsequently used for IM. Fig. 5.17A shows the setup and inserts placement in the IM machine, while Fig. 5.17B shows the fabricated parts in both cavities. In the polymer insert, cracks occurred in different regions, as shown in Fig. 5.17C. The blue (gray in printed version) arrow on the right side of the insert shows the gate location as well as the injection direction. Five areas were considered to be prone to cracks. The inserts had a tendency to crack on the opposite walls from the gate and in a direction parallel to the injection direction. It was noticed on the top corner of the bricks that cracking happened in most of the inserts and then on the wall between two features on each side of

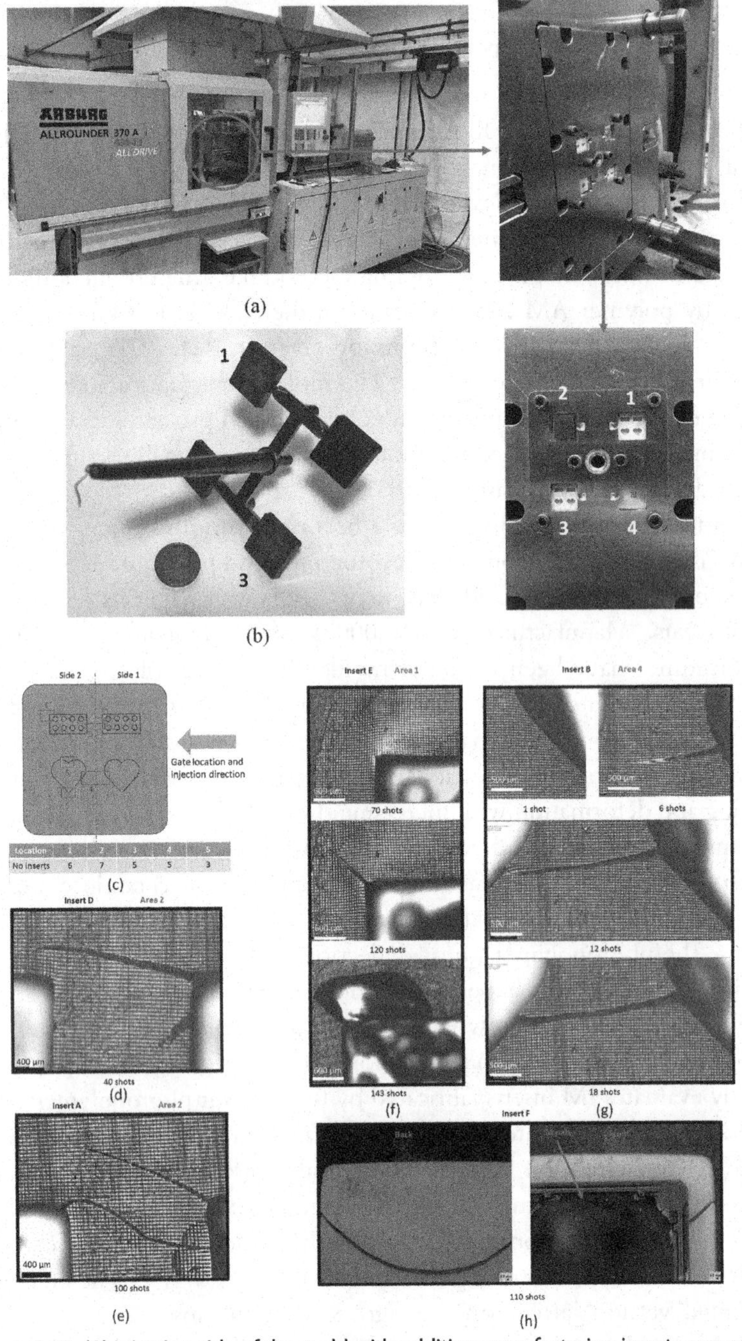


**Figure 5.17** (A) Injection side of the mold with additive manufacturing inserts mounted in two cavities for injection molding (IM); (B) IM parts; (C) crack locations on the insert; and (D–H) molded part cracks and crack propagation on the IM parts in different shots. *(Reproduced with kind permission of Springer Nature from Reference A. Davoudinejad, M. Bayat, D. B. Pedersen, Y. Zhang, J. H. Hattel, and G. Tosello, "Experimental investigation and thermo-mechanical modelling for tool life evaluation of photopolymer additively manufactured mould inserts in different injection moulding conditions," Int. J. Adv. Manuf. Technol., Jan. 2019, PP 403–420, and with kind permission of Elsevier from Reference A. Davoudinejad, M. R. Khosravani, D. B. Pedersen, and G. Tosello, "Influence of thermal ageing on the fracture and lifetime of additively manufactured mold inserts," Eng. Fail. Anal., vol. 115, p. 104694, 2020).*

the insert [32]. Another study investigates the AM tool inserts and injection molded parts in terms of surface roughness and geometrical accuracy prior and after injection molding. The results revealed slight difference in surface roughness at the left and right side of the mold in both 3D-printed inserts and IM parts. The variation, in terms of inserts, might be due to the nonuniform quality of the 3D printer vat that replicates the insert with the same uneven quality. Concerning the IM parts, the final quality was affected by the surface of the insert and the IM parameters as well. The best surface was achieved with the first batch of the IM parts that shows the lowest surface roughness of the insert and the combination of the IM parameters [33].

Therefore, particular attention has to be paid to develop the soft tooling process chain and the IM experimental procedure. In an investigation with soft tooling, the parts were analyzed and evaluated by the measurements of different features and the influence of the IM process. The variation of the dimensional features of parts were mainly due to the IM setting parameters and also might be due to the shrinkage of ABS material and the crack of the insert that affects the feeding of the mold [30,34,35].

In addition to the IM, soft tooling was applied in the extrusion process. Dies manufacturing by conventional processes for extrusion are time- and cost-consuming in comparison with AM processes, which have the capability of producing complex geometries, both external and internal components. In a study, different experimental tests were carried out with the additively manufactured dies made of photopolymer material. The failure of the die was investigated in several processing conditions, and numerical analysis were applied to predict the weakest point of the die for further consideration during the die design phase. The model helped the improvement of the AM design for optimized die manufacturing. The SLA AM process was used for fabrication of the die with photopolymer material in a plastic extrusion process. After proper tool setup and machine warm up for a stable process condition, the extrusion started with low screw speed that was then progressively increased. The failure of the die occurred after a short run of approximately 5 min, when the die pressure reached a peak of 6 bar. The obtained extruded product was acceptable only for a small production since the die failed shortly after running; however, with the modified die design, much longer extrudate was fabricated with the AM die [36]. Fig. 5.18 shows the die fabricated with SLA method for a U-form profile, process setup and failure analyses, respectively.

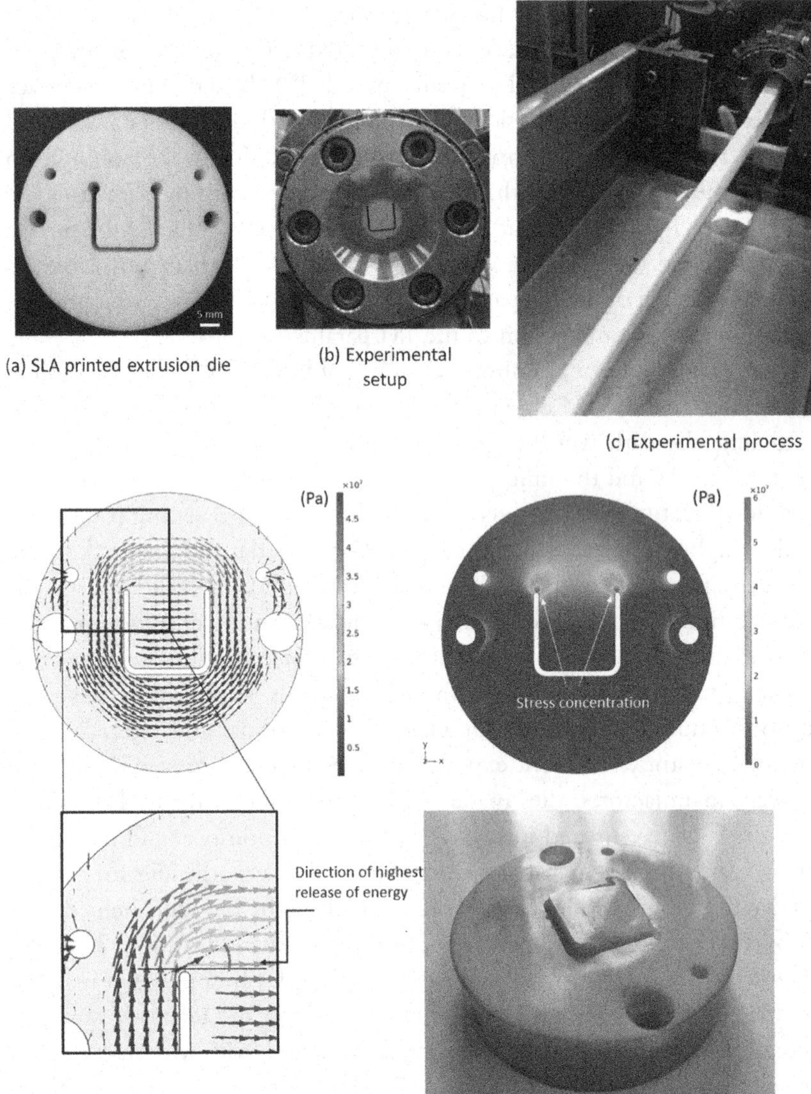


**Figure 5.18** (A) Additively manufactured (AM) die and (B) experimental setup. (C) Extrusion process with AM die, and (D) vectors of first principal stresses. The colors show the magnitude of the principal stresses. *(Reproduced with kind permission of American Institute of Physics from Reference A, Davoudinejad; Bayat, M.; Larsen, A.; Pedersen, D. B.; Hattel, J. H.; Tosello, G. Mechanical properties of additively manufactured die with numerical analysis in extrusion process. In: AIP Conference Proceedings Series, 2019, PP 1–5).*

## References

[1] A. Davoudinejad, D.B. Pedersen, G. Tosello, Additive manufacturing for micro tooling and micro Part Rapid prototyping, in: Micro Injection Molding, Carl Hanser Verlag GmbH & Co. KG, 2018, pp. 289–313.
[2] C.M. González-Henríquez, M.A. Sarabia-Vallejos, J. Rodriguez-Hernandez, Polymers for additive manufacturing and 4D-printing: materials, methodologies, and biomedical applications, Prog. Polym. Sci. 94 (July 2019) 57–116.
[3] S.A.M. Tofail, E.P. Koumoulos, A. Bandyopadhyay, S. Bose, L. O'Donoghue, C. Charitidis, Additive manufacturing: scientific and technological challenges, market uptake and opportunities, Mater. Today 21 (1) (January–February 2018) 22–37.
[4] V. Khare, S. Sonkaria, G.Y. Lee, S.H. Ahn, W.S. Chu, From 3D to 4D printing – design, material and fabrication for multi-functional multi-materials, Int. J. Precis. Eng. Manuf. - Green Technol. 4 (2017) 291–299.
[5] A. Davoudinejad, M. Annoni, L. Rebaioli, Q. Semeraro, Improvement of surface flatness in high precision milling, in: Conference Proceedings - 14th International Conference of the European Society for Precision Engineering and Nanotechnology, EUSPEN 2014, vol. 2, 2014.
[6] A. Davoudinejad, G. Tosello, M. Annoni, Influence of the worn tool affected by built-up edge (BUE) on micro end-milling process performance: a 3D finite element modeling investigation, Int. J. Precis. Eng. Manuf. (2017) 1321–1332.
[7] ASTM International, F2792-12a - Standard Terminology for Additive Manufacturing Technologies, Rapid Manuf. Assoc, 2013, pp. 10–12.
[8] J. V Crivello, E. Reichmanis, Photopolymer materials and processes for advanced technologies, Chem. Mater. 26 (1) (January, 2014) 533–548.
[9] I. Gibson, D. Rosen, B. Stucker, Vat photopolymerization processes, in: Additive Manufacturing Technologies: 3D Printing, Rapid Prototyping, and Direct Digital Manufacturing, Springer New York, New York, NY, 2015, pp. 63–106.
[10] P.F. Jacobs, Rapid Prototyping & Manufacturing: Fundamentals of Stereolithography, Society of Manufacturing Engineers, 1992.
[11] D.W. Rosen, Stereolithography and rapid prototyping, in: P.J. Hesketh (Ed.), Bio-NanoFluidic MEMS, Springer US, Boston, MA, 2008, pp. 175–196.
[12] A. Davoudinejad, et al., Geometrical and feature of size design effect on direct stereolithography micro additively manufactured components, Proc. Struct. Integr. 13 (January, 2018) 1250–1255.
[13] A. Davoudinejad, M.M. Ribo, D.B. Pedersen, A. Islam, G. Tosello, Direct fabrication of bio-inspired gecko-like geometries with vat polymerization additive manufacturing method, J. Micromech. Microeng. 28 (8) (2018) 1–10.
[14] A. Davoudinejad, A.K. Jessen, S.D. Farahani, N. Franke, D.B. Pedersen, G. Tosello, Geometrical shape assessment of additively manufactured features by Continuous Liquid Interface Production, vat photopolymerization method, in: Proceedings of the 19th International Conference and Exhibition (EUSPEN 2019), The European Society for Precision Engineering and Nanotechnology, 2019, pp. 116–117.
[15] A. Davoudinejad, Y. Cai, D.B. Pedersen, X. Luo, G. Tosello, Fabrication of micro-structured surfaces by additive manufacturing, with simulation of dynamic contact angle, Mater. Des. 176 (August, 2019) 107839.
[16] A. Charalambis, A. Davoudinejad, G. Tosello, D.B. Pedersen, Cost estimation of a specifically designed direct light processing (DLP) additive manufacturing machine for precision printing, in: 17th Euspen International Conference & Exhibition, 2017.

[17] A. Davoudinejad, D.B. Pedersen, G. Tosello, Evaluation of polymer micro parts produced by additive manufacturing processes by using vat photopolymerization method, in: Joint Special Interest Group Meeting between Euspen and ASPE Dimensional Accuracy and Surface Finish in Additive Manufacturing, 2017, pp. 1–3.

[18] A. Davoudinejad, D.B. Pedersen, G. Tosello, Characterization of additive manufacturing processes for polymer micro parts productions using direct light processing ( DLP ) method, in: 33rd Conference of the Polymer Processing Society, at Cancun, Mexico, 2017, pp. PP1–5.

[19] A. Davoudinejad, M.M. Ribo, D.B. Pedersen, G. Tosello, A. Islam, Biological features produced by additive manufacturing processes using vat photopolymerization method, in: Joint Special Interest Group Meeting between Euspen and ASPE Micro/Nano Manufacturing, 2017, pp. 3–5.

[20] A. Davoudinejad, D.B. Pedersen, G. Tosello, Direct fabrication of microstructured surfaces by additive manufacturing, in: euspen's 19th International Conference & Exhibition, 2019.

[21] A. Davoudinejad, et al., Additive manufacturing with vat polymerization method for precision polymer micro components production, Proc. CIRP 75 (January, 2018) 98–102.

[22] A. Davoudinejad, et al., Geometric and fature size design effect on vat photopolymerization micro additively manufactured surface deatures, in: Special Interest Group Meeting: Additive Manufacturing: Advancing Precision in Additive Manufacturing, 2018.

[23] L.C.D. Péreza, et al., Geometrical shape assessment of additively manufactured features by direct light processing vat polymerization method, in: 18th International Conference of the European Society for Precision Engineering and Nanotechnology (Euspen 18), 2018. PP1-3.

[24] D.C. Aduba, et al., Vat photopolymerization 3D printing of acid-cleavable PEG-methacrylate networks for biomaterial applications, Mater. Today Commun. 19 (June, 2019) 204–211.

[25] M. Kain, et al., Quality assurance of reference specimens manufactured by continuous liquid interface production using coordinate metrology, in: euspen's 19th International Conference & Exhibition, 2019.

[26] S.O. Onuh, Y.Y. Yusuf, Rapid prototyping technology: applications and benefits for rapid product development, J. Intell. Manuf. 10 (3) (1999) 301–311.

[27] C.K. Chua, K.F. Leong, Z.H. Liu, Rapid tooling in manufacturing, in: A.Y.C. Nee (Ed.), Handbook of Manufacturing Engineering and Technology, Springer London, London, 2015, pp. 2525–2549.

[28] M. Mischkot, Advanced Process Chains for Prototyping and Pilot Production Based on Direct Rapid Soft Tooling, vol. ISBN 978-8, p. PhDthesis, Department of Mechanical Engineering, 2018.

[29] G. Tosello, et al., Value chain and production cost optimization by integrating additive manufacturing in injection molding process chain, Int. J. Adv. Manuf. Technol. (2019) 783–795.

[30] A. Davoudinejad, M.R. Khosravani, D.B. Pedersen, G. Tosello, Influence of thermal ageing on the fracture and lifetime of additively manufactured mold inserts, Eng. Fail. Anal. 115 (2020) 104694.

[31] M. Mischkot, A. Davoudinejad, A. Charalambis, G. Tosello, D.B. Pedersen, H. Nørgaard, Dimensional accuracy of Acrylonitrile Butadiene Styrene injection molded parts produced in a pilot production with an additively manufactured insert, in: 33rd Conference of the Polymer Processing Society, at Cancun, Mexico, 2017.

[32] A. Davoudinejad, M. Bayat, D.B. Pedersen, Y. Zhang, J.H. Hattel, G. Tosello, Experimental investigation and thermo-mechanical modelling for tool life evaluation of photopolymer additively manufactured mould inserts in different injection moulding conditions, Int. J. Adv. Manuf. Technol. (January, 2019) 403–420.
[33] A. Davoudinejad, A. Charalambis, D.B. Pedersen, G. Tosello, Evaluation of surface roughness and geometrical characteristic of additive manufacturing inserts for precision injection moulding, in: AIP Conference Proceedings, vol. 2139, 2019, p. 190006, 1.
[34] A. Davoudinejad, et al., Assessment of sub-mm features replication capability in injection moulding using a multi-cavity tool produced by additive manufacturing, in: 18th International Conference of the European Society for Precision Engineering and Nanotechnology (Euspen 18), 2018, pp. 1–3.
[35] A. Davoudinejad, A. Charalambis, Y. Zhang, M. Calaon, G. Tosello, H.N. Hansen, Evaluation of part consistency with photopolymer inserts in different injection moulding process parameters, in: 18th International Conference of the European Society for Precision Engineering and Nanotechnology (Euspen 18), 2018, pp. 1–3.
[36] A. Davoudinejad, M. Bayat, A. Larsen, D.B. Pedersen, J.H. Hattel, G. Tosello, Mechanical properties of additively manufactured die with numerical analysis in extrusion process, in: AIP Conference Proceedings Series, 2019, pp. 1–5.

CHAPTER 6

# Polymer and composites additive manufacturing: material extrusion processes

**Vidya Kishore, Ahmed Arabi Hassen**
Oak Ridge National Laboratory (ORNL), Oak Ridge, TN, United States

## 6.1 Introduction

Extrusion-based additive manufacturing (AM) process, termed as "material extrusion" by ASTM International (American Society for Testing and Materials) and the International Organization for Standardization (ISO), is defined as "an additive manufacturing process in which material is selectively dispensed through a nozzle or orifice" [1]. This process, often referred to as fused deposition modeling (FDM), or fused filament/freeform fabrication (FFF), is currently the most widely used polymer AM technology globally. Material extrusion enables process scalability, design freedom, the use of a wide variety of feedstock materials, and rapid qualification. Ever since the late-1980s, several extrusion-based AM systems have been developed for polymeric materials, ranging in build volumes from a typical 200 × 200 × 200 mm desktop-scale platform, to a 30 × 6.7 × 3 m large-scale system, widely referred to as large format additive manufacturing (LFAM) systems (Fig. 6.1) [2–4]. Concurrently, several materials have also been developed, encompassing thermoplastics, thermosets, and their composites, specifically for AM applications [5–8]. As industries and researchers started exploring extrusion AM as a technique for fabricating functional components, instead of just prototyping, new thermoplastic and thermoset-based composites have also been developed, involving a variety of reinforcements such as carbon fiber (CF), glass fiber (GF), clay, metal powder, and so on. Starting with the commonly used thermoplastic matrices such as acrylonitrile butadiene styrene (ABS), polylactic acid (PLA), polycarbonate (PC), and nylons, other polymeric systems using high-performance thermoplastics such as poly(aryl ether ketones), polysulfones, polyetherimide (PEI), and polyphenylene sulfide (PPS) have also been developed for extrusion AM processes. In the case of thermosets,

*Additive Manufacturing*
ISBN 978-0-12-818411-0
https://doi.org/10.1016/B978-0-12-818411-0.00021-5

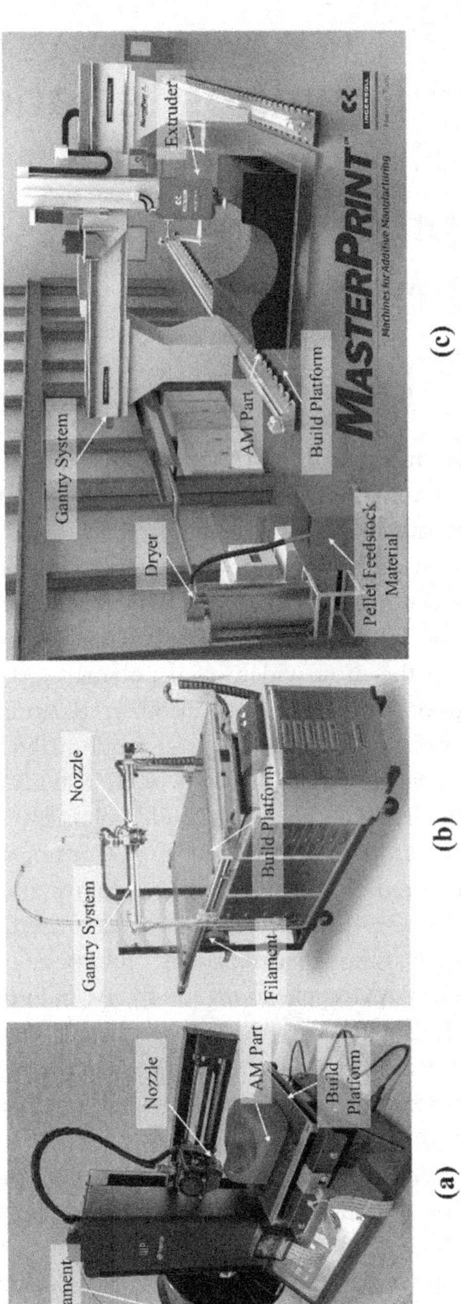


**Figure 6.1** Examples of extrusion AM systems with different build volumes. (A) Desktop FFF printer (Build volume: 120 × 120 × 120 mm), (B) Midscale 3D Platform printer (Build volume: 1000 × 1500 × 700 mm), and (C) A rendering of Ingersoll's MasterPrint 3D Printer (Build volume: 30 × 6.7 × 3 m). *AM*, additive manufacturing. *(A) Photo courtesy of Oak Ridge National Laboratory, U.S. Dept. of Energy; (B) Photo courtesy of 3D Platform, US; (C) Photo courtesy of Ingersoll Machine Tools Inc., USA).*

matrices such as epoxies and silicones have been extensively explored with nanofillers and short fiber reinforcements. In addition to developments in feedstock materials, complementary multifunctional fabrication technologies, including wire embedding, multimaterial extrusion, coextrusion, 3D printing electronics, and smart manufacturing via embedded sensors, are also evolving [9,10].

The polymer-based AM industry is the fastest growing segment among all other AM technologies, with an average annual growth of 39% between 2015 and 2020 [11,12]. There is a global active interest from industry, academia, and government to advance such novel technologies for making manufacturing faster, more cost-effective, and environmentally sustainable. For example, in the United States, industries in this area are typically heavily involved in materials, process, and application development. Government and academic research entities, in collaboration with industries, are involved in new process development, smart manufacturing, materials characterization, and other novel developments. Public—private partnership organizations such as America Makes leverage the knowledge from researchers and industrial experts to accelerate qualification and certification. With material extrusion AM technology maturing aggressively, it is being adopted for several applications in areas such as medicine, automotive, energy, tooling, and aerospace, for cost, energy, and lead time benefits.

This chapter will provide a comprehensive discussion on systems development for extrusion AM processes and platforms. This will be followed by discussions on advances in feedstock materials, including thermoplastics, thermosets, and their composites, used in extrusion-based system and their printability criteria. Finally, a discussion on the different process and material challenges, limitations, applications, automation, and paths forward will be covered.

## 6.2 Extrusion additive manufacturing processes

Similar to other AM techniques, the extrusion AM process first starts with a three-dimensional (3D) computer-aided design (CAD) file that is converted into a stereolithography (STL) file. The STL file is then imported into a slicer program that converts it into G-code (numerical control programming language), with commands for printer motion and customized build parameters such as number of layers, layer height, print speed, and part orientation. Finally, the G-code is loaded into the printer to fabricate parts, layer-by-layer. In general, all material extrusion platforms can be considered to comprise of three primary sections based on their functionality, namely, feedstock material loading system, extrusion mechanism, and build space and motion system (i.e., motion control system, build platform, and print enclosure).

## 6.2.1 Feedstock material loading system

The possibility of utilizing a wide variety of feedstock materials available in different forms, such as filaments, pellets, and pastes, makes material extrusion AM a highly versatile, flexible, and a vastly adopted process. For thermoplastics, feedstock material is in the form of either extruded filaments (~1.75 mm diameter) or pellets, as shown in Fig. 6.2. In filament-fed printers or FFF systems, the feedstock material is loaded by directing the filament from the spool into the extruder head, with a torque and pinch system controlling filament movement. For a multimaterial FFF process with multiple extruders, different filaments can be fed into each of the extruders for simultaneous printing.

On the other hand, systems operating with pelletized feedstock, such as LFAM platforms, have a more complex material loading and shuttling mechanism. The feedstock loading system for such printers comprises of a dryer for moisture removal, vacuum-driven transfer lines to transport pellets from the dryer to the hopper, and a hopper that feeds the extruder. This

**Figure 6.2** Different forms of thermoplastic feedstock materials used in extrusion AM processes. (A) Polylactic acid (PLA) filament feedstock (1.75 mm in diameter), and (B) Pelletized feedstock material (ABS with 20% by weight short carbon fiber). *AM*, additive manufacturing. *(A) Photo courtesy of Oak Ridge National Laboratory, U.S. Dept. of Energy; (B) Reproduced with kind permission of SAGE Publishing from Reference Ajinjeru C, Kishore V, Chen X, Hershey C, Lindahl J, Kunc V, et al. Rheological survey of carbon fiber-reinforced high-temperature thermoplastics for Big Area Additive Manufacturing tooling applications. J. Thermoplast. Compos. Mater. 2019:0892705719873941).*

arrangement can be extended for multimaterial printing as well, with the addition of multiple dryers and a precision blender system that can mix different materials in user-defined proportions. The dryers and the blender should be designed to match the minimum drying time and the deposition rate of the extrusion system. For example, for a system with a deposition rate of 45 kg/h and a minimum material drying time of 4 h, a dryer of 272 kg capacity and a blender capable of handling pellets up to 59 kg/h should be used. After drying, solid pellets are then transferred to the hopper. In most cases, a single hopper is used for processing one single material for a print. When two different materials are used in a single print, pellets are transferred to two different hoppers, with a material switching mechanism attached between the hopper and the extruder. Oak Ridge National Laboratory (ORNL) has developed a large-scale multimaterial print system with such material switching capability [13]. This allows for multimaterial extrusion through a standard single screw extruder without the need for an additional extrusion mechanism (weighing about 90 kg).

In the case of extrusion AM of thermoset materials, i.e., reactive extrusion AM, feed material typically involves a resin and a liquid catalyst or hardener. Direct-write (DW) AM technology is a commonly used extrusion AM technique for processing thermoset polymers on a small scale. Here, the resin or the paste is premixed with the hardener and loaded into luer lock syringes mounted on a gantry [15]. However, in large-scale reactive extrusion processes, such as the ThermoBot system developed by ORNL and Magnum Venus Products (MVP), United States, separate resin and catalyst delivery mechanisms are employed. The resin is transferred using pumping equipment located next to the printer and directly connected to the deposition nozzle via lightweight hoses (Fig. 6.3) [8]. The pumping system consists of transfer pumps (with 19 or 208 L capacities) to convey the noncatalyzed resin to the metering pumps. The catalyst is gravity fed into another metering pump from the catalyst reservoir. The two metering pumps are synchronized by a drive assembly arm controlled by a servo hydraulic actuator to ensure a controlled initiator-to-resin ratio, as shown in Fig. 6.3. These design considerations ensure consistent resin flow to and fluid pressure on the deposition head. Such decoupling of heavy pumping and metering systems from the gantry provides the added benefit of decreased weight in comparison with large-scale thermoplastic extrusion systems. This provides the opportunity to increase print speed and production rate [8].

**Figure 6.3** Large-scale thermoset polymer additive manufacturing system showing the build platform, resin pumping station, and the deposition head system. *(Reproduced with kind permission of the Society for the Advancement of Material and Process Engineering (SAMPE) from Reference Lindahl J, Hassen A, Romberg S, Hedger B, Hedger Jr P, Walch M, et al. Large-scale additive manufacturing with reactive polymers. In: Proc. 2018 CAMX Conference, Dallas, TX, USA; 2018).*

### 6.2.2 Extrusion mechanism

In FFF systems, the extruder consists of a heater block with a coil or cartridge heater that softens or melts the filament, which then passes through the extruder nozzle. Most FFF nozzles are made from brass, aluminum, or stainless steel, with diameters ranging from 0.2 to 1.2 mm. For the material to extrude out of the nozzle, the extrusion force provided by the filament should be adequate to overcome the pressure drop across the entire system. FFF printers can print parts with a good resolution (typically 100–300 microns) and highly detailed features; however, their deposition rates are low (typically about 80 g/h), thereby leading to long build times.

Large-scale AM or LFAM systems that process thermoplastics employ a single screw extruder for material extrusion and deposition, with capacities ranging from 0.5 kg/h to over 100 kg/h. Fig. 6.4 shows a setup for an extrusion mechanism with a dual-material deposition capability in an LFAM process. The extruder has three primary units: a screw, barrel, and a nozzle. The screw conveys feed pellets to the barrel, where the pellets are melted using heating elements and then dispensed through the nozzle. The rotational speed of the screw is controlled by an electric motor drive unit and a gearbox. Screws are often customized by varying parameters such as flight shape and width, channel depth and length, zone lengths, and compression ratios, for desired output. These factors primarily depend on the desired throughput and the material's thermal and rheological properties. Nozzles used in LFAM processes have a diameter ranging from 2.5 to 12.7 mm. As the material exits the nozzle, the deposited layer or bead often undergoes an approximate 50% reduction in height due to a reciprocating force applied by a vertically vibrating metal plate called the "tamper" that is attached right by the nozzle. The tamper was developed by Cincinnati Incorporated, United States, for the big area additive manufacturing (BAAM) system (Fig. 6.5A) [16]. Another similar mechanism was developed by Thermwood Corporation, United States, for the large-scale additive manufacturing (LSAM) system in which a controlled, articulated roller that followed the nozzle was used to force the extruded bead down to a consistent height. The main objective of using a tamper or a roller is to increase the area of contact between the deposited layers (or beads) for better bonding and to reduce macrovoids between them (Fig. 6.5B) [5].

On the other hand, for thermoset systems, the syringes used in the DW processes are first loaded with the resin–catalyst mixture. These syringes are then mounted onto a three-axis gantry-style stage fitted with a high-pressure adaptor that drives, pneumatically controls, and pushes the material through

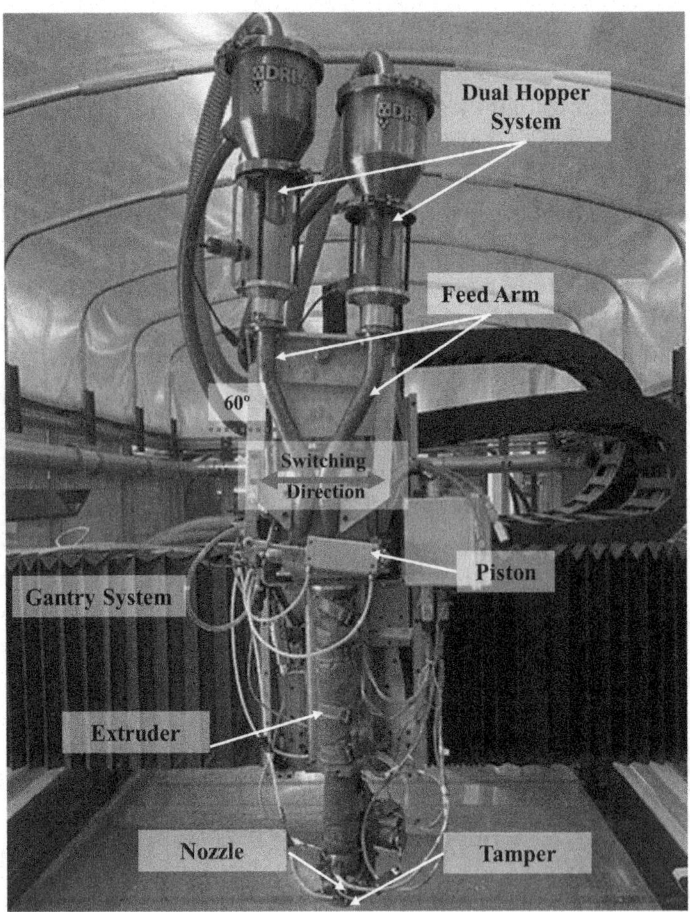


**Figure 6.4** Single screw extruder with dual-material deposition capability for the BAAM system located at ORNL. *BAAM*, big area additive manufacturing; *ORNL*, Oak Ridge National Laboratory. *(Reproduced with kind permission of the Society for the Advancement of Material and Process Engineering (SAMPE) from Reference Smith T, Hassen AA, Lind R, Lindahl J, Chesser P, Roschli A, etal Dual material system for polymer large scale additive manufacturing. In: Proc. 2020 SAMPE Conference and Exhibition, Seattle, WA, USA; 2020).*

the nozzle at the desired throughput [15]. In the case of large-scale reactive extrusion systems, the resin and catalyst are first pumped separately to a gun block where they are combined into a single fluid channel. The combined fluid then goes through a static mixing section (with a length of about 23 cm) for adequate mixing. Mixing just prior to deposition helps overcome challenges that may arise due to material thickening and curing, especially for long print times. After mixing, the catalyzed

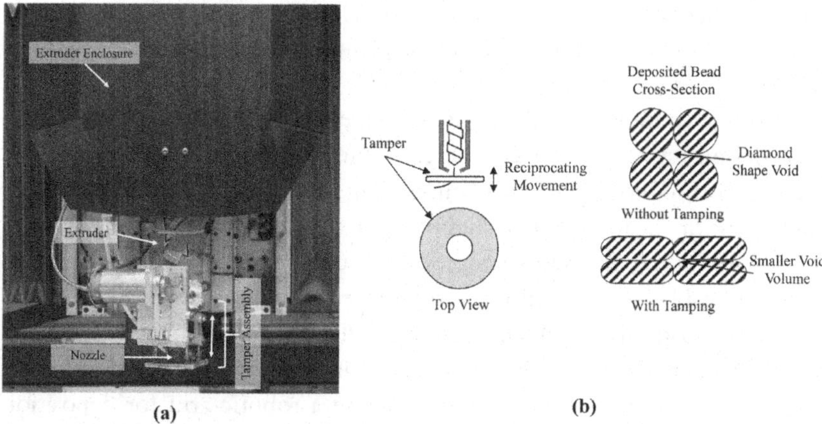


**Figure 6.5** Tamping mechanism used in LFAM to increase the area of contact and reduce macrovoids between the deposited layers. (A) Typical tamper assembly integrated into BAAM systems, and (B) Schematic representing macrovoids formation during deposition and the effect of tamping on void geometry. *BAAM*, big area additive manufacturing; *LFAM*, large format additive manufacturing. *(A) Photo courtesy of Cincinnati Incorporated, United States; (B) Reproduced with kind permission of the American Society for Nondestructive Testing Inc. from Reference Hassen AA, Kirka MM. Additive Manufacturing: the rise of a technology and the need for quality control and inspection techniques. Mater. Eval: 2018; 76(4):438–453; Copyright [2018]).*

resin flows to the air-controlled, fast-acting pneumatic deposition nozzle. Nozzle diameters for the large-scale reactive extrusion process range from 1.27 to 10.16 mm [8].

## 6.2.3 Build space and motion system

Most material extrusion AM platforms consist of, at the least, a motion system, build platform, and a print enclosure. Irrespective of the build volume, a majority of widely adopted AM systems for polymers and composites use a three-axis gantry-based motion platform. Gantry systems are easier to control, more economical, and possess the most robust deposition accuracy among the existing motion systems. Travel speed of the deposition head typically varies depending upon the desired material throughput, melt pressure, and rheological characteristics of the material. Desktop-scale gantry systems are capable of speeds up to 60 mm/s. In LFAM systems, travel speed is constrained by the weight of the extruder, preventing excess inertia during gantry movement. As an example, the maximum gantry speed for safe operation of BAAM systems is constrained to 0.28 m/s. Although gantry-based systems are reliable and easy to operate, with their limited degrees of freedom (three), they are primarily suited only for planar structures and not for out-of-plane printing applications without an additional support structure.

Another motion system employed in extrusion AM processes is the cable-driven system [18–20]. Here, the extrusion head is moved throughout the workspace by manipulating the lengths of the cables used. The extrusion head is usually lightweight in comparison with a gantry-based system, and the work area or the build space is also much larger. However, cable-driven platforms are slower, have more complex controls, and are less stiff, all of which affect their deposition accuracy. To enable out-of-plane printing, several companies and researchers are developing extrusion AM systems for polymers that use a robotic arm for deposition [20–22]. With six degrees of freedom, robotic arms make printing more flexible and enable novel deposition strategies for out-of-plane printing and printing over curved surfaces. This greatly reduces the need for support structures, especially for applications that demand structures with overhangs at angles greater than 45 degrees. Currently, industrial adoption of this technology is still limited due to high capital cost, limited reach of the robotic arm (build volume), limited maximum weight capacity of the arm, and control system complexity.

Build platforms, commonly referred to as the print bed, used for thermoplastics-based systems are all usually heated to temperatures ranging between 80 and 110°C. Heating helps reduce warping and distortion of the deposited beads by maintaining the initial few layers at elevated temperatures for the entire print duration. In LFAM processes, for both thermoplastics and reactive extrusion, the build platform consists of precision-machined aluminum platens with a stainless steel substructure. The platform is also equipped with a vacuum port to hold down the build sheets on which the parts are printed, facilitating ease of removal.

The third element that makes up the build space is the print enclosure. Most material extrusion AM systems are out-of-oven deposition systems, devoid of heated enclosures. However, there are a few midscale systems that employ a heated enclosure (up to 120°C) for reducing thermal gradients in printed structures. A heated enclosure enables fabrication of parts with better interlayer strength and increases the ability to print structures with a longer layer time without encountering significant delamination or cracking. However, these enclosures add to the overall energy consumption and limit process scalability. Although LFAM systems do not employ heated enclosures, other approaches to reduce part warping and deformation, such as a combination of high throughput and printing with CF—reinforced systems, have been demonstrated to be effective [5,23].

## 6.3 Hybrid systems

One of the distinct advantages of LFAM systems for thermoplastics is the use of pelletized feedstock which costs lower than filament-based feedstock used in small-scale systems. However, these processes also employ nozzles with a large diameter (up to 12.7 mm) for faster deposition, thereby creating parts with low surface resolution [3]. Typically, post-processing methods such as milling, coating, or polishing are required to obtain surfaces with the desired quality of finish. In cases where machining is the desired postprocessing method, parts are first printed such that the deposited beads are oversized, i.e., about 1.5 times the desired bead width. These parts are then machined down to the desired tolerances to obtain a smooth surface. On the other hand, while employing the surface coating method of postprocessing, printed parts are generally undersized by a factor equal to the thickness of the coating layer (ranging from 2.25 to 12.7 mm), and then a self-leveled coating layer is applied to the part surface. Having two separate systems, one for additive and the other for postprocessing, can significantly add to the capital cost of the manufacturing process, as well as to the complexity of the postprocessing operations. Locating, registering, and aligning the printed part for accurate machining is quite challenging when two separate systems are used.

To address this, several LFAM systems are currently adopting a hybrid approach, where both additive and subtractive manufacturing processes can take place in one single system simultaneously. The system would have two gantries on a single frame, with one gantry for an AM extrusion head, and the other for a milling head for subtractive processing. Examples of hybrid systems include the large-scale additive manufacturing (LSAM) system developed by Thermwood Cooperation, United States, and the Master-Print system developed by Ingersoll Machine Tools, United States. Utilizing this hybrid technique enhances part location and aligning accuracy since both the additive and the subtractive process heads use the same axis of motion relative to the location of the printed parts. The only major limitation of a hybrid system is the shared build volume. When parts that take up the entire build space are to be printed or machined, simultaneous processing would not be possible. This would reduce the overall manufacturing throughput as the machine would either be used for printing or machining. However, such scenarios are less frequent and the machine is typically in the hybrid processing mode most of the time.

## 6.4 In-line monitoring and automation for smart manufacturing

Since AM processes are highly computerized and digital in nature, they can be easily integrated into smart manufacturing, i.e., Industry 4.0. Integration of sensors, automation systems, and continuous in-line monitoring of AM processes can help minimize losses in material and time while maintaining effective processing to enable lean manufacturing. Currently, for LFAM platforms, several in-line monitoring systems such as infrared (IR) imaging and laser profilometer are in use [17]. The data obtained from these monitoring systems can be used as a feedback loop system for "lights-out operation." Lights-out operation refers to operation of an equipment without any human intervention. IR cameras mounted on the printer are used for monitoring the thermal profile of the printed structure to maintain and control the substrate (previously deposited layer or bead) temperature such that it remains close to or above the glass transition temperature ($T_g$) of the material throughout the print time (Fig. 6.6). Maintaining the temperature above $T_g$ during the print helps with better interlayer adhesion, especially for large parts with long layer times [24,25]. On the other hand, depositing material while the substrate is too hot (close to the melt temperature), due to short layer time, can lead to the collapse of the printed structure. In this case, the IR camera is used to ensure the deposited layer has cooled sufficiently to the desired temperature before continuing the print. The other in-line monitoring system, laser profilometer, is used to scan the surface of the deposited beads to create height maps (over fill or under fill), which help determine whether an optimal volume of material is being deposited. Based on the data obtained from this system, deposition speed and material throughput can be adjusted for subsequent deposition. For screw speed and throughput adjustments, LFAM systems are also equipped with an automated purge and weigh system, which helps measure the mass flow rate of the material being processed at different screw speeds. The extruder deposits (purges) the material at a predefined screw speed onto a scale where the extrudate is collected and weighed. This process is repeated for different screw speeds.

Besides in-line process monitoring, the incorporation of sensors into printed structures during the print has also been possible on LFAM platforms

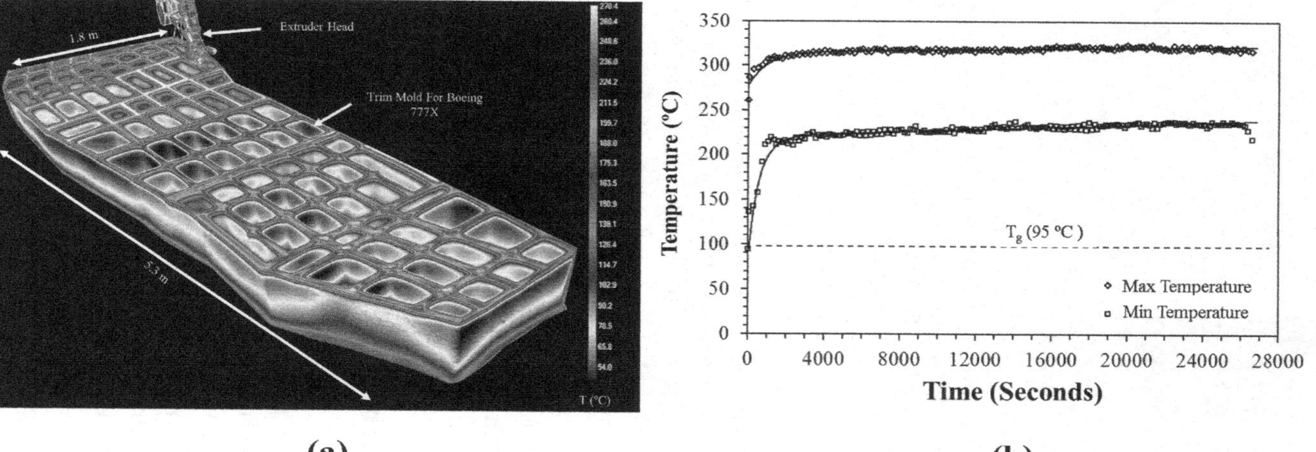


**Figure 6.6** In situ IR thermography data for LFAM parts; (A) IR thermogram for ABS/CF 20% by weight AM trim tool for Boeing 777X passenger jet (mold dimensions: 5.3 m, 1.8 m, and 1.2 m), and (B) Interface temperature during printing of polyphenylene sulfide (PPS)/CF 50% by weight. *ABS*, acrylonitrile butadiene styrene; *AM*, additive manufacturing; *CF*, carbon fiber; *IR*, infrared; *LFAM*, large format additive manufacturing. *(A) Reproduced with kind permission of the American Society for Nondestructive Testing Inc. from Reference Hassen AA, Kirka MM. Additive Manufacturing: the rise of a technology and the need for quality control and inspection techniques. Mater. Eval. 2018; 76(4):438–453; Copyright [2018]; (B) Reproduced with kind permission of the Society for the Advancement of Material and Process Engineering (SAMPE) from Reference Kunc V, Lindahl J, Dinwiddie R, Post B, Love L, Matlack M, etal Investigation of in-autoclave additive manufacturing composite tooling. In: Proc. of CAMX Conference, Anaheim, CA, USA; 2016).*

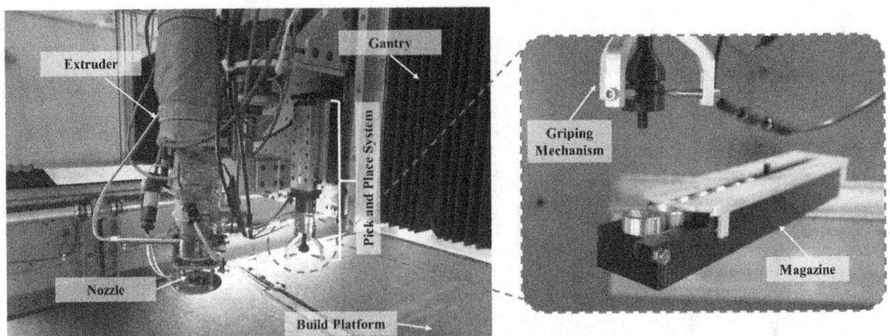


**Figure 6.7** Pick and place system installed on the BAAM system for integrating sensors or alignment fiducials into additively manufactured components during the printing process. *BAAM*, big area additive manufacturing. *(Reproduced with kind permission of Solid Freeform Fabrication (SFF) Symposium from Reference Boulger AM, Chesser PC, Post BK, Roschli AC, Hilton JS, Welcome CJ, etal Pick and place robotic actuator for big area additive manufacturing. In: Proc. of 29th Annual International Solid Freeform Fabrication Symposium, Austin, TX, USA; 2018).*

with the development of a pick and place system, demonstrated by researchers at ORNL (Fig. 6.7). The system consists of a robotic arm with a pneumatic griping mechanism and a tray (magazine) where sensors are stored and picked from. This can be used to integrate thermocouples, cartridge heaters, radiofrequency identification (RFID) tags, pressure sensors, and fiducials, for machining alignment. The primary objective of integrating a pick and place system with an LFAM platform is to digitize the manufacturing, shipping, and inventory process, especially with manufacturing technologies moving toward automation and the industrial Internet of things (IIoT).

## 6.5 Materials development

The development of materials for extrusion-based AM platforms started with FFF systems, with the most common ones being, acrylonitrile butadiene styrene (ABS), polylactic acid (PLA), polycarbonate (PC), polyphenylsulfone (PPSU), polyamides (PA), and PEI [28]. Since then, the quest for better properties and multifunctional printed components has led to the development of filament feedstock with reinforcements such as carbon fiber, metal powders such as copper and iron, nanoparticles, ceramic fillers, and so on, for enhanced mechanical properties [29–36]. The loading levels in these composite filaments mostly vary between 2% and 20% by weight. The development of single screw extrusion-based systems operating with pelletized feedstock led to faster deposition rates, along with the flexibility of using a wide range of thermoplastics and composite materials even with much higher filler loadings (up to 50%–60% by weight) [7]. In addition, switching from filament to pellets lowered the feed material cost by a factor of about 20 times in several cases [5]. Besides these advantages, such processes have also opened opportunities for printing components with blended and functionally graded properties [37–39] using LFAM systems with multimaterial capability. Some of the thermoplastic matrices developed for pellet-based feedstock include low-temperature thermoplastics such as ABS, PLA, polyethylene terephthalate (PET), elastomers such as thermoplastic polyurethane (TPU), and high temperature matrices such as PPSU, PEI, PPS, poly(ether ether ketone) (PEEK), poly(ether ketone ketone) (PEKK), and polyethersulfone (PESU). The most commonly used reinforcements for these processes are short carbon fibers and glass fibers.

Another class of pelletized material being developed for AM is bio-based thermoplastic composites, with one of the most explored matrices being bio-derived PLA. Some of the bio-derived reinforcements used in large-scale printing include bamboo fibers, poplar fibers, wood flour, flax, microcellulose, and nanocellulose [40]. Such bio-derived matrices and reinforcements serve as a greener alternative to petroleum-based materials with a reduced cost. Polymer-bonded rare earth magnets, consisting of NdFeB powder and polyamide (Nylon 12), have also been developed for applications such as electric motors, sensors, and vehicles [41]. Using AM techniques to fabricate magnets that incorporate such critical rare earth materials significantly reduces waste and provides a pathway to recycle printed parts back to feedstock pellets for reprinting. In addition, syntactic foams, and thermoplastic foams with expandable microspheres as foaming agents, have also been developed for large-scale platforms to fabricate lightweight components with densities as low as $0.2\ g/cm^3$ [42,43]. Fig. 6.8A–C represents some of the components printed on LFAM systems using different thermoplastics as single material components, multimaterial components, and permanent magnets, respectively.

On the other hand, for thermoset materials, unlike thermoplastics, printing starts with feedstock material in the form a viscoelastic liquid (resin or paste), which is then dispensed through a nozzle onto the print bed. The deposited layers are then allowed to undergo a curing process, often facilitated by heat (thermal-cure) or ultraviolet radiation (UV-cure), to obtain structures with the desired properties. For small-scale processes such

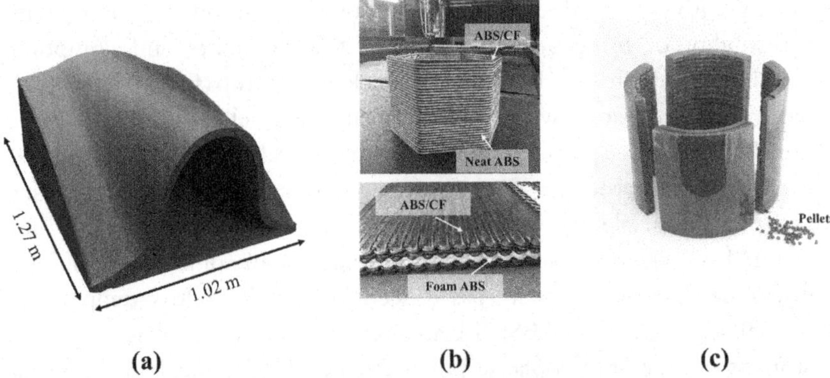


**Figure 6.8** Components manufactured using LFAM processes. (A) Autoclave AM mold fabricated using PPS/CF (50% CF by weight), (B) Multimaterial parts manufactured using ABS/CF (20% CF by weight) and neat ABS or ABS foam, and (C) Isotropic and near-net-shape neodymium–iron–boron (NdFeB)–bonded AM magnets. *ABS*, acrylonitrile butadiene styrene; *AM*, additive manufacturing; *CF*, carbon fiber; *LFAM*, large format additive manufacturing; *PPS*, polyphenylene sulfide. *(Photos courtesy of Oak Ridge National Laboratory, U.S. Dept. of Energy).*

as DW AM, epoxies have been the most extensively used resin systems. Numerous epoxy-based composites containing short fibers, silica, clays, NdFeB magnetic particles, and other functional materials have also been developed to enhance mechanical properties and obtain multifunctional parts [6,44–46]. Other thermoset matrices used in the DW process include cyanate ester resins and silicones [47–49]. For the recently developed large-scale reactive extrusion system, polyester and vinyl ester–based resin systems with peroxide catalysts have been developed [8,50].

## 6.5.1 Printability criteria

The development of a variety of new materials that can be successfully printed for structural and multifunctional applications involves a fundamental understanding of "what makes a material printable?" Although several of these materials are often processed using traditional manufacturing techniques such as injection and compression molding, extrusion, thermoforming, and so on, processing them through an extrusion AM platform, followed by deposition on to a print bed in ambient environment, requires knowledge of the relevant process–structure–property relationship for each material system. For a successful print, the material should first be dispensed out of the extruder nozzle, then form a stable deposited bead, and finally be able to form a stable structure, which makes up the desired component.

As a methodical approach toward characterizing the printability of a material, Duty et al. developed a simple, practical viscoelastic model for evaluating the printability of various polymer feedstock materials [51,52]. The model describes a fundamental set of print conditions based on certain viscoelastic and thermomechanical properties of polymers measured using small quantities of materials in a controlled test environment. The primary conditions for successful printing of a material have been classified into four different print criteria: material extrusion, bead formation (geometry), bead functionality, and component functionality, as shown in Fig. 6.9 [52].

For material extrusion, pressure-driven extrusion of the material must occur through an orifice of a given geometry at a certain specified flow rate (criteria Ia and Ib in Fig. 6.9). The pressure required by the material to flow must be below the system limit. For fiber-reinforced systems, there can also be potential fiber entanglement leading to restricted flow and clogging of the nozzle during the print. For these systems, the fiber volume fraction is also accounted for while estimating the pressure drop to achieve a desired throughput, and this value must again be lower than the maximum system pressure.

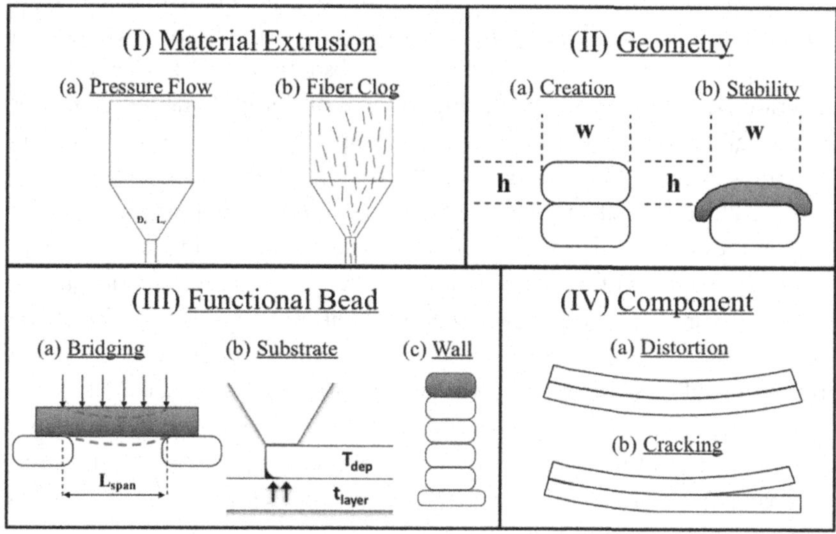


**Figure 6.9** Primary conditions for successful printing, classified into four different print criteria; (I) material extrusion, (II) bead geometry, (III) bead functionality, and (IV) component functionality. *(Reproduced with kind permission of Solid Freeform Fabrication (SFF) Symposium from Reference Duty CE, Ajinjeru C, Kishore V, Compton B, Hmeidat N, Chen X, et al. A viscoelastic model for evaluating extrusion-based print conditions. In: Proc. of the 28thAnnual International Solid Freeform Fabrication Symposium, Austin, TX, USA; 2017[52]).*

Once the material is dispensed through the nozzle, the next criterion requires the material to form a stable bead of a semirectangular geometry, having a consistent height and width (criteria IIa in Fig. 6.9). The free-standing height of the deposited bead should be at least the spacing between the substrate and the nozzle (i.e., the bead height). This free-standing height depends on melt density of the deposited material and the surface energy of the material, which is dependent upon the material in contact. If the free-standing height of a material is lower than the desired bead height, then the material will not be able to form a consistent bead, i.e., the material would flow too freely from the nozzle without achieving the desired bead height. In addition to the formation of a semirectangular bead, the deposited bead must also remain geometrically stable over time, typically for the duration of the processing time (criterion IIb in Fig. 6.9). The material should be able to support the entire weight of the bead (can be calculated as a simple hydrostatic pressure) during the print layer time and at the deposition temperature. The extent of deformation of the material subjected to this hydrostatic pressure over time is dependent on the viscoelastic characteristics of the material, ranging between that of an elastic solid and a viscous liquid.

The third criterion, bead functionality, calls for the extruded material to be able to successfully bridge an unsupported gap (often required while printing structures with sparse in-fill patterns and complex geometries with overhangs) (criterion IIIa), serves as a substrate for subsequent layers (criterion IIIb), and provides a strong foundation for building tall structures with a significant number of layers (criterion IIIc). The successful bridging of a gap is evaluated by determining the time required to achieve a certain permitted level of deflection in the bead. For substrate support functionality (IIIb), it is essential to have the total stress on the substrate material be lower than the yield stress of the extruded material at $T_g$ for thermoplastics, and in a partially cured state for thermosets. In addition, the elastic strain of the substrate should also be below a certain defined limit. For successful printing of tall structures (IIIc), the criteria are similar to that of substrate support but also account for the number of deposited layers in the structure. Overall, for bead functionality criteria, factors such as bead geometry, print conditions, and extruded material's viscoelastic and thermomechanical properties play a critical role.

The fourth criterion in the printability model describes the functionality of the entire printed component. For each substrate layer, the printed component should maintain some degree of dimensional stability and integrity (criterion IV in Fig. 6.9). Deformation and residual stresses induced in the part due to thermal contractions during printing can become more significant as the size of the printed structure increases. The coefficient of thermal expansion (CTE) of the material and the thermal profiles encountered during printing play a critical role in determining dimensional changes to the part. For printing a part without significant distortion, the printability model defines a quantity, distortion ratio, which is the ratio of structure deformation to the layer height. A distortion ratio greater than unity implies that the previously deposited layers have deflected more than the layer height, thereby physically preventing subsequent material deposition. The printability criterion for cracking (criterion IVb) is defined based on stresses at the bead interface relative to the interfacial bond strength.

## 6.5.2 Key material properties influencing printing

### *6.5.2.1 Composition and physical properties*

Factors influencing the ability to successfully print a given polymer or a composite material first begin with the feed material composition. Some of these material-related factors include molecular weight of the polymer, chemical structure, polymer architecture (linear, branched, or cross-linked), crystallization kinetics (for semicrystalline thermoplastics), cure kinetics and resin—catalyst ratio (for thermoset polymers), filler aspect ratio and loading, and presence of additives such as viscosity modifiers. These parameters influence appropriate processing conditions, melt rheological properties prior to and after deposition, bead surface quality, and thermomechanical properties of printed components [6,7,53—61]. In addition, properties such as pellet geometry and surface quality (for LFAM of thermoplastics), filament stiffness (for FFF processes), and material density are also important parameters that influence the ability of the material to feed into the extruder for continuous printing.

### *6.5.2.2 Thermophysical properties*

For thermoplastics, the first step to processing involves determining a suitable processing temperature window. For amorphous polymers, typically the processing temperature is at least 100—120°C above the $T_g$ of the material, while for semicrystalline polymers, temperature at the nozzle is set to at least 15°C above the melting temperature of the polymer, ensuring melting of crystallites existing in the feed material [53,59]. For materials wherein rheological properties such as viscosity are affected by temperature changes, processing temperatures are often varied slightly to ensure printability at the desired deposition rates. However, the upper bound of processing temperatures for thermoplastics is limited by the degradation temperature of the matrix material, i.e., the temperature at which there are significant chemical changes in the material, thereby altering melt viscosity, releasing volatiles, and so on [59]. For thermoset materials, extrusion deposition takes place at ambient temperature in most cases. In the case of thermoplastics, CTE, thermal conductivity, and heat capacity of the material play a significant role in determining the dimensional stability of the printed structure. The addition of fillers such as carbon fibers has been shown to lower the CTE of the material, which minimizes the induced strain as the printed part cools down from the deposition temperature to ambient temperature, leading to a significant reduction in warping (as shown in Fig. 6.10) [23]. This makes filled systems to be the preferred candidates for printing large structures on LFAM platforms. Also, increasing the thermal conductivity of the material reduces thermal gradients throughout the part, thereby reducing part warpage [23].

**Figure 6.10** Effect of carbon fiber in thermoplastic matrix on part distortion during LFAM process. Carbon fiber reduces the materials' CTE which minimizes the induced strain as the printed part cools down from the deposition temperature to ambient temperature. *CTE*, coefficient of thermal expansion; *LFAM*, large format additive manufacturing. *(Reproduced with kind permission of Cambridge University Press from Reference Love L, Kunc V, Rios O, Duty C, Elliott A, Post B, et al. The importance of carbon fiber to polymer additive manufacturing. J. Mater. Res. 2014;29(17):1893–1898[23]).*

### 6.5.2.3 Rheological properties

The rheological properties of polymers in shear and extension, such as the viscosity, elastic and viscous moduli, yield stress, and so on, play an important role in determining the success of a print. In the case of LFAM of thermoplastics, as the material passes through the extruder and gets deposited, it experiences different shear rates in various regions of the process ranging from about 5000 $s^{-1}$ in the screw and 10–100 $s^{-1}$ in the nozzle, to much lower rates ($<0.1\ s^{-1}$) for postdeposition conditions. Understanding the effect of different process parameters such as shear rates (determined by throughput, nozzle geometry), temperature, and processing environment, on the viscoelastic properties of the chosen materials, helps determine suitable processing conditions, optimize the process and design, and ease troubleshooting.

For example, fiber-filled systems typically show much higher viscosities (even up to an order of magnitude in some cases) when compared with their unreinforced counterparts, as observed for unreinforced and carbon fiber–reinforced PEKK in Fig. 6.11A. However, the addition of fillers has also shown to increase the shear thinning behavior for several thermoplastic systems (such as PEKK in Fig. 6.11A), which is preferred for easier processability [53,59]. Hence, for a given polymer, if the viscoelastic properties tend to indicate that shear rate alters viscosity more than temperature changes (as seen for PEKK in Fig. 6.11A where the drop in viscosity is greater with increasing shear rate or frequency than an increase in temperature), this knowledge would help the operator vary screw speed or nozzle geometry for the print to obtain optimal flow, instead of varying the processing temperature. Dynamic moduli (elastic or storage modulus and viscous or loss modulus) at different shear rates and temperatures help determine printability for postdeposition scenarios, such as bead stability, substrate support, and part functionality. In addition to this, characterization of printed components using techniques such as dynamic mechanical analysis (DMA) helps with the understanding of thermomechanical performance of printed components. For example, in Fig. 6.11B, DMA analysis of 3D-printed ABS with 20% CF by weight has been tested in both parallel to (x) and perpendicular to (z) the print direction. The results indicate that the storage modulus significantly drops at temperatures above $T_g$, as expected for amorphous polymers.

For thermoset polymers, rheological properties such as viscosity are important to determine the resin pumpability, i.e., the ability to pump, extrude, and deposit the material at the desired volumetric flow rates. Fig. 6.11C shows the effect of increasing nanoclay content in an epoxy

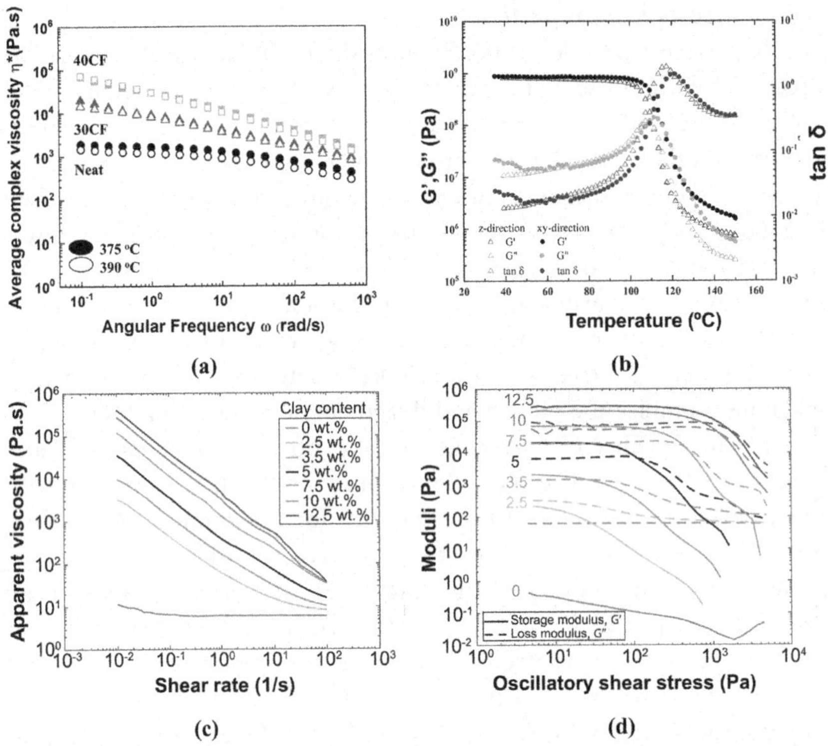


**Figure 6.11** Rheological properties indicating (A) the effect of shear rate, temperature, and carbon fiber addition on viscosity of PEKK composites, (B) the effect of temperature on dynamic moduli of printed ABS/CF composite (20% by weight CF) along x- and z-directions, (C) effect of nanoclay loading on apparent viscosity of epoxy at different shear rates, and (D) effect of nanoclay loading on dynamic moduli of epoxy across a range of oscillatory shear stress. *(A) Reproduced with kind permission of John Wiley & Sons from Reference Kishore V, Ajinjeru C, Hassen AA, Lindahl J, Kunc V, Duty C. Rheological behavior of neat and carbon fiber-reinforced poly (ether ketone ketone) for extrusion deposition additive manufacturing. Polym. Eng. Sci.' 2020; 60(5): 1066–1075; (B) Reproduced with kind permission of Sage Publishing from Reference Ajinjeru C, Kishore V, Chen X, Hershey C, Lindahl J, Kunc V, et al. Rheological survey of carbon fiber-reinforced high-temperature thermoplastics for Big Area Additive Manufacturing tooling applications. J. Thermoplast. Compos. Mater.' 2019:0892705719873941; (C) Reproduced with kind permission of Elsevier from Reference Hmeidat NS, Kemp JW, Compton BG. High-strength epoxy nanocomposites for 3D printing. Compos. Sci. Technol.'2018;160:9–20; (D) Reproduced with kind permission of Elsevier from Reference Hmeidat NS, Kemp JW, Compton BG. High-strength epoxy nanocomposites for 3D printing. Compos. Sci. Technol.'2018;160:9–20).*

on the viscosity at different shear rates. Increasing filler loading increased the apparent shear viscosity of the composite, as well as the shear thinning behavior. At a shear rate typically expected during deposition (about 50 $s^{-1}$), viscosity values ranged between 10 and 100 Pa s, indicating that a high degree of shear thinning narrowed down the range considerably [6]. For postdeposition scenarios, the desired rheological properties for bead stability and build stability are obtained either by controlling feed material properties or through cure reaction kinetics. For example, if the resin–catalyst ratio is set such that the part begins to gel or cure during the print, although curing is expected to increase the viscoelastic moduli, which is in favor of bead stability, the exothermic nature of this process would lead to heating up of the printed structure [62]. This can lead to a significant decrease in the modulus and stability of the printed part, especially when the resin is still a viscoelastic liquid during the print. Therefore, rheological properties of thermosets are controlled by a combination of modifications to the feed material, such as adding fillers, as well as carefully controlling cure kinetics. Fig. 6.11D shows the effect of fillers on the dynamic moduli of epoxy nanoclay composites across a range of shear stress. At low filler loadings, loss modulus is greater than storage modulus, indicating that the material predominantly behaves as a viscous fluid. As the nanoclay content increases, the material starts to display more solid-like behavior. Yield stress (shear) values are also seen to increase with increasing clay content [6].

### 6.5.3 Material-related challenges

#### *6.5.3.1 Porosity*

Although AM enables property optimization of printed structures through design, one of the fundamental material challenges that limits these structures from attaining properties close to their molded counterparts is the intrinsic porosity. Porosity in printed parts can be of two types: voids within a bead, or intrabead porosity (microvoids), and voids between the printed beads, or interbead porosity (macrovoids). Optical microscopy or computed tomography (CT) X-ray scans are used to map and characterize porosity in printed components, as shown in Fig. 6.12. The images can further be analyzed to provide statistical information about the shape and size distribution of the pores in printed parts. The levels of porosity can be more magnified for components printed on large-scale platforms.

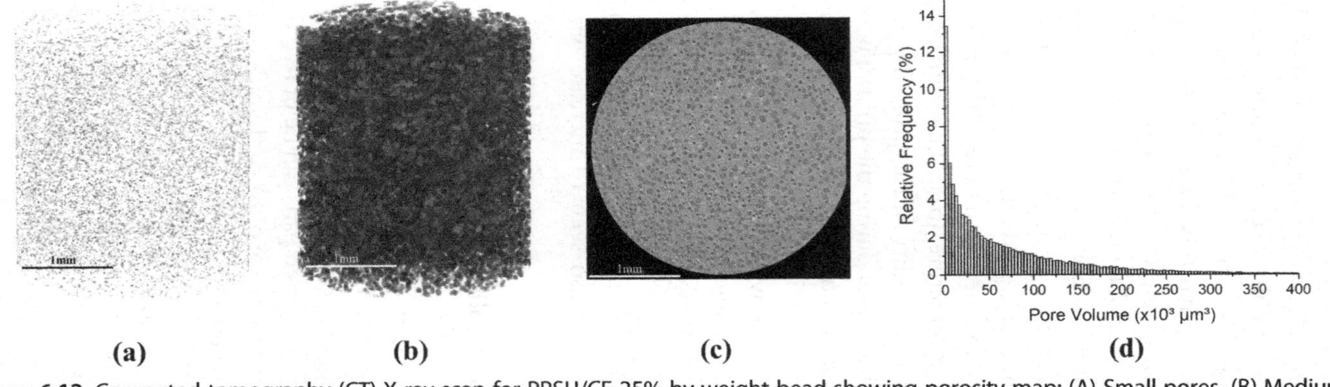


**Figure 6.12** Computed tomography (CT) X-ray scan for PPSU/CF 25% by weight bead showing porosity map: (A) Small pores, (B) Medium and large pores, (C) Top-down view cross section for a scanned CT slice, and (D) Relative frequency of occurrence of the pore volume showing the size distribution of the pores in the bead. *CF*, carbon fiber; *CT*, computed tomography; *PPSU*, polyphenylsulfone. *(Data courtesy of Oak Ridge National Laboratory, U.S. Dept. of Energy).*

For FFF parts made from short fiber-reinforced systems, sources of porosity could be factors such as interbead voids, cavitation, and fiber pull-out [33,63]. All these factors are heavily dependent on interactions between the material and processing. Ning et al. found that initial addition of carbon fibers up to 3 wt.% in ABS matrix decreased overall porosity from about 2.5% to less than 1%. However, samples with 10 wt.% CF showed the highest mean value of porosity, about 9%, which influenced tensile properties as well [33]. Tekinalp et al. reported that for fiber-reinforced ABS, intrabead voids contributed more significantly to porosity than interbead voids. Voids within the beads potentially created stress concentration regions, leading to reduced mechanical performance. Although interbead voids were also present in the samples, they were all aligned in the loading direction and were not expected to significantly affect the mechanical properties. Also, increasing CF loading led to better packing of the deposited beads, thereby creating smaller voids between them (interbead). On the other hand, an increase in the number of fiber ends caused more voids within the bead itself [36].

For parts printed on thermoplastic LFAM systems, intrabead porosity can arise due to factors such as air entrapped in pellets during the compounding process, entrapped air in the feed section of the screw, and material shrinkage upon solidification. In some cases, polymer degradation and off-gassing can also lead to void formation within a bead. Interbead porosity levels depend on the contact area between the printed beads, as well as the postdeposition thermal profile of the part. The primary bonding mechanism for thermoplastics is thermal fusion and polymer interdiffusion, both of which are favored when the surface temperature of the substrate bead (previously deposited bead) is well above the material's $T_g$ [64,65]. In general, for LFAM, fiber-reinforced materials have shown higher degrees of both intrabead and interbead porosity when compared with their unreinforced counterparts [5]. Mechanisms such as tamping, as discussed in Section 6.2.2, have shown to improve the area of contact between the beads. Currently, there are only a few reported works on porosity studies for parts printed on large-scale platforms, and therefore, this remains an active topic of current research.

#### *6.5.3.2 Anisotropic properties*

Additively manufactured parts often tend to have different mechanical and thermal properties along the print direction (or x-direction) and across successive layers (or z-direction), as shown in Fig. 6.13. This effect is more

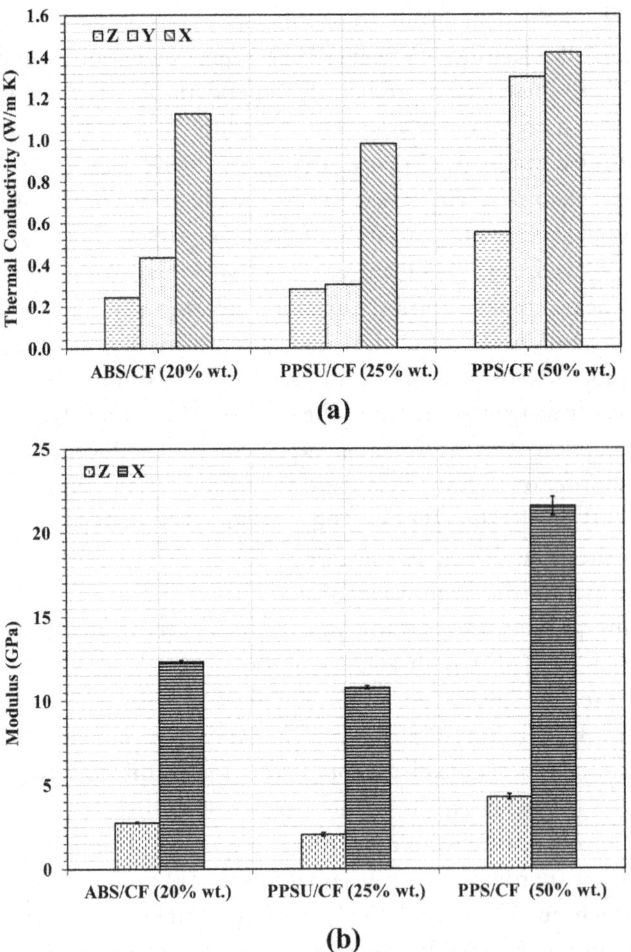


**Figure 6.13** Anisotropic properties in carbon fiber–reinforced thermoplastics printed on LFAM system. (A) Thermal conductivity (at room temperature) along the three print axes, and (B) Tensile moduli measured along directions parallel to the print (x-direction) and perpendicular to the print (z-direction). *LFAM*, large format additive manufacturing. *(Data courtesy of Oak Ridge National Laboratory, U.S. Dept. of Energy).*

predominant in fiber-filled systems wherein the fibers tend to align along the print direction. Anisotropy in properties is primarily influenced by fiber content and orientation, and bonding between the printed layers. Fibers are highly aligned in the flow direction near the nozzle wall, but this alignment is gradually disrupted while approaching the bead core (i.e., bead center). This is attributed to the shear flow in a nozzle, where shear flow is dominant near the nozzle walls, but the shear stress gradually decreases such that the flow becomes extension dominated at the center of the bead. Smaller nozzles result in a higher fiber orientation in the deposition direction, thereby leading to more anisotropic properties in the final printed structure.

In thermoplastics, for a good bond to be established between the printed beads, the bead surface should be at a temperature above the material's glass transition temperature long enough to allow for intermolecular diffusion across the interface, thereby strengthening the bond [64,65]. This depends upon the material's thermal properties and the cooling characteristics of the build as each layer gets deposited. Several physical processes and material modification techniques have been explored to reduce the mechanical anisotropy of printed components. Some of the methods developed for FFF-based systems include exposing printed components to ionizing radiation to create cross-links between the printed beads, using microwaves to locally heat composite materials, using a "z-pinning" approach to deposit material into intentionally aligned voids across multiple layers, developing a core—shell structure with different peak crystallization temperatures for semi—crystalline polymer filaments, and creating bimodal blends using low-molecular-weight additives [66—71]. For large-scale systems, process-based approaches such as infrared preheating of the substrate bead prior to depositing the next bead have been explored [24,25]. For thermosets, the source of anisotropic properties is primarily from aspect ratio and orientation of the fillers. Since bonding between the printed layers can take place by curing during the print or postprocess curing, anisotropy is not significantly affected by interfacial bonding, unlike thermoplastics.

#### *6.5.3.3 Part distortion and cracking*

The other notable material-related challenge for extrusion AM parts printed in ambient conditions is part distortion and cracking. In thermoplastics, the printed part undergoes repetitive heating and cooling as the hot polymer melt gets deposited onto a relatively cold substrate layer. This causes a residual stress build-up during deposition, resulting in distortion, or warpage and microcracking of the printed part [72]. In addition, for semicrystalline polymers, effects of nonisothermal crystallization of the deposited beads also add to the stress build-up. On the other hand, thermoset polymers do experience shrinkage during the curing process. However, the differential shrinkage of individual layers can be limited through material modification and the control of curing process for cases where parts begin to cure during the print. In other cases where the part is cured separately after printing, this effect of differential shrinkage would not be a significant issue.

## 6.6 Current applications and path forward

Applications of polymers and composite parts fabricated using extrusion AM processes range from prototyping for the architecture, medical, art, and automotive industries, to fabrication of large structures, functional molds, and dies. In medicine, this technology has been adopted for dental implants, prosthetics, customized fixtures, and surgical tools. FFF parts are also widely being used in consumer products such as customized footwear, electronics, accessories, and sporting goods. LFAM-printed thermoplastic and composite components have been used for demonstrating 3D printing of large-scale structures such as a full-sized vehicle (Fig. 6.14B), in integrated energy endeavors such as project Additive Manufacturing Integrated Energy (AMIE), aerospace tooling, wind turbine blades, and so on [73–75]. LFAM systems have been used to build architectural structures such as the 2017 Design Miami pavilion, which was built using about 4535 kg of bio-derived PLA reinforced with bamboo [40]. Besides demonstrations, functional applications of composite structures printed on LFAM have been successful in tooling. For example, Boeing's 777X trim and drill tool for the aircraft's composite wing skin was successfully printed using an LFAM platform (Fig. 6.14D) [76]. Autoclave tooling using CF reinforced composites of high-temperature materials such as PPS and PPSU was successfully fabricated and tested in a production autoclave [4,26]. In 2018, LFAM-printed thermoplastic molds were successfully used for precast concrete panels that covered the textured façade of a 42-story building in New York [77]. LFAM-printed structures using thermoset materials can also be used in composite tooling, ranging from out-of-autoclave processes such as vacuum-assisted resin transfer molding (VARTM) and hand lay-up molds, to autoclave processing. Fig. 6.14 shows some of the applications of parts printed on LFAM systems.

Although the applications of extrusion AM of polymeric composites are constantly evolving, it is evident that this technology is uniquely well suited for low volume production of highly customized and complex composite components. Growth in manufacturing is primarily being driven by product customization in the automotive and aerospace sectors, as well as an increase in the consumption of consumer electronics and renewable energy

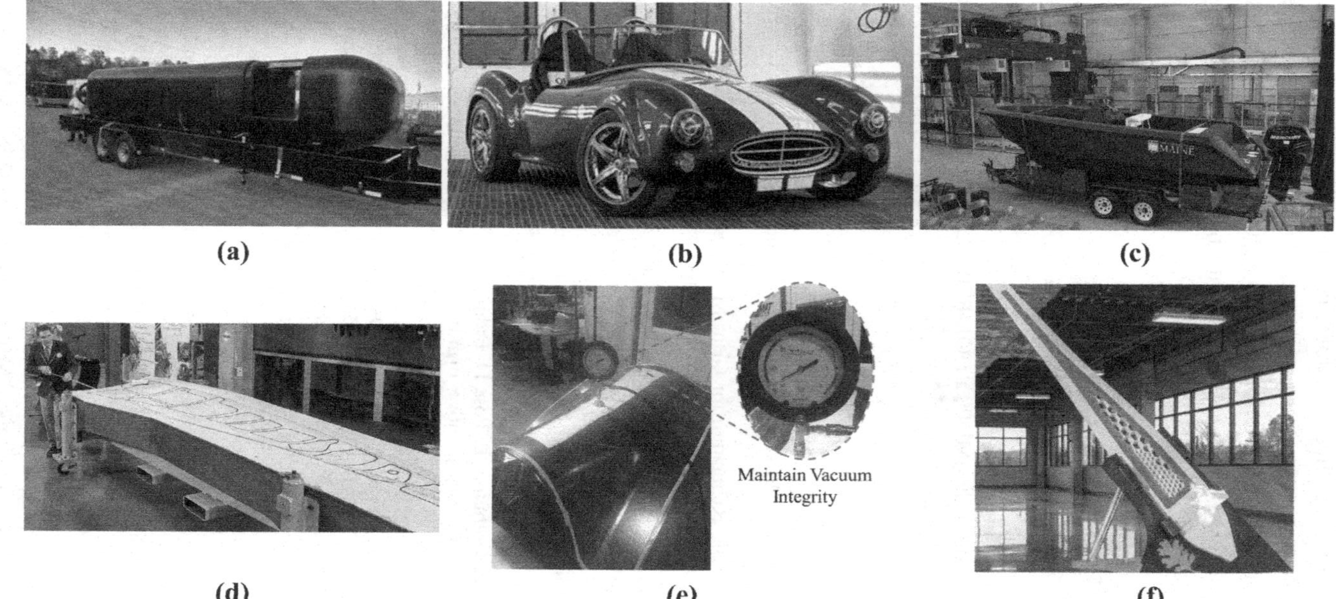


**Figure 6.14** Applications of structures printed on LFAM systems, (A) 9.14-m-long proof-of-concept submersible hull 3D-printed using carbon fiber composite material, (B) 3D-printed Shelby Cobra manufactured using ABS/CF thermoplastic composite, (C) 3D-printed boat by University of Maine (7.62 m long, weighing 2.2 tons), (D) AM trim and drill tool for Boeing 777X passenger jet winglet (Mold dimensions: 5.3 m, 1.8 m, and 1.2 m), (E) Autoclave AM mold printed using PPS/CF, and (F) A demonstration of wind turbine blade with 3D-printed PLA foam core having a property optimized graded density honeycomb structure. *ABS*, acrylonitrile butadiene styrene; *AM*, additive manufacturing; *CF*, carbon fiber; *LFAM*, large format additive manufacturing; *PLA*, polylactic acid; *PPS*, polyphenylene sulfide. *(A), (B), (D), and (E) (F) Photo courtesy of Oak Ridge National Laboratory, U.S. Dept. of Energy; (C) Photo courtesy of University of Maine, United States).*

systems. Tooling, encompassing molds, dies, and fixtures, forms the basis for the manufacture of high-quality, durable components. For material extrusion AM technology, composite tooling remains to be an attractive area, wherein both cost and lead time for tool fabrication can be reduced by an order of magnitude by switching from traditional composite tooling techniques to AM [3,4,26,78].

## Acknowledgments

Resources used by the authors for preparing this chapter were supported by the U.S. Department of Energy, Office of Energy Efficiency and Renewable Energy, Advanced Manufacturing Office, under contract DE-AC05-00OR22725 with UT-Battelle, LLC, and Manufacturing Demonstration Facility (MDF), a DOE-EERE User Facility at Oak Ridge National Laboratory.

## References

[1] ISO/ASTM52900-15, Standard Terminology for Additive Manufacturing – General Principles–Terminology, ASTM International, West Conshohocken, PA, 2015. www.astm.org.
[2] I. Wolff, Ingersoll Has Big Reputation for Giant Machines, 2018 accessed 04.15.20, https://www.sme.org/technologies/articles/2018/october/ingersoll-has-big-reputation-for-giant-machines/.
[3] A.A. Hassen, R. Springfield, J. Lindah, B.K. Post, L.J. Love, C. Duty, et al., The durability of large-scale additive manufacturing composite molds, in: Proc. 2016 CAMX Conference, Anaheim, CA, USA, 2016.
[4] V. Kunc, A.A. Hassen, J. Lindahl, S. Kim, B. Post, L. Love, Large scale additively manufactured tooling for composites, in: Proc. 2017 International SAMPE Symposium and Exhibition, Japan, 2017.
[5] C.E. Duty, V. Kunc, B. Compton, B. Post, D. Erdman, R. Smith, et al., Structure and mechanical behavior of big area additive manufacturing (BAAM) materials, Rapid Prototyp. J. 23 (1) (2017).
[6] N.S. Hmeidat, J.W. Kemp, B.G. Compton, High-strength epoxy nanocomposites for 3D printing, Compos. Sci. Technol. 160 (2018) 9–20.
[7] A. Hassen, J. Lindahl, X. Chen, B. Post, L. Love, V. Kunc, Additive manufacturing of composite tooling using high temperature thermoplastic materials, in: Proc. 2016 SAMPE Conference and Exhibition, Long Beach, CA, USA, 2016.
[8] J. Lindahl, A. Hassen, S. Romberg, B. Hedger, P. Hedger Jr., M. Walch, et al., Large-scale additive manufacturing with reactive polymers, in: Proc. 2018 CAMX Conference, Dallas, TX, USA, 2018.
[9] K.M. Billah, J.L. Coronel, M.C. Halbig, R.B. Wicker, D. Espalin, Electrical and thermal characterization of 3D printed thermoplastic parts with embedded wires for high current-carrying applications, IEEE Access 7 (January 31, 2019) 18799–18810.
[10] C. Shemelya, F. Cedillos, E. Aguilera, E. Maestas, J. Ramos, D. Espalin, et al., 3D printed capacitive sensors, in: Sensors, 2013 IEEE, IEEE, November 3, 2013, pp. 1–4.
[11] K.D. Migler, R.E. Ricker, Measurement Science Roadmap for Polymer-Based Additive Manufacturing, 2016. No. Advanced Manufacturing Series (NIST AMS)-100-5.

[12] I. Campbell, O. Diegel, J. Kowen, T. Wohlers, Wohlers Report 2018: 3D Printing and Additive Manufacturing State of the Industry: Annual Worldwide Progress Report, Wohlers Associates, 2018.
[13] T. Smith, A.A. Hassen, R. Lind, J. Lindahl, P. Chesser, A. Roschli, et al., Dual material system for polymer large scale additive manufacturing, in: Proc. 2020 SAMPE Conference and Exhibition, Seattle, WA, USA, 2020.
[14] C. Ajinjeru, V. Kishore, X. Chen, C. Hershey, J. Lindahl, V. Kunc, et al., Rheological survey of carbon fiber-reinforced high-temperature thermoplastics for big area additive manufacturing tooling applications, J. Thermoplast. Compos. Mater. (2019), 0892705719873941.
[15] B.G. Compton, J.A. Lewis, 3D-printing of lightweight cellular composites, Adv. Mater. 26 (34) (2014) 5930–5935.
[16] Cincinnati Incorporated. Operation, Safety, and Maintenance Manual for Cincinnati® BAAM-100. OH, Cincinnati Incorporated, USA, 2016 accessed 04.15.20, http://wwwassets.e-ci.com/PDF/Preinstallation/Additive/EM-565-BAAM-100_Operation-Safety-and-Maintenance-Manual.pdf.
[17] A.A. Hassen, M.M. Kirka, Additive Manufacturing: the rise of a technology and the need for quality control and inspection techniques, Mater. Eval. 76 (4) (2018) 438–453.
[18] E. Barnett, C. Gosselin, Large-scale 3D printing with a cable-suspended robot, Addit. Manufact. 7 (2015) 27–44.
[19] S.M. Chamberlain, R.T. Eberheim, Cable Driven Manipulator for Additive Manufacturing, Google Patents, 2017.
[20] P. Urhal, A. Weightman, C. Diver, P. Bartolo, Robot assisted additive manufacturing: a review, Robot. Comput. Integrat. Manuf. 59 (2019) 335–345.
[21] X. Li, Q. Lian, D. Li, H. Xin, S. Jia, Development of a robotic arm based hydrogel additive manufacturing system for in-situ printing, Appl. Sci. 7 (1) (2017) 73.
[22] G.Q. Zhang, A. Spaak, C. Martinez, D.T. Lasko, B. Zhang, T.A. Fuhlbrigge, Robotic additive manufacturing process simulation-towards design and analysis with building parameter in consideration, in: 2016 IEEE International Conference on Automation Science and Engineering (CASE), IEEE, 2016.
[23] L. Love, V. Kunc, O. Rios, C. Duty, A. Elliott, B. Post, et al., The importance of carbon fiber to polymer additive manufacturing, J. Mater. Res. 29 (17) (2014) 1893–1898.
[24] V. Kishore, C. Ajinjeru, A. Nycz, B. Post, J. Lindahl, V. Kunc, et al., Infrared preheating to improve interlayer strength of big area additive manufacturing (BAAM) components, Addit. Manufact. 14 (2017) 7–12.
[25] A. Nycz, V. Kishore, J. Lindahl, C. Duty, C. Carnal, V. Kunc, Controlling substrate temperature with infrared heating to improve mechanical properties of large-scale printed parts, Addit. Manufact. 33 (2020) 101068.
[26] V. Kunc, J. Lindahl, R. Dinwiddie, B. Post, L. Love, M. Matlack, et al., Investigation of in-autoclave additive manufacturing composite tooling, in: Proc. of CAMX Conference, Anaheim, CA, USA, 2016.
[27] A.M. Boulger, P.C. Chesser, B.K. Post, A.C. Roschli, J.S. Hilton, C.J. Welcome, et al., Pick and place robotic actuator for big area additive manufacturing, in: Proc. of 29th Annual International Solid Freeform Fabrication Symposium, Austin, TX, USA, 2018.
[28] I. Gibson, D. Rosen, B. Stucker, Additive Manufacturing Technologies: 3D Printing, Rapid Prototyping, and Direct Digital Manufacturing, Springer, New York, 2014.
[29] M. Nikzad, S. Masood, I. Sbarski, Thermo-mechanical properties of a highly filled polymeric composites for fused deposition modeling, Mater. Des. 32 (6) (2011) 3448–3456.

[30] D. Roberson, C.M. Shemelya, E. MacDonald, R. Wicker, Expanding the applicability of FDM-type technologies through materials development, Rapid Prototyp. J. 21 (2) (2015) 137–143.
[31] S.J. Kalita, S. Bose, H.L. Hosick, A. Bandyopadhyay, Development of controlled porosity polymer-ceramic composite scaffolds via fused deposition modeling, Mater. Sci. Eng. C 23 (5) (2003) 611–620.
[32] S. Masood, W. Song, Development of new metal/polymer materials for rapid tooling using fused deposition modelling, Mater. Des. 25 (7) (2004) 587–594.
[33] F. Ning, W. Cong, J. Qiu, J. Wei, S. Wang, Additive manufacturing of carbon fiber reinforced thermoplastic composites using fused deposition modeling, Compos. B Eng. 80 (2015) 369–378.
[34] M. Shofner, K. Lozano, F. Rodríguez-Macías, E. Barrera, Nanofiber-reinforced polymers prepared by fused deposition modeling, J. Appl. Polym. Sci. 89 (11) (2003) 3081–3090.
[35] W. Zhong, F. Li, Z. Zhang, L. Song, Z. Li, Short fiber reinforced composites for fused deposition modeling, Mater. Sci. Eng. 301 (2) (2001) 125–130.
[36] H.L. Tekinalp, V. Kunc, G.M. Velez-Garcia, C.E. Duty, L.J. Love, A.K. Naskar, et al., Highly oriented carbon fiber–polymer composites via additive manufacturing, Compos. Sci. Technol. 105 (2014) 144–150.
[37] Z. Sudbury, C. Duty, V. Kunc, V. Kishore, C. Ajinjeru, J. Failla, et al., Characterizing material transition for functionally graded material using Big Area Additive Manufacturing, in: Proc. of the 27thAnnual International Solid Freeform Fabrication Symposium, Austin, TX, USA, 2016.
[38] J. Brackett, Y. Yan, D. Cauthen, V. Kishore, J. Lindahl, T. Smith, et al., Development of functionally graded material capabilities in large-scale extrusion deposition additive manufacturing, in: Proc. of the 30thAnnual International Solid Freeform Fabrication Symposium, Austin, TX, USA, 2019.
[39] T.Z. Sudbury, C. Ajinjeru, V. Kishore, C. Duty, P. Liu, V. Kunc, Blending of fiber reinforced materials using big area additive manufacturing (BAAM), in: Proc. 2017 SAMPE Conference and Exhibition, Seattle, WA, USA, 2017.
[40] Oak Ridge National Laboratory, Exploring Bioderived Composite Materials: ORNL 3D Prints Design Miami Pavilions, 2017 accessed 04.15.20, https://www.ornl.gov/research-highlight/exploring-bioderived-composite-materials-ornl-3d-prints-designmiami-pavilions.
[41] L. Li, B. Post, V. Kunc, A.M. Elliott, M.P. Paranthaman, Additive manufacturing of near-net-shape bonded magnets: prospects and challenges, Scripta Mater. 135 (2017) 100–104.
[42] P. Liu, J. Lindahl, A.A. Hassen, V. Kunc, Rheology of acrylonitrile butadiene styrene with hollow glass microspheres for extusion process, in: Proc. 2017 SPE Antec Conference, Anaheim, CA, USA, 2017.
[43] S. Kim, G.D. Dreifus, B.T. Beard, A. Glick, A.K. Messing, A.A. Hassen, et al., Graded infill structure of wind turbine blade accounting for internal stress in Big Area Additive Manufacturing, in: Proc. of CAMX Conference, Dallas, TX, USA, 2018.
[44] B.G. Compton, N.S. Hmeidat, R.C. Pack, M.F. Heres, J.R. Sangoro, Electrical and mechanical properties of 3D-printed graphene-reinforced epoxy, J. Miner. Met. Mater. Soc. 70 (3) (2018) 292–297.
[45] B.G. Compton, J.W. Kemp, T.V. Novikov, R.C. Pack, C.I. Nlebedim, C.E. Duty, et al., Direct-write 3D printing of NdFeB bonded magnets, Mater. Manuf. Process. 33 (1) (2018) 109–113.
[46] J.P. Lewicki, J.N. Rodriguez, C. Zhu, M.A. Worsley, A.S. Wu, Y. Kanarska, et al., 3D-printing of meso-structurally ordered carbon fiber/polymer composites with unprecedented orthotropic physical properties, Sci. Rep. 7 (1) (2017) 1–14.

[47] S. Chandrasekaran, E.B. Duoss, M.A. Worsley, Lewicki, 3D printing of high performance cyanate ester thermoset polymers, J. Mater. Chem. 6 (3) (2018) 853–858.
[48] A.S. Wu, I.V.W. Small, T.M. Bryson, E. Cheng, T.R. Metz, S.E. Schulze, et al., 3D printed silicones with shape memory, Sci. Rep. 7 (1) (2017) 1–6.
[49] L-y Zhou, Q. Gao, J-z Fu, Q-y Chen, J-p Zhu, Y. Sun, et al., Multimaterial 3D printing of highly stretchable silicone elastomers, ACS Appl. Mater. Interface. 11 (26) (2019) 23573–23583.
[50] C. Hershey, J. Lindahl, S. Romberg, A. Roschli, B. Hedger, M. Kastura, et al., Large-Scale reactive extrusion deposition of sparse infill structures with solid perimeters, J. Adv. Mater. (2019) 76–82.
[51] C. Duty, C. Ajinjeru, V. Kishore, B. Compton, N. Hmeidat, X. Chen, et al., What makes a material printable? A viscoelastic model for extrusion-based 3D printing of polymers, J. Manuf. Process. 35 (2018) 526–537.
[52] C.E. Duty, C. Ajinjeru, V. Kishore, B. Compton, N. Hmeidat, X. Chen, et al., A viscoelastic model for evaluating extrusion-based print conditions, in: Proc. of the 28thAnnual International Solid Freeform Fabrication Symposium, Austin, TX, USA, 2017.
[53] C. Ajinjeru, V. Kishore, P. Liu, J. Lindahl, A.A. Hassen, V. Kunc, et al., Determination of melt processing conditions for high performance amorphous thermoplastics for large format additive manufacturing, Addit. Manufact. 21 (2018) 125–132.
[54] C. Ajinjeru, V. Kishore, J. Lindahl, Z. Sudbury, A.A. Hassen, B. Post, et al., The influence of dynamic rheological properties on carbon fiber reinforced polyetherimide for large scale extrusion-based additive manufacturing, Int. J. Adv. Manuf. Technol. 99 (1–4) (October 1, 2018) 411–418.
[55] V. Kishore, Melt Processability and Post-processing Treatment of High Temperature Semi-crystalline Thermoplastics for Extrusion Deposition Additive Manufacturing [dissertation on the Internet], Knoxville TN: University of Tennessee Knoxville, 2018 [cited 2020 Apr 15]. Available from: https://trace.tennessee.edu/cgi/viewcontent.cgi?article=6775&context=utk_graddiss.
[56] V. Kishore, C. Ajinjeru, A.A. Hassen, J. Lindahl, P. Liu, V. Kunc, et al., Rheological characteristics of fiber reinforced Poly(ether ketone ketone) (PEKK) for melt extrusion additive manufacturing, in: Proc. 2017 SAMPE Conference and Exhibition, Seattle, WA, USA, 2017.
[57] C. Ajinjeru, V. Kishore, X. Chen, J. Lindahl, T.Z. Sudbury, A.A. Hassen, et al., The influence of rheology on melt processing conditions of amorphous thermoplastics for Big Area Additive Manufacturing (BAAM), in: Proc. of the 27thAnnual International Solid Freeform Fabrication Symposium, Austin, TX, USA, 2016.
[58] C. Ajinjeru, V. Kishore, Z. Sudbury, C. Duty, J. Lindahl, A.A. Hassen, et al., The influence of rheology on melt processing conditions of carbon fiber reinforced polyetherimide for Big Area Additive Manufacturing, in: Proc. 2017 SAMPE Conference and Exhibition, Seattle, WA, USA, 2017.
[59] V. Kishore, C. Ajinjeru, A.A. Hassen, J. Lindahl, V. Kunc, C. Duty, Rheological behavior of neat and carbon fiber-reinforced poly (ether ketone ketone) for extrusion deposition additive manufacturing, Polym. Eng. Sci. 60 (5) (2020) 1066–1075.
[60] J. Lindahl, C. Hershey, G. Gladysz, V. Mishra, K. Shah, V. Kunc, Extrusion deposition additive manufacturing utilizing high glass transition temperature latent cured epoxy systems, in: Proc. 2019 SAMPE Conference and Exhibition, Charlotte, NC, USA, 2019.
[61] M.Q. Ansari, M.J. Bortner, D.G. Baird, Generation of polyphenylene sulfide reinforced with a thermotropic liquid crystalline polymer for application in fused filament fabrication, Addit. Manufact. 29 (2019) 100814.

[62] S. Romberg, C. Hershey, J. Lindahl, W. Carter, B.G. Compton, V. Kunc, Large-scale additive manufacturing of highly exothermic reactive polymer systems, in: Proc. 2019 SAMPE Conference and Exhibition, Charlotte, NC, USA, 2019.
[63] K.R. Hart, E.D. Wetzel, Fracture behavior of additively manufactured acrylonitrile butadiene styrene (ABS) materials, Eng. Fract. Mech. 177 (2017) 1–3.
[64] Q. Sun, G. Rizvi, C. Bellehumeur, P. Gu, Effect of processing conditions on the bonding quality of FDM polymer filaments, Rapid Prototyp. J. 14 (2) (2008) 72–80.
[65] N. Turner B, R. Strong, A. Gold S, A review of melt extrusion additive manufacturing processes: I. Process design and modeling, Rapid Prototyp. J. 20 (3) (2014) 192–204.
[66] S. Shaffer, K. Yang, J. Vargas, M.A. Di Prima, W. Voit, On reducing anisotropy in 3D printed polymers via ionizing radiation, Polymer 55 (23) (2014) 5969–5979.
[67] C.B. Sweeney, B.A. Lackey, M.J. Pospisil, T.C. Achee, V.K. Hicks, A.G. Moran, et al., Welding of 3D-printed carbon nanotube–polymer composites by locally induced microwave heating, Sci. Adv. 3 (6) (2017) e1700262.
[68] C. Duty, J. Failla, S. Kim, T. Smith, J. Lindahl, V. Kunc, Z-Pinning approach for 3D printing mechanically isotropic materials, Addit. Manufact. 27 (2019) 175–184.
[69] J.K. Mikulak, C.R. Deckard, R.L. Zinniel, Core-shell Consumable Materials for Use in Extrusion-Based Additive Manufacturing Systems, Google Patents, 2012.
[70] N.P. Levenhagen, M.D. Dadmun, Bimodal molecular weight samples improve the isotropy of 3D printed polymeric samples, Polymer 122 (2017) 232–241.
[71] N.P. Levenhagen, M.D. Dadmun, Interlayer diffusion of surface segregating additives to improve the isotropy of fused deposition modeling products, Polymer 152 (September 12, 2018) 35–41.
[72] M.R. Talagani, S. DorMohammadi, R. Dutton, C. Godines, H. Baid, F. Abdi, et al., Numerical simulation of big area additive manufacturing (3D printing) of a full size car, SAMPE J. 51 (4) (2015) 27–36.
[73] Oak Ridge National Laboratory. 3D printed Shelby Cobra, https://www.ornl.gov/content/3d-printed-shelby-cobra [accessed 04.15.20].
[74] Oak Ridge National Laboratory, ORNL Integrated Energy Demo Connects 3D-Printed Building, Vehicle, 2015 accessed 04.15.20, https://www.ornl.gov/news/ornl-integrated-energy-demo-connects-3d-printed-building-vehicle.
[75] Oak Ridge National Laboratory, 3D Printed Tooling Holds Promise for Appliance Manufacturing, 2017 accessed 04.15.20, https://www.ornl.gov/blog/3d-printed-tooling-holds-promise-appliance-manufacturing.
[76] Oak Ridge National Laboratory, 3D Printed Tool for Building Aircraft Achieves Guinness World Records Title, 2016 accessed 04.15.20, https://www.ornl.gov/news/3d-printed-tool-building-aircraft-achieves-guinness-world-records-title.
[77] Oak Ridge National Laboratory, 3D Printing Shapes Building Industry, Creates Rapid Construction Potential, 2019 accessed 04.15.20, https://www.ornl.gov/news/3d-printing-shapes-building-industry-creates-rapid-construction-potential-1.
[78] A.A. Hassen, M. Noakes, P. Nandwana, S. Kim, V. Kunc, U. Vaidya, et al., Scaling up metal additive manufacturing process to fabricate molds for composite manufacturing, Addit. Manufact. 32 (March 1, 2020) 101093.

CHAPTER 7

# Introduction to fused deposition modeling

**Przemysław Siemiński**
Academy of Fine Arts in Warsaw, Faculty of Design and Warsaw University of Technology, The Faculty of Automotive and Construction Machinery Engineering, Warsaw, Poland

## 7.1 Historical outline and used labels

According to the 3D Hubs company report for 2020 [1], the most popular additive manufacturing method worldwide is a process based on the melting, extrusion, and deposition of thermoplastic polymers. This method is known as **FDM** (fused deposition modeling), although some companies and authors of publications call it differently—a matter discussed further in this chapter.

The FDM 3D printing method was submitted as a patent application on September 30, 1989 in the United States by the inventor S. Scott Crump, and the patent was granted on July 9, 1992, with the number 5121329 [2]. In the patent, there was described a numerically controlled device which melts a thermoplastic polymer and deposits it in the form of fibers in layers on the surface of a table (Fig. 7.1). A detailed description of the technology is provided in one of the following subchapters.

The inventor of the FDM technology, S. Scott Crump, along with his wife Lisa, established the Stratasys company in 1989 [3]. The company was the first to sell 3D printers using the FDM process. It has introduced to the market a number of devices with interesting and innovative designs. Patent protection for FDM expired in 2009 and since then, other companies started to sell their own 3D printers using the FDM process.

The "FDM" acronym was reserved for the Stratasys company in the United States (in the "United States Patent and Trademark Office" under No. 1663961 [4]) as its trademark. Therefore, other manufacturers of 3D printers cannot use it for their devices. The Stratasys company used the FDM abbreviation to denote their early device models (and reserved the right to their names); for example, the series of office (so-called desktop) printers were designated as "FDM 2000/3000/8000" (Fig. 7.2), while the industrial devices were designated as "FDM Vantage" or "FDM Maxum"

*Additive Manufacturing*
ISBN 978-0-12-818411-0
https://doi.org/10.1016/B978-0-12-818411-0.00008-2

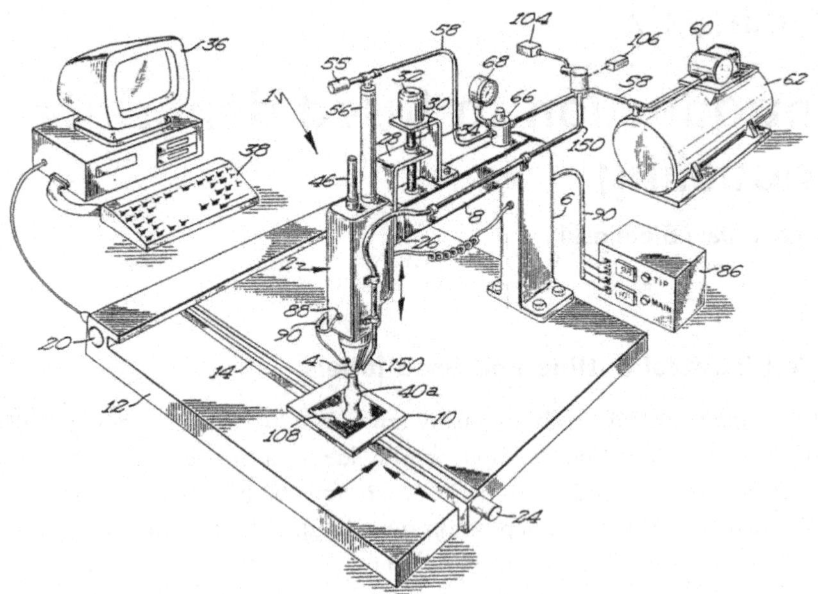

**Figure 7.1** Drawing from the description of the patent No. 5121329, showing a schematic representation of a numerically controlled device, the head of which melts and dispenses thermoplastic fibers on a base, building a three-dimensional, physical object [2].

(Fig. 7.3). For this reason, other manufacturers of FDM 3D printers typically use different ways to denote this 3D printing method, even though the underlying principle is the same. For example, the America company "3D Systems", known for introducing the 3D printing stereolithography method, uses the acronym **PJP** (plastic jet printing) for their thermoplastic extruding 3D printers, while the Polish Zortrax company uses **LPD** (layer plastic deposition), and the largest Chinese manufacturer of 3D printers, Tiertime, uses the acronym **MEM** (melted extrusion modeling).

However, a large number of other 3D printer manufacturers using the FDM additive technology describe it using the acronym **FFF** (fused filament fabrication). This solution is particularly prevalent among companies using design methods derived from the RepRap project (described in the following subchapter). Such companies are, for example, American companies widely known in the 3D printing market, such as "Ulitmaker", "Makerbot" (recently acquired by Stratasys), or "Markforged"; the Chinese "Flashforge"; Taiwanese "XYZPRINTING"; Spanish "BCN3D"; Dutch "Builder 3D Printers"; and Czech "Prusa." However, the following chapter will most often refer to Stratasys 3D

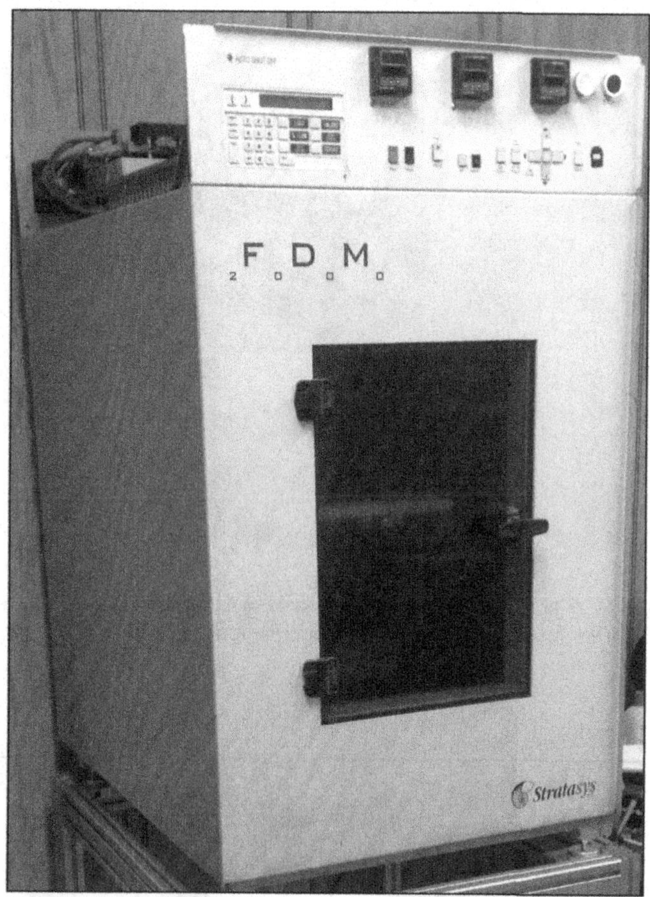


**Figure 7.2** A view of the older models of Stratasys FDM 3D printers: "FDM 2000" [5]. *FDM*, fused deposition modeling.

printers, as the company has significantly developed the FDM technology since its first patent and has introduced a number of rapid prototyping device models to the market—a number of them with innovative design solutions—and has set a standard in the FDM 3D printer construction.

An example of the currently produced and sold Stratasys FDM 3D printers is the F123 series and the Fortus series. The F123 series consists of machines for office use (desktop devices), designed for prototyping and small-scale production of parts. They are dedicated for design studios and agencies, educational centers, and small tool shops. The F123 series consists of several device models: F120, F170, F270, and F370, with the F120 (Fig. 7.4A) being an office (desktop) device, while the other models having

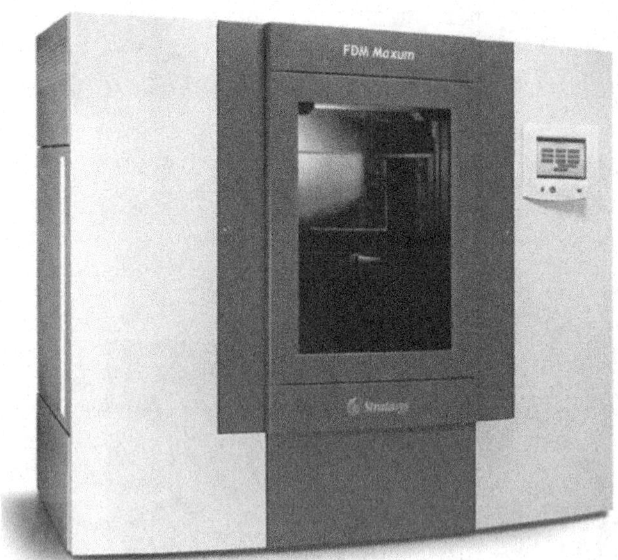


**Figure 7.3** A view of the older industrial models of Stratasys FDM 3D printers: "FDM Maxum" [3]. *FDM*, fused deposition modeling. *(Reproduced with kind permission from Stratasys.)*

**Figure 7.4** A view of the currently produced Stratasys FDM 3D printer models: F120 (A) [3] and multimaterial F370 (B) [3]. *FDM*, fused deposition modeling. *(Reproduced with kind permission from Stratasys.)*

a) 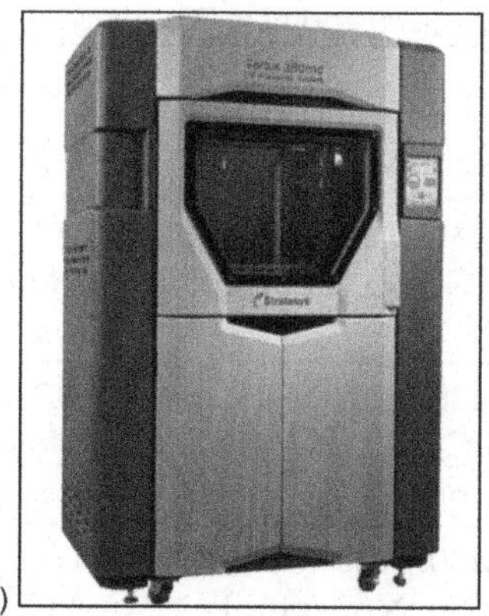
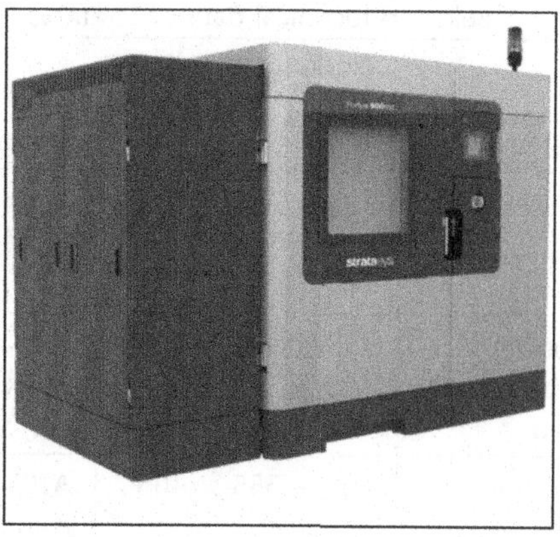 b)

**Figure 7.5** A view of the currently produced Stratasys FDM 3D printer models: "Fortus 380mc" model (A) [3] and "F900" the largest model [3]. *FDM,* fused deposition modeling. *(Reproduced with kind permission from Stratasys.)*

larger dimensions, e.g., model F370 (Fig. 7.4B). Industrial devices offered by the Stratasys company are the Fortus line and the large-scale model F900 (Fig. 7.5). These are multimaterial 3D printers, dedicated to continuous part production. Hence, the Fortus series was indicated with the "mc" abbreviation—from "manufacturing center". Working chamber sizes for both FDM 3D printer groups and their respective model materials are shown in Tables 7.1 and 7.2. A more detailed description of the model materials is included in the following subchapters.

**Table 7.1** Selected technical data of Stratasys F123 Series printers [3].

| Model name | Device dimensions:H x W x D [mm] | Working chamber size X × Y × Z [mm]: | Available model materials |
|---|---|---|---|
| F120 | 889 × 870 × 721 | 254 × 254 × 254 | ABS-M30, ASA |
| F170 | 1626 × 864 × 711 | 254 × 254 × 254 | ABS-M30, PLA, ASA, TPU 92A |
| F270 | | 305 × 254 × 305 | ABS-M30, PLA, ASA, TPU 92A |
| F370 | | 355 × 254 × 355 | ABS-M30, PLA, ASA, TPU 92A, PC-ABS |
| Available layer thickness [mm] | ABS-M30: 0.127/0.178/0.254/0.330<br>ASA: 0.127/0.178/0.254/0.330<br>PC-ABS: 0.127/0.178/0.254/0.330<br>PLA: 0.254<br>TPU 92A: 0.254 | | |

Reproduced with kind permission from Stratasys.

**Table 7.2** Selected technical data of Stratasys F123 Series printers [3].

| Model name | Working chamber size X × Y × Z [mm]: | Available model materials |
|---|---|---|
| Fortus 380mc | 355 × 305 × 305 | ABS-M30, ABS-M30i, ABS-ESD7, ASA, PC-ISO, PC, PC-ABS, Nylon 12 |
| Fortus 380mc Carbon Fiber edition | 355 × 305 × 305 | ABS-M30, ABS-M30i, ABS-ESD7, ASA, PC-ISO, PC, PC-ABS, Nylon 12, Nylon 12CF |
| Fortus 450mc | 406 × 355 × 406 | ABS-M30, ABS-M30i, ABS-ESD7, Antero 800NA, ASA, PC-ISO, PC, PC-ABS, Nylon 12, Nylon 12CF, ST-130, ULTEM 9085, ULTEM 1010 |
| F900 | 914 × 610 × 914 | ABS-M30, ABS-M30i, ABS-ESD7, Antero 800NA, ASA, PC-ISO, PC, PC-ABS, PPSF, FDM Nylon 12, Nylon 12CF, Nylon 6, ST-130, ULTEM 9085, ULTEM 1010 |
| Available layer thickness [mm] | ABS-M30: 0.127/0.178/0.254/0.33<br>ABS-ESD7: 0.178/0.254<br>ABS-M30i: 0.127/0.178/0.254/0.33<br>ASA: 0.127/0.178/0.254/0.33/0.508<br>PC-ABS: 0.127/0.178/0.254/0.33<br>PC: 0.127/0.178/0.254/0.33<br>PC-ISO: 0.178/0.254/0.33<br>Antero 800NA: 0.254<br>ULTEM 1010: 0.254/0.33/0.508<br>ULTEM 9085: 0.254/0.33<br>PPSF: 0.254/0.33<br>ST-130: 0.33<br>Nylon 6: 0.254/0.33<br>Nylon 12: 0.178/0.254/0.33<br>Nylon 12CF: 0.254 | |

Reproduced with kind permission from Stratasys.

## 7.2 The RepRap project—history and models of 3D printers

The RepRap project has been particularly significant in the history of the development and popularization of 3D printing. The **RepRap** (replicating rapid prototyper) abbreviation is the name given to the idea of building a machine capable of creating its own copy. The project was initiated in 2004 by Dr. Adrian Bowyer [6,7] from the University of Bath in the United Kingdom. In 2007, within the scope of the RepRap project, a design of a 3D printer (named Darwin) using the FDM method (Figs. 7.6 and 7.7) was developed at several universities. Standard and commercially available threaded rods were used to build this prototyping machine. To connect the

**Figure 7.6** Design of the first RepRap 3D printer (Darwin) for which technical documentation was made available; visible threaded rods connected with 3D printed connectors and standardized screws [9]. *(Reproduced with kind permission from RepRap Wiki.)*

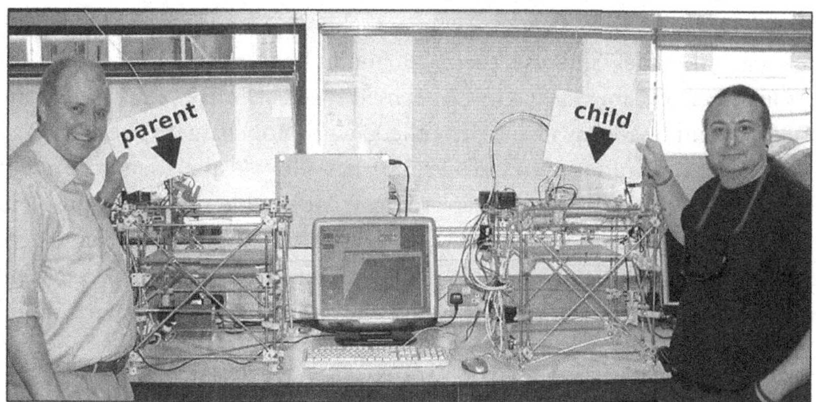


**Figure 7.7** Two Darwin 3D printer models—the device on the left (labeled "parent") was used to make the rod connectors for the device on the right (labeled as "child"); on the left side of the picture, Dr. Adrian Bowyer [7]. *(Reproduced with kind permission from RepRap Wiki.)*

rods, custom connectors were designed to be 3D-printed (e.g., using FDM) (Fig. 7.5), similar to the rest of the project-specific parts, such as bearing brackets, electronics and power supply holders, extruder gears, and so on. All other necessary components, i.e., nuts, bearings, guides, electric wires, stepper motors, and drive belts were also standard and then easily available and relatively inexpensive. This allowed for a fully functional 3D printer for domestic (amateur) use to be built at a highly affordable price (around 500 euros) [8].

To distinguish from the more advanced, but significantly more expensive FDM 3D printers, authors in the RepRap project used the acronym FFF to refer this technology. In consequence, it has been decided to use the FFF acronym in this chapter to denote the amateur and budget 3D printers (usually smaller in size and mostly intended for domestic use), which have their roots in the RepRap project. Conversely, the FDM acronym will be used as a general description of the method and for the Stratasys 3D printers. In some cases, a double abbreviation "FDM/FFF" may be used. The abbreviation is supposed to indicate that given design solutions or model materials are used in both of the aforementioned groups of devices. The designations "FDM/FFF" or "FDM (FFF)" are used by a number of authors in their publications, especially when they study FDM technology, but use machines derived from the RepRap project [10–12].

In 2009, after patent protection for the FDM 3D printing method expired, the RepRap project initiators made available the full technical documentation of their 3D printers on their website [6], along with assembly instructions and source code for electronics. The project files were based on GNU-free license [13], which allows everyone to legally download the files and use them to build their own device either for personal use or for sale. GNU-licensed projects may also be legally modified and made available to others with no charge. The RepRap website forum (online discussion group) has allowed the users to present their solutions, comment on their progress, and enable support in problem solving among the users. Over the years, it has generated an active community to form and work on the development of different FDM 3D printer designs, improving their controls, testing new model materials, and finishing methods.

Around the year 2010, one of the most popular amateur FDM 3D printer designs, taking its roots in the RepRap project, was the Mendel model (Fig. 7.8A). These 3D printer's 3D CAD files were downloaded from the RepRap Wiki website [14]. Similar to the Darwin model, its

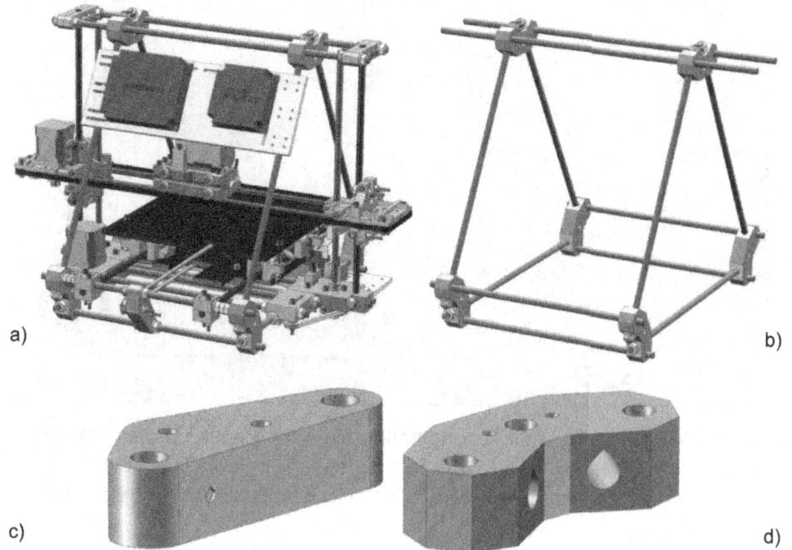


**Figure 7.8** The 3D CAD objects of the Mendel 3D printer: fully assembled (A); main threaded rod and connector frame (B); to be 3D printed: idler bracket (C) and rod connector (D); all 3D CAD models were downloaded from the RepRap Wiki website. *CAD*, computer-aided design.

design was based on threaded rods, although fewer than its predecessor (Fig. 7.8B). The newer model was also easier to assemble and less expensive. As the rod connectors were easily 3D-printed using the FDM method (Fig. 7.8C and D), every person owning a simple 3D printer could make the necessary components and sell them at cost to an acquaintance or on the Internet. This served to further popularize 3D printers among amateurs, as well as in schools and at universities.

The Mendel model was improved and modified in many ways. An example of such a modification is the MendelMax (Fig. 7.9A), the arm of which was made from aluminum profiles. This made assembly easier and served to increase the structural stiffness of the frame [15]. Another approach is visible in the design of the Wallace model (Fig. 7.9B), the design of which was significantly simplified. Such a device was considerably less expensive, though less stiff as well—so in the printing process, reduced feed velocity had to be accounted for.

Among RepRap 3D printer models popular at the time, the "Prusa Mendel" is particularly worth noting. The design of this model derives from the Mendel model. It was modified by the Czech inventor Josef Prusa. The inventor designed it as "Prusa i3" (i3 means 'iteration 3) model. This is

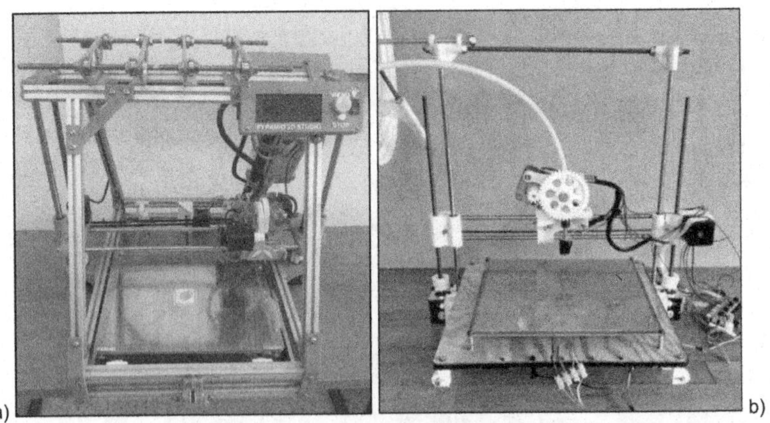

**Figure 7.9** FDM 3D printer models originating from the RepRap project: MendelMax (A) and Wallace model (B) [16]. *FDM*, fused deposition modeling. *(On figure from Pyramid 3D Studio Company; Reproduced with kind permission from RepRap Wiki.)*

still popular and sold by many companies (Fig. 7.10). It is built with several laser-cut plates or sheets, which makes the assembly even easier and contributes to an increase in structural stiffness [17]. The RepRap project has also led to the development of a number of other designs, with different kinematics and employing other additive manufacturing methods—not only FDM [18].

The fact that projects available on the RepRap websites were free (and legal) to download and modify, as well as serve to build devices both for personal use and for sale, has contributed to the rise of the number of companies in the current 3D printing industry. Some of them did not survive, but others have developed significantly. Among the best known companies are, among others, "Bits From Bytes" (acquired by "3D Systems"), Makerbot (acquired by Stratasys several years ago), and Ultimaker. An example of such a device, directly descended for the RepRap Darwin model, is the "Rapman" 3D printer, in which a part of the threaded rods constituting the frame was exchanged for laser-cut plastic panels.

Some of the first 3D printer models, e.g., by Makerbot and Ultimaker, had been made from laser-cut plywood (Fig. 7.11A), but the generations that followed were constructed with precise molds (Fig. 7.11B), plastic panels, or metal sheets. These first devices were sold as DIY (do it yourself) sets to be assembled by the user, to which they owed their lower price and even greater availability. Currently, most machines are sold already assembled and precalibrated—practically ready for work.

**Figure 7.10** One of the most popular 3D printer of RepRap machines to this day—"Prusa i3"; figures below show model names "Prime 3D" with one extruder. *(By Monkeyfab Company.)*

**Figure 7.11** Budget FDM 3D printer models: the historical "Thing-O-Matic" by Makerbotwith a laser-cut frame (A) [19] and the contemporary "Replicator" by Makerbot (B). *FDM*, fused deposition modeling.

As a result of the RepRap project, the number of companies has risen, manufacturing model material filaments as well as producing electronics and parts dedicated for 3D printers. Their increased output has also allowed for a reduction in component pricing, as well as the consumables for FDM 3D printers.

A few years after the RepRap project had ended (about 5–7 years ago), the number of 3D printing users has significantly grown in comparison with the period before that. A drop in the prices of both the devices and the prints has caused a rapid increase in the number of applications of additive manufacturing technologies, which in turn has caused a spike in media interest in the topic. At the time, a portion of journalists and futurists were inclined to believe that 3D printing will serve as the basis of a new industrial revolution, in which the people will start producing goods by and for themselves, thereby limiting mass production [20]. Presently, the interest of the popular media in 3D printing has dwindled, and 3D printer sales (especially FDM/FFF) are lower than several years ago [21]. This is explained by market saturation and a more rational approach of the users to additive technologies.

## 7.3 Model and support materials

In the FDM/FFF additive technology, the model material (also known as building material) is a thermoplastic polymer, which is a hard or an elastic plastic which, after heating to the melting point, becomes ductile and can be shaped as desired, retaining the formed shape after cooling. In the case of thermoplastics, the process of shaping objects through melting and cooling may be repeated many times, each time causing a certain degree of polymer degradation.

Most thermoplastics used as model materials in FDM/FFF 3D printers are produced from pellets used in traditional plastic processing methods, i.e., injection molding, free blowing, or thermoforming. However, for FDM/FFF 3D printing, those polymers are modified in such a way as to minimize their shrinkage or to improve layer bonding, thus making the prints less prone to delamination.

According to a report in 2018 of one of the most popular 3D printing industry services (3D Hubs), 66% of its users used FDM/FFF 3D printing, 15% used SLA + DLP, and 12% used SLS and MJF (Multi Jet Fusion) [22]. Taking into account all 3D printing methods, the most frequently chosen model materials were FDM/FFF 3D printing polymers: PLA (34%), ABS

(16%), and others more than 12%. Resins were used by 10% of users, and polyamide powders for SLS + MJF 9%.

According to a 2019 press release from the Polish chemical industry company "Grupa Azoty" [23], approximately 5000 tons of polymer filaments for FDM/FFF 3D printers were sold in Europe, while approximately 750 tons of polymer powders, mainly polyamide (PA), were sold for SLS + MJF 3D printers. Such a large difference is due to the large popularity of FDM/FFF 3D printing, which is linked to affordable 3D printer prices (particularly budget devices) and model materials. Moreover, thermoplastics used in FDM/FFF are durable and safe to use, particularly in comparison with light-cured resins used in SLA + DLP + LCD + 3SP or PolyJet Modeling or MultiJet Printing.

In the FDM method, thermoplastic model material is supplied to the printhead in the form of so-called filament, or line (sometimes called a wire) with a circular cross-sectional area (Fig. 7.12B). Currently, the most widely used FDM/FFF filament has a diameter of 1.75 mm (less often, filaments with 2.85 mm or 3 mm in diameter are also used), with diameter consistency being important for the dimensional accuracy of the prints. The most popular filaments currently in the market are those with a diameter of 1.75 mm, manufactured with a diameter tolerance of ±0.05 mm. The commercially available filament is wound on a spool (Fig. 7.12A), so it does

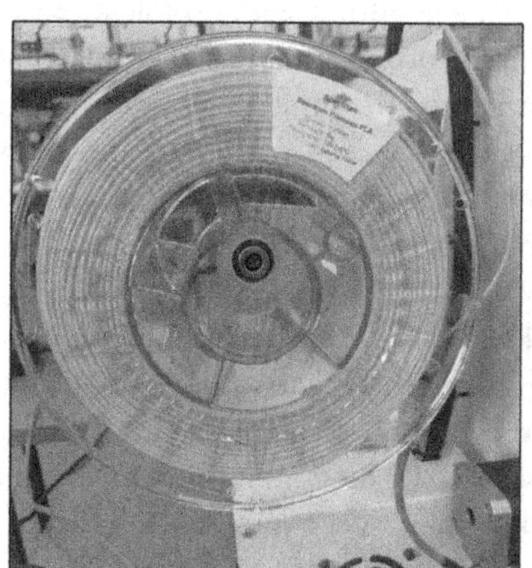

**Figure 7.12** View of the filament on the spool and placed on a 3D printer derived from the RepRap Prusa i3 project (A); a close-up on the shape of the filament and information label: manufacturer logo, filament trade name (Spectrum Premium PLA), wire diameter (1.75 mm), weight of the spool wound material (1 kg), color (Bahama Yellow [light gray in print version]), printing temperature (185–215°C) (B).

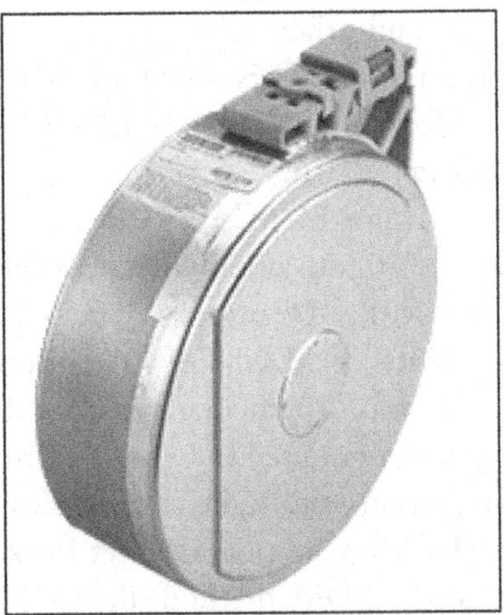

**Figure 7.13** Filament cartridges for Stratasys FDM 3D printers: for Dimension series devices (A) and for Fortus series devices (B) [3]. *FDM*, fused deposition modeling. *(Reproduced with kind permission from Stratasys.)*

not get entangled (which could hinder the printing process) and makes it easy to quickly install in the device.

The manufacturers usually deliver the filament tightly enclosed in a plastic bag, ensuring it is not susceptible to environmental humidity. Some of 3D printer manufacturers sell filaments in special plastic (Fig. 7.13A) or metal containers (cartridges) (Fig. 7.13B), sometimes with a built-in electronic memory system. The system allows for the monitoring of material consumption level, so there is no risk of starting the printing process with an insufficient amount of model material in the device; however, this prevents the operator from winding their own material. The electronic chip can also hold information on the type and color of the filament, or the recommended 3D printing parameters.

Some manufacturers of FDM/FFF 3D printers recommend brand filaments, which are usually more expensive than widely available materials. However, an advantage of these brand filaments is that the user practically does not have to do any test prints. The significant competition between filament manufacturers makes them very affordable, but also highly diverse in quality. An inadequate filament may vary in diameter or circularity, as it can have gas bubbles or contaminants causing the nozzle to clog. Therefore, for FDM/FFF 3D printing, quality of the model material used in the process and its established processing parameters are particularly important.

The FDM/FFF 3D printing method is based on the layer-on-layer deposition of plasticized polymer. This method of fabrication explains the anisotropic mechanical properties of the 3D-printed structures. During stretching, it is easier to delaminate the layers than separate the fibers from each other. Therefore, it is important for the filament manufacturing companies to provide strength parameters for prints, not for bulk material (e.g., through testing of injected samples). In the case of maximum tensile stress, it is important to provide information on the sample orientation in the working chamber. Most often, three basic sample orientations are taken into account (Fig. 7.14): flat (XY), on edge (XZ), and upright (ZX). Prints exhibit a significantly larger tensile strength along the layers (XY) than in the case of layer delamination (ZX). In the case of stretching of dumbbell specimens, the fiber structure is more favorable in the XZ direction than in XY, which makes XZ somewhat more robust. In XY, contour fibers replicate the outer shape of the dumbbell; therefore, they do not carry the full load. Unfortunately, most of filament manufacturers provide only one tensile strength value—maximum tensile stress, which is most probably the value for the orientation with the highest tensile strength, which is XZ. Of course, tensile properties also depend on printing parameters, particularly the melting point of the extruded polymer, so the filament manufacturer should provide them in their documentation.

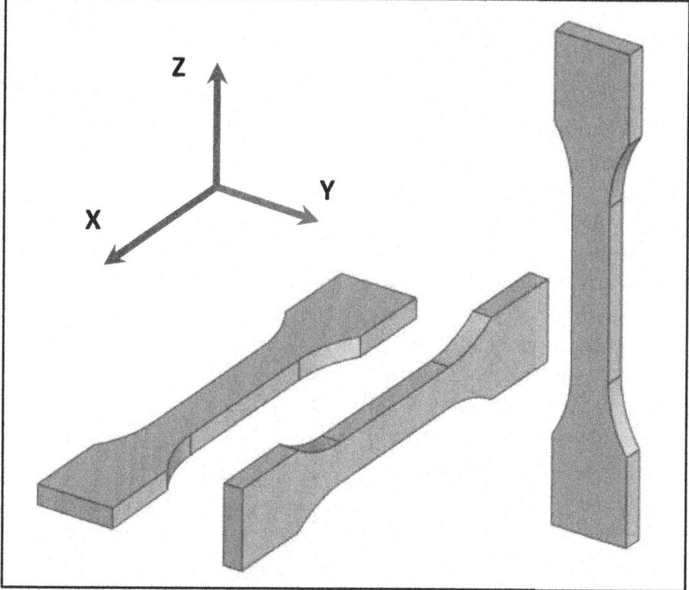


**Figure 7.14** Basic orientations of a dumbbell specimen during printing in the device working chamber: flat (XY), on edge (XZ), upright (ZX).

### 7.3.1 Model materials used in stratasys 3D printers

In industrial applications of FDM 3D printing, the most commonly used polymer is ABS and its blends, e.g., PC-ABS (abbreviations used for those polymers are shown in Table 7.3, in which characteristics, applications, and

**Table 7.3** Properties, applications, and chosen parameters (HDT stands for heat deflection temperature) of the most common model materials used in Stratasys FDM 3D printers [3].

| Model material | Properties and applications of a given model material | Tensile strength according to ASTM D638 [MPa] (axes labeled as per Fig. 7.14) | HDT at 0.46 MPa, acc. to ASTM D648 |
|---|---|---|---|
| PLA | PLA (polylactide) is a biodegradable thermoplastic obtained from renewable sources. It is the least expensive model material sold by Stratasys. As it is less resistant to temperature, it is recommended for use in models and prototypes. The material is easily applied for 3D printing purposes due to its low processing shrinkage. Approved for use in contact with food. | 45 (XZ)<br>26 (ZX) | 53°C |
| ASA | ASA (acrylonitrile styrene acrylate) is a thermoplastic with tensile strength similar to ABS, but with high resistance to UV radiation, which makes it proper for use in prints situated outdoors and directly exposed to sunlight; it is suitable for prototypes and less mechanically loaded parts, such as housings, guards, and casings. | 29 (XZ)<br>27 (ZX) | 98°C |

**Table 7.3** Properties, applications, and chosen parameters (HDT stands for heat deflection temperature) of the most common model materials used in Stratasys FDM 3D printers [3].—cont'd

| Model material | Properties and applications of a given model material | Tensile strength according to ASTM D638 [MPa] (axes labeled as per Fig. 7.14) | HDT at 0.46 MPa, acc. to ASTM D648 |
|---|---|---|---|
| ABS-M30 | ABS is a copolymer of acrylonitrile, butadiene, and styrene. ABS-M30 is an ABS modified to provide better temperature resistance and mechanical strength, particularly bonding between layers and fibers. It is suitable for the production of models and fully functional prototypes, e.g., power tool housings with elastic hooks. | 31 (XZ)<br>26 (ZX) | 96°C |
| ABS-M30i | ABS-M30i is the same polymer as ABS-M30, which is also compliant with the ISO 10993 standard, which defines biocompatibility. It is suitable for the production of medical models, surgical tools, and templates, which can be sterilized with gamma radiation or ethylene oxide. | 31 (XZ)<br>26 (ZX) | 96°C |
| ABS-ESD7 | The material dissipates static electricity, so it is dedicated to produce components used next to electronic systems, e.g., housings; the material does not attract dust, powders, and fog, which makes it get less prone to dirt. | 30 (XZ)<br>24 (ZX) | 96°C |

*Continued*

**Table 7.3** Properties, applications, and chosen parameters (HDT stands for heat deflection temperature) of the most common model materials used in Stratasys FDM 3D printers [3].—cont'd

| Model material | Properties and applications of a given model material | Tensile strength according to ASTM D638 [MPa] (axes labeled as per Fig. 7.14) | HDT at 0.46 MPa, acc. to ASTM D648 |
|---|---|---|---|
| Nylon 12 | Nylon 12 is a polyamide (PA) polymer, characterized by the highest resistance to bending, impact, fatigue, and breaking from all the materials listed here. It is also characterized in a good chemical resistance. | 32 (XZ)<br>28 (ZX) | 97°C |
| Nylon 12CF | For applications requiring high tensile strength. Nylon 12CF has the highest stiffness and tensile strength among all FDM Stratasys materials. It is used mainly for the manufacturing of production equipment and tools. | 63 (XZ)<br>29 (ZX) | 143°C |
| Nylon 6 | This material is suitable for the production of durable tools or parts and prototypes, which allow for a precise determination of the functionality of the end product. | 49 (XZ)<br>29 (ZX) | 93°C |
| PC-ABS | PC-ABS is a blend of two polymers: PC (polycarbonate) and ABS; it has good mechanical properties, particularly impact strength; it is resistant to scratching and allows for the production of esthetic prints with small details. It is particularly suitable for covers and bumpers. | 36 (XZ)<br>25 (ZX) | 110°C |

**Table 7.3** Properties, applications, and chosen parameters (HDT stands for heat deflection temperature) of the most common model materials used in Stratasys FDM 3D printers [3].—cont'd

| Model material | Properties and applications of a given model material | Tensile strength according to ASTM D638 [MPa] (axes labeled as per Fig. 7.14) | HDT at 0.46 MPa, acc. to ASTM D648 |
|---|---|---|---|
| PC-ISO | Biocompatible material (ISO 10993 USP Class VI) is suitable for medical and food grade products, can be sterilized using gamma rays and ethylene oxide. | 57 (no data) | 133°C |
| PC | PC (polycarbonate) is highly resistant to bending (nearly does not break), and abrasion; it is more thermally and mechanically resilient than ABS. It is highly resistant to scratching. | 40 (XZ)<br>30 (ZX) | 138°C |
| ST-130 | Material with the trade name of ST-130 is an auxiliary material, which enables the production of hollow composite parts. After the desired shape is printed, it is covered by the desired material, e.g., epoxy resin with fiberglass mat. The leftover ST-130 can be easily dissolved after the process. | 14 (XZ)<br>28 (ZX) | 108°C |
| ULTEM 9085 | This polymeric material is a PEI (polyetherimide), it has an FST certificate of limited flammability, smoke generation, and combustion products toxicity; it is used for the production of responsible machine parts and devices used in aviation or military vehicles; it has a high thermal resistance; ULTEM 9085 CG has a certificate of conformity for direct contact with food and the human body. | 69 (XZ)<br>33 (ZX) | 153°C |

*Continued*

**Table 7.3** Properties, applications, and chosen parameters (HDT stands for heat deflection temperature) of the most common model materials used in Stratasys FDM 3D printers [3].—cont'd

| Model material | Properties and applications of a given model material | Tensile strength according to ASTM D638 [MPa] (axes labeled as per Fig. 7.14) | HDT at 0.46 MPa, acc. to ASTM D648 |
|---|---|---|---|
| ULTEM 1010 | Prints from this filament are characterized with a very high mechanical strength, as well as thermal and chemical resistance. They are suitable for the production of heavily loaded components of machines, vehicles, and devices; ULTEM 1010 CG has a certificate of conformity for direct contact with food and the human body, making it suitable for the production of e.g., surgical instruments. | 81 (XZ)<br>29 (ZX) | 216°C |
| PPSF/ PPSU | PPSF (polyphenylsulfone) has high mechanical strength and a very high resistance to chemical agents and high temperatures; it is suitable for use in corrosive media; details can be subject to plasma or chemical sterilization, as well as with the use of gamma radiation, ethylene oxide, and steam autoclave. | 55 (no data) | 189°C |
| Antero 800NA | A material based on the PEEK (polyetherketoneketone) polymer, with high tensile strength and abrasion resistance; it is resistant to higher temperatures and chemical agents; it is also dimensionally stable. | 93 (XZ)<br>46 (ZX) | 150°C |

**Table 7.3** Properties, applications, and chosen parameters (HDT stands for heat deflection temperature) of the most common model materials used in Stratasys FDM 3D printers [3].—cont'd

| Model material | Properties and applications of a given model material | Tensile strength according to ASTM D638 [MPa] (axes labeled as per Fig. 7.14) | HDT at 0.46 MPa, acc. to ASTM D648 |
|---|---|---|---|
| TPU 92A | TPU is a thermoplastic elastomer, belonging to a group of polymers called polyurethanes (PU), with rubber-like properties. It has a high resistance to abrasion, cracking, and tearing, as well as to atmospheric conditions and chemical agents, i.e., oils, fats, a range of solvents; elongation at break is 552% (XY) and 482% (XZ); Shore A hardness is 92. | 15 (XZ) (No data) | 40°C |

Reproduced with kind permission from Stratasys.

chosen parameters of model materials used in Stratasys FDM 3D printers currently on the market are shown). Stratasys was the first company to implement FDM on the market and to set a standard in FDM 3D printer construction, as well as in the model and support materials used. For this reason, specific materials manufactured by this company are described in the following; then, filaments for open FDM/FFF 3D printing systems will be described.

For F123 Stratasys 3D printers (Table 7.1), thermoplastics with a lower strength than those for the Fortus series are available (Table 7.2). The F123 series is used, among others, in design offices or tool shops, where there are often built models or prototypes (e.g., form PLA) and less responsible parts, i.e., guards, casings (from ABS or ASA polymers), medical models (from ABS-M30i or PC-ISO polymers; Fig. 7.15). These devices also allow for 3D printing with a TPU elastomer with rubber-like properties (Fig. 7.17). However, the Fortus series and the F900 Stratasys model are dedicated for small-scale production of thermally and mechanically durable parts fitted in end products or for the manufacturing of production assembly instrumentation. A good example of FDM 3D printing application are mechanically resistant pipes or profiles with variable cross-sectional area and

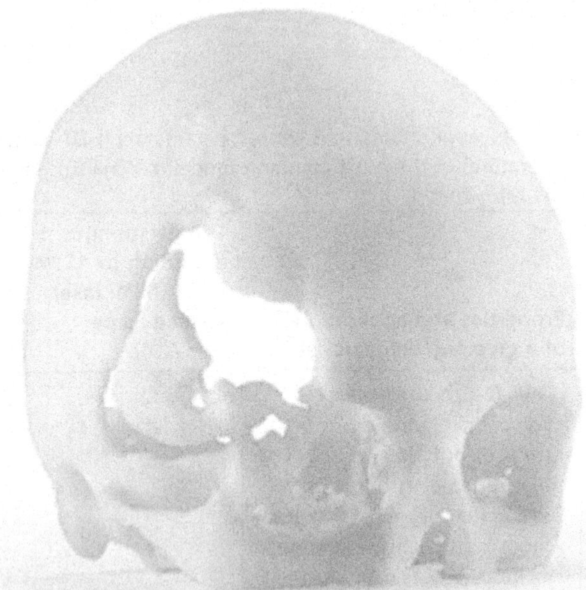

**Figure 7.15** A medical model of skull bones made from ABS-M30i [3]. *(Reproduced with kind permission from Stratasys.)*

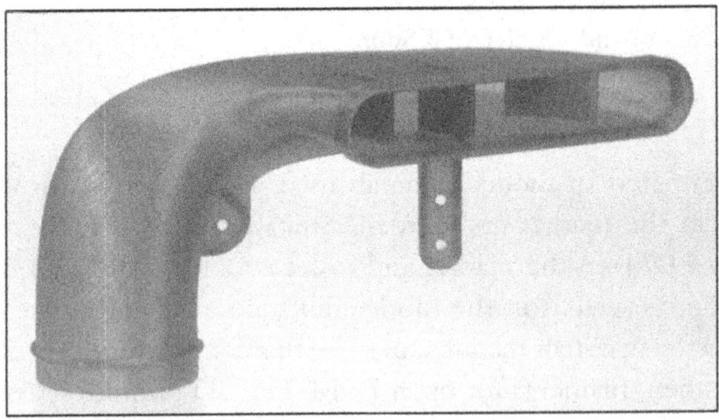

**Figure 7.16** A profile with a varying cross-sectional area and internal ribs, 3D printed using PEEK polymer [3]. *PEEK*, polyetherketoneketone. *(Model material Antero 800NA by Stratasys; Reproduced with kind permission from Stratasys.)*

**Figure 7.17** Elastic pipe 3D printed using TPU 92A elastomer [3]. *(Reproduced with kind permission from Stratasys.)*

with internal reinforcing ribs (Fig. 7.16), which are extremely difficult (or practically impossible) to produce using traditional polymer processing methods (i.e., injection molding, flexible casting, rotational casting).

For filaments (summarized in Table 7.3), Stratasys does not provide any processing parameters in their Technical Data Sheets (TDS) [3]. This is because in the CAM software (e.g., CatalystEx), dedicated to 3D printers made by this company, extrusion temperature or working chamber temperature is set permanently and cannot be changed by the user. Stratasys has determined their processing parameters on an experimental basis; therefore, any user using these brand materials in a correctly working device should obtain good prints and with the expected tensile parameters.

For the model materials shown in Table 7.3, chosen filaments are used for building overhang support structures and highly inclined overhang walls. Manufacturers of these materials provide information on the model materials to be used for these applications. Some of the model materials are soluble in specific aqueous solutions (Fig. 7.18A and B), and others have to be manually broken off, which is time-consuming and may cause damage to the printed object. Soluble support material allows for the production of prints with significantly more complex shapes, such as spatial trusses, deep curved holes, or pockets, which would not be possible to clear from breakaway supports. However, small traces of support structures are always left on the supported surfaces—therefore, finishing operations, e.g., grinding with sandpaper is usually applied.

#### 7.1.3.1.1 Model materials used in open 3D printing systems

In the case of open FDM/FFF 3D printing systems (most often used in budget, amateur and laboratory devices, often originating from the RepRap project), PLA (polylactic acid) is the most commonly used model material. For example, in Belgium in 2016, PLA filaments constituted 73% of the general sales, with PET-G at 10%, ABS at 9%, and others at 7% [24].

Due to its particularly low processing shrinkage, PLA is the model material easiest to use for FDM/FFF 3D printing and is therefore often chosen by amateurs. It is not necessary to have a 3D printer with a closed, heated working chamber for PLA. A heated working table is also not necessary, but it is sometimes useful for larger components (to prevent models from curling and to make them easier to remove from the table). PLA is a lactic acid–based polymer, thanks to which it is safe to use in 3D printing and fully biodegradable [25]. PLA is obtained from natural resources, such as corn, which makes it generally safe for contact with food.

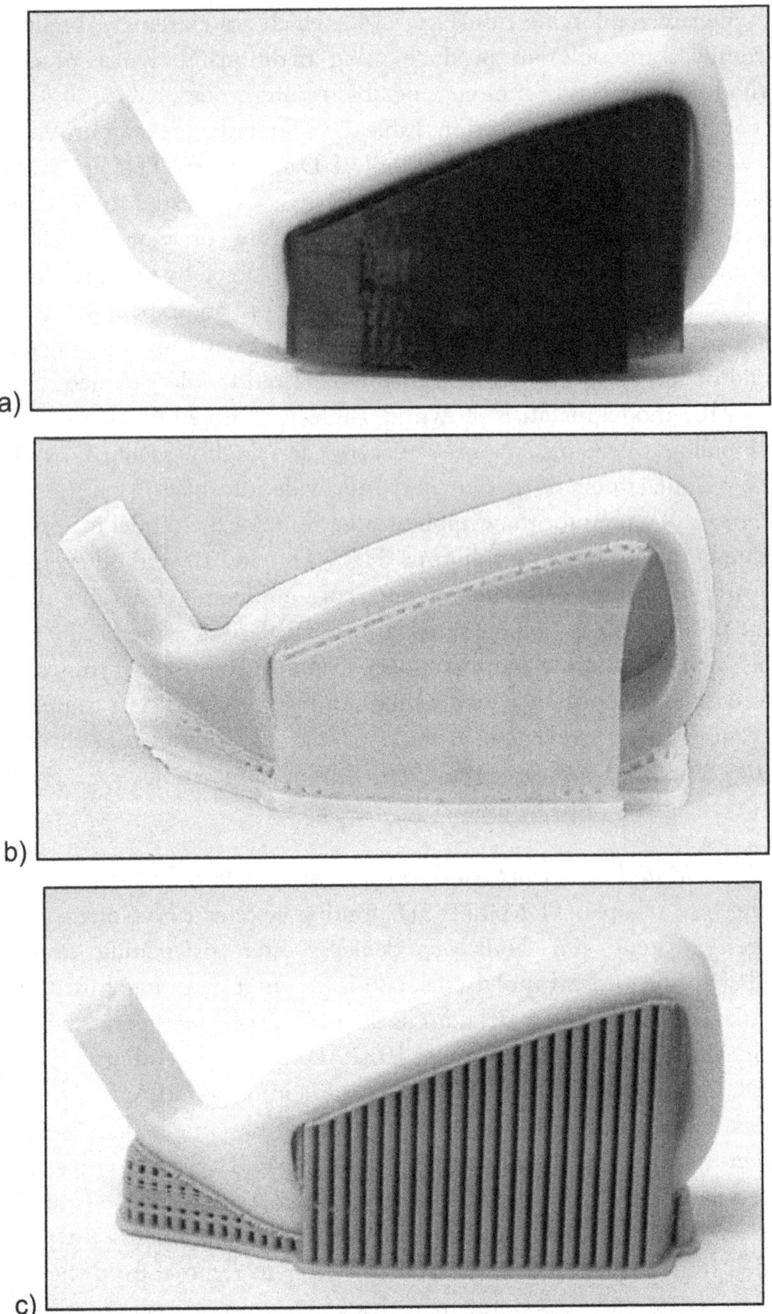


**Figure 7.18** Golf clubhead 3D printed using different model materials based on the ABS polymer; a view on the support structures made from a polymer soluble in alkaline aqueous solutions: SR-20 (A) and SR-30 (B) (*SR*, soluble release) and nonsoluble support structures for manual removal (C) (*BASS*, breakaway support structures) [3]. *(Reproduced with kind permission from Stratasys.)*

Medical-grade PLA is used, among others, for the production of resorbable surgical thread. Due to its low heat deflection temperature (HDT), typically around 50°C, PLA is not suitable for the printing of elements, e.g., kept inside cars in hot summer days or washed in a dishwasher. However, there are special PLA-based filaments which need to be properly heat-processed after the 3D printing process. This postprocessing operation causes their HDT to rise up to 85°C. Similarly, prints from modified ABS filaments do not undergo deformation even up to 95°C.

In Table 7.4, we have described the most popular model materials used in open FDM/FFF 3D printing systems, including printers based on the RepRap project. Most important properties and applications of these

**Table 7.4** Properties, applications, and chosen parameters of the most commonly used model materials used in open FDM/FFF 3D printing systems (including budget and amateur devices originating from the RepRap project) [25–27]; [29,32,33].

| Model material | Properties and applications of a given model material | HDT at 0.45 MPa, according to ASTM D648 |
|---|---|---|
| PLA | PLA polymer (polylactic acid) is suitable for the production of models, test prototypes, and because of its high hardness (higher than that of ABS, ASA, HIPS, and PET-G), also suitable for casting patterns or core boxes. PLA can be easily dyed, thanks to which an incredibly large range of filament colors are commercially available. Moreover, PLA filaments with the addition of glitter, wood particles, lignin, gypsum, minerals, and metal powders (aluminum, copper). Pure PLA is more mechanically resistant than pure ABS, but it is also slightly heavier and stiffer and more brittle. Pure PLA extrusion temperature for FDM/FFF 3D printer hot ends is approximately 200°C. Thanks to low shrinkage, it is possible to print using a heated working table or working chamber, though some manufacturers recommend heating the table to 45°C. In two head 3D printers, PVA or BVOH are recommended as support filaments for PLA. | Approx. 55°C |

*Continued*

**Table 7.4** Properties, applications, and chosen parameters of the most commonly used model materials used in open FDM/FFF 3D printing systems (including budget and amateur devices originating from the RepRap project) [25–27]; [29,32,33].—cont'd

| Model material | Properties and applications of a given model material | HDT at 0.45 MPa, according to ASTM D648 |
|---|---|---|
| PET-G | PET-G is a polymer based on poly(ethylene terephthalate) subjected to glycolization. PET-G prints are highly crystalline, which gives them highly transparent, especially with a low number of external contours. PET-G prints are odorless. An advantage of PET-G is low processing shrinkage and low humidity absorption in comparison to other model materials. PET-G is resistant to creep under constant load. There are modified kinds of PET-G resistant to temperatures up to 100°C. The prints can be disinfected by spraying with concentrated ethyl or isopropyl alcohol, UV-C radiation, or hydrogen peroxide [31]. | Approximately 70°C |
| ABS | The ABS (acrylonitrile-butadiene-styrene) polymer is a durable construction material with high resilience and stiffness; ABS is less brittle than PLA; ABS has lower resistance to UV radiation than ASA; after printing, it has a more glossy sheen and is more resistant to scratching; it can be smoothed in acetone vapor; ABS prints can be metalized and sanded, mudded, and painted; they can also be glued with acetone or cyanoacrylate; 3D printing hot end temperature is roughly 240°C; due to significant processing shrinkage it is recommended to use a heated working chamber with a temperature of 80°C - in exceptional cases, a heated working table with a temperature of 90°C may be used, in which case it is best to cover it with a layer of an ABS/acetone solution (so-called ABS juice). During 3D printing from ABS, styrene odor is perceptible. For two head 3D printers, HIPS is the dedicated support material for ABS. | Approx. 85°C |

**Table 7.4** Properties, applications, and chosen parameters of the most commonly used model materials used in open FDM/FFF 3D printing systems (including budget and amateur devices originating from the RepRap project) [25–27]; [29,32,33].—cont'd

| Model material | Properties and applications of a given model material | HDT at 0.45 MPa, according to ASTM D648 |
|---|---|---|
| HIPS | The HIPS polymer (high Impact Polystyrene) can be used as a model or a support material (for ABS prints). HIPS has lower shrinkage than common ABS, which makes it easier to 3D print. HIPS has the same density as pure ABS, but it is more stiff and has lower tensile strength. HIPS prints are characterized with a high impact resistance, have a matt surface, can be mechanically processed, sanded, and painted. The recommended hot end temperature for 3D printing is approx. 230°C; it is recommended to use a heated working chamber with a temperature of 80°C or a heated working table with a temperature of 90°C. HIPS is used as a support material for ABS. HIPS is dissolved in a concentrated aqueous solution of D-limonene, which is a natural substance, safe for the skin, and available in stores with cleaning products. HIPS is not suitable as a support material for PLA prints. | Approx. 85°C |
| ASA | The ASA (acrylonitrile-styrene-acrylate) polymer has a tensile strength similar to ABS, but it is significantly more resistant to UV radiation, which makes it suitable for outdoor components exposed to sunlight; the polymer is easier to 3D print than ABS; ASA is a good choice for the production of prototypes, functional production tools, components exposed to changing atmospheric conditions, and even everyday objects. | Approx. 85°C |

*Continued*

**Table 7.4** Properties, applications, and chosen parameters of the most commonly used model materials used in open FDM/FFF 3D printing systems (including budget and amateur devices originating from the RepRap project) [25–27]; [29,32,33].—cont'd

| Model material | Properties and applications of a given model material | HDT at 0.45 MPa, according to ASTM D648 |
|---|---|---|
| PA (Nylon) | The PA (polyamide) polymer, with the trade name nylon, is a model material with a significantly higher tensile strength than ABS or ASA. PA filaments are problematic to 3D-print due to their high hygroscopicity, causing them to be best used shortly after unpacking. 3D printing of damp PA tends to fail. PA is suitable for the production of gear wheels and racks, slide bearings, scrapers, and guide rollers working in higher temperatures, reaching 100°C in the short term and 150°C in the long term. The recommended extrusion temperature is approximately 260°C, and the working table temperature is approximately 90°C. | Approx. 100°C |
| TPU | TPU is a thermoplastic polyurethane with rubber-like mechanical properties. The prints are highly extensible, up to 500%. Depending on the TPU composition, it is printed using hot end with a temperature of approximately 200–230°C. TPU has a relatively low processing shrinkage, which makes it possible for the material to be printed without a heated working chamber; however, a working table heated to approx. 50°C is often recommended; it is possible to print on Kapton, masking tape, or clean glass. Due to filament extensibility and frequent problems with pressing it into the nozzle, the recommended feed speed is low (up to 0–30 mm/s). TPU is resistant to compression, impact, abrasion, and hydrocarbons. TPU in a range of hardness is commercially available. | From 50 to 80°C |

filaments, as well as recommended hot end temperatures during the printing process or heated working chamber or heated working table, are given. HDT is given as a separate parameter, as it is often the deciding factor in the possibility of application of a given polymer. In Table 7.4, no data on mechanical strength of prints from these polymers are given, as there are many filament manufacturers, who make a large number of model materials and provide different processing parameters for their use (most often extrusion temperature ranges and fiber feed rate range). Moreover, users of open FDM/FFF 3D printing systems can usually set their own processing parameters in their CAM software and even modify them in the 3D printer control system. The used processing parameters influence strength properties of the prints; therefore, they are not given in Table 7.4, except for HDT.

Usually, amateur FDM/FFF 3D printers have one head; then, they are only able to print one material at a time. It is not possible to use another material for the support structures as is standard in Stratasys devices. Supports in single head devices are made from the model material (see Fig. 7.19); therefore, it is necessary to remove them by manually breaking, which is time-consuming and causes a risk of damage of thin walls or particularly small print features. To facilitate the removal of supports, CAM software generates them as lattice structures surrounding the proper model, slightly shifted in relation to its surface, with the lattices as delicate walls built of single fibers. Such walls are relatively easy to remove manually or with the use of stork beak pliers.

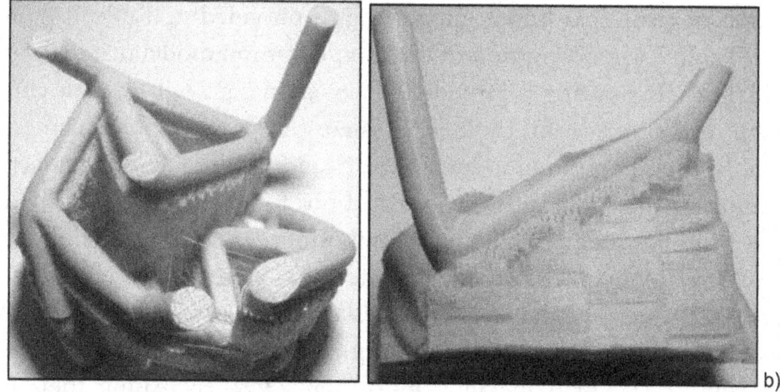


**Figure 7.19** A view from two sides (A and B) of a model of spatially connected rods made on a single-head FDM/FFF 3D printer from green PLA; visible support structure made from the model material, which needs to be broken away manually. *FDM*, fused deposition modeling; *FFF*, fused filament fabrication; *PLA*, polylactic acid.

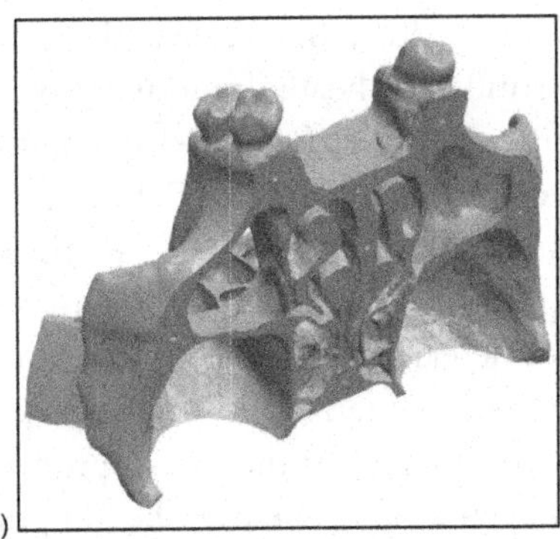

**Figure 7.20** Virtual 3D (STL mesh from computed tomography scan) model of human jaw (A); the same model made from white polymer PLA with light yellow PVA support structures (B); the FDM/FFF method is not the best for such shapes; the powder methods such as SLS, MJF, and CJP or photopolymer methods such as PJM, MJP, LCD, DLP, and 3SP are definitely better. *FDM*, fused deposition modeling; *FFF*, fused filament fabrication; *MJF*, Multi Jet Fusion; *PLA*, polylactic acid; *PVA*, polyvinyl alcohol; *SLS*, selective laser sintering; *CJP*, color jet printing; *PJM*, polyJet modeling; *MJM*, multiJet modeling; *LCD*, liquid-crystal display; *DLP*, digital light processing; *3SP*, scan, spin and selectively photocure.

Soluble support structure materials need to be selected with the model material in mind. In the case of prints from PLA-based filaments, support structures are built from PVA (polyvinyl alcohol) or BVOH (a copolymer of butenediol and PVA), which are soluble in clean water (see Fig. 7.20). However, when ABS is the model material, HIPS (high impact polystyrene) is used for support structures. HIPS dissolves in a concentrated solution of citric acid (more information is provided in Table 7.4).

In Table 7.4, a description of the most common model materials used in FDM/FFF 3D printing is provided. They are designated by their chemical names, in contrast to in Table 7.3, where the polymer trade names were given. In the 3D printing industry, there are many filament manufacturers, and most of them use different designations for their products. Their materials may also differ in chemical composition and properties. These materials may contain additives designed to increase the layer-to-layer bonding, plasticizers, stabilizers (e.g., UV radiation protection), pigments, or fillers changing their appearance (e.g., glitter), giving them a wood-like look (Fig. 7.21), or a metallic shine (Fig. 7.22), increasing their tensile strength (e.g., cut carbon fibers), or thermal conductivity (e.g., copper or bronze powder) or electrical conductivity. There are also commercially available fluorescent (glow-in-the-dark) [26] variants of PLA filament and

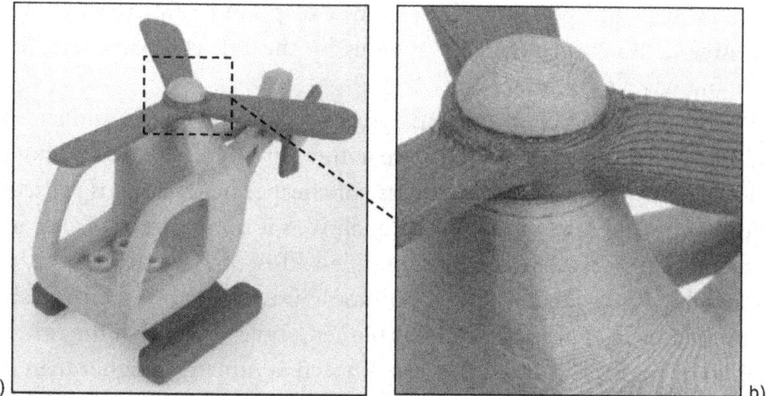

**Figure 7.21** A toy which looks as if it was made of wood—3D printed with FDM/FFF from several types (colors) of FiberWood filaments by Fiberlogy (A); visible print layers emphasize the similarity to wood grain (B) [27]. *(Reproduced with kind permission from Fiberlogy.)*

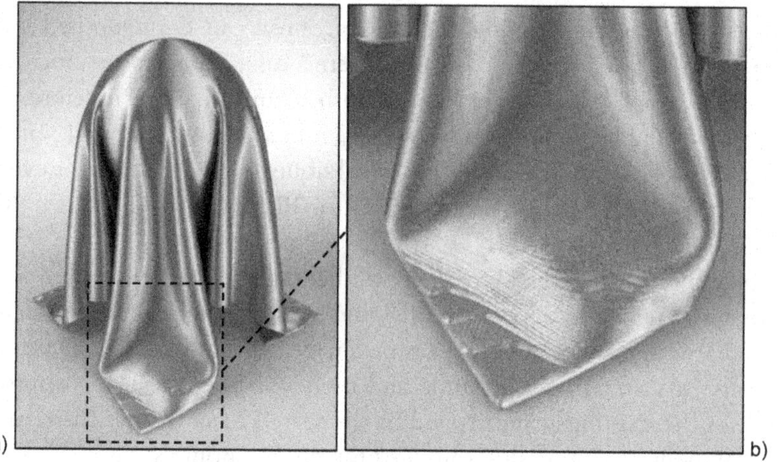

**Figure 7.22** Artistic INVISIBALL print—designed by J. Krejcira (Extreme3Dprint), 3D printed from FiberSilk Metallic filament by Fiberlogy [27] (A); a 3D model available on Thingiverse under the name INVISIBALL; a view on the 3D printing layers (B) [28]. *(Reproduced with kind permission from Fiberlogy.)*

those changing their color with temperature [26]. A number of colors and shades are available for PLA and ABS filaments, which makes them more attractive for FDM/FFF 3D printing. In other polymer additive manufacturing technologies, there is either one basic color of the model material (SLS, SLA, DLP, MJP), or the manufacturer provides few basic colors (LOM—Solido, PJM). If a possibility exists to color the prints, the colors are pale and poorly saturated. An exception to this rule are (highly expensive to buy and run) MJF systems by the HP company and PolyJet Modeling Matrix by Stratasys.

Price of FDM/FFF model materials is currently quite similar; non-modified PLA, PET-G, or ABS are rather affordable and cost about 20 EUR per 750−1000 g [29]. In the case of cheaper materials, it is particularly important to take into account their quality, as it is a significant factor in the 3D printing process. The cost of modified PLA or ABS filaments is often considerably higher than that of commonly used PLA, but their use allows for more visually appealing effects or higher tensile strength of the final part. Price for average ASA and TPU (also known as rubber) is higher than PLA, while PVA is three to four times more expensive than common PLA [29]. Technical filaments, e.g., PA (polyamide, also known as nylon), or PC (polycarbonate), are also several times more expensive than PLA.

It is also important to protect the filament from humidity. Polymers are hygroscopic, and during high-temperature extrusion, water evaporates, causing problems in printing quality, e.g., breaks in the deposited fibers. The prints become fragile and, in extreme cases, take on an openwork structure, which makes it possible to crush them in hand. Therefore, new filament should be vacuum-packed with a moisture absorbing pack, and the unpacked rolls should be dried for several hours in special driers, or even in an ordinary oven, not exceeding 45°C for PLA and 80°C [30] for ABS, PET-G, or PA.

New polymers and additives are still being developed and tested in the laboratories of chemical companies. Some of them, after a time, are adapted for 3D printing as model or support materials. Currently, in the 3D printing industry, it seems that more time and money is invested in the improvement of the existing materials and in the search of new ones than on the development of new design solutions for the 3D printers.

## 7.4 Extrusion head structure

The 3D printing method known as FDM is based on melting thermoplastic polymer in a hot end and its extrusion in the form of a thin fiber. The fiber is deposited in appropriate places on the 3D printer working platform, to which it adheres and hardens during cooling. After the first fiber layer is deposited, the nozzle is raised by the width of the layer (or, in 3D printers with different kinematics, the platform is lowered), and the next layer is built. The process is repeated until the last layer programmed in the 3D model is deposited. A basic scheme of the FDM 3D printing process is shown and described in Fig. 7.23.

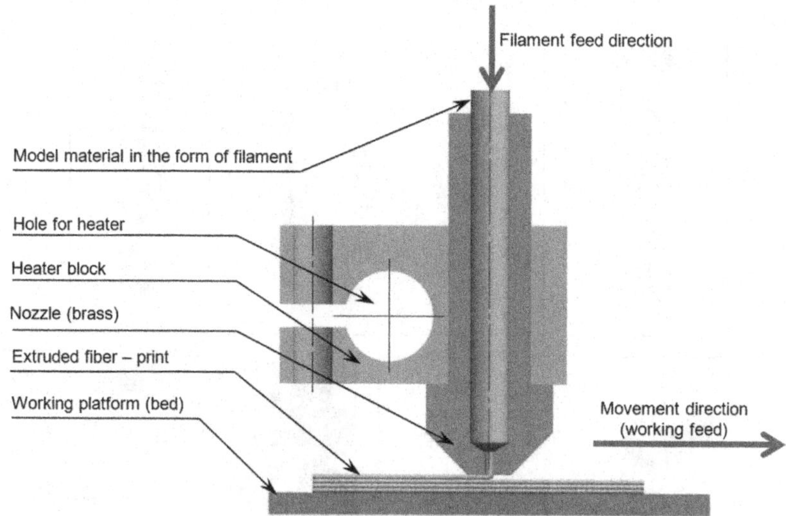


**Figure 7.23** A basic FDM/FFF 3D printing scheme: a thermoplastic in the form of filament is pressed into a hot nozzle, extruded as a fiber, and deposited on the 3D printer working platform; after one step is finished, the head is moved away from the platform and the fibers are deposited on the previous layer; these steps are repeated until the whole object is built. *FDM*, fused deposition modeling.

An example of a printing head used in simple FDM/FFF 3D printers is shown in Fig. 7.24E. The head consists of a brass nozzle, fitted with a box-shaped aluminum heater block with holes. An electric heater and temperature sensor (typically a thermistor) are mounted using these holes. Both these elements are connected to the 3D printer, which maintains a given temperature in the set range.

To assure that in the process of 3D printing a fiber is extruded through the small opening in the nozzle, the filament is pushed in from the other side. This is realized through a specific injecting mechanism called extruder. It is typically made of a textured roll, powered by a stepper motor (Fig. 7.25D). The filament is pressed from the other side with a smooth roll (usually, a bearing) (Fig. 7.26A), supported by a spring element (Fig. 7.26A). The extruder can be placed directly next to the extrusion head—the so-called direct drive system, or removed from it—the so-called Bowden system (Fig. 7.25B). In the Bowden system, the filament must be supplied to the printing head through a plastic tube (for example, Teflon tube).

The main advantage of the Bowden systems is that the extruder is separated from the extrusion head, so the head has a significantly lower mass.

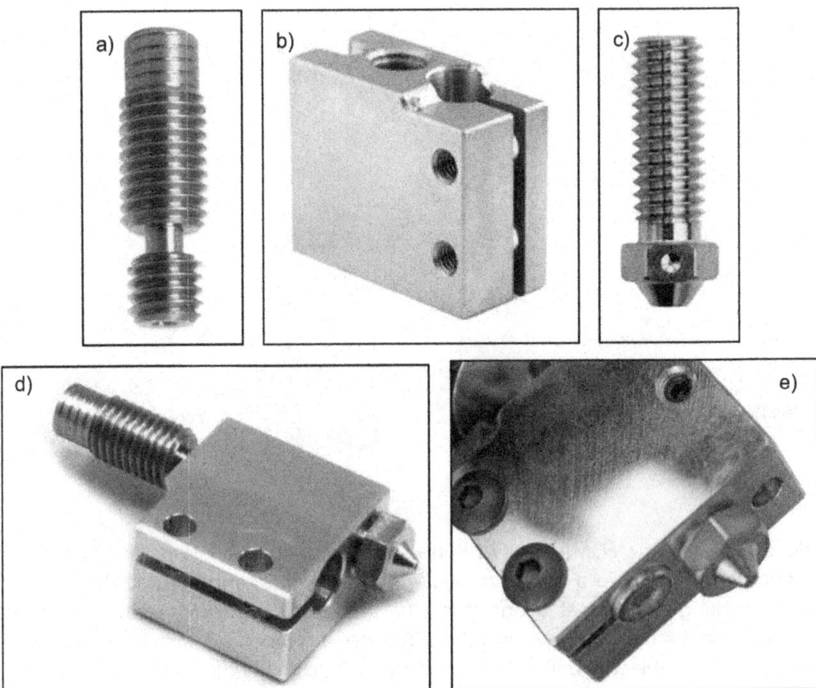


**Figure 7.24** Sample hot end—Volcano from E3D-Online: (A) steel heat break [34], (B) aluminum heater block [34], (C) brass nozzle [34], (D) head preassembled [34], (E) head fully assembled with heater (securing with two socket dome screws—on left side) and thermistor (securing with grub screw—on right side) [34]. *(Reproduced with kind permission from E3D-online Ltd.)*

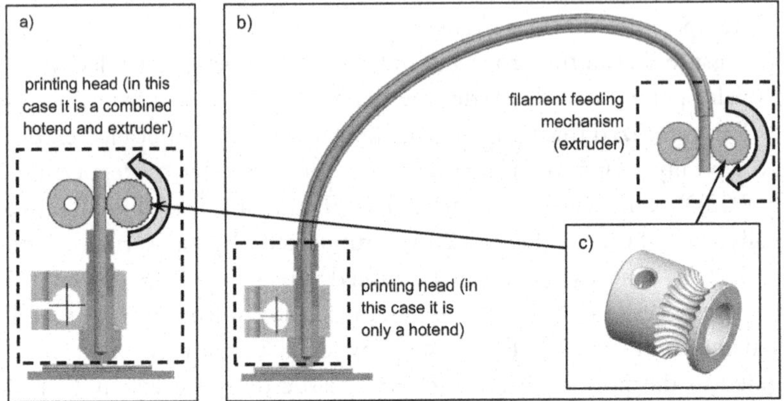


**Figure 7.25** There are two methods for injecting the filament into the printing head: (A) direct, also known as direct drive and (B) indirect, also known as the Bowden system, through a tube; gray arrow indicates rotational direction of the motor-driven knurl (C), feeding the filament.

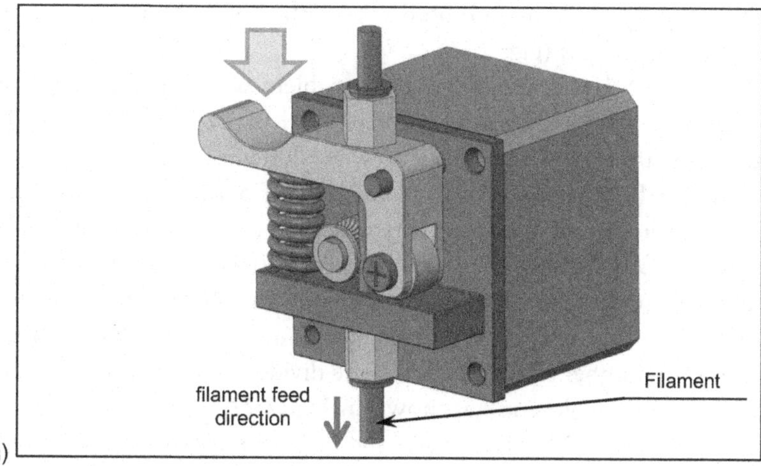


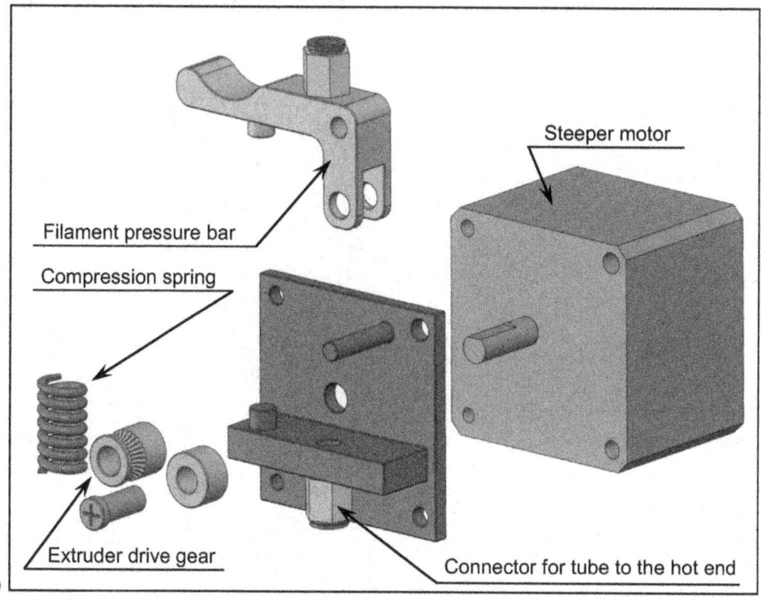


**Figure 7.26** Scheme of extruder as an example of a filament feeding mechanism used in open FDM/FFF 3D printing systems with Bowden type filament feeding systems: (A) assembly; (B) mechanism in exploded view without filament; pressing the beam indicated with a big gray arrow with a finger moves away the press roller and enables free insertion or removal of the filament from the head. *FDM*, fused deposition modeling; *FFF*, fused filament fabrication.

This is because the quite heavy stepper motor needs to be directly near the filament feeding mechanism. 3D printers with a light extrusion head, movable in the X and/or Y axes, allow for larger feed speed, which gives a faster printing. In Bowden systems, filament is fed through a plastic tube, which is usually connected on both sides with special connectors, e.g., used in pneumatic systems (Fig. 7.26).

In extrusion heads, there may be a problem with pressing the filament into the hot nozzle. This is a problem particularly noticeable for elastic filaments with rubber-like properties. A difficulty may also arise with pressing the hard thermoplastic filament with a large force, particularly when it has a low softening temperature, such as PLA. This thermoplastic softens at approximately 50°C, which means that if it exceeds this temperature at the distance between the rolls and the channel entrance, it will undergo bending, making it impossible to press it into the hot nozzle. Therefore, a number of printing heads are divided into a hot zone (hot end) and a cold zone (cold end), as shown in Fig. 7.27.

The built of FDM extruder was shown in the patent No. 5121329 on FDM 3D printing [2]. A scheme for the head design with a mechanism

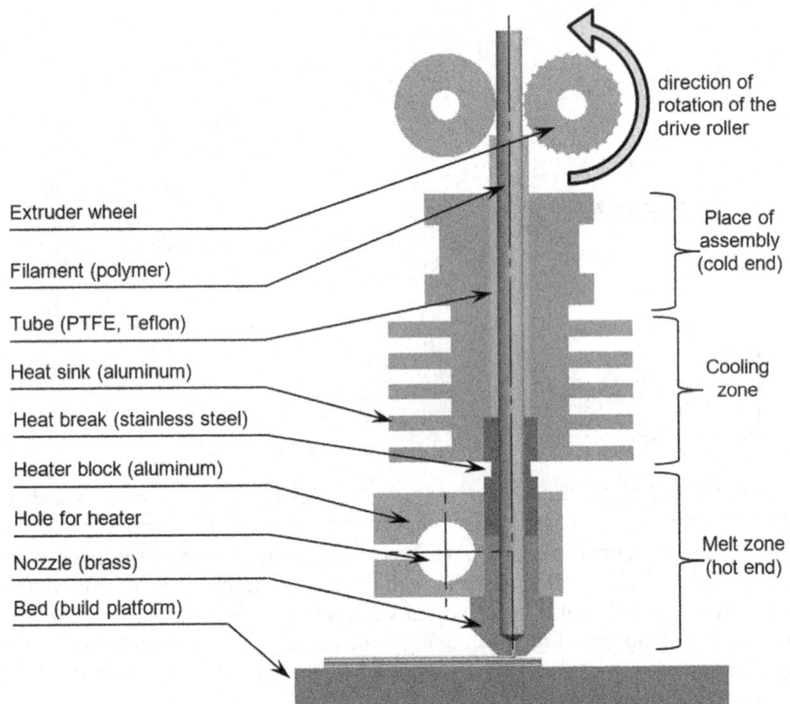


**Figure 7.27** An example of a printing head with a direct drive; the same heads are used e.g., in open FDM/FFF 3D printing systems; the gray arrow shows the rotational direction of the motor-driven roll with a knurl, pushing the filament into the head. *FDM*, fused deposition modeling; *FFF*, fused filament fabrication.

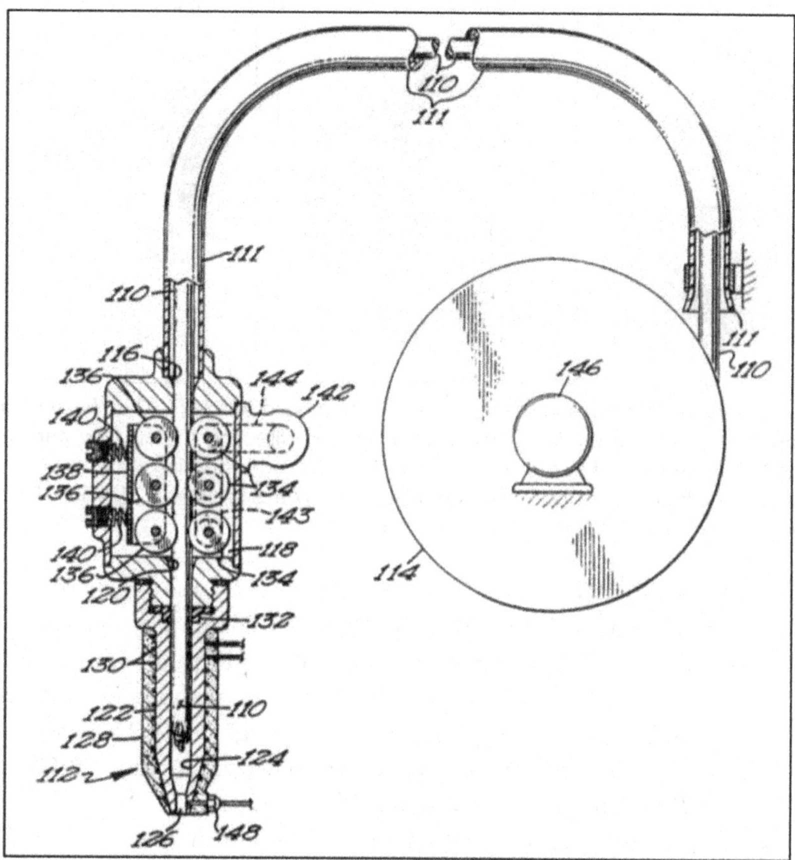


**Figure 7.28** Drawing from patent description No. 5121329 [2]; filament (110) is unwound from a spool (114), and then fed to the head (112) through a tube (111); the filament is pressed with a system of drive rollers (134) and press rollers (136); rollers (134) are driven by a motor (142) through a transmission (144); the filament is melted in the hot end (122) and extruded in the form of a fiber through an opening (126).

feeding the filament into the hot end, where it is melted and extruded in the form of a fiber, and finally deposited on the working platform of the device or on the previously deposited layers, is shown in Fig. 7.28. A scheme of the fiber deposition during the 3D printing process, shown in the patent description, is depicted in Fig. 7.1.

In 3D printers originating from the RepRap project, hot zone (hot end) and a cold zone (cold end) are typically separated with a heat break, e.g., a steel neck bushing, which allows for a limited heat transfer between these zones. Additionally, a radiator, usually mounted with a fan (as shown in Fig. 7.29B), is used to increase the cooling efficiency of the filament

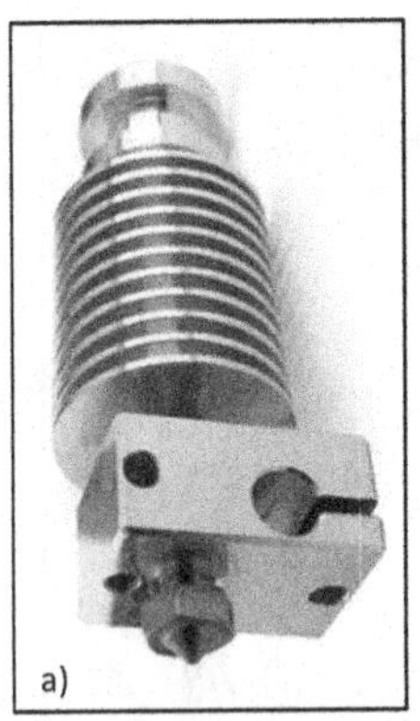

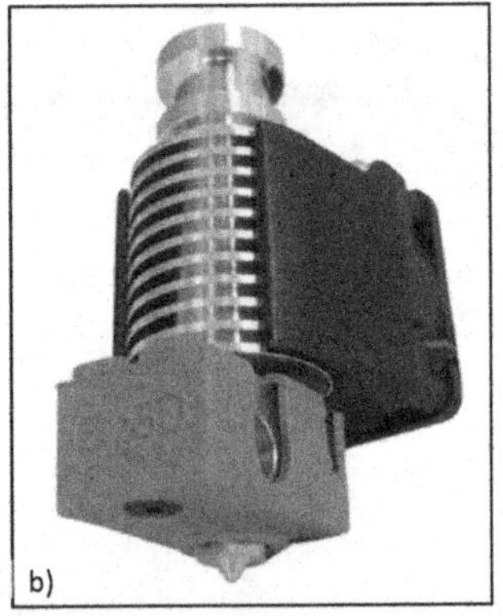

**Figure 7.29** Metal elements of the E3D V6 printing head (A) [34]; E3D V6 head with a mounted electric heater and a fan with a housing directing the air flow onto the radiator (B) [34]; E3D head mounted on a direct extruder housing (front cover and filament feeding mechanism were removed to show the interior) (C) [34,35]. *(Reproduced with kind permission from E3D-online Ltd.)*

entrance zone (filament inlet nozzle). In Fig. 7.27, the schematic of a printing head with a radiator is shown, which is similar in shape to the E3D V6 model (Fig. 7.29), popular among users of RepRap 3D printers. E3D series printing heads are typically sold separately, without the filament feeding mechanism (i.e., the extruder) as seen in Fig. 7.29B. These are able to fit extruders with direct injection systems (Figs. 7.25A and 7.29C) and indirect or Bowden type systems (Figs. 7.25B and 7.26).

Most 3D printers using the FDM/FFF method (including open systems originating from the RepRap project) use integrated printing heads—the filament feeding system is direct (Fig. 7.25A). This means that a filament feeding mechanism (extruder) is mounted on the part responsible for the filament melting (hot end). Thanks to this solution, problems typical for Bowden systems can be avoided, e.g., filament friction inside the tube, filament folding, etc. An example of an integrated printing head with one extruder is shown in Fig. 7.30.

Vast majority of amateur FDM/FFF printers have printing heads with one nozzle. However, professional or industrial devices are always equipped with two nozzle heads (Fig. 7.31). This enables the use of two materials in one printing process, e.g., model and support material (Fig. 7.20), which allows for the production of much more complex shapes than it is possible using one-head 3D printers, where the support structures need to be made from the same material as the model (Fig. 7.19). In a basic version, two-nozzle 3D printers allow for the use of two model materials of the same type but in different colors.

**Figure 7.30** E3D Hemera Direct Kit form E3D-online—an example of a printing head build from a metal type extruder with stepper motor [34]. *(Reproduced with kind permission from E3D-online Ltd.)*

**Figure 7.31** HotEnd Chimera+ from E3D-online—this is the example of printing head (without filament feeding mechanism) with two nozzles at constant height, so their leveling is required [34]. *(Reproduced with kind permission from E3D-online Ltd.)*

The easiest method to equip a 3D printer with two heads is to add a second hot end-extruder set, in which the nozzle is constantly at the same height (Fig. 7.30). Such a solution requires precise leveling of both nozzles and precise measurement of the distance between them. During a two-nozzle head printing process, the unused nozzle scratches against the print, and should the print rise slightly, the nozzle might collide with it and break it away. For this reason, in professional 3D printers, two-nozzle heads are used, but with a variable height. FDM Stratasys 3D printers described in Section 7.1 use this type of head construction. It is also worth mentioning that when the head is changed to a two-nozzle version in the amateur devices, the available working space along the axis determined by the nozzles will be lower by double the distance between the nozzles.

New types of printing heads and extruders have been developed for many years. Printing heads with one nozzle but, capable of printing from several materials, are an interesting solution. The integrated dual-feed extruder by the Dutch company Builder has this type of construction. The head has one nozzle with a heater block, to which materials are supplied using two extruders. Generally, two operational modes are available. The first is when one extruder is stopped, and the other presses the material. The required materials need to be polymers printed in the same temperatures, most preferably of the same type. In this case, it is possible to print just as using a two-nozzle device, but without the aforementioned limitations of the working surface and leveling. The second mode allows for the mixing the model materials in specific proportions and extrusion of the resulting fiber. If colors of the same polymer are used,a smooth tonal transition between the layers can be obtained. The best visual effects are obtained with the use of transparent PET-G filaments.

A more advanced example of a one-nozzle extrusion head is Diamond HotEnd (Fig. 7.32A). Such a head allows for printing using two or three model materials in one printing process. One heater (Fig. 7.32B) is mounted in the head, so one melting temperature is set. Using the filaments of the same polymer but with different colors, it is possible to obtain multicolored prints in one single process (Fig. 7.33A). If those colors are the three basic colors, intermediate colors may be obtained by their extrusion in proper proportions (Fig. 7.33B).

## 7.5 Selected details about heads in open systems

To extrude a fiber from a hot nozzle, it is necessary to press the filament into the printing head channel with a large enough force. Therefore, the

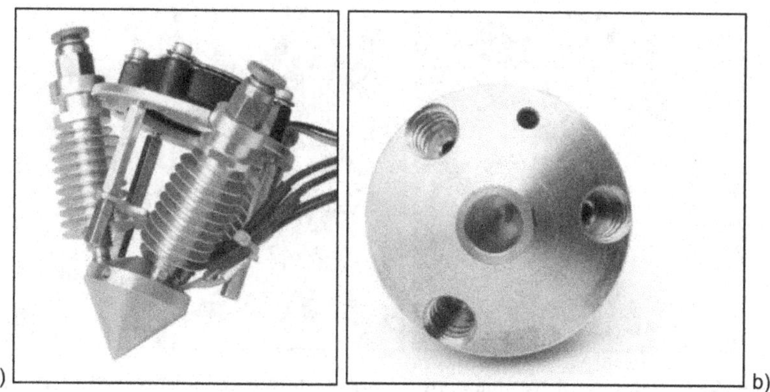

**Figure 7.32** Diamond HotEnd—a printing head with a special nozzle allowing three filaments to be mixed; quick-connect fittings allow the connection with three separate Bowden extruders via tubes (A); top view on the printing head with a centrally visible electric heater socket and a small opening for the temperature sensor at the top (B) [36]. *(Reproduced with kind permission from RepRap Wiki.)*

**Figure 7.33** Examples of prints obtained using the Diamond HotEnd head: (A) from three filaments with specific colors (visible on the prints); (B) from three filaments in basic colors and mixed in proper proportions, obtaining a number of intermediate colors [37]. *(Reproduced with kind permission from RepRap Wiki.)*

pressed filament needs to be stiff enough to avoid bending just after the driving fluted roller (Fig. 7.34A). Filament folding before the head channel is a common problem in 3D printing from elastic model materials (with rubber-like properties, e.g., TPU described in Table 7.4), which is shown

**Figure 7.34** Examples of two filament feeding system mechanisms (extruders): made using ABS injection (A); made using FDM/FFF 3D printing (B); folding of the elastic filament around the driven roll, which blocks the 3D printing process visible on (A); the shaping of the filament channel inlet, preventing its bending, marked on (B) with a red [light black in print version] *arrow* [38]. *FDM*, fused deposition modeling; *FFF*, fused filament fabrication. *(Reproduced with kind permission from Gyrobot Limited.)*

in Fig. 7.34A. Therefore, Bowden systems are not recommended for these materials.

FDM/FFF 3D printers are usually not equipped with fiber flow blockage sensors, so the device does not stop the 3D printing process and remains idling. To prevent the low filament stiffness problem in direct drives, the printing head channel is specifically shaped (indicated with a red [light black in printed version] arrow in Fig. 7.34B). Alternatively, a tube is placed between the channel entrance and the driving rollers, and an incision is made to place it as close to them as possible (Fig. 7.35B). Tube ending in the Bowden Extruder mechanism is treated similarly.

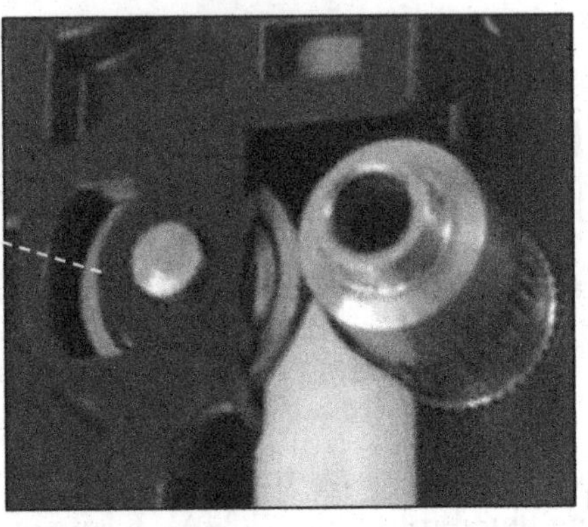

**Figure 7.35** A E3D Titan-type printing head (A); a view on the specifically shaped end of a Teflon tube, placed directly after the rollers feeding the filament into the channel leading to the nozzle (B) [34,35]. *(Reproduced with kind permission from E3d-online Ltd.)*

**Figure 7.36** A belt transmission with a toothed belt of the filament feeding mechanism (extruder), mounted on a Monkeyfab Prime 3D printer head; the gears are cut from transparent plastic plates; cool end zone radiator cooling fan visible at the bottom.

In filament feeding mechanisms (extruders), the driving roller can be equipped with a knurl pressing the filament in the notch (Figs. 7.25D and 7.35B) or resemble a cylindrical gear in shape (Figs. 7.29B and 7.34A,B). Moreover, such a roll can be mounted directly on the stepper motor axis (Fig. 7.29B) or be driven by it through a transmission, e.g., gear (Fig. 7.28C) or toothed belt gear (Fig. 7.36). In such case, a stepper motor with a lower torque, and therefore, lower mass, may be used.

During the FDM/FFF 3D printing process, it sometimes happens that short fiber fragments do not attach to the working platform or the previous layer. In such cases, they tend to remain at the nozzle, which either moves them to different areas of the prints or they melt, sticking to the nozzle and the heater block. A view of a hot end covered with degraded and charred thermoplastic is shown in Fig. 7.37B. To prevent this, silicon casings may be used (Fig. 7.37C), which are resistant to melting temperatures of the thermoplastics used in FDM/FFF 3D printers (given in Table 7.4). Nozzle and heater block casing also protects the printed layers from excessive

**Figure 7.37** E3D printing heads; unused (A); used many times—visible nozzle and heater block stains of degraded and charred thermoplastic (B) [34,39]; E3D head with a heater block covered with a silicone sock (V6 Silicone Socks from E3D-online) (C) [34]. *(Reproduced with kind permission from E3d-online Ltd.)*

heating due to the infrared radiation emitted by hot parts of the head. Similar role is played by Kapton sheet cover of the heater block (Figs. 7.29A and 7.30) or nozzles with Teflon tips, used in Stratasys 3D printers (shown in Fig. 7.42B).

For the print to have a smooth surface (Fig. 7.39B) and correct dimensions, the extruded fibers need to be deposited on initially cooled layers. This problem is particularly visible when a single tall element with a low surface area is printed, e.g., a vertical column. If there is no special time gap between the subsequent layers, the nozzle deposits the new fibers in the still hot and plastic previous fibers. This increases their cohesion but causes bending or deformation of the entire print. Therefore, proper cooling of the deposited fibers is required. Separate fans mounted on the printing head (Fig. 7.38) are used for this purpose. Efficient cooling systems are used to improve the prints surface quality and increase the possible printing speed. An example of this can be the change of standard cooling (Fig. 7.38A) to a two-head version (Fig. 7.38B). For some prints, the direction of cooling airflow may be an issue, so different head designs are used, e.g., omnidirectional.

## 7.6 An example of head construction in a Stratasys device

As the Stratasys company was the first manufacturer of FDM 3D printers, it has had the possibility to develop device designs for the longest time. Therefore, in this chapter, the construction of a printing head of an exemplary Stratasys device is presented.

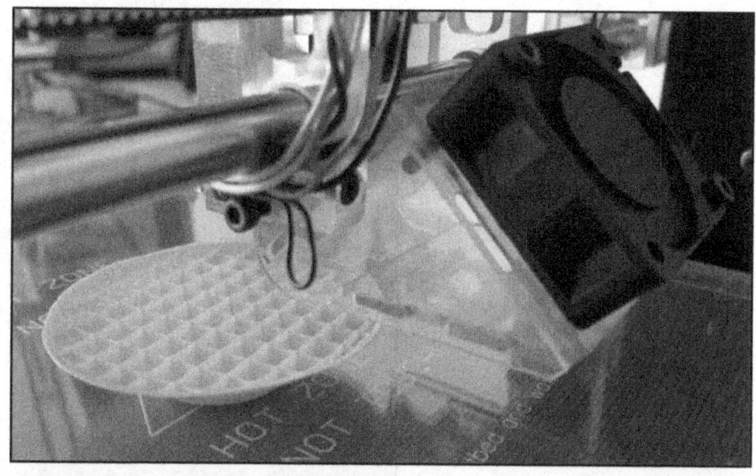

a)

b)

**Figure 7.38** Examples of cooling systems in a "Monkeyfab Prime 3D" 3D printer (originating form the Prusa I3 project—Fig. 7.10): (A) standard—one fan with a slot nozzle; (B) extended—two fans with large heads and Teflon cover [40]. *(Reproduced with kind permission from 3D w praktyce—Łukasz Długosz.)*

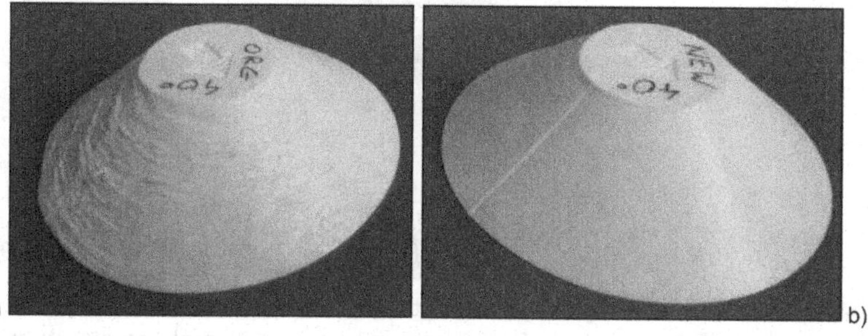

a) b)

**Figure 7.39** Examples of two test prints—truncated cones inclined at an angle of 40 degrees and 3D printed from the side of the smaller flat wall (see Fig. 7.38B) on a "Monkeyfab Prime 3D" printer; (A) with standard cooling—one fan with a straight nozzle (see Fig. 7.38A); (B) with extended cooling—i.e., two fans with large heads (see Fig. 7.38B) [40]. *(Reproduced with kind permission from 3D w praktyce—Łukasz Długosz.)*

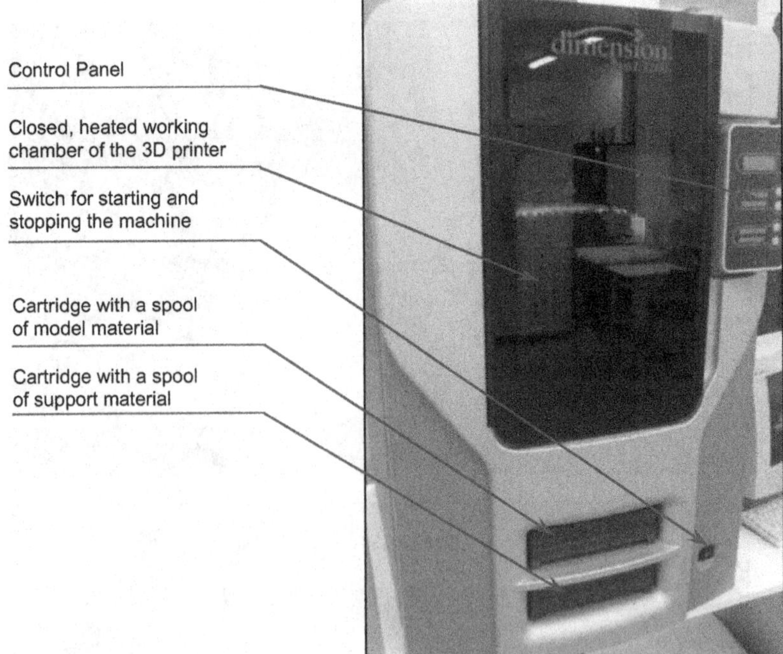


**Figure 7.40** Stratasys dimension BST 1200 3D printer.

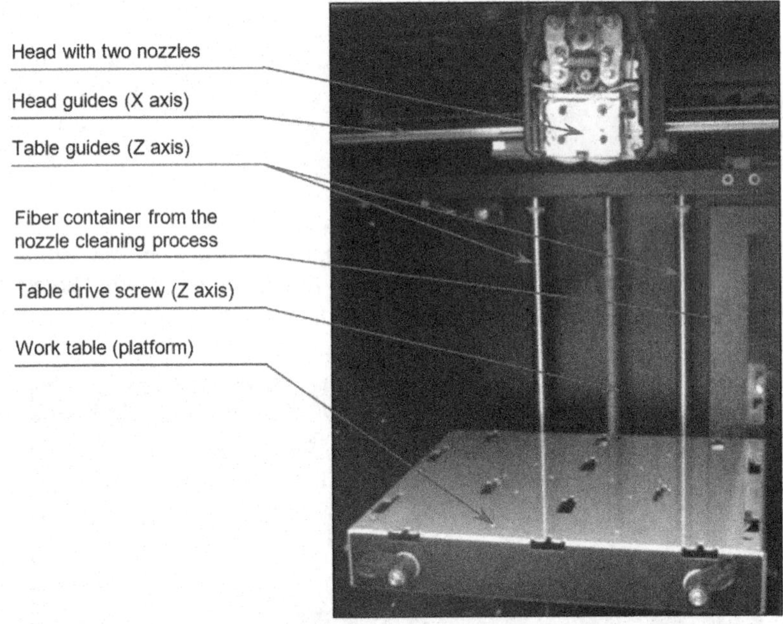


**Figure 7.41** Printer working chamber of Stratasys dimension BST 1200: visible steel table, on which the working platform is mounted— an ABS molding (which assures better adhesion of the first ABS fiber layer, firmly fixing the whole print); the table is driven by two rollers, and lifted and lowered in the Z axis using a with trapezoidal thread screw connected to the motor; the head is moved along guides in the X and Y axes; the head is moved by stepper motors using toothed belts.

a)

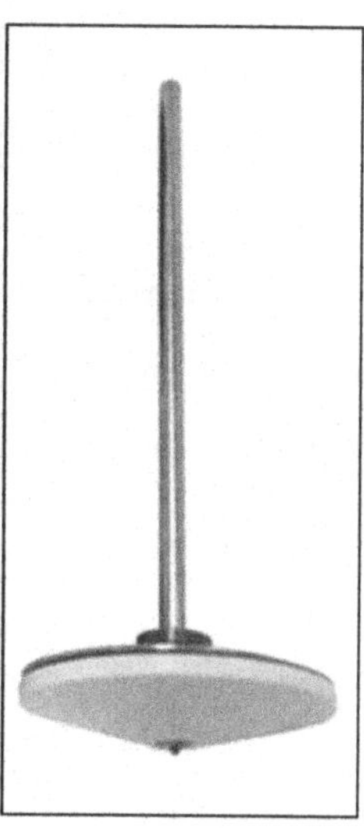
b)

**Figure 7.42** Printing head of the Stratasys Dimension BST 1200 device after the front cover is removed (A); two nozzles are visible: for support material (left) and model material (right); nozzle with a Teflon casing protecting it from the adhesion of free material fibers (B).

The view of the 3D printer (Dimension BST 1200) from Stratasys Company, which is typically dedicated for design offices and educational purposes, is shown in Fig. 7.40. The external view from the inside of working chamber of a printer is shown in Fig. 7.41. The device can produce prototypes and functional parts, as well as simple tools (mounting brackets for mechanisms, laminating and thermoforming forms, form sockets for PET blowing preforms, etc.) from pure ABS polymer.

This 3D printer has a two-nozzle head (Fig. 7.42A) for model and support materials. Nozzles can be lifted and lowered with a simple mechanism. Its operation is based on a press beam (Fig. 7.42A), which is moved by the head when it reaches a wall of the working chamber. Thanks to this, no separately controlled mechanism is required. Two nozzles with Teflon casings are mounted on the bottom side of the head (Fig. 7.42B), preventing free material fibers from sticking to it. Internal structure of the head is shown in Figs. 7.43 and 7.44.

Fig. 7.44 shows the shape of the air supply nozzle and airflow outlets. The air is cooling the nozzle switch mechanism, the filament pressing rollers, and filament between rollers and inlets to the printing nozzles. Due to the cooling, the filament should be rigid enough, so as to be pressed into the nozzle channel. If the cooling is not sufficient, the filament may collapse

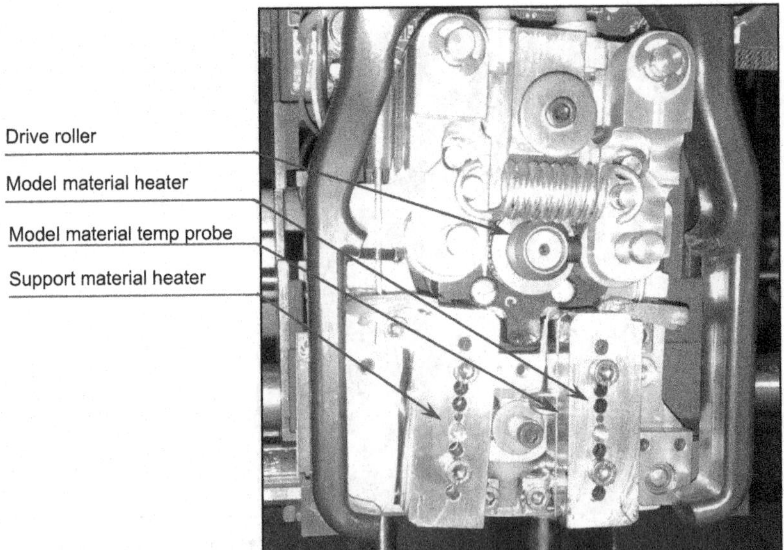


**Figure 7.43** A view on the head of a Stratasys 3D dimension BST 1200 3D printer after thermal cover was removed along with both printing nozzles; text in red (gray in printed version) describes two heating blocks for the printing nozzles.

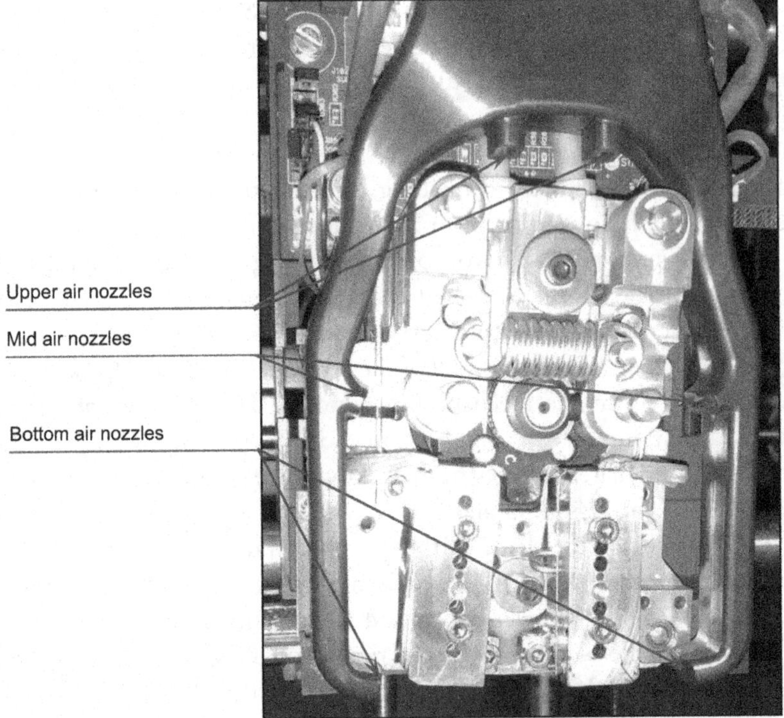


**Figure 7.44** A view on the shape of the air supply nozzle, and airflow outlets (from the top) to cool the nozzle switch mechanism, to cool the filament pressing rollers and its inlets to the printing nozzles, to cool the prints.

or melt in front of the channel and block the possibility of it being pressed in. This would interrupt the fiber extrusion process. The 3D printer does not have any nozzle clogging sensor, so the machine control system will continue the pointless 3D printing process.

## 7.7 Fiber deposition strategy and finishing process

To precisely describe the FDM/FFF 3D printing process, it is necessary to describe the axes along which the head moves in respect to the working platform (table), on which the extruded fibers are deposited. 3D printer manufacturers have established that the 3D printer build platform surface determines the XY plane for nearly every additive manufacturing method. The subsequent layers are built in the Z axis direction, so raising the printer head or lowering the table (depending on the device kinematics) is always performed in the Z axis. Fig. 7.45A shows a schematic representation of a flat XZ axis system, with 3D build platform and an extrusion head moving over it. Melted model or support material fibers are deposited on this on this working platform and the previously built layers. In cross-section perpendicular to the path, the extruded fiber has a shape shown in Fig. 7.45B, and its parametric description is shown in Fig. 7.46 (after [41]).

In the FDM/FFF additive technology, layer thickness within the 0.1–0.5 mm is used. Dedicated commercial CAM software of Stratasys devices allows it to be set to the following values: 0.178, 0.254, or 0.33 mm, depending on the device model and the diameter of the extruder nozzle.

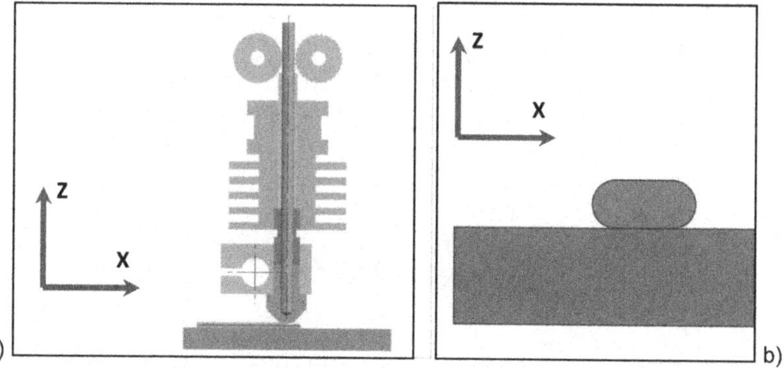


**Figure 7.45** 3D printer axis system (XZ) scheme—visible platform and moving head (A), a close-up on an extruded fiber deposited on the table (B).

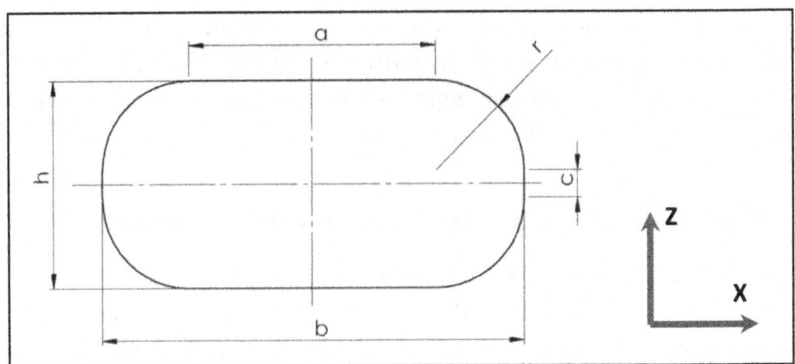


**Figure 7.46** Parametric description of the extruded fiber shape.

In the case of printers originating from the RepRap project, values of 0.2 or 0.25 mm are commonly set (for a 0.4 mm diameter nozzle). 0.3 mm layer thickness may be set if shorter print time is desired. For 3D models with only horizontal and vertical walls, a change in thickness from 0.1 to 0.3 mm will not significantly influence the initial shape reproduction quality, but it will shorten the printing process three times. However, if the model is printed in such an orientation that most of the sides are at an incline from the Z axis, it is better to choose a lower deposited layer thickness. The staircase effect is most visible in the print sides which are "almost" horizontal, regardless of them being flat or free surfaces. An example of this phenomenon is presented in Fig. 7.47.

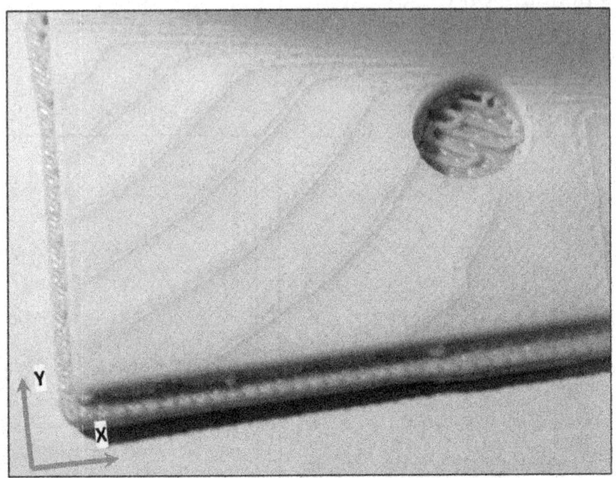


**Figure 7.47** A view on an example of a print made using the FDM method from ABS polymer, with a dimension BST 1200 by Stratasys; under the model you can see support structures; view on fiber deposition on the wall of the model, forming the so-called staircase effect. *FDM*, fused deposition modeling.

There are several ways to eliminate the staircase effect and finish the outer walls of printouts. The first way is traditional manual modeling work, i.e., multiple sanding and putting, and then painting. The second method of smoothing the surface requires using of a solvent that will chemically soften slightly and gently spread the staircase. Unfortunately, this can degrade fine bumps in printouts and thin walls. Shapes printed from ABS polymer require using acetone as a solvent. ABS objects are usually immersed in acetone vapors for about 30 s. When it comes about PLA printouts, ethyl acetate is required as a solvent. Sadly, this product is less commercially available than acetone. These methods of smoothing the surface of prints using the solvents are not completely safe, so they are dedicated to professionals users. The third method of smoothing the prints (regardless the model material and the selected additive technology) is circular vibratory surface finishing machine (mostly ceramic). Due to high cost, it is usually cost-effective in industrial applications only.

Going back to the fiber deposition strategy typically, at the beginning of the 3D printing process, so-called contour fibers are placed on the working platform. These fibers define the outer shape of the produced model in a given layer (Fig. 7.48A). This serves to maintain the highest possible dimensional accuracy of the model.

After contour deposition, layer fill is performed. For the first layer of the model, to ensure all outer walls of the model are closed, so-called full filling with fibers is conducted. In 3D CAM software for 3D printers, fill is usually set to 100%, but in practice small gaps are always left next to the contour. This is shown in Figs. 7.49B,C, 7.50 and 7.51. If feed rate of the material

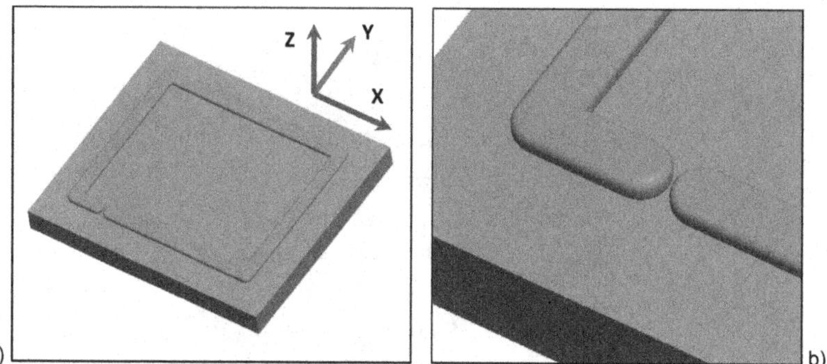


**Figure 7.48** 3D CAD model showing an example of extrusion, on the working platform, of a single fiber in the form of a rectangular frame, which is a contour of the deposited layer (A); close-up at the place, where its deposition started and ended (B). *CAD*, computer-aided design.

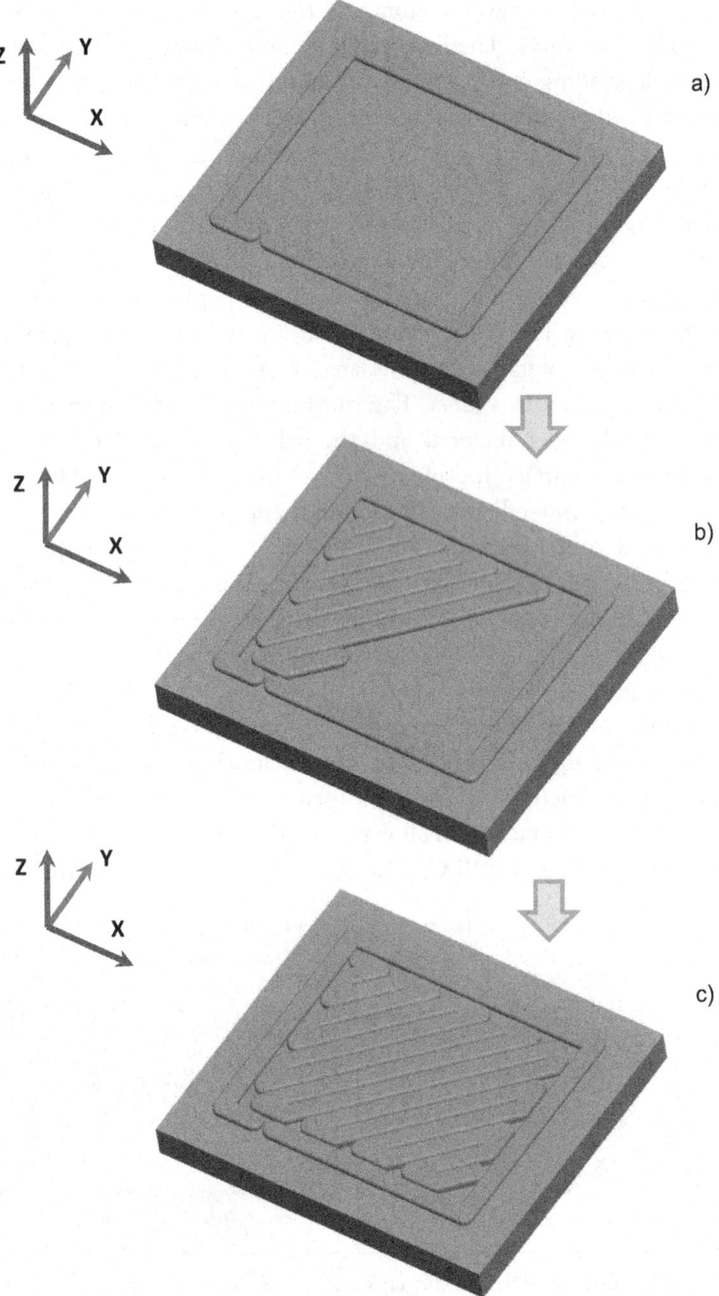


**Figure 7.49** An example of a strategy for fiber deposition on the working platform (table) for the first layer (A) contour fiber deposition, (B) a visible part of the extruded and already deposited fibers, which fill the inside of a given layer; those fibers are at an angle of 45 degrees to the X axis; full filling was used; (C) fiber fill of the whole inside of the layer.

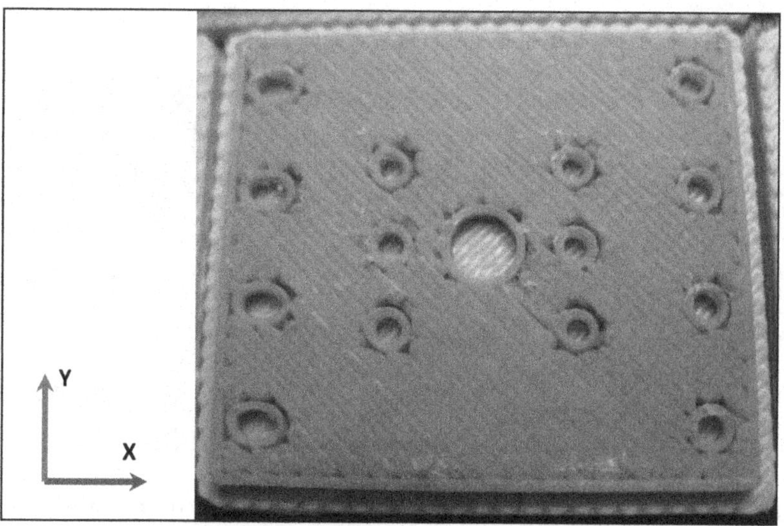


**Figure 7.50** View of the last layer of an exemplary, small object made by means of FDM 3D printing from ABS polymer; you can see the places of incomplete filling of a given layer. *FDM*, fused deposition modeling.

**Figure 7.51** An example of an object (cup holder) made using a 3D FDM printer from ABS polymer (A); a close-up on a full infill with fibers in the topmost layer (B); visible contour fibers and parallel fibers of the internal fill. *FDM*, fused deposition modeling.

through the extruder is not increased, it is difficult to avoid such gaps in the prints. Lack of fillings lowers mechanical strength of the prints and causes them to leak fluids, should, e.g., a bottle prototype be printed.

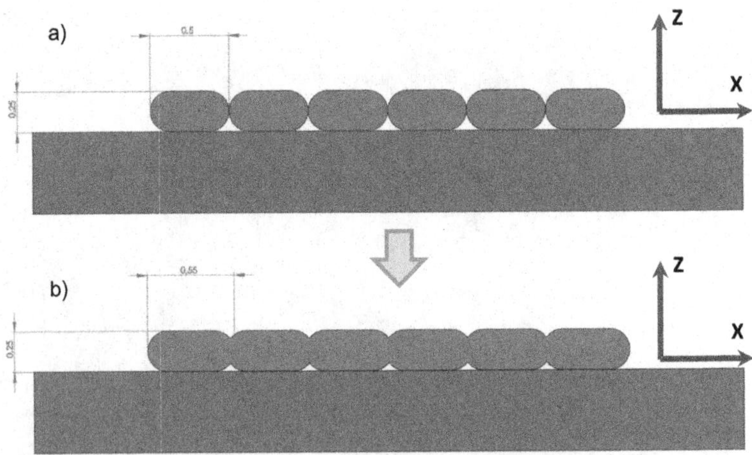


**Figure 7.52** A cross section through parallel fibers deposited on the 3D printer working platform; (A) the fibers practically do not connect; (B) an increase fiber width from 0.5 to 0.55 mm without path changes causes the fibers to connect more tightly.

For mechanical strength of the prints it is also important for the deposited fibers to come into contact at the sides. To show this, a cross section of a 3D CAD model shown in Fig. 7.48 was done, and its view is shown in Fig. 7.52, where parallel fibers deposited on the 3D printer working platform are visible. Initially, the fibers are 0.5 mm in diameter. After increasing their thickness to 0.55 mm (but without changes to distance between paths), the fibers connect more closely. Therefore, the whole print will be more solid (have less pores) and have higher mechanical strength. To change fiber diameter without modifying their paths and with a constant layer thickness, feed rate of the material through the extruder is increased. This means that the material is pressed into the nozzle a slightly faster than it was initially calculated as nominal.

An increase in the material flow rate causes an increase in the diameter of extruded and deposited fibers. Should the flow rate be changed without changing the shape and location of the deposited fiber paths, the layer full infill may be adjusted in such a way as not to cause gaps between fibers and to prevent overmelts (Fig. 7.53). A correctly selected flow rate parameter allows for a proper infill of a given layer.

a) 

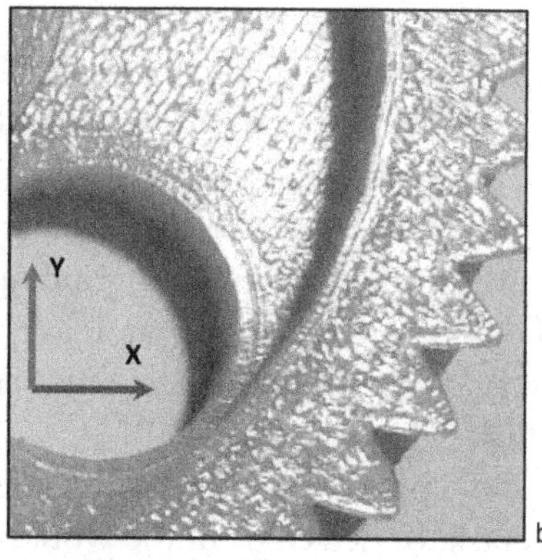 b)


**Figure 7.53** View of the last layer of a spur gear made by 3D FFF printing from PLA polymer (A); the close-up shows the exact layer filled with model material, no gaps (B). *FFF*, fused filament fabrication; *PLA*, polylactic acid.

## 7.8 Conclusions

As mentioned earlier, FDM/FFF 3D printing is currently the most common additive manufacturing method, thanks to its many advantages, such as follows:

**(a)** FDM/FFF 3D printers have a relatively simple design, thanks to which they are trouble-free, easy to transport, relatively inexpensive to buy and upkeep; FDM/FFF devices may be used in offices: they do not emit dusts and odors, are not loud, and do not heat up significantly; the devices may be easily prepared for the printing process, the model materials are easy to replace; no time-consuming cleaning of the working chamber is required after the 3D printing process.

**(b)** In comparison with other additive methods, the FDM/FFF method is easy and safe for everyday use, so it is most commonly chosen by people who want to begin using 3D printing as the first to learn or buy.

**(c)** A wide range of nontoxic and durable model materials are available for FDM/FFF 3D printers at low prices, an incredibly wide range of colors, which faithfully reflect those of traditionally produced objects; most model materials used in standard FDM/FFF are more thermally resistant than light curing resins used in SLA, PJM, DLP, and LCD methods.

**(d)** Lattice fillings can be used in the prints, which drastically reduces the amount of model material needed and shortens the process (if the printed objects do not need to have high mechanical strength); it is possible to pause the FDM/FFF process, thanks to which metal inserts may be placed inside the model and built over with the model material.

**(e)** The produced models are suitable for the production of mechanisms (latches, locks, gearboxes, etc.); the staircase effect can be smoothed out mechanically or chemically; the prints may be subjected to reductive machining (milling, grinding), mudding, and painting.

Like any other technology, FDM/FFF 3D printing additive manufacturing method has its limits and disadvantages, main of which are as follows:

**(a)** Large processing shrinkage of model materials causes the prints to curl, crack, or detach from the working platform; it is therefore often necessary to use a heated working chamber or working table.

**(b)** The need to use support structures for overhang walls (inclined more than 45 degrees from the vertical), which prolongs the 3D printing process, increases the amount of material needed, and raises the price of the method; walls with supports have a lower quality, support removal requires additional time, particularly when the supports are made from the same material as the model; in the case of breakaway support material (however, not in the case of chemically removed material), there are significant problems with its removal from narrow gaps or deep and small holes; therefore, FDM/FFF is not suitable for the production of lattice structures, space trusses, or other more complex shapes.

**(c)** High cost of high-strength model materials (PEEK) and materials dedicated for particular 3D printer types; most model materials are hygroscopic, which makes it necessary to provide specific packaging and storage, or placing them in airtight cassettes.

**(d)** In devices with one extruder, there is no possibility to use different materials in one 3D printing process; apart from exceptions described before, it is not possible to obtain multicolored prints in one process.

**(e)** Significantly lower mechanical strength of the prints between the layers (i.e., along the Z axis) than along the fibers (i.e., in the XY plane); this causes a need for specific design (particularly mechanical analysis) of parts produced using this method.

**(f)** Print stratification is visible, particularly in walls inclined at a low angle; so-called stairs are visible and finishing is required—which is often a time-consuming and expensive process.

**(g)** Dimensional accuracy of most prints is approximately ±0.2 mm; it depends on the used layer thickness, extruder nozzle diameter, feed rate, and model fill strategy.

To sum up the advantages and limits of the FDM/FFF additive method, it is safe to say that this method will still be the most popular 3D printing technique in the next 5–10 years, especially in amateur applications. In industrial applications, this can change due to the MJF method of the Hewlett–Packard Company.

It can be definitely said that FDM 3D printing along with the RepRap project has led to the international popularity of all the additive manufacturing technologies, not only FDM. The information which had spread, however, caused a number of people to identify the limitations of the FDM 3D printing method (and often low-quality FFF devices and printouts) with additive technologies in general. It is also important to note that those outside the 3D printing industry tend to call the FDM method simply "printing from plastic" and often have little knowledge on the topic. For this reason, a portion of business representatives and industrial users of additive technologies, i.e., SLM, EBM, MJF, take a highly critical view on the FFF method, considering it rather fun than a serious manufacturing method. Therefore, publications describing and comparing particular additive technologies are still very much needed and necessary.

## References

[1] T. Roberts, A. Bournias Varotsis, 3D Printing Trends, 2020.
[2] S.S. Crump, "Apparatus and Method for Creating Three-Dimensional Objects," 5,121,329, 1992.
[3] Stratasys Ltd. [Online]. Available: www.stratasys.com. [Accessed: 02-Feb-2020].
[4] Trademark - FDM, 1991 [Online]. Available, http://tmsearch.uspto.gov/. (Accessed 2 February 2020).
[5] Have Blue [dot org]. [Online]. Available: https://haveblue.org/?p=1600. [Accessed: 20-Jul-2020].
[6] RepRap project official website, RepRap. [Online]. Available: www.reprap.com. [Accessed: 02-Feb-2020].
[7] RepRap About." [Online]. Available: https://reprap.org/wiki/About. [Accessed: 20-May-2020].
[8] R. Jones, et al., Reprap - the replicating rapid prototyper, Robotica 29 (1) (2011) 177–191, https://doi.org/10.1017/S026357471000069X. SPEC. ISSUE.
[9] RepRap Darwin. [Online]. Available: https://reprap.org/wiki/Darwin. [Accessed: 20-May-2020].
[10] V.E. Kuznetsov, A.N. Solonin, A. Tavitov, O. Urzhumtsev, A. Vakulik, Increasing of strength of FDM (FFF) 3D printed parts by influencing on temperature-related parameters of the process, Rapid Prototyp. J. (2018) 1–32, https://doi.org/10.1108/RPJ-01-2019-0017, vol. ahead-of-p, no. ahead-of-print.

[11] G. Ćwikła, C. Grabowik, K. Kalinowski, I. Paprocka, P. Ociepka, The influence of printing parameters on selected mechanical properties of FDM/FFF 3D-printed parts, IOP Conf. Ser. Mater. Sci. Eng. 227 (1) (August, 2017) 012033, https://doi.org/10.1088/1757-899X/227/1/012033.
[12] B. Brenken, E. Barocio, A. Favaloro, V. Kunc, R.B. Pipes, Fused filament fabrication of fiber-reinforced polymers: a review, Addit. Manuf. 21 (February) (2018) 1–16, https://doi.org/10.1016/j.addma.2018.01.002.
[13] A. Bowyer, RepRap GPL Licence, 2016 [Online]. Available, https://reprap.org/wiki/RepRapGPLLicence. (Accessed 2 February 2020).
[14] RepRap Mendel. [Online]. Available: https://reprap.org/wiki/Mendel. [Accessed: 02-Feb-2020].
[15] RepRap MendelMax. [Online]. Available: https://reprap.org/wiki/MendelMax. [Accessed: 02-Feb-2020].
[16] RepRap Wallace. [Online]. Available: https://reprap.org/wiki/Wallace. [Accessed: 20-May-2020].
[17] RepRap Prusa I3. [Online]. Available: https://reprap.org/wiki/Prusa_i3. [Accessed: 02-Feb-2020].
[18] RepRap Family Tree. [Online]. Available: https://reprap.org/wiki/RepRap_Family_Tree. [Accessed: 02-Feb-2020].
[19] Wikipedia Makerbot. [Online]. Available: https://en.wikipedia.org/wiki/MakerBot. [Accessed: 20-May-2020].
[20] R. Smith, 3D Printing Is about to Change the World Forever, *Forbes*, 2015 [Online]. Available, https://www.forbes.com/sites/ricksmith/2015/06/15/3d-printing-is-about-to-change-the-world-forever/. (Accessed 2 February 2020).
[21] T. Roberts, A.B. Varotsis, 3D Printing Trends 2020. Industry Highlights and Market Trends, 2020. Chicago, USA.
[22] 3D Hubs, 3D printing trends Q1/2018, 3D Hubs 14 (2018).
[23] Grupa Azoty Wchodzi Na Rynek Materiałów Do Druku 3D, 05 March, 2020 [Online]. Available, https://grupaazoty.com/aktualnosci/grupa-azoty-wchodzi-na-rynek-materialow-do-druku-3d.
[24] S. Sarah, Filaments.directory Releases Results of 2018 3D Printing Filament Survey, 2018 [Online]. Available, https://3dprint.com/210552/filaments-directory-2018-survey/.
[25] Z. Foltynowicz, P. Jakubiak, Poli(kwas mlekowy) - biodegradowalny polimer otrzymywany z surowców roślinnych, Polimery 47 (2002) 769–774.
[26] Spectrum Filaments. High quality 3D printing filaments for each application, Filament Catalog (2019).
[27] Fiberlogy [Online]. Available, fiberlogy.com, 2020. (Accessed 2 May 2020).
[28] Facebook Fiberlogy [Online]. Available, https://www.facebook.com/fiberlogy/, 2020. (Accessed 20 May 2020).
[29] Simplify3D. Filament Properties Table, 2020 [Online]. Available, https://www.simplify3d.com/support/materials-guide/properties-table/. (Accessed 2 May 2020).
[30] P. Ślusarczyk, Kilka Słów Na Temat Domowego Suszenia Wilgotnych Filamentów, 2020 [Online]. Available, https://centrumdruku3d.pl/kilka-slow-na-temat-domowego-suszenia-wilgotnych-filamentow/.
[31] Prusa Face Shield Disinfection, 2020 [Online]. Available, https://help.prusa3d.com/pl/article/prusa-face-shield-disinfection_125459/. (Accessed 2 May 2020).
[32] Spectrum Filaments, 2020 [Online]. Available, https://spectrumfilaments.com/. (Accessed 2 May 2020).
[33] P. Siemiński, G. Budzik, Techniki Przyrostowe. Druk 3D. Drukarki 3D, Warsaw University of Technology Press, Warsaw, 2015.
[34] E3D-Online [Online]. Available, e3d-online.com, 2020. (Accessed 20 May 2020).

[35] Ekstruder E3D Titan 1.75 Mm I 3 Mm, 2020 [Online]. Available, https://blackfrog.pl/ekstruder-e3d-titan-175-mm-i-3-mm-p-606.html. (Accessed 20 May 2020).

[36] 3 In1 Out Nozzle Brass Extruder Diamond Hot End 0.4mm for 1.75mm 3D Printer Part, 2020 [Online]. Available, https://www.ebay.com/itm/3-in1-out-Nozzle-Brass-Extruder-Diamond-Hot-End-0-4mm-for-1-75mm-3D-Printer-Part-/253655171427. (Accessed 20 May 2020).

[37] Triple Filament Hotend (Diamond Hotend), 2020 [Online]. Available, https://reprap.org/wiki/Diamond_Hotend. (Accessed 20 May 2020).

[38] S. Wood, How to 3D Print with Flexible Filaments, 2014 [Online]. Available, http://www.gyrobot.co.uk/blog/how-to-3d-print-with-flexible-filaments. (Accessed 20 May 2020).

[39] K. Stevenson, Socks for Your 3D Printer Nozzle?, 2016 [Online]. Available, https://www.fabbaloo.com/blog/2016/9/1/socks-for-your-3d-printer-nozzle. (Accessed 20 May 2020).

[40] Nowe rozwiεązanie do chłodzenia wydruków 3D, 2015 [Online]. Available, https://3dwpraktyce.pl/2015/03/nowe-rozwiazanie-do-chlodzenia-wydrukow-3d/. (Accessed 20 May 2020).

[41] R. Szczesiak, M. Kowalik, M. Cader, P. Pyrzanowski, Parametryczny model numeryczny do predykcji właściwości mechanicznych struktur wytwarzanych W technologii FDM Z materiałów polimerowych vol. 9, 2018, pp. 626–632.

CHAPTER 8

# Electron beam melting process: a general overview

**Manuela Galati**
Department of Management and Production Engineering, Torino, Italy

## 8.1 Introduction

Electron beam melting (EBM) or electron beam powder-bed fusion (EB-PBF) [1] is a powder-bed additive manufacturing (AM) process. The process is based on the selective melting of metallics powder by an electron beam [2]. The EBM process is used for mass production in aerospace and medical fields for a wide variety of components, which are expensive or difficult to make by other conventional manufacturing or AM based processes [3]. Examples of industrial successes include turbine blades, turbocharger wheels, valves for combustion engines [4–6], and customized implants with high biocompatibility and structures with osseointegration properties [7]. Because of that, the EBM process is attracting attention and especially for the development of new materials at high performance such as intermetallic titanium aluminide alloys [8]. This chapter provides a foundation on the EBM process and an overview of the current state of the art on the research in the field. The chapter is organized into six sections. A detailed description of the process is provided in the section "Process description," which includes a resume of the processed materials and their applications. The second section "EBM physical mechanism" explains in detail the dynamics of the process. Then the section "Process control and process parameters" gives an outline of the main process parameters that are used for the process control and optimization. The next section, "Part features," is focused on the main characteristics an EBM printed part in terms of surface roughness, internal defects, part delamination, and variation of the chemical composition of the material. The section "Process monitoring" presents the tools and the methodologies developed for EBM process for the process monitoring and defects detection. The last section "Numerical simulation" provides a complete overview of the current state of the art of the EBM process models, which has been developed to explore "what-is scenario," speed up the process optimization and improve the understanding of the process.

*Additive Manufacturing*
ISBN 978-0-12-818411-0
https://doi.org/10.1016/B978-0-12-818411-0.00014-8

### 8.1.1 Process description

An EBM system is based on the operating principle of an electron microscope while the main hardware is similar to a welding system. Arcam AB (Sweden), founded in 1997, has developed the first EBM technology [2]. The machine hardware (Fig. 8.1) consists of the top column and the processing chamber. The top column contains the electron gun and the magnetic lenses. In the electron gun, a cathode emits electrons. The cathode can be a LaB crystal or a tungsten filament, which is heated to emit electrons. The electrons are accelerated by an anodic potential of 60 kV up to 40% the speed of light [9]. Magnetic lenses [10] focus and deflect the electron beam [9,11]. To avoid the deflection of the electrons, the whole process occurs under high vacuum. In the processing chamber, the typical pressure of residual gasses is around $10^{-3}$ Pa and $10^{-5}$ Pa in the electron gun [3]. A flow of helium gas ($10^{-1}$ Pa) is also used during the process to avoid the accumulation of electrical charges in the powder bed, which facilitates the cooling phases and guarantee a high thermal stability [12]. The processing chamber consists of the steel build tank, the powder hoppers, and the raking system [13]. The steel build tank contains the start plate that represents the building plane of the layer and can translate along building

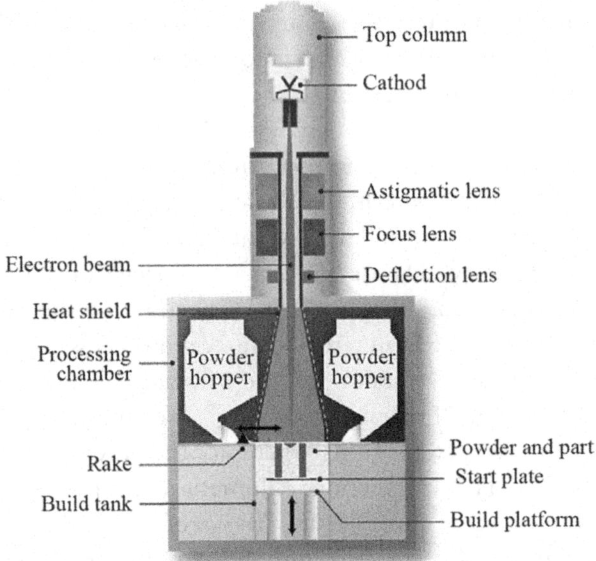


**Figure 8.1** Architecture of an electron beam melting machine.

direction [13]. The two hoppers are located in the left and right corners of the chamber and store the powders. The powder falls in the chamber through an outlet located on the bottom part of the hopper and is distributed on the start plate or the previous layer by a rake system. The rake system consists of a series of thin and flexible stainless-steel teeth.

The build process begins with the heating of the start plate by the electron beam. The material of the start plate and the temperature to be achieved before to start the process depend on the processed powder [13]. Before the melting, each deposited layer is also completely preheated by a series of passages of the defocused beam at low power and high speed [12]. After the melting, the layer may be further heated. The start plate is then lowered, and new powder is delivered from the powder hoppers and then raked. The process is repeated until the part is completed. After the building is finished, the job cools down in the chamber with a helium flow. The term "job" is referred to the set of the parts with the soft agglomeration of powders that adheres to the fabricated parts and covered them completely [3,12]. This agglomerate also named breakaway powder [12] and is removed by a blasting process [3,12,14,15] in which the same powders of the job is used. After the cleaning phase, the unused powder is sieved and completely recycled for the next job [3]. Differently from other AM processes for metal components, EBM is identified as a hot process because of the high working temperature in the chamber. Thanks to this, the final parts do not need any stress-relieving treatment and material with a high melting point can be processed.

## 8.1.2 Materials and applications

The metallic powders for EBM process are produced by the gas atomization process [16–18]. The powder shape is spherical [19], and the distribution of the size of virgin powders ranges between 45 and 100 μm [17,18,20] while the size of the sieve mesh is usually 150 μm. As mentioned, the powder can be recycled without appreciable modification in chemical composition or physical properties [3,12,21]. EBM systems can work with many classes of materials that include steels (17–4 and H13), Ni-based superalloys (625 and 718), Co-based superalloys (Stellite 21), low-expansion alloys (Invar), hard metals (NiWC), intermetallic compounds, aluminum, copper, beryllium, and niobium [3]. γ-TiAl alloys are central for EBM applications [3] especially the intermetallic alloys because to date they cannot be processed with any other process. Applications include automotive components [22,23] and medical implants [24–29] that are customized with structure at high biocompatibility and osseointegration [14,24,26,28,30–32] and aerospace components [14,22] at high performance, lightweight, complex shapes [3,12,33] and with a low buy-to-fly ratio [3,34].

## 8.2 Electron beam melting physical mechanisms

Fig. 8.2 shows the main physical mechanisms that occur thanks to the energy provided by the electrons during the EBM process. The kinetic energy of the electrons is transformed into thermal energy. According to the amount of energy which is provided into the powder bed, the material could be sintered, melted, or vaporized [35]. Then the heat transfer takes places mainly by heat conduction between the powder particles and between the powders and the bulk substrate; irradiation from the powder bed to the chamber; and convection between the powder bed and the surrounding environment due to the helium flow. The phase change from powder to the liquid and then to solid occurs over a short time [35]. The typical lifetime of the melt pool is typically a few tens of milliseconds [36]. The generation of high-temperature gradients between the melted and the surrounding areas involves a high dynamicity of the melt pool and the generation of a material flow due to the Marangoni convection and the capillary forces [35].

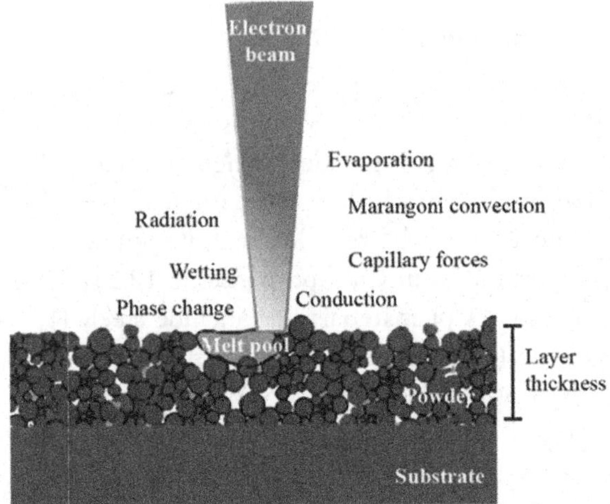


**Figure 8.2** Electron beam melting physical mechanism during the melting phase.

At the beginning of the interaction between the electron beam and the powders a spread of powders in the build chamber, also called smoke, may occur like an explosion [10,37]. The causes are still not clear. Qi, Yan [19] found that the momentum transferred by electrons would cause the spreading because the generated impact force is greater than the force of the cohesion between the powder. This effect is increased by the fluidity of the powders and is reduced by increasing the beam current and particle size. The analytic studies developed by J. Milberg [10] and Sigl, Lutzmann [37] have detected the repulsion force due to the electrostatic charge of powder particles as the main cause of the smoke. The powder particles would be negatively charged because the charge distributed by electrons during the hitting. The generated repulsion forces would result far higher than the weight of a particle. The preheating step, which allows the formation of circular necks (Fig. 8.3) [38,39] that connect the particles, would help to increase the apparent total weight the powder particles and to reduce the process because produce a strong enough cohesion between the powder particles that increases the thermal conductivity of the powder bed. Moreover, the strength of the sintered powders allows reducing the number of structures need to support the overhangs of the component to be produced [3,12,14,15].

During the melting of powders, the formation of the melt pool leads to rapid sintering of the surrounding material and mass transportation. The movement of the melt pool wets the surrounding material and may increase the surface tension. Higher surface tensions lead to soaking more powder into the melt pool [19,35] and a particle rearrangement [40]. Because of

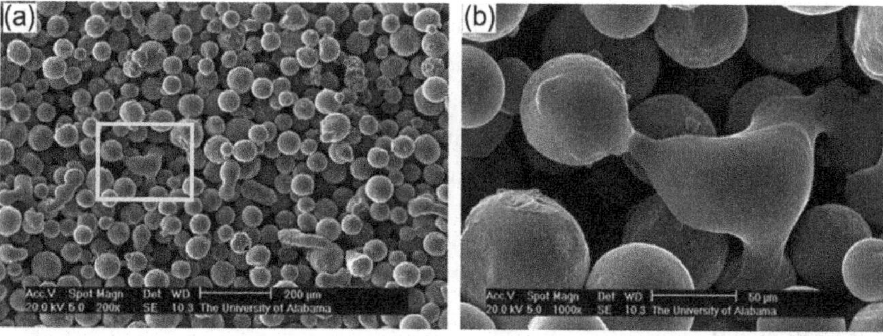


**Figure 8.3** Neck formation: (A) raw powder and (B) sintered powder. *(Reproduced with kind permission of EDP Sciences from Reference Gong, X., T. Anderson, and K. Chou, Review on powder-based electron beam additive manufacturing technology. Manuf. Rev., 2014. 1.)*

that, the control of molten material flows is extremely important [35,41] and is closely correlated to the wetting characteristics of the solid phase by the liquid phase [35]. The wetting characteristics depend on the material temperature, impurities, and contaminations present in the powders [40,42]. If the melted powder and the solid substrate have similar temperatures the wetting will not occur [35] and additional beam energy may be required to guarantee the adhesion between the actual layer and previous one [42]. However, the formation of the melt pool and the required excess of energy increase the heat transfer rapidly, producing high thermal gradients that generate numerous thermal flows. These flows increase the heat transfer, which causes the drop in the temperature of the melt pool and as a consequence, a further increase of the thermal gradient. The surface temperature gradient leads to turbulent flows, and the temperature gradient throughout the melt pool depth causes the generation of flotation forces. These flows are known as thermocapillarity flows or Marangoni convection flows [43]. The flow instability can lead to Rayleigh instabilities [43] and the balling effect for which the melt pool breaks into small droplets [35,40,43,44]. When this effect is evident, the roughness of top surface may not be smooth enough to enable the spreading of the new powder uniformly [45], causing marked defects or even the failure of the process. The excess of energy may also provoke the evaporation of some alloying element [12]. This effect is also greatest because of the vacuum environment that decreases the evaporation temperature of the element [3].

## 8.3 Process control and process parameters

A set of process parameters is called "build theme" and contains hundreds of process parameters. Most of them are optimization functions that can be set for a specific condition. Arcam AB standard themes include a specific theme for the preheating phase and melting phase, which considers different process parameters for the melting of the lattice structure, supports, and solid bulk. The main process parameters can be considered:

- Layer thickness
- Scanning strategy
- Line offset
- Scan speed
- Beam current
- Focus offset

Experimental and numerical investigations have also shown that parameters, such as the energy density [46] and line energy [9], are relevant to measure the quality of the process. These parameters are calculated by a combination of the abovementioned ones. The energy density is defined as the ratio between the beam power and the section of the beam. The line energy is calculated by the ratio between the beam power and the scan speed. The beam power is calculated by the product between the acceleration voltage and the beam current. The section of the beam depends on the beam diameter, which is set by the focus offset and the beam current [47], and it is controlled by the electron magnetic lenses [48].

The layer thickness is ranging between 0.050 and 0.200 mm depending on the powder material [49]. Nonuniform layer thickness can increase the smoke effect [13]. The beam current and the scan speed values could be set as a function of the value of layer thickness [50]. The set layer thickness corresponds to the lowering of the start plate. However, the deposited layer of powder is greater the set layer thickness to account to the consolidation of the powder during the sintering and the melting phase.

The scanning strategy includes the scanning process and the scanning mode [51]. These process parameters set affects the beam impact on the powder, the heat transfer, the temperature distribution, the final microstructure of the part, and the shrinkage after the solidification. As mentioned, the scanning process includes the preheating steps [13], the melting phase, and postheating [13]. During the preheating step, the beam performs a full scanning of the layer to ensure high-dimensional precision and a fine surface. A second preheating may take place only in the area, which corresponds to an offset of the area to be melted. The whole preheating phase aims to have a uniform temperature on the powder bed equal approximately 60%–70% the melting point of the material. Usually, two melting strategies are used during the melting step: the contour is melted using the MultiBeam strategy [48,52] while the inner area using a hatching strategy. The contour is defined by the perimeter of the section to be melted. Up to five concentric offsets of this perimeter can be used in the contour [13,48]. These additional contours are called inner contours [48]. During MultiBeam mode, the control system rapidly moves the beam according to a discontinuous pattern [48,53]. In this way, separate melt pools are activated in different points of the contour that is molten quasi simultaneously [44,48,52,53]. The process parameters to define a MultiBeam strategy are the number of beam spots, the spot time, the scan speed, and the minimum distance between two consecutive jumping points [47].

This kind of strategy is used to better define the contour of the melted part and avoid a high heat dissipation from the melt pool to the surrounding area typical of the use a continuous melting. The roughness of the surface is also improved. During the hatching, parallel continuous lines are melted at a constant beam current and speed for a certain distance according to the area to be melted [47]. The distance between two parallel lines is called hatch and is settable by the line offset parameter [50]. The scan path is similar to a snaking movement [48]. To avoid anisotropy of material, the hatching direction should be rotated by 90 degrees between each layer [48,51]. For a job with multiparts, it is necessary to define the behavior of the beam. Usually, to increase the productivity and to uniform the temperature distribution, the parts that are built in the same job are considered simultaneously. Therefore, both the contours and the inner of the parts are considered as a part of a unique part. The hatching is adapted to have single continuous lines, which are melted over the parts according to the reference length parameter. During the postheating phase, the layer can be either cooled down or further heated depending on the total amount of energy supplied during the previous steps [54]. The process parameters for this phase are set equal to the first preheating phase [51]. The preheating and the postheating keep the chamber warm and help to reduce the total shrinkage [51].

The scanning speed can set using the speed function (SF) [50]. The SF index is a parameter setting [55], which is used to dynamically control the translation of electron beam during the process [13,55] and, as consequence, the melt pool size [50]. Negative SF values indicate a linear relationship between the scan speed and beam current value. Positive SF values indicate the use of SF algorithms, which are not available in the literature and outside the Arcam [13,50,56]. Mahale [13] observed that the instant scan speed value increase by increasing the SF index [13].

Since the acceleration voltage is constant during the process, the beam current is the only parameter that affects the power of the beam. The beam current value can be constant during the scanning [56] or can be varied according to the line to be melted. The beam current can be expressed as a function of the imposed maximum current value, the part height, and the layer thickness [13,39]. In general, the beam current can be ranging between 1 and 50 mA [5]. The beam current and the focus offset, jointly control the spot size.

The focus offset is the additional current that the electromagnetic lenses can provide to translate the focal plane from its zero position and adjust the beam diameter [57]. The focus offset values can range between negative and positive values. Studies on the effect of the variation of the focus offset were mainly focused on the roughness of the surface [50,58]. As mentioned, the beam current and the focus offset control the beam diameter but this relation is not available outside of Arcam AB [56]. For constant beam current, the relation between focus offset and beam size is not linear [56]. Differently from what is reported in the literature, the increase of the focus offset at constant beam current does not always mean a bigger diameter and vice versa [56]. In the range of higher-focus offset values, increasing the focus offset at constant beam the spot size rises [56]. On the contrary, for lower-focus offset values (negative values) the spot size increases by decreasing the beam current [56]. Therefore, for a given value of beam current, the morphology of melt pool changes varying the focus offset. Larger beam diameters lead to a greater width of melt pool but a lower depth [50]. Because the spot is less concentred, an increased beam section means a lower value of energy density [50]. However, during hatching, at constant line offset values, the overlap area increases when the spot size increases. In the same way, a smaller beam section leads to a smaller overlap area, but a higher energy concentration increases the melt depth. Generally, in a standard procedure, the beam is used at its most focused during contouring and less focused during the hatching [48]. A procedure to measure the beam diameter is reported in Ref. [47]. Recently, Arcam AB introduced a control system, called xQam [49], which is an X-ray detection for in-situ monitoring and can be used to calibrate the beam characteristics automatically using an autocalibration algorithm [44].

Table 8.1 [59] reports the main process parameters of the Arcam theme for processing Ti6Al4V powder into an A2x model equipped with the EBM control 5.2.

## 8.4 Part features

### 8.4.1 Surface roughness

Because of their dimensional accuracy and surface finish [3,22,31], the parts produced by the EBM process are only near-net-shape. The morphology of the surface changes with the orientation of the surface. Fig. 8.4 [59] shows the typical morphology of a specimen produced by EBM process and which include upward and downward surfaces. Vertical and upward surfaces are characterized by an exterior appearance rippled, with visible sintered powder particles [22,60]. Although the surfaces of the parts are

**Table 8.1** Arcam standard theme for processing Ti6Al4V powder in A2X model equipped with EBM control 5.2 version. Layer thickness is set equal to 0.050 mm.

| ***Preheating 1*** | | | | | | |
|---|---|---|---|---|---|---|
| Scan speed (mm/s) | Focus offset (mA) | Beam current (mA) | Max current (mA) | Number of repetition | Line offset (mm) | |
| 10,000 | 70 | 16 | 30 | 2 | 1,2 | |
| ***Preheating 2*** | | | | | | |
| Scan speed (mm/s) | Focus offset (mA) | Beam current (mA) | Max current (mA) | Number of repetition | Line offset (mm) | |
| 13,000 | 70 | 19 | 38 | 3 | 1,2 | |
| ***Outer contour*** | | | | | | |
| Melting strategy | Scan speed (mm/s) | Focus offset (mA) | Beam current (mA) | Hatch contours | Number of contours | |
| MultiBeam | 850 | 6 | 5 | 0,29 | 3 | |
| ***Hatching*** | | | | | | |
| Melting strategy | Speed function | Focus offset (mA) | Max beam current (mA) | Reference length (mm) | Reference current (mA) | Line offset (mm) |
| Continuous | 45 | 25 | 20 | 45 | 12 | 0,2 |

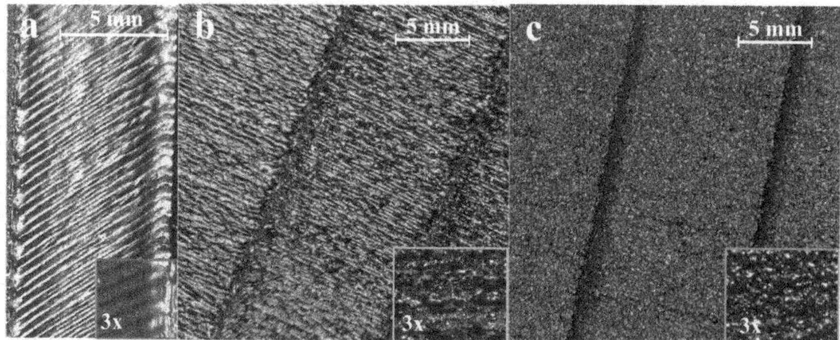


**Figure 8.4** Surface roughness of an electron beam melting part with the use of MultiBeam strategy: (A) top surface, average Ra equal to 6 μm; (B) upward surfaces, average Ra value equal to 21 μm; (C) downward surfaces, average Ra value equal to 23 μm. *(Reproduced with kind permission of MDPI from Reference Galati, M., P. Minetola, and G. Rizza, Surface roughness characterisation and analysis of the electron beam melting (EBM) process. Materials, 2019. 12(13): p. 2211.)*

rough, usually the surfaces are well formed and continuous [31]. In Ref. [60], the external ripples were highlighted by building a vertical cylindrical surface without the use of the MultiBeam strategy [32]. Generally, the average surface roughness, in terms of Ra, ranges between 15 μm [59] and 50 μm [60,61] with a minimum for the top surfaces, which is around 6 μm [59].

Several studies have shown that the main cause of the surface roughness is the combination of the process parameters, which can be lead an only partial melting of the powder particles that adhere to the solidified surface [37,61–65]. Consequently, the surface roughness is closely correlated to the powder particle size [16,63,66]. A higher number of small particles in the powder batch increases the possibility to find sintered particles attached to the surface [66]. Safdar, He [63] detected that the roughness of surface decreases for high values of focus offset and scan speed because a smaller-focus offset concentrates the beam and generates a larger temperature gradient. In Ref. [58] the focus offset has been found to have the greatest effect on the surface roughness over the beam current and the beam speed. Jamshidinia and Kovacevic [62] showed that the relation between the surface roughness and the hatching distance is inversely proportional. Klingvall Ek, Rännar [61] noticed a relationship between the surface roughness and the combination between the beam current used during the contour and the distance between two adjacent contours. For the upward surfaces, the roughness is affected by the staircase effect and varies linearly with the slope angle [59]. For the downward surface, the surface roughness is mainly by heat distribution [59].

The morphology of the surface changes when the line offset increase up to the appearance of voids, porosities, and the balling effect [44,50]. Porosity appears when the line offset values do not allow a correct overlap between two adjacent lines [50,67]. Rough and bumpy surfaces such as the ones shown in Fig. 8.5 may inhibit the process from continuation [44] because they preclude the correct wetting and adhesion between layers.

As mentioned, the balling is due to the break of the melt pool and the subsequent solidification of melted material into balls. Physically the surface tensions drive the melt balls formation and transform the melted volume into a cylinder [43]. When the cylinder length exceeds to circumference length (or $\pi$ times to diameter), the melted cylinder breaks into droplets [68,69]. The first studies about the balling effect concerned the selective laser melting process [70–72]. For the EBM process, Zäh, Lutzmann [46] have shown that the limit can be largely exceeded without the balling effect occurs. The authors noticed that the surface tensions are lower for the EBM process, thanks to the preheating phase and vacuum environment. Qi, Yan [51] found that the instability can be reduced by reducing the length of the inner lines and thus of the melted cylinder. A smaller length difference between the two adjacent hatching lines can also help to reduce the balling effect [51]. Experiments on the single-line scan by Korner, Attar [9] have shown that the formation of the droplets begins independently during the

**Figure 8.5** Balling effect. *(Reproduced with kind permission of Springer Nature from Reference Zäh, M.F. and S. Lutzmann, Modelling and simulation of electron beam melting. J. Inst. Eng. Prod., 2010. 4(1): p. 15–23.)*

melt phase due to the local spread of powder, wetting, gravity, and capillarity forces. Lower values of line energy cause an increase of the balling effect.

## 8.4.2 Internal defects

Powder defects (Fig. 8.6) and process parameters could cause local defects and zones of unmelted powder [12,50,73].

Incorrect process parameters cause macroscopic density gradient [3,12,31,74,75] due to unmelted powder (Fig. 8.7A–F) and the balling effect. unmelted powders cause localized unconsolidated regions (Fig. 8.7A) and unmelted pockets (Fig. 8.7C–F [12]; [76]). The numbers and the wideness of the unmelted regions could be reduced by increasing the line energy and the energy density. The line energy increases when the scan speed decrease [12], and the beam current increases. However, the beam current variation leads to change of a beam size that may lead to a reduction of the energy density [56]. Gong, Rafi [67] and in the subsequent work [50] investigated the residual porosity varying the focus offset and the line offset. The porosity dramatically increases over a certain value of line offset because of the smaller overlap between adjacent lines. Unmelted pockets are localized under the top surface between two

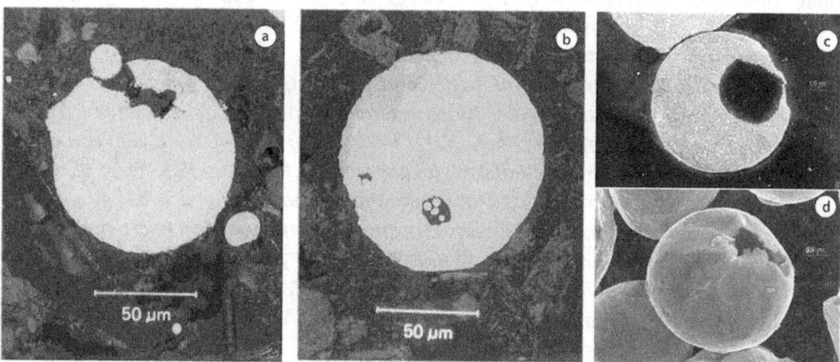


**Figure 8.6** Powder defects. Light micrographs of two powder particles from the size fraction 63–90 mm: (A) Particle with an open pore (B) powder particle with a closed pore of Argon (C) Optical metallographic view of the ground, polished and etched powder particle and (D) SEM view of gas bubble breaking powder particle surface. *(A), (B) Reproduced from Reference Kahnert, M., S. Lutzmann, and M. Zaeh. Layer formations in electron beam sintering. in Solid Freeform Fabrication Symposium. 2007; Reproduced with kind permission of Taylor and Francis Group from Reference Gaytan, S.M., et al., Advanced metal powder based manufacturing of complex components by electron beam melting. Mater. Technol., 2009. 24(3): p. 180–190.)*

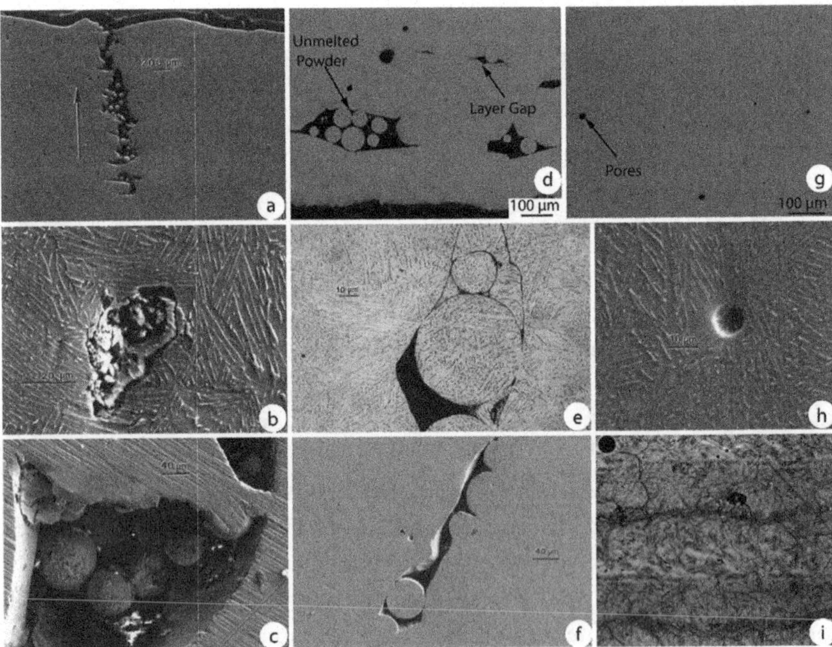


**Figure 8.7** (A) Continuous beam tripping resulting in unconsolidated region progressing with build (*arrow*) (B) build flaw revealed after grinding and polishing sample (C) unmelted pocket (D) OM image of lack-of-fusion defects observed in EBM-built, difference between unmelted pocket and layer gap Ti–6Al–4V (E) and (F) unmelted and unconsolidated zone (G) Optical micrograph of micropores (H) spherical gas void and (I) oxide deposits in H11. *(A), (B), (C), (E), (F) and (H) Reproduced with kind permission of Taylor and Francis Group from Reference Gaytan, S.M., et al., Advanced metal powder based manufacturing of complex components by electron beam melting. Mater. Technol., 2009. 24(3): p. 180–190; (D) and (G) Reproduced with kind permission of Emerald Publishing Limited from Reference Kok, Y., et al., Fabrication and microstructural characterisation of additive manufactured Ti-6Al-4V parts by electron beam melting: this paper reports that the microstructure and micro-hardness of an EMB part is thickness dependent. Virtual Phys. Prototyp., 2015. 10(1): p. 13–21; (I) Reproduced with kind permission from Reference Sigl, M., S. Lutzmann, and M. Zäh. Transient physical effects in electron beam sintering. in Solid Freeform Fabrication Symposium Proceedings. Austin, TX. 2006.)*

adjacent lines until the melt pool does not fail to contact the previous hatch line. After that, the unmelted powder is visible through the pores of the top surface, causing a gap between the hatch lines. Lower values of focus offset cause profound melting depth while higher values increase the overlap but reduce the melting depth and create an unmelted pocket and an unstable melt pool. In last, inconsistent raking could also cause unmelted pocket and layer gaps [76].

Fig. 8.7I [37] shows the typical defects in the microstructure due to impurities such as oxide or nitrates. The impurities may be deposited in the virgin or recycled powder [12,37] and have lower density [73], which causes a discontinuity of the beam path [12]. The residual impurities can be avoided producing and storing the powders in an inert environment [37] with helium or argon gasses [77]. During the powder production by gas atomization process, the rapid solidification of the particles may cause the entrapment of gas and a residual porosity into the powder particles [3,12,77,78] as gas bubbles or spherical voids (Fig. 8.7B and G [12]; [76]). These gas bubbles persist after the melting (Fig. 8.7H) [12]. To eliminate such voids, EBM parts can be hot isostatically pressed [3,12,77].

## 8.4.3 Delamination

Delamination is the separation of two or more adjacent layers within the part due to an incomplete melting between layers [79]. This defect [73] is macroscopic and cannot be corrected by postprocessing [79]. High thermal gradient and different cooling rate between the melted areas generate differentiated shrinkage and internal tensions. When these tensions exceed the limits of cohesion of the layers, the layers split and forming gaps. Highest residual stresses emerge close to free edges, causing the maximum delamination and bending of the borders of the part [73]. Higher energy density and tuned scanning mode improve the connection between layers and reduced residual stresses [44,73]. Since the most uniform energy input causes the fewest amount of residual stresses, the best results can be achieved avoiding the unidirectional scanning raster.

## 8.4.4 Chemical composition of the material

Due to the thermal nature of the EBM process, the final chemical composition of the material may be slightly different from the original composition of the powder. In the case of Ti48Al2Cr2Nb, overheating causes the losses of aluminum content [12]. Experimental studies [3,75] highlighted that the evaporation of lightweight elements is also affected by the high vapor pressures of some alloying elements at high temperature and under vacuum. An experimental study performed by Cormier, Harrysson [75] on H13 stainless steel as shown that as-built chemistry differs slightly from that of the initial powder. A similar study of Biamino, Penna [3] on Ti48Al2Cr2Nb has quantified the aluminum losses to be lower than 1.0 wt%, and in particular, the total maximum variation of aluminum content was 0.2 wt% both along the axial length of 300 mm long samples built along with the vertical position and through the build area of 200 × 200 $mm^2$. Similar studies have been performed also to verify the chemical composition of the powder after the process highlighting that the powder can be recycled many times without appreciable modification of its chemical composition and its physical properties [3]. Gaytan, Murr [12] compared the content of alloy elements in virgin powder, in breakaway powder mixed with recycled powder, and in the fabricated part made by Ti6Al4V. After 40 cycles, the differences between chemistry of the powder and the part has been quite constant even using reused powders. Considering a less number of reuse cycles, Tang, Qian [21] shown that the aluminum content decreased from 6%, 47% to 6%, 37% by weight, while oxygen content increased from 0%, 08% to 0%, 19%.

## 8.5 Process monitoring

All Arcam systems monitor and record continuously all sensors with which the machine is equipped. These data are available during the process and at the end of the job to provide a certain qualification of the produced part. Among these data, the temperature of the start plate is measured by a thermocouple located under the start plate. This temperature can be taken as indicative of the chamber temperature. Additionally, the Arcam systems are also equipped with a high-resolution camera, called LayerQam, which takes an image of each layer after the melting. At the end of the process, an image-processing software compares the nominal section to be melted and the acquired picture for each layer to detect any melting defect [7]. To monitor the local temperature of the melt pool a near-infrared camera has been used in the work Ref. [80]. The camera was installed out of the machine and in front of the door window. For this reason, the camera needed a calibration. Inside the chamber, an infrared camera was used in the works [57,81]. The camera was placed almost coaxially to the electron gun and was used to acquire images at the end of the melting of each layer. In Ref. [81], a closed-loop feedback control was implemented to automatically adjusts the beam in the case of melting process defects. However, the system was slower than the process to allow a prompt process correction. Yoder, Nandwana [82] examined the in-situ data collected from LayerQam camera of Arcam Q10pLus machines to correlate the mechanical properties of the part to its internal defects due to the part proximity, melt order, and cross-sectional area. In the works [83–85], an electronic imaging system that uses the backscattered electrons (BSEs) was developed and installed in an Arcam A1 to investigate the performance of the systems and the effect of the high temperature on the quality of the images. An image of a bulk part was acquired at room temperature and approximately 650°C. A small effect of the temperature on the quality of the images was detected. In Ref. [86], a BSE coaxial detector has been developed to acquired electron-optical (ELO) images during the whole process. To validate the system, ELO images have been acquired for a whole job and compared with the tomography of the part. Although the limited spatial resolution of the BSE detector, ELO-images showed to have a great potential for monitoring the EBM process to reduce the time to determine the process window from weeks to hours the time [87]. An in-situ quantitative inspection technique for assessing the powder quantity spread by the rake has been developed by Liu, Blunt [88]. The system used a fringe projection profilometry to inspect the powder bed before and after the melting to assure a proper powder distribution and part geometry.

## 8.6 Numerical simulation

Process simulation has been recognized as a powerful tool to speed up the process optimization, the material development, and the process understanding. Since the process simulation is a description of a real process, the modeling phase includes a careful analysis of the mechanism, which are involved in the physics of the process. That means that the model will be an abstraction from the real process and will be as simple or complex as is necessary to predict and explain a specific aspect of the process. As far as the EBM process is concerned, the modeling phase must consider three main aspects: the powder modeling, the heat source model, and the modeling of the boundary conditions. According to the level of detail [89], the models for the EBM process can be grouped into three categories: white, gray, and black-box models. The black-box model solves a numerical problem with a high level of approximation. A gray model considers some of the physics of the process and the complexity and the flexibility of the model depends on the number of approximations, which have been introduced at the modeling phase. The white models involve much information, they reproduce the process in a real small scale, and the level of approximation is low or insignificant. The greater the level of detail with which the process is modeled, the greater the physical meaning and the complexity of the model are. A high level of detail involves high computational cost. White models are the most complex kind of model. There will be no purely black box or white models, but models whose structure and characteristics aimed toward the limit conditions. A further classification of the gray models groups the models according to the scale of the analyzed phenomenon. Microscale models consider EBM phenomena at the microscale such as the interaction between the electron beam and the powder. Such effect is mainly investigated to introduce fewer approximations such as in Ref. [90]. Mesoscale models consider the powder particle to particle and involve many physical mechanisms. Examples of these models are reported in Refs. [9,91–94]. The level of approximation is low, and some physical phenomena such as powder and the solid shrinkage are neglected. The resulting model is complex, and expensive and thus not suitable for large-scale simulations. For these reasons, those simulations are rarely at the 3D scale. To decrease the computation cost, macroscale models have been developed. The models consider the powder as a continuum and are solved using the finite difference method, the finite element method (FEM), or the finite volume method. The macroscale models can be further classified according to the number of physics that have been considered. Therefore, the models can be solved using a unique solver (uncoupled models) or multiple solvers simultaneously (coupled) or separately (weakly coupled models). Uncoupled models, which are also called pure thermal models considered only thermal phenomena, have been developed by Qi, Yan [51], Mahale [13], Zäh and Lutzmann [44], Neira Arce [16], Professor Chou's research team [39,55,95–97], Galati et al. [47,90,98] as well as Riedlbauer, Scharowsky [99]. The weakly coupled models are those in which the result of thermal analysis is used as input into a subsequent analysis. An example of the application of this kind of model for the EBM process is provided in Ref. [100] in which the temperature distribution is used as input to simulate the final microstructure and the grain growth. Coupled models aggregate different physics, which are solved simultaneously. A thermo-fluid single-track model has been developed in Ref. [101] using an FEM analysis coupled with a CFD analysis. Thermomechanical models for the melting of a single track have been developed in the works [102,103] to investigate the residual thermal stresses. A thermomechanical model for the melting of a single layer with both MultiBeam and hatching strategies has been developed in Ref. [54] in which the effect porosity reduction of the powder bed on the melt pool size has been also included. A simplified simulation of thermomechanical effect has been implemented in Ref. [104] to simulate a multilayer-crossed raster scan. This model has been also used to analyze the

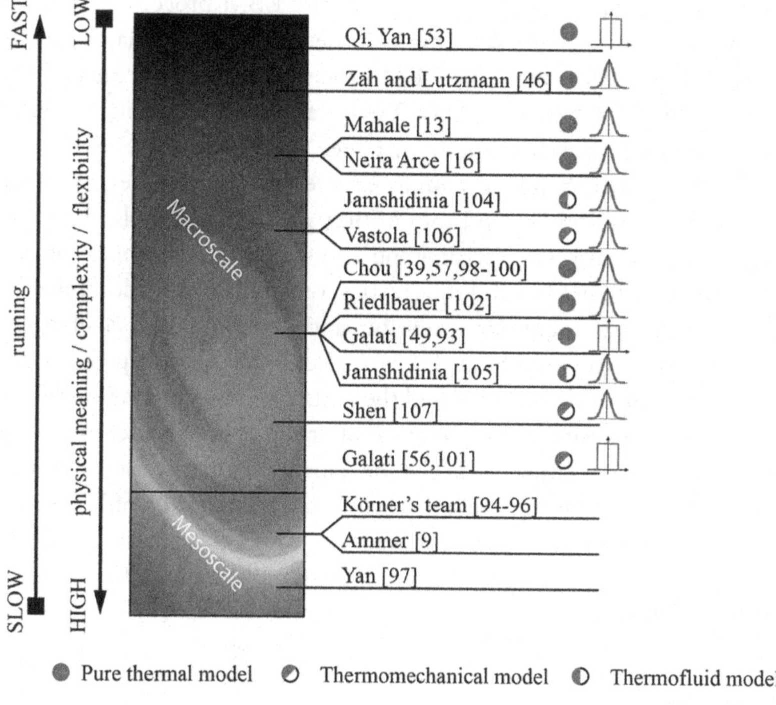


**Figure 8.8** Classification of the gray model for the electron beam melting process simulation. *(Reproduced with kind permission of Elsevier from Reference Galati, M. and L. Iuliano, A literature review of powder-based electron beam melting focusing on numerical simulations. Additive Manufacturing, 2018. 19: p. 1–20.)*

thermal stresses, which are associated with part overhang [105,106]. Fig. 8.8 classifies the gray models according to their level of approximation. The models differ from each other in the number of solvers, the modeling of the heat source, and material properties. In most of the studies, the heat source is modeled with a Gaussian distribution in which the heat flux is calibrated by using proper coefficients. A detailed review of these models can be found in Ref. [89].

## 8.7 Summary and scientific and technological challenges

This chapter provided a complete overview of the current state of the art on the research about the EBM process. The process has been presented from a physical and technological point of view. The chapter resumes scientific literature, which was focused on the EBM process and provides numerous references to explore more in detail the topics that have been addressed. The full vision of the EBM process revealed its great potential, which is still locked by the current issues in terms of part quality and efforts to come with an optimized solution. Chief among the presented issues is the need for a better understanding of the process. Numerical process models and in-situ sensing and monitoring are the only tools, which can support both the process understanding and the process optimization. The development of these tools and the creation of new tools that combine their results represent the current scientific and technological challenge to eliminate build errors, improve the part quality and speed up the qualification and testing of the material and the part. The implementation of these tools and their outputs in a framework of artificial intelligence may make easier the implementation of real-time closed feedback control for the process optimization and the exploring what-if scenario for offline process optimization. All these activities aimed to improve the part quality and process efficiency, in turn, will result in greater adoption and proliferation of EBM systems.

## References

[1] ISO/ASTM, ISO/ASTM 52900: 2015 Additive Manufacturing—General Principles—Terminology, International Organization for Standardization, Geneva, 2015.
[2] L.U. Larsson Morgan, O. Harrysson, Rapid manufacturing with electron beam melting (EBM)—a manufacturing revolution?, in: Solid Freeform Fabrication Symposium, 2003 (Austin, TX).
[3] S. Biamino, et al., Electron beam melting of Ti-48Al-2Cr-2Nb alloy: microstructure and mechanical properties investigation, Intermetallics 19 (6) (2011) 776—781.
[4] H. Clemens, H. Kestler, Processing and applications of intermetallic γ-TiAl-based alloys, Adv. Eng. Mater. 2 (9) (2000) 551—570.
[5] D. Dimiduk, Gamma titanium aluminide alloys—an assessment within the competition of aerospace structural materials, Mater. Sci. Eng. 263 (2) (1999) 281—288.
[6] S. Knippscheer, G. Frommeyer, Intermetallic TiAl (Cr, Mo, Si) alloys for lightweight engine parts, Adv. Eng. Mater. 1 (3-4) (1999) 187—191.
[7] F. Calignano, et al., Design of additively manufactured structures for biomedical applications: a review of the additive manufacturing processes applied to the biomedical sector, J. Healthcare Eng. 2019 (2019).
[8] Wartbichler, R., H. Clemens, and S. Mayer, Electron beam melting of a β-Solidifying intermetallic titanium aluminide alloy. Adv. Eng. Mater. 21 (12) (n.d).
[9] C. Korner, E. Attar, P. Heinl, Mesoscopic simulation of selective beam melting processes, J. Mater. Process. Technol. 211 (6) (2011) 978—987.
[10] J. Milberg, M.S., Electron beam sintering of metal powder, J. Inst. Eng. Prod. 2 (2008).

[11] X. Gong, T. Anderson, K. Chou, Review on powder-based electron beam additive manufacturing technology, in: ASME/ISCIE 2012 International Symposium on Flexible Automation, American Society of Mechanical Engineers, 2012.
[12] S.M. Gaytan, et al., Advanced metal powder based manufacturing of complex components by electron beam melting, Mater. Technol. 24 (3) (2009) 180–190.
[13] T. Mahale, Electron Beam Melting of Advanced Materials and Structures, Mass Customization, Mass Personalization, North Carolina State University, Raleigh, NC, 2009.
[14] P. Heinl, et al., Cellular titanium by selective electron beam melting, Adv. Eng. Mater. 9 (5) (2007) 360–364.
[15] L.H. Shen, Q.J. Li, Q.L. Cui, Correlation between the growth direction, morphology and defect luminescence in aluminum nitride nanowires Synthesized through direct nitridation method, Sci. Adv. Mater. 3 (5) (2011) 725–729.
[16] A. Neira Arce, Thermal Modeling and Simulation of Electron Beam Melting for Rapid Prototyping on Ti6Al4V Alloys, 2012.
[17] Arcam, ASTM F75 CoCr Alloy, 2016. Available from: http://www.arcam.com/wp-content/uploads/Arcam-ASTM-F75-Cobalt-Chrome.pdf.
[18] Arcam, Ti6Al4V ELI Titanium Alloy, 2016. Available from: http://www.arcam.com/wp-content/uploads/Arcam-Ti6Al4V-ELI-Titanium-Alloy.pdf.
[19] H.B. Qi, et al., Direct metal part forming of 316L stainless steel powder by electron beam selective melting, Proc. IME B J. Eng. Manufact. 220 (11) (2006) 1845–1853.
[20] Arcam, Grade 2 Titanium, 2016. Available from: http://www.arcam.com/wp-content/uploads/Arcam-Titanium-Grade-2.pdf.
[21] H. Tang, et al., Effect of powder reuse times on additive manufacturing of Ti-6Al-4V by selective electron beam melting, J. Occup. Med. 67 (3) (2015) 555–563.
[22] L.E. Murr, et al., Characterization of titanium aluminide alloy components fabricated by additive manufacturing using electron beam melting, Acta Mater. 58 (5) (2010) 1887–1894.
[23] D. Cormier, et al., Freeform fabrication of titanium aluminide via electron beam melting using prealloyed and blended powders, Adv. Mater. Sci. Eng. 2007 (2008).
[24] P. Heinl, et al., Cellular Ti–6Al–4V structures with interconnected macro porosity for bone implants fabricated by selective electron beam melting, Acta Biomaterialia 4 (5) (2008) 1536–1544.
[25] X. Li, et al., Fabrication and characterization of porous Ti6Al4V parts for biomedical applications using electron beam melting process, Mater. Lett. 63 (3) (2009) 403–405.
[26] P. Thomsen, et al., Electron beam-melted, free-form-fabricated titanium alloy implants: material surface characterization and early bone response in rabbits, J. Biomed. Mater. Res. B Appl. Biomater. 90 (1) (2009) 35–44.
[27] L. Murr, et al., Microstructure and mechanical properties of open-cellular biomaterials prototypes for total knee replacement implants fabricated by electron beam melting, J. Mech. Behavior Biomed. Mater. 4 (7) (2011) 1396–1411.
[28] A. Palmquist, et al., Long-term biocompatibility and osseointegration of electron beam melted, free-form–fabricated solid and porous titanium alloy: experimental studies in sheep, J. Biomater. Appl. (2011), 0885328211431857.
[29] G. Chahine, et al., Design optimization of a customized dental implant manufactured via electron beam melting®, in: International Solid Freefrom Fabrication SymposiumAustin, Texas, USA, 2009, pp. 631–640.
[30] O.L. Harrysson, et al., Direct metal fabrication of titanium implants with tailored materials and mechanical properties using electron beam melting technology, Mater. Sci. Eng. C 28 (3) (2008) 366–373.

[31] J. Parthasarathy, et al., Mechanical evaluation of porous titanium (Ti6Al4V) structures with electron beam melting (EBM), J. Mech. Behavior Biomed. Mater. 3 (3) (2010) 249–259.
[32] O. Harrysson, et al., Evaluation of titanium implant components directly fabricated through electron beam melting technology, in: Materials and Processes for Medical Devices Conference, Boston, 2006.
[33] L.E. Murr, et al., Characterization of Ti-6Al-4V open cellular foams fabricated by additive manufacturing using electron beam melting, Mater. Sci. Eng. a-Struct. Mater. Propert. Microstruct. Process. 527 (7–8) (2010) 1861–1868.
[34] A. Antonysamy, J. Meyer, P. Prangnell, Effect of build geometry on the β-grain structure and texture in additive manufacture of Ti 6Al 4V by selective electron beam melting, Mater. Char. 84 (2013) 153–168.
[35] E. Attar, Simulation of Selective Electron Beam Melting Processes, Dr.-Ing., University of Erlangen, Nuremberg, Germany, 2011.
[36] T. Scharowsky, et al., Melt pool dynamics during selective electron beam melting, Appl. Phys. A 114 (4) (2014) 1303–1307.
[37] M. Sigl, S. Lutzmann, M. Zäh, Transient physical effects in electron beam sintering, in: Solid Freeform Fabrication Symposium Proceedings. Austin, TX, 2006.
[38] X. Gong, T. Anderson, K. Chou, Review on powder-based electron beam additive manufacturing technology, Manuf. Rev. 1 (2014).
[39] B. Cheng, et al., On process temperature in powder-bed electron beam additive manufacturing: model development and validation, J. Manufact. Sci. Eng.-Trans. Asme 136 (6) (2014).
[40] M. Agarwala, et al., Direct selective laser sintering of metals, Rapid Prototyp. J. 1 (1) (1995) 26–36.
[41] J.-P. Kruth, et al., Binding mechanisms in selective laser sintering and selective laser melting, Rapid Prototyp. J. 11 (1) (2005) 26–36.
[42] S. Das, Physical aspects of process control in selective laser sintering of metals, Adv. Eng. Mater. 5 (10) (2003) 701–711.
[43] A.V. Gusarov, et al., Heat transfer modelling and stability analysis of selective laser melting, Appl. Surf. Sci. 254 (4) (2007) 975–979.
[44] M.F. Zäh, S. Lutzmann, Modelling and simulation of electron beam melting, J. Inst. Eng. Prod. 4 (1) (2010) 15–23.
[45] S. Sih, J.W.B., Emissivity of powder beds, in: Solid Freeform Fabrication Symposium Proceedings, Kluwer, Boston, Austin (Texas, USA), 1995.
[46] M. Zäh, et al., Determination of process parameters for electron beam sintering (EBS), in: Excerpt from the Proceedings of the COMSOL Conference Hannover, 2008.
[47] M. Galati, A. Snis, L. Iuliano, Experimental validation of a numerical thermal model of the EBM process for Ti6Al4V, Comput. Math. Appl. 78 (7) (2019) 2417–2427.
[48] S. Tammas-Williams, et al., XCT analysis of the influence of melt strategies on defect population in Ti-6Al-4V components manufactured by Selective Electron Beam Melting, Mater. Char. 102 (2015) 47–61.
[49] AB, A. http://www.arcam.com/. [cited 2016 November 2016].
[50] H. Gong, R., K., T. Starr, B. Stucker, The Effects of Processing Parameters on Defect Regularity in Ti-6Al-4V Parts Fabricated by Selective Laser Melting and Electron Beam Melting, 2013.
[51] H. Qi, et al., Scanning method of filling lines in electron beam selective melting, Proc. IME B J. Eng. Manufact. 221 (12) (2007) 1685–1694.
[52] C. Körner, Additive manufacturing of metallic components by selective electron beam melting—a review, Int. Mater. Rev. (2016) 1–17.
[53] J. Karlsson, et al., Digital image correlation analysis of local strain fields on Ti6Al4V manufactured by electron beam melting, Mater. Sci. Eng. 618 (2014) 456–461.

[54] M. Galati, A. Snis, L. Iuliano, Powder bed properties modelling and 3D thermo-mechanical simulation of the additive manufacturing Electron Beam Melting process, Additive Manufact. 30 (2019) 100897.
[55] B. Cheng, et al., Speed function effects in electron beam additive manufacturing, in: ASME 2014 International Mechanical Engineering Congress and Exposition, American Society of Mechanical Engineers, 2014.
[56] Arcam. Personal Communication, June, 2016.
[57] J. Schwerdtfeger, R.E. Singer, C. Korner, In situ flaw detection by IR-imaging during electron beam melting, Rapid Prototyp. J. 18 (4) (2012) 259−263.
[58] D.H. Abdeen, et al., Effect of processing parameters of electron beam melting machine on properties of Ti-6Al-4V parts, Rapid Prototyp. J. 22 (3) (2016) 609−620.
[59] M. Galati, P. Minetola, G. Rizza, Surface roughness characterisation and analysis of the electron beam melting (EBM) process, Materials 12 (13) (2019) 2211.
[60] M. Koike, et al., Evaluation of titanium alloy fabricated using electron beam melting system for dental applications, J. Mater. Process. Technol. 211 (8) (2011) 1400−1408.
[61] R. Klingvall Ek, et al., The effect of EBM process parameters upon surface roughness, Rapid Prototyp. J. 22 (3) (2016) 495−503.
[62] M. Jamshidinia, R. Kovacevic, The influence of heat accumulation on the surface roughness in powder-bed additive manufacturing, Surf. Topogr. Metrol. Prop. 3 (1) (2015) 014003.
[63] A. Safdar, et al., Effect of process parameters settings and thickness on surface roughness of EBM produced Ti-6Al-4V, Rapid Prototyp. J. 18 (5) (2012) 401−408.
[64] S. Ponader, et al., Effects of topographical surface modifications of electron beam melted Ti-6Al-4V titanium on human fetal osteoblasts, J. Biomed. Mater. Res. 84A (4) (2008) 1111−1119.
[65] O. Cansizoglu, et al., Properties of Ti-6Al-4V non-stochastic lattice structures fabricated via electron beam melting, Mater. Sci. Eng. a-Struct. Mater. Propert. Microstruct. Process. 492 (1−2) (2008) 468−474.
[66] J. Karlsson, et al., Characterization and comparison of materials produced by Electron Beam Melting (EBM) of two different Ti−6Al−4V powder fractions, J. Mater. Process. Technol. 213 (12) (2013) 2109−2118.
[67] H. Gong, et al., Defect morphology in Ti−6Al−4V parts fabricated by selective laser melting and electron beam melting, in: 24rd Annual International Solid Freeform Fabrication Symposium—An Additive Manufacturing Conference, Austin, TX, August, 2013.
[68] L. Rayleigh, XX, *On the theory of surface forces.—II. Compressible fluids.* the London, Edinburgh, and Dublin, Philosoph. Magaz. J. Sci. 33 (201) (1892) 209−220.
[69] P.-G. De Gennes, F. Brochard-Wyart, D. Quéré, Capillarity and Wetting Phenomena: Drops, Bubbles, Pearls, Waves, Springer Science & Business Media, 2013.
[70] M. Van Elsen, F. Al-Bender, J.-P. Kruth, Application of dimensional analysis to selective laser melting, Rapid Prototyp. J. 14 (1) (2008) 15−22.
[71] J.P. Kruth, et al., Selective laser melting of iron-based powder, J. Mater. Process. Technol. 149 (1−3) (2004) 616−622.
[72] C. Hauser, Selective Laser Sintering of a Stainless Steel Powder, University of Leeds, 2003.
[73] M. Kahnert, S. Lutzmann, M. Zaeh, Layer formations in electron beam sintering, in: Solid Freeform Fabrication Symposium, 2007.
[74] L. Facchini, et al., Microstructure and mechanical properties of Ti-6Al-4V produced by electron beam melting of pre-alloyed powders, Rapid Prototyp. J. 15 (3) (2009) 171−178.

[75] D. Cormier, A. Harrysson, H. West, Characterization of H13 steel produced via electron beam melting, Rapid Prototyp. J. 10 (1) (2004) 35–41.
[76] Y. Kok, et al., Fabrication and microstructural characterisation of additive manufactured Ti-6Al-4V parts by electron beam melting: this paper reports that the microstructure and micro-hardness of an EMB part is thickness dependent, Virtual Phys. Prototyp. 10 (1) (2015) 13–21.
[77] R. Gerling, R. Leitgeb, F.-P. Schimansky, Porosity and argon concentration in gas atomized γ-TiAl powder and hot isostatically pressed compacts, Mater. Sci. Eng. 252 (2) (1998) 239–247.
[78] B. Choi, et al., Densification of rapidly solidified titanium aluminide powders—I. Comparison of experiments to hiping models, Acta Metall. Mater. 38 (11) (1990) 2225–2243.
[79] W. Sames, et al., The metallurgy and processing science of metal additive manufacturing, Int. Mater. Rev. (2016) 1–46.
[80] S. Price, et al., Temperature measurements in powder-bed electron beam additive manufacturing, in: ASME 2014 International Mechanical Engineering Congress and Exposition, American Society of Mechanical Engineers, 2014.
[81] J. Mireles, et al., Closed-loop automatic feedback control in electron beam melting, Int. J. Adv. Manuf. Technol. 78 (5–8) (2015) 1193–1199.
[82] S. Yoder, et al., Approach to qualification using E-PBF in-situ process monitoring in Ti-6Al-4V, Addit. Manufact. 28 (2019) 98–106.
[83] H. Wong, et al., Pilot investigation of feedback electronic image generation in electron beam melting and its potential for in-process monitoring, J. Mater. Process. Technol. 266 (2019) 502–517.
[84] H. Wong, et al., Pilot capability evaluation of a feedback electronic imaging system prototype for in-process monitoring in electron beam additive manufacturing, Int. J. Adv. Manuf. Technol. 100 (1–4) (2019) 707–720.
[85] H. Wong, et al., Benchmarking spatial resolution in electronic imaging for potential in-situ Electron Beam Melting monitoring, Addit. Manufact. 29 (2019) 100829.
[86] C. Arnold, et al., Layerwise monitoring of electron beam melting via backscatter electron detection, Rapid Prototyp. J. 24 (8) (2018) 1401–1406.
[87] C.R. Pobel, et al., Immediate development of processing windows for selective electron beam melting using layerwise monitoring via backscattered electron detection, Mater. Lett. 249 (2019) 70–72.
[88] Y. Liu, et al., In-situ areal inspection of powder bed for electron beam fusion system based on fringe projection profilometry, Addi. Manufact. 31 (2020) 100940.
[89] M. Galati, L. Iuliano, A literature review of powder-based electron beam melting focusing on numerical simulations, Addit. Manufact. 19 (2018) 1–20.
[90] M. Galati, et al., Modelling energy source and powder properties for the development of a thermal FE model of the EBM additive manufacturing process, Addi. Manufact. 14 (2017) 49–59.
[91] R. Ammer, et al., Simulating fast electron beam melting with a parallel thermal free surface lattice Boltzmann method, Comput. Math. Appl. 67 (2) (2014) 318–330.
[92] Ammer, R., et al., Modeling of Thermodynamic Phenomena with Lattice Boltzmann Method for Additive Manufacturing Processes.
[93] T. Scharowsky, et al., Observation and numerical simulation of melt pool dynamic and beam powder interaction during selective electron beam melting, in: Proceedings of the Solid Freeform Fabrication Symposium, 2012.
[94] W. Yan, et al., Meso-scale modeling of multiple-layer fabrication process in selective electron beam melting: inter-layer/track voids formation, Mater. Des. 141 (2018) 210–219.

[95] N. Shen, K. Chou, Thermal modeling of electron beam additive manufacturing process: powder sintering effects, in: ASME 2012 International Manufacturing Science and Engineering Conference Collocated with the 40th North American Manufacturing Research Conference and in Participation with the International Conference on Tribology Materials and Processing, American Society of Mechanical Engineers, 2012.
[96] B. Cheng, K. Chou, Melt pool geometry simulations for powder-based electron beam additive manufacturing, in: 24th Annual International Solid Freeform Fabrication Symposium-An Additive Manufacturing Conference, Austin, TX, USA, 2013.
[97] N. Shen, Y. Chou, Numerical thermal analysis in electron beam additive manufacturing with preheating effects, in: Proceedings of the 23rd Solid Freeform Fabrication Symposium, Austin, TX, 2012.
[98] M. Galati, O. Di Mauro, L. Iuliano, Finite element simulation of multilayer electron beam melting for the improvement of build quality, Crystals 10 (6) (2020) 532.
[99] D. Riedlbauer, et al., Macroscopic simulation and experimental measurement of melt pool characteristics in selective electron beam melting of Ti-6Al-4V, Int. J. Adv. Manuf. Technol. (2016) 1–9.
[100] J. Koepf, et al., Numerical microstructure prediction by a coupled finite element cellular automaton model for selective electron beam melting, Comput. Mater. Sci. 162 (2019) 148–155.
[101] M. Jamshidinia, F. Kong, R. Kovacevic, Numerical modeling of heat distribution in the electron beam melting® of Ti-6Al-4V, J. Manuf. Sci. Eng. 135 (6) (2013) 061010.
[102] M. Jamshidinia, F. Kong, R. Kovacevic, The coupled CFD-FEM model of electron beam melting (EBM), Mech. Eng. Res., Paper 4 (2013).
[103] G. Vastola, et al., Controlling of Residual Stress in Additive Manufacturing of Ti6Al4V by Finite Element Modeling, Additive Manufacturing, 2016.
[104] N. Shen, K. Chou, Simulations of thermo-mechanical characteristics in electron beam additive manufacturing, in: ASME 2012 International Mechanical Engineering Congress and Exposition, American Society of Mechanical Engineers, 2012.
[105] B. Cheng, K. Chou, Thermal stresses associated with part overhang geometry in electron beam additive manufacturing: process parameter effects, in: Proc. Annu. Int. Solid Freeform Fabr. Symp, 2014.
[106] B. Cheng, K. Chou, Geometric consideration of support structures in part overhang fabrications by electron beam additive manufacturing, Comput. Aided Des. 69 (2015) 102–111.

CHAPTER 9

# Introduction to 4D printing: methodologies and materials

**Xiao Kuang**
The George W. Woodruff School of Mechanical Engineering, Georgia Institute of Technology, Atlanta, GA, United States

## 9.1 Introduction

In 2013, Tibbits first introduced the concept of 4D printing [1]. He combined 3D printing with smart (or active) materials to obtain printed objects whose shape could change after treating in water. In the same year, Ge et al. [2] published the first research paper on 4D printing using printed active composites (PACs). Since then, 4D printing attracts great attention to both research communities of smart materials and 3D printing as well as industry.

In the initial definition, 4D printing was "3D printing + time," with the fourth dimension being time. Fig. 9.1A shows the comparison of concepts from 1D to 4D. There are at least two stable states in 4D-printed materials and structures, which can shift from one state to another under the predetermined stimulus. The concept itself has evolved with time in the past few years. Up to date, a popular definition of 4D printing is that the shape, property, and functionality of a 3D-printed structure could evolve with time when it is exposed to an environmental stimuli, such as exposure to heat [3,4], water [5,6], light [7,8], or pH [9]. Fig. 9.1B shows that the 4D printing techniques involve the use of proper design and programming strategies in 3D printing of smart materials for predetermined stimulus-responsive dynamic structures. During the past years, significant efforts have been undertaken in 4D printing using different materials and mechanisms. The advances in 4D printing involve the fast growth and interdisciplinary research of 3D printing techniques, stimulus-responsive smart materials, design, and modeling-based programming technology [10–14].

Currently, most of the 4D printing refers to the shape and architecture transformation of 3D-printed structures. 4D printing for dynamic structure changes involves at least two stable shapes with some mechanical deformation during the shape-shifting process. Fig. 9.2 shows that the

*Additive Manufacturing*
ISBN 978-0-12-818411-0
https://doi.org/10.1016/B978-0-12-818411-0.00004-5

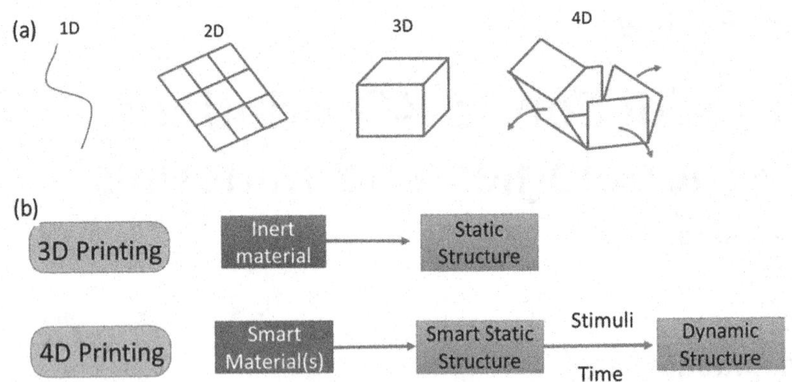


**Figure 9.1** (A) Schematics show the concept form 1D to 4D. (B) Flowchart demonstrating the difference between 3D and 4D printing.

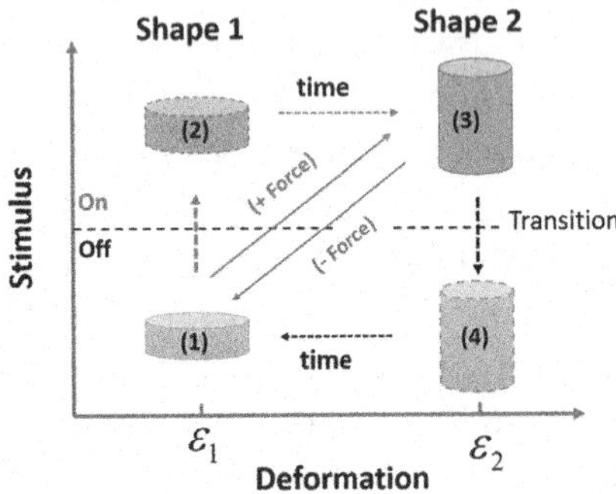


**Figure 9.2** The scheme of 4D printing of shape-shifting structure with an initial configuration of *Shape 1* that can transform into another configuration of *Shape 2* with different strain state. State 1 and state 3 are stable states at corresponding conditions, whereas state 2 and state 4 are intermediate metastable states.

4D-printed shape-shifting structures can change the shapes under an external stimulus. The shape deformation from the initial shape (Shape 1) to the end shape (Shape 2) indicates some global or local strain gradients in the system. The stable and metastable shapes present different strain states at different or even the same environment condition after a programming process. The underlining principle of stable shapes with different deformation can be understood by introducing an inelastic strain or eigenstrain distribution in printed smart materials. After a shape-programming procedure under the external stimulus, eigenstrain usually exists in the material, and the metastable shapes (state 3) can be stabilized at an external stress-free state (state 4). For different printing techniques, however, the eigenstrain was encoded at a different stage either during the printing process or postprinting step [13–16].

With the added dimension of time, 4D printing has a few notable advantages: (1) allowing the fabrication of intelligent devices, which can respond to environmental by changing its shape or function; (2) saving printing time and materials, especially in the case of printing thin-walled or lattice structures [17,18]; and (3) saving space for storage and transportation of the printed parts. The pursuing of materials, structures, or devices that can change their shape, properties, or functionalities has been a focus in the smart material research community for several decades [19–22]. 4D printing has become a new and exciting research field and attracts enormous interest in different disciplines, especially in smart material and advanced manufacturing communities.

Various materials, including polymers [2,23,24], metals [25], and ceramics [26,27], have been used for 4D printing. Comparing with smart metal and ceramic, smart polymers have the advantages of low-cost, diverse stimulus-responsivities, and large deformability for shape programming [25]. Thus, most of the 4D printing to-date is related to polymers. In this chapter, we will comprehensively review the 4D printing polymers and composites. We first introduce the basic elements including 3D printing techniques, shape-programmable materials, and underlying mechanisms. The state-of-the-art progress in 4D printing using different materials and printing techniques is reviewed in detail.

## 9.2 Fundamentals of 4D printing

As abovementioned, the development of 4D printing involves the interdisciplinary research of 3D printing techniques, stimulus-responsive smart materials, design, and modeling-based programming. Compared with the static objects created by 3D printing, 4D printing has the additional dimension of time. A wide range of physical and chemical stimuli, including temperature, water, solvent/chemical composition, light, and magnetic field, have been used to trigger different stimulus-responsive polymers for shape shifting (Fig. 9.3). We will summarize four most extensively studied programmable materials: shape-changing capability are shape memory polymers (SMPs), hydrogel composites, liquid crystal elastomers (LCEs), and magnetoactive materials for 4D printing. Meanwhile, the primary 3D printing techniques, including extrusion-based printing, inkjet printing, and vat photopolymerization–based printing, are discussed.

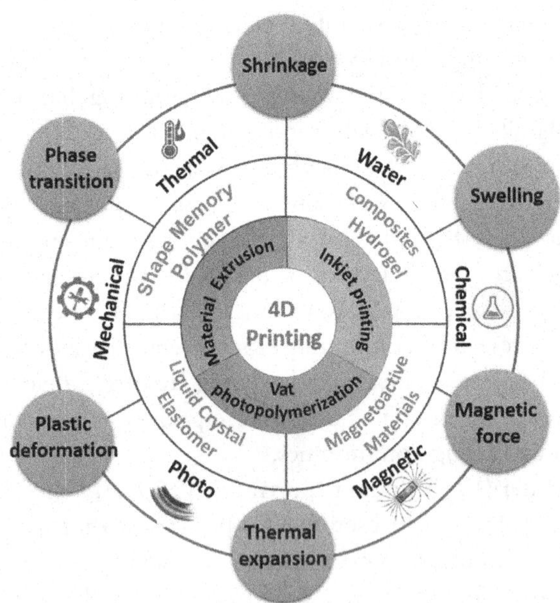


**Figure 9.3** Primary printing techniques, stimulus-responsive programmable materials, conventional stimulus, and shape change mechanism for shape transformation in 4D printing materials.

For the 4D printing of shape-shifting materials, the eigenstrain usually derived from phase transformation, plastic strain, swelling/thermal expansion, or shrinkage. For SMP, the elastic deformation by the external load can be frozen and stored as internal energy leading to a stress-free state by phase evolution, such as glass transition and crystallization. For example, the deformed molecular chain at high temperature with elastic strain can be immobilized by cooling down below the glass transition temperature. SMP can be also programmable using the mechanical programming by proceeding with a plastic deformation [28–30]. The shape actuation of LCEs also lies in the phase transition between isotropic (I) state and nematic (N) state at different conditions [31–33]. In hydrogel, the shape programming is enabled by the swelling and deswelling under the environmental stimulus. In the case of magnetoactive materials, a microtorque is exerted on the material for shape change due to the high magnetization of magnetic particles in the soft matrix. In the following context, we will provide more details about the mechanism of different shape-shifting material and the corresponding printing techniques.

## 9.2.1 Shape-programmable materials

### *9.2.1.1 Shape memory polymers*

SMPs are a kind of smart materials that can maintain a temporary shape and recover their initial shape in the presence of an external stimulus such as heat [34–38]. The widely investigated thermally triggered SMP can be achieved by direct heating or indirect heating, such as Joule heating and photothermal effect. In the past 5–10 years, SMPs were witnessed significant progress with a focus on implanting new mechanisms to extend functionality and new materials to enhance the thermal and mechanical properties of SMPs for specific application requirements.

From the standpoint of modeling, SMPs consist of a "fixing phase" and "switch phase." The fixing phase holds the permanent shape of the material, which is usually made of either chemical/physical cross-linking or inert microphase [34,37]. To get better shape memory performance, chemically cross-linked materials were widely used for SMP. The switching phase can be in both phase level, such as amorphous and crystalline, and molecular levels, such as supramolecular entities and reversible molecular units. In the case of phase switch, an amorphous phase with a $T_g$ and a semicrystalline phase with a $T_m$ as phase transition temperature ($T_t$) are employed for shape programming (shape fixing and shape recovery). Fig. 9.4 shows the shape memory effect of SMP. The molecular chains of an SMP at the initial shape usually adopt conformations with the highest entropy, which is a thermodynamically stable state. Upon heating above the $T_t$, chain mobility is significantly activated. Upon the loading of an external deformation, the chain conformations are changed to a lower entropy state as indicated by the macroscopic shape change. Meanwhile, mechanical

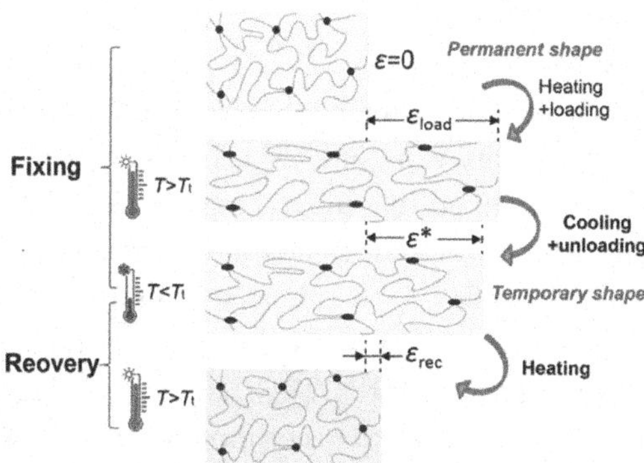


**Figure 9.4** Schematic illustration of the strain evolution during fixing and recovery step for thermally triggered SMP using a phase switch. *SMP*, shape memory polymer.

work can also be stored as internal energy in the material. This high energy state is not stable, which will lead to strain recovery after the instant force removal. However, when cooling down than $T_t$ and proceeding the phase transition such as crystallization, the decrease of enthalpy as indicated by the exothermic heat leads to a metastable state with a relatively low energy state. The reduction in free volume significantly reduces the mobility of macromolecular chains, and the conformational change of the network becomes increasingly difficult. The intermolecular barrier to the entropic motions leads to the viscoelastic/viscoplastic behaviour. As a result, the temporary shape is thus kinetically trapped due to the freezing of the molecular chain segments. The eigenstrian enabled by the phase transition results in a stress-free state for SMP. However, upon reheating up above $T_t$ under a constraint-free condition, the molecular mobility is reactivated, which allows the chains to transform from the programmed nonequilibrium configuration (temporary shape) to the recovered equilibrium configuration (permanent shape) with both low energy and highest entropy. It should be noted that the programming of SMPs can be also an enthalpy controlled process in some cases, such as for compression cold programming [39].

A variety of polymers, including semicrystalline polymer (such as polyolefin [40,41], polyester [42–45], and polyurethane (PU) [46–49]) and amorphous polymers (polyacrylate [50–52] and epoxy resin), have been used for SMP with different $T_t$, shape memory performance, and mechanical properties. Both the shape fixing step and shape recovery step of SMP can be used for shape programming. For example, the eigenstrain of SMP can be generated during the material printing process, where the followed shape recovery by heating would drive the shape shifting of the SMP composites [8,53].

#### 9.2.1.2 Liquid crystal elastomers

Liquid crystals are a class of materials that demonstrate both the order of solid crystals and the fluidity of liquids [31]. LCEs, with the mesogens of a liquid crystal covalently linked to a polymer chain or backbone, are further loosely chemically cross-linked that can exhibit large changes in order when subject to an external stimulus, such as heat and photo. Thermal-responsive LCEs make this transition as a result of passing through a nematic-to-isotropic transition temperature ($T_{NI}$) as seen in Fig. 9.5A. Photo-responsive LCE-based azobenzene compound has also been widely studied. As seen in Fig. 9.6B, the rod-like trans-azobenzene mesogens stabilize the LC alignment, whereas the bent cis forms reduce the LC order parameter. Upon UV light irradiation, the LCE mesogens change from trans to cis configurations resulting in a mechanical shrinkage response.

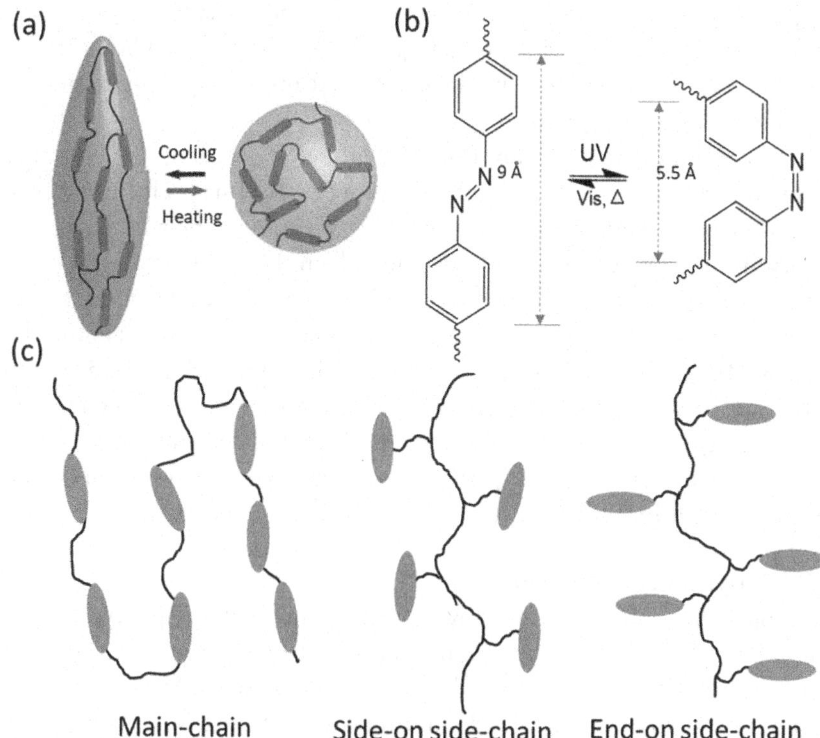


**Figure 9.5** (A) The change in order at or above the nematic-to-isotropic phase transition results in anisotropic deformation with shrinkage of the sample parallel to the orientation direction. (B) Schematics show the photo characteristics of photo-responsive azobenzene isomerization. (C) Chain conformation of main-chain, side-on side-chain, and end-on side-chain LCEs. *LCEs*, liquid crystal elastomers.

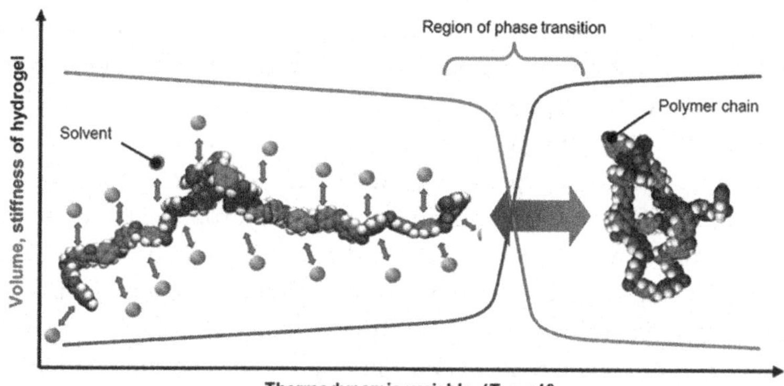


**Figure 9.6** Phase transition behavior of stimuli-responsive hydrogels for programmable properties. *(Reproduced with kind permission of IntechOpen from Reference M. Bahram, N. Mohseni, M. Moghtader, In Emerging Concepts in Analysis and Applications of Hydrogels, IntechOpen 2016.)*

Shape actuation using light is attractive due to broad availability, convenient manipulation, and the ability to be applied remotely. In addition, some thermal-responsive LCE can be light-activated utilizing the photothermal effect of nanoparticle additives [55,56].

LCE can be divided into three categories depending on how the mesogens are connected to the polymer network. In a main-chain LCE, the mesogens are linked to the polymer directly through their long axes, as shown in Fig. 9.5C. In a side-on side-chain LCE, the mesogens are connected to the polymer backbone by their sides, with a short spacer linking the two together. In the end-on side-chain configuration, the mesogen backbones are linked by a spacer. LCE needs to align the mesogen before use. Researchers have developed numerous methods for aligning the LCE mesogens using techniques, such as mechanical alignment [57–60], polarized light patterning [61], shearing during extrusion [4,62–64], and magnetic fields [65] for hand-free reversible shape actuation.

#### 9.2.1.3 Hydrogel composites

Hydrogels were first reported by Wichterle in 1960 [66]. Hydrogel is composed of a three-dimensional cross-linked hydrophilic polymer network, which is capable of swelling or deswelling reversibly in water and solution. By definition, hydrogel constitutes at least 10% water of the total weight (or volume). The hydrophilicity and stimulus-responsive capability of the network are attributed to the presence of hydrophilic functional groups, such as $-NH_2$, $-COOH$, $-OH$, $-CONH_2$, $-CONH-$, and $-SO_3H$. Various stimulus-responsive hydrogels were fabricated by copolymerization of functional monomers [5,67–69]. Uptake and release of water molecules are the main driving forces of shape actuation for hydrogel and composites. Fig. 9.6 shows the hydrogel has two states [54]. One state is a gel-like structure dominated by polymer–polymer interactions featured by the maximum hydrophobicity leading to the shrinkage of the hydrogel. In the second state, interactions between the solvent and the polymer network create a mixed phase in which the polymer and the aqueous solution are well mixed. The maximal value of hydrophilicity and swelling occurs in this state. Hydrogels can undergo a significant volume phase transition in response to certain physical and chemical stimuli, including temperature, electric, solvent composition, light, pH, ions, and so on. Exposure to one or multiple stimuli causes reversible swelling/shrinkage of hydrogels due to a change of interactions of the polymer molecules with water. The extent of swelling is influenced by any factor that alters the intermolecular interaction between the media and the polymer network, which is influenced by the nature of the functional groups in the monomer. In addition, the chain architectures, and the degree of cross-linkage, as well as the applied external stimulus, have a significant effect on hydrogel actuation.

Besides the simple deformation of expanding and contraction, more complex and high-order deformation, such as bending and twisting as well as snap-bulking can be achieved by either alternating the stimulus field or using heterogeneous material structure [70–74]. The multimaterial or hydrogel composite enables more complex and well-controlled shape programming. Hydrogels are similar to natural tissue due to their significant water content, which is also a good candidate for tissue engineering and biomedical applications [75–79].

#### 9.2.1.4 Magnetoactive materials

The magnetic responsive materials have drawn significant interest in shape-shifting materials using remote magnetic fields for actuation. Magnetoactive soft material (MSM) composed of magnetic particles in a soft polymer matrix can undergo rapid and large deformation via the magnetic field, offering a wide range of applications such as soft actuators [80], soft robots [81], and drug release [82]. There are two types of magnetic particles: soft magnetic particles and hard magnetic particles based on the coercivity and residual magnetic flux density (B). Fig. 9.7A shows the magnetic hysteresis loops for these two magnetic materials. Both of these ferromagnetic magnetic particles can develop strong magnetic flux density when exposed

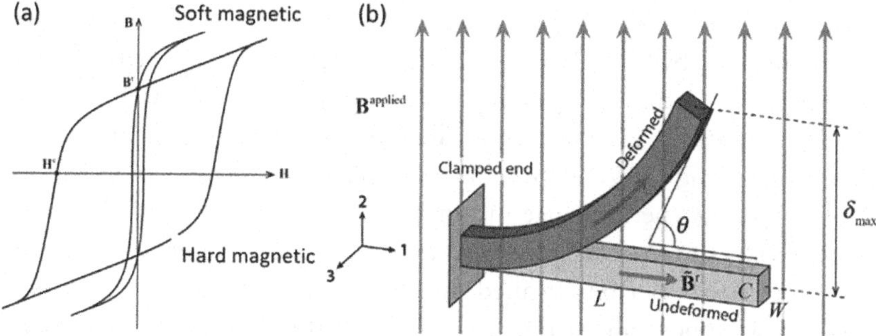


**Figure 9.7** (A) Magnetic hysteresis loops and B–H curves of soft magnetic and hard magnetic materials. (B) Schematic actuation of a beam whose initial residual magnetic flux density Br is aligned with the one direction. *(Reproduced with kind permission of Elsevier from Reference R. Zhao, Y. Kim, S. A. Chester, P. Sharma, X. Zhao 2019.Mechanics of hard-magnetic soft materials J. Mech. Phys. Solid., 124 244.)*

to an external magnetic field strength (H). Soft magnetic materials form a sharp and narrow magnetic B—H curve due to the low coercive field ($H_c$). Therefore, soft magnetic particles cannot sustain high residual magnetic flux density $B_r$. Typically, MSMs such as magnetorheological elastomers [84] and ferrogels [82] use magnetically soft particles characterized by the low coercivity such as iron and iron oxides [85]. By contrast, hard magnetic materials possess much higher coercivity, which can retain high residual magnetic flux density ($B_r$) once magnetized to saturation [83].

Hard particles, such as neodymium-iron-boron (NdFeB) featured with high-coercivity, have been embedded in soft materials to achieve fast, untethered, and remote-controlled shape shifting [81,83,86]. Upon applying an external magnetic field using permanent magnets or electric coils, the embedded hard magnetic particles with programmed domains exerted microtorques on the materials creating internal stresses leading to controlled macroscale shape changes. Complicated shape transformation and different modes of locomotion with high spatial resolutions can be controlled by designing the magnetization profile in the material and using a time-varying magnetic field for actuation. As shown in Fig. 9.7B, during the actuation, a cantilever can bend under a uniform magnetic field applied in the direction perpendicularly to the beam length (magnetization profile direction). The bending deflection is influenced by the magnetic field.

### 9.2.2 3D printing techniques

3D printing, or additive manufacturing (AM), is defined by a range of technologies that are capable of translating virtual solid model data into physical models in a rapid process [87]. 3D printing enables decentralized fabrication of customized, personalized objects with on-demand complex shapes in a lay-by-layer manner. 3D printing has become an advanced manufacturing technique gaining tremendous progress during the past decade. It finds broad applications ranging from visual prototypes [88] to tissue engineering [76,78,79,89], biomedicine [75,78,90], electronic devices [91,92], and high-performance materials/metamaterials [93—99]. According to the ASTM International criteria, there are over 50 different polymer AM technologies that can be classified into seven different categories: material extrusion, vat photopolymerization, powder bed fusion, energy deposition, binder jetting, material jetting, and sheet lamination [100—102]. 3D printing has become a hype in both the academy and industry. Based on how the raw material (or ink) is deposited, there are three main 3D printing techniques: extrusion-based printing, inkjet printing, and vat photopolymerization.

Various stimulus-responsive materials can be formulated as functional ink and resins to satisfy different printing requirements.

Material extrusion-based 3D printing is the computer-controlled layer-by-layer deposition of molten and semimolten polymers, pastes, polymer solutions, and dispersions through a movable nozzle serving as the extrusion print head (Fig. 9.8A). Extrusion-based printing including fused filament fabrication (FFF; or fused deposition modeling, FDM) and direct ink writing (DIW; or direct-write, DW), which is applicable to a wide range of range materials such as engineering thermoplastics and thermosetting polymers assisted by a postcure [93,103–106]. The resolution of FDM and DIW is limited by the nozzle diameter, which is typically in the range hundreds of micrometers. FDM requires the materials to be melted and softened when heated during extrusion and then rapidly solidified upon cooling. A wide range of engineering thermoplastic polymers, including polylactic acid (PLA), acrylonitrile butadiene styrene (ABS), thermoplastic polyurethane (TPU), polycarbonate (PC), Nylon, and so on, have been printed [100]. During the printing process, the polymer chain can be partially aligned by the shearing force and moving nozzle. There is some built-in residual stress in the parts, which has been well used for shape shifting. Upon heating above its $T_g$ of the printed materials, the internal strain energy imparted in the polymer releases and drives the shape change. DIW, also known as "solid free-form fabrication" or "robocasting," employed a computer-controlled

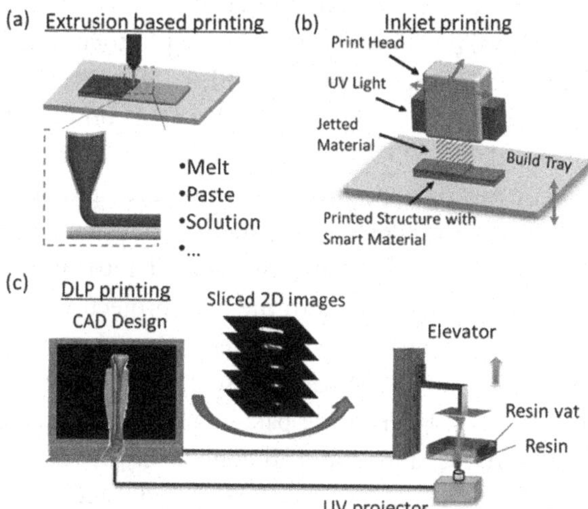


**Figure 9.8** Three main 3D printing techniques: extrusion-based printing (A), inkjet printing (B), and vat photopolymerization (C) based on how the ink is deposited.

translation stage for depositing viscous inks by a pneumatic pressure from a nozzle to create materials with controlled architecture and composition in a filament-by-filament and then layer-by-layer manner [107]. To achieve successful printing, a viscous ink with a shear-thinning effect can be formulated. Various functional materials in responsive to stimulus, such as heat [4,104,108–111], photo [7], water [73,112], and magnetic field [83,86], have been used for DIW printing shape-programmable materials.

Inkjet 3D printing is a high-resolution, multimaterial AM technique. The most widely used Polyjet 3D printing utilizes multiple inkjet printheads to jet different photopolymer inks to fabricate multimaterial structures that can have mostly varying material properties (called digital materials) from point to point within the same structure (Fig. 9.8B). The low-viscosity (usually less than 20 mPs) photocurable liquid resins were sprayed from multiple printheads on the printing bed, followed by UV curing to printing one layer [113,114]. Typical in-plane resolution of an inkjet printer is ~30–40 μm for single material printing and 200–400 μm for multimaterials printing [114,115]. Recently, Polyjet technology (Connex series, by Stratasys) has been widely used to printed multimaterial-based shape-programmable structures for self-assembly [116], and on-demand shape shifting [2,6,23,24,117–120].

Vat photopolymerization used a photopolymerizable chemical moiety, which can be cured under light exposure. Free radical photopolymerizations of (meth)acrylates and (meth)acrylamides are most commonly employed due to the rapid reaction rate. A laser beam or a UV light can be used as a light source to induce the photocuring. When a laser beam is used, it is called stereolithography (SLA). When a projector is used, it is called digital light processing (DLP) [88]. DLP printing based on the digital micromirror device (DMD) as a dynamic mask for photocuring has been becoming increasingly popular since its development 20 years ago [121]. Fig. 9.8C shows the typical procedures of bottom-up DLP printing. The designed object with CAD files was sliced into 3D monochromic grayscale images, which were passed to the light projector. The light source positioned below the resin vat is directed upward to the transparent window made of low adhesion material (such as TEFLON and PDMS). During printing, the first layer of the part is cured and adheres firmly onto the build stage surface, and the subsequent layers were cured onto the previous layer. The process was repeated, and the object was reconstructed in a layer-by-layer manner. The recent advance of using an oxygen-permeable window [122,123] or mobile liquid interface [124] for easy separation of the cured part enables a fast printing speed comparable with industrial production. In addition, high-resolution (micro- and nanoscale) DLP can be achieved with optical lens systems (call projection microstereolithography, PμSL) [123,125,126]. In addition, a two-stage curing approach has been combined with DLP to make engineering polymers [94,95,104,127,128] Moreover, multimaterial DLP is also gaining some progress using switchable vats [129], grayscale light [73,130–134], and selective wavelength curing [135] for DLP printing of functional materials with stimulus responsible capability. The application of the high-resolution projection microstereolithography (PμSL) technique allows for producing shape-changing materials at the microscale, which is favored for some biomedical applications. Alternatively, the direct laser writing by two photonic polymerizations (TPP) enables high-resolution fabrication of nano- and submicrostructures [136–140].

## 9.3 Material extrusion-based 4D printing

### 9.3.1 Fused deposition modeling–based 4D printing

FDM requires the materials to be melted and softened when heated during extrusion and then rapidly solidified upon cooling. During the printing process and subsequent cooling, there is built-in residual stress in the parts due to the shearing-induced chain alignment and volume shrinkage. The residual stress in the printed parts can be utilized for shape shifting after heating. Upon heating above its $T_g$ of the polymer, the internal strain energy imparted in the polymer release drives the shape change.

Heat shrinkage polymer can be printed by FDM printing for controlled shape shifting. For example, Manen et al. [141] reported a 2D/3D shape-shifting structure using a single-step FDM printing a heating shrinkage polymer (PLA). The percentage of the shrinkage of the filament was controlled by the adjustment of the printing parameters (extrusion temperature, layer thickness) and the actuation condition (such as triggering temperatures). The multiply panels design provides more freedom to achieve widely tunable shape-shifting structures including self-folding origami, pop-up, and "sequential" recovery structure (Fig. 9.9A). Fig. 9.9B shows a printed multiply flower structure can self-fold from the initially flat petals to create a tulip. The heat shrinkage of aligned polymer filament can be combined with geometry control to enable more complex shape changing. Zhang et al. [142] used FDM to print PLA filament for

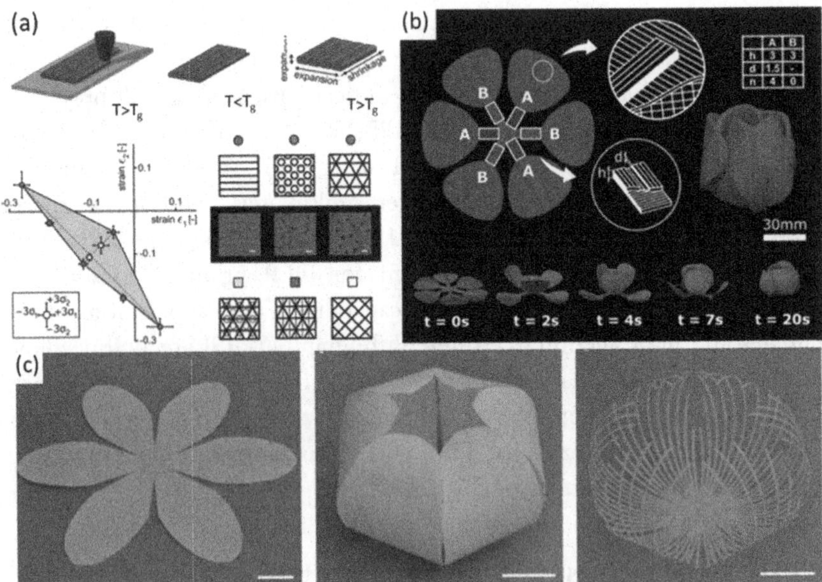


**Figure 9.9** (A) 3D-printed PLA filaments simultaneously decrease in length and thickens once heated above their glass transition temperature. Two-step folding of the initially flat petals to create a tulip. (B) The initial planar shape of the 3D-printed composite sheet self-folded into the final flower-like 3D shape after a process of heating and cooling (scale bars = 2 cm). *PLA*, polylactic acid. *(A) Reproduced with kind permission of Royal Society of Chemistry from Reference T. van Manen, S. Janbaz, A. A. Zadpoor 2017.Programming 2D/3D shape-shifting with hobbyist 3D printers Mater. Horiz., 4 1064; Copyright 2017; (B) Reproduced with kind permission of Nature Publishing Group from Reference Q. Zhang, K. Zhang, G. Hu 2016.Smart three-dimensional lightweight structure triggered from a thin composite sheet via 3D printing technique Sci. Rep., 6 22431; Copyright 2016.)*

thermally triggering pattern transformation in a controllable way. Shape-shifting lattice metamaterial can be obtained. Fig. 9.9C shows a layer of polymer was printed on a piece of paper (passive layer). The bilayer composites can fold like a flower. After shape shifting, the paper layer was removed to obtain the skeleton of the printed structure.

Multimaterial FDM printing has also been used for 4D printing to achieve more complex shape actuation. An et al. [18] used a desktop FDM printer to printing bilayer of PLA (active layer)/TPU (constrain rubber layer) composite for shape shifting, which is called Thermorph. A printed flat thermoplastic bilayer composites can be triggered to bend above $T_g$ of the PLA. With the aid of software editor, complex geometries (up to 70 faces), like a rose, can be designed, printed, and self-folded with the readily available

**Figure 9.10** (A) Blooming and color-shifting flower printed by a multimaterial FDM printer. The green [dark black in print version] and orange [light black in print version] flower change from the temporary bud shape to a yellow [light gray in print version] bloom shape [111]. (B) The curled artificial octopus tentacles were colorfully stretched out in chronological order while matching different color change speeds on a hot plate at 80°C. *FDM*, fused deposition modeling. *(Reproduced with kind permission of John Wiley & Sons from Reference J. Wang, Z. Wang, Z. Song, L. Ren, Q. Liu, L. Ren 2019. Biomimetic shape—color double-responsive 4D Printing Adv. Mater. Technol., 4 1900293; Copyright 2019.)*

printing materials from a flat sheet. Besides shape shifting, Wang et al. [111] reported the shape—color double-responsive 4D printing. The PLA as an SMP was mixed with thermochromic pigments for FDM printing. By regulating the geometric thickness, layer thicknesses, process parameters (nozzle temperature), and stimulus conditions, the speed of shape changing and color transition can be controlled simultaneously. Fig. 9.10 shows a sequential shape and color-shifting blooming flower by FDM printing.

Various nanoparticles are mixed with engineering polymer for FDM printing of functional polymer nanocomposite for 4D printing [7,143]. Yang et al. [7] reported the 3D printing of photoresponsive SM devices by combining FDM printing technology and photothermal effect of carbon black-reinforced polyurethane (PU). Owning to the high photothermal conversion efficiency, natural sunlight can be used to trigger the SM behavior of 3D-printed devices. Fig. 9.11A shows the resulting shape change of a printed cube under direct sunlight. Similarly, Hua et al. [143] reported the 3D printing of photo-responsive shape changing composites based on polylactic acid (PLA) and multiwalled carbon nanotubes (MWCNTs) on paper substrates with FDM printing for the construction of flexible photothermal-responsive actuators. In Fig. 9.11B, the 3D-printed flower

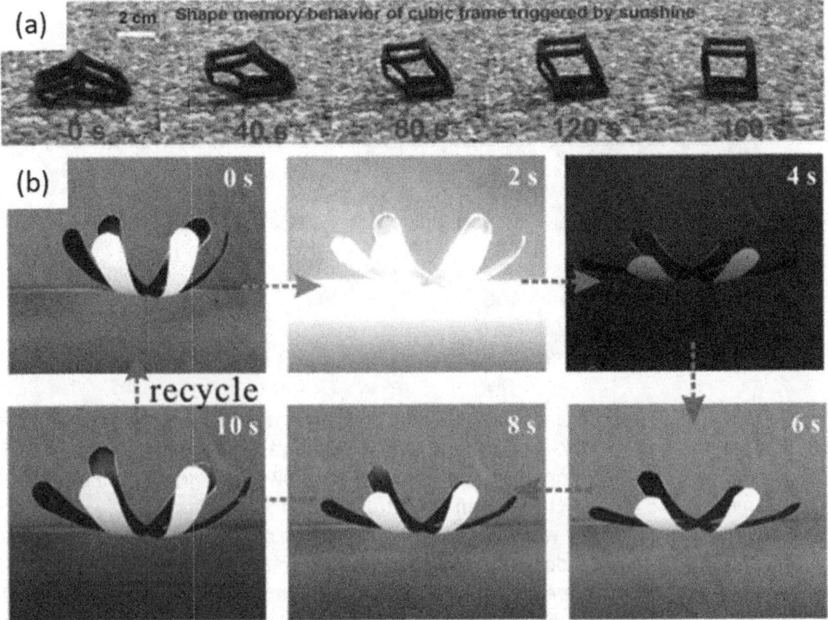


**Figure 9.11** (A) Sunlight-activated recovery of an SMP/carbon black composite. (B) Photo-activated shape-changing the behavior of a 3D-printed flower from the closed state to the open state. *SMP*, shape memory polymer. *(A) Reproduced with kind permission of John Wiley & Sons from Reference H. Yang, W. R. Leow, T. Wang, J. Wang, J. Yu, K. He, D. Qi, C. Wan, X. Chen 2017. 3D printed photoresponsive devices based on shape memory composites Adv. Mater., 29 1701627; (B) Reproduced with kind permission of Royal Society of Chemistry from Reference D. Hua, X. Zhang, Z. Ji, C. Yan, B. Yu, Y. Li, X. Wang, F. Zhou 2018. 3D printing of shape changing composites for constructing flexible paper-based photothermal bilayer actuators J. Mater. Chem. C, 6 2123.)*

shifted from the closed to open state like the blooming of flowers. During the illumination, the flower turned from bud to bloom and then recovered to its shape in the form of closed petals upon turning off the light source.

## 9.3.2 4D printing by direct ink writing printing

### *9.3.2.1 Direct ink writing printing of shape memory polymer*

As a most versatile printing technique, DIW printing can accommodate various different inks, including polymer paste, solution, and colloid for 3D and 4D printing [107]. For example, Kuang et al. [110] developed novel resins and used it for UV-assisted DIW printing of high-strain shape memory and self-healing materials. A composite ink containing semicrystalline polymer, acrylate-based photopolymer, and fumed silica was extruded at an elevated temperature (70°C) followed by photocuring. As a result, a loosely cross-linked semiinterpenetrating polymer network contains a percolation network of nanoscale crystals (Fig. 9.12A). Fig. 9.12B shows that a printed Archimedean spiral structure recovered its original shape first and then healed the crack at 80°C. Inserted pictures are the optical micrograph showing the surface morphology (scale bar is1 mm).

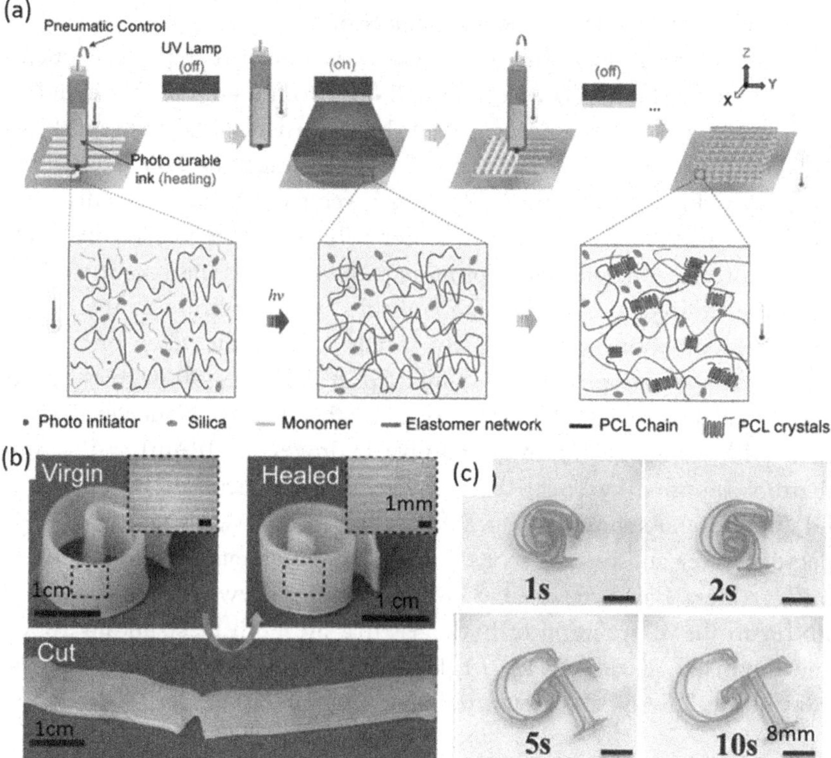


**Figure 9.12** (A) DIW-based 3D printer equipped with heating elements prints each layer of the filament followed by shining UV light to cure the resin. (B) Pictures showing SM-assisted SH process for printed Archimedean spiral structure. Inserted pictures are the optical micrograph showing the surface morphology. (B) Shape memory-assisted self-healing process for printed Archimedean spiral structure. (C) The results shape the recovery of printed epoxy SMP in a "GT" logo in an oil bath. *DIW*, direct ink writing; *SH*, self-healing; *SMP*, shape memory polymer. *(B) Reproduced with kind permission of American Chemistry of Society from Reference X. Kuang, K. Chen, C. K. Dunn, J. Wu, V. C. Li, H. J. Qi 2018.3D printing of highly stretchable, shape-memory and self-healing elastomer toward novel 4D Printing ACS Appl. Mater. Interfaces, 10 7381; Copyright 2018; (C) Reproduced with kind permission of Royal Society of Chemistry from Reference K. Chen, X. Kuang, V. Li, G. Kang, H. J. Qi 2018.Fabrication of tough epoxy with shape memory effects by UV-assisted direct-ink write printing Soft Matter, 14 1879.)*

Using the same methods, Chen et al. [104] used UV-assisted DIW to print interpenetrating epoxy resin with tunable $T_g$ and modulus via two-stage curing for SMPs (Fig. 9.12C). This kind of two-stage cured epoxy thermosets can be potentially used for high-performance SMPs.

#### *9.3.2.2 Direct ink writing printing of hydrogel*

The shearing forces generated during DIW printing process can also be exploited to align microparticulates/fillers or polymeric chains to achieve unique shape transformations. For example, Gladman et al. [5] developed a hydrogel and nanofibrillated cellulose (NFC) composite ink that could be DIW-printed. As shown in Fig. 9.13A, NFC particulates underwent shear-induced alignment as the ink was extruded through the DIW deposition nozzle. By doing so, the printed hydrogel architectures were encoded with localized, anisotropic swelling behavior which was controlled by NFC alignment along the prescribed printing pathways. This anisotropic swelling could be utilized to allow precise control of the curvature in the bilayer hydrogel composite structures. By locally tuning the fiber orientation during DIW printing, a flower comprised of 0 degrees/90 degrees bilayer lattice that could yield folding after swelling (Fig. 9.13B). Naficy et al. [109] used DIW printing a thermoresponsive hydrogel (PNIPAm) and a non-thermoresponsive hydrogel that could realize bidirectional bending after DIW printing. As shown in Fig. 9.13C, the hydrogel swells and bends in 20 degrees water and returns to its initial, flat configuration after heating in 60°C water. Bakarich et al. [144] designed a new ink composed of PNIPAm for 3D printing of hydrogels that are both mechanically robust and thermally actuating. Fig. 9.13D shows that a printed smart valve can control the flow of water with different temperatures.

#### *9.3.2.3 Direct ink writing printing of liquid crystal elastomer*

DIW printing has also been used to print LCE with reversible shape actuation [4,62,63,145–150]. The extrusion-based printing process can not only deposit the ink but also align the long liquid crystalline chains due to high shear. Yuan et al. [150] first developed a methodology for using electrical current, or Joule heating, to actuate the LCE above its $T_{NI}$ in a multimaterial structure. In 2018, Ambulo et al. [63] utilized the shear forces generated during the DIW 3D printing process and then fixed the alignment using a photo-cross-linking reaction (Fig. 9.14A). Central to this printing scheme is the use of elevated temperatures to reduce the LCE to a printable viscosity. The printed LCE actuation direction is along the

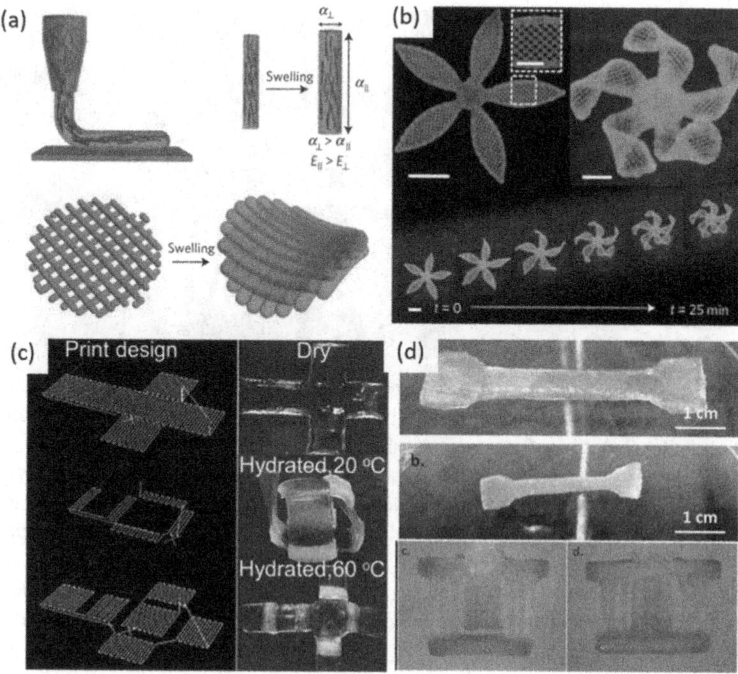


**Figure 9.13** (A) Schematic of the shear-induced alignment of cellulose fibrils during DIW and subsequent effect on anisotropic stiffness and swelling for controlled bending (scale bars = 15 mm). (B) 3D-printed bilayers oriented with respect to the long axis of each petal, with time-lapse sequences of the flowers during the swelling process. (C) Hydrogel multimaterial printing pattern used for a folding box and the resulting print at various temperatures. (D) 3D-printed N-isopropylacrylamide, alginate/PNIPAAm hydrogel tensile specimen swollen in water at 20°C (top) and 60°C (bottom). *(A) and (B) Reproduced with kind permission of Nature Publishing Group from Reference A. Sydney Gladman, E. A. Matsumoto, R. G. Nuzzo, L. Mahadevan, J. A. Lewis 2016.Biomimetic 4D printing Nat. Mater., 15 413; (C) Reproduced with kind permission of Nature Publishing Group from John Wiley & Sons from Reference S. Naficy, R. Gately, R. Gorkin, H. Xin, G. M. Spinks 2017.4D printing of reversible shape morphing hydrogel structures Macromol. Mater. Eng., 302; (D) Reproduced with kind permission of John Wiley & Sons from Reference S. E. Bakarich, R. Gorkin, M. i. h. Panhuis, G. M. Spinks 2015.4D printing with mechanically robust, thermally actuating hydrogels macromol. Rapid. Comm., 36 1211.)*

printing direction. By locally controlling the alignment direction, complex actuation, from 2D to 3D and 3D to 3D' can be achieved [4,145]. The actuation temperature of the above LCE was very high (around 200°C) for full actuation, limiting its potential for practical applications.

Later, researchers modified synthesis chemistry by using soft thiol spacer to tune the actuation temperature and achieve room temperature printing of LCE [62,151]. Roach et al. [62] synthesized an LCE oligomer with

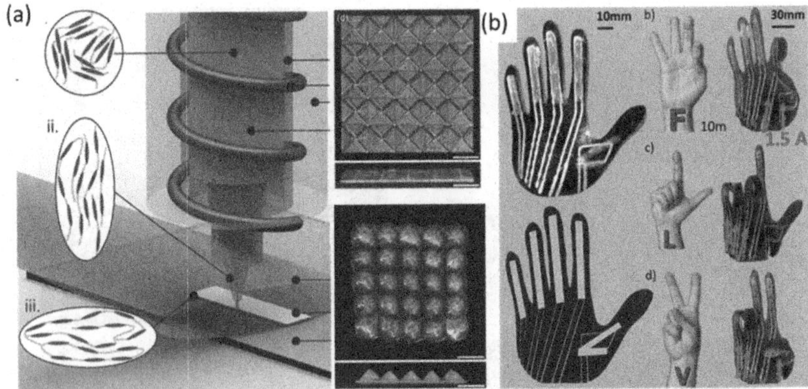


**Figure 9.14** (A) DIW 3D printing of LCE by aligning the LCE mesogens in the printing direction using DIW shear forces. (B) A fully 3D-printed hand where each finger can be activated separately to form American sign language letters. *DIW*, direct ink writing; *LCE*, liquid crystal elastomer. *(A) Reproduced with kind permission of John Wiley & Sons from Reference A. Kotikian, R. L. Truby, J. W. Boley, T. J. White, J. A. Lewis 2018.3D printing of liquid crystal elastomeric actuators with spatially programed nematic order Adv. Mater., 30 1706164; (B) Reproduced with kind permission of IOP Publishing from Reference D. J. Roach, X. Kuang, C. Yuan, K. Chen, H. J. Qi 2018.Novel ink for ambient condition printing of liquid crystal elastomers for 4D Printing Smart Mater. Struct., 27 125011.)*

lower $T_{NI}$ (40°C) by copolymerizing with a flexible thiol spacer. This facilitates room temperature printing and mild temperature actuation of LCE. LCE is to be printed into multimaterial structures to create novel actuators and soft robotics for practical applications [145,146,148,150]. Fig. 9.14B shows a 3D-printed hand with each finger controlled separately by Joule heating to produce letters from the American sign language alphabet. More recently, Kotikian et al. [145] printed a soft robotic structure composed of liquid LCE bilayers with orthogonal director alignment and different nematic-to-isotropic transition temperatures ($T_{NI}$) to form active hinges that interconnect polymeric tiles. The actuation response of different LCEs is programmed by varying their chemistry and printed architecture. The integration of design and AM approaches leads to passively controlled, untethered soft robotic matter with on-demand configurations. Based on this, a self-twisting origami polyhedron and soft robot with programmed sequential folding and deformation can be obtained.

#### *9.3.2.4 Direct ink writing printing of magnetoactive material*

Magnetic field—responsive soft active materials can be 3D-printed to achieve remote-controlled actuation devices for soft robotics and biomedical devices [80,81,86,152,153]. Roh et al. [152] reported 3D-printed silicone soft architectures with programmed magnetocapillary reconfiguration. Fig. 9.15A shows the 3D printing process using home composite capillary pastes for ultrasoft actuators. The ultrasoft actuators easily deform by the magnetic force exerted on carbonyl iron particles embedded in the silicone, as well as lateral capillary forces. In Fig. 9.15B, the water droplet is held on top of the mesh when it is closed by the applied magnetic field. Upon removing the magnetic field, the mesh is opened, and the water droplet is

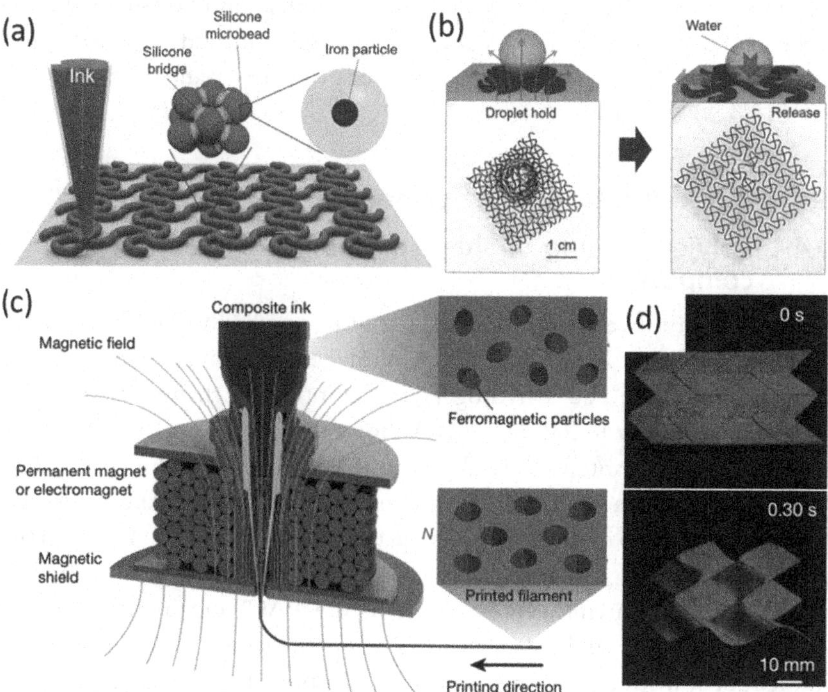


**Figure 9.15** (A) Schematic of 3D printing via homocomposite capillary pastes. (B) The droplet is held on top of the mesh when it is closed by the applied magnetic field upon removing the magnetic field, the mesh is opened and the water droplet is released and penetrates through the mesh. (C) DIW printing of untethered fast-transforming magnetoactive soft materials. (D) The printing planar soft material can self-fold into origami rapidly. *DIW*, direct ink writing. *(A) and (B) Reproduced with kind permission of John Wiley & Sons from Reference S. Roh, L. B. Okello, N. Golbasi, J. P. Hankwitz, J. A.-C. Liu, J. B. Tracy, O. D. Velev 2019.3D-Printed silicone soft architectures with programmed magneto-capillary reconfiguration Adv. Mater. Technol., 4 1800528; (D) Reproduced with kind permission of Nature Publishing Group from Reference Y. Kim, H. Yuk, R. Zhao, S. A. Chester, X. Zhao 2018. Printing ferromagnetic domains for untethered fast-transforming soft materials Nature, 558 274.)*

released and penetrates through the mesh. To achieve more complicated shape actuation, hard magnetic-based magnetoactive materials with a controlled magnetization profile can be used. Kim et al. [86] reported magnetic field—assisted DIW printing of MSMs for untethered, fast shape transforming. During the extrusion process, a magnetic field was applied to align the magnetic polarity in the elastomer composite containing NdFeB particles (Fig. 9.15C). By controlling the filaments infilling direction, the magnetization profile in complex 3D-printed soft materials can be programmed. Fig. 9.15D shows a printed flat soft material can self-fold into origami rapidly. The remote, untethered, complex, and fast shape-shifting under a mild magnetic field of the soft materials can find potential applications in flexible electronics, biomedical devices, and soft robotics [80,81,84,86,153—159].

## 9.4 4D printing by polyjet printing

### 9.4.1 Polyjet printing of shape memory polymer composites

One of the first investigations into 4D printing was by the Tibbits using the Connex 500 inkjet 3D printer [23]. Multimaterial structures consisting of a rigid polymer base and a hydrogel were printing to change its shape on immersion in water [6,23]. Fig. 9.16A shows that the hydrogel (red springs: gray springs in printed version) expansion induced the hinge bending deformation, which can be precisely controlled by changing the spatial distribution of the two materials (the rigid plastic and the hydrogel). Based on this principle, a printed multimaterial strand could self-fold into a 3D cube [23] (Fig. 9.16B). Meanwhile, Ge et al. [2] used a PAC containing a laminate composed of a pure rubber lamina and a composite lamina with parallel fibers in the rubber matrix. After the laminate was printed, it was heated, stretched, cooled, and released for shape transforming into a complex temporary shape (Fig. 9.16C). Various complex 3D shapes including bending twisting, coiling, and folding shapes could be obtained depending on the fiber distribution and orientation. The SMP composites could also be used as a smart hinges to enable active origami for creating complex 3D architectures [24].

One of the advantages of Polyjet printing is digital material fabrication. Multimaterial SMPs can be printed to achieve sequential shape actuation. Mao et al. [117] combined the printed elastomer/SMP composite containing a sandwiched hydrogel layer for shape-shifting materials with enhanced mechanical properties and shape locking capability. The actuation speeds and

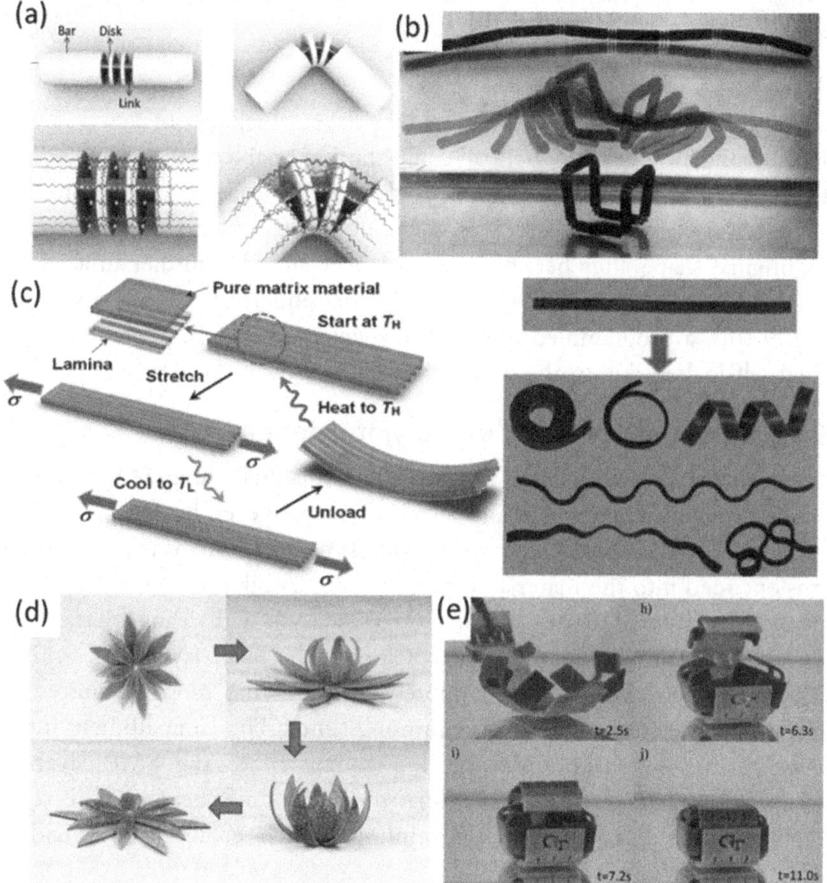


**Figure 9.16** 4D printing of multimaterial hydrogel and composites activated by water. (A) Renderings of an initial joint and its folding, with their corresponding spring–mass systems and 4D printing example that shows the deformation of a grid into a hyperbolic surface. (B) The schematic graph of the reversible actuation component by 3D printing where the hydrogel is confined by the SMP and the elastomer layers and 4D printing examples showing a shape memory petal-like structure for sequential actuation. (C) Schematic demonstration of the laminate architecture and the programming process and representative images for folding with different fiber architectures. (D) Sequential actuation of hydrogel that is confined by the SMP and the elastomer layers showing a shape memory petal-like structure for sequential actuation. (E) 3D-printed sheet folds into a box with a self-locking mechanism. *(A) Reproduced with kind permission of Nature Publishing Group from Reference D. Raviv, W. Zhao, C. McKnelly, A. Papadopoulou, A. Kadambi, B. Shi, S. Hirsch, D. Dikovsky, M. Zyracki, C. Olguin, R. Raskar, S. Tibbits 2014.Active printed materials for complex self-evolving deformations Sci. Rep., 4 7422; Copyright 2014; (B) Reproduced with kind permission of John Wiley & Sons from Reference S. Tibbits 2014.4D printing: multi-material shape change Architect. Des, 84 116; (C) Reproduced with kind permission of AIP Publishing from Reference Q. Ge, H. J. Qi, M. L. Dunn 2013. Active materials by four-dimension printing Appl. Phys. Lett., 103 131901; Copyright 2013; (D) Reproduced with kind permission of Nature Publishing Group from Reference Y. Mao, Z. Ding, C. Yuan, S. Ai, M. Isakov, J. Wu, T. Wang, M. L. Dunn, H. J. Qi 2016.3D printed reversible shape changing components with stimuli responsive materials Sci. Rep., 6 24761; (E) Reproduced with kind permission of Nature Publishing Group from Reference Y. Mao, K. Yu, M. S. Isakov, J. Wu, M. L. Dunn, H. J. Qi 2015.Sequential self-folding structures by 3D printed digital shape memory polymers Sci. Rep., 5 13616.)*

curvature of actuators can be tuned by layer thickness of active layers. Fig. 9.16D shows the sequential shape-shifting enabled by controlling the thickness of SMP layers at high temperatures. The subsequent cooling and drying of hydrogel would lock the bending shape, which would recover to its original shape after heating again. Mao et al. [117] further utilized the printed digital material with tunable $T_g$ for sequential shape recovery. In Fig. 9.16E, a programmed 3D printed sheet can self-fold into a box with then self-locked due to the sequential recovery of the panel.

### 9.4.2 Direct 4D printing by polyjet printing

Most of the previous shape-programmable material using SMPs needs a shape-programming step before shape shifting. Recently, Ding et al. [3] reported an SMP-based direct 4D printing process, where the eigenstrain was encoded into the material during printing. A bilayer laminate strip was printed by using Connex3 Objet500 (Stratasys) with TangoBlack+ (or Tango+) as the elastomer and VeroClear as the SMP layer (Fig. 9.17A). The elastomer layer exists a compressive strain that can be adjusted by printing parameters (such as layer printing time). The laminate bent into a new permanent shape upon heating above $T_g$ of the SMP layer. A buckyball can transform from a planar rod mesh upon heating. This result highlights the advantage of 4D printing, which enables a tremendous

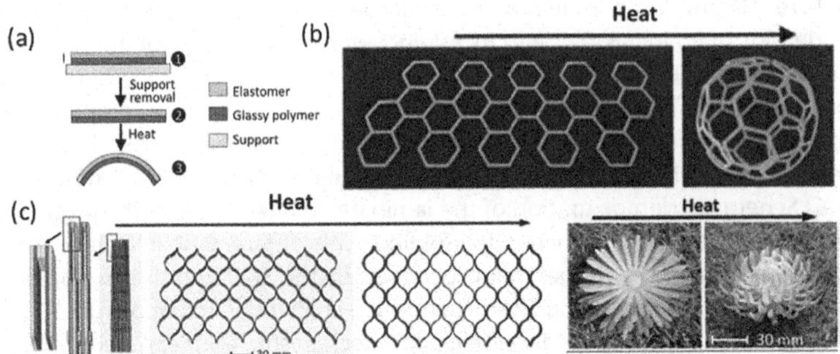


**Figure 9.17** (A) Schematics of bilayer composite for direct 4D printing. (B) A buckyball directly transformed from a planar rod mesh and helix directly transformed from a flat rod upon heating. (C) Lattice structure printed in a collapsed configuration that deploys into an open configuration and printed flower consisting of multiple petals at multiple layers that blooms into a configuration with petals showing different curvatures upon heating. *(A) and (B) Reproduced with kind permission of Elsevier from Reference Z. Ding, O. Weeger, H. J. Qi, M. L. Dunn 2018.4D rods: 3D structures via programmable 1D composite rods Mater. DES., 137 256; (C) Reproduced with kind permission of AAAS from Reference Z. Ding, C. Yuan, X. Peng, T. Wang, H. J. Qi, M. L. Dunn 2017.Direct 4D printing via active composite materials Sci. Adv., 3 e1602890.)*

reduction of both material consumption and printing time (70%–90%) of the thin-walled structure printing (Fig. 9.17B). More complex structure was also achieved with this method. As shown in Fig. 9.17C, a printed thin-walled lattice in a collapsed configuration deployed into an open configuration after heating. The bending curvature can be tuned by the layer printing time and geometrical thickness. Fig. 9.17C also shows that a printed flower can bloom with petals showing different curvatures upon heating. Using the same mechanism, Ding et al. [17] further developed the so-called 4D rod where a rod with composite cross sections could be transformed into a prescribed 3D shape simply by heating.

## 9.5 Vat photopolymerization-based 4D printing

### 9.5.1 Digital light processing

DLP using multivat or grayscale curing enables printing multimaterial with location-specific material properties for SMPs. Kuang et al. [134] developed a new two-stage curing resin for grayscale DLP printing of functionally graded materials with widely tunable mechanical and thermomechanical properties. Fig. 9.18A shows the design and print part of a multimaterial SMP as an artificial arm with sequential recovery capability [134]. Ge et al. [129] used the projection microstereolithography (PμSL) to printing SMPs with widely tunable mechanical and thermomechanical properties. Fig. 9.18B shows a shape recovery of PμSL-printed high-resolution Eiffel Tower. Shape memory stents with different geometric parameters can be also printed, which can find potential biomedicine applications. Using the same approach, Yang et al. [160] developed the reconfigurable, deployable, and mechanically tunable metamaterials by digital micro-3D printing of an SMP. As shown in Fig. 9.18C, an octet-truss microlattice in its original and deformed shape can load the same mass, which can further recover its unique shape upon heating up. Dramatic and reversible changes in the stiffness, geometry, and functions of the metamaterials were achieved.

Recently, Zhang et al. [161] reported a double-network self-healing SMP (SH-SMP) system for high-resolution self-healing 4D printing. In the SH-SMP system, the semicrystalline linear polymer polycaprolactone (PCL) is incorporated into a methacrylate-based SMP system which has good compatibility with the DLP–based 3D printing technology and can be used to fabricate complex 4D printing structures with high resolution (up to 30 μm). Fig. 9.18D shows that a shape memory Kelvin foam lattice structure and a printed chess structure can self-heal after the cut.

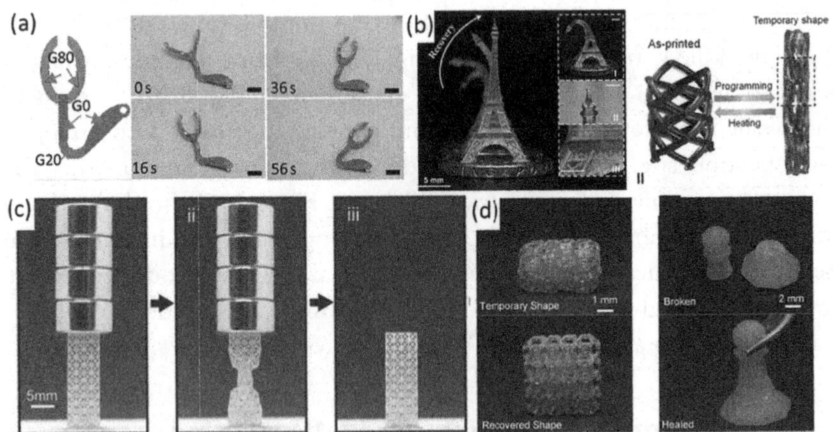


**Figure 9.18** 4D printing by digital light processing printing. (A) Grayscale DLP printed multimaterials SMP for artificial arms with sequential shape recovery. (B) Projection microstereolithography (PμSL) printing of high-resolution shape memory Eiffel Tower and stent with tunable dimension parameters. (C) Reconfigurable octet-truss microlattice in its original shape and deformed geometry can bear the same load and then returns to its original shape again upon heating. Scale bar is 5 mm. (D) DLP printing of self-healing and shape memory semiinterpenetrating polymer network: 3D lattice and a chess structure. *DLP*, digital light processing; *SMP*, shape memory polymer. *(A) Reproduced with kind permission of AAAS from Reference X. Kuang, J. Wu, K. Chen, Z. Zhao, Z. Ding, F. Hu, D. Fang, H. J. Qi 2019.Grayscale digital light processing 3D printing for highly functionally graded materials Sci. Adv., 5 eaav5790; (B) Reproduced with kind permission of Nature Publishing Group from Reference Q. Ge, A. H. Sakhaei, H. Lee, C. K. Dunn, N. X. Fang, M. L. Dunn 2016.Multimaterial 4D printing with tailorable shape memory polymers Sci. Rep., 6 31110; Copyright 2016; (C) Reproduced with kind permission of Royal Society of Chemistry from Reference C. Yang, M. Boorugu, A. Dopp, J. Ren, R. Martin, D. Han, W. Choi, H. Lee 2019.4D printing reconfigurable, deployable and mechanically tunable metamaterials Mater. Horiz., 6 1244; (D) Reproduced with kind permission of ACS Publications from Reference B. Zhang, W. Zhang, Z. Zhang, Y. Zhang, H. Hingorani, Z. Liu, J. Liu, Q. Ge 2019.Self-healing four-dimensional printing with ultraviolet curable double network shape memory polymer system ACS Appl. Mater. Interfaces, 11 10328.)*

Besides printing SMP, DLP also enables digital manufacturing multimaterial for desolvation-based shape shifting [70,71,112,130,133,163,164]. The reaction conversion and cross-linking density can be tuned by controlling the light dose for photocuring either by the light attenuation or using grayscale light. The gradient in multimaterial desolvation and swelling can be utilized for shape shifting. Zhao et al. [165] utilized the graded stress by the nonuniform volume shrinkage during frontal photopolymerization to fold thin films into 3D origami structures. As shown in Fig. 9.19A, using the method of two-side illumination by two different grayscale patterns

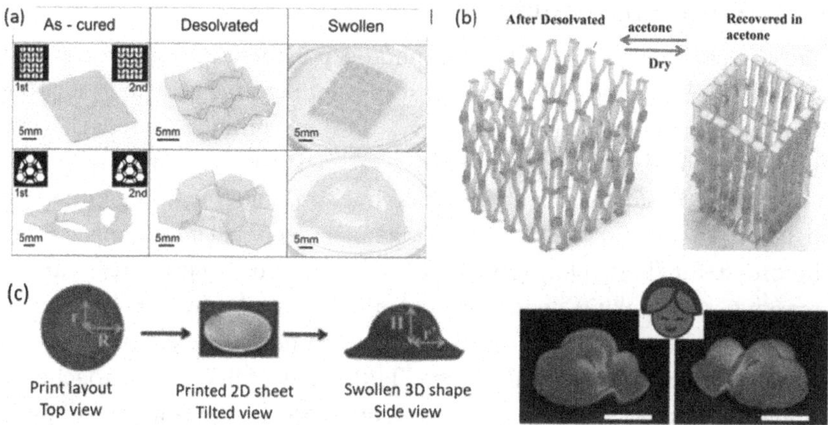


**Figure 9.19** (A) The as-cured flat pattern, dissolved shape, and swollen flat shape of different 3D origami structures. (B) Grayscale DLP printed shape-shifting structure for reversible expansion/shrinkage via swelling/drying in acetone. (C) 3D cartoon face mask viewed from two different angles right after photopolymerization and the corresponding free bending of spatial, differently cured sheet. *DLP*, digital light processing. *(A) Reproduced with kind permission of John Wiley and Sons from Reference Z. Zhao, J. Wu, X. Mu, H. Chen, H. J. Qi, D. Fang 2017.Desolvation induced origami of photocurable polymers by digit light processing macromol. Rapid. Comm., 38 1600625; (B) Reproduced with kind permission of IOP Publishing from Reference W. Jiangtao, Z. Zeang, K. Xiao, M. H. Craig, F. Daining, H. J. Qi 2018.Reversible shape change structures by grayscale pattern 4D printing Multifunctional Materials, 1 015002; (C) Reproduced with kind permission of John Wiley and Sons from Reference L. Huang, R. Jiang, J. Wu, J. Song, H. Bai, B. Li, Q. Zhao, T. Xie 2017.Ultrafast digital printing toward 4D shape changing materials Adv. Mater., 29 1605390.)*

followed by releasing the mold, the cured sheet folded into origami Crane. Wu et al. [133] used the grayscale light pattern to control the local light intensity distribution so as to tune the cross-linking densities at different locations of the bilayer pattern. The desolvation of residual monomers in the solvent leads to local volume shrinkage and shape shifting after drying. The subsequent swelling in a good solvent (such as acetone) drove the material to recovery. Using this method, a reversible shape-shifting structure with self-expanding/shrinking structures can be achieved (Fig. 9.19B). Huang et al. [162] reported an ultrafast digital 4D printing by controlling the light exposure time at different locations of a 2D flat sheet. Fig. 9.19C shows that the polymer network with spatially variable degrees of monomer conversion and cross-linking density led to different monomer desolvation and swelling ratios in water. Internal stress created by the swelling difference between the pixels drove the flat sheet into a prescribed 3D architecture.

## 9.5.2 Direct laser writing

Direct laser writing (DLW) by two-photon polymerization (TPP) offers a powerful tool to fabricated microscale shape-shifting structure. Jin et al. [166] develop a microscale 4D printing via TPP to printing the degradable superparamagnetic hydrogel composite as a swimming microrobot for constructing 3D-to-3D shape-morphing micromachines in a single-material single-step mode. Fig. 9.20A shows a schematic representation using acrylate photoresist for TPP. During the process, cuboid microstructures with an increasing cross-linking density were printed from the hydrogel precursor solution by elevating the laser power. By programming the exposure dosage of femtosecond laser pulses in TPP-based DLW, heterogeneous stimulus-responsive hydrogels with location-specific cross-linking density, stiffness, and swelling/shrinkage ratios can be distributed spatially into arbitrary 3D shapes with submicrometer features. A printed microumbrella

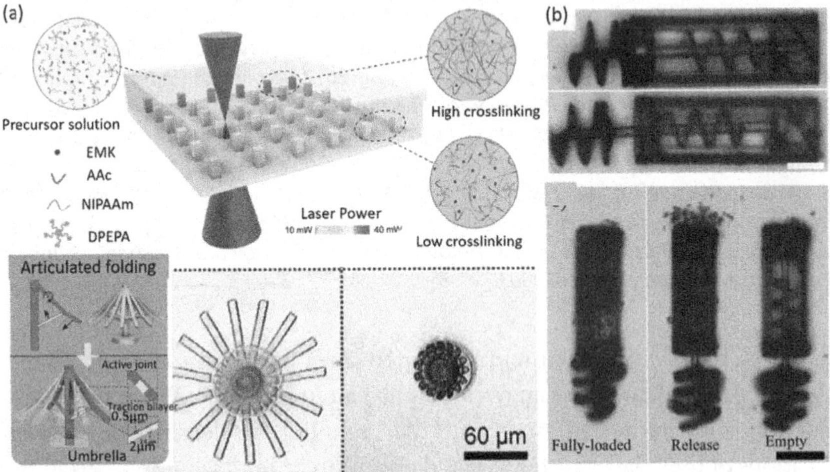


**Figure 9.20** (A) Schematic of the 4D-DLW process. The color bar of the laser power ranges from 10 to 40 mW. The printed microumbrella can be constructed to achieve rapid, precise, and reversible articulated-lever 3D-to-3D folding. (B) 3D-printed micromachine with a microcapsule and microsyringe functioning two different modes of the reciprocating mechanism (closed and open) for the particle release process. The scale bars are 20 μm. *DLW*, direct laser writing. *(A) Reproduced with kind permission of Elsevier from Reference D. Jin, Q. Chen, T.-Y. Huang, J. Huang, L. Zhang, H. Duan 2019. Four-dimensional direct laser writing of reconfigurable compound micromachines Mater. Today Off., 32 19; (B) Reproduced with kind permission of John Wiley and Sons from Reference T. Y. Huang, M. S. Sakar, A. Mao, A. J. Petruska, F. Qiu, X. B. Chen, S. Kennedy, D. Mooney, B. J. Nelson 2015.3D printed microtransporters: compound micromachines for spatiotemporally controlled delivery of therapeutic agents Adv. Mater., 27 6644.)*

showed precise and reversible 3D-to-3D shape transformation in response to multiple external stimuli (such as pH), which can be used as smart and multifunctional micromachine for biomedical applications.

Similarly, Huang et al. [167] used the TPP technique in a combination of selective magnetic film deposition to fabricate the magnetic-responsive microtransporter devices with our assembling step for targeted and triggered delivery of particles, biological materials, and smaller micromachines. As shown in Fig. 9.20B, inspired by the Archimedean screw-pump, a micromachine containing microcapsule and microsyringe with two different reciprocating modes were directly printed without an assembling step. Under a rotating magnetic field of 9 mT at several frequencies between 1 and 10 Hz, the micromachine-loaded 3- and 6-μm-diameter polystyrene microspheres can gradually release. The vat photopolymerization-based 3D printing of shape-programmable materials in multiscale finds great application potentials as flexible sensors, actuators, biomedical devices, and tissue engineering.

## 9.6 Summary

Since the first conceptualization in 2013, 4D printing has realized unprecedented advances. 4D printing technique derived from 3D printing, smart materials science. The future development of 4D printing requires interdisciplinary research advances in new 3D printing techniques and novel functional materials as well as advanced simulation/modeling tools to realize well-controlled and precise shape shifting (Fig. 9.21).

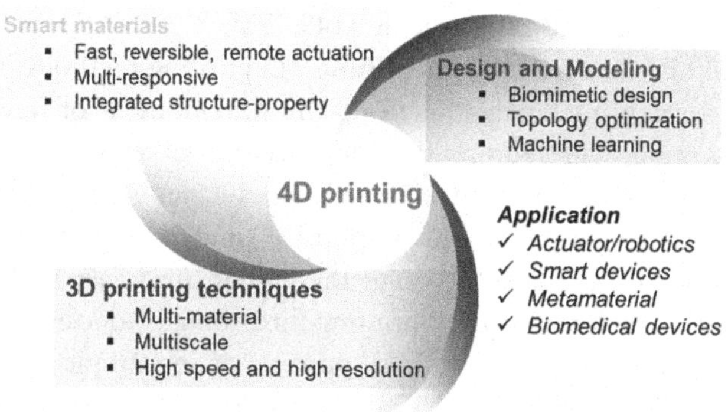


**Figure 9.21** The further development of 4D printing, including smart materials, design/modeling methods, and new 3D printing techniques, can find broad applications.

New smart materials for 4D printing. The properties of stimulus-responsive materials determine the stimulus and method for actuation. The majority of SMPs used for 4D printing is thermal-responsive one-way (irreversible) SMPs. Other stimulus-responsive SMPs, especially photo-responsive SMP [168,169] and magnetic field–responsive SMP [155,170,171], are highly desirable for remote-controlled actuation of 4D printing. 4D printing of either single or multimaterial SMP usually involves a manual shape-programming step for shape shifting. Recently, direct 4D printing was achieved by encoding the eigenstrain in the multimaterial structure during printing. Despite this advance, 4D printing of SMP still suffers from one-way actuation. By contrast, hydrogel composites enable reversible shape shifting due to stimulus-responsive swelling and deswelling process. The actuation of hydrogels involves diffusion and mass transport of water or other chemicals, which is a time-consuming process. This issue can be tackled by decreasing the sample dimension and using porous material design. However, a hydrogel is a mechanically weak soft material, which limited broad application. By contrast, LCE also enables reversible shape actuation with relatively enhanced mechanical properties. Both photo- and thermal-responsive LCEs have been used for 4D printing to allow large strain reversible actuation. Recently, magnetoactive materials are used for 4D printing with remote-controlled, untethered, and fast actuation. The development of advanced smart materials with integrated shape actuation and multistimulus-responsive capability is challenging and highly desired in the future. Other multifunctional materials, such as self-healing material and biomaterials, can be developed for 4D-printed structures [110,172]. In addition, the material shape shifting can be combined with other functional properties, such as wettability and surface adhesion for broad applications in robotics and optics and biomedicine [149,173].

It should be noted that most existing 3D printing methods may not be applicable to new materials; therefore, the development of new materials for 4D printing will require the modification of existing 3D printing methods. External field–assisted 3D provides a new method to fabricate high-performance and stimulus-responsive materials. For example, an electric field can be applied to align the carbon materials, such as carbon nanotube and graphite, for DLP printing functional composites [174–176]. Magnetic field was used to assist DLP printing of anisotropic composites for controlled shape shifting [177,178]. In addition, high-resolution, high-speed 3D printing technologies can be developed to fabricate materials and devices with complex multiscale geometries [179]. More sophisticated multimaterial 3D printing technologies are especially desirable to fabricate more complex structure and devices with shape shifting and other functional properties.

New structural design and advanced simulation and modeling tools can further advance 4D printing. For example, the porous materials with multiscale pore size enable faster diffusion and mass transport for water and chemical-based actuation. As multimaterials are extensively used for programmable matter, bilayers structure and biomimetic concept can provide a lot of inspiration [175,180–182]. For shape actuation involving multiphysics problems, such as hydrogel composites with large swelling deformation, an intuition-based design method may be difficult to capture all of the physical processes. Multiple finite-element modeling can be used to accurately predict the shape transformation of the printed shape under the external stimulus [8,118]. In addition, the simulation and modeling approach provides a nonintuitive solution for the inverse design problem [183,184]. To this end, genetic algorithms and machine learning offer a powerful approach to optimize material distribution for the rational design of active materials for controlled shape shifting. Topology optimization can be used to optimize the design of smart structures based on the target architecture with reduced weight [183,185].

In this chapter, we introduced basic elements and the advance of 4D printing with a focus on materials and printing techniques. 4D printing enables the targeted evolution of 3D-printed structures' shape, property, and functionality as a function of time in response to external stimulus. 4D printing is very young and still needs a significant amount of effort for future development. With continuous progress in the field of new stimulus-responsive materials, new printing techniques, structural design, and modeling tools/software, 4D printing will inevitably realize transformative applications in smart devices, soft robotics, metamaterial, and biomedical devices.

## References

[1] S. Tibbits, The Emergence of "4D Printing", https://www.ted.com/talks/skylar_tibbits_the_emergence_of_4d_printing, (accessed).

[2] Q. Ge, H.J. Qi, M.L. Dunn, Active materials by four-dimension printing, Appl. Phys. Lett. 103 (2013) 131901.

[3] Z. Ding, C. Yuan, X. Peng, T. Wang, H.J. Qi, M.L. Dunn, Direct 4D printing via active composite materials, Sci. Adv. 3 (2017) e1602890.

[4] A. Kotikian, R.L. Truby, J.W. Boley, T.J. White, J.A. Lewis, 3D printing of liquid crystal elastomeric actuators with spatially programed nematic order, Adv. Mater. 30 (2018) 1706164.

[5] A. Sydney Gladman, E.A. Matsumoto, R.G. Nuzzo, L. Mahadevan, J.A. Lewis, Biomimetic 4D printing, Nat. Mater. 15 (2016) 413.
[6] D. Raviv, W. Zhao, C. McKnelly, A. Papadopoulou, A. Kadambi, B. Shi, S. Hirsch, D. Dikovsky, M. Zyracki, C. Olguin, R. Raskar, S. Tibbits, Active printed materials for complex self-evolving deformations, Sci. Rep. 4 (2014) 7422.
[7] H. Yang, W.R. Leow, T. Wang, J. Wang, J. Yu, K. He, D. Qi, C. Wan, X. Chen, 3D printed photoresponsive devices based on shape memory composites, Adv. Mater. 29 (2017) 1701627.
[8] O. Kuksenok, A.C. Balazs, Stimuli-responsive behavior of composites integrating thermo-responsive gels with photo-responsive fibers, Mater. Horiz. 3 (2016) 53.
[9] M. Nadgorny, Z. Xiao, C. Chen, L.A. Connal, Three-dimensional printing of pH-responsive and functional polymers on an affordable desktop printer, ACS Appl. Mater. Interface. 8 (2016) 28946.
[10] J. Choi, O.C. Kwon, W. Jo, H.J. Lee, M.-W. Moon, 4D printing technology: a review, 3D Print. Addit. Manuf. 2 (2015) 159.
[11] J.-J. Wu, L.-M. Huang, Q. Zhao, T. Xie, 4D printing: history and recent progress *Chin*, J. Polym. Sci. 36 (2017) 563.
[12] D.-G. Shin, T.-H. Kim, D.-E. Kim, Review of 4D printing materials and their properties, Int. J. Precis. Eng. Manufact.-Green Technol. 4 (2017) 349.
[13] F. Momeni, S.M. Mehdi HassaniN, X. Liu, J. Ni, A review of 4D printing, Mater. Des. 122 (2017) 42.
[14] Z.X. Khoo, J.E.M. Teoh, Y. Liu, C.K. Chua, S. Yang, J. An, K.F. Leong, W.Y. Yeong, 3D printing of smart materials: a review on recent progresses in 4D printing, Virt. Phys. Prototyp. 10 (2015) 103.
[15] J. Lee, H.-C. Kim, J.-W. Choi, I.H. Lee, A review on 3D printed smart devices for 4D printing, Int. J. Precis. Eng. Manufact.-Green Technol. 4 (2017) 373.
[16] B. Shen, O. Erol, L. Fang, S.H. Kang, Programming the time into 3D printing: current advances and future directions in 4D printing, Multifunct. Mater. 3 (2020) 012001.
[17] Z. Ding, O. Weeger, H.J. Qi, M.L. Dunn, 4D rods: 3D structures via programmable 1D composite rods, Mater. Des. 137 (2018) 256.
[18] B. An, Y. Tao, J. Gu, T. Cheng, X.A. Chen, X. Zhang, W. Zhao, Y. Do, S. Takahashi, H.-Y. Wu, in: Presented at Proceedings of the 2018 CHI Conference on Human Factors in Computing Systems, 2018.
[19] R. Bogue, Smart materials: a review of capabilities and applications, Assemb. Autom. 34 (2014) 16.
[20] D.J. Leo, Engineering Analysis of Smart Material Systems, John Wiley & Sons, 2007.
[21] A. Lendlein, V.P. Shastri, Stimuli-sensitive polymers, Adv. Mater. 22 (2010) 3344.
[22] D. Roy, J.N. Cambre, B.S. Sumerlin, Future perspectives and recent advances in stimuli-responsive materials, Prog. Polym. Sci. 35 (2010) 278.
[23] S. Tibbits, 4D printing: multi-material shape change, Architect. Des 84 (2014) 116.
[24] G. Qi, K.D. Conner, H.J. Qi, L.D. Martin, Active origami by 4D printing, Smart Mater. Struct. 23 (2014) 094007.
[25] J. Ma, B. Franco, G. Tapia, K. Karayagiz, L. Johnson, J. Liu, R. Arroyave, I. Karaman, A. Elwany, Spatial control of functional response in 4D-printed active metallic structures, Sci. Rep. 7 (2017) 46707.
[26] F.L. Bargardi, H. Le Ferrand, R. Libanori, A.R. Studart, Bio-inspired self-shaping ceramics, Nat. Commun. 7 (2016) 13912.
[27] G. Liu, Y. Zhao, G. Wu, J. Lu, Origami and 4D printing of elastomer-derived ceramic structures, Sci. Adv. 4 (2018).
[28] G. Li, A. Wang, Cold, warm, and hot programming of shape memory polymers, J. Polym. Sci. B Polym. Phys. 54 (2016) 1319.

[29] E.D. Rodriguez, X. Luo, P.T. Mather, Linear/network poly(ε-caprolactone) blends exhibiting shape memory assisted self-healing (SMASH), ACS Appl. Mater. Interfaces 3 (2011) 152.
[30] X. Gu, P.T. Mather, Entanglement-based shape memory polyurethanes: synthesis and characterization, Polymer 53 (2012) 5924.
[31] M. Warner, E.M. Terentjev, Liquid Crystal Elastomers, OUP, Oxford, 2003.
[32] S.W. Ula, N.A. Traugutt, R.H. Volpe, R.R. Patel, K. Yu, C.M. Yakacki, Liquid crystal elastomers: an introduction and review of emerging technologies, Liquid Cryst. Rev. 6 (2018) 78.
[33] T.J. White, D.J. Broer, Programmable and adaptive mechanics with liquid crystal polymer networks and elastomers, Nat. Mater. 14 (2015) 1087.
[34] A. Lendlein, O.E.C. Gould, Reprogrammable recovery and actuation behaviour of shape-memory polymers, Nat. Rev. Mater. 4 (2019) 116.
[35] J. Hu, Y. Zhu, H. Huang, J. Lu, Recent advances in shape—memory polymers: structure, mechanism, functionality, modeling and applications, Prog. Polym. Sci. 37 (2012) 1720.
[36] Y. Liu, J. Genzer, M.D. Dickey, "2D or not 2D": shape-programming polymer sheets, Prog. Polym. Sci. 52 (2016) 79.
[37] H. Xie, K.-K. Yang, Y.-Z. Wang, Photo-cross-linking: a powerful and versatile strategy to develop shape-memory polymers, Prog. Polym. Sci. 95 (2019) 32.
[38] Q. Zhao, H.J. Qi, T. Xie, Recent progress in shape memory polymer: new behavior, enabling materials, and mechanistic understanding, Prog. Polym. Sci. 79 (2015) 49—50.
[39] J. Fan, G. Li, High enthalpy storage thermoset network with giant stress and energy output in rubbery state, Nat. Commun. 9 (2018) 642.
[40] Y. Gao, W. Liu, S. Zhu, Polyolefin thermoplastics for multiple shape and reversible shape memory, ACS Appl. Mater. Interface. 9 (2017) 4882.
[41] J. Zhao, M. Chen, X. Wang, X. Zhao, Z. Wang, Z.-M. Dang, L. Ma, G.-H. Hu, F. Chen, Triple shape memory effects of cross-linked polyethylene/polypropylene blends with cocontinuous architecture, ACS Appl. Mater. Interface. 5 (2013) 5550.
[42] T. Gong, W. Li, H. Chen, L. Wang, S. Shao, S. Zhou, Remotely actuated shape memory effect of electrospun composite nanofibers, Acta Biomaterialia 8 (2012) 1248.
[43] S. Pandini, F. Baldi, K. Paderni, M. Messori, M. Toselli, F. Pilati, A. Gianoncelli, M. Brisotto, E. Bontempi, T. Riccò, One-way and two-way shape memory behaviour of semi-crystalline networks based on sol—gel cross-linked poly(ε-caprolactone), Polymer 54 (2013) 4253.
[44] M. Huang, X. Dong, L. Wang, J. Zhao, G. Liu, D. Wang, Two-way shape memory property and its structural origin of cross-linked poly([varepsilon]-caprolactone), RSC Adv. 4 (2014) 55483.
[45] Y. Zhang, N. Zheng, Y. Cao, F. Wang, P. Wang, Y. Ma, B. Lu, G. Hou, Z. Fang, Z. Liang, Climbing-inspired twining electrodes using shape memory for peripheral nerve stimulation and recording, Sci. Adv. 5 (2019) eaaw1066.
[46] M. Behl, K. Kratz, J. Zotzmann, U. Nöchel, A. Lendlein, Reversible bidirectional shape-memory polymers, Adv. Mater. 25 (2013) 4466.
[47] J. Zotzmann, M. Behl, D. Hofmann, A. Lendlein, Reversible triple-shape effect of polymer networks containing polypentadecalactone- and poly(ε-caprolactone)-segments, Adv. Mater. 22 (2010) 3424.
[48] M.K. Jang, A. Hartwig, B.K. Kim, Shape memory polyurethanes cross-linked by surface modified silica particles, J. Mater. Chem. 19 (2009) 1166.
[49] A. Lendlein, R. Langer, Biodegradable, elastic shape-memory polymers for potential, Biomed. Applicat. Sci. 296 (2002) 1673.

[50] W. Voit, T. Ware, R.R. Dasari, P. Smith, L. Danz, D. Simon, S. Barlow, S.R. Marder, K. Gall, High-strain shape-memory polymers, Adv. Funct. Mater. 20 (2010) 162.
[51] C.M. Yakacki, R. Shandas, D. Safranski, A.M. Ortega, K. Sassaman, K. Gall, Strong, tailored, biocompatible shape-memory polymer networks, Adv. Funct. Mater. 18 (2008) 2428.
[52] Y. Luo, Y. Guo, X. Gao, B.-G. Li, T. Xie, A general approach towards thermoplastic multishape-memory polymers via sequence structure design, Adv. Mater. 25 (2013) 743.
[53] Q. Zhang, K. Zhang, G. Hu, Smart three-dimensional lightweight structure triggered from a thin composite sheet via 3D printing technique, Sci. Rep. 6 (2016) 22431.
[54] M. Bahram, N. Mohseni, M. Moghtader, In Emerging Concepts in Analysis and Applications of Hydrogels, IntechOpen, 2016.
[55] X. Lu, H. Zhang, G. Fei, B. Yu, X. Tong, H. Xia, Y. Zhao, Liquid-crystalline dynamic networks doped with gold nanorods showing enhanced photocontrol of actuation, Adv. Mater. 30 (2018) 1706597.
[56] R.R. Kohlmeyer, J. Chen, Wavelength-selective, IR light-driven hinges based on liquid crystalline elastomer composites, Angew. Chem. Int. Ed. 52 (2013) 9234.
[57] M.O. Saed, A.H. Torbati, D.P. Nair, C.M. Yakacki, Synthesis of programmable main-chain liquid-crystalline elastomers using a two-stage thiol-acrylate reaction, J. Vis. Exp. 0 (2016) 53546.
[58] C.M. Yakacki, M. Saed, D.P. Nair, T. Gong, S.M. Reed, C.N. Bowman, Tailorable and programmable liquid-crystalline elastomers using a two-stage thiol-acrylate reaction, RSC Adv. 5 (2015) 18997.
[59] M. Barnes, R. Verduzco, Direct shape programming of liquid crystal elastomers, Soft Matter 15 (2019) 870.
[60] Y. Li, Y. Zhang, O. Rios, J. Keum, M. Kessler, Liquid crystalline epoxy networks with exchangeable disulfide bonds, Soft Matter 13 (2017) 5021.
[61] T.H. Ware, M.E. McConney, J.J. Wie, V.P. Tondiglia, T.J. White, Voxelated liquid crystal elastomers, Science 347 (2015) 982.
[62] D.J. Roach, X. Kuang, C. Yuan, K. Chen, H.J. Qi, Novel ink for ambient condition printing of liquid crystal elastomers for 4D, Print. Smart Mater. Struct. 27 (2018) 125011.
[63] C.P. Ambulo, J.J. Burroughs, J.M. Boothby, H. Kim, M.R. Shankar, T.H. Ware, Four-dimensional printing of liquid crystal elastomers, ACS Appl. Mater. Interface. 9 (2017) 37332.
[64] M.O. Saed, A. Gablier, E.M. Terentejv, Liquid crystalline vitrimers with full or partial boronic-ester bond exchange, Adv. Funct. Mater. 30 (2019) 1906458.
[65] M. Tabrizi, T.H. Ware, M.R. Shankar, Voxelated molecular patterning in 3-dimensional freeforms, ACS Appl. Mater. Interface. 11 (2019) 28236.
[66] O. Wichterle, D. Lim, Hydrophilic gels for biological use, Nature 185 (1960) 117.
[67] P.K. Annamalai, K.L. Dagnon, S. Monemian, E.J. Foster, S.J. Rowan, C. Weder, Water-responsive mechanically adaptive nanocomposites based on styrene—butadiene rubber and cellulose nanocrystals—processing matters, ACS Appl. Mater. Interface. 6 (2014) 967.
[68] C. Zhu, C.J. Bettinger, Photoreconfigurable physically cross-linked triblock copolymer hydrogels: photodisintegration kinetics and structure—property relationships, Macromolecules 48 (2015) 1563.
[69] M. Ma, L. Guo, D.G. Anderson, R. Langer, Bio-inspired polymer composite actuator and generator driven by water gradients, Science 339 (2013) 186.

[70] Z.L. Wu, M. Moshe, J. Greener, H. Therien-Aubin, Z. Nie, E. Sharon, E. Kumacheva, Three-dimensional shape transformations of hydrogel sheets induced by small-scale modulation of internal stresses, Nat. Commun. 4 (2013) 1586.
[71] J. Kim, J.A. Hanna, M. Byun, C.D. Santangelo, R.C. Hayward, Designing responsive buckled surfaces by halftone gel lithography, Science 335 (2012) 1201.
[72] D. Han, Z. Lu, S.A. Chester, H. Lee, Micro 3D printing of a temperature-responsive hydrogel using projection micro-stereolithography, Sci. Rep. 8 (2018) 1963.
[73] Z. Zhao, X. Kuang, C. Yuan, H.J. Qi, D. Fang, Hydrophilic/Hydrophobic composite shape-shifting structures, ACS Appl. Mater. Interface. 10 (2018) 19932.
[74] Q. Zhao, X. Yang, C. Ma, D. Chen, H. Bai, T. Li, W. Yang, T. Xie, A bioinspired reversible snapping hydrogel assembly, Mater. Horiz. 3 (2016) 422.
[75] C. Colosi, S.R. Shin, V. Manoharan, S. Massa, M. Costantini, A. Barbetta, M.R. Dokmeci, M. Dentini, A. Khademhosseini, Microfluidic bioprinting of heterogeneous 3D tissue constructs using low-viscosity bioink, Adv. Mater. 28 (2016) 677.
[76] S. Miao, N. Castro, M. Nowicki, L. Xia, H. Cui, X. Zhou, W. Zhu, S.-j. Lee, K. Sarkar, G. Vozzi, 4D printing of polymeric materials for tissue and organ regeneration *Mater*, Today Off. 20 (2017) 577.
[77] S.V. Murphy, A. Atala, 3D bioprinting of tissues and organs, Nat. Biotechnol. 32 (2014) 773.
[78] B. Derby, Printing and prototyping of tissues and scaffolds, Science 338 (2012) 921.
[79] G. Villar, A.D. Graham, H. Bayley, A tissue-like printed material, Science 340 (2013) 48.
[80] J. Kim, S.E. Chung, S.-E. Choi, H. Lee, J. Kim, S. Kwon, Programming magnetic anisotropy in polymeric microactuators, Nat. Mater. 10 (2011) 747.
[81] W. Hu, G.Z. Lum, M. Mastrangeli, M. Sitti, Small-scale soft-bodied robot with multimodal locomotion, Nature 554 (2018) 81.
[82] J. Qin, I. Asempah, S. Laurent, A. Fornara, R.N. Muller, M. Muhammed, Injectable superparamagnetic ferrogels for controlled release of hydrophobic drugs, Adv. Mater. 21 (2009) 1354.
[83] R. Zhao, Y. Kim, S.A. Chester, P. Sharma, X. Zhao, Mechanics of hard-magnetic soft materials, J. Mech. Phys. Solid. 124 (2019) 244.
[84] J.A. Jackson, M.C. Messner, N.A. Dudukovic, W.L. Smith, L. Bekker, B. Moran, A.M. Golobic, A.J. Pascall, E.B. Duoss, K.J. Loh, C.M. Spadaccini, Field responsive mechanical metamaterials, Sci. Adv. 4 (2018) eaau6419.
[85] V.Q. Nguyen, A.S. Ahmed, R.V. Ramanujan, Morphing soft magnetic composites, Adv. Mater. 24 (2012) 4041.
[86] Y. Kim, H. Yuk, R. Zhao, S.A. Chester, X. Zhao, Printing ferromagnetic domains for untethered fast-transforming soft materials, Nature 558 (2018) 274.
[87] I. Gibson, D. Rosen, B. Stucker, Additive Manufacturing Technologies, Springer, Boston, MA, 2015.
[88] J.P. Kruth, M.C. Leu, T. Nakagawa, Progress in additive manufacturing and rapid prototyping, CIRP Ann. - Manuf. Technol. 47 (1998) 525.
[89] B. Gao, Q. Yang, X. Zhao, G. Jin, Y. Ma, F. Xu, 4D bioprinting for biomedical applications, Trends Biotechnol. 34 (2016) 746.
[90] D.B. Kolesky, R.L. Truby, A.S. Gladman, T.A. Busbee, K.A. Homan, J.A. Lewis, 3D bioprinting of vascularized, heterogeneous cell-laden tissue constructs, Adv. Mater. 26 (2014) 3124.
[91] J.H. Cho, J. Lee, Y. Xia, B. Kim, Y. He, M.J. Renn, T.P. Lodge, C.D. Frisbie, Printable ion-gel gate dielectrics for low-voltage polymer thin-film transistors on plastic, Nat. Mater. 7 (2008) 900.

[92] J.M. Whiteley, P. Taynton, W. Zhang, S.-H. Lee, 3D printing of customized Li-ion batteries with thick electrodes, Adv. Mater. 27 (2015) 6922.
[93] B.G. Compton, J.A. Lewis, 3D-Printing of lightweight cellular composites, Adv. Mater. 26 (2014) 5930.
[94] M. Hegde, V. Meenakshisundaram, N. Chartrain, S. Sekhar, D. Tafti, C.B. Williams, T.E. Long, 3D printing all-aromatic polyimides using mask-projection stereolithography: processing the nonprocessable, Adv. Mater. 29 (2017) 1701240.
[95] X. Kuang, Z. Zhao, K. Chen, D. Fang, G. Kang, H.J. Qi, High-speed 3D printing of high-performance thermosetting polymers via two-stage curing, Macromol. Rapid Commun. 39 (2018) 1700809.
[96] R. Matsuzaki, M. Ueda, M. Namiki, T.-K. Jeong, H. Asahara, K. Horiguchi, T. Nakamura, A. Todoroki, Y. Hirano, Three-dimensional printing of continuous-fiber composites by in-nozzle impregnation, Sci. Rep. 6 (2016) 23058.
[97] F. Kotz, K. Arnold, W. Bauer, D. Schild, N. Keller, K. Sachsenheimer, T.M. Nargang, C. Richter, D. Helmer, B.E. Rapp, Three-dimensional printing of transparent fused silica glass, Nature 544 (2017) 337.
[98] Z.C. Eckel, C. Zhou, J.H. Martin, A.J. Jacobsen, W.B. Carter, T.A. Schaedler, Additive manufacturing of polymer-derived ceramics, Science 351 (2016) 58.
[99] X. Zheng, H. Lee, T.H. Weisgraber, M. Shusteff, J. DeOtte, E.B. Duoss, J.D. Kuntz, M.M. Biener, Q. Ge, J.A. Jackson, S.O. Kucheyev, N.X. Fang, C.M. Spadaccini, Ultralight, ultrastiff mechanical metamaterials, Science 344 (2014) 1373.
[100] S.C. Ligon, R. Liska, J. Stampfl, M. Gurr, R. Mülhaupt, Polymers for 3D printing and customized additive manufacturing, Chem. Rev. 117 (2017) 10212.
[101] D.L. Bourell, Perspectives on additive manufacturing, Annu. Rev. Mater. Res. 46 (2016) 1.
[102] A. Zhakeyev, P. Wang, L. Zhang, W. Shu, H. Wang, J. Xuan, Additive manufacturing: unlocking the evolution of energy materials, Adv. Sci. 4 (2017) 1700187.
[103] C.J. Hansen, R. Saksena, D.B. Kolesky, J.J. Vericella, S.J. Kranz, G.P. Muldowney, K.T. Christensen, J.A. Lewis, High-throughput printing via microvascular multi-nozzle arrays, Adv. Mater. 25 (2013) 96.
[104] K. Chen, X. Kuang, V. Li, G. Kang, H.J. Qi, Fabrication of tough epoxy with shape memory effects by UV-assisted direct-ink write printing, Soft Matter 14 (2018) 1879.
[105] M. Invernizzi, G. Natale, M. Levi, S. Turri, G. Griffini, UV-assisted 3D printing of glass and carbon fiber-reinforced dual-cure polymer composites, Materials 9 (2016) 583.
[106] L.L. Lebel, B. Aissa, M.A.E. Khakani, D. Therriault, WILEY-VCH Verlag, Patent, 2010, pp. 1521–4095.
[107] J.A. Lewis, Direct ink writing of 3D functional materials, Adv. Funct. Mater. 16 (2006) 2193.
[108] K. Chen, L. Zhang, X. Kuang, V. Li, M. Lei, G. Kang, Z.L. Wang, H.J. Qi, Dynamic photomask-assisted direct ink writing multimaterial for multilevel triboelectric nanogenerator, Adv. Funct. Mater. (2019) 0 1903568.
[109] S. Naficy, R. Gately, R. Gorkin, H. Xin, G.M. Spinks, 4D printing of reversible shape morphing hydrogel structures *Macromol*, Mater. Eng. 302 (2017).
[110] X. Kuang, K. Chen, C.K. Dunn, J. Wu, V.C. Li, H.J. Qi, 3D printing of highly stretchable, shape-memory and self-healing elastomer toward novel 4D, Print. ACS Appl. Mater. Interface. 10 (2018) 7381.
[111] J. Wang, Z. Wang, Z. Song, L. Ren, Q. Liu, L. Ren, Biomimetic shape–color double-responsive 4D, Print. Adv. Mater. Technol. 4 (2019) 1900293.
[112] A. Nojoomi, H. Arslan, K. Lee, K. Yum, Bioinspired 3D structures with programmable morphologies and motions, Nat. Commun. 9 (2018) 3705.

[113] M.W. Barclift, C.B. Williams, in: Presented at International Solid Freeform Fabrication Symposium, 2012.
[114] A. Cazón, P. Morer, L. Matey, PolyJet technology for product prototyping: tensile strength and surface roughness properties, Proc. Institut. Mech. Eng., Part. B: J. Eng. Manufact. 228 (2014) 1664.
[115] M. Jochen, C. Diana, S. Manuel, S. Ralph, S. Kristina, Mechanical properties of interfaces in inkjet 3D printed single- and multi-material parts, 3D Print. Addit. Manuf. 4 (2017) 193.
[116] E. Pei, 4D Printing: dawn of an emerging technology cycle, Assemb. Autom. 34 (2014) 310.
[117] Y. Mao, K. Yu, M.S. Isakov, J. Wu, M.L. Dunn, H.J. Qi, Sequential self-folding structures by 3D printed digital shape memory polymers, Sci. Rep. 5 (2015) 13616.
[118] Y. Mao, Z. Ding, C. Yuan, S. Ai, M. Isakov, J. Wu, T. Wang, M.L. Dunn, H.J. Qi, 3D printed reversible shape changing components with stimuli responsive materials, Sci. Rep. 6 (2016) 24761.
[119] K. Liu, J. Wu, G.H. Paulino, H.J. Qi, Programmable deployment of tensegrity structures by stimulus-responsive polymers, Sci. Rep. 7 (2017) 3511.
[120] J. Wu, C. Yuan, Z. Ding, M. Isakov, Y. Mao, T. Wang, M.L. Dunn, H.J. Qi, Multi-shape active composites by 3D printing of digital shape memory polymers, Sci. Rep. 6 (2016) 24224.
[121] L.J. Hornbeck, The DMD TM projection display chip: a MEMS-based technology, MRS Bull. 26 (2001) 325.
[122] R. Janusziewicz, J.R. Tumbleston, A.L. Quintanilla, S.J. Mecham, J.M. DeSimone, Layerless fabrication with continuous liquid interface production, Proc. Natl. Acad. Sci. U. S. A 113 (2016) 201605271.
[123] J.R. Tumbleston, D. Shirvanyants, N. Ermoshkin, R. Janusziewicz, A.R. Johnson, D. Kelly, K. Chen, R. Pinschmidt, J.P. Rolland, A. Ermoshkin, E.T. Samulski, J.M. DeSimone, Continuous liquid interface production of 3D objects, Science 347 (2015) 1349.
[124] D.A. Walker, J.L. Hedrick, C.A. Mirkin, Rapid, Large-Volume, Thermally Controlled 3D Printing Using a Mobile Liquid Interface *Science* vol. 366, 2019, p. 360.
[125] J.-W. Choi, E. MacDonald, R. Wicker, Multi-material microstereolithography, Int. J. Adv. Manuf. Technol. 49 (2010) 543.
[126] C. Sun, N. Fang, D. Wu, X. Zhang, Projection micro-stereolithography using digital micro-mirror dynamic mask, Sensor Actuat. Phys. 121 (2005) 113.
[127] J. Herzberger, V. Meenakshisundaram, C.B. Williams, T.E. Long, 3D printing all-aromatic polyimides using stereolithographic 3D printing of polyamic acid salts, ACS Macro Lett. 7 (2018) 493.
[128] C. Lu, C. Wang, J. Yu, J. Wang, F. Chu, Two-step 3 d-printing approach toward sustainable, repairable, fluorescent shape-memory thermosets derived from cellulose and rosin, ChemSusChem 12 (2020) 893.
[129] Q. Ge, A.H. Sakhaei, H. Lee, C.K. Dunn, N.X. Fang, M.L. Dunn, Multimaterial 4D printing with tailorable shape memory polymers, Sci. Rep. 6 (2016) 31110.
[130] Z. Zhao, J. Wu, X. Mu, H. Chen, H.J. Qi, D. Fang, Desolvation induced origami of photocurable polymers by digit light processing *macromol*, Rapid. Comm. 38 (2017) 1600625.
[131] Z. Zhao, X. Kuang, J. Wu, Q. Zhang, G.H. Paulino, H.J. Qi, D. Fang, 3D printing of complex origami assemblages for reconfigurable structures, Soft Matter 14 (2018) 8051.

[132] G.I. Peterson, J.J. Schwartz, D. Zhang, B.M. Weiss, M.A. Ganter, D.W. Storti, A.J. Boydston, Production of materials with spatially-controlled cross-link density via vat photopolymerization, ACS Appl. Mater. Interface. 8 (2016) 29037.
[133] W. Jiangtao, Z. Zeang, K. Xiao, M.H. Craig, F. Daining, H.J. Qi, Reversible shape change structures by grayscale pattern 4D printing, Multifunct. Mater. 1 (2018) 015002.
[134] X. Kuang, J. Wu, K. Chen, Z. Zhao, Z. Ding, F. Hu, D. Fang, H.J. Qi, Grayscale digital light processing 3D printing for highly functionally graded materials, Sci. Adv. 5 (2019) eaav5790.
[135] J.J. Schwartz, A.J. Boydston, Multimaterial actinic spatial control 3D and 4D printing, Nat. Commun. 10 (2019) 791.
[136] A. Tudor, C. Delaney, H. Zhang, A.J. Thompson, V.F. Curto, G.-Z. Yang, M.J. Higgins, D. Diamond, L. Florea, Fabrication of soft, stimulus-responsive structures with sub-micron resolution via two-photon polymerization of poly(ionic liquid) s, Mater. Today Off. 21 (2018) 807.
[137] B.H. Cumpston, S.P. Ananthavel, S. Barlow, D.L. Dyer, J.E. Ehrlich, L.L. Erskine, A.A. Heikal, S.M. Kuebler, I.-Y.S. Lee, D. McCord-Maughon, Two-photon polymerization initiators for three-dimensional optical data storage and microfabrication, Nature 398 (1999) 51.
[138] T. Gissibl, S. Thiele, A. Herkommer, H. Giessen, Two-photon direct laser writing of ultracompact multi-lens objectives, Nat. Photon. 10 (2016) 554.
[139] S. Juodkazis, Manufacturing: 3D printed micro-optics, Nat. Photon. 10 (2016) 499.
[140] S.K. Saha, D. Wang, V.H. Nguyen, Y. Chang, J.S. Oakdale, S.-C. Chen, Scalable submicrometer additive manufacturing, Science 366 (2019) 105.
[141] T. van Manen, S. Janbaz, A.A. Zadpoor, Programming 2D/3D shape-shifting with hobbyist 3D printers, Mater. Horiz. 4 (2017) 1064.
[142] Q. Zhang, D. Yan, K. Zhang, G. Hu, Pattern transformation of heat-shrinkable polymer by three-dimensional (3D) printing technique, Sci. Rep. 5 (2015) 8936.
[143] D. Hua, X. Zhang, Z. Ji, C. Yan, B. Yu, Y. Li, X. Wang, F. Zhou, 3D printing of shape changing composites for constructing flexible paper-based photothermal bilayer actuators, J. Mater. Chem. C 6 (2018) 2123.
[144] S.E. Bakarich, R. Gorkin, M.i. h. Panhuis, G.M. Spinks, 4D printing with mechanically robust, thermally actuating hydrogels *macromol*, Rapid. Comm. 36 (2015) 1211.
[145] A. Kotikian, C. McMahan, E.C. Davidson, J.M. Muhammad, R.D. Weeks, C. Daraio, J.A. Lewis, Untethered soft robotic matter with passive control of shape morphing and propulsion, Sci. Robot. 4 (2019) eaax7044.
[146] D.J. Roach, C. Yuan, X. Kuang, V.C.-F. Li, P. Blake, M.L. Romero, I. Hammel, K. Yu, H.J. Qi, Long liquid crystal elastomer fibers with large reversible actuation strains for smart textiles and artificial muscles, ACS Appl. Mater. Interface. 11 (2019) 19514.
[147] R.H. Volpe, D. Mistry, V.V. Patel, R.R. Patel, C.M. Yakacki, Dynamically crystalizing liquid-crystal elastomers for an expandable endplate-conforming interbody fusion cage advanced healthcare materials 9 (2020) 1901136.
[148] Q. He, Z. Wang, Z. Song, S. Cai, Bioinspired design of vascular artificial muscle, Adv. Mater. Technol. 4 (2019) 1800244.
[149] M. López-Valdeolivas, D. Liu, D.J. Broer, C. Sánchez-Somolinos, 4D printed actuators with soft-robotic functions *macromol*, Rapid. Comm. (2018), https://doi.org/10.1002/marc.2017007101700710.
[150] C. Yuan, D.J. Roach, C.K. Dunn, Q. Mu, X. Kuang, C.M. Yakacki, T.J. Wang, K. Yu, H.J. Qi, 3D printed reversible shape changing soft actuators assisted by liquid crystal elastomers, Soft Matter 13 (2017) 5558.

[151] L. Yu, H. Shahsavan, G. Rivers, C. Zhang, P. Si, B. Zhao, Programmable 3D shape changes in liquid crystal polymer networks of uniaxial orientation, Adv. Funct. Mater. 28 (2018) 1802809.
[152] S. Roh, L.B. Okello, N. Golbasi, J.P. Hankwitz, J.A.-C. Liu, J.B. Tracy, O.D. Velev, 3D-Printed silicone soft architectures with programmed magneto-capillary reconfiguration, Adv. Mater. Technol. 4 (2019) 1800528.
[153] J. Cui, T.-Y. Huang, Z. Luo, P. Testa, H. Gu, X.-Z. Chen, B.J. Nelson, L.J. Heyderman, Nanomagnetic encoding of shape-morphing micromachines, Nature 575 (2019) 164.
[154] S. Tottori, L. Zhang, F. Qiu, K.K. Krawczyk, A. Franco-Obregón, B.J. Nelson, Magnetic helical micromachines: fabrication, controlled swimming, and cargo transport, Adv. Mater. 24 (2012) 811.
[155] Q. Ze, X. Kuang, S. Wu, J. Wong, S.M. Montgomery, R. Zhang, J.M. Kovitz, F. Yang, H.J. Qi, R. Zhao, Magnetic shape memory polymers with integrated multifunctional shape manipulation, Adv. Mater. 32 (2020) 1906657.
[156] H.-W. Huang, M.S. Sakar, A.J. Petruska, S. Pané, B.J. Nelson, Soft micromachines with programmable motility and morphology, Nat. Commun. 7 (2016) 12263.
[157] J.A.-C. Liu, J.H. Gillen, S.R. Mishra, B.A. Evans, J.B. Tracy, Photothermally and magnetically controlled reconfiguration of polymer composites for soft robotics, Sci. Adv. 5 (2019) eaaw2897.
[158] Y. Kim, G.A. Parada, S. Liu, X. Zhao, Ferromagnetic soft continuum robots, Sci. Robot. 4 (2019) eaax7329.
[159] T. Xu, J. Zhang, M. Salehizadeh, O. Onaizah, E. Diller, Millimeter-scale flexible robots with programmable three-dimensional magnetization and motions, Sci. Robot. 4 (2019) eaav4494.
[160] C. Yang, M. Boorugu, A. Dopp, J. Ren, R. Martin, D. Han, W. Choi, H. Lee, 4D printing reconfigurable, deployable and mechanically tunable metamaterials, Mater. Horiz 6 (2019) 1244.
[161] B. Zhang, W. Zhang, Z. Zhang, Y. Zhang, H. Hingorani, Z. Liu, J. Liu, Q. Ge, Self-healing four-dimensional printing with ultraviolet curable double network shape memory polymer system, ACS Appl. Mater. Interface. 11 (2019) 10328.
[162] L. Huang, R. Jiang, J. Wu, J. Song, H. Bai, B. Li, Q. Zhao, T. Xie, Ultrafast digital printing toward 4D shape changing materials, Adv. Mater. 29 (2017) 1605390.
[163] Y. Klein, E. Efrati, E. Sharon, Shaping of elastic sheets by prescription of non-Euclidean metrics, Science 315 (2007) 1116.
[164] M. Jamal, A.M. Zarafshar, D.H. Gracias, Differentially photo-crosslinked polymers enable self-assembling microfluidics, Nat. Commun. 2 (2011) 527.
[165] Z. Zhao, J. Wu, X. Mu, H. Chen, H.J. Qi, D. Fang, Origami by frontal photopolymerization, Sci. Adv. 3 (2017) e1602326.
[166] D. Jin, Q. Chen, T.-Y. Huang, J. Huang, L. Zhang, H. Duan, Four-dimensional direct laser writing of reconfigurable compound micromachines, Mater. Today Off. 32 (2019) 19.
[167] T.Y. Huang, M.S. Sakar, A. Mao, A.J. Petruska, F. Qiu, X.B. Chen, S. Kennedy, D. Mooney, B.J. Nelson, 3D printed microtransporters: compound micromachines for spatiotemporally controlled delivery of therapeutic agents, Adv. Mater. 27 (2015) 6644.
[168] G. Li, S. Wang, Z. Liu, Z. Liu, H. Xia, C. Zhang, X. Lu, J. Jiang, Y. Zhao, 2D-to-3D shape transformation of room-temperature-programmable shape-memory polymers through selective suppression of strain relaxation, ACS Appl. Mater. Interface. 10 (2018) 40189.

[169] J.W. Cho, J.W. Kim, Y.C. Jung, N.S. Goo, Electroactive shape-memory polyurethane composites incorporating carbon nanotubes *macromol*, Rapid. Comm. 26 (2005) 412.
[170] H. Wei, Q. Zhang, Y. Yao, L. Liu, Y. Liu, J. Leng, Direct-write fabrication of 4D active shape-changing structures based on a shape memory polymer and its nanocomposite, ACS Appl. Mater. Interface. 9 (2017) 876.
[171] M. Razzaq, M. Behl, A. Lendlein, Magnetic memory effect of nanocomposites, Adv. Funct. Mater. 22 (2012) 184.
[172] D.L. Taylor, Self-healing hydrogels, M. in het Panhuis, Adv. Mater. 28 (2016) 9060.
[173] Y. Yang, Z. Pei, Z. Li, Y. Wei, Y. Ji, Making and remaking dynamic 3D structures by shining light on flat liquid crystalline vitrimer films without a mold, J. Am. Chem. Soc. 138 (2016) 2118.
[174] A. Velasco-Hogan, J. Xu, M.A. Meyers, Additive manufacturing as a method to design and optimize bioinspired structures, Adv. Mater. (2018) 0 1800940.
[175] Y. Yang, Z. Chen, X. Song, Z. Zhang, J. Zhang, K.K. Shung, Q. Zhou, Y. Chen, Biomimetic anisotropic reinforcement architectures by electrically assisted nanocomposite 3D printing, Adv. Mater. 29 (2017) 1605750.
[176] Y. Yang, X. Li, M. Chu, H. Sun, J. Jin, K. Yu, Q. Wang, Q. Zhou, Y. Chen, Electrically assisted 3D printing of nacre-inspired structures with self-sensing capability, Sci. Adv. 5 (2019) eaau9490.
[177] R.M. Erb, J.S. Sander, R. Grisch, A.R. Studart, Self-shaping composites with programmable bioinspired microstructures, Nat. Commun. 4 (2013) 1712.
[178] J.J. Martin, B.E. Fiore, R.M. Erb, Designing bioinspired composite reinforcement architectures via 3D magnetic printing, Nat. Commun. 6 (2015) 8641.
[179] X. Chen, W. Liu, B. Dong, J. Lee, H.O.T. Ware, H.F. Zhang, C. Sun, High-speed 3D printing of millimeter-size customized aspheric imaging lenses with sub 7 nm surface roughness, Adv. Mater. 30 (2018) 1705683.
[180] D. Kokkinis, F. Bouville, A.R. Studart, 3D printing of materials with tunable failure via bioinspired mechanical gradients, Adv. Mater. 30 (2018) 1705808.
[181] Y. Yang, S. Xuan, L. Xiangjia, C. Zeyu, Z. Chi, Z. Qifa, C. Yong, Recent progress in biomimetic additive manufacturing technology: from materials to functional structures, Adv. Mater. 0 (2018) 1706539.
[182] A.R. Studart, Additive manufacturing of biologically-inspired materials, Chem. Soc. Rev. 45 (2016) 359.
[183] G. Sossou, F. Demoly, H. Belkebir, H.J. Qi, S. Gomes, G. Montavon, Design for 4D printing: a voxel-based modeling and simulation of smart materials, Mater. Des. 175 (2019) 107798.
[184] G.Z. Lum, Z. Ye, X. Dong, H. Marvi, O. Erin, W. Hu, M. Sitti, Shape-programmable magnetic soft matter, Proc. Natl. Acad. Sci. U. S. A 113 (2016) E6007.
[185] M.J. Geiss, N. Boddeti, O. Weeger, K. Maute, M.L. Dunn, Combined level-set-XFEM-density topology optimization of four-dimensional printed structures undergoing large deformation, J. Mech. Des. 141 (2019) 051405.

CHAPTER 10

# Laser polishing of additive-manufactured Ti alloys and Ni alloys

**Yingchun Guan[1,2,3], Yuhang Li[1,2,3], Huaming Wang[1,2,3]**
[1]School of Mechanical Engineering and Automation, Beihang University, Beijing, China; [2]National Engineering Laboratory of Additive Manufacturing for Large Metallic Components, Beihang University, Beijing, China; [3]International Research Institute for Multidisciplinary Science, Beihang University, Beijing, China

## 10.1 Introduction

Additive manufacturing (AM) or 3D printing is based on an incremental layer-by-layer manufacturing technology for producing high-complexity components with intricate internal and external structures [1–4]. SLM and LMD techniques have been typically explored to produce near-net-shaped components for various applications in aerospace [5], marine and offshore [6], nuclear and fossil fuel power plants [7], and biomedical industries [8]. However, poor surface integrity and geometric tolerances usually occur at the surface of AM components due to the staircase effect [9], balling effect [10], and molten pool splashing during the AM process [11], which requires additional post-treatment to meet surface smoothness and in-service tolerance requirements [12].

Compared with traditional surface post-treatment technologies, such as mechanical polishing [13–15], chemical mechanical polishing [16], and electrochemical polishing [9,17], laser polishing is effective to reduce the surface roughness of complex and freeform surface components with high efficient, precise, selective, mechanical damage free and environmentally friendly, which has become one of the most promising processes in surface smoothening of AM components [18–20]. During laser polishing processing, smooth metallic surfaces can be obtained by melting and evaporation rather than removal [21–23]. When the laser energy converts into heat energy through the interaction between laser and the metal lattice, the surface material will be melted rapidly and the melted surface will produce an extremely shallow molten pool, in which the liquid phase is driven by

*Additive Manufacturing*
ISBN 978-0-12-818411-0
https://doi.org/10.1016/B978-0-12-818411-0.00002-1

surface tension and gravity rapidly flowing into the valley from the convex peaks [19,24,25].

Bourell et al. [26] reported that laser polishing could produce a smooth finish on selective laser sintered (SLS) iron—copper parts by fast melting and resolidification with the percent reduction in surface roughness Ra up to 31%. Lamikiz et al. [27] proposed that laser energy density was the most important parameter during polishing process, and the surface roughness ($R_a$) of SLS parts (60% sintered AISI 420 stainless steel and 40% infiltrated bronze) could be reduced by 80% at optimized conditions. Rosa and Mognol et al. [28,29] concluded that laser polishing could reduce the surface roughness $S_a$ from 21 to 0.79 μm for 316 L parts manufactured via direct metal deposition. Bhaduri et al. [10] used a fiber laser with a maximum average power of 50 W and wavelength (λ) of 1064 nm to polish 3D-printed stainless steel (SS316L) components, in which the surface roughness reduced by 94%. Obeidi et al. [30] using a $CO_2$ laser working under a continuous-wave (CW) mode polished AM 316L stainless steel cylindrical components, and the surface roughness was decreased from 10.4 to 2.7 μm. In particular, the laser polishing parameters were firstly optimized to polish additively manufactured CoCr components under atmospheric and shielding gas environments by Wang et al. [31—33], and the surface roughness was reduced by 93% compared with the as-received components. After that, Richter et al. used a CW laser to polish powder-bed fusion Co-Cr—Mo alloy, which demonstrated the suitability of theoretical frequency and capillary smoothing prediction model [34]. Moreover, Černašėjus et al. [35] assessed the properties of SLS steel 1.2709 surfaces after laser polishing, in which the hardness was increased up to ~88% and the wear resistance was improved up to about four times compared with the as-received components.

Note that Ti alloys and nickel-based superalloys have been widely used to various components by AM technology, such as aircraft gas turbines, nuclear and fossil fuel fastenings, turbine discs of aircraft engines and biomedical implant, in which the average surface roughness is usually more than 10 μm [36—42]. In this chapter, we report our recent study and development in laser polishing of LMD TC11, SLM TC4, and SLM Inconel 718 superalloy in corresponding sections. The surface roughness and microstructure evolution are carefully investigated in the laser-irradiated zone. The thermodynamics and rapid solidification during the laser polishing process are analyzed by finite element simulation. Finally, the mechanical properties of laser-polished layer are examined by comparing those of the as-received layer, including microhardness, wear resistance, tensile property, and corrosion resistance.

## 10.2 Laser polishing LMD TC11

### 10.2.1 Surface morphology

A high-finish titanium alloy surface can be obtained by laser polishing on the macrolevel [43–45]. In this study, a nanosecond-pulsed fiber laser was used to polish the as-received surface of LMD TC11 specimens (the chemical compositions are given in Table 10.1) with the dimension of 50 mm (length) × 20 mm (width) × 10 mm (thickness) [19]. The optimized laser polishing parameters such as power density, scanning speed, and overlapping ratio were set as $1.2 \times 10^7$ W/cm$^2$, 200 mm/s, 50% respectively. Macroscale photographs of LMD TC11 before and after laser polishing were shown in Fig. 10.1A. It is noticed that machined ripples on the as-received surface were significantly reduced via laser polishing, meanwhile a reflective surface was acquired. It involves rescanning and remelting solidified layers, in which an extremely shallow molten pool is produced in a very short time interval. After that, the molten phase is driven to move away from the convex peaks and flow into the valleys by gravity and surface tension. Eventually, the liquid material will be solidified with high cooling rate, which leads to a significant reduction of the surface roughness [44,46,47]. From the SEM morphology in Fig. 10.1B, liquid spheroidizing and powder bonding were existing on the as-received surface leading to rough surface and undesirable surface quality. Meanwhile, slight laser scanning track was observed after laser polishing as shown in Fig. 10.1C. Surface profiles before and after laser polishing were measured as shown in Fig. 10.1C. The peak-to-valley height of TC11 surface decreased from 80 to 4.5 μm as shown in Fig. 10.1D and E. Eventually, the average roughness is decreased from over Ra 7.21 μm to less than 0.73 μm.

**Table 10.1** The chemical compositions of TC11.

| Material | Al | V | Mo | Si | Zr | Fe | Ti |
|---|---|---|---|---|---|---|---|
| TC11 | 5.8 −7.0 | – | 2.80 −3.80 | 0.20 −0.35 | 0.8 −2.0 | < 0.25 | Balance |

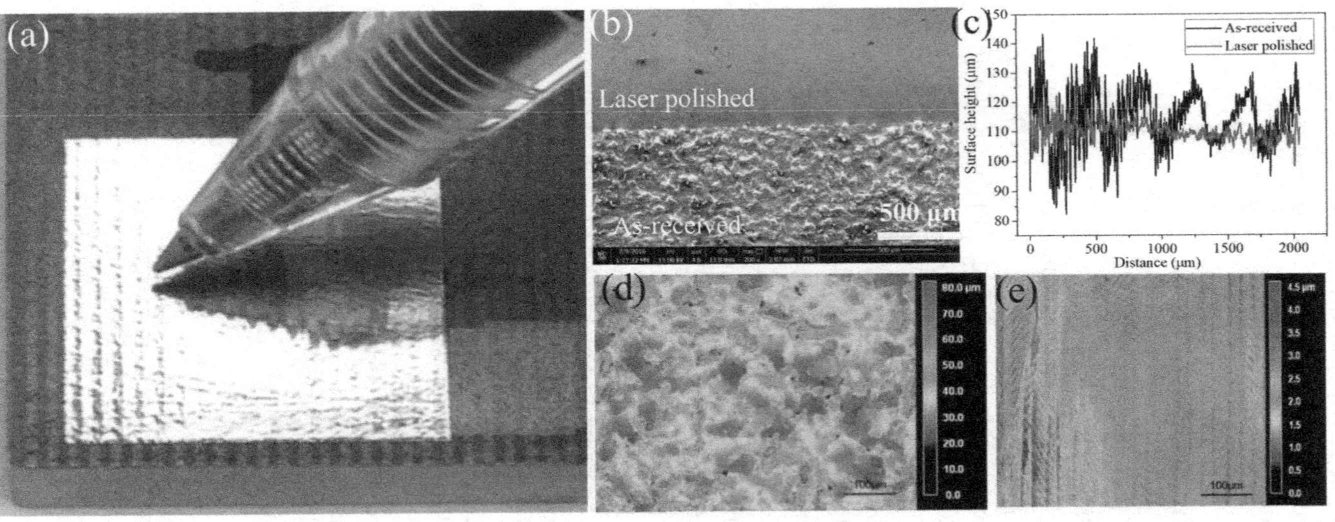


**Figure 10.1** Effects of laser polishing on LMD TC11: (A) the polished region and the as-received surface; (B) SEM morphology of the as-received and laser-polished surface; (C) the profiles of the as-received and optimized laser-polished regions; (D) topographic image from laser scanning confocal microscope (LSCM) of the as-received surface; and (E) topographic image from LSCM of laser-polished surface. *(Reproduced with kind permission of Elsevier from Reference C.P. Ma, Y.C. Guan, W. Zhou, Laser polishing of additive manufactured Ti alloys, Optic Laser. Eng., 93 (2017) 171—177.)*

## 10.2.2 Numerical simulation and microstructure

To calculate the temperature effect in the polishing process, the thermal field and thermal cycle were solved in TC11 plate model with the size of 4 × 2 × 1.5 mm by using the finite element solver ABAQUS. The upper surface was defined as laser polishing zone, which consists of minor gridding size (12 × 12 × 12 μm). According to practical needs, the model was divided into two parts, which were adopted with different mesh density to speed up the computation time. Besides, the physical parameters of the materials were temperature-dependent, and the processing parameters were identical to the experiments. Heat transport through the bottom walls was neglected (adiabatic walls), convective was dominant in energy transmission during the laser polishing process. The calculation used linear elements for solving the heat equation. The energy distribution was defined as Gaussian heat source [48]:

$$q = \frac{2P}{\pi r^2} \times eff \times e^{-2\times\frac{[(x1-x0)^2+(y1-y0)^2]}{r^2}} \tag{10.1}$$

where q is the heat flux density; P (85 W) is the laser power; $r$ (25 μm) is the spot radius; *eff* (46.5%) is the laser absorption rate; $x_0$ and $y_0$ are the instantaneous positions of the spot center; and $x_1$, $y_1$ are the variables of the position.

Fig. 10.2A shows the temperature distribution of the surface with laser radiation. It is known that liquid temperature and β-transus temperature of TC11 were approximately 1660°C and 1000°C, respectively. The results of numeric simulation reveal laser polishing pool was around 80°μm in diameter, 25°μm in depth. The dimensions of phase transition area were

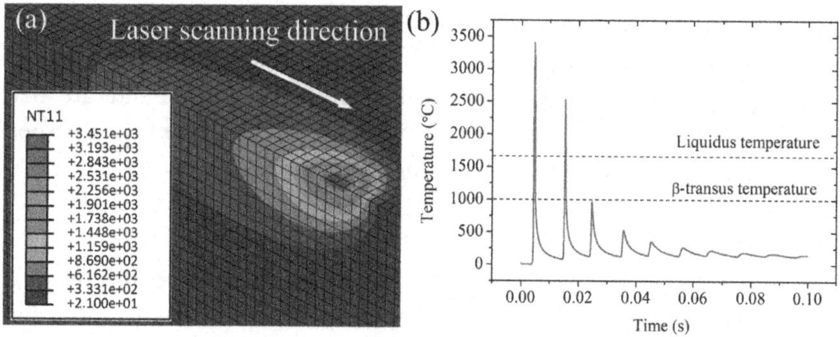


**Figure 10.2** Numerical simulation and thermal cycle: (A) temperature field simulation and (B) change curve of surface temperature at a certain point.

100 μm in diameter, 36 μm in depth. Compared with the experimental results, the thermal field of the material surface can be simulated by FEM. The thermal cycle of midpoint in laser polishing trace was shown in Fig. 10.2B, this position experienced phase transition three times during the laser polishing process. Besides, the multiple scan tracks can result in different peaks appear in it. While both of maximum temperature can be cooled to 250°C within 0.002 sec since the laser radiation area is small and heat dissipates rapidly to the surrounding. In the first cycle, material was heated above the melting temperature, then molten metal generated β columnar crystal during solidification. Because the cooling rate ($3.7 \times 10^5$°C/s) was higher than the critical cooling rate of martensite transformation (410°C/s), the primary phase (β phase) cannot be converted into the secondary phase (α phase) by alloy element diffusion. Therefore, bcc β phase transforms completely into metastable hpc martensite α′ phase by diffusionless and shear-type transformation process. Martensite grows inside the original columnar crystal, several parallel primary acicular martensite formed firstly, then a series of relatively small secondary acicular martensite formed, which stopped by grain boundary or primary martensite, resulting in the formation of the typical staggered distribution of acicular tissue in the polished layer.

To investigate the influence of laser polishing on the subsurface materials, cross section was examined by an optical microscope (OM) and a scanning electron microscope (SEM). Fig. 10.3A shows that the thickness of the polished layer was 90 μm; moreover, refined grains existed in the polished layer owing to rapid melting and solidification process. After heat treatment, the deposited basket-weave microstructure of LMD TC11 changed from ultrafine needle-shaped α phase (the average width was 0.4 μm) to the thick lamellae α phase (the average width was 2~3 μm) because of grain growth coarsening. Fig. 10.3B and C show the typical microstructure of coarsened α + β phase in matrix and acicular martensite α' in the polished zone. XRD profiles in Fig. 10.3D further indicate that as-received TC11 consists of α phase and β phase, while the laser-polished surface mainly consists of α' martensitic phase without β phase.

### 10.2.3 Mechanical properties

The surface properties are caused by significant changes on the material surface due to the complex melting, vaporization, solidification, and other non-equilibrium phase transformation as well as thermal stress distribution

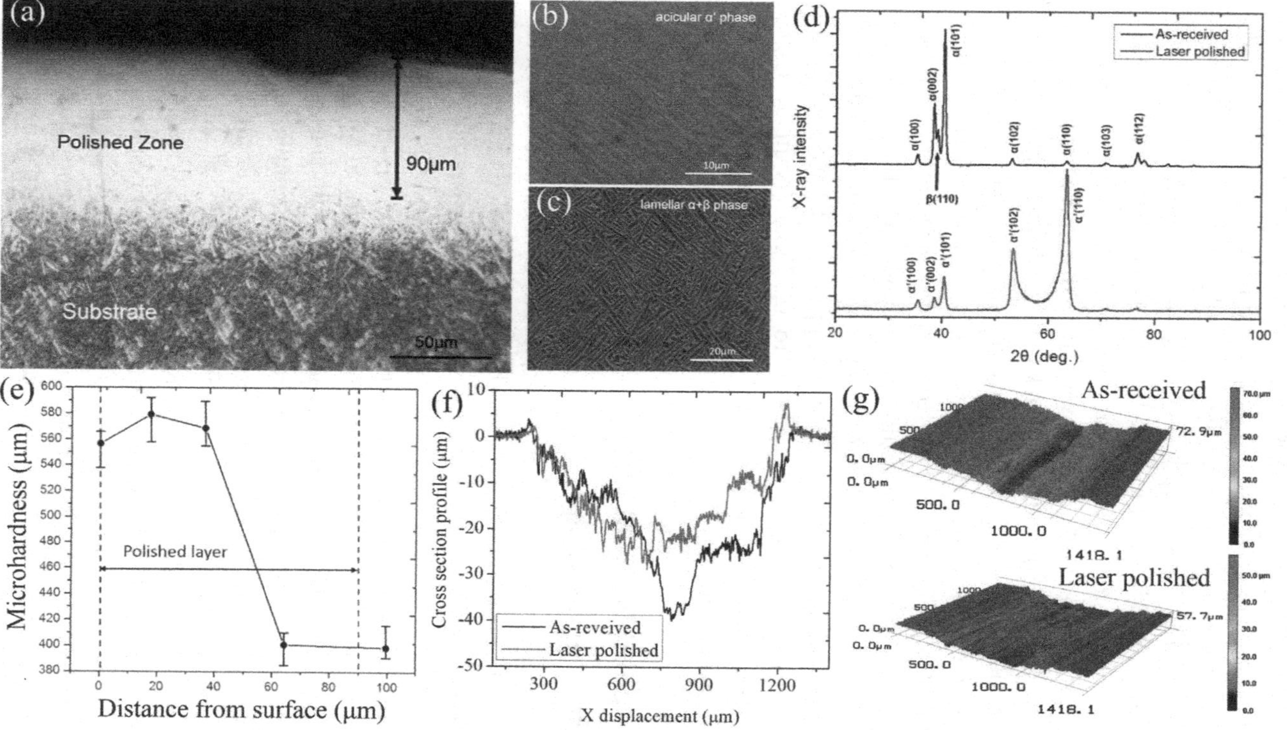


**Figure 10.3** Microstructure and properties after laser polishing LMD TC11: (A) microstructure in cross section; (B) microstructure of the polished layer; (C) microstructure of the as-received layer; (D) XRD patterns for the as-received and laser-polished surface; (E) microhardness distributions; (F) cross-section profile of wear tracks on the as-received and laser-polished surface; (G) laser scanning confocal microscope of wear tracks. *(Reproduced with kind permission of Elsevier from Reference C.P. Ma, Y.C. Guan, W. Zhou, Laser polishing of additive manufactured Ti alloys, Optic Laser. Eng., 93 (2017) 171–177.)*

on the surface of materials during the polishing process [18,19]. To examine the improvement of mechanical properties, microhardness test, and dry sliding friction test were conducted. The average microhardness of initial substrate TC11 is about 400HV. In the laser-polished layer, the microhardness was 560 HV increasing by 42% comparing with that of the as-received layer, which can be seen in Fig. 10.3E. Microhardness is enhanced because of microstructure transformation, where hcp structure ($\alpha'$ martensitic) has higher bulk modulus value than that of the bcc structure ($\beta$ phase). Cross-section profile of wear track of TC11 before and after laser polishing was shown in Fig. 10.3F. Because of the presence of $\alpha'$ martensitic, greater wear resistance was attained in the laser-polished zone. From the LSCM morphology in Fig. 10.3G, the conclusion TC11 have poor wear resistance at room temperature can be drawn, which is consistent with the previous studies. From LSCM morphology, we can see adhesive traces and furrows along the sliding direction appear on the worn surface. Wear rate decreased from 0.41 $mm^3/N\cdot m$ to 0.27 $mm^3/N\cdot m$ by laser polishing, which means laser polishing will be a promising surface strengthening technology.

## 10.3 Laser polishing SLM TC4

### 10.3.1 Surface morphology

In this work, an SLM TC4 alloy was used to polish by a nanosecond fiber laser. It was annealed at 1003 K for 2 h and then slow cooled in a vacuum furnace to 803 K. Macroscale photograph and surface topography of the laser-polished SLM TC4 sample was shown in Fig. 10.4. The complex as-received surface contained a high surface roughness due to the staircase effect, bonding powder, and molten pool splashing in SLM process (Fig. 10.4B) [49]. After laser polishing, smoother surface with laser melting tracks was observed and the defects were removed, which implied a clear improvement on surface roughness (Fig. 10.4C). Surface profiles of as-received and laser-polished surfaces were measured, as shown in Fig. 10.4D. After laser polishing, Ra was reduced from 6.53 to 0.32 μm, Rz was reduced from 140.29 to 6.42 μm. The basic principle of laser polishing is that the surface layer melts or vaporizes due to the absorption of high energy on the material surface. It demonstrated that the morphological apexes and valleys of the as-received surface were remelted during laser irradiation process, and the molten material was redistributed by surface tension and gravity. Then in following cooling process, the molten materials solidified and formed the smoother surface morphology.

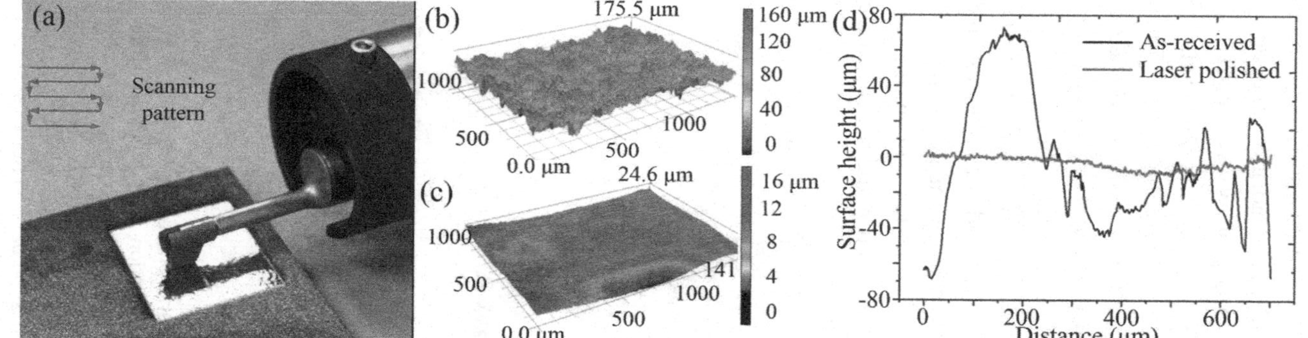


**Figure 10.4** Effects of laser polishing on the SLM TC4: (A) the polished region and the as-received surface; (B) topographic image from laser scanning confocal microscope (LSCM) of the as-received surface; (C) topographic image from LSCM of the laser-polished surface; and (D) the profiles of the as-received and optimized laser-polished regions. *(Reproduced with kind permission of MDPI from Reference Y.H. Li, B. Wang, C.P. Ma, Z.H. Fang, L.F. Chen, Y.C. Guan, S.F. Yang, Material characterization, thermal analysis, and mechanical performance of a laser-polished Ti alloy prepared by selective laser melting, Metals, 9 (2019) 112.)*

### 10.3.2 Numerical simulation and microstructure

Combining the characteristics of the initial surface absorption and the non-equilibrium phase transition under laser radiation, the temperature field and thermal cycle during laser polishing process was calculated by finite element simulation method. ABAQUS was used to simulate the temperature field within the TC4 surface during laser polishing process [50]. The relationship between melting pool depth, cooling rate, and melting duration during laser polishing was predicted. Calculation was done on a rectangle substrate with dimension of 4 × 2 × 0.5 mm, and smallest element size was 30 μm. The specific heat and thermal conductivity of TC4 were temperature-dependent. The heat input of CW laser spot was modeled as a Gaussian distribution with same intensity and scan strategy as experiment (Section 2.2).

As shown in Fig. 10.5A, it is clear that the depth of area heated above β-transus temperature was about 160 μm and the melting depth was about

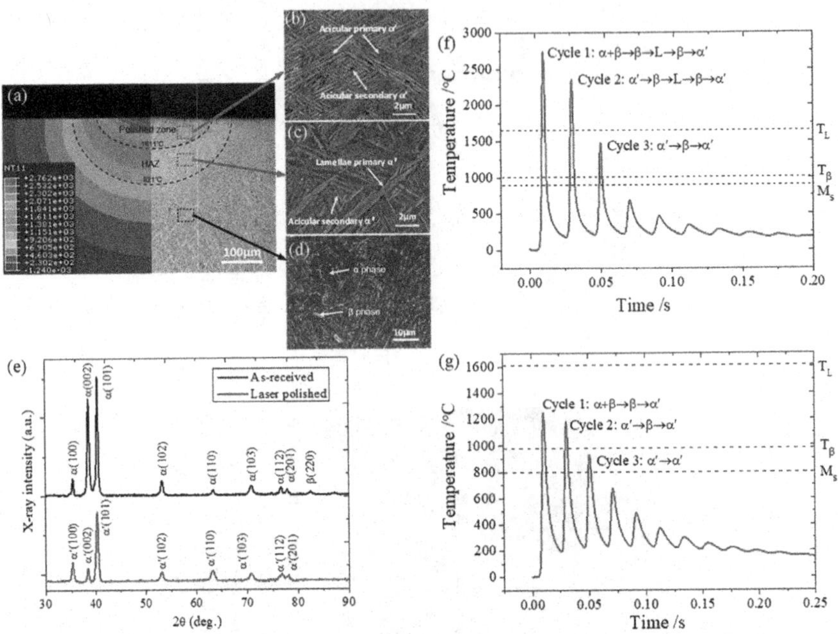


**Figure 10.5** Numerical simulation and phase diagram of SLM TC4 alloy by laser polishing: (A) temperature field simulation and microstructure in cross section; (B) microstructure on the polished zone; (C) microstructure on the heat-affected zone (HAZ); (D) microstructure on the as-received layer; (E) XRD patterns for the as-received and laser-polished surface; (F) thermal cycle in the polished layer; and (G) thermal cycle in HAZ. *(Reproduced with kind permission of MDPI from Reference Y.H. Li, B. Wang, C.P. Ma, Z.H. Fang, L.F. Chen, Y.C. Guan, S.F. Yang, Material characterization, thermal analysis, and mechanical performance of a laser-polished Ti alloy prepared by selective laser melting, Metals, 9 (2019) 112.)*

60 μm, which meant a good agreement with the microstructure analysis obtained from experiment. Also, the typical microstructure evolution of laser-polished sample observed under SEM and OM is shown in Fig. 10.5A–D. In the cross section of the near surface, a melted layer with thickness of 60 μm was found, which fully consisted of fine acicular martensite α'. A heat-affected zone (HAZ) with thickness of 90 μm was found under melted layer; it is mainly composed of a coarser martensite α' than the melted layer, which may be caused by reheating and cooling by lower peak temperature than melted layer [25,51]. Needle-shaped α phase and β phase were observed in the substrate produced by SLM processing [52,53]. The microstructure evolution was also agreed with XRD profile in Fig. 10.5E, which indicated that as-received microstructure contained α phase and tiny amounts of β phase, while the polished surface mostly consists of α' martensitic phase. Generally, martensitic transformation of TC4 is related to high cooling rate [54]. During the laser polishing process, rapid melting occurred on the surface due to the high laser energy density. When laser spot left, high temperature gradient induced by self-quenching caused high cooling rate. In the cooling process from liquidus temperature to room temperature, TC4 alloy would undergo nonequilibrium phase transition $\beta \rightarrow \alpha'$ due to the high cooling rates and multiple thermal cycles.

The temperature cycle of one fixed point on the polished layer and HAZ were shown in Fig. 10.5F and G, respectively. During the laser scanning process, material at that point was heated above liquidus temperature in very short time and melted several times. When laser left, it was also rapidly cooled to below 250°C in less than 0.01 s, the cooling rate was about $3 \times 10^{6o}$C/s in the first heating and cooling process. Because of the much higher cooling rate than the critical cooling rate of 410 K/s, martensitic $\alpha'$ would formed in the polished zone after laser irradiation.

We further used transmission electron microscope (TEM) to observe the martensite α', which was shown in Fig. 10.6. It can be seen that a large number of dislocations pile-up are contained in the martensite α'. The stacking of dislocations causes lattice distortion, which leads to the decrease in toughness, and increment of brittleness and strength.

### 10.3.3 Mechanical properties

The evolution of the structural properties of the microstructure before and after polishing was studied, in which the relationship among initial roughness, laser parameters, microstructure, and mechanical properties was

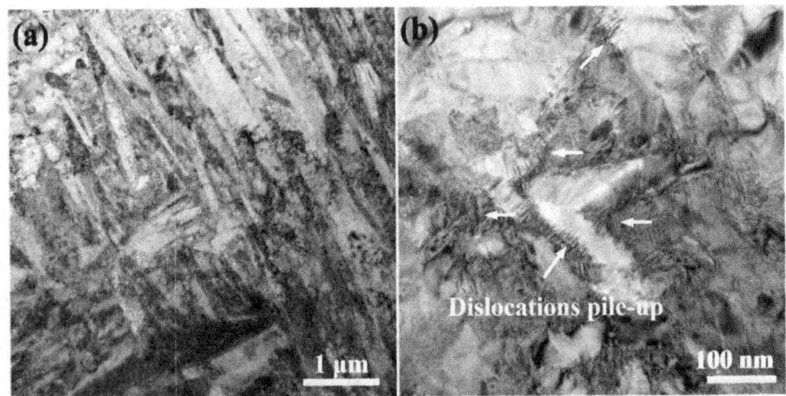


**Figure 10.6** Typical TEM observations of the polished layer: (A) $\alpha'$ martensite and (B) a large number of dislocations pile-up in $\alpha'$ martensite.

established by experiments. The microhardness distributions on cross section of laser-polished TC4 is shown in Fig. 10.7A. It is clear the average hardness of substrate is about 340 HV. In the range of polished layer, microhardness increased to about 426 HV in average, which was 25% higher than that of as-received material. The improvement of microhardness could be attributed to the formation of α' martensitic phase in laser-polished layer. Because α' martensitic of TC4 has a hexagonal closed-packed (hcp) structure and higher bulk modulus value than α and β phase [55].

With the improvement of microhardness, wear resistance of polished surface was also enhanced. Fig. 10.7B shows cross-section profiles of wear scars of TC4 before and after laser polishing. Wear scar of the laser-polished sample was shallower than that of as-received sample Fig. 10.7C and D. Wear rate could be calculated for quantitative comparison of wear resistance of the samples, and it is defined as:

$$W = \frac{\Delta V}{L \times D} \tag{10.2}$$

where $\Delta V$ ($\mu m^3$) is the volume of the wear trace, $D$ (μm) is the sliding distance and $L$ (N) is the applied load in the experiment [56]. According to the calculation, wear rate of laser-polished surface was 39% less than the as-received surface.

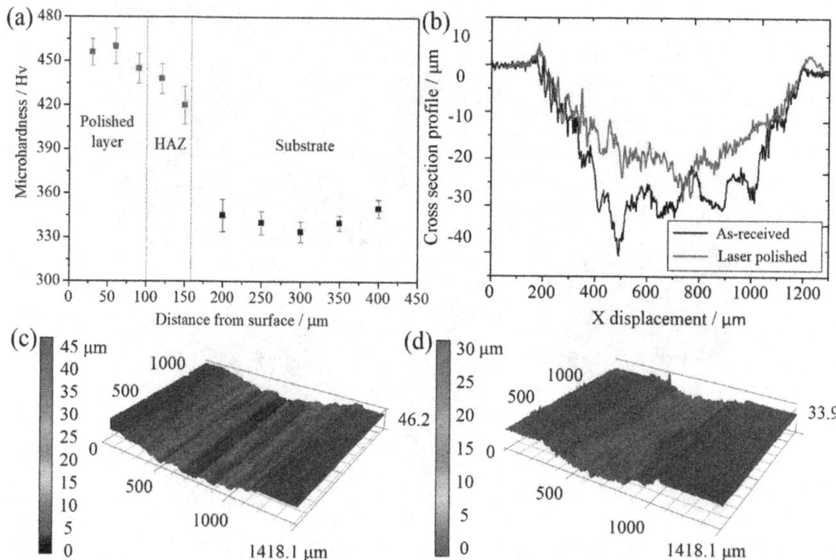


**Figure 10.7** Properties after laser polishing SLM TC4: (A) microhardness distributions; (B) Cross-section profile of wear tracks on the as-received and laser-polished surface; (C) laser scanning confocal microscope (LSCM) of wear tracks on the as-received layer; and (D) LSCM of wear tracks on the polished layer. *(Reproduced with kind permission of MDPI from Reference Y.H. Li, B. Wang, C.P. Ma, Z.H. Fang, L.F. Chen, Y.C. Guan, S.F. Yang, Material characterization, thermal analysis, and mechanical performance of a laser-polished Ti alloy prepared by selective laser melting, Metals, 9 (2019) 112.)*

### 10.3.4 Examples of laser polishing on large-area SLM Ti components

Fig. 10.8 shows typical Ti components after laser polishing in our group. In Fig. 10.8A and B, a thin-wall cylindrical part was polished with average surface roughness Ra reduction from 21 to 0.25 μm. Moreover, a complex biomedical device was polished with the minimum surface roughness Ra up to 0.1 μm, which was decreased by 68.6% comparing with that of the as-received surface. In Fig. 10.8D, the surface roughness of ultrathin Ti mesh structure with average thickness as 1.5 mm could reach Ra 0.08 μm after laser polishing with the efficiency as 150–350 $cm^2/h$.

## 10.4 Laser polishing SLM inconel 718 superalloy

In terms of nickel-based superalloys, non-AM surface has been reported to reduce the roughness by pulsed laser [57], CW Nd:YAG laser [58], and

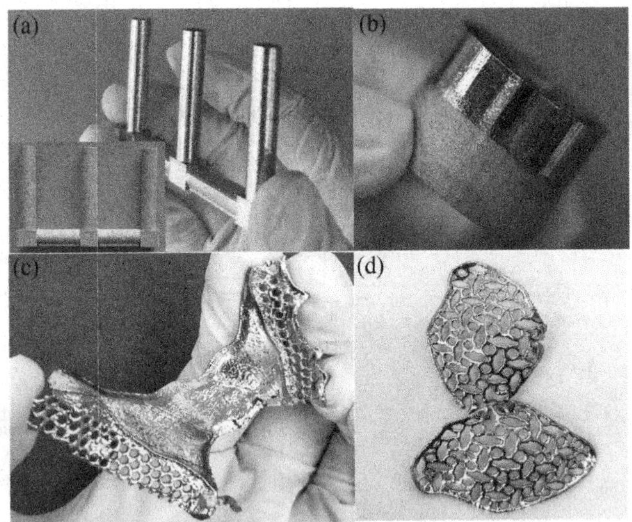


**Figure 10.8** Examples of practical applications using laser polishing techniques on SLM TC4 alloys: (A) thin-wall three-way tube; (B) cylindrical device; (D) removable partial denture framework; and (E) titanium mesh (thickness: 1.5 mm).

picosecond laser [59]. It should be noted that the above initial surfaces are either casted or forged with average roughness Ra less than 5 μm, which is lower than that of the AM surface with Ra more than10 μm [13,14,17]. Therefore, the aim of this part is to explore how laser polishing for AM rough surfaces.

## 10.4.1 Surface morphology

In this work, the rough surface of SLM IN718 superalloy was polished using a nanosecond fiber laser (wavelength = 1064 nm, spot size = 50 μm) [12,60]. The chemical composition of the IN718 samples is summarized in Table 10.2. The laser parameters such as power, pulse width, repetition frequency, scanning speed, and overlapping ratio were set as 90W, 220 ns, 20 kHz, 70 mm/s, and 90%, respectively. A macroscale photograph of the as-received IN718 was shown in Fig. 10.9A. The SEM image of as-received surface was shown in Fig. 10.9B, and the LSCM topographic images as shown in Fig. 10.9C. The average roughness Ra and Rz were 8 and 33 μm, respectively, which were correspondingly decreased to 0.1 and 0.8 μm after laser polishing (Fig. 10.9D and E). During laser irradiation, a certain amount of heat accumulation will form on the surface, which leads to the remelting or evaporation on the material surface. When the continuous

**Table 10.2** Chemical composition of gas atomization Inconel 718 powder (wt.%).

| Elements | Ni | Cr | Nb | Mo | Si | Mn | Cu | Ti | Al | C | P | S | B | Fe |
|---|---|---|---|---|---|---|---|---|---|---|---|---|---|---|
| Compositions | 55.53 | 19.6 | 5.3 | 3.1 | 0.25 | 0.35 | 0.3 | 1.15 | 0.3 | 0.08 | 0.015 | 0.015 | 0.006 | Bal |

Reproduced with kind permission of Elsevier from Reference Y. Li, Z. Zhang, Y. Guan, Thermodynamics analysis and rapid solidification of laser polished Inconel 718 by selective laser melting, Appl. Surf. Sci., 2020, 511, 145423.

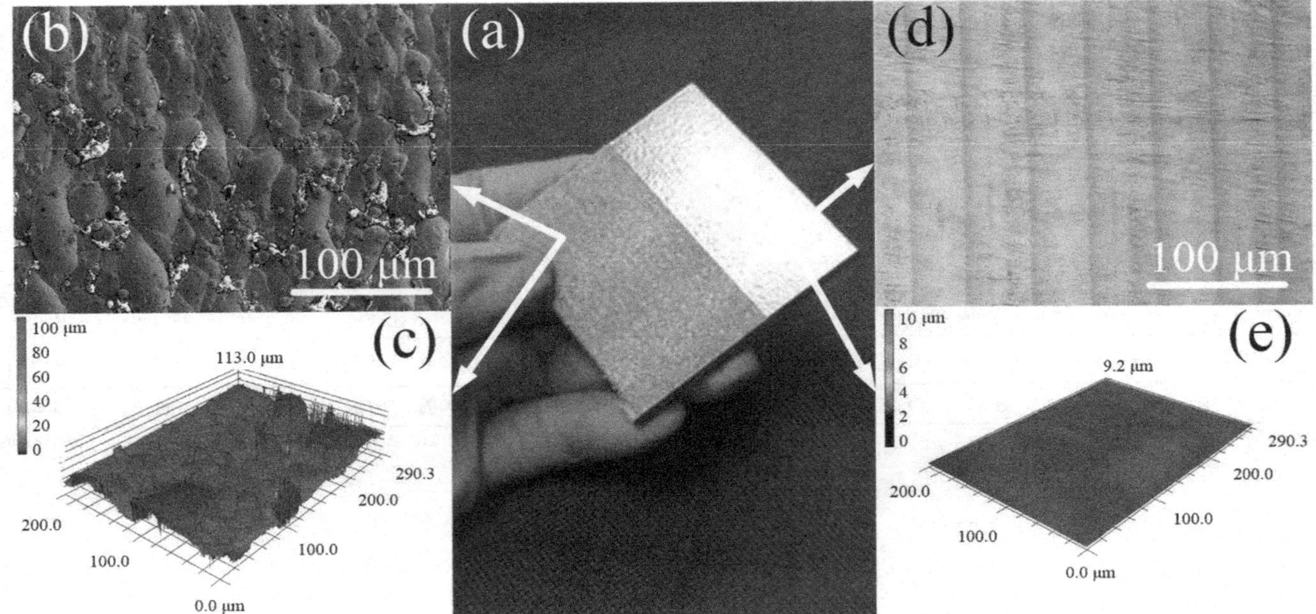


**Figure 10.9** Effects of laser polishing on SLM IN718: (A) the polished region and the as-received surface; (B) as-received surface; (C) topographic image from laser scanning confocal microscope (LSCM) of as-received surface; (D) laser-polished surface; (E) topographic image from LSCM of laser-polished surface. *(Reproduced with kind permission of KeAi from Reference G. Hu, Y. Song, Y. Guan, Tailoring metallic surface properties induced by laser surface processing for industrial applications, Nanotechnology and Precision Engineering, 2 (2019) 29–34.)*

laser has a long wavelength and a long pulse width, the heat effect will be greater than that of the short pulse laser. This phenomenon can remove the structure of the peak on the surface and reduce the surface roughness. Therefore, the results show that laser polishing reduced the roughness reduction by up to 97.5%.

## 10.4.2 Numerical simulation and microstructure

The temperature field during the laser polishing process has been calculated by the finite element simulation method. The commercial software ABAQUS (ABAQUS 2018, SIMULIA, Rhode Island, RI, USA) has been used to simulate the temperature field at the surface of the SLM IN718 superalloy. The thermal model was established by solving the general heat transfer equation for conduction [50]. The calculation of the model have been conducted on a rectangle substrate with dimensions of $2 \times 1 \times 0.5$ mm, and the minimum differential volume element is a cube with a size of $30 \times 30 \times 30$ μm. The density of the input heat flow rate of each facula is calculated by the Gaussian function, and it is consistent with the laser heat source used in laser polishing experiment. Besides, the exhaustive explanation of its convergence criteria, boundary conditions, and governing equations was provided in previous work [61].

Since laser polishing process has great thermal variation [27,32,33], the short interaction times of high energy laser with small spot size and rapid solidification of thin layers by high cooling rate ($9.238 \times 10^3$ to $1.209 \times 10^4$ K/s) results in directional grain growth [8,62]. To analyze the microstructural evolution by thermal distribution and phase transition, the numerical simulation and phase diagram during laser polishing process is shown in Fig. 10.10A [63]. High-temperature contours extend inside the polished layer, and they extend further away from the center of the irradiated spot due to the movement of the laser beam along the laser polishing direction (Fig. 10.10A). The temperature in the polished zone is over 1410°C, and the maximum temperature can reach 2491°C, which is greater than the temperature of liquid phase formation (1167°C) that the NbC entirely disappears and Laves coexistent $\gamma$ phase and liquid phase [64].

The cross section of the polished sample is shown in Fig. 10.10B, and the thickness of the remelted layer is estimated as 110 μm. XRD patterns in Fig. 10.10F indicate that both the as-received and polished Inconel 718 layer consists of $\gamma$, $\gamma'$, and $\gamma''$ phases, but the (111), (200), and (220) peaks are significantly changed due to rapid melting and solidification at uneven

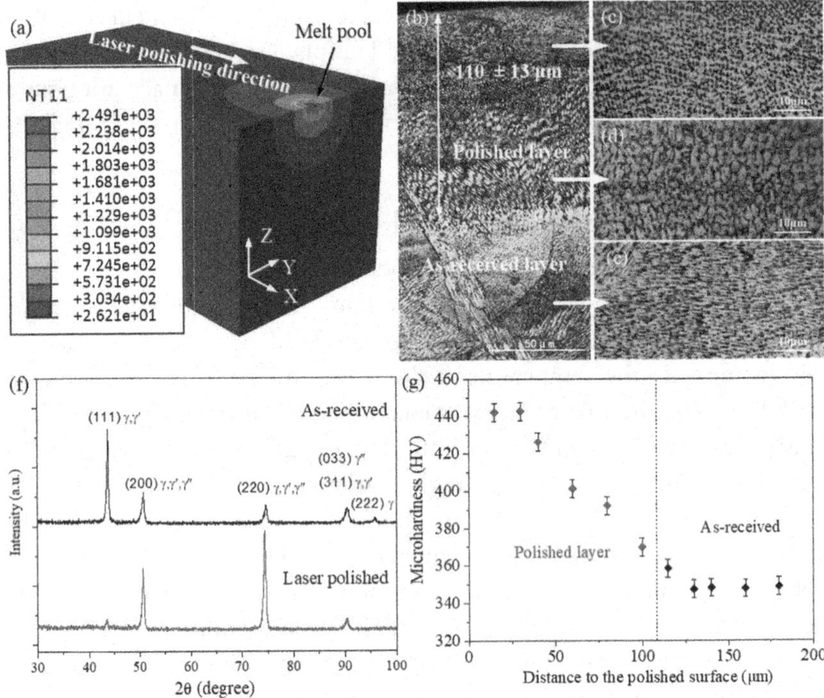


**Figure 10.10** Numerical simulation and microstructure of SLM IN718 alloy by laser polishing: (A) temperature field simulation; (B) microstructure in cross section; (C) microstructure in the polished layer; (D) microstructure in the heat-affected zone; (E) microstructure in the as-received layer; (F) XRD patterns for the as-received and laser-polished surface; and (G) microhardness distributions. *(Reproduced with kind permission of Elsevier and KeAi from References G. Hu, Y. Song, Y. Guan, Tailoring metallic surface properties induced by laser surface processing for industrial applications, Nanotech. Precis. Eng., 2 (2019) 29–34; Y. Li, Z. Zhang, Y. Guan, Thermodynamics analysis and rapid solidification of laser polished Inconel 718 by selective laser melting, Appl. Surf. Sci., 2020, 511, 145423.)*

high temperature [12,63]. The matrix Inconel 718 alloy comprises γ phase (Ni–Cr solid solution), and the γ'' phase (bct-Ni3Nb) is the main strengthening phase, forming a coherent precipitate with the γ phase in the IN718. On the top of the polished zone, the grains are fine for the high cooling rate as shown in Fig. 10.10C, while the grains at the bottom of the polished zone are columnar (Fig. 10.10E), but the grain size increases compared with those at the top zone. The substrate microstructure revealed typical segregation patterns, resulting from the dendritic/cellular growth during SLM process. The presence of more secondary phases and carbides contributes to grain boundary pinning, hence restricting the grain growth during the cooling process of laser polishing. There are more γ'' phases along the grain boundaries on the top of the polished zone because of the smaller grain size compared with the size of the substrate microstructure.

### 10.4.3 Mechanical properties

The evolution of microstructures and phase composition will inevitably lead to changes of surface properties. Therefore, we research the microhardness distributions, coefficient of friction under reciprocating sliding, potentiodynamic polarization, and tensile stress–strain properties after laser polishing.

Fig. 10.10G shows the microhardness distributions of polished and as-received IN718 superalloy. The average hardness of the as-received layer is about 345 HV. After laser polishing, the hardness of the polished layer increases by 27.5% (about 440 HV) than that of the as-received layer. Obviously, laser polishing improves the surface hardness of as-received IN718 superalloy. Combined with the analysis of microstructures (Section 4.2), the high surface hardness is caused by the precipitation strengthening of γ'' on the polished layer. Besides, laser rapid heating and cooling lead to the grain refinement, which is beneficial for improving the hardness of the polished layer [7].

The tensile testing was performed in accordance with ASTM standard E8/E8M-16a [65], and laser polishing is carried out on all sides of the plate-like tensile specimens. Uniaxial tensile test was conducted on Zwick Z100/SN5A electronic universal testing machine (loading rate: 0.5 mm/min, 5 kN load cell), in which five measurements were conducted, and the results were averaged. The typical tensile properties of the as-received and polished samples at room temperature are shown in Fig. 10.11A. The polished sample exhibits a higher ultimate tensile strength (1093.74 MPa) than the as-received sample (1079.68 MPa). In the later part of the stress–strain curve in Fig. 10.11A, it is demonstrated that the stress declines with increasing strain, which verifies the existence of necking and plastic deformation in tensile failure. As a surface treatment process, laser polishing has can realize the problem of "control performance" on high-performance AM components.

The corrosion resistance of the laser-polished surface has been improved, which can be seen in Fig. 10.11B. To outstanding the effectiveness of laser polishing, we also compared three different specimens: as-received,

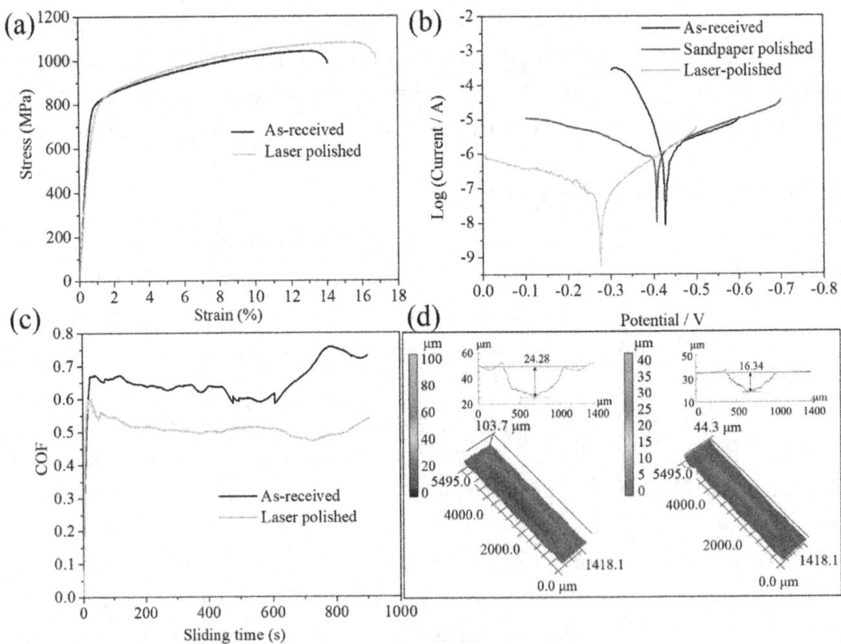


**Figure 10.11** Properties after laser polishing SLM IN718 alloy: (A) tensile stress–strain properties of as-received and laser-polished sample; (B) potentiodynamic polarization results; (C) coefficient of friction under reciprocating sliding for as-received and optimized laser-polished surface; and (D) Cross-section profile of wear tracks on as-received and laser-polished surface. *(Reproduced with kind permission of KeAi from Reference G. Hu, Y. Song, Y. Guan, Tailoring metallic surface properties induced by laser surface processing for industrial applications, Nanotechnology and Precision Engineering, 2 (2019) 29–34.)*

sandpaper-polished, and laser-polished surface. The results are listed in Table 10.3. The corrosion potential of the polished surface increased from −0.43 to −0.27 V, and the corrosion current decreased from −5.90 to −7.13 (Log A). Besides, the result of laser polishing is better than that of sandpaper polishing (corrosion potential −0.41 V, corrosion current −6.06). The improvement is attributed to the fine-grain strengthening, γ'' phase precipitation strengthening, and smoother surface morphology in the polished layer.

**Table 10.3** Electrochemical corrosion parameters of the three surfaces.

| | As-received | Sandpaper polished | Laser polished |
|---|---|---|---|
| Potential (V) | −0.43 | −0.41 | −0.27 |
| Current (log A) | −5.90 | −6.06 | −7.13 |

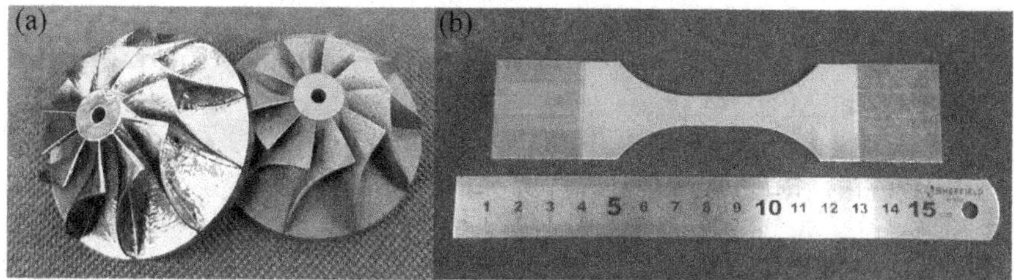

**Figure 10.12** Examples of practical applications using laser polishing techniques on SLM IN718 alloys: (A) aeroturbine engine blade and (B) plate fatigue test components.

Furthermore, the variation of friction coefficient with sliding time of the as-received and polished surface is shown in Fig. 10.11C. It can be seen that the average friction coefficient of polished surface decreases by 0.15 compared with that of the as-received surface. It indicates that the improvement of surface hardness and finish can decrease the friction coefficient of SLM IN718 alloy. The wear scar of the polished surface is smaller than that of the as-received surface as shown in Fig. 10.11D. The maximum depth of profile on cross-section is decreased from 24.28 to 16.34 μm after laser polishing. Results showed that the wear resistance of the laser-polished surface was significantly improved.

### 10.4.4 Examples of laser polishing on large-area SLM IN718 components

Fig. 10.12 shows two typical samples of Ni components after laser polishing in our group. In Fig. 10.12A, the mini turbine engine blade of SLM IN718 was polished with the surface roughness Ra reduction from 17.5 to 0.15 μm. In Fig. 10. 12B, the average roughness Ra of the plate tensile specimen was decreased from 6.4 μm to less than 0.1 μm, which is decreased by 98.4% comparing to that of the as-received surface.

## 10.5 Conclusions

In this chapter, we have presented the laser polishing method to improve the surface quality on LMD TC11, SLM TC4 and SLM Inconel 718 superalloy. The main results are summarized as follows:

**(1)** After laser-polished, the average roughness is decreased from over Ra 7.21 μm to less than 0.73 μm on LMD TC11. In addition, the microhardness increased from 400 HV to 560 HV, and wear rate decreased from $0.41\ mm^3/N\cdot m$ to $0.27\ mm^3/N\cdot m$ due to the acicular martensite $\alpha'$ in polished zone.

**(2)** For laser-polished SLM TC4, the average roughness was reduced from Ra 6.53 −0.32 μm. Similar to LMD TC11, the typical microstructure of coarsened $\alpha + \beta$ phase in as-received layer transform into acicular martensite $\alpha'$ after laser polishing. Meanwhile, the microhardness was increasing by 25% comparing to that of the as-received layer, and wear rate of laser-polished surface was decreased by 39% less than as-received layer.

**(3)** For laser-polished SLM Inconel 718 superalloy, the average roughness was reduced from over Ra 8−0.1 μm. Both the as-received and laser-polished layers consist of $\gamma$, $\gamma'$ and $\gamma''$ phases. The microhardness of the polished layer increases by 27.5% than that of the as-received layer. Besides, the ultimate tensile strength on the polished sample is about 1093.74 MPa, which is higher than that of the as-received sample. For corrosion resistance, the corrosion potential of polished surface increased from −0.43 to −0.27 V, and the corrosion current decreased from −5.90 to −7.13(Log A). Besides, the average friction coefficient of polished surface decreases by 0.15 compared to that of as-received surface.

## Acknowledgments

This work was financially supported by National Natural Science Foundation of China under Grants 51705013 and 51875313; National Key Research and Development Program of China under Grants 2018YFB1107400, 2018YFB1107700, and 2016YFB1102503.

## References

[1] E.S. Huilong Hou, T. Ma, N.S. Johnson, S. Qian, C. Cissé, S. Drew, N.Al Hasan, L. Zhou, Y. Hwang, R. Radermacher, V.I. Levitas, M.J. Kramer, M.A. Zaeem, A.P. Stebner, R.T. Ott, J. Cui, I. Takeuchi, Fatigue-resistant high-performance elastocaloric materials made by additive manufacturing, Science 366 (2019) 1116−1121.

[2] D. Zhang, D. Qiu, M.A. Gibson, Y. Zheng, H.L. Fraser, D.H. StJohn, M.A. Easton, Additive manufacturing of ultrafine-grained high-strength titanium alloys, Nature 576 (2019) 91−95.

[3] P. Barriobero-Vila, J. Gussone, A. Stark, N. Schell, J. Haubrich, G. Requena, Peritectic titanium alloys for 3D printing, Nat. Commun. 9 (2018) 3426.

[4] D.C. Hofmann, S.N. Roberts, H. Kozachkov, Infrared thermal processing history of a Ti-based bulk metallic glass matrix composite manufactured via semi-solid forging, Acta Mater. 95 (2015) 192−200.

[5] E. Cakmak, M.M. Kirka, T.R. Watkins, R.C. Cooper, K. An, H. Choo, W. Wu, R.R. Dehoff, S.S. Babu, Microstructural and micromechanical characterization of IN718 theta shaped specimens built with electron beam melting, Acta Mater. 108 (2016) 161−175.

[6] X. Tan, Y. Kok, Y.J. Tan, M. Descoins, D. Mangelinck, S.B. Tor, K.F. Leong, C.K. Chua, Graded microstructure and mechanical properties of additive manufactured Ti–6Al–4V via electron beam melting, Acta Mater. 97 (2015) 1–16.
[7] Z. Wang, K. Guan, M. Gao, X. Li, X. Chen, X. Zeng, The microstructure and mechanical properties of deposited-IN718 by selective laser melting, J. Alloys Compd. 513 (2012) 518–523.
[8] N. Li, S. Huang, G. Zhang, R. Qin, W. Liu, H. Xiong, G. Shi, J. Blackburn, Progress in additive manufacturing on new materials: a review, J. Mater. Sci. Technol. 35 (2019) 242–269.
[9] Z. Baicheng, L. Xiaohua, B. Jiaming, G. Junfeng, W. Pan, S. Chen-nan, N. Muiling, Q. Guojun, W. Jun, Study of selective laser melting (SLM) Inconel 718 part surface improvement by electrochemical polishing, Mater. Des. 116 (2017) 531–537.
[10] D. Bhaduri, P. Penchev, A. Batal, S. Dimov, S.L. Soo, S. Sten, U. Harrysson, Z. Zhang, H. Dong, Laser polishing of 3D printed mesoscale components, Appl. Surf. Sci. 405 (2017) 29–46.
[11] S. Marimuthu, A. Triantaphyllou, M. Antar, D. Wimpenny, H. Morton, M. Beard, Laser polishing of selective laser melted components, Int. J. Mach. Tool Manufact. 95 (2015) 97–104.
[12] L.L. Fang Zhihao, L. Chen, Y. Guan, Laser polishing of additive manufactured superalloy, Proc. CIRP 71 (2018) 150–154.
[13] S. Bagehorn, J. Wehr, H.J. Maier, Application of mechanical surface finishing processes for roughness reduction and fatigue improvement of additively manufactured TC4 parts, Int. J. Fatig. 102 (2017) 135–142.
[14] L.E. Murr, E. Martinez, K.N. Amato, S.M. Gaytan, J. Hernandez, D.A. Ramirez, P.W. Shindo, F. Medina, R.B. Wicker, Fabrication of metal and alloy components by additive manufacturing: examples of 3D materials science, J. Mater. Res. Technol. 1 (2012) 42–54.
[15] T.M. Mower, M.J. Long, Mechanical behavior of additive manufactured, powder-bed laser-fused materials, Mater. Sci. Eng., A 651 (2016) 198–213.
[16] E. Łyczkowska, P. Szymczyk, B. Dybała, E. Chlebus, Chemical polishing of scaffolds made of Ti–6Al–7Nb alloy by additive manufacturing, Archiv. Civil Mech. Eng. 14 (2014) 586–594.
[17] V. Urlea, V. Brailovski, Electropolishing and electropolishing-related allowances for powder bed selectively laser-melted TC4 alloy components, J. Mater. Process. Technol. 242 (2017) 1–11.
[18] Y.H. Li, B. Wang, C.P. Ma, Z.H. Fang, L.F. Chen, Y.C. Guan, S.F. Yang, Material characterization, thermal analysis, and mechanical performance of a laser-polished Ti alloy prepared by selective laser melting, Metals 9 (2019) 112.
[19] C.P. Ma, Y.C. Guan, W. Zhou, Laser polishing of additive manufactured Ti alloys, Optic Laser. Eng. 93 (2017) 171–177.
[20] A. Krishnan, F. Fang, Review on mechanism and process of surface polishing using lasers, Front. Mech. Eng. 14 (2019) 299–319.
[21] T. Deng, J. Li, Z. Zheng, Fundamental aspects and recent developments in metal surface polishing with energy beam irradiation, Int. J. Mach. Tool Manufact. 148 (2020) 103472.
[22] L. Jiang, A.D. Wang, B. Li, T.H. Cui, Y.F. Lu, Electrons dynamics control by shaping femtosecond laser pulses in micro/nanofabrication: modeling, method, measurement and application, Light Sci. Appl. 7 (2018) 17134.
[23] L.B. Boinovich, E.B. Modin, A.R. Sayfutdinova, K.A. Emelyanenko, A.L. Vasiliev, A.M. Emelyanenko, Combination of functional nanoengineering and nanosecond laser texturing for design of superhydrophobic aluminum alloy with exceptional mechanical and chemical properties, ACS Nano 11 (2017) 10113–10123.

[24] M.V. Chao Ma, N.A. Duffie, F.E. Pfefferkorn, X. Li, Melt pool flow and surface evolution during pulsed laser micro polishing of Ti6Al4V, J. Manuf. Sci. Eng. 135 (2013), 061023-061021-061028.

[25] J. Yang, J. Han, H. Yu, J. Yin, M. Gao, Z. Wang, X. Zeng, Role of molten pool mode on formability, microstructure and mechanical properties of selective laser melted TC4 alloy, Mater. Des. 110 (2016) 558–570.

[26] D.L.B.J.A. Ramos-Grez, Reducing surface roughness of metallic freeform-fabricated parts using non-tactile finishing methods, Int. J. Mater. Prod. Technol. 21 (2004) 297–315.

[27] A. Lamikiz, J.A. Sánchez, L.N. López de Lacalle, J.L. Arana, Laser polishing of parts built up by selective laser sintering, Int. J. Mach. Tool Manufact. 47 (2007) 2040–2050.

[28] P.M. Benoit Rosa, J.-yves Hascoët, Laser polishing of additive laser manufacturing surfaces, J. Laser Appl. 27 (2015) S29102.

[29] B. Rosa, P. Mognol, J.-Y. Hascoët, Modelling and optimization of laser polishing of additive laser manufacturing surfaces, Rapid Prototyp. J. 22 (2016) 956–964.

[30] M.A. Obeidi, E. McCarthy, B. O'Connell, I. Ul Ahad, D. Brabazon, Laser polishing of additive manufactured 316L stainless steel synthesized by selective laser melting, Materials (2019) 12.

[31] W.J. Wang, K.C. Yung, H.S. Choy, T.Y. Xiao, Z.X. Cai, Effects of laser polishing on surface microstructure and corrosion resistance of additive manufactured CoCr alloys, Appl. Surf. Sci. 443 (2018) 167–175.

[32] K.C. Yung, T.Y. Xiao, H.S. Choy, W.J. Wang, Z.X. Cai, Laser polishing of additive manufactured CoCr alloy components with complex surface geometry, J. Mater. Process. Technol. 262 (2018) 53–64.

[33] K.C. Yung, W.J. Wang, T.Y. Xiao, H.S. Choy, X.Y. Mo, S.S. Zhang, Z.X. Cai, Laser polishing of additive manufactured CoCr components for controlling their wettability characteristics, Surf. Coating. Technol. 351 (2018) 89–98.

[34] B. Richter, N. Blanke, C. Werner, F. Vollertsen, F.E. Pfefferkorn, Effect of initial surface features on laser polishing of Co-Cr-Mo alloy made by powder-bed fusion, J. Occup. Med. 71 (2018) 912–919.

[35] O. Černašėjus, J. Škamat, V. Markovič, N. Višniakov, S. Indrišiūnas, Surface laser processing of additive manufactured 1.2709 steel parts: preliminary study, Adv. Mater. Sci. Eng. (2019) 1–9, 2019.

[36] J. Sun, Y. Yang, D. Wang, Mechanical properties of a Ti6Al4V porous structure produced by selective laser melting, Mater. Des. 49 (2013) 545–552.

[37] J. Vaithilingam, E. Prina, R.D. Goodridge, R.J.M. Hague, S. Edmondson, F. Rose, S.D.R. Christie, Surface chemistry of Ti6Al4V components fabricated using selective laser melting for biomedical applications, Mater. Sci. Eng. C Mater. Biol. Appl. 67 (2016) 294–303.

[38] C. Qiu, H. Chen, Q. Liu, S. Yue, H. Wang, On the solidification behaviour and cracking origin of a nickel-based superalloy during selective laser melting, Mater. Char. 148 (2019) 330–344.

[39] K.N. Amato, S.M. Gaytan, L.E. Murr, E. Martinez, P.W. Shindo, J. Hernandez, S. Collins, F. Medina, Microstructures and mechanical behavior of Inconel 718 fabricated by selective laser melting, Acta Mater. 60 (2012) 2229–2239.

[40] M.J. Anderson, C. Panwisawas, Y. Sovani, R.P. Turner, J.W. Brooks, H.C. Basoalto, Mean-field modelling of the intermetallic precipitate phases during heat treatment and additive manufacture of Inconel 718, Acta Mater. 156 (2018) 432–445.

[41] Y. Chen, F. Lu, K. Zhang, P. Nie, S.R. Elmi Hosseini, K. Feng, Z. Li, Dendritic microstructure and hot cracking of laser additive manufactured Inconel 718 under improved base cooling, J. Alloys Compd. 670 (2016) 312–321.

[42] S. Sui, H. Tan, J. Chen, C. Zhong, Z. Li, W. Fan, A. Gasser, W. Huang, The influence of Laves phases on the room temperature tensile properties of Inconel 718 fabricated by powder feeding laser additive manufacturing, Acta Mater. 164 (2019) 413–427.
[43] H.-L.T. Chen Chen, Fundamental study of the bulge structure generated in laser polishing process, Optic Laser. Eng. 107 (2018) 54–61.
[44] L. Giorleo, E. Ceretti, C. Giardini, Ti surface laser polishing: effect of laser path and assist gas, Proc. CIRP 33 (2015) 446–451.
[45] F.E. Pfefferkorn, N.A. Duffie, X. Li, M. Vadali, C. Ma, Improving surface finish in pulsed laser micro polishing using thermocapillary flow, CIRP Annal. 62 (2013) 203–206.
[46] C.-S. Chang, T.-H. Chen, T.-C. Li, S.-L. Lin, S.-H. Liu, J.-F. Lin, Influence of laser beam fluence on surface quality, microstructure, mechanical properties, and tribological results for laser polishing of SKD61 tool steel, J. Mater. Process. Technol. 229 (2016) 22–35.
[47] Q. Wang, J.D. Morrow, C. Ma, N.A. Duffie, F.E. Pfefferkorn, Surface prediction model for thermocapillary regime pulsed laser micro polishing of metals, J. Manuf. Process. 20 (2015) 340–348.
[48] L. Chen, Y. Yang, F. Jiang, C. Li, Experimental investigation and FEM analysis of laser cladding assisted by coupled field of electric and magnetic, Mater. Res. Exp. 6 (2018) 016516.
[49] Y. Wang, R. Chen, X. Cheng, Y. Zhu, J. Zhang, H. Wang, Effects of microstructure on fatigue crack propagation behavior in a bi-modal TC11 titanium alloy fabricated via laser additive manufacturing, J. Mater. Sci. Technol. 35 (2019) 403–408.
[50] A. Foroozmehr, M. Badrossamay, E. Foroozmehr, S.i. Golabi, Finite element simulation of selective laser melting process considering optical penetration depth of laser in powder bed, Mater. Des. 89 (2016) 255–263.
[51] J. Yang, H. Yu, J. Yin, M. Gao, Z. Wang, X. Zeng, Formation and control of martensite in TC4 alloy produced by selective laser melting, Mater. Des. 108 (2016) 308–318.
[52] A.R. Nassar, E.W. Reutzel, Additive manufacturing of TC4 using a pulsed laser beam, Metall. Mater. Trans. 46 (2015) 2781–2789.
[53] E. Brandl, D. Greitemeier, Microstructure of additive layer manufactured Ti–6Al–4V after exceptional post heat treatments, Mater. Lett. 81 (2012) 84–87.
[54] S.Q. Wu, Y.J. Lu, Y.L. Gan, T.T. Huang, C.Q. Zhao, J.J. Lin, S. Guo, J.X. Lin, Microstructural evolution and microhardness of a selective-laser-melted Ti–6Al–4V alloy after post heat treatments, J. Alloys Compd. 672 (2016) 643–652.
[55] C.N.L.K. Iyakutti, S. Anuratha, S. Mahalakshmi, Pressure-induced electronic phase transitions and superconductivity in titanium, Int. J. Mod. Phys. B 23 (2009) 723–741.
[56] G. Guo, G. Tang, X. Ma, M. Sun, G.E. Ozur, Effect of high current pulsed electron beam irradiation on wear and corrosion resistance of Ti6Al4V, Surf. Coating. Technol. 229 (2013) 140–145.
[57] D. W., T.L. Perry, X. Li, F.E. Pfefferkorn, N.A. Duffie, The effect of laser pulse duration and feed rate on pulsed laser polishing of microfabricated nickel samples, J. Manuf. Sci. Eng. 131 (2009), 031002-031001-031007.
[58] J. Lambarri, J. Leunda, C. Soriano, C. Sanz, Laser surface smoothing of nickel-based superalloys, Phy. Proc. 41 (2013) 255–265.
[59] A.M.K. Hafiz, E.V. Bordatchev, R.O. Tutunea-Fatan, Experimental analysis of applicability of a picosecond laser for micro-polishing of micromilled Inconel 718 superalloy, Int. J. Adv. Manuf. Technol. 70 (2013) 1963–1978.
[60] G. Hu, Y. Song, Y. Guan, Tailoring metallic surface properties induced by laser surface processing for industrial applications, Nanotech. Precis. Eng. 2 (2019) 29–34.

[61] T.M.Y.L.F. Guo, H.C. Man, A finite element method approach for thermal analysis of laser cladding of magnesium alloy with preplaced Al–Si powder, J. Appl. Phys. 16 (2004) 229–235.
[62] M.L.S.N. Quy Bau Nguyen, Z. Zhu, C.N. Sun, J. Wei, W. Zhou, Characteristics of Inconel powders for powder-bed additive manufacturing, Engineering 3 (2017) 695–700.
[63] Y. Li, Z. Zhang, Y. Guan, Thermodynamics analysis and rapid solidification of laser polished Inconel 718 by selective laser melting, Appl. Surf. Sci. 511 (2020) 145423.
[64] H. Wang, K. Ikeuchi, M. Takahashi, A. Ikeda, Microstructures of Inconel 718 alloy subjected to rapid thermal and stress cycle – joint performance and its controlling factors in friction welding of Inconel 718 alloy, Weld. Int. 23 (2009) 662–669.
[65] L.R. Lise Sandnes, Ø. Grong, F. Berto, T. Welo, Assessment of the mechanical integrity of a 2 mm AA6060-T6 butt weld produced using the hybrid metal extrusion & bonding (HYB) process – Part II: tensile test results, Proc. Struct. Integrit. 17 (2019) 632–642.

CHAPTER 11

# On surface quality of engineered parts manufactured by additive manufacturing and postfinishing by machining

M. Pérez[1,2], A. García-Collado[1], D. Carou[1,3,4], G. Medina-Sánchez[1], R. Dorado-Vicente[1]

[1]Department of Mechanical and Mining Engineering, EPS de Jaén, University of Jaén, Campus Las Lagunillas, Jaén, Spain; [2]Department of Manufacturing Engineering, Universidad Nacional de Educación a Distancia (UNED), Madrid, Spain; [3]Centre for Mechanical Technology and Automation (TEMA), University of Aveiro, Aveiro, Portugal; [4]Departamento de Deseño na Enxeñaría, Universidade de Vigo, Campus As Lagoas, Ourense, Spain

## 11.1 Introduction

Additive manufacturing (AM) is one of the major trends in manufacturing in the past decades and is recognized as one key emerging technology by the international Organisation for Economic Co-operation and Development [1]. Conventionally, the 3D printing term is used as synonym of AM, although it is properly related to low-capacity/cost-printing machines [2,3]. The manufacturing process, which currently includes several technologies, is considered one of the enabling technologies in the latest industrial revolution, Industry 4.0 [4], term that is the translation of the German *Industrie* 4.0 concept [5]. This new movement is encouraging the integration of intelligent production systems and advanced information technologies [6]. AM can support Industry 4.0 by connecting production lines with customers in real time, thus reducing supply chains and increasing the speed of production [5,7].

AM, particularly stereolithography, was firstly developed as a technique for rapid prototyping during the 1980s [94]. But, in a few years, new techniques were developed [8]. In 2005, the 3D printing revolution began with the RepRap project, whose goal was to create a machine that could replicate itself [9]. The process offers advantages such as no restrictions in design, no special tooling needed, and near net shape parts (reduced waste) [10]. Besides, thanks to the support by amateur users and governments, expiration of AM

*Additive Manufacturing*
ISBN 978-0-12-818411-0
https://doi.org/10.1016/B978-0-12-818411-0.00015-X

patents, and to the advances that appeared in fields such as computer technology, electronics, and software; nowadays, AM is in full development [11].

The development of new technologies and improvement of the existing ones are making possible to increase the number of industrial applications. A large number of sectors are benefiting from these advances; among them, it is possible to cite the aerospace, biotechnology, electronics, and medical sectors [12]. Noteworthy examples of the use of AM include the fuel nozzle developed by General Electric [13] for their LEAP engines. Other interesting examples are those that are appearing in biomedical applications, in which it is possible to appreciate the nature of the impact that can occur. For instance, Singh and Ramakrishna [14] reviewed a wide range of applications under research and development in biomedical engineering that can lead to breakthroughs for patients and society. To name a few, the authors listed custom designs in orthopedics, heart implants, retinal implants, liver printing, heterogeneous bioprinting for in vitro drug toxicology testing, and 3D printing biodegradable vascular stents. The technology is not just important for industrial or research activities but becoming of great social importance. As an example of this, the OECD [15] identifies 3D printers as one of the symbolic products for the Generation Z (>1995), as compared with car, television, computer, and mobile for previous generations.

Nowadays, the evolution of AM is also opening new opportunities for the apparition of new technologies that, despite sharing the basic principles, include novel features that make them innovative compared with the previous ones. For instance, 4D printing, firstly introduced in 2013, represents an evolution of "conventional" 3D printing, relying on the development of the smart materials [16]. The technology, developed by the Self-Assembly Lab at the Massachussetts Institute of Technology (MIT), takes as baseline the principles of 3D printing, but it needs materials with the capacity to be stimulated. The pieces manufactured by this technology evolve as a function of time, presenting intelligent behaviors [17].

While every day new interesting applications for AM technologies are appearing, they still have to face some challenges for their adoption by industries. Improving the finishing of printed parts is one of the most challenging issues. Surface quality is of great importance for engineering applications [18]. For instance, surface roughness is a widely used index of product quality and, in most cases, a technical requirement for mechanical products due to its relation with the functional behavior of a part [19]. For example, the relation of surface quality (i.e., surface roughness) with fatigue strength should be also considered for AM parts [20]. But, as the number of applications is being increased, the requirements for the manufactured parts are becoming more demanding, such as applications in biomedicine demand surface specifications (i.e., roughness and structures) to improve the biological interaction of the part with the body [21]. Surface characteristics can be also modified depending on the needs, as for instance proposed Dinis et al. [22]. The authors presented a software, called POMES (Porous and Modifications for Engineering Surfaces), that allows modifying the surfaces by creating a specific roughness pattern and porous structures in any geometry based on AM. Surface modification can have a wide range of applications for manufacturing suitable orthopedic and dental implants in terms of osseointegration.

### 11.1.1 Materials

For any manufacturing process, the raw material selected must be compatible with the process in question, besides having acceptable properties to withstand the manufacturing process and offering the expected results during the predicted service life of the part. The type of material directly affects the shape, dimensions, applications, durability, and cost. Moreover, it should be considered that, in industry, the manufacture of products can be made from one or multiple materials [31].

AM has still a limited number of materials available. However, the range of materials is being constantly increased in parallel to the development of the different technologies. Depending on the state of the raw material, it is possible to classify them in three main categories: liquid, solid, and powder [9]. The type of material to be used limits the number of suitable technologies. So, for instance, liquid-based materials are mainly thermoplastic and thermosetting polymers to be used in stereolithography [9].

The materials can be also classified depending on their nature in ceramics, composites, metals, and polymers [11]. Because of the availability of economical solutions for printing, polymers are the most commonly used material. Besides, polymers provide adequate properties such as low electrical and thermal conductivity, low density, and high strength-to-weight ratio. Typical polymers include polylactic acid (PLA), acrylonitrile butadiene styrene (ABS), polycarbonate, and polyamides. Metals such as steel [23], titanium alloys [24,25], aluminum alloys [26]), and nickel-based superalloys [27] can be used for taking advantage of properties such as strength, toughness, thermal and electrical conductivity, and machinability. Applications for other materials such as ceramics ($Al_2O_3$, $SiO_2$, and $ZrO_2$)

can be also found based on their high strength. Composite materials are also finding applications in AM. Their properties depend on the base materials (ceramic, metal, or polymer-based material) and their composition [9]. Besides, the use of advanced novel materials such as digital and smart materials (i.e., shape memory polymers), electronic (i.e., conductive polymers), and biomaterials (i.e., hydrogels) is also being studied as reported by Lee et al. [32].

For years, the use of single materials has been the standard; however, the advance of the technology allows moving forward to the multimaterial AM. By combining different materials, parts or products can take advantage of the different properties of the materials to manufacture enhanced products: functionality, hardness, wear resistance, thermal, performance, and osteoconduction [28].

### 11.1.2 Main technologies and principles of additive manufacturing

The basic principle of AM consists of building 3D geometries by material addition, generally layer by layer. Therefore, the technique used for depositing the material in each layer becomes critical, considering that all materials have different behavior regarding adhesion and temperature (i.e., melting point). Several principles can be used to create the parts in AM, namely polymerization, ink/binder, melting, freezing, and joining [29].

AM is no longer just a prototyping technique as it was firstly introduced. Currently, it includes a wide range of technologies, resulting in an increase in complexity to deal with the topic. Besides, it is gaining a lot of attention and becoming a popular technology with an important number of users and, more important, potential users. Therefore, efforts were carried out to organize the knowledge by the ASTM International [2] to deliver a comprehensive standard with terminology and classifications, the "Standard Terminology for Additive Manufacturing Technologies." Lately, this standard was withdrawn when the new ISO/ASTM 52900:2015 "Additive manufacturing. General principles. Terminology" standard was launched [3]. In this way, communication and understanding on AM can be improved for both technical and nontechnical users.

The ISO/ASTM 52900 standard [3] establishes a general classification of the process: direct (the parts can obtain the material properties and geometry in only one step) or indirect (the parts acquire the final geometry in the first step and the material properties in the second step). In all cases, both types of processes may require postprocessing operations such as finishing and thermal treatments. Moreover, the standard establishes a classification in process categories: binder jetting, directed energy deposition, material extrusion, material jetting, powder-bed fusion, sheet lamination, and vat polymerization. These terms are destined to group the machines that share processing similarities.

Within each process category, there are different AM approaches that have in common the technology and the state of the material that they use. Some of the most used technologies include fused deposition modeling (FDM), laminated object manufacturing (LOM), stereolithography (SLA), selective laser melting (SLM), and selective laser sintering (SLS) [30]. Different authors have used the seven ISO/ASTM categories to sort different AM processes. In the same way, the seven categories are presented in Fig. 11.1 along with different processes and the state of the raw material that they use.

AM processes have in common the following characteristics: a computer to store, process geometric information, and guide the user, and a deposition material that is processed by points, lines, or areas to create parts [11]. All parts manufactured with the current technologies follow several common steps during the manufacturing process. First, a solid or surface model, created by means of Computer-Aided Design (CAD) software, is converted into a file format readable by the software of the machine. This format is usually the Standard Triangle Language (STL), although the ISO 10303 "Automation systems and integration—Product data representation and exchange" standard promotes the Standard for the Exchange of Products (STEP) [34]. This conversion transforms a complex CAD surface into an approximate surface created with triangles, allowing dividing them with ease in two-dimensional cuts that are then reproduced physically with the AM process [35]. Next, the piece is electronically sectioned into layers with a certain thickness, and the path of the tool is generated. The information of each section in layers is electronically transmitted to the machine, processing layer by layer until the piece is finally completed [9,33]. The final step is the postprocessing for improving surface finish and dimensional accuracy [36].

### 11.1.3 Additive versus subtractive manufacturing processes

In the ISO/ASTM 52900 standard [3], AM is defined as "process of joining materials to make parts from 3D model data, usually layer upon layer, as opposed to subtractive manufacturing and formative manufacturing methodologies" [3].

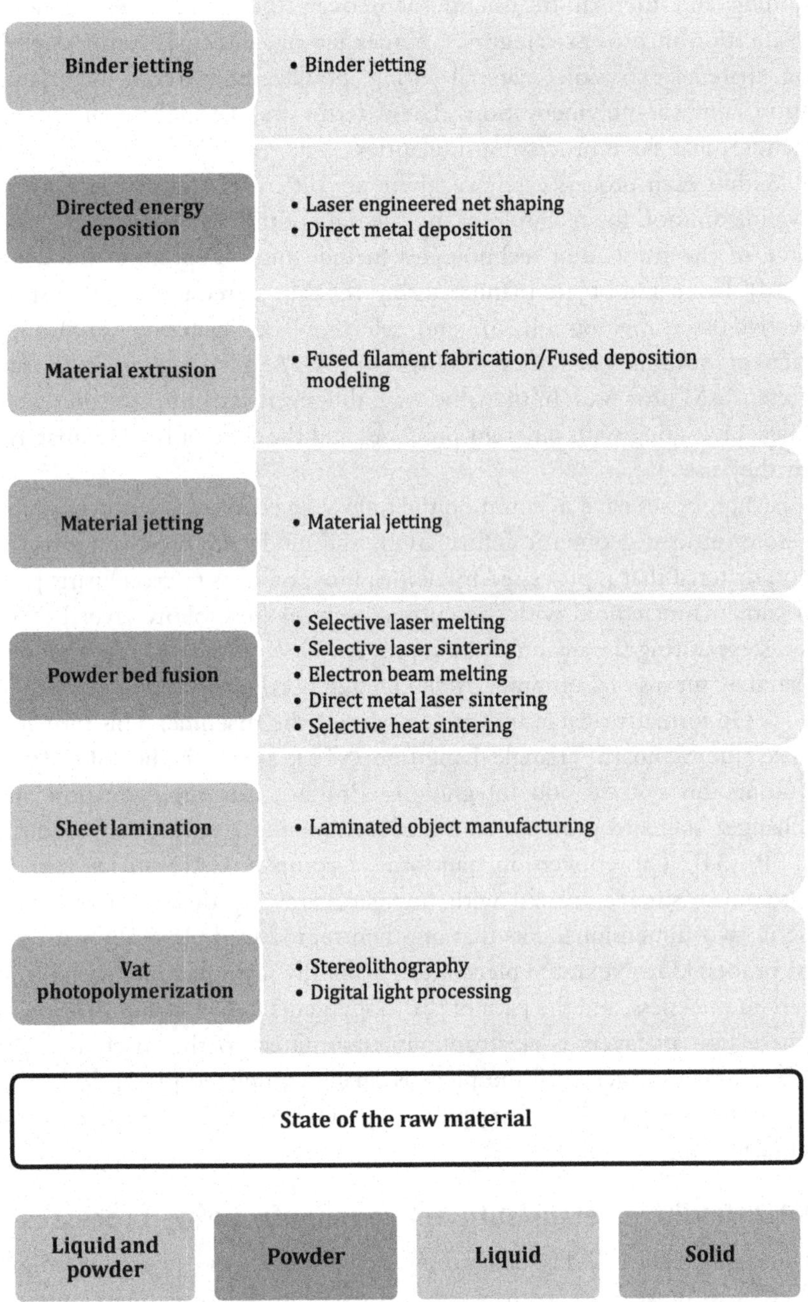


**Figure 11.1** Classification of different additive manufacturing processes. *(Based on [3,32,29,93,31,33].)*

The aforementioned definition helps to compare the process against subtractive processes such as machining. Although the fundamentals of these processes are different, one (additive, AM) can be imagined as the opposite of the other (subtractive, machining). The opposite nature of these two processes makes possible to image them just as completely nonrelated processes, but it could be also possible to imagine them as complementary, diminishing the limitations of one by using also the other.

Noorani [9] identified the key characteristics of additive and subtractive machining processes. AM offers advantages such as its adequacy for manufacturing entire assemblies, where high complexity can be obtained, including small holes of unlimited length. However, it is a still relatively expensive technology, and part size is a limitation. Subtractive manufacturing offers advantages such as no part size limitations and cost because it is a conventional and widespread technology. Moreover, it allows achieving high accuracy and excellent surface finish, which is still a limitation for AM.

## 11.2 Surface roughness

Surface quality conventionally determines the credibility of the manufactured components [37]. The evaluation of the surface quality of a part includes the analysis of the surface integrity; thus, it requires a comprehensive study of surface roughness, residual stresses, work hardening, and microstructure alterations after a manufacturing process [38]. Conventionally, surface topography has been one of the most studied characteristics to assess the quality of the parts. Particularly, these studies have long tradition since the 1930s [39].

### 11.2.1 Measurement methodology

The long tradition in the evaluation of surface topography has resulted in a wide range of achievements. For instance, experimental methods for measuring surface roughness such as stylus instruments, scanning probe microscopy, and optical instruments [39–41] have been developed and used for measuring numbers of different parts in all types of manufacturing processes.

The study of the surface roughness refers to deviations (from third to sixth order) on the surface of parts or products characterized by irregularities resulting in poor quality [19]. Several criteria can be used, being preferred the quantitative methods for assessing the roughness of a specific surface [42]. However, the determination of the roughness presents a complex technical problem because it has to be measured in a 3D geometry. This challenge has been traditionally overcome by using 2D profiles but, with the advance of the measurement technology, the analysis of 3D topographies is gaining importance and becoming more usual.

Surface roughness can be evaluated using a wide number of parameters (more than 100 and 40 for 2D and 3D, respectively) [43]. As there are a wide range of methods for the study of surface texture, ISO deals with the classification of methods to measure surface quality. For the analysis of 2D profiles, the ISO 4287 [44] "Geometrical product specification (GPS) Surface texture: Profile method; Terms, definitions and surface texture parameters" standard has been the main reference for years. Methods for assessing 3D topographies are conveniently analyzed in ISO 25178 [45] "Geometric Product Specifications (GPS)—Surface texture: areal" collection of standards to analyze 3D areal surface textures [46]. For 2D topographies, some parameters such as the arithmetical mean roughness average (*Ra*) and the mean roughness depth ($R_z$) are commonly accepted [47] and have been widely adopted by industry and researchers. However, it should be noted that industry is increasing the range of surface texture parameters used as reported by an international survey conducted in 2016 aimed at CIRP (*Collège International pour la Recherche en Productique*) industry contacts [48].

It should be also taken into account that the values obtained for the parameters should be also validated. This is because each instrument has different data storage formats and software for analyzing the results [49]. So the results can vary depending on the measuring instrument. As consequence, efforts are being carried out for coming up with a set of master algorithms that allow comparisons and traceability between measurements using common software such as the ones developed by the National Institute of Standards and Technology (NIST) [50] and the *Physikalisch-Technische Bundesanstalt* (PTB) [49].

The conventional methodology for evaluating 2D surface textures uses statistical sampling in different sections, making a comparison between the theoretical and real profile of the part surface. This real profile is not determined exhaustively, as it would be costly and unnecessary. For obtaining this profile, a common way is to traverse it with a feeler that, with changes in height due to surface irregularities of a section of the piece, generates electrical signals proportional to these changes. The profile obtained from the real surface by this method is an effective profile that is close to the real one. To eliminate ripples (low-frequency or long-wave signals) and roughness (high-frequency or short-wave signals), as well as other more specific irregularities, the effective profile passes through different filters, which separate this profile into both long and short wavelength components [42,44,51].

Several measurement methods have been used for measuring surface quality in AM processes, particularly for metals, as reported by Townsend et al. [52]. Among them, it is possible to cite: contact stylus, confocal microscopy, atomic force microscopy, scanning force microscope (SEM), X-ray computed tomography, and Raman spectroscopy. The authors identified the development of the AM technology as early-stage considering their findings on the analysis of surface quality. According to the authors, quantitative measurement of surface texture has been predominantly achieved by stylus-based profile measurements, using the ISO 4287 standard [44] and the average roughness (*Ra*) parameter. Thus, there is still the need of analyzing in detail the 3D topography of the surface.

## 11.2.2 Surface roughness in additive manufacturing

AM can be still considered a novel process because of the constant development of new technologies and the ongoing improvement of the existing ones. As a result, process improvement and optimization was a key research issue at the initial development of the process and topics such as surface quality were scarcely studied. Although several research works were presented relating the surface quality of AM parts, a limited number of works deal with the modeling of surface roughness, as reviewed by Ahn et al. [53], up to 2009. However, some noteworthy efforts can be highlighted. For instance, Paul and Voorakarnam [54] deduced an expression of surface roughness for LOM based on centerline average roughness. Luis-Pérez et al. [55] proposed an analytical model to estimate a roughness average value of layered manufactured parts. Similarly, Pandey et al. [56] suggested an analytical equation to predict surface roughness for FDM by approximating the layer edge profile as a parabolic curve. Ahn et al. [57] proposed a new methodology to predict surface roughness in layered manufacturing processes. The theoretical and actual characteristics of the surface roughness distributions of the layer manufacturing parts were investigated to represent the actual roughness predictions.

The surface quality of additively manufactured pieces is related to their strength and the dimensional accuracy [58]. The analysis of the surface roughness must focus on the specific technology used and its main characteristics but, because of the various types of materials that could be used, the properties of the materials are also of great importance. For instance,

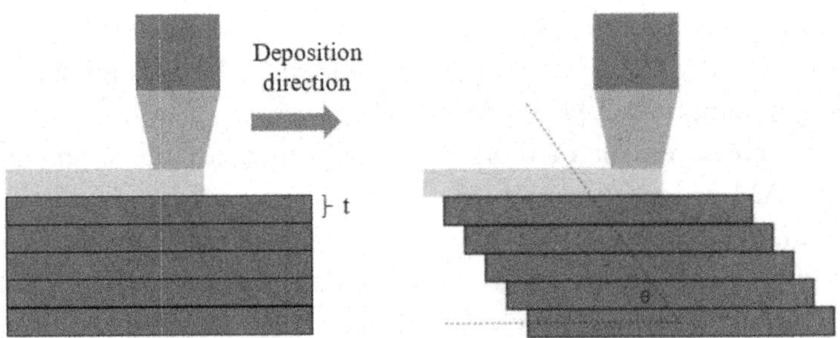


Figure 11.2 Staircase effect on additive manufacturing.

polymers are one of the most extended materials in processes such as fused modeling deposition. When using polymers as process material, one of the problems that could affect the surface roughness is related to the thermal and mechanical aspects. In this sense, the material cools quickly leading to distortions and stresses, which could cause eventually problems due to shrinkages in unpredictable areas of the piece [59].

The process has also a big responsibility on the poor surface quality obtained. In this sense, the layer-by-layer process produces a staircase effect that leads to high surface roughness, especially on curved and inclined surfaces [60]. The staircase effect can be notably limited when properly selecting the manufacturing parameters, especially the layer thickness and part orientation. Several studies have analyzed the influence of printing parameters such as workpiece orientation, layer thickness, and the orientation of the material deposition on surface roughness [94]. For instance, the influence of the layer height on the surface roughness has been highlighted by Pérez et al. [61] on FDM processes and by Strano et al. [62] on SLM processes. In Fig. 11.2, it is possible to appreciate how surface roughness can be affected by the layer thickness ($t$) and the printing angle ($\theta$).

The staircase effect markedly affects FDM parts as it employs thick filaments, generating layers of 0.254 mm thickness in most cases and, only for some materials, of 0.127 mm [63,64]. These values of layer thickness are higher than the ones used in stereolithography (0.05 mm), SLS (0.02 mm), and polymer material jetting (0.016 mm) [59]. As a result, the surface quality of additively manufactured parts is poor. For instance, in Fig. 11.3, the results of surface roughness[1] obtained for various technologies along its minimum

[1] All the results of surface roughness are presented using the arithmetical mean roughness ($Ra$) parameter, which is the most used parameter in research as reported by Sanz et al. [50], or conveniently indicated when needed.

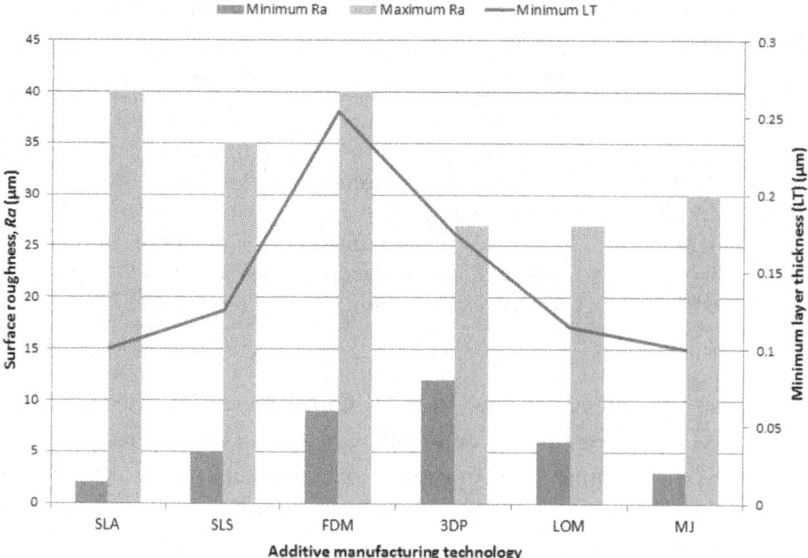


**Figure 11.3** Surface roughness for various additive manufacturing technologies along minimum layer thickness. *3DP*, 3D printing; *FDM*, fused deposition modeling; *LOM*, laminated object manufacturing; *MJ*, material jetting; *SLA*, stereolithography; *SLS*, selective laser sintering. *(Based on Campbell, R.I., Martorelli, M., and Lee, H.S. (2002). Surface roughness visualization for rapid prototyping models. Comput. Aided Des., 34, pp. 717–725; Kumbhar, N.N., and Mulay, A.V. (2018). Post processing methods used to improve surface finish of products which are manufactured by additive manufacturing technologies: a review. J. Inst. Eng.: Series C, 99(4), pp. 481–487.)*

layer thickness are presented [65]; [94]. The graph allows identifying wide ranges for surface roughness for all of these technologies and, particularly, it is possible to see how maximum values are higher than 25 µm in all cases.

Examples on the study of the surface roughness for different materials can be found in the literature (see Table 11.1). For instance, an experimental study by Pérez [61] analyzed the FDM process for manufacturing PLA samples. The authors evaluated the influence of various printing parameters such as layer height, wall thickness, printing speed, printing temperature, and printing (filling) strategy. The values obtained for surface roughness depended greatly on the selected layer height (0.15–0.25 mm) and were in the range of 10.698–26.045 µm.

Tseng et al. [66] proposed the use of a screw extrusion method to overcome the existing problems of the filament-feeding methods to be used with polyether ether ketone (PEEK). The studied system includes three main subsystems: extruder, thermal control, and traversing subsystems. The

**Table 11.1** Surface roughness measured in additively manufactured metal and polymeric parts.

| Material | Process | *Ra* (μm) |
|---|---|---|
| PLA [61] | Fused deposition modeling | From 10.698 to 26.045 |
| PEEK [66] | Extrusion | 0.945 and 3.46 ($10^{-3}$) |
| AISI 316L [62] | Selective laser melting | From 8 to 16 |
| Stainless steel 420 [67] | Binder jetting | From 6.22 to 9.01 |
| AlSi10Mg [68] | Direct metal laser sintering | From 14.35 to 24.71 |
| Inconel 625 [69] | Selective laser melting | From 2 to 7 |
| Inconel 625 [70] | Selective laser melting | From 4 to 37 |

authors recommended to use nozzle temperatures in the range of 370–390°C and a heated plate up to 280°C with an extrusion rate of 1.5 g/min. Surface roughness was evaluated in both single-layer samples of 123 μm in thickness and five-layer samples of 1.5 mm, obtaining values of 0.945 and 3.46 nm, respectively.

The study of the surface roughness obtained in the AM of metal parts is gaining attention. In this sense, Strano et al. [62] analyzed surface roughness in parts of AISI 316L generated by SLM. The authors found an important influence of the sloping angle, resulting in a minimum surface roughness for the 0 degree angle. Approximately, surface roughness varied between 8 and 16 μm. Similarly, Do et al. [67] analyzed the surface roughness of another stainless steel, 420 stainless steel, using an X1-Lab 3D printer and SS420 stainless steel powder with various boron-based additives. The samples obtained were sintered at 1150, 1200, and 1250°C. Surface roughness was measured for the sintered samples, depending their values on the used additives. The authors reported surfaces roughness of 9.01 μm for pure stainless steel, 8.2 μm for the samples with 0.5% BN, and 6.22 μm for the samples with 0.5% BC.

Aluminum pieces of AlSi10Mg alloy were manufactured by AM by Calignano et al. [68]. The process studied was direct metal laser sintering using an EOSINT M270 Xtended machine. Several parameters, such as scan speed, laser power, and hatching distance, were varied to address their influence on the process. Surface roughness values were measured in the range of 14.35–24.71 μm. The authors also proved how shot blasting could be used to reduce the surface roughness between 30% and 83% from the original value depending on the pressure used.

Li et al. [69] studied the SLM process for manufacturing Inconel 625 samples performed on a lab-made SLM facility, consisting of a fiber laser system, a CNC XYZ platform, and an inert shielding gas system. Laser power and scan speed were varied to study their influence. The authors identified how, in general, surface roughness decreases as the laser power is increased. The surface roughness range varied between approximately 2 and 7 μm. Similarly, Mumtaz and Hopkinson [70] studied SLM for producing Inconel 625 samples using a GSI Lumonics JK701H Nd:YAG laser. Samples consisted of multilayer parts of 25 mm in length and of four 0.1 mm powder layers. The authors analyzed top/side surface roughness at variable scan speeds and spot overlaps. Top surface roughness was notably lower than side surface roughness, and a clear influence of spot overlap was found. So, when the spot overlap increases, the side surface roughness increased, while the opposite occurred for the top surface roughness. The surface roughness varied approximately between 4 and 37 μm for all experiments.

### 11.2.3 Postfinishing: conventional machining

The previous results clearly indicate that there is a great potential for improving surface quality of additively manufactured pieces. Although the surface roughness can be improved by the adequate selection of the manufacturing parameters: layer thickness, part orientation, or raster angle/width [61,71,72], for the production of parts that require tight tolerances, a postprocessing is usually done to improve its microstructure and reduce porosity and roughness, so that they comply with the necessary geometric and structural tolerances [11]. For example, the pieces produced by SLM have a dimensional tolerance of 40–80 μm due to the thickness of the fusion layers, and the sintered particles stuck to the edge of the pieces. For this reason, a postfinishing by machining is necessary so that the product meets the dimensional and structural requirements of an industrial part [73].

The postprocessing procedures are time-consuming and reduce the productivity of the whole manufacturing process [74]. This is especially important when manual and laborious postprocessing procedures are used to reduce the surface roughness and improve dimensional accuracy [74,75]. Because of that, several postprocessing techniques are being tested to optimize the quality of the parts and also reduce the cost and time of the whole manufacturing process. Two approaches can be used as postprocessing techniques depending on their principle: by addition or removal of material [71].

A large number of postprocessing techniques are already available for AM. Kumbhar and Mulay [94] presented a review on postprocessing techniques, focusing on laser surface finishing techniques. But the authors

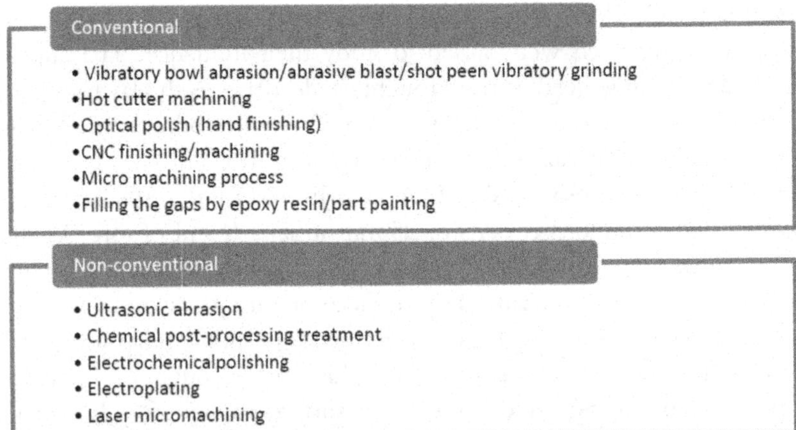


**Figure 11.4** Main types of conventional and nonconventional postprocessing processes for additive manufacturing.

have also identified other techniques, which were successfully used in AM. In addition, they have classified these techniques in conventional and nonconventional techniques and, also, by the type of material (metal, polymer and ceramic). In Fig. 11.4, the classification of conventional and nonconventional processes is shown. By means of these techniques, it is possible to notably reduce the surface roughness of the manufactured parts. For instance, examples on the use of vibratory bowl and ultrasonic abrasion, hot cutter machining, epoxy filling, and laser polishing let attain reductions in surface roughness of 74%, 87%, 83.85%, and 80.1%, respectively.

Machining is a conventional and mature process that has been used extensively for decades. The quality of the parts obtained by machining is excellent, with surface roughness notably lower than the typical values obtained in AM (previously shown in Fig. 11.3). Moreover, the evolution of surface roughness in machining can be predicted with high accuracy depending on the manufacturing parameters, knowing that it is mainly related to factors such as feed rate and tool geometry [76,77].

Conventional machining processes such as turning and milling can be used as postprocess, but also grinding can be used as a finishing and ultraprecision process. With grinding, values of surface roughness as small as 0.3 nm can be obtained [78]. Besides, it is a widespread technology, so conventional and also new modern machines are available practically in almost any workshop around the world. Thus, machining can be a suitable candidate as postprocessing technique to improve the quality of AM processes. However, it is important to take into account the specific characteristics of the manufacturing process. For instance, shrinkage can play an important role on the dimensional accuracy of the part and, so, the machining strategy should consider it [75]. Moreover, proper studies on the selection of the parameters to be used for machining additively manufactured pieces are needed because the optimum parameters could be different from those required to finish pieces obtained by other methods such as wrought materials.

## 11.3 Experimental studies on additive manufacturing and machining

The importance of surface quality, particularly surface roughness, encourages carrying out specific studies for AM. However, nowadays, the number of publications on the topic is still limited. On contrary, during decades, surface quality has been thoroughly researched for machining processes. These published studies can serve as baseline for developing similar investigations on AM to boost new applications but also to provide insights for selecting the best technology or for the integration of both technologies.

Comparative studies of these two technologies were carried out as, for instance, the one presented by Hällgren [79] in which AM (powder bed fusion) and high-speed machining were compared based on cost. As guidance, the authors suggest to accept printed quality avoiding postprocessing via machining due to the increase of costs that could make advisable to manufacture the piece just using machining. Moreover, the authors identified how the used material or shape complexity to be produced can help AM to be competitive against machining. So, specific cost analyses are needed for identifying whether machining is a more economical option against AM followed by postprocessing via machining or not. The present work is developed based on the assumption that machining can be used as a postprocessing method for improving surface roughness, without considering an economic analysis of the different solutions.

Despite the current diffusion of AM and the wide range of existing industrial applications, the technology still presents important limitations regarding the surface quality of the printed parts as it was previously stated.

This is a well-known problem for conventional plastic parts, but it is also evident when dealing with new applications for manufacturing metal parts, which are gaining importance for sectors such as aeronautics and automotive. In the past years, machining processes are being analyzed for improving surfaces generated with AM, particularly for metal parts. In the case of metals, typical values of surface roughness presented in Section 11.2 can be greatly reduced when using suitable machining processes.

In the following sections, some of the latest studies on surface roughness of both additively manufactured and postprocessed via machining parts are presented, serving as a tool for comparing results. The studies are conveniently presented in (Tables 11.2 and 11.3) for both metals and polymers, respectively. Widely used conventional machining operations such as grinding, milling, and turning are the ones analyzed.

### 11.3.1 Metals

Bagehorn et al. [80] studied the surface roughness generated by AM and compared the results with several postprocessing techniques. Particularly, Ti−6Al−4V parts were generated by direct metal laser sintering from Ti−6Al−4V powder with an averaged particle size of 44 μm (Dp50), using an EOS M270 machine. Surface roughness measured was $17.9 \pm 2.0$ μm. The plates manufactured were machined on a Deckel FP3NC milling machine, obtaining a reduction in surface roughness higher than 95%. Surface roughness was reduced to $0.3 \pm 2.0$ μm ($Ra$), with $Rz$ and $Rt$ values of $1.9 \pm 0.8$ μm and $2.9 \pm 1.5$ μm, respectively. Recently, Mostafaei et al. [81] studied the binder-jet 3D printing process (M-Flex ExOne binder jet printer) of the nickel-based 625 superalloy. The authors analyzed the surface roughness of the parts before and after grinding using a stylus profilometer. Grinding was performed using a rotary tool (Dremel 4000). The surface roughness obtained before the grinding was $1.10 \pm 0.16$ μm and after grinding $0.38 \pm 0.02$ μm.

Guo et al. [86] analyzed the surface roughness after machining of AISI 316L samples produced by direct laser deposition, using a prototype coaxial powder-feeding HP DLD machine. The authors obtained surface roughness between 0.2 and 0.4 μm after dry finish milling using a Deckel Maho DMU 70V CNC milling machine. Cutting conditions included cutting speed between 60 and 150 m/min, feed rate of 0.05 mm/tooth, and depth of cut of 0.1 mm. The authors did not provide the values of the surface roughness before machining but, regarding the postprocess, they showed how by increasing the cutting speed, the surface roughness can be reduced.

Bruschi et al.[82] analyzed the influence of several machining parameters on the surface roughness obtained after semifinishing wrought and additively manufactured samples of Ti−6Al−4V turned on a Mori Seiki CNC lathe. Cryogenic and dry machining conditions were studied along with the influence of cutting speed and feed rate. No significant influence of the Ti−6Al−4V as-delivered condition was found. Surface roughness was found to be mainly dependent on the feed rate. The additively manufactured samples were produced using electron beam melting, obtained from cylindrical billets manufactured by using an ARCAM Q10 machine. The values obtained after the machining of the additively manufactured parts were lower than $2.34 \pm 0.05$ μm, while for wrought samples $2.29 \pm 0.07$ μm.

Balachandramurthi et al. [83] produced pieces of Inconel 718 by using electron beam melting and SLM. The Arcam A2X machine with powder having a size range of 40−105 μm was used. The surface roughness of the pieces was measured in terms of various parameters; among them, the arithmetical mean height (Sa) of the surface was used. The obtained values were approximately 45 and 15 μm for electron beam melting and SLM, respectively. A grinding process was used to finish these parts to a surface roughness as low as 0.2 μm. Inconel 718 pieces were also produced by SLM by Kaynak and Tascioglu [84]. A postprocessing process by turning was carried out allowing reducing the surface roughness to a great extent: from 19 to 24 μm to approximately 0.5−1.85 μm.

Direct metal laser sintering was used by Morel et al. [85] to produce pieces of 300 maraging steel. The authors used a EOSINT M 280 machine and a Romi D600 machining center for postprocessing by milling the pieces. When analyzing the surface roughness, an important influence of the build orientation was found for the additively manufactured pieces, being the lowest values obtained for 0°. The obtained range for surface roughness was 3−11.36 μm. Milling reduced greatly surface roughness for all the conditions tested, reaching values from 0.37 to 0.6 μm.

**Table 11.2** Surface roughness measured in both additively manufactured and postprocessed by machining metal pieces.

| Material | Process | Postprocess | *Ra* (μm) | |
|---|---|---|---|---|
| | | | Before | After |
| Ti–6Al–4V [80] | Direct metal laser sintering | Milling | 17.9 ± 2.0 | 0.3 ± 2.0 |
| Ti–6Al–4V [82] | Electron beam melting | Turning | – | 0.83 ± 0.09 –2.34 ± 0.05 |
| Inconel 718 [83] | Electron beam melting<br>Selective laser melting | Grinding | Approximately 45[a]<br>Approximately 15[a] | 0.2[a] |
| Inconel 718 [84] | Selective laser melting | Turning | From 19 to 24 | From 0.5 to 1.85 |
| AISI 316L [86] | Direct laser deposition | Milling | – | From 0.2 to 0.4 |
| Maraging steel 300 [85] | Direct metal laser sintering | Milling | From 3 to 11.36 | From 0.37 to 0.6 |

[a]These values (*Sa*) are the equivalent to *Ra* for a surface.

### 11.3.2 Polymers

Boschetto et al. [35] produced parts of ABS-P400 using a Stratasys Dimension Bst 768 machine, layer thickness of 0.254 mm, solid model filling, and break-away support generation method. Several deposition angles were studied from 0 to 90 degrees. The authors studied the surface roughness obtained using the FDM process and postmachining using several processing conditions. When studying nonmachined surfaces, values of 17 and 50 μm are identified for deposition angles of 60 and 9 degrees, respectively. The parts were finished using a three-axis CNC milling. Surface roughness was notably reduced in the aforementioned cases when using a cutting depth of 0.1 mm, concretely to 2.5 and 20 μm, respectively. However, low values of cutting depth are incapable of eliminating the whole geometry generated by the FDM process. When increasing the cutting depth, the surface roughness can be reduced, but the evolution of the surface roughness depends also on the deposition angle. The range of the values is really high, approximately, from 1-2 to 35 μm. It is interesting to point out that very high peak to valley heights, as high as 250–270 μm, were found when using low depths of cut.

Huang et al. [87] studied the polymer material jetting process using a Stratasys Connex 500 machine to print using a polyjet system optical lenses. The pieces were subsequently machined by means of a precise diamond turning machine, Precitech Freeform705XG. The material used was VeroClear 810. Three printing configurations, namely, horizontal, vertical, and oblique were studied, and three different thickness were produced (10, 15, and 20 mm), obtaining values of surface roughness between 0.14 and 1.07 μm. After that, the samples were finished by diamond turning without analyzing the influence of the machining parameters. However, surface roughness was notably decreased, remaining in the range 0.14–0.32 μm.

Guo et al. [72] studied the postfinishing of selective laser-sintered polyamide 12, PA12 powders (PA2200), by precision grinding and magnetic field–assisted finishing. A EOS P395 machine with a $CO_2$ laser was used for the manufacturing process. A SMART N10 KOMBI precision grinding machine with an electroplated diamond wheel with a diameter of 300 mm and width of 8 mm was used for finishing. Though magnetic field–assisted finishing allowed reducing surface roughness to a greater extent, by grinding, surface roughness was reduced to 2.85 μm from over 15 μm.

Li et al. [88] studied a hybrid manufacturing process, including both additive and subtractive processes. Specifically, the authors manufactured PLA pieces using a FDM printing head and a milling head. The measured surface roughness varied depending on the used stratification angle (from 20 to 90 degrees). The hybrid process, the postprocessing by milling, allowed reducing the surface roughness from a range of 17.332–56.021 μm to 4.870–24.511 μm.

**Table 11.3** Surface roughness measured in both additively manufactured and postprocessed by machining polymeric pieces.

| Material | Process | Post-process | *Ra* (μm) | |
|---|---|---|---|---|
| | | | Before | After |
| ABS-P400 [35] | Fused deposition modeling | Milling | From 17 to 50 | From 1 to 2 to 35 |
| VeroClear 810 [87] | Material jetting | Turning | From 0.14 to 1.07 | From 0.14 to 0.32 |
| Polyamide 12 [72] | Selective laser sintering | Grinding | Over 15 | 2.85 |
| PLA [88] | Fused deposition modeling | Milling | From 17.332 to 56.021 | From 4.870 to 24.511 |

## 11.4 Challenges and opportunities

Surface roughness is still one of the drawbacks of AM. By recognizing the problem, the technology can be further improved by refining the manufacturing conditions. In this sense, reductions on the layer thickness used in the process and the optimization of other processing parameters would be of great importance for improving the surface roughness of the manufactured parts. In this regard, the research community is developing new experimental studies that will contribute to the development of more accurate and precise printing machines.

Despite the improvement of the technology, it is unlikely that AM could compete anytime soon with other manufacturing processes in terms of surface quality without using postprocessing. Among these processes, machining, which is a mature and widespread technology, offers excellent surface finish, being a suitable postprocess for improving surface quality of additively manufactured pieces. However, the cost of the combined process (AM plus machining) can be higher than the cost of machining alone, so it would be advisable to manufacture the part just using machining, abandoning the benefits of AM.

The advantage of machining in terms of the quality of the machined surface is one of the drivers for the development of new hybrid-multitask machines. These new machines produce complex and very close to the final shape geometries by means of AM, improving part roughness and dimensions through machining. Hybrid machines are recognized as an adequate solution for manufacturing parts of aerospace alloys, high hardness materials, manufacturing tools, and high precision parts such as medical devices [89]. This is evidenced by the increase in the number of hybrid machine tools coming to market, which merge powder bed/machining capabilities or direct energy deposition (DED)/machining capabilities [90,91]. Latest developments in hybrid machines are conveniently reviewed by Cortina et al. [92]. However, several issues that can affect the integration of both AM and machining should be considered. One is the current state of the art and technology for the machining of some materials as it was stated, which could be also solved by combining nonconventional machining processes. But it should be also considered the particularities of AM in terms of complexity. AM allows generating complex geometries that, for instance, current milling technology cannot reproduce such as shapes including undercuts or inner structures [80].

## 11.5 Conclusions

AM is gaining importance as a process for manufacturing prototypes, small batches, and final parts in serial production in a wide range of materials such as ceramics, composites, metals, and polymers. The technology is being used in advanced industrial sectors. So, the specifications required to manufacture the parts are increasingly demanding. However, issues related to the technology itself such as the staircase effect lead to poor surface quality, particularly to high surface roughness.

To improve surface roughness of additively manufactured parts, several postprocessing methods have been successfully proved in the past, among them conventional machining. The reviewed experimental investigations provide interesting results on surface roughness, proving how machining can be an adequate method to greatly improve surface roughness after AM. The results proved how the surface of both metal and polymer parts can be further improved by machining. In this sense, machining is offering opportunities for improving parts produced by AM and, in this way, increase the number of applications of the technology to the more advanced industrial sectors.

The study of surface quality of additively manufactured parts is still limited. However, the number of studies is increasing for parts produced using AM and postprocessed by various methods. More comprehensive studies are still needed, particularly, dealing with surface integrity and studying in detail, not only surface roughness, but also residual stresses, defects, hardness, and microstructure. Besides, the improvement of the technology would benefit from studies on the 3D surface topography of the surface.

## References

[1] OECD, The Next Production Revolution: Implications for Governments and Business, OECD Publishing, Paris, 2017.

[2] ASTM, ASTM F2792-12A: Standard Terminology for Additive Manufacturing Technologies, ASTM International, West Conshohocken, PA, 2012.

[3] ISO/ASTM, ISO/ASTM 52900: Additive Manufacturing. General Principles. Terminology, 2015.

[4] A. Sartal, D. Carou, R. Dorado-Vicente, L. Mandayo, Facing the challenges of the food industry: might additive manufacturing be the answer? Proc. IME B J. Eng. Manufact. 233 (8) (2018) 1902–1906.

[5] M. Ghobakhloo, The future of manufacturing industry: a strategic roadmap toward Industry 4.0, J. Manuf. Technol. Manag. 29 (6) (2018) 910–936.

[6] U.M. Dilberoglu, B. Gharehpapagh, U. Yaman, M. Dolen, The role of additive manufacturing in the era of Industry 4.0, Proc. Manufact. 11 (2017) 545–554.

[7] H. Lasi, P. Fettke, H.-G. Kemper, T. Feld, M. Hoffmann, Industry 4.0, Business Informat. Syst. Eng. 6 (4) (2014) 239–242.

[8] G.N. Levy, R. Schindel, J.P. Kruth, Rapid manufacturing and rapid tooling with layer manufacturing (lm) technologies, state of the art and future perspectives, CIRP Ann. 52 (2) (2003) 589–609.

[9] R. Noorani, 3D Printing. Technology, Applications and Selection, CRC Press, Boca Ratón, Florida (US), 2018.

[10] F. Cabanettes, A. Joubert, G. Chardon, V. Dumas, J. Rech, C. Grosjean, Z. Dimkovski, Topography of as built surfaces generated in metal additive manufacturing: a multi scale analysis from form to roughness, Precis. Eng. 52 (2018) 249–265.

[11] D. Bourell, J.P. Kruth, M. Leu, G. Levy, D. Rosen, A.M. Beese, A. Clare, Materials for additive manufacturing, CIRP Ann. Manuf. Technol. 66 (2017) 659–681.

[12] O.F. Beyca, G. Hancerliogullari, I. Yazici, Additive manufacturing technologies and applications (chapter 13), in: A. Ustundag, E. Cevikcan (Eds.), Industry 4.0: Managing the Digital Transformation, Springer, Switzerland, 2018.

[13] General Electric (last accessed September 2018), https://www.ge.com/reports/epiphany-disruption-ge-additive-chief-explains-3d-printing-will-upend-manufacturing/, 2018.

[14] S. Singh, S. Ramakrishna, Biomedical applications of additive manufacturing: present and future, Curr. Opin. Biomed. Eng. 2 (2017) 105–115.

[15] OECD, OECD Science, Technology and Innovation Outlook 2016: Megatrends Affecting Science, Technology and Innovation, 2016.

[16] J. Choi, O.-C. Kwon, W. Jo, H.J. Lee, M.-W. Moon, 4D printing technology: a review, 3D Print. Addit. Manuf. 2 (2015) 159–167.

[17] F. Momeni, S.M.M. Hassani, X. Liu, J. Ni, A review of 4D printing, Mater. Des. 122 (2017) 42–79.

[18] M. Mahesh, Y.S. Wong, J.Y.H. Fuh, H.T. Loh, Benchmarking for comparative evaluation of RP systems and processes, Rapid Prototyp. J. 10 (2004) 123–135.

[19] P.G. Benardos, G.-C. Vosniakos, Predicting surface roughness in machining: a review, Int. J. Mach. Tool Manufact. 43 (2003) 833–844.

[20] J. Pegues, M. Roach, R.S. Williamson, N. Shamsaei, Surface roughness effects on the fatigue strength of additively manufactured Ti-6Al-4V, Int. J. Fatig. 116 (2018) 543–552.

[21] F. Bartolomeu, M.M. Costa, J.R. Gomes, N. Alves, C.S. Abreu, F.S. Silva, G. Miranda, Implant surface design for improved implant stability – a study on Ti–6Al–4V dense and cellular structures produced by Selective Laser Melting, Tribol. Int. 129 (2019) 272–282.

[22] J.C. Dinis, T.F. Moraes, P.H.J. Amorim, M.R. Moreno, A.A. Nunes, J.V.L. Silva, POMES: an open-source software tool to generate porous/roughness on surfaces, Proc. CIRP 49 (2016) 178–182.

[23] D. Herzog, V. Seyda, E. Wycisk, C. Emmelmann, Additive manufacturing of metals, Acta Mater. 117 (2016) 371–392.

[24] L.S. Bertol, W.K. Júnior, F.P. da Silva, C. Aumund-Kopp, Medical design: direct metal laser sintering of Ti–6Al–4V, Mater. Des. 31 (2010) 3982–3988.

[25] L.E. Murr, S.M. Gaytan, A. Ceylan, E. Martinez, J.L. Martínez, D.H. Hernández, B.I. Machado, D.A. Ramirez, F. Medina, S. Collins, Characterization of titanium aluminide alloy components fabricated by additive manufacturing using electron beam melting, Acta Mater. 58 (2010) 1887–1894.

[26] E.O. Olakanmi, R.F. Cochrane, K.W. Dalgarno, A review on selective laser sintering/melting (SLS/SLM) of aluminium alloy powders: processing, microstructure, and properties, Prog. Mater. Sci. 74 (2015) 401–477.
[27] A. Basak, S. Das, Microstructure of nickel-base superalloy MAR-M247 additively manufactured through scanning laser epitaxy (SLE), J. Alloys Compd. 705 (2017) 806–816.
[28] A. Bandyopadhyay, B. Heer, Additive manufacturing of multi-material structures, Mater. Sci. Eng. R 129 (2018) 1–16.
[29] S.A.M. Tofail, E.P. Koumoulos, A. Bandyopadhyay, S. Bose, L. O'Donoghue, C. Charitidis, Additive manufacturing: scientific and technological challenges, market uptake and opportunities, Mater. Today 21 (1) (2018) 22–37.
[30] L. Jin, K. Zhang, T. Xu, T. Zeng, S. Cheng, The fabrication and mechanical properties of SiC/SiC composites prepared by SLS combined with PIP, Ceram. Int. 44 (2018) 20992–20999.
[31] H. Bikas, P. Stavropoulos, G. Chryssolouris, Additive manufacturing methods and modelling approaches: a critical review, Int. J. Adv. Manuf. Technol. 83 (2016) 389–405.
[32] J.-Y. Lee, J. An, C.K. Chua, Fundamentals and applications of 3D printing for novel materials, Appl. Mater. Today 7 (2017) 120–133.
[33] D. Rietzel, M. Friedrich, T.A. Osswald, Additive manufacturing, in: E.T.A. Osswald (Ed.), Understanding Polymer Processing (147-169), Editorial: Hanser, 2017.
[34] ISO, ISO 10303: Automation Systems and Integration — Product Data Representation and Exchange, 2014.
[35] A. Boschetto, L. Bottini, F. Veniali, Finishing of fused deposition modeling parts by CNC machining, Robot. Comput. Integrated Manuf. 41 (2016) 92–101.
[36] M. Taufik, P.K. Jain, Role of build orientation in layered manufacturing: a review, Int. J. Manuf. Technol. Manag. 27 (1/2/3) (2013) 47–73.
[37] A. Haridas, M.V. Matham, A. Crivoi, P. Patinharekandy, T.M. Jen, K. Chan, Surface roughness evaluation of additive manufactured metallic components from white light images captured using a flexible fiberscope, Optic Laser. Eng. 110 (2018) 262–271.
[38] S.S. Sarnobat, H.K. Raval, Experimental investigation and analysis of the influence of tool edge geometry and work piece hardness on surface residual stresses, surface roughness and work-hardening in hard turning of AISI D2 steel, Measurement 131 (2019) 235–260.
[39] P.M. Lonardo, D.A. Lucca, L.De Chiffre, Emerging trends in surface metrology, CIRP Ann. Manuf. Technol. 51 (2) (2002) 701–723.
[40] T.V. Vorburger, J.A. Dagata, G. Wilkening, K. Lizuka, G. Thwaite, P. Lonardo, Industrial uses of STM and AFM, Ann. CIRP 46 (2) (1997) 597–620.
[41] R.J. Hocken, N. Chakraborty, C. Brown, Optical metrology of surfaces, CIRP Ann. Manuf. Technol. 54 (2) (2005) 169–183.
[42] G.G. Lakić, D. Kramar, J. Kopač, Metal Cutting. Theory and Applications. Editorial Universidad de Banja Luka, Facultad de Ingeniería Mecánica, 2014.
[43] Q. Qi, X. Jiang, X. Liu, P.J. Scott, An unambiguous expression method of the surface texture, Measurement 43 (2010) 1398–1403.
[44] ISO, ISO 4287: Geometrical Product Specification (GPS) Surface Texture: Profile Method; Terms, Definitions and Surface Texture Parameter, 1997.
[45] ISO, ISO 25178-6: Geometrical Product Specifications (GPS) – Surface Texture: Areal - Part 6: Classification of Methods for Measuring Surface Texture, ISO, 2010.
[46] Q. Qi, T. Li, P.J. Scott, X. Jiang, A correlational study of areal surface texture parameters on some typical machined surfaces, Proc. CIRP 27 (2015) 149–154.
[47] N.K. Myshkin, A.Y. Grigoriev, S.A. Chizhik, K.Y. Choi, M.I. Petrokovets, Surface roughness and texture analysis in microscale, Wear 254 (2003) 1001–1009.

[48] L.D. Todhunter, R.K. Leach, S.D.A. Lawes, F. Blateyron, Industrial survey of ISO surface texture parameters, CIRP J. Manufact. Sci. Technol. 19 (2017) 84–92.

[49] S.H. Bui, T.B. Renegar, T.V. Vorburger, J. Raja, M.C. Malburg, Internet-based surface metrology algorithm testing system, Wear 257 (2004) 1213–1218.

[50] A. Sanz, A.A. Negre, R. Fernández, F. Calvo, Comparative study about the use of two and three-dimensional methods in surface finishing characterization, Proc. Eng. 63 (2013) 913–921.

[51] P.P. Company, M. Vergara, S. Mondragón, Dibujo Industrial, Editorial Universidad de Jaume I, 2007.

[52] A. Townsend, N. Senin, L. Blunt, R.K. Leach, J.S. Taylor, Surface texture metrology for metal additive manufacturing: a review, Precis. Eng. 46 (2016) 34–47.

[53] D. Ahn, J.-H. Kweon, S. Kwon, J. Song, S. Lee, Representation of surface roughness in fused deposition modeling, J. Mater. Process. Technol. 209 (2009a) 5593–5600.

[54] B.K. Paul, V. Voorakarnam, Effect of layer thickness and orientation angle on surface roughness in laminated object manufacturing, J. Manuf. Process. 3 (2001) 94–101.

[55] C.J. Luis-Pérez, J. Vivancos-Calvet, M.A. Sebastián-Pérez, Geometric roughness analysis in solid free-form manufacturing processes, J. Mater. Process. Technol. 119 (2001) 52–57.

[56] P.M. Pandey, N. Venkata Reddy, S.G. Dhande, Improvement of surface finish by staircase machining in fused deposition modeling, J. Mater. Process. Technol. 132 (2003) 323–331.

[57] D. Ahn, H. Kim, S. Lee, Surface roughness prediction using measured data and interpolation in layered manufacturing, J. Mater. Process. Technol. 209 (2009b) 664–671.

[58] A. Kantaros, D. Karalekas, Fiber Bragg grating based investigation of residual strains in ABS parts fabricated by fused deposition modeling process, Mater. Des. 50 (2013) 44–50.

[59] A. Boschetto, L. Bottini, Roughness prediction in coupled operations of fused deposition modeling and barrel finishing, J. Mater. Process. Technol. 219 (2015) 181–192.

[60] W. Oropallo, L.A. Piegl, Ten challenges in 3D printing, Eng. Comput. 32 (2016) 135–148.

[61] M. Pérez, G. Medina-Sánchez, A. García-Collado, M. Gupta, D. Carou, Surface quality enhancement of fused deposition modeling (FDM) printed samples based on the selection of critical printing parameters, Materials 11 (2018) 1382.

[62] G. Strano, L. Hao, R.M. Everson, K.E. Evans, Surface roughness analysis, modelling and prediction in selective laser melting, J. Mater. Process. Technol. 213 (2013) 589–597.

[63] C.K. Chua, K.F. Leong, C.S. Lim, Rapid Prototyping: Principles and Applications, third ed., World Scientific, River Edge, 2010.

[64] S. Singh, S. Ramakrishna, R. Singh, Material issues in additive manufacturing: a review, J. Manuf. Process. 25 (2017) 185–200.

[65] R.I. Campbell, M. Martorelli, H.S. Lee, Surface roughness visualization for rapid prototyping models, Comput. Aided Des. 34 (2002) 717–725.

[66] J.-W. Tseng, C.-Y. Liu, Y.-K. Yen, J. Belkner, T. Bremicker, B.H. Liu, T.-J. Sun, A.-B. Wang, Screw extrusion-based additive manufacturing of PEEK, Mater. Des. 140 (2018) 209–221.

[67] T. Do, C.S. Shin, D. Stetsko, G. VanConant, A. Vartanian, S. Pei, P. Kwon, Improving structural integrity with boron-based additives for 3D printed 420 stainless steel, Proc. Manufact. 1 (2015) 263–272.

[68] F. Calignano, D. Manfredi, E.P. Ambrosio, L. Iuliano, P. Fino, Influence of process parameters on surface roughness of aluminum parts produced by DMLS, Int. J. Adv. Manuf. Technol. 67 (2013) 2743–2751.

[69] C. Li, Y.B. Guo, J.B. Zhao, Interfacial phenomena and characteristics between the deposited material and substrate in selective laser melting Inconel 625, J. Mater. Process. Technol. 243 (2017) 269–281.
[70] K. Mumtaz, N. Hopkinson, Top surface and side roughness of Inconel 625 parts processed using selective laser melting, Rapid Prototyp. J. 15 (2) (2009) 96–103.
[71] M. Adel, O. Abdelaal, A. Gad, A.B. Nasr, A.M. Khalil, Polishing of fused deposition modeling products by hot air jet: evaluation of surface roughness, J. Mater. Process. Technol. 251 (2018) 73–82.
[72] J. Guo, J. Bai, K. Liu, J. Wei, Surface quality improvement of selective laser sintered polyamide 12 by precision grinding and magnetic field-assisted finishing, Mater. Des. 138 (2018) 39–45.
[73] G.L. Coz, M. Fischer, R. Piquard, A. D'Acunto, P. Laheurte, D. Dudzinski, Micro cutting of Ti-6Al-4V parts produced by SLM Process, Proc. CIRP 58 (2017) 228–232.
[74] W.P. Syam, R. Leach, K. Rybalcenko, A. Gaio, J. Crabtree, In-process measurement of the surface quality for a novel finishing process for polymer additive manufacturing, Proc. CIRP 75 (2018) 108–113.
[75] M. Vispute, N. Kumar, P.K. Jain, P. Tandon, P.M. Pandey, Shrinkage compensation study for performing machining on additive manufactured parts, Mater. Today: Proc. 5 (2018) 18544–18551.
[76] G. Boothroyd, W.A. Knight, Fundamentals of Machining and Machine Tools, third ed., CRC Press, Taylor & Francis Group, 2006.
[77] D. Carou, E.M. Rubio, C.H. Lauro, J.P. Davim, Experimental investigation on surface finish during intermittent turning of UNS M11917 magnesium alloy under dry and near dry machining conditions, Measurement 56 (2014) 136–154.
[78] K. Wegener, F. Bleicher, P. Krajnik, H.-W. Hoffmeister, C. Brecher, Recent developments in grinding machines, CIRP Ann. - Manuf. Technol. 66 (2017) 779–802.
[79] S. Hällgren, L. Pejryd, J. Ekengren, Additive Manufacturing and High Speed Machining -Cost comparison of short lead time manufacturing methods, Proc. CIRP 50 (2016) 384–389.
[80] S. Bagehorn, J. Wehr, H.J. Maier, Application of mechanical surface finishing processes for roughness reduction and fatigue improvement of additively manufactured Ti-6Al-4V parts, Int. J. Fatig. 102 (2017) 135–142.
[81] A. Mostafaei, S.H.V.R. Neelapu, C. Kisailus, L.M. Nath, T.D.B. Jacobs, M. Chmielus, Characterizing surface finish and fatigue behavior in binder-jet 3D-printed nickel-based superalloy 625, Addit. Manufact. 24 (2018) 200–209.
[82] S. Bruschi, R. Bertolini, A. Bordin, F. Medea, A. Ghiotti, Influence of the machining parameters and cooling strategies on the wear behavior of wrought and additive manufactured Ti–6Al–4V for biomedical applications, Tribol. Int. 102 (2016) 133–142.
[83] A.R. Balachandramurthi, J. Moverare, N. Dixit, R. Pederson, Influence of defects and as-built surface roughness on fatigue properties of additively manufactured Alloy 718, Mater. Sci. Eng. 735 (2018) 463–474.
[84] Y. Kaynak, E. Tascioglu, Finish machining-induced surface roughness, microhardness and XRD analysis of selective laser melted Inconel 718 alloy, Proc. CIRP 71 (2018) 500–504.
[85] C. Morel, V.V. Cioca, S. Lavernhe, A. Jardini, E.G. Conte, Part surface roughness on laser sintering and milling of maraging steel 300, in: 14th International Conference on High Speed Manufacturing, Apr 2018, San-Sebastian, Spain. 14th International Conference on High Speed Manufacturing, 2018.
[86] P. Guo, B. Zou, C. Huang, H. Gao, Study on microstructure, mechanical properties and machinability of efficiently additive manufactured AISI 316L stainless steel by high-power direct laser deposition, J. Mater. Process. Technol. 240 (2017) 12–22.

[87] Y. Huang, C.-M. Chang, C.-F. Ho, T.-W. Lee, P.-H. Lin, W.-Y. Hsu, The research on surface characteristics of optical lens by 3D printing technique and precise diamond turning technique, in: Proc. SPIE 10449, Fifth International Conference on Optical and Photonics Engineering, 2017, pp. 1–6, 1044910.
[88] L. Li, A. Haghighi, Y. Yang, Theoretical modelling and prediction of surface roughness for hybrid additive–subtractive manufacturing processes, IISE Trans. 51 (2) (2019) 124–135, https://doi.org/10.1080/24725854.2018.1458268.
[89] T. Yamazaki, Development of A Hybrid multi-tasking machine tool:integration of additive manufacturing technology with CNC machining, Proc. CIRP 42 (2016) 81–86.
[90] J.M. Flynn, A. Shokrani, S.T. Newman, V. Dhokia, Hybrid additive and subtractive machine tools –Research and industrial developments, Int. J. Mach. Tool Manufact. 101 (2016) 79–101.
[91] O. Oyelola, P. Crawforth, R. M'Saoubi, A.T. Clare, On the machinability of directed energy deposited Ti–6Al–4V, Addit. Manufact. 19 (2018) 39–50.
[92] M. Cortina, J.I. Arrizubieta, J.E. Ruiz, E. Ukar, A. Lamikiz, Latest developments in industrial Hybrid Machine Tools that combine additive and subtractive operations, Materials 11 (12) (2018) 2583.
[93] A. Mitchell, U. Lafont, M. Hołyńska, C. Semprimoschnig, Additive manufacturing — a review of 4D printing and future applications, Addit. Manufact. 24 (2018) 606–626.
[94] N.N. Kumbhar, A.V. Mulay, Post processing methods used to improve surface finish of products which are manufactured by additive manufacturing technologies: a review, J. Inst. Eng.: Series C 99 (4) (2018) 481–487.

CHAPTER 12

# Standards for additive manufacturing technologies: structure and impact

**Asunción Martínez-García[1], Mario Monzón[2], Rubén Paz[2]**
[1]Innovative Materials and Manufacturing Area, AIJU, Ibi (Alicante), Spain; [2]Departamento de Ingeniería Mecánica, ULPGC, Las Palmas de Gran Canaria, Spain

## 12.1 Introduction

The International Organization for Standardization (ISO) is an independent, non-governmental international organization with a membership of 164 national standard bodies [1]. It currently has 780 technical committees and subcommittees (SCs) devoted to the development of standards and related documents for almost all industrial sectors. According to the ISO, the foremost aim of international standardization is to facilitate the exchange of goods and services through the elimination of technical barriers to trade.

To ensure that global work is carried out with high quality in the additive manufacturing (AM) processes and be able to offer reliable parts while innovating in the market among other values, the ISO/TC 261 Additive Manufacturing standardization committee was created in 2011. It was created with the scope of standardization in the field of AM concerning their processes, terms and definitions, process chains (Hard- and Software), test procedures, quality parameters, supply agreements, and all kinds of fundamentals [2].

According to the definition given in ISO/ASTM 52900:2017 standard [3], AM is the general term for those technologies that, based on a geometrical representation, create physical objects by successive addition of material.

AM processes are therefore characterized by the direct manufacture of parts from 3D CAD data without the need to use molds or tools to produce series of parts, with advantages of cost and time savings.

The momentum of AM in recent years has been caused by various factors, such as the improvement in existing technologies and the development of new AM equipment, the greater experience of users in the

*Additive Manufacturing*
ISBN 978-0-12-818411-0
https://doi.org/10.1016/B978-0-12-818411-0.00013-6

management of technologies, the expansion of applications in which AM is being used, the development of new AM-specific design systems, and so on. All this increasingly leads to a trend of using AM technologies as final part manufacturing processes and not only to obtain mere prototypes or tooling [4].

In these cases, the parts need some requirements and certifications not provided so far by suppliers and customers. In addition, the materials must meet precise specifications, and the processes must be perfectly controlled to ensure the reproducibility of the parts and achieve adequate productivity to be able to consider these technologies at the height of other manufacturing processes.

Although standards as such are voluntary, using them proves that your products and services reach a certain level of quality, safety, and reliability [5]. The development of industry standards is a requirement to adopt AM for the production [6] because it is difficult to compete with traditional techniques. This is where a set of standards can help guarantee a level of reproducibility and give business and manufacturers the needed assurance that AM processes, materials, and technologies are safe and reliable.

According to the Business Plan developed by the ISO/TC 261 committee [7], thanks to the standardization, companies and users can reach or expect a set of benefits through availability of the standards. These benefits comprise the following:

- systematic development, modification, and use of processes of joining materials from 3D model data (AM) resulting in innovative products;
- assistance to users within the assessment of different additive processes resulting in using the appropriate technology for the specified product demands;
- specification of quality parameters of different processes needed for standardized test procedures;
- specification of appropriate test procedures, thereby ensuring uniform interpretation and evaluation of quality parameters;
- standardization of process chains of AM technologies securing functionality and compatibility;
- standardization of data formats, data structures, and metrics for AM models; and
- standardization of vocabulary required to define the product and to find a common speech.

In September 2011, ISO and ASTM signed a cooperative agreement to govern the ongoing collaborative efforts between the two organizations to adopt and jointly develop international standards that serve the global marketplace in the field of AM. The purpose of this agreement is to eliminate duplication of effort while maximizing resource allocation within the AM industry.

ISO and ASTM International have jointly crafted the Additive Manufacturing Standards Development Structure [8], a framework that will help meet the needs for new technical standards in this fast-growing field.

To develop the standards, a new structure was proposed, as commented in the following section, for among other reasons, helping the improvement of usability and acceptance among the AM community, including manufacturers, entrepreneurs, consumers, and so on. According to this structure, all the standards are classified under three levels, depending on the specificity of the document: feedstock materials, process/equipment, and finished parts.

In addition, to achieve that European countries adopt the international standards and to ensure consistency and harmonization with ISO, the Technical committee TC438—Additive Manufacturing of the European Committee for Standardization (CEN) was created in 2015 [9]. By applying the Vienna agreement with ISO/TC 261, CEN publishes the ISO standards as EN ISO, ensuring its alignment with European regulations. Therefore, the corresponding national standardization bodies will adopt all European standards through identical national standards, withdrawing any national standards that conflict with them.

## 12.2 Structure of additive manufacturing standardization working groups

### 12.2.1 ISO committee

The Technical Committee ISO/TC 261 Additive Manufacturing Technologies was established in 2011. It has currently 25 participating members that have influence on ISO standards development and strategy by participating and voting in ISO technical and policy meetings. The committee has also eight observer countries, with no voting participation but who act as observers of the development of ISO standards and strategy by attending ISO meetings.

Table 12.1 includes the participating and observer members and their corresponding national standardization body, which represents the interests of companies and society of their countries in European and international

**Table 12.1** Participating and observer countries for ISO/TC 261.

| Participating P-members | | Observer O-members |
|---|---|---|
| Australia (SA) | Jordan (JSMO) | Austria (ASI) |
| Belgium (NBN) | Korea, Republic of (KATS) | Czech Republic (UNMZ) |
| Brazil (ABNT) | Netherlands (NEN) | Iran, Islamic Republic of (ISIRI) |
| Canada (SCC) | Norway (SN) | New Zealand (NZSO) |
| China (SAC) | Poland (PKN) | Portugal (IPQ) |
| Denmark (DS) | Russian Federation (GOST R) | Romania (ASRO) |
| Finland (SFS) | Singapore (SSC) | South Africa (SABS) |
| France (AFNOR) | Spain (UNE) | Turkey (TSE) |
| Germany (DIN) | Sweden (SIS) | |
| Ireland (NSAI) | Switzerland (SNV) | |
| Israel (SII) | United Kingdom (BSI) | |
| Italy (UNI) | United States (ANSI) | |
| Japan (JISC) | | |

Reproduced with kind permission of ISO from Reference ISO/TC 261Chairman statement, https://committee.iso.org/home/tc261, (accessed 02.03.2020).

standardization organizations. Specifically, the German Institute for Standardization, DIN, is currently chairing the secretariat of the ISO/TC 261 since its creation.

## 12.2.2 Working groups

The ISO/TC 261 committee is structured in working groups (WGs) with the objective of developing standards and other technical documents related to certain topics. The members of the WG are experts who have been nominated by their national standards body and should be aware of their national point of view; however, they act in a personal capacity [10].

ISO/TC 261 is divided into seven technical WGs working on the following topics:

- ISO/TC 261/WG 1 Terminology
- ISO/TC 261/WG 2 Processes, systems and materials
- ISO/TC 261/WG 3 Test methods and quality specifications
- ISO/TC 261/WG 4 Data and design
- ISO/TC 261/JWG 5 Joint ISO/TC 261−ISO/TC 44/SC 14 WG: Additive manufacturing in aerospace applications
- ISO/TC 261/WG 6 Environment, health and safety
- ISO/TC 261/JWG 11 Joint ISO/TC 261−ISO/TC 61/SC 9 WG: Additive manufacturing for plastics

### 12.2.3 Joint groups

As indicated before, ISO/TC 261 and ASTM F42 developed an ASTM/ISO Partner Standards Developing Organization (PSDO) cooperative agreement to collaborate in a fruitful way to create AM-related standards maximizing the effort of the experts and avoiding risks of duplication or competing standards.

Both organizations have nominated dedicated experts for joint standards development through the creation of joint groups (JGs), where about three to five members from each organization work together. The JG will ensure the visibility of its work to the corresponding internal WG structure of ISO/TC 261 and the corresponding internal sub-subcommittee structure of ASTM F42.

The WGs, in turn, have to ensure that the information flows between the experts of both organizations, ISO/TC 261 and ASTM F 42, coordinating the work of the ISO/TC 261 experts in the JGs, monitoring of progress, efficiency, and effectiveness of the work of the JGs from ISO point of view and developing proposals for further joint ISO/ASTM activities and possible JGs.

The last proposed structure for the ISO/TC 261 Committee is presented in Fig. 12.1 [11]. As it can be appreciated, there are a steering group on JG activities (JAG) and a chairman's advisory group (CAG) organizing and coordinating these WG. The main role of the CAG Committee is to oversee the WG scopes, to appoint the JG Convenor; the allocation of standards work, the allocation of each ISO/ASTM JG to a WG of ISO/TC 261, to define timeliness of standards production; the approval and maintenance of existing standards; reviewing the need for new standard projects; and maintenance of liaisons. This group comprises a Chair and the WG Convenor of ISO/TC 261.

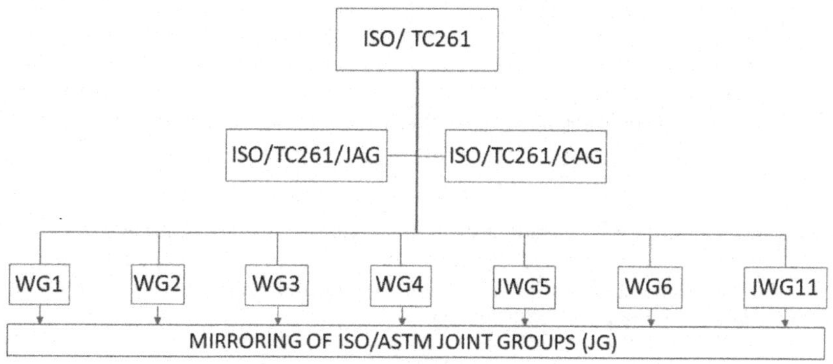


**Figure 12.1** Structure of the ISO/TC 261 committee.

On the other hand, the Joint Steering Group (JAG) reviews progress reports on the JG activities, manages JGs, resolves any problems, presents proposals for joint activities to the respective plenary groups, and maintains a 3-year plan for joint standards development. The ISO/ASTM Steering may also decide upon the provisional nomination of ISO experts to JGs (final nomination to be performed by ISO/TC 261).

Regarding the proposals for joint activities, during the committee meetings, the experts propose (depending on the interest of the market, stakeholders, new regulations, etc.) a series of initiatives to develop, which, depending on the subject, are placed under the responsibility of the most appropriate WG. As it can be seen in Fig. 12.2, 25 JGs have been constituted for the developments of specific standards in a cooperative

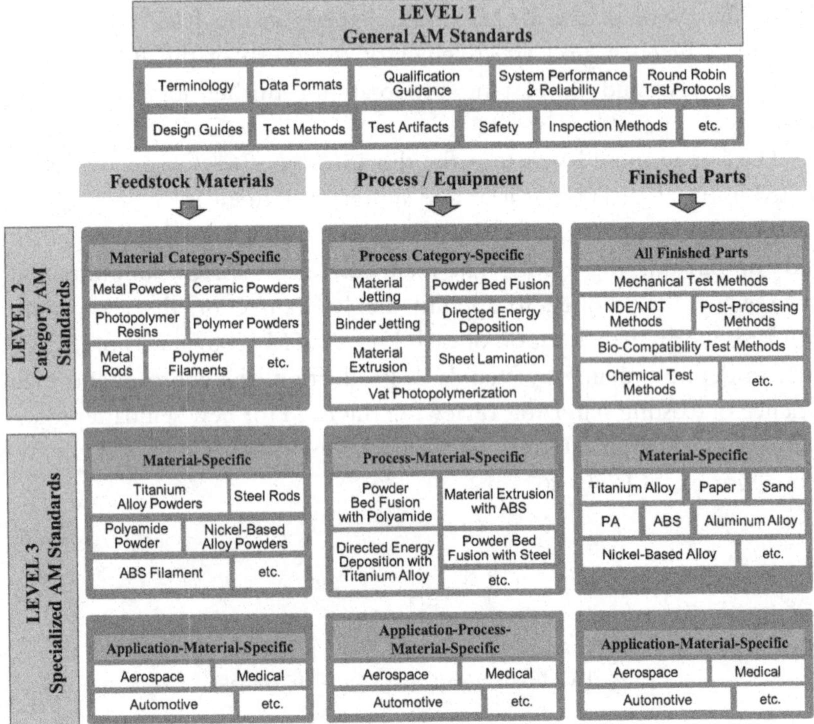


**Figure 12.2** Structure of the standards on AM developed by ISO/TC261-ASTM F42. *AM*, additive manufacturing. *(Reproduced with kind permission of ISO from Reference ISO/ASTM Additive Manufacturing Standards Structure, https://committee.iso.org/sites/tc261/home/projects.htm, (accessed 02.03.2020).)*

way. Although it takes a long period of time, as soon as the standard is finished, voted and approved according to the Committee procedure, the JG stops its activity, and the experts can focus on new projects. Consequently, the standards developed by these JGs have the terminology of both organizations, i.e., ISO/ASTM 52901—Additive Manufacturing—General principles—Requirements for Purchased AM Parts (see Table 12.4).

Table 12.2 includes some examples of JGs that are currently working in standards and the topic in which they work.

**Table 12.2** Examples of active joint groups ISO/TC 261-ASTM F 42 and the topic.

| Joint ISO/261-ASTM F 42 group | References |
|---|---|
| JG 51 | Terminology |
| JG 55 | Standard specification for extrusion-based additive manufacturing of plastic materials |
| JG 57 | Process-specific design guidelines and standards |
| JG 58 | Qualification, quality assurance, and postprocessing of powder bed fusion metallic parts |
| JG 59 | NDT (nondestructive tests) for AM parts |
| JG 64 | Additive manufacturing file format (AMF) |
| JG 67 | Technical report for the design of functionally graded additive manufactured parts |
| JG 69 | EH&S for use of metallic materials |
| JG 71 | Powder quality assurance |
| JWG 10 | Additive manufacturing in aerospace applications |
| JWG 11 | Additive manufacturing for plastics |

Reproduced with kind permission of ISO.

**Table 12.3** Examples of liaisons of ISO TC 261 with different TC.

| References | Title of the committee | ISO/IEC |
|---|---|---|
| IEC/TC 76 | Optical radiation safety and laser equipment | IEC |
| ISO/IEC JTC 1 | Information technology | ISO/IEC |
| ISO/TC 24/SC 4 | Particle characterization | ISO |
| ISO/TC 61/SC 9 | Thermoplastic materials | ISO |
| ISO/TC 119 | Powder metallurgy | ISO |
| ISO/TC 135 | Nondestructive testing | ISO |
| ISO/TC 150 | Implants for surgery | ISO |
| ISO/TC 150/SC 1 | Materials | ISO |

Reproduced with kind permission of ISO.

**Table 12.4** Some published AM standards.

| |
|---|
| ISO 17296-2:2015 |
| Additive manufacturing—General principles—part 2: overview of process categories and feedstock |
| ISO 17296-3:2014 |
| Additive manufacturing—General principles—part 3: main characteristics and corresponding methods |
| ISO/ASTM 52901:2017 |
| Additive manufacturing—General principles—requirements for purchased AM parts |
| ISO/ASTM 52904:2010 |
| Additive manufacturing—process characteristics and performance—practice for metal powder bed fusion process to meet critical applications |
| ISO/ASTM 52907:2019 |
| Additive manufacturing—feedstock materials—methods to characterize metal powders |
| ISO/ASTM 52910:2018 |
| Additive manufacturing—design—requirements, guidelines and recommendations |
| ISO/ASTM 52911–1:2019 |
| Additive manufacturing—design—part 1: Laser-based powder bed fusion of metals |
| ISO/ASTM 52911–2:2019 |
| Additive manufacturing—design—part 2: Laser-based powder bed fusion of polymers |
| ISO/ASTM 52915:2016 |
| Specification for additive manufacturing file format (AMF) version 1.2 |

Reproduced with kind permission of ISO.

### 12.2.4 Liaisons with other groups

Apart from working with ASTM F42, ISO TC 261 has liaison agreements with other organizations that are related to the AM field, such as CECIMO (European Association of the Machine Tool Industries), EPMA (European Powder Metallurgy Association), and EWF (European Federation for Welding, Joining and Cutting). These organizations can provide input on the standard development, focusing on the impact and alignment with the needs of their associated companies that are working in the AM sector.

Additionally, other standardization committees (ISO, IEC, and related organizations) have reached liaison agreements with ISO/TC 261 allowing the access to the developed documents, taking both parts advantage of the preliminary results of the ongoing standardization activities and the knowledge of experts of complementary fields, and avoiding duplication of efforts, among others.

Some examples of current liaisons are presented in Table 12.3.

### 12.2.5 CEN committee

In 2015 the committee CEN/TC 438—Additive Manufacturing was created. The relevance of this CEN/TC is that those approved standards are automatically approved in all the EU countries, removing the local/national standards of AM.

The main objectives of CEN/TC 438 are as follows [9]:

- To provide a complete set of European standards on processes, test procedures, quality parameters, supply agreements, fundamentals and vocabulary based, as far as possible, on international standardization work. The aim is to apply the Vienna Agreement with ISO/TC 261 "Additive Manufacturing" to ensure consistency and harmonization.
- To strengthen the link between European Research programs and standardization in additive manufacturing.
- To ensure visibility to the European standardization in additive manufacturing by centralizing standardization initiatives in Europe on additive manufacturing.

Currently, there are 3 EN ISO and 9 EN ISO/ASTM published standards and 27 projects in progress. CEN/TC 438 will develop new projects that are related to aeronautic, medical, 3D manufacturing and data protection.

## 12.3 Published AM standards in ISO/ASTM

As indicated before, the standards could help the industry to improve the quality and reliability of the manufactured parts and AM processes. The extended use of these documents contributes to enlarge the common knowledge of users in this field by adopting the same terminology associated with AM (ISO/ASTM 52900), by following specific AM guidelines for designing (ISO/ASTM 52910) that could lead to better design opportunities both in metal or plastic parts, by clarifying the communication between AM providers and clients (ISO/ASTM 529019) to define the most important requirements, or by manufacturing and/or testing in a harmonized way regardless of the machine, material batch, user, etc. Depending on the evolution of technologies, there will be standards, as the terminology one, which will be continuously reviewed and updated.

Different sectors such as automotive, aerospace, medicine and consumer products are implementing the use of the applicable AM standards in different steps of their product development and manufacturing cycles [2,12].

Currently, there are 14 published ISO standards and 23 ISO standards under development, under the direct responsibility of ISO/TC 261. Some of them are included in Table 12.4.

How these standards are organized was decided by ISO and ASTM International proposing an Additive Manufacturing Standards Development Structure [11], to meet the needs for new technical standards in this fast-growing field.

The new structure, shown in Fig. 12.2, will help to

- guide the work of global experts and standard development organizations involved in AM standardization;
- identify standard-related gaps and needs in the AM industry;
- prevent overlap and duplicative efforts in AM standard development;
- ensure cohesion among AM standards;
- prioritize AM standard areas; and
- improve usability and acceptance among the AM community, including manufacturers, entrepreneurs, consumers, and others.

Based on this structure, all standards are classified under three levels and three subgroups: feedstock materials, process/equipment, and finished parts.

The first or top level, General standards, includes general concepts, common requirements, guidelines, and so on, and the standards are generally applicable. The second level is category AM standards, related to material category or process category. And the third level is for specialized AM standards, specific to material (e.g., aluminum alloy powders), process (e.g., material extrusion with ABS), or application (e.g., aerospace, medical).

Some examples to understand how the developed standards fit with such a structure are the following:

- Level 1: ISO/ASTM 52900:2015. Additive manufacturing—General principles—Terminology
- Level 2: ISO/ASTM 52911-1:2019 Additive manufacturing—Design—Part 1: Laser-based powder bed fusion of metals
- Level 3: ISO/ASTM 52942:2020 Additive manufacturing—Qualification principles—Qualifying machine operators of laser metal powder bed fusion machines and equipment used in aerospace applications

## 12.4 Impact of standards for additive manufacturing

The success in the implementation of any standard in the industry depends on several factors: (1) to respond to a need, (2) the standard is easy to apply, (3) the standard is realistic and fits very well with the sector where is applied, and (4) in some cases, the standard facilitates a process of certification/regulation. However, this is not always the general rule, and many standards do not achieve a relevant impact among the industrial community. Standards in AM are starting to take off, and the dissemination of them is a key factor for further implementation. One indicator to study the impact or level of knowledge of the approved standards is to analyze the number of publications in different databases. If we are more focused on the scientific impact of these standards, a good reference is to study the papers published in scientific journals. The first standard approved by ASTM (ASTM F2792) and ISO/ASTM some years later (ISO/ASTM 52900) is the one of terminologies for AM [3]. The number of scientific papers, according to the database of the library of the University of Las Palmas de Gran Canaria (ULPGC) (source RSS 2.0), during the past years is shown in Fig. 12.3.

In Fig. 12.4 is observed that from 2015 (the year of approval) the number of references to ISO/ASTM 52900 was continuously increased in papers and ASTM F2792 started to decrease in 2017. The reason of this is that the authors have progressively replaced the first ASTM standard by the joint standard ISO/ASTM, showing a good knowledge of the latest versions of the standards and the collaboration ASTM-ISO. Comparing the impact with the impact of publications with the general topic of AM, the

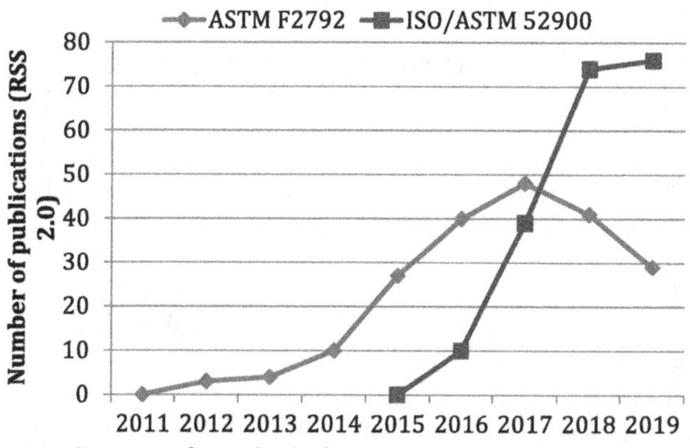


**Figure 12.3** Citations of standard of terminology (library of ULPGC, RSS2.0).

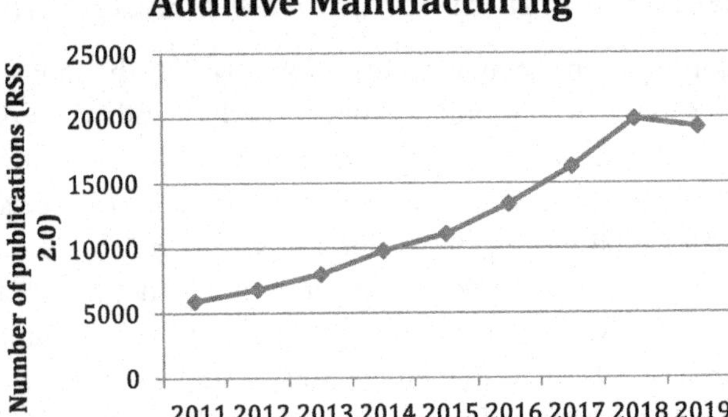


**Figure 12.4** Citations of the concept of AM (library ULPGC, RSS2.0). *AM*, additive manufacturing.

difference is still very high because most of the papers do not mention any specific standard for AM. In some cases, despite the use of standards for testing, design, etc., most of them are from other ISO or ASTM committees, showing the need of more specific standards for AM.

Beyond the research activity, published in scientific papers, the other source of documents where not only scientific articles but also other kinds of documents such as technical papers, dissemination documents, books, thesis, and so on are considered is Google Scholar. Google Scholar can show another measurement of the impact of the developed standards. In this source, the comparison of ASTM F2792 and ISO/ASTM 5200 (Fig. 12.5) follows a similar trend to the one shown in Fig. 12.4, but the decrease in the number of publications referencing F2792 is not so remarkable, probably because many documents are not so updated, or the rigor is not so relevant as formal papers. In any case, the trend in the use of F2792 started to increase slowly in 2016.

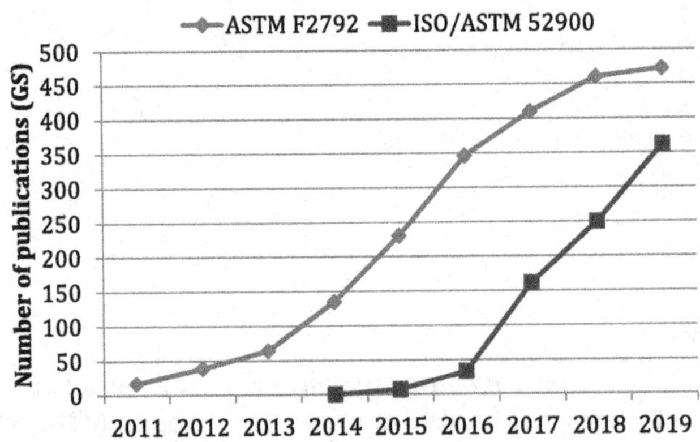


**Figure 12.5** Citations of the standards of terminology (Google Scholar).

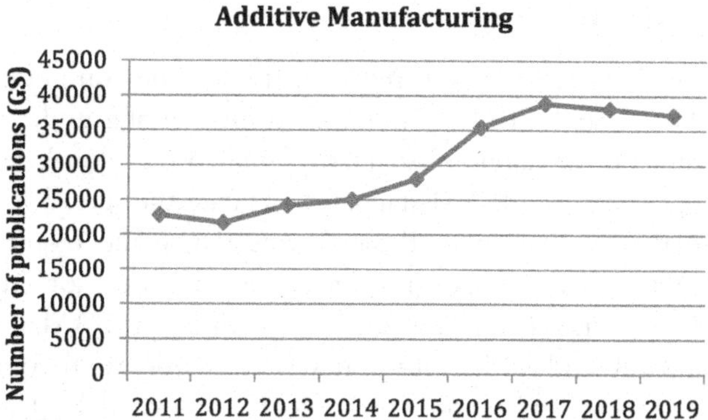


**Figure 12.6** Citations of the concept of AM (Google Scholar). *AM*, additive manufacturing.

In terms of the topic of AM in Google Scholar, the trend is more constant from 2017, as shown in Fig. 12.6, which is a sign that the information about AM standards are reaching the general public.

Another relevant standard for the AM community is the Additive Manufacturing File Format (AMF), which is the alternative to STL (not formal standard). The citations of this AMF standard, ISO/ASTM 5295 [13], have a continuous growth but are still quite far from a relevant impact (see Fig. 12.7). It is clear that the real impact in the use of the new AMF is the general implementation in the commercial software and AM equipment, mainly taking into account the competition of the traditional STL, or other recent formats such as 3 MF [12].

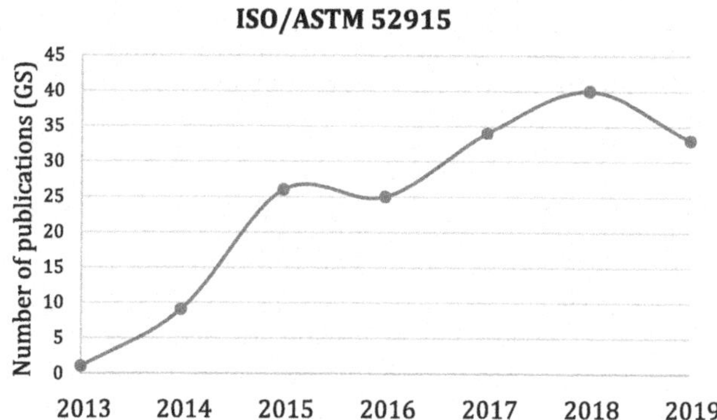


**Figure 12.7** Citations of the standard of AM file, AMF (Google Scholar). *AM*, additive manufacturing.

## 12.5 Conclusions

Standards for AM are a need in the industry and the committees ASTM F42, ISO TC261 and CEN TC438 have been working under a collaborative agreement, developing standards applied at global level: One world—one standard [10]. AM standards facilitate the spread of the technology and the application in those sectors where the requirements of quality or regulations are mandatory or at least necessary for the competitiveness of the companies. The different roadmaps for AM include this need of standards, and in the future, newer standards are to come, mainly focused on specific applications, technologies or materials. The dissemination of the developed standard in ISO TC261 and F42 is a key factor, and it is also important to send out the message that the AM technology requires specific standards in all the stages of the product development: design, materials, manufacturing, and final part.

## References

[1] ISO Committee≤www.iso.org> (accessed 02.03.2020).
[2] M. Monzón, et al., Standardisation in additive manufacturing: activities carried out by international organizations and projects, Int. J. Adv. Manuf. Technol. 76 (5–8) (2014) 1111–1121.
[3] ISO/ASTM 52900, Additive Manufacturing — General Principles — Terminology, 2015.
[4] Terry Wholers Report, 2019. wohlersassociates.com.
[5] European standards≤https://europa.eu/youreurope/business/product-requirements/standards/standards-in-europe/index_en.htm> (accessed 02.03.2020).
[6] ISO news≤https://www.iso.org/news/2015/05/Ref1956.html> (accessed 027.03.2020).
[7] ISO business Plan ≤https://committee.iso.org/files/live/sites/tc261/files/ISO_TC_261_Related_Links/ISO_TC%20261%20_Additive%20manufacturing_Business%20Plan%202015.pdf> (accessed 27.03.2020).
[8] ISO news≤https://www.iso.org/news/2016/10/Ref2124.html> (accessed 02.03.2020).
[9] CEN/TC 438, Additive Manufacturing ≤https://standards.cen.eu/dyn/www/f?p=204:7:0::::FSP_ORG_ID:1961493&cs=1725A335494BA95FA4CC9FE85A6F6B4B1> (accessed 02.03.2020).
[10] ISO/TC 261Chairman statement≤https://committee.iso.org/home/tc261> (accessed 02.03.2020).
[11] ISO/ASTM Additive Manufacturing Standards Structure ≤https://committee.iso.org/sites/tc261/home/projects.htm> (accessed 02.03.2020).
[12] E. Pei, et al., Investigating the impact of additive manufacturing data exchange standards for re-distributed manufacturing, Prog. Addit. Manufact. 4 (2019) 331–344.
[13] ISO/ASTM 52915, Specification for Additive Manufacturing File Format (AMF) Version 1.2, 2016.

CHAPTER 13

# Metal matrix composites processed by laser additive manufacturing: microstructure and properties

**Anne I. Mertens**
University of Liège, Faculty of Applied Science, Aerospace and Mechanical Engineering Department, Metallic Materials Science (MMS), Liège, Belgium

## 13.1 Introduction

Metal matrix composites (MMCs), in bulk or coatings, have been attracting more and more attention as they offer the possibility to enhance the properties of metallic materials and meet increasingly stringent requirements in terms, e.g., of higher service temperatures or mechanical loads [1]. Moreover, their use is not restricted to structural applications [2,3], and significant efforts have been put over the past few years into the development of MMCs with new functionalities such as improved biocompatibility [4–7], or enhanced tribological behavior [7–10].

In recent years, laser-based additive manufacturing (AM) processes have demonstrated their potential as powerful methods for the production of MMCs [3,11]. Laser-based additive processes can be classified in two main categories, each with its own advantages and specificities. On the one hand, powder bed fusion processes such as laser powder bed fusion (LPBF), selective laser melting (SLM), direct metal laser sintering (DMLS), or laser beam melting (LBM) rely on a focused energy beam to melt selected zones of a powder bed, following a CAD (computer-aided design) pattern. Powder bed fusion processes allow for the production of MMCs with complex geometries, including lattice structures, but they generally require a premixing of the matrix and reinforcement powders. Powder bed processes are thus limited to the production of MMCs with a constant volume fraction of reinforcement. Moreover, the correct premixing of the powders may prove particularly challenging when the matrix and reinforcement powders exhibit significant differences in densities and thus a high tendency to segregation [12,13]. On the other hand, beam deposition

*Additive Manufacturing*
ISBN 978-0-12-818411-0
https://doi.org/10.1016/B978-0-12-818411-0.00005-7

processes as direct metal deposition (DMD), laser engineering net shaping (LENS), or laser cladding (LC) use a high-energy laser beam to fuse a powder or mixture of powders while it is being projected onto a substrate. Besides, the term laser cladding is sometimes used as a synonym to laser surface alloying, i.e., the laser-induced melting of a preplaced layer of powders, a process suitable to fabricate a thin MMC coating [14,15]. This second category of processes has demonstrated its usefulness to repair damaged components and to produce coatings or local inserts. By feeding the matrix and reinforcement powders from two separate tanks, it is possible to vary gradually the volume fraction of reinforcements, thus producing functionally graded materials (FGMs). However, beam deposition processes are not suitable for the fabrication of MMCs with complex shapes.

For both powder bed fusion and beam deposition processes, the reinforcing or second phase may be produced ex situ then mixed directly with the powder of the matrix alloy (Fig. 13.1A). As opposed to this rather conventional approach, the actual second phase of the composite may also be produced in situ, as the product of chemical reactions taking place inside the starting mixture of powders during processing (Fig. 13.1C) [3,16]. Yet, in reality, the distinction between these two processing routes may be quite

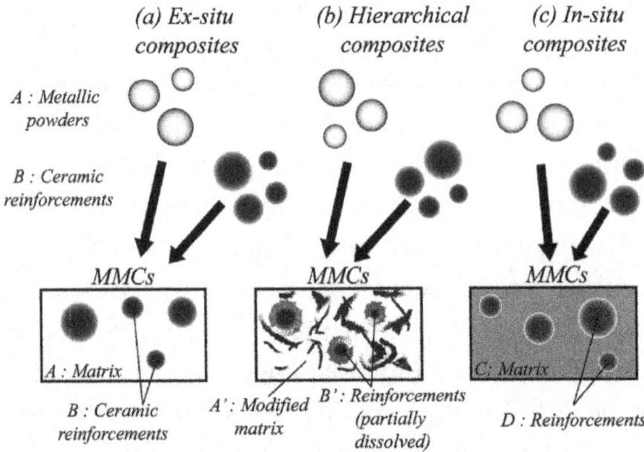


**Figure 13.1** Schematic illustration of the three possible routes for the laser additive manufacturing of MMCs; (A) pure ex situ MMCs where reactions between the reinforcements and the metallic matrix are avoided or limited as much as possible; (B) hierarchical MMCs in which partial dissolution of the reinforcements and interfacial reactions between the reinforcement and metallic matrix give birth to a complex hierarchical microstructure with a metallic matrix that is modified with respect to the original matrix alloy, and (C) in situ synthesized MMCs whose reinforcing phase is the product of chemical reactions taking place during fabrication [T. Maurizi Enrici (ULiège)]. *MMCs*, metal matrix composites.

blurred. Indeed, ex situ production of MMCs by laser additive processes is often accompanied by a (partial) dissolution of the second phase in the melted metallic matrix and by interfacial reactions between the second phase and the matrix. These phenomena lead to the development of complex hierarchical microstructures combining the remaining original reinforcements with a modified metallic matrix that is itself potentially reinforced by solute elements dissolved from the original reinforcement and by new phases produced by reactions between the metallic matrix and the original reinforcements (Fig. 13.1B). Ex situ production of MMCs *stricto sensu* actually requires specific precautions for the conservation of the intended second phase. These three different cases—i.e., (1) in situ synthesis of MMCs, (2) pure ex situ fabrication of MMCs, and (3) synthesis of mixed hierarchical structures—will be reviewed in detail in the following three subsections.

Besides, both powder bed fusion and beam deposition processes are characterized by extremely fast heating and cooling rates, resulting in out-of-equilibrium microstructures that exhibit significant chemical supersaturation and size refinement effects. The last two subsections of this chapter will thus focus more particularly on the challenges posed by the characterization of the complex out-of-equilibrium microstructures of MMCs produced by laser AM and on their specific usage properties and applications.

## 13.2 In situ synthesis of metal matrix composites by laser additive manufacturing

The reinforcement or second phase particles of MMCs may be synthesized in situ during laser AM [3,16]. This method is particularly interesting for the production of MMCs reinforced with very small particles, i.e., in the ~1 μm or nanosize range, that are prone to agglomerate during processing by ex situ methods. Indeed, in situ synthesis allows for obtaining a fine and uniform distribution of second-phase particles, as well as a better wetting and improved cohesion between the particles and the metallic matrix [17–19].

Reactants may take the form of blended pure elemental powders that will react to form a metallic matrix reinforced by finely distributed intermetallic or ceramic particles. For instance, Al-matrix composite coating reinforced with $TiB_2$, $Al_3Ti$, and $Al_3Fe$ intermetallics have been produced by beam deposition from blended Al, Ti, and Fe-coated B powders. Fe coating was applied on the B particles to protect them from burning under direct interaction with the laser, and the ratio of Fe-coated B with respect to Ti was set to reach the stoichiometry of $TiB_2$ [17]. Similarly, TiC particles have been formed in situ during the beam deposition of a Fe-based coating from blended pure Fe, Ti and C powders [20,21]. In this case, the pure elemental powders were premixed by ball milling with an atomic C:Ti ratio of 45:55 to promote the formation of TiC as predicted by the Ti–C phase diagram [20]. TiC and $Ti_3SiC_2$ reinforcing particles have also been synthesized in situ during the beam deposition of a composite Ti–Si–C powder that had itself been obtained through ball milling of pure elemental Ti, Si, and C powders followed by sintering by means of high-frequency thermal plasma technique [22]. Alternatively, alloyed powders in appropriate proportions may also be used as reactants, e.g., Fe-matrix composite coating reinforced with in situ $TiB_2$ particles have been produced by laser cladding from a mixture of 60.73 wt.% ferrotitanium with 39.27 wt.% ferroboron alloy powders [18]. Choosing alloy powders instead of pure elemental powders as precursors was found beneficial by reducing the cost and by lowering the melting temperature of the material.

Metallic powders may also be reacted with the gaseous protective atmosphere and/or the carrier gas during beam deposition. This method is an efficient way to synthesize various nitrides. TiN reinforcement particles have thus been produced during the beam deposition of commercial pure Ti powder under an atmosphere of $Ar + N_2$ [23]. Similarly, TiN has been synthesized during the beam deposition of a $CoCr_2FeNiTi_{0.5}$ high-entropy alloy [24]. Al/AlN MMCs have also been obtained via the beam deposition of an Al–Si eutectic alloy using $N_2$ as carrier gas [25]. Moreover, it is found that the volume fraction of nitride particles can be modified by adjusting the reaction time, e.g., by tuning the laser scan speed [23,25].

The third variant of in situ synthesis of MMCs relies on reactions between metallic and ceramic powders in which the ceramic fully decomposes to give birth to new reinforcing phase(s). Ceramic reactants include oxides as well as carbides, but the use of oxide has met with mixed results [3]. Dadbakhsh and Hao [26] have attempted to produce Al matrix composites reinforced with a fine dispersion of oxides via SLM of a mixture of $Fe_2O_3$ particles with various Al alloys. They succeeded in producing Al matrix composites reinforced with finely distributed $Al_2O_3$ and Al–Fe intermetallics with a size inferior to 400 nm. However, the mechanical properties of these composites were compromised due to their relatively high volume fraction of porosities. Attempts at producing Fe matrix composites reinforced by finely dispersed $Al_2O_3$ particles and mixed carbides were more successful, based on the thermite reaction of $Fe_2O_3$ with Al [27]. $Al_2O_3$ particles have been successfully added to produce an oxide dispersion–strengthened (ODS) AlSi10Mg matrix by promoting the formation of finely distributed aluminum silicates ($Al_2Si_4O_{10}$) [19]. Carbides as SiC or WC, on the other hand, have already been widely used as reactant for the production of steel matrix composites reinforced with various types of finely distributed complex alloyed carbides [14,28,29].

## 13.3 Laser additive manufacturing for the production of "pure" ex situ metal matrix composites

In favorable cases, i.e., when it is stable with respect to interactions with the laser beam and with the molten metallic matrix, the reinforcing second phase may be added ex situ in the MMCs. The $TiB_2$/Al alloys system is a good example as $TiB_2$ is thermally stable and chemically inert with respect to Al alloys [30]. Since $TiB_2$ nanoparticles are known as potent inoculants, acting as efficient heterogeneous nuclei during the solidification of Al alloys, this characteristic has recently given way to an interesting application. $TiB_2$ nanoparticles have been used to modify the columnar solidification structure typical of AM Al alloys into a more equiaxed and refined structure resulting in more isotropic mechanical properties [31,32].

However, the production of ex situ MMCs frequently requires taking specific precautions to prepare the feedstock powders to preserve the second phase during fabrication and protect it from undesired interactions with the laser or with the metallic matrix. While these methods may be of general use for the laser AM of MMCs, they are often a necessity for the production of ex situ MMCs and are thus presented in this section.

The first strategy to overcome these difficulties consists in premixing the matrix and second-phase powders into a composite or "satellited" powder [10,13,33,34,71]. This strategy is particularly recommended for processing small-size second-phase particles whose low intrinsic flowability may prove detrimental during the fabrication of the MMCs [13,71]. Moreover, it has the advantage of promoting a more uniform distribution of the second phase in the final microstructure [13]. Due to its low cost and simplicity, ball milling is the most commonly used method for mixing the matrix and second-phase powders. However, it is worth noting that the milling parameters (duration, rotation speed, choice of milling medium, etc.) must be controlled very carefully. Indeed, excessive milling may result in extensive damage to the powder particles such as particles breakage leading to the presence of small debris or to loss of sphericity, which may in turn impair the powder flowability [13,71]. Chemical method for the production of tailored composite feedstock powder are also receiving increasing attention. The feasibility of using electroless plating to synthesize a Ni/nano-$Al_2O_3$ composite powder suitable as feedstock for the SLM process has thus recently been demonstrated [35].

The second strategy for the production of tailored feedstock for the AM of MMCs consists in encapsulating the second phase with a metallic shell or coating layer. Ball milling can be used to this end, for example, to encapsulate h-BN particles inside a shell of Cu nanoparticles [9]. Electroless plating of a Ni-based layer has also been used to successfully protect $CaF_2$ lubricating particles from decomposition and evaporation under the laser beam [36]. However, if not chosen carefully, the encapsulating material may sometimes be considered as an impurity with detrimental effect on the properties of the final MMCs. This consideration has triggered investigations into alternative and well-targeted chemical methods for the preparation of specific feedstock for the AM of MMCs. Zhao et al. [37] have thus recently reported encouraging results on the SLM of AlSi10Mg/graphene MMCs using Al-precoated graphene obtained by means of an organic Al chemical reduction process.

## 13.4 Laser additive manufacturing of hierarchical metal matrix composites

As mentioned in Section 13.1, attempts at ex situ production of MMCs by laser additive processes are often accompanied by a (partial) dissolution of the second phase in the melted metallic matrix and by interfacial reactions between the second phase and the matrix [3,38,72]. These phenomena lead to the development of complex hierarchical microstructures combining the remaining original reinforcements with a modified metallic matrix as illustrated in Fig. 13.2 for a laser clad stainless steel 316L + WC composite. An original WC particle (top of the picture) has partially dissolved in the stainless steel 316L melt pool, and upon solidification, the dissolved W and C have given way to the formation of new complex alloyed carbides with a refined lamellar structure.

The extent of the dissolution and reaction processes depends on a number of variables. The thermal history experienced by the composite during its fabrication, which is controlled by processing parameters such as the laser power or the scan speed, play an important role [40–42]. Besides,

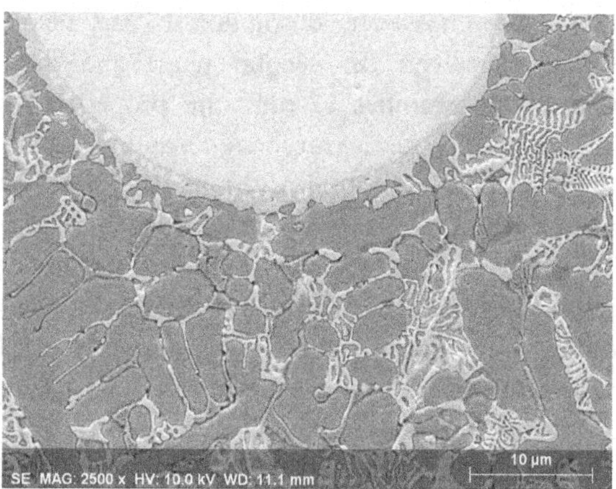


**Figure 13.2** SEM micrograph of a stainless steel 316L composite coating with 16% vol. of tungsten carbides produced by beam deposition [39].

the size of the second-phase particles has an obvious effect, as smaller particles dissolve more rapidly due to their higher surface-to-volume ratio [42–44]. The nature of the particles and their chemical stability with respect to the molten metallic matrix also has an influence. For instance, TiC is known for its high stability with respect to steel matrix, resulting in very low dissolution rates. SiC, on the contrary, has a high tendency to dissolve in steel, while WC exhibits an intermediate behavior between those of SiC and TiC [45].

Such reactions may sometimes be undesirable, when their products are detrimental to the usage properties of the composite. The formation of $Al_4C_3$ in Al matrix composites reinforced with carbides or other carbonaceous reinforcements as graphene or carbon nanotubes is a known example. Indeed, $Al_4C_3$ is brittle and susceptible to hydrolysis and corrosion in humid environment, and significant efforts have been made to limit or avoid its formation during the processing of Al matrix composites [3,46–49]. Successful strategies include the use of a very high energy density of 666 $J/mm^3$ in the LPBF of AlSi10Mg + SiC composites, to promote the formation of the less detrimental mixed $Al_4SiC$ carbide that is favored at temperatures higher than 1350–1400°C [50]. Alloy modification is another promising method to avoid the formation of $Al_4C_3$ in Al matrix composites. Additions of Ti in Al–Si alloys have been found to promote the formation of TiC over $Al_4C_3$ [48,51]. Additions of high amounts of Si (above 40 wt.%) have also been found to inhibit the formation of $Al_4C_3$ during the beam deposition of Al–Si alloy + SiC composites [48,49].

On a positive side, however, dissolution of the second phase and interfacial reactions between the second phase and the matrix offer practically limitless opportunities to tune the properties of the MMCs processed by laser AM. Indeed, the surviving second phase particles may act as inoculants during the solidification of the metallic matrix, leading to an enhanced grain refinement in comparison with the laser processed matrix alloy in the absence of reinforcement [41,52–54]. In favorable cases, interfacial reactions between the matrix and the reinforcing particles may prove beneficial by improving the interfacial cohesion [1,41]. Furthermore, the modified metallic matrix may benefit from (1) the solid solution strengthening due to the dissolved elements released by the second phase [42,53,55] and (2) the precipitation hardening by the new phases formed by reactions between the matrix and the reinforcing particles [40,52,53]. By optimizing the synergy of the aforementioned strengthening mechanisms and of the conventional composite strengthening effect, it is indeed possible to enhance the usage properties of MMCs in view of their target applications. A recent and interesting example of such optimization lies with the beam deposition of Al–W composites, which open the possibility to combine the high electrical conductivity of aluminum with the refractory behavior of tungsten [55]. While alloying of Al and W by more conventional methods is known to be challenging, beam deposition offers the possibility to combine the effects of solid solution strengthening, precipitation hardening by $Al_{12}W$ and $Al_4W$, and composite strengthening by retained W particles that have a beneficial effect on the resistance to pitting and grain boundary corrosion.

More examples of hierarchical MMCs produced by laser AM are reviewed in Refs. [3,16].

## 13.5 Microstructural characterization of metal matrix composites produced by laser additive manufacturing

As exposed in Sections 13.2–13.4, MMCs processed by laser AM exhibit complex and out-of-equilibrium microstructures. The extremely high cooling rates experienced by these materials during solidification may result in a significant microstructural refinement—an effect that is even enhanced in laser AM MMCs [41,52–54]—and in a chemical supersaturation of the metallic matrix [72]. The latter effect has been widely documented in AM Al-Si [56–58], Al-Mg-Zr [59], and Al–Mg-Sc alloys [60]. Supersaturated solute W content as high as 0.4 at. %—i.e., well in excess of the equilibrium concentration—have also been reported in Al–W MMCs processed by direct laser metal deposition [55]. These specificities of the microstructures of laser AM MMCs contribute to make their characterization particularly challenging. Optical microscopy (OM) can obviously be used to characterize the macrostructure of laser AM MMCs and evaluate their overall metallurgical health. Other conventional characterization techniques as scanning electron microscopy (SEM) (that is sometimes combined with energy-dispersive spectrometry [EDS]), electron backscattered diffraction (EBSD), and X-Ray diffraction (XRD) allow for microstructural analysis and phase identification. However, EBSD and XRD face limitations due to the peculiar nature of laser AM microstructures exhibiting chemical supersaturation and residual stresses, both of which are susceptible of inducing significant modifications of crystal lattice parameters [41,61,72]. At a finer scale, transmission electron microscopy (TEM) is a powerful technique to investigate nanosize reinforcements or the nature and condition of interfaces between the reinforcements and the metallic matrix [10,37,51]. Raman spectroscopy, on the other hand, is dedicated in particular to the study of boron nitride nanoplatelets [62] and of carbonaceous reinforcements as graphene or carbon nanotubes [34,63].

Considering this, the careful cross-consideration of several techniques is generally needed to reach a deep knowledge and understanding of the microstructure of laser AM MMCs. The cross-consideration of differential thermal analysis (DTA) with SEM + EDS and EBSD observations is a nice example in that respect [61,72]. This method, which had originally been designed to elucidate the solidification sequence of complex graphitic high speed steel under the fast cooling conditions obtained under centrifugal casting, is also well suited for laser AM MMCs, as illustrated in Fig. 13.3. Fig. 13.3A shows the DTA heating curve relative to a 316L stainless steel + WC MMC fabricated by laser cladding. Indeed, heating curves of thermal tests carried out at low scans (e.g., from 1 to 10°C/min) allow considering the thermodynamic stability of phases formed at high temperature. Considered simultaneously, classical SEM + EDS and EBSD analyses (Fig. 13.3B) allow to identify the precise condition of a phase structure (out-of-equilibrium or supersaturated crystal lattice). The cross-consideration of this information is particularly helpful to clarify the nature of the intermetallic phases. As an example, EBSD analysis in Fig. 13.3B allows to index both solidification carbides ($M_6C$ and $M_{23}C_6$) formed around the original WC reinforcement as FCC. Chemical analyses

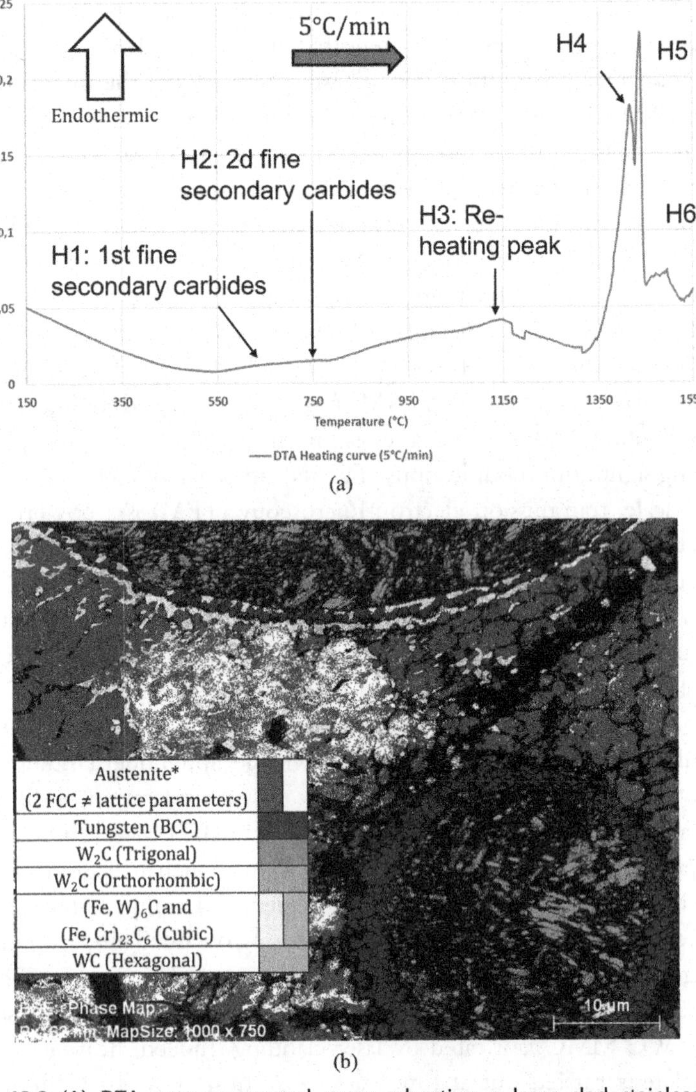


**Figure 13.3** (A) DTA curve measured upon reheating a laser clad stainless steel 316L + WC composite at a scan rate of 5°C/min; (B) EBSD map of the same laser clad stainless steel 316L + WC composites [T. Maurizi Enrici (ULiège)].

by EDS then allow to distinguish these two types of carbides and to associate them to a specific melting peak of the DTA curve. Finally, the analysis of the whole DTA curve can lead to the restoration of the complete solidification sequence.

## 13.6 Properties and applications of metal matrix composites produced by laser additive manufacturing

As already stated in Section 13.1, the development of functional MMCs has been receiving an increasing attention over the past few years. As far as laser AM MMCs are concerned, the current state-of-the-art reveals a clear focus on biomedical and on tribological applications.

MMCs for biomedical applications may be designed to meet several diverse requirements as a high load-bearing capacity, improved wear resistance to minimize the risks of wear-induced osteolysis, and improved biocompatibility [7]. To this end, Ti-based matrix composites have been developed with TiN additions [23,64], $Si_3N_4$, or calcium phosphate tribasic (TCP) additions [6]. Hydroxyapatite additions have also been used in combination with a nitinol [5] or a stainless steel 316L matrix [4,65].

Interesting examples of MMCs for tribological applications include self-lubricating composites, as well as coatings with ultrafine microstructures for improved resistance to wear by abrasion, erosion, or cavitation/erosion. Indeed, the combined beneficial effects of the strongly refined microstructure, solid solution, and precipitation strengthening of in situ synthesized and hierarchical laser AM MMCs have been well demonstrated for applications that involve handling hard particles or slurries [8,14,28,29,66]. Laser AM of self-lubricating MMCs has also been a hot topic, targeting applications where conventional lubrication does not work, e.g., at high temperatures or under vacuum [3,62]. A large variety of self-lubricating reinforcements have already been investigated, such as hexagonal boron nitride (h-BN) [9,37,62], $CaF_2$ [15,36,67], graphene platelets [10,63,68], or the newly developed $Ti_3SiC_2$ compound whose layered hexagonal structure makes into a very good solid lubricant [69].

In contrast, the corrosion properties of laser AM MMCs have received less attention than their tribological behavior. Indeed, it is feared that the precipitation of reaction carbides in stainless steel matrix MMCs might result in the sensitization of the material due to local Cr depletion. Yet, some reports suggest that the corrosion behavior of such stainless steel - + carbides composite may remain acceptable under the condition that the extent of the precipitation of reaction carbides is limited [52,70]. In a few instances, laser AM MMCs have been reported to possess an excellent corrosion resistance that was considered a beneficial effect of the strong grain refinement characteristic of those materials [24].

These few examples illustrate the diversity and versatility of laser AM MMCs, and the large number of applications that can be envisaged for these materials whose potential has not yet been fully exploited.

## 13.7 Concluding remarks

Laser AM has imposed itself as a powerful method for the production of MMCs, with a number of advantages and specificities. MMCs may be produced either in complex shapes by laser powder bed fusion processes, or alternatively as localized insert or coating with constant or graded composition by direct deposition processes.

Moreover, three distinct processing routes may be distinguished based on the method for introducing the reinforcing phase in the material. Indeed, the second phase of the composite may be produced in situ, i.e., as the product of a chemical reaction that occurs during fabrication. Alternatively, the reinforcing phase as a powder may be mixed directly with the powder of the matrix alloy (ex situ processing). In this case, reactions between reinforcing phase and the metallic matrix may be limited or sometimes even completely avoided although this may require specific precautions regarding, e.g., feedstock preparation. In many cases, however, a (partial) dissolution of the second phase in the matrix and interfacial reactions between the second phase and the matrix are observed, leading to the formation of a hierarchical composite whose complex microstructure combines some features of both the in situ and the ex situ MMCs.

Since the extremely high cooling rates experienced during laser AM generally result in the formation of out-of-equilibrium microstructures (exhibiting strong chemical supersaturation and significant grain size refinement), their microstructural characterization may prove challenging. The careful cross-consideration of several different techniques is thus shown as a powerful way to reach a deep understanding of the genesis of the complex microstructures of laser AM MMCs. Finally, some examples of specific properties and applications of laser AM MMCs are reviewed, illustrating the diversity and versatility of these materials whose many potentialities yet remain to be explored.

## References


[1] K.U. Kainer, Basics of Metal matrix composites, in: K.U. Kainer (Ed.), Metal Matrix Composites. Custom-Made Materials for Automotive and Aerospace Engineering, WILEY-VCH Verlag, Weinheim, 2006.

[2] A.D. Mogahdam, B.F. Schultz, J.B. Ferguson, E. Omrani, P.K. Rohatgi, N. Gupta, Functional metal matrix composites: self-lubricating, self-healing, and nano-composites – an outlook, J. Miner. Met. Mater. Soc. 66 (2014) 872–881.

[3] A.I. Mertens, J. Lecomte-Beckers, On the role of interfacial reactions, dissolution and secondary precipitation during the laser additive manufacturing of metal matrix composites: a review, in: I.V. Shishkovsky (Ed.), New Trends in 3D Printing, InTech, Rijeka, 2016.

[4] L. Hao, S. Dadbakhsh, O. Seaman, M. Felstead, Selective laser melting of a stainless steel and hydroxyapatite composite for load-bearing implant development, J. Mater. Process. Technol. 209 (2009) 5793–5801.

[5] I.V. Shishkovskii, I.A. Yadroitsev, I.Y. Smurov, Theory and technology of sintering, thermal and chemicothermal treatment: selective laser sintering/melting of nitinol-hydroxyapatite composite for medical applications, Powder Metall. Met. C+ 50 (2011) 275–283.

[6] X. Xu, J. Han, C. Wang, A. Huang, Laser cladding of composite bioceramic coatings on titanium alloy, J. Mater. Eng. Perform. 25 (2016) 656–667.

[7] Y. Hu, W. Cong, A review on laser deposition-additive manufacturing of ceramics and ceramic reinforced metal matrix composites, Ceram. Int. 44 (2018) 20599–20612.

[8] Z. Zhang, R. Kovacevic, Laser Cladding of iron-based erosion resistant metal matrix composites, J. Manuf. Process. 38 (2019) 63–75.

[9] Y. Zhao, K. Feng, C. Yao, P. Nie, J. Huang, Z. Li, Microstructure and tribological properties of laser cladded self-lubricating nickel-base composite coatings containing nano-Cu and h-BN solid lubricants, Surf. Coating. Technol. 359 (2019) 485–494.

[10] L. Wu, Z. Zhao, P. Bai, W. Zhao, Y. Li, M. Liang, H. Liao, P. Huo, J. Li, Wear resistance of graphene nano-platelets (GNPs) reinforced AlSi10Mg matrix composite prepared by SLM, Appl. Surf. Sci. 503 (2020) 144156.

[11] S. Kumar, J.P. Kruth, Composites by rapid prototyping technology, Mater. Des. 31 (2010) 850–856.

[12] R.M. Mahamood, E.T. Akinlabi, Laser metal deposition of functionally graded Ti6Al4V/TiC, Mater. Des. 84 (2015) 402–410.

[13] E. Fereiduni, A. Ghasemi, M. Elbestawi, Characterization of composite powder feedstock from powder bed fusion additive manufacturing perspective, Materials 12 (2019), https://doi.org/10.3390/ma12223673.

[14] G. Thawari, G. Sundararajan, S.V. Joshi, Laser surface alloying of medium carbon steel with $SiC_{(p)}$, Thin Solid Films 423 (2003) 41–53.

[15] H. Yan, J. Zhang, P. Zhang, Z. Yu, C. Li, Y. Lu, Laser cladding of Co-based alloy/TiC/$CaF_2$ self-lubricating composite coatings on copper for continuous casting mold, Surf. Coating. Technol. 232 (2013) 362–369.

[16] S. Dadbakhsh, R. Mertens, L. Hao, J. Van Humbeeck, J.P. Kruth, Selective laser melting to manufacture "in situ" metal matrix composites: a review, Adv. Eng. Mater. 21 (2019) 1801244.

[17] J. Xu, W. Liu, Wear characteristic of in situ $TiB_2$ particulate-reinforced Al matrix composite formed by laser cladding, Wear 260 (2006) 486–492.

[18] B. Du, Z. Zou, X. Wang, S. Qu, Laser cladding of in situ $TiB_2$/Fe composite coating on steel, Appl. Surf. Sci. 254 (2008) 6489–6494.

[19] L. Wang, J. Jue, M. Xia, L. Guo, B. Yan, D. Gu, Effect of the thermodynamic behavior of selective laser melting on the formation of in situ oxide dispersion-strengthened aluminum-based composites, Metals 6 (2016), https://doi.org/10.3390/met6110286.

[20] A. Emamian, M. Alimardani, A. Khajepour, Correlation between temperature distribution and in situ formed microstructure of Fe-TiC deposited on carbon steel using laser cladding, Appl. Surf. Sci. 258 (2012) 9025–9031.

[21] A. Emamian, M. Alimardani, A. Khajepour, Effect of cooling rate and laser process parameters on additive manufactured Fe-Ti-C metal matrix composites microstructure and carbide morphology, J. Manuf. Process. 16 (2014) 511–517.

[22] N. Li, W. Liu, H. Xiong, R. Qin, S. Huang, G. Zhang, C. Gao, In-situ reaction of Ti-Si-C composite powder and formation mechanism of laser deposited Ti6Al4V/($TiC+Ti_3SiC_2$) system functionally graded material, Mater. Des. 183 (2019) 108155.

[23] H.C. Man, S. Zhang, F.T. Cheng, X. Guo, In situ formation of a TiN/Ti metal matrix composite gradient coating on NiTi by laser cladding and nitriding, Surf. Coating. Technol. 200 (2006) 4961–4966.

[24] Y.X. Guo, Q.B. Liu, X.J. Shang, In situ TiN-reinforced $CoCr_2FeNiTi_{0.5}$ high-entropy alloy composite coating fabricated by laser cladding, Rare Met. (2019), https://doi.org/10.1007/s12598-018-1194-8.

[25] A. Riquelme, P. Rodrigo, M.D. Escalera-Rodriguez, J. Rams, Effect of the process parameters in the additive manufacturing of in situ Al/AlN samples, J. Manuf. Process. 46 (2019) 271–278.

[26] S. Dadbakhsh, L. Hao, Effect of Al alloys on selective laser melting behaviour and microstructure of in situ formed particle reinforced composite, J. Alloys Compd. 541 (2012) 328–334.

[27] H. Tan, Z. Luo, Y. Li, F. Yan, R. Duan, Y. Huang, Effect of strengthening particles on the dry sliding wear behavior of $Al_2O_3$-$M_7C_3$/Fe metal matrix composite coatings produced by laser cladding, Wear 324–325 (2015) 36–44.

[28] K.H. Lo, F.T. Cheng, C.T. Kwok, H.C. Man, Improvement of cavitation erosion resistance of AISI 316 stainless steel by laser surface alloying using fine WC powder, Surf. Coating. Technol. 165 (2003) 258–267.

[29] G. Abbas, U. Ghazanfar, Two-body abrasive wear studies of laser produced stainless steel+SiC composite clads, Wear 258 (2003) 258–264.

[30] R. Anandkumar, A. Almeida, R. Vilar, Wear behavior of Al-12Si/$TiB_2$ coatings produced by laser cladding, Surf. Coating. Technol. 205 (2011) 3824–3832.

[31] X. Wen, Q. Wang, Q. Mu, N. Kang, S. Sui, H. Yang, X. Lin, W. Huang, Laser solid forming additive manufacturing $TiB_2$ reinforced 2024Al composite: microstructure and mechanical properties, Mater. Sci. Eng. A 745 (2019) 319–325.

[32] Y.K. Xiao, Z.Y. Bian, Y. Wu, G. Ji, Y.Q. Li, M.J. Li, Q. Lian, Z. Chen, A. Addad, H.W. Wang, Effect of nano-$TiB_2$ particles on the anisotropy in an AlSi10Mg alloy processed by selective laser melting, J. Alloys Compd. 798 (2019) 644–655.

[33] P.K. Farayibi, T.E. Abioye, A. Kennedy, A.T. Clare, Development of metal matrix composites by direct energy deposition of 'satellited' powders, J. Manuf. Process. 45 (2019) 429–437.

[34] D. Gu, X. Rao, D. Dai, C. Ma, L. Xi, K. Lin, Laser additive manufacturing of carbon nanotubes (CNTs) reinforced aluminum matrix nanocomposites: processing optimization, microstructure evolution and mechanical properties, Additive. Manufact. 29 (2019) 100801.

[35] M. Li, A. Fang, E. Martinez-Franco, J.M. Alvarado-Orozco, Z. Pei, C. Ma, Selective laser melting of metal matrix composites : feedstock powder preparation by electroless plating, Mater. Lett. 247 (2019) 115–118.
[36] X.B. Liu, S.H. Shi, J. Guo, G.F. Yu, M.D. Wang, Microstructure and wear behavior of γ-TiAl intermetallic alloy prepared by Nd:YAG laser cladding, Appl. Surf. Sci. 255 (2009) 5662–5668.
[37] Z. Zhao, P. Bai, R.D.K. Misra, M. Dao, M. Dong, R. Guan, Y. Li, J. Zhang, L. Tan, J. Gao, T. Ding, W. Du, Z. Guo, AlSi10Mg alloy nanocomposites reinforced with aluminum-coated graphene: selective laser melting, interfacial microstructure and property analysis, J. Alloys Compd. 792 (2019) 203–214.
[38] A. Mertens, T. L'Hoest, J. Magnien, R. Carrus, J. Lecomte-Beckers, On the elaboration of metal-ceramic coatings by laser cladding, Mater. Sci. Forum 879 (2017) 1288–1293.
[39] T. L'Hoest, Production of Metal/ceramic Composite Coating by Laser Cladding, Master Thesis, University of Liège, Belgium, 2015.
[40] J.C. Betts, B.L. Mordike, M. Grech, Characterisation, wear and corrosion testing of laser-deposited AISI316 reinforced with ceramic particles, Surf. Eng. 26 (2010) 21–29.
[41] B. Li, B. Qian, Y. Xu, Z. Liu, J. Zhang, F. Xuan, Additive manufacturing of ultrafine-grained austenitic stainless steel matrix composite via vanadium carbide reinforcement addition and selective laser melting: formation mechanisms and strengthening effect, Mater. Sci. Eng. A 745 (2019) 495–508.
[42] F. Fazliana, S.N. Aqida, I. Ismail, Effect of tungsten carbide partial dissolution on the microstructure evolution of a laser clad surface, Optic Laser. Technol. 121 (2020) 105789.
[43] K. Das, T.K. Bandyopadhyay, S. Das, A review on the various synthesis routes of TiC reinforced ferrous based composites, J. Mater. Sci. 37 (2002) 3881–3892.
[44] S.K. Ghosh, P. Saha, S. Kishore, Influence of size and volume fraction of SiC particulates on properties of ex situ reinforced Al-4.5Cu-3Mg matrix composite prepared by direct metal laser sintering process, Mater. Sci. Eng. A 527 (2010) 4694–4701.
[45] F.T. Cheng, C.T. Kwok, H.C. Man, Laser surfacing of S31603 stainless steel with engineering ceramics for cavitation erosion resistance, Surf. Coating. Technol. 139 (2001) 14–24.
[46] R. Anandkumar, A. Almeida, R. Vilar, Microstructure and sliding wear resistance of an Al-12 wt.% Si/TiC laser clad coating, Wear 282–283 (2012) 31–39.
[47] A.I. Mertens, J. Delahaye, J. Lecomte-Beckers, Fusion-based additive manufacturing for processing aluminum alloys: state-of-the-art and challenges, Adv. Eng. Mater. 19 (2017) 1700003.
[48] A. Riquelme, M.D. Escalera-Rodriguez, P. Rodrigo, E. Otero, J. Rams, Effect of alloying elements added on microstructure and hardening of Al/SiC laser clad coating, J. Alloys Compd. 727 (2017) 671–682.
[49] A. Riquelme, P. Rodrigo, M.D. Escalera-Rodriguez, J. Rams, Characterisation and mechanical properties of Al/SiC metal matrix composite coatings formed on ZE41 magnesium alloys by laser cladding, Result. Phy. 13 (2019) 102160.
[50] F. Chang, D. Gu, D. Dai, P. Yuan, Selective laser melting of in-situ Al4SiC4 + SiC hybrid reinforced Al matrix composites: influence of starting SiC particle size, Surf. Coating. Technol. 272 (2015) 15–24.
[51] F. Li, Z. Gao, Y. Zhang, Y. Chen, Alloying effect of titanium on WCp/Al composite fabricated by coincident wire-powder laser deposition, Mater. Des. 93 (2016) 370–378.
[52] J.D. Majumdar, A. Kumar, L. Li, Direct laser cladding of SiC dispersed AISI 316L stainless steel, Tribol. Int. 42 (2009) 750–753.

[53] Q. Li, Y. Lei, H. Fu, Laser cladding in-situ NbC particle reinforced Fe-based composite coatings with rare earth oxide addition, Surf. Coating. Technol. 239 (2014) 102–107.
[54] B. Song, S. Dong, C. Coddet, Rapid in situ fabrication of Fe/SiC bulk nanocomposites by selective laser melting directly from a mixed powder of microsized Fe and SiC, Scripta Mater. 75 (2014) 90–93.
[55] A. Ramakrishnan, G.P. Dinda, Microstructural control of an Al-W aluminum matrix composite during direct laser metal deposition, J. Alloys Compd. 813 (2020) 152208.
[56] X.P. Li, X.J. Wang, M. Saunders, A. Suvorova, L.C. Zhang, Y.J. Liu, M.H. Fang, Z.H. Huang, T.B. Sercombe, A selective laser melting and solution heat treatment refined Al–12Si alloy with a controllable ultrafine eutectic microstructure and 25% tensile ductility, Acta Mater. 95 (2015) 74–82.
[57] S. Marola, D. Manfredi, G. Fiore, M.G. Poletti, M. Lombardi, P. Fino, L. Battezzati, A comparison of Selective Laser Melting with bulk rapid solidification of AlSi10Mg alloy, J. Alloys Compd. 742 (2018) 271–279.
[58] J. Delahaye, J.T. Tchuindjang, J. Lecomte-Beckers, O. Rigo, A.M. Habraken, A. Mertens, Influence of Si precipitates on fracture mechanisms of AlSi10Mg parts processed by selective laser melting, Acta Mater. 175 (2019) 160–170.
[59] J.R. Croteau, S. Griffiths, M.D. Rossell, C. Leinenbach, C. Kenel, V. Jansen, D.N. Seidman, D.C. Dunand, N.Q. Vo, Microstructure and mechanical properties of Al-Mg-Zr alloys processed by selective laser melting, Acta Mater. 153 (2018) 35–44.
[60] A.B. Spierings, K. Dawson, T. Heeling, P.J. Uggowitzer, R. Schaublin, F. Palm, K. Wegener, Microstructural features of Sc- and Zr-modified Al-Mg alloys processed by selective laser melting, Mater. Des. 115 (2017) 52–63.
[61] T. Maurizi Enrici, A. Mertens, M. Sinnaeve, J.T. Tchuindjang, Elucidation of the solidification sequence of a complex graphitic HSS alloy under a combined approach of DTA and EBSD analysis, J. Therm. Anal. Calorim. 141 (2020) 1075–1089.
[62] Y. Song, G. He, Y. Wang, Y. Chen, Tribological behavior of boron nitride nanoplatelet reinforced $Ni_3Al$ intermetallic matrix composite fabricated by selective laser melting, Mater. Des. 165 (2019) 107579.
[63] G. Lu, X. Shi, X. Liu, H. Zhou, Y. Chen, Z. Yang, Y. Huang, Tribological performance of functionally gradient structure of graphene nanoplatelets reinforced $Ni_3Al$ metal matrix composites prepared by laser melting deposition, Wear 428–429 (2019) 417–429.
[64] V.K. Balla, A. Bhat, S. Bose, A. Bandyopadhyay, Laser processed TiN Ti6Al4V composite coatings, J. Mech. Behav. Biomed. 6 (2012) 9–20.
[65] Q. Wei, S. Li, Q. Han, W. Li, L. Cheng, L. Hao, Y. Shi, Selective laser melting of stainless steel/nano-hydroxyapatite composites for medical applications. Microstructure, element distribution, crack and mechanical properties, J. Mater. Process. Technol. 222 (2015) 444–453.
[66] R. Kumar, M. Antonov, U. Beste, D. Goljandin, Assessment of 3D printed steels and composites intended for wear applications in abrasive, dry or slurry erosive conditions, Int. J. Refract. Met. H. 86 (2020) 105126.
[67] H. Yan, P. Zhang, Z. Yu, Q. Lu, S. Yang, C. Li, Microstructure and tribological properties of laser-clad Ni-Cr/$TiB_2$ composite coatings on copper with the addition of $CaF_2$, Surf. Coating. Technol. 206 (2012) 4045–4053.
[68] S. Wen, K. Chen, W. Li, Y. Zhou, Q. Wei, Y. Shi, Selective laser melting of reduced graphene oxide/S136 metal matrix composites with tailored microstructures and mechanical properties, Mater, Design 175 (2019) 107811.

[69] X. Li, C.H. Zhang, S. Zhang, C.L. Wu, Y. Liu, J.B. Zhang, M. Babar Shahzad, Manufacturing of $Ti_3SiC_2$ lubricated Co-based alloy coatings using laser cladding technology, Optic Laser. Technol. 114 (2019) 209–215.
[70] J.D. Majumdar, L. Li, Studies on direct laser cladding of SiC dispersed AISI316L stainless steel, Metall. Mater. Trans. 40 (2009) 3001–3008.
[71] O. Ertugrul, T. Maurizi Enrici, H. Paydas, E. Saggionetto, F. Boschini, A. Mertens, Laser cladding of TiC reinforced 316L stainless steel composites: feedstock powder preparation and microstructural evaluation, Powder Technol. 375 (2020) 384–396.
[72] T. Maurizi Enrici, O. Dedry, F. Boschini, J.T. Tchuindjang, A. Mertens, Microstructural and thermal characterization of 316L + WC composite coatings obtained by laser cladding, Adv. Eng. Mater. 22 (12) (2020), 2000291.

CHAPTER 14

# Laser aided metal additive manufacturing and postprocessing: a comprehensive review

**Rajkumar Velu[1,4], Arun V. kumar[2], A.S.S. Balan[3], Jyoti Mazumder[1]**
[1]Centre for Laser Aided Intelligent Manufacturing, University of Michigan, Ann Arbor, MI, United States; [2]Department of Mechanical Engineering, VIT University, Vellore, Tamil Nadu, India; [3]Department of Mechanical Engineering, National Institute of Technology Karnataka, Mangaluru, Karnataka, India; [4]Department of Mechanical Engineering, Indian Institute of Technology Jammu, Jammu & Kashmir, India

## 14.1 Introduction

3D printing or additive manufacturing (AM) has now become a reality in the industrial production. It has proven itself foremost in prototyping and small batch manufacturing, and now it is propelling toward a wide range of applications such as medical, military, manufacturing, and utilities or fashion [1]. Due to cost benefits gain in production and time savings too, it has been a classic example of where innovation has been nipping at the heels of conventional manufacturing process. It allows to fabricate accurately, with high quality and reliability [2]. Moreover, AM shows its strengths where conventional manufacturing reaches its limits; for example, at commencing January 2020, the pandemic coronavirus (COVID-19) appeared. Many patients manifested severe symptoms of the respiratory illness, requiring specialist respirators to take over the lungs; however, respirators were in short supply apparently due to the rapid spread of the disease. To alleviate the respirator shortage, AM providers, manufacturers, and designers began to respond the global crisis by volunteering their appropriate skills to fabricate 3D-printed respirators. 3D-printed respirators lent a hand to the affected country governments to save many human lives [3,4]. Apart from respirators, 3D printing services further fabricated face shield, face mask, and ventilator components. They were fabricated using different 3D printing technologies based on materials and processing compatibility [5–7].

The primary motivation of AM processing techniques is to build the parts with good accuracy and mechanical properties for functional prototypes

*Additive Manufacturing*
ISBN 978-0-12-818411-0
https://doi.org/10.1016/B978-0-12-818411-0.00023-9

and products also known as solid freeform fabrication (SFF). The main classifications of AM based on processing techniques are vat photopolymerization, material extrusion, material jetting, binder jetting, powder bed fusion, direct energy deposition, and sheet lamination. They use different energy sources such as laser light, light projection, UV light, heat, bonding agents, or an electron beam. However, these processing techniques also tend to depend significantly on different materials such as polymer, metal, ceramics, and its composites [8,9]. However, laser-based processing techniques have several advantages compared with conventional processing techniques; particularly, they can produce relatively more accurate results with inducing much less heat [10]. Amid all of these techniques, this chapter is focused on the development of fabricating metal parts using AM techniques, in particular, by using laser-based processing methods. Several AM techniques have been developed to produce metal parts, such as 3D welding [11], microcasting [12], selective laser sintering (SLS) [13], laser-engineered net shaping [14], shape deposit manufacturing, directed light fabrication, direct metal deposition, selective laser melting [15,16], and also some other hybrid methods. As mentioned earlier, laser-based AM techniques have noteworthy advantages for producing metal parts. In laser melting, quite a few number of parameters govern the layer-by-layer building process, and they are sensitive and influence on each other. There are two main sets of significant parameters: one is based on the processing equipment, and the other one is related to the phenomena occurring in the interaction zone. The equipment-based parameters are the laser type, beam diameter or beam shaping optics, the nozzle, the powder and gas delivery system, and the substrate on the part bed [17,18]. The main interaction zone parameters are thermal conductivity, reflectivity (or conversely absorptivity), and the flowability of the powder material. In AM technologies, critical laser parameters are related to the laser—material interaction.

Moreover, all additively manufactured parts require some postprocessing stages to achieve the exact dimension and certain final mechanical properties. Production of a particular 3D-printed metal part involves certain steps, such as powder removal, stress relief, part removal, machining, advanced heat treatments, surface treatments, and final inspection. This chapter addresses all the aforementioned steps related to the fabrication of metal parts by using AM laser-based techniques, including the postprocessing methods. Initially, principles of the interaction of laser radiation with metallic particles and its classification are discussed. Furthermore, we have discussed the most common laser-based AM process such as selective laser melting (SLM) and directed energy deposition (DED). The working principle, types of laser used, and significant parameters influencing the part quality are reviewed. Eventually, different types of postprocessing techniques are addressed with respect to critical parameters influencing the end use of the product.

## 14.2 Laser additive metal manufacturing

### 14.2.1 History and classification of lasers

A laser is a device that emits a beam of coherent light generated by an optical amplification process. A laser beam is highly energetic, nearly monochromatic, and unidirectional because of its coherence. These characteristics confer it with a significant potential to produce very high power densities and make it suitable in many metallurgical applications, like welding, cutting, drilling, and heat treatment [19–21]. In all of these applications, the laser is utilized as a controlled heat source, which is produced by the thermalization of the laser light energy in the surface of the materials. Power densities in the focused laser spot around $10^3$–$10^{11}$W/mm$^2$ are easily obtained; these are quite higher than that emitted by conventional light and other heat sources utilized, for example, in welding. In laser processing, the power density can be controlled within the aforementioned range to achieve different conditions (heating, melting, or vaporization) required in different processes such as cutting, drilling, welding, and heat treatment.

In the early days, both solid-state and gas lasers have been used for material processing, and in particular for AM. However, most common lasers used for metallurgical processing are solid-state lasers (e.g., fiber, disk, or Nd:YAG lasers), gas lasers (e.g., $CO_2$ laser), dye lasers, diode lasers, and excimer lasers [22,23]. They all have their own set of components on their utilization in different applications [24]. The most relevant laser sources used in material processing and its wavelength are listed in Fig. 14.1. Some solid-state lasers, such as the Nd:YAG laser and (low power) fiber lasers, are mostly used for processing thin components, while others, such as the $CO_2$, disk, or fiber lasers, are used for processing thick materials. The choice of the laser for any particular application depends on the thickness of the material and on its physical, thermal, and optical properties. It may be noted that there are some differences in the performance of lasers, especially when Nd:YAG laser is compared with $CO_2$, fiber, or disk lasers [25,26]. The comparison between Nd:YAG laser, fiber laser, and $CO_2$ laser is summarized in Table 14.1.

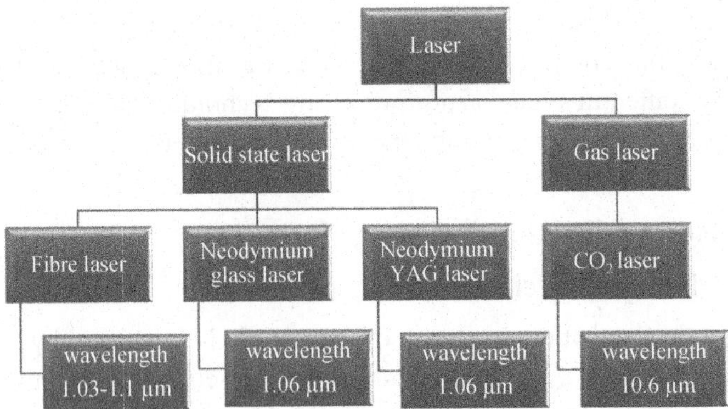


**Figure 14.1** Classification of most relevant lasers (and wavelength) used in laser materials processing, in particular, for AM. *AM*, additive manufacturing.

**Table 14.1** Comparison of Nd:YAG, fiber, and $CO_2$ lasers.

| | Criteria | Nd:YAG laser | $CO_2$ laser | Fiber laser |
|---|---|---|---|---|
| 1 | Wavelength | 1.06 μm | 10.6 μm | 1.03–1.1 μm |
| 2 | Maximum power | <1 kW | 100,000 W | 100,000 W |
| 3 | Laser efficiency | 2%–3% | 10%–20% | 70%–80% |
| 4 | Minimum spot size | Small | Big | Small |
| 5 | Depth of focus | Large | Low | Large |
| 6 | Reflectivity (metals) | Low | High | Low |
| 7 | Gas consumption | No gas | $CO_2$, $N_2$, He needed | No gas |
| 8 | Optics | Fused silica | ZnSe, GaAs | Fused silica |
| 9 | View during operation | Direct viewing | Special viewing | Direct viewing |
| 10 | Size | Small | Big | Small |
| 11 | Cost per watt | Cheaper | Costly | Cheaper |

## 14.2.2 Critical laser parameters in additive manufacturing

In general, when a laser beam strikes any material, the process of optical absorption is mainly driven by the free electrons, and the thermalization of the energy occurs by phonon/electron interaction which raises the energy of electrons in the conduction band. It may be mentioned that photon energy for the laser radiation emitted by a Nd:YAG or fiber laser ($E = hc/\lambda = 1.2$ eV, where h and c are Plank's constants and velocity of light, respectively, and λ is the radiation wavelength ≈ 1 μm) is too low; therefore, X-rays are not generated during the laser processing [27,28].

For dense materials, the absorption lengths are smaller than for the same materials in form of powder; however, for powder materials, only part of the incident radiation is absorbed on the outer surface of particles, and the rest of the radiation penetrates into the interparticle spaces of the loose powder, interacting with the underlying particles. Jehnming et al. studied the attenuation of the laser radiation by 304 L stainless steel powder jets with different powder stream focus and proved that 50% of the laser energy could be lost in the powder stream; further focused streams attain less powder catchment than the nonfocused streams during coaxial laser cladding process [29]. The working material gets heated due to the thermal conduction effect, and when the beam power density is very high, evaporation occurs as temperature rises significantly.

It may be noted that high power density and low interaction time result in the evaporation of materials, while low power density and high interaction time result in heating. Intermediate power density and interaction times are required in welding or joining where controlled melting is needed [30].

The absorptivity (or absorptance) of materials mainly depends on the laser wavelength and nature of materials to be processed. In AM process, powder materials are undergone laser processing. The laser interaction (absorption) in powder particles is highly influenced on the particle shape and size distribution. Due to this, the pore structure of the powder bed changes drastically in the course of laser processing. In AM, the morphological changes in materials due to the laser irradiation depend on the following:

**(a)** The materials properties, e.g., absorptivity, thermal conductivity, specific heat, etc. [31]

**(b)** Laser parameters, e.g., beam power, wavelength, beam diameter, pulse width [32], and pulsing rate [33] for pulsed laser processing

**(c)** The process parameters such as processing speed, shielding gas, and so on

Absorptivity is one of the most important parameters determining the laser interaction as the efficiency of the laser processing critically depends on the absorption and conversion of the light energy by the workpiece [34]. The absorptivity percentage (or absorptance) of different metallic powder materials typically used in laser-based AM for different laser wavelengths is shown in Fig. 14.2 [35]. As observed, the absorption to 1 μm laser radiation is higher as compared with 10.6 μm radiation emitted by the $CO_2$ laser (emitting 10.6 μm laser radiation). The difference is especially relevant for highly reflective materials, such as Cu and alloys. This, in combination with

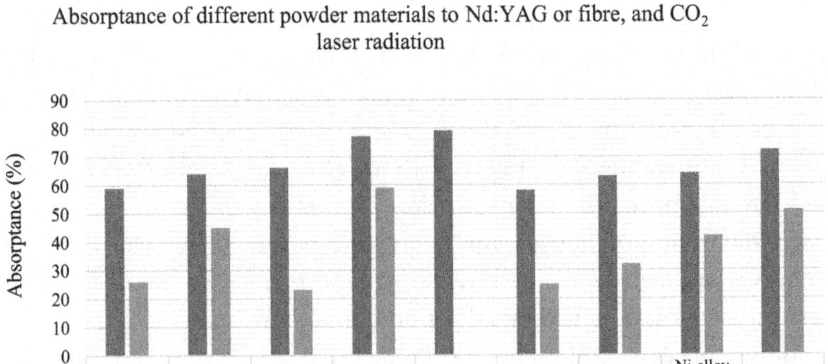

Absorptance of different powder materials to Nd:YAG or fibre, and $CO_2$ laser radiation

Absorptance (%)

| | Cu | Fe | Sn | Ti | Pb | Co-alloy (1% C; 28% Cr; 4% W) | Cu-alloy (10% Al) | Ni-alloy (13% Cr; 3% B; 4% Si; 0.6% C) | Ni-alloy (15% Cr; 3.1 % Si; 0.8% C) |
|---|---|---|---|---|---|---|---|---|---|
| Nd:YAG/Fibre | 59 | 64 | 66 | 77 | 79 | 58 | 63 | 64 | 72 |
| CO2 | 26 | 45 | 23 | 59 | | 25 | 32 | 42 | 51 |

**Figure 14.2** Absorptivity of different compositions of metallurgical powder materials to Nd:YAG or fiber and $CO_2$ laser radiation. *(Data extracted from Reference N. K. Tolochko, Y. V Khlopkov, S. E. Mozzharov, M. B. Ignatiev, T. Laoui, and V. I. Titov, "Absorptance of powder materials suitable for laser sintering," Rapid Prototyp. J: 2000. Vol. 6 No. 3, pp. 155–161. https://doi.org/10.1108/13552540010337029.)*

the high laser power available, and the better focusability of the 1 μm laser radiations, has made fiber lasers the preferred option as laser source in laser-based AM systems used for producing metallic parts.

Absorptivity of a surface depends upon the nature of material its roughness, state of oxidation, temperature, wavelength and power density of laser radiation, and also shape and size distribution of the powder. Nikolay et al. determined the normal spectral absorptance of powders for two laser wavelengths of 1.06 and 10.6 m (using Nd:YAG and $CO_2$ lasers, respectively). The result reveals that the absorptance depends on the laser wavelength. The absorptance for metal and carbide powders is lower for higher wavelengths, and vice versa was observed for oxides and polymers [35].

Laser parameters, such as beam power, beam diameter, and the interaction time, are important parameters for materials processing [36]. The first two can be combined into one single parameter, the power density or irradiance ($I = 8P/D_b^2$) (where P is total power and $D_b$ is beam diameter). This can be controlled by varying the power and diameter of the beam. It has also been observed that a minimum threshold power is required for a particular type of process and material [37–39]. For metal AM, the build rate increases with the laser power; however, the feature quality at a high build rate may be affected. Therefore, the laser energy must be higher than certain threshold (depending on the material), and it should be selected taking into account the build rate and feature quality [40].

### 14.2.3 Laser-based additive manufacturing techniques

Recently, metal-based AM processes have experienced a huge growth and interest in different industries. Based on this evolution, new and complex shapes and designs are entirely viable, whereas they could not be previously conceived [41]. There are diverse set of processes used to form feedstock into 3D objects. In the metal AM processes, the feedstock is bonded together into a dense part. Eventually, the metals are melted to attain the final dense product form [42]. In accordance with the interaction energy of these processing techniques on the feedstock, a classification of metal AM processing technologies is shown in Fig. 14.3. Each process has its own unique applications, strengths, and challenges. However, in this chapter, we have focused our attention into the laser-based metal AM techniques. In Fig. 14.3, the red (dark gray in printed version)-colored shapes denoted the laser processing techniques used in metal AM; they are selective laser melting

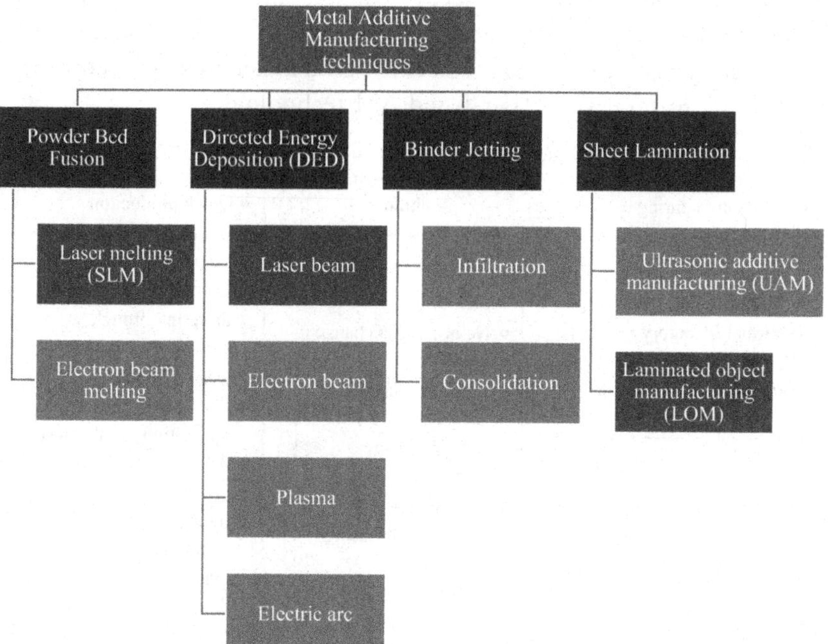


**Figure 14.3** Classification of metal additive manufacturing techniques.

(SLM), laser-assisted directed energy deposition (laser-DED), and laminated object manufacturing (LOM). These processing techniques use the laser as a heat source [43–45]. As mentioned earlier (in Fig. 14.2), the absorptivity of metals is higher for 1 μm lasers. In consequence, fiber lasers (and more marginally, Nd:YAG lasers) are used for metal AM. On the contrary, $CO_2$ lasers are recommended for polymer-based materials as their absorptivity is higher for this laser wavelength as compared with fiber lasers [46,47].

Across many industries, laser-based metal AM is identified as a potential alternative to conventional manufacturing process because it is cost-effective (even at low volumes), also with a shortened lead time and reduced part count, and offers many more advantages as shown in Fig. 14.4 [48,49]. Laser-based processing techniques are quite worthy because of the higher precision and accuracy than in other processing techniques. Based on the laser parameters, a good surface finish can be achieved, and the complete consolidation of metals is also possible by optimizing the laser power density [50]. The components can be produced quickly and directly from the CAD design. Mainly, this method eliminates support processes such as expensive tooling. Laser-based processing techniques allow the manufacturing of almost any metal material. From ferrous metals such as stainless steel to copper, aluminum, brass, and even other metal alloys or composites, laser-based AM manufacturing can be successfully performed [51,52]. Recently, gold materials were identified as challenging to process by laser-based AM technology.

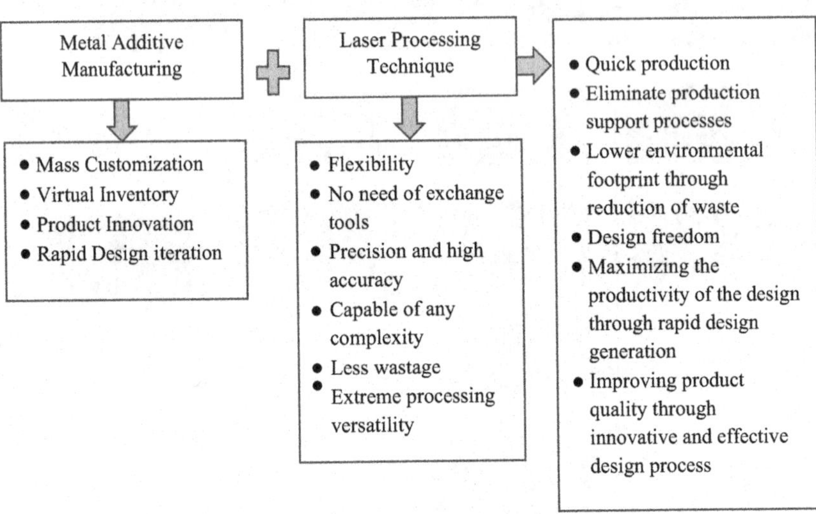


**Figure 14.4** Advantages of laser-based metal additive manufacturing techniques.

The most common laser-based metal AM processes are SLM [53] and DED [54]; both are discussed in detail in the forthcoming sections. Though AM process, in particular SLM process, is highly recommended for fabrication of highly complex metal parts, some challenges were identified. SLM process has been successfully applied using Ni-based alloys, Fe-based alloys, or Ti-based alloys; however, fabrication of aluminum alloy parts is still challenging because of its poor flowability and high reflectivity along with high thermal conductivity. Moreover, apart from fabrication of aluminum alloy parts, it is quite challenging to attain a high densification in other alloys, which significantly influences the mechanical properties of the parts produced by SLM [55].

#### 14.2.3.1 Selective laser melting

In early 1990s, selective laser melting (SLM) process was developed and used for producing metallic components; later, in 1999, the first SLM machine was developed by Fockele and Schwarze (F&S) with the support of the Fraunhofer Institute of Laser Technology ILT (Aachen). Initially, only steel materials were used as precursor materials in the machine. Subsequently, due to the huge attraction by different industrial sectors and researchers, the first SLM machine was commercialized in 2004, and in 2005, a high-resolution SLM machine (Realizer 100) was introduced. The huge interest of the market over AM systems promoted the development of other systems, such as those developed by Concept Laser GmbH and EOS GmbH that headed toward manufacturing metal-based AM machines such as laser curing and direct metal laser sintering. With further development on EOS in 2003, they established EOSINT M 270 DMLS and aimed for direct metal fabrication. In 2008, MCP Tooling Technologies Ltd. introduced SLM 250 and SLM 125 machines in the market. Recently, many other companies have developed SLM and DMLS machines for industrial and research purposes such as 3D Systems Corporation, Lumex, Renishaw, and so on. The recent manufacturers focused to develop advanced machines by incorporating CNC operation and advanced processing software. Apparently, scientists and researchers focused on optimizing processing parameters, developing in situ monitoring systems, and more recently on the application of artificial intelligence to fuse with SLM machines to achieve the best product quality [56,57].

Powder bed fusion (PBF) processes involve the selective melting or sintering of a layer of a powder bed using concentrated energy sources such as a laser or an electron beam [58]. In SLS process (SLS is discussed in detail in Chapter 2), the laser selectively hit the powder particles, and the surface of the powder particles is sintered to form a layer; in this case, polymer-based materials are used. However, metals must completely melt instead of sintering to achieve the proper coalescence in a dense layer. Remelting the previous layers during the melting of the current layer allows for a good adhesion between layers, and the production of a solid part [59,60]. A schematic of an SLM machine is shown in Fig. 14.5 (SLM is discussed in detail in Chapter 3). As depicted in this figure, the laser beam selectively melts the metal powders in a layer-wise fashion in contrast to SLS where the powder materials are only sintered. The hardware setup used to manage the optical energy in SLM machines is quite simple, as compared with the analogous one in other powder-bed processes such as in electron beam melting (EBM) machines. These include beam expanders, lenses, and scanning mirrors or galvanometers to deviate the laser beam along the powder bed. The powder bed system used in SLM consists of a dispersing piston and a roller to spread the powder layer by layer, as shown in Fig. 14.5. Technically, SLM process involves certain operation steps such as machine setup, operation powder recovery, and substrate removal. The operation of an SLM machine is governed by optimized scanning strategies and processing parameters [61].

Apart from the hardware setup, the type of laser is a significantly important parameter to deal when processing metal powders. As discussed earlier, metallic powders have higher absorptivity to the 1 μm laser radiation than to that emitted for a $CO_2$ laser (see Fig. 14.2). Initially, Nd:YAG

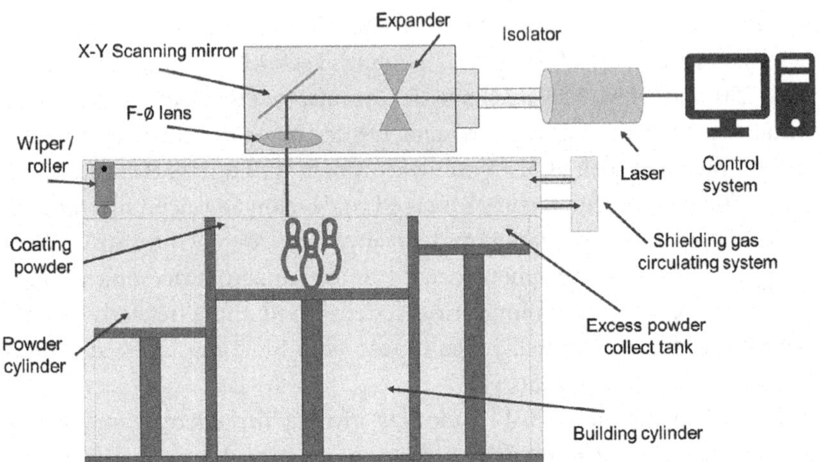


**Figure 14.5** General schematic of selective laser melting (SLM) process.

lasers were considered for metal melting processes. However, the better beam quality of fiber and disk lasers made these laser sources as the preferred option for the manufacturers of SLM machines [62–64].The laser power for these sources typically ranges from 50 W to 2 kW.

The fast melting and cooling occurring during laser processing leads to the formation of small grains, nonequilibrium phases, and new chemical compounds. These phenomena are also related to the composition of the metal powders and have an impact on mechanical properties such as yield strength, ultimate tensile strength, and ductility of the fabricated part. SLM process produces high-strength parts when compared with laser sintering process. The main metal powders used in SLM are steel and iron-based alloys, titanium and titanium-based alloys, Inconel625, aluminum-based alloys, cobalt-based alloys, gold, copper, and so on. Although SLM process has been identified as a versatile technology, there are some common issues occurring in the melting process. The main defect is the porosity, which is relevant in metal AM process. Other defects are mainly related to the scan strategy, process temperature, feedstock, and build chamber atmosphere. The observation of the defects during the laser melting process improves the process reliability and the quality of the fabricated parts [65,66]. The detailed discussion of defects and need for postprocessing techniques are discussed in Section 14.3.

#### 14.2.3.2 Laser-assisted directed energy deposition

Laser-DED is based on the well-established technology of laser cladding, which is used in aerospace, marine, defense, oil, and gas industries to repair components and reconditioning systems such as aircraft frames and structures, refractory metal components, ballistic materials, or marine propulsion components (laser-based DED is discussed in detail in Chapter 4) [67]. This process is quite popular due to its unique features which allows for a wide range of new applications. DED technologies are used to repair (adding material to existing components) or to create new metal parts [68]. The main advantages are high build rates, production of dense and strong parts, and near net shaping; it can be used for repairing components, multimaterial possibility, production of large parts, easy material change, and reduced material waste. DED process deposits the powder particles as per the need to build the components, whereas in the SLM system, the build platform has to be filled with metal powder, which leads to a large material wastage [69]. However, there are some challenging limitations occurring during DED processing; mainly, they are a low build resolution and no support structures.

After processing, the parts exhibit a poor surface finishing, and overhangs are not possible. To overwhelm the poor surface finish of the product, secondary processes or postprocessing is highly recommended (these are detailed in Section 14.3 of this chapter).

DED technology not only is limited to metal materials but also suits for polymer and ceramic material processing; however, DED process is predominantly used for producing metal parts. DED processes are classified into four types based on the energy source: laser beam, electron beam, plasma, and electric arc. Furthermore, in all of these classes, they can be either powder-based or wire-based feedstock DED systems. In this chapter, as discussed earlier, we are only going to consider the laser-based processing systems. Fig. 14.6 shows the schematic of a laser-based DED processing system using a wire feedstock hardware setup. The DED system uses an inert gas injection system for laser processing to avoid undesirable reactions with the ambient; since the laser-assisted DED processing involves in creating a molten metal pool for layer deposition, this inert gas plays an important role in the product quality, in particular for reactive metals like titanium. As shown in Fig. 14.6, the main part of the system is the laser head, which transfers the laser energy source and the wire feedstock (or the powder injection); this is supplied near to the laser beam focus, where

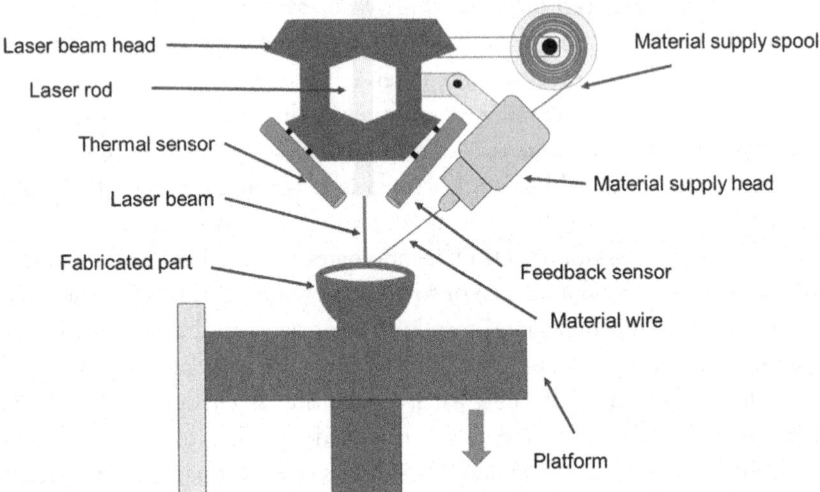


**Figure 14.6** Schematic of directed energy deposition (DED) using wire feedstock (powder material, instead of wire is also used as feedstock).

the part is build. In powder-based systems, the powder jet supplied by the nozzles (coaxially or laterally) converges at the point of deposit. The laser beam head is generally mounted in multiaxis CNC head, but in some cases, an articulated robot arm is also used. The process starts with the movement of the laser head and build platform to generate the layer-by-layer process. The laser beam melts and creates the molten pool of the metal materials to deposit on the substrate at the origin point along the build path. The feeding system supplies the wire or powder feedstock to fabricate dense metal parts. The materials used are titanium and alloys, Inconel718 or 625, Hastelloy, tantalum, tungsten, niobium, stainless steels, aluminum alloys, zinc alloys, and copper—nickel alloys. Similar to SLM process, fiber and disk lasers (and more marginally Nd:YAG lasers) are used in current commercial machines which are mainly manufactured by Optomec, POM, and Accufusion Inc. [70—73].

DED laser-based systems significantly depends on certain process parameters to attain the best part quality; typically, they are laser power, laser beam spot size, powder or wire feed rate, scanning speed, gas flow rate, clad angle, feedstock properties, and layer dimension [74]. Consequently, a wide range of process parameters integrated with complicated transport phenomena, such as conduction of heat into the substrate, convention due to Marangoni effects, and scattering of the laser radiation by the powder particles, lead to a high complexity to understand the effect of these individual process parameters on the overall DED process [75]. Due to this complexity, some defects such as porosity, changes in chemical composition due to solute segregation and loss of alloying elements, and printability of alloys can hinder the process. All of the alloys are not suitable to be processed by laser-based DED. The printability of alloys is defined using a dimensionless parameter known as thermal strain. Thermal strains arise due to volume changes caused by both temperature differences and phase transformations, including solidification and solid-state phase changes. Lower values of thermal strain lead to lesser residual stress in the material, which increases the printability of the alloy by laser-based AM process. Moreover, there are lot of limitations and challenges in DED process due to its extreme heating and cooling rates. It is expected that future investigations, experimentally and/or computationally conducted, on printability of other alloy materials and composites expands the applicability limits and reduce these challenges [76,77]. Based on these defects, postprocessing of laser-based metal AM is required. This is discussed in Section 14.3.

## 14.3 Postprocessing techniques for additive manufactured components

### 14.3.1 Need for postprocessing

Components that are built using AM techniques require postprocessing operations to prepare the part for the intended form or function. Post-processing operations are implemented for improving the surface texture reliability of the components. As mentioned earlier, AM is being widely accepted by industrial designers due to its enormous advantages. However, AM is still not able to implement up to a full extend, even though AM is being extensively used in the industry. Feuerhahn et al. [78] studied the microstructure and properties of selective laser-melted high hardened tool steel. The results reveal low hardness values due to irregular microstructure, though the samples were crack-free and with high density. Furthermore, to increase the hardness, the samples were heat-treated and attained a homogenous microstructure with ultrafine carbide precipitation. Functionality of a mechanical component depends on their microstructure, surface characteristic, and residual stress developed during printing. Fukuo Hashimoto et al. [79] investigated the influence of surface property on the product functionality. Different finishing operations such as grinding, superfinishing, hard turning, and isotropic finishing were done to evaluate the change in fatigue life of the components with changing surface properties. Improved surface property helps to increase the fatigue life and improves the frictional parameters. As a result, the torque and heat generated may reduce and also increases the wear resistance of the material. C Li et al. [80] and Protasov et al. [81] predicted the residual stresses in selective laser melting with help of a multiscale modeling approach, and it was found that maximum tensile residual stresses (TRSs) appears on the top portion of the part and further decreases as the depth increases. The changes in the residual stress between the boundary of the part and the substrate due to tensile and compressive stresses (in the part and substrate respectively) were observed. Molaei et al. [82] studied about the fatigue design with additive manufactured metals, and it was found that the most common defects for fatigue crack growth is the air entrapped in the form of irregular voids. Residual stresses initiate fatigue cracks, but their effect can be reduced with help of suitable heat treatments. Greitemeier et al. [83] studied the effect of the surface roughness on the fatigue life of Ti–6 Al–4V parts that were manufactured using DMLS and EBM, which have then undergone a heat treatment to relieve residual stress. Fatigue test and fatigue crack growth tests along with roughness measurement were done, and it was found that surface roughness significantly influences the fatigue performance due to multiple stress concentration. Roughness found in the material was due to the solidification of the melt pool and the adhesion of partially melted powder particles. During fractographic analyses, the appearance of multiple crack initiation at the surface was found, and it was concluded that high cycle fatigue properties were dominated by surface roughness. Most of the components that are used in the industry undergo cyclic loading states through their life cycle. Multiaxial stresses are common in many additively manufactured components. Even for uniaxial loading conditions, stress might be multiaxial due to the geometry complexity, notches, and residual stresses. Pegues et al. [84] studied about the effect of the surface roughness on the fatigue life of additively manufactured Ti–6Al–4V parts. It was found that crack initiation was highly sensitive to surface roughness where all cracks initiated from the rougher down the skin surface. So, it was evidently found that poor surface finishing obtained during the manufacturing of a product with AM results in poor mechanical property such as fatigue life.

Table 14.2 clearly explains the need of postprocessing due to the effects caused as a result of the printing that it will adversely affect the functionality of the final part. Suitable postprocessing operations will help to overcome these drawbacks. Postprocessing techniques are certain mechanical or thermal operations done on additively manufactured components to ensure a proper reliability. Effect of different postprocessing operations on various materials is shown in Table 14.3.

## 14.3.2 Classification of postprocessing methods

Postprocessing of additively manufactured components is mainly categorized into two categories. Postprocessing methods adopt thermal methods

**Table 14.2** Additive manufacturing limitations and effects.

| Defects in additively manufactured parts | Effects on printed parts |
|---|---|
| 1. Irregular microstructure | • Lower hardness, reduced strength |
| 2. Poor surface texture | • Reduction in fatigue life<br>• High friction in mating components which result in heat generation |
| 3. Residual stress | • Tensile residual stress reduces the fatigue life |

**Table 14.3** Variation of parameter with different postprocessing operation.

| Postprocessing method | Material | Parameter influenced | Remarks |
|---|---|---|---|
| Heat treatment (Feuerhahn et al. [78]) | X110CrMoVAl 8-2 | • Microstructure<br>• Hardness | • Fine and homogenous microstructure<br>• Increase in hardness |
| Laser shock peening (Kalentics et al. [85]; Sun et al. [86]) | SS 316L | • Residual stress | • TSR converted into CSR<br>• Smaller spot size induces maximum CRS<br>• Larger spot size increase depth of CRS |
| | 2319 aluminum alloy | • Residual stress<br>• Grain size<br>• Microhardness | • Improvement in surface hardness due to high density dislocations<br>• Refined microstructure<br>• Improved tensile properties |
| Laser polishing (Zhihao et al. [87]; Guan et al. [88]) | Inconel 178<br>Ti–6Al–4V<br>Inconel 625 | • Surface roughness | • Reduction in surface roughness<br>• Increase in hardness<br>• Increased wear resistance |
| Burnishing (Zhang et al. [89]) | Cr–Ni based stainless steel | • Surface quality<br>• Hardness<br>• Residual stress | • Increase in surface quality and hardness<br>• Plastic deformation resulted in CRS<br>• Improvement in porosity and bonding strength. |
| Shot peening (Maamoun et al. [90]; Uzan et al. [91]) | AlSi10Mg<br>17-4 stainless steel | • Hardness<br>• Residual stress<br>• Wear resistance | • Depth of fatigue crack initiation increased<br>• Improved wear resistance and fatigue life |

to modify the surface texture of properties; this includes heat treatment, laser shock peening (LSP), and laser remelting. In other cases, mechanical finishing operations are adopted. Fig. 14.7 shows different postprocessing methods adopted to enhance the properties of additively manufactured components.

#### *14.3.2.1 Laser shock peening*

In LSP, a laser beam is used to create a plasma shock wave which is reflected on the surface of the component inducing a compressive residual stress (CRS) which deeply penetrates into the material, as shown in Fig. 14.8. Kalentics et al. [85] introduced3D LSP—a new method for the 3D control of residual stresses in parts produced by selective laser melting. This work describes a hybrid manufacturing process in which LSP and selective laser melting were combined to process austenitic steel parts. TRSs are easily converted into CRSs. The laser spot size defines the depth and

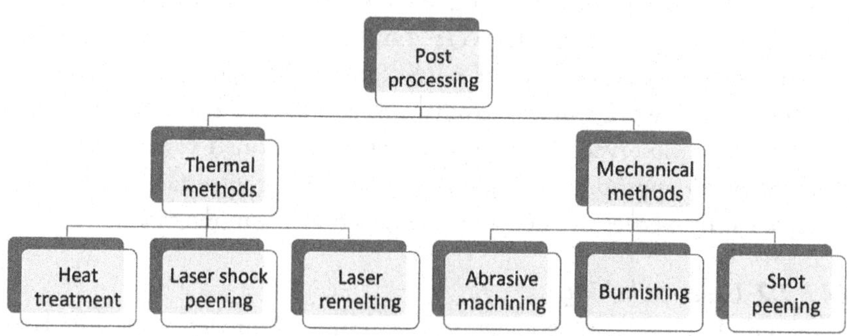


**Figure 14.7** Different postprocessing methods.

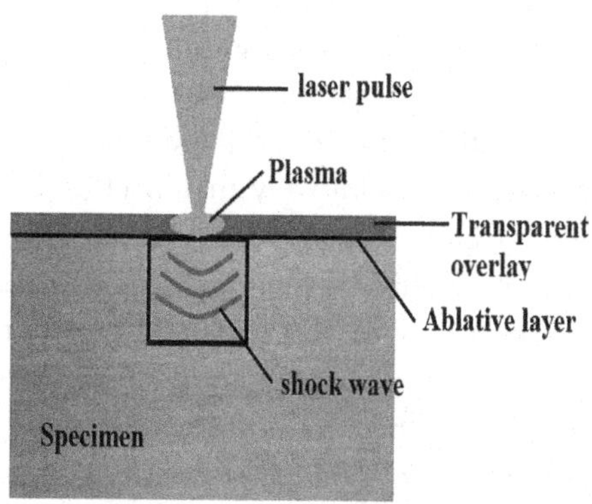


**Figure 14.8** Schematic diagram of laser shock peening (LSP).

area of the induced CRSs. It was also found that increasing the number of SLM layers between LSP treatments leads to an increase of CRS depth. In this study, LSP of SLM-printed SS 316L enabled to convert the TRSs induced in the subsurface region due to the high temperature gradient between the layers into CRSs. TRSs have a detrimental effect on the fatigue life. CRSs induced as a result of LSP will compress the TRS, which will increase the fatigue life. Sun et al. [86] studied the microstructural and mechanical properties of wire arc additively manufactured 2319 aluminum alloy subjected to LSP. High density dislocation caused due to LSP increased the microhardness at the surface level. LSP was able to impart a maximum compressive stress of 100 MPa up to a depth of 0.75 mm, which significantly improved the tensile properties. The result revealed that LSP significantly improved the microstructure, and microhardness as the TRSs was converted into CRSs. Furthermore, it was observed that the yield strength was increased by 72%, and no influence on the ultimate tensile stress (UTS) was detected. Hackel et al. [92] discussed LSP as a post-processing operation for AM. Fatigue strength of unpenned, shot peened, and laser-shock peened samples were compared and observed that LSP was able to impart a deeper level of CRS. Since LSP is apparently having a minimum cold work effect, no changes in material hardness and yield strength were observed. Based on the aforementioned review, it shows that LSP is an appropriate method for improving the surface quality and mechanical properties of additively manufactured components.

#### *14.3.2.2 Laser polishing*

Laser polishing is a noncontact surface finishing process that uses laser irradiation to achieve subsequent surface smoothening, as shown in Fig. 14.9 [93]. In the process, a laser beam is used to melt the surface of the work piece to reduce the surface roughness. The remelting temperature must be chosen according to the initial surface roughness and material which is being used. Dadbakhsh et al. [94] and Guan et al. [88] studied the surface finish improvement of additively manufactured samples after using laser polishing. Laser polishing of Ti−6Al−4V parts helps to reduce the surface roughness by the ability of remelted liquid material to uniformly spread due to the surface tension and gravity. Formation of α-martensite during remelting of titanium alloy resulted in higher hardness. Reduction in surface roughness also helped to improve the wear resistance of the components. It was noted that due to laser polishing, the surface roughness was reduced to 80% than the initial surface roughness. Zhihao et al. [87]

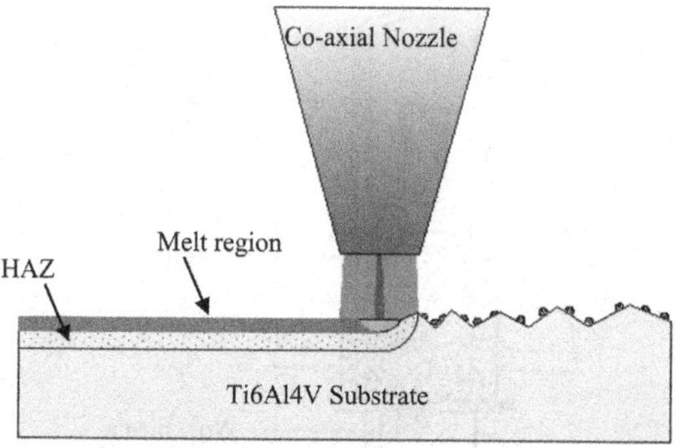


**Figure 14.9** Schematic diagram of laser polishing. *(Reproduced with kind permission of Elsevier from Reference S. Marimuthu, A. Triantaphyllou, M. Antar, D. Wimpenny, H. Morton, and M. Beard, "Laser polishing of selective laser melted components," Int. J. Mach. Tools Manuf., vol. 95, pp. 97–104, 2015, doi: https://doi.org/10.1016/j.ijmachtools.2015.05.002.)*

studied the laser polishing of an additively manufactured superalloy parts and found that laser polishing improves the surface property of parts manufactured by SLM. There was a sufficient decrease in the average surface roughness (Ra), from 5 μm to less than 0.1 μm, and mean roughness depth (Rz) values from 31 to 0.6 μm. Furthermore, it was observed the reduction in the grain size along with a decrease in the value of coefficient of friction and wear resistance.

#### *14.3.2.3 Abrasive machining*

To attain a high dimensional accuracy, a small amount of material from the surface can be removed with the help of abrasive particles which are small, and hard, having sharp edges and an irregular shape. A hydraulic ram forces the abrasive medium through the work piece, as shown in Fig. 14.10. Abrasive particles may be quartz, zirconium oxide, tungsten carbide, or aluminum oxide. Wang et al. [95] studied the finishing of additively manufactured metals after abrasive flow machining and found that the surface roughness of AM parts was about 10–50 μm, whereas for convectional manufacturing, they had values of roughness about 2.5 μm. In the experiment, abrasive with hybrid grit was adopted to implement the polishing of the material, and after machining, the surface quality improved significantly by reducing the surface roughness up to a value of 0.94 μm. Tan et al. [96] investigated on a postprocessing technique called ultrasonic cavitation abrasive finishing (UCAF). In this investigation, the abrasive particles were fed amid of the transducer and horn of an ultrasonic

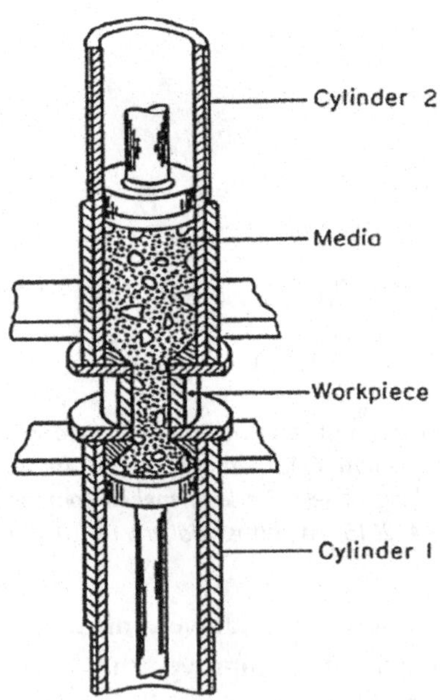


**Figure 14.10** Schematic diagram of abrasive flow machining. *(Reproduced with kind permission of Elsevier from Reference A. Maamoun, M. Elbestawi, and S. Veldhuis, "Influence of shot peening on AlSi10Mg parts fabricated by additive manufacturing," J. Manuf. Mater. Process., vol. 2, no. 3, p. 40, 2018, doi: https://doi.org/10.3390/jmmp2030040.)*

wave of 20 Hz. Abrasive machining on Inconel parts resulted in a high-quality surface finishing. The process involved in the removal of large sized surface peaks due to the rubbing action of abrasive particles (as shown in Fig. 14.11) results in a homogenized surface texture, thereby reducing the fatigue crack growth from the surface. This process potentially removes the irregularities of additively manufactured components. The initial surface roughness of the material was between 6.5 and 7.5 μm which was efficiently reduced up to 3.65 μm after the postprocessing operation.

### *14.3.2.4 Burnishing*

Burnishing is a superfinishing operation in which the plastic deformation of the surface, due to sliding contact with another object, is produced (as shown in Fig. 14.12). This results in a mirrorlike finishing. As the stress in the contact area increases, the material is deformed. Salmi et al. [97] studied the

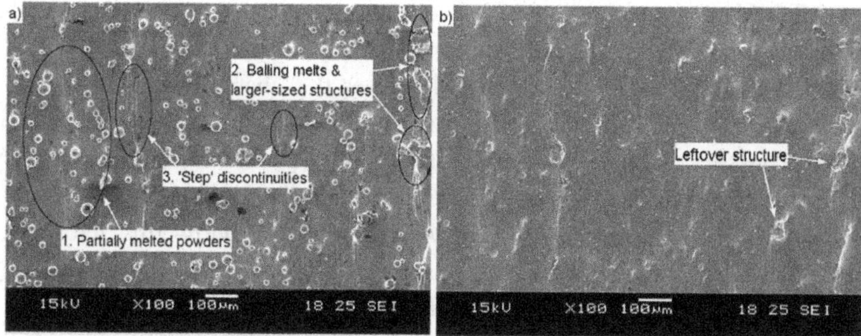


**Figure 14.11** (A) Typical side surface characteristics of as-manufactured DMLS Inconel 625 component. (B). Surface topography after 30 min of UCAF processing. *UCAF*, ultrasonic cavitation abrasive finishing. *(Reproduced with kind permission of Elsevier from Reference K. L. Tan and S. H. Yeo, "Surface modification of additive manufactured components by ultrasonic cavitation abrasive finishing," Wear, vol. 378–379, pp. 90–95, 2017, doi: https://doi.org/10.1016/j.wear.2017.02.030.)*

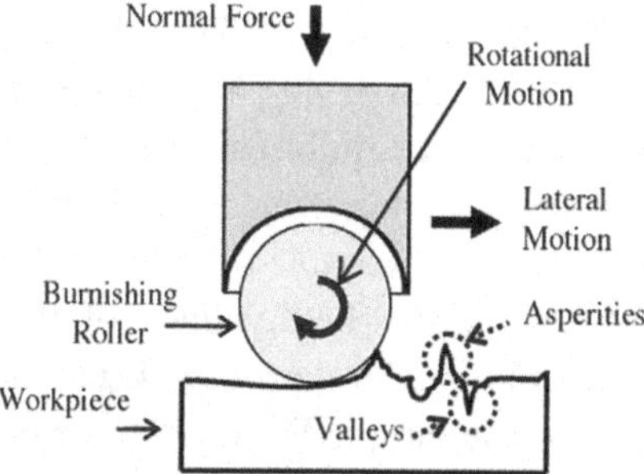


**Figure 14.12** Schematic diagram of ball burnishing. *(Reproduced with kind permission of Elsevier from Reference N. E. Uzan, S. Ramati, R. Shneck, N. Frage, and O. Yeheskel, "On the effect of shot-peening on fatigue resistance of AlSi10Mg specimens fabricated by additive manufacturing using selective laser melting (AM-SLM)," Addit. Manuf., vol. 21, no. 2010, pp. 458–464, 2018, doi: https://doi.org/10.1016/j.addma.2018.03.030.)*

effect of ultrasonic burnishing on parts made by AM. As a result, it improves the surface quality and surface hardness. Zhang et al. [89] investigated the effect of sequential turning and burnishing on the surface integrity of Cr–Ni-based stainless steel formed by laser cladding. After sequential turning and burnishing operations, it was observed that there was an improvement in the surface topography, roughness, residual stresses, and microhardness after the machining of the samples. Porosity and bonding strength were also found to be improved due to the formation of CRSs during the machining process.

#### 14.3.2.5 Shot peening

Shot peening is a cold working process which is used to impart a CRS, which will suppress the TRSs that decreases the chances of crack propagation. In shot peening, solid objects such as metal shots, glass, or beads are used. This shot is made to impact on the solid part at certain velocity as shown in Fig. 14.13; as a result of the impact, a plastic deformation occurs on the surface of the material, and the material just below this deformed area resists this deformation. In consequence, a CRS is induced in the material. AlMangour et al. [98] studied the improvement of the surface quality and mechanical properties of additively manufactured stainless steel parts fabricated using DMLS. In general, shot peening effectively enhances the roughness, hardness, compressive yield strength, and wear resistance of the part. Accordingly, it was noted that due to shot peening, grain size refinement occurred in the material; apparently, it is associated with the activity of the high dislocation density and the formation of smaller shear bands. In this study, the initial residual stress (with a value of −119MPa) was increased up to a value of −700 MPa. Maamoun et al. [90] studied the influence of shot peening on AlSi10Mg parts manufactured by AM. It was observed that shot peening successfully eliminated surface defects. Author also compared the effect of shot peening and LSP on the CRSs in the machined surface. Shot peening was able to induce more CRS than LSP, as shown in Fig. 14.14 [99]; also, the hardness of the material increased significantly due to the cold working process. Uzan et al. [91] studied the effect of shot peening on the fatigue life of AlSi10Mg manufactured with selective laser melting. It shows that nanoindentations were present on the

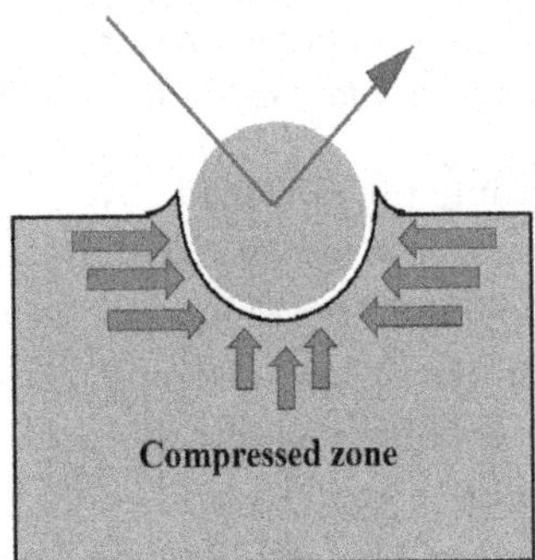


**Figure 14.13** Schematic diagram of shot peening.

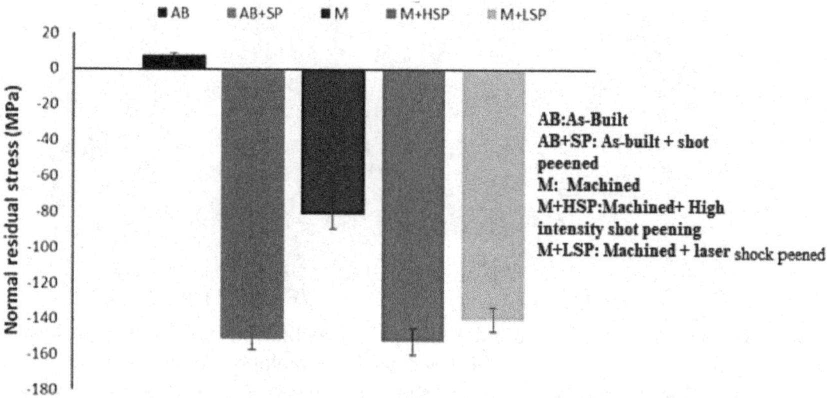


**Figure 14.14** Variation of residual stress with different peening process.

surface of the material and an increase in surface hardness was also identified. In shot peened samples, the depth of fatigue crack initiation was found to be lower than the one which is observed in additively manufactured samples [100].

#### 14.3.2.6 Heat treatment

To enhance the mechanical and microstructural properties, Feuerhahn et al. studied the heat treatment of X110CrMoVAl 8–2 parts produced by SLM [78]. Selectively laser-melted X110CrMoVAl 8–2 parts show an irregular microstructure with few voids observed along with limited carbide precipitations which resulted in low hardness value. After a full heat treatment, a fine and homogenous microstructure is produced (as shown in Fig. 14.15B) along with carbide precipitation, which resulted in high hardness values.

After the postprocessing operation, a better microstructure and hardness were attained than in the conventional manufactured components. DMLS-printed Ti–6Al–4V parts were found to be imparted with residual stresses due to the high cooling rates and steep temperature gradient, which can affect the fatigue life. TRS imparted in the material can initiate the fatigue crack growth. Stress relief annealing, at suitable temperatures, helps to reduce the residual stresses imparted due to printing. Imparting CRS will help to increase the fatigue life of the material.

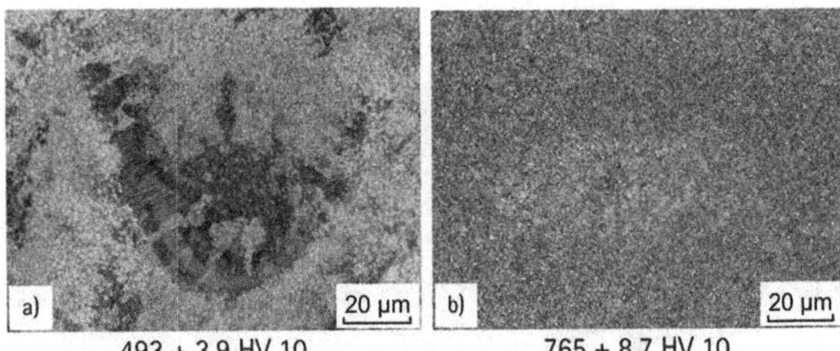


**Figure 14.15** SLM-printed X110CrMoVAl 8–2. (A) As-built microstructure; (B) Microstructure after heat treatment. *SLM*, selective laser melting. *(Reproduced with kind permission of Elsevier from Reference F. Feuerhahn, A. Schulz, T. Seefeld, and F. Vollertsen, "Microstructure and properties of selective laser melted high hardness tool steel," Phys. Procedia, vol. 41, pp. 843–848, 2013, doi: https://doi.org/10.1016/j.phpro.2013.03.157.)*

## 14.4 Future scope of laser-based additive manufacturing

AM of metal components is currently a noteworthy manufacturing approach based on its potential growth in a wide range of industries. Industries such as aerospace, automotive, electronics, defense, and medical are highly motivated to use AM processes. Evidently, AM processes proved their potential with respect to traditional methods; however, cost-effective based on equipment, material, and design standards are still in a limited range. In particular, laser-based equipments are quite costlier than other processing techniques. The SLM and DED machine cost or hardware setup margin rate is highly influenced by the selection of the laser system. Therefore, reduction of cost in equipment and material will provide an incentive for any upcoming new industries to potentially use this technology. Apparently, to achieve this low-cost investment, in the near future, AM process must work together with subtractive or forming technologies. Currently, commercial machines based on laser-assisted metal AM process incorporate laser sources with an average power around 1 −6 kW. Then, it is foreseeable that new laser sources, with a higher power, more efficient, and better performance control, could contribute in the near future to address the existing AM process limitations in terms of repeatability, part reproducibility, layer thickness/deposition rate, and high energy use.

## Acknowledgment

We would like to show our gratitude to the Center for Laser-Aided Intelligent Manufacturing (CLAIM), University of Michigan, United States, and Department of

Mechanical Engineering, Indian Institute of Technology Jammu, VIT University and NIT Surathkal, India, for their support and encouragement. Also, we thank to Mr. Murali Krishnan researcher at SUTD, Singapore, for assistance with designing and modeling of process schematic diagram as well as manuscript consolidations.

## References

[1] J. Savolainen, M. Collan, Additive manufacturing technology and business model change—a review of literature, Addit. Manuf. (2020) 101070.

[2] P. Sahoo, S.K. Das, 1 Emerging trends in additive and subtractive manufacturing, Addit. Subtract. Manuf. Emerge. Technol. 4 (2020).

[3] D.C.Y. Liu, et al., "Adapting Reusable Elastomeric Respirators to Utilise Anaesthesia Circuit Filters Using a 3D-printed Adaptor; a Potential Alternative to Address N95 Shortages during the COVID-19 Pandemic, *Anaesthesia*, 2020, https://doi.org/10.1111/anae.15108.

[4] R. Tino, et al., COVID-19 and the Role of 3D Printing in Medicine, Springer, 2020, https://doi.org/10.1186/s41205-020-00064-7, 3D Print Med 6, 11 (2020).

[5] J.M. Zuniga, A. Cortes, "The role of additive manufacturing and antimicrobial polymers in the COVID-19 pandemic." Taylor & Francis, Expet Rev. Med. Dev. 17 (6) (2020) 477–481, https://doi.org/10.1080/17434440.2020.1756771.

[6] S.T. Flanagan, D.H. Ballard, 3D Printed Face Shields: A Community Response to the COVID-19 Global Pandemic, *Acad. Radiol*, 2020.

[7] N.J. Rowan, J.G. Laffey, Challenges and solutions for addressing critical shortage of supply chain for personal and protective equipment (PPE) arising from Coronavirus disease (COVID19) pandemic—Case study from the Republic of Ireland, Sci. Total Environ. (2020) 138532.

[8] R. Velu, F. Raspall, S. Singamneni, 3D printing technologies and composite materials for structural applications, Green Compos. Automot. Appl. (Jan. 2019) 171–196, https://doi.org/10.1016/B978-0-08-102177-4.00008-2.

[9] R. Velu, T. Calais, A. Jayakumar, F. Raspall, A comprehensive review on bio-nanomaterials for medical implants and feasibility studies on fabrication of such implants by additive manufacturing technique, Materials 13 (1) (Dec. 2019) 92, https://doi.org/10.3390/ma13010092.

[10] Z.C. Oter, et al., Benefits of laser beam based additive manufacturing in die production, Optik 176 (2019) 175–184.

[11] K. Zhang, M. Yan, T. Huang, J. Zheng, Z. Li, 3D reconstruction of complex spatial weld seam for autonomous welding by laser structured light scanning, J. Manuf. Process. 39 (2019) 200–207.

[12] C. Shao, S. Zhao, X. Wang, Y. Zhu, Z. Zhang, R.O. Ritchie, Architecture of high-strength aluminum—matrix composites processed by a novel microcasting technique, NPG Asia Mater. 11 (1) (2019) 1–12.

[13] J. Chen, G. Wu, Y. Xie, Z. He, N. Wan, Y. Wu, Study on performance of metal and polymer composites parts based by additive manufacturing, in: 2019 20th International Conference on Electronic Packaging Technology, ICEPT), 2019, pp. 1–4.

[14] M. Izadi, A. Farzaneh, M. Mohammed, I. Gibson, B. Rolfe, A review of laser engineered net shaping (LENS) build and process parameters of metallic parts, Rapid Prototyp. J. 26 (6) (2020) 1059–1078, https://doi.org/10.1108/RPJ-04-2018-0088.

[15] S. Pandey, R. Srivastava, R. Narain, Study on the effect of powder feed rate on direct metal deposition layer of precipitation hardened steel, Mater. Today Proc. 26 (2020) 2272–2276, part 2, Issue n/a.

[16] A. Khorasani, I. Gibson, U.S. Awan, A. Ghaderi, The effect of SLM process parameters on density, hardness, tensile strength and surface quality of Ti-6Al-4V, Addit. Manuf. 25 (2019) 176–186.
[17] L. Moniz, et al., Additive manufacturing of an oxide ceramic by laser beam melting—comparison between finite element simulation and experimental results, J. Mater. Process. Technol. 270 (2019) 106–117.
[18] E. Alexander, Residual Stresses in Laser Beam Melting (LBM)–Critical Review and Outlook of Activities at BAM, 2019.
[19] S. Asadi, T. Saeid, A. Valanezhad, J. Khalil Allafi, Dissimilar laser welding of NiTi shape memory alloy to austenitic stainless steel archwires, J. Weld. Sci. Technol. Iran 5 (2) (2020) 141–152.
[20] P. Hirsch, et al., Effect of thermal properties on laser cutting of continuous glass and carbon fiber-reinforced polyamide 6 composites, Mach. Sci. Technol. 23 (1) (2019) 1–18.
[21] T. Gilboa, E. Zvuloni, A. Zrehen, A.H. Squires, A. Meller, Automated, ultra-fast laser-drilling of nanometer scale pores and nanopore arrays in aqueous solutions, Adv. Funct. Mater. (2019) 1900642.
[22] H.J. Eichler, H. Fritsche, O. Lux, S.G. Strohmaier, High power diode and solid state lasers, in: XXI International Symposium on High Power Laser Systems and Applications 2016, vol. 10254, 2017, p. 102540R.
[23] M. Endo, R.F. Walter, Gas Lasers, CRC Press, 2018.
[24] D. Haley, O. Pratt, Basic principles of lasers, Anaesth. Intens. Care Med. 18 (12) (2017) 648–650.
[25] N.A. Saud, S.A. Faris, The role of Nd-YAG laser with A wavelength of 1064 Nm for the treatment of skin wounds in laboratory mice, J. Educ. Pure Sci. Thi-Qar 9 (2) (2019) 16–24.
[26] J. Conroy, Investigation of improved label cutting by $CO_2$ lasers with wavelength optimization, in: International Congress on Applications of Lasers & Electro-Optics, vol. 2016, 2016, p. 2004, 1.
[27] M. V Shugaev, et al., Fundamentals of ultrafast laser–material interaction, MRS Bull. 41 (12) (2016) 960–968.
[28] M. V Shugaev, et al., "Advances in the application of lasers in materials science, chapter 5, insights into laser-materials interaction through modeling on atomic and macroscopic scales," oak ridge national laboratory, oak ridge leadership computing facility OLCF, in: Springer Series in Materials Science, vol. 274, Springer, Cham, 2019, https://doi.org/10.1007/978-3-319-96845-2_5.
[29] J. Lin, Laser attenuation of the focused powder streams in coaxial laser cladding, J. Laser Appl. 12 (1) (2000) 28–33, https://doi.org/10.2351/1.521910.
[30] J. Dutta Majumdar, I. Manna, Laser processing of materials, Sadhana Acad. Proc. Eng. Sci. 28 (3–4) (2003) 495–562, https://doi.org/10.1007/BF02706446.
[31] A. Rubenchik, I. V Golosker, M.M. LeBlanc, S.C. Mitchell, S.S. Wu, System and Method for the Direct Calorimetric Measurement of Laser Absorptivity of Materials, Google Patents, 19 March, 2019.
[32] R. Sundar, et al., Laser shock peening and its applications: a review, Lasers Manuf. Mater. Process 6 (4) (2019) 424–463.
[33] S. Miller, C.G. Jones, K. Mitra, Modeling and experimental analysis of thermal therapy during short pulse laser irradiation, in: Modeling of Microscale Transport in Biological Processes, Elsevier, 2017, pp. 243–259.
[34] S.J. Wolff, Laser-Matter Interactions in Directed Energy Deposition, Northwestern University, 2018.

[35] N.K. Tolochko, Y. V Khlopkov, S.E. Mozzharov, M.B. Ignatiev, T. Laoui, V.I. Titov, Absorptance of powder materials suitable for laser sintering, Rapid Prototyp. J. 6 (3) (2000) 155–161, https://doi.org/10.1108/13552540010337029.
[36] A.K. Dubey, V. Yadava, Experimental study of Nd: YAG laser beam machining—an overview, J. Mater. Process. Technol. 195 (1–3) (2008) 15–26.
[37] J. Ready, Effects of High-Power Laser Radiation, Elsevier, 2012.
[38] E.C. Honea, et al., 115-W Tm: YAG diode-pumped solid-state laser, IEEE J. Quant. Electron. 33 (9) (1997) 1592–1600.
[39] S. Backus, C.G. Durfee III, M.M. Murnane, H.C. Kapteyn, High power ultrafast lasers, Rev. Sci. Instrum. 69 (3) (1998) 1207–1223.
[40] H. Lee, C.H.J. Lim, M.J. Low, N. Tham, V.M. Murukeshan, Y.J. Kim, Lasers in additive manufacturing: a review, Int. J. Precis. Eng. Manuf. - Green Technol. 4 (3) (2017) 307–322, https://doi.org/10.1007/s40684-017-0037-7.
[41] Y. Huang, M.C. Leu, J. Mazumder, A. Donmez, Additive manufacturing: current state, future potential, gaps and needs, and recommendations, J. Manuf. Sci. Eng. 137 (1) (Feb. 2015) 014001, https://doi.org/10.1115/1.4028725.
[42] G. Tapia, A. Elwany, A review on process monitoring and control in metal-based additive manufacturing, J. Manuf. Sci. Eng. 136 (6) (2014).
[43] D. Gu, Laser additive manufacturing (am): classification, processing philosophy, and metallurgical mechanisms, in: Laser Additive Manufacturing of High-Performance Materials, Springer, 2015, pp. 15–71.
[44] C.B. Williams, F. Mistree, D.W. Rosen, A functional classification framework for the conceptual design of additive manufacturing technologies, J. Mech. Des. 133 (12) (Dec. 2011) 121002, https://doi.org/10.1115/1.4005231.
[45] L. Scime, J. Beuth, Anomaly detection and classification in a laser powder bed additive manufacturing process using a trained computer vision algorithm, Addit. Manuf. 19 (2018) 114–126.
[46] P.A. Kobryn, S.L. Semiatin, The laser additive manufacture of Ti-6Al-4V, J. Occup. Med. 53 (9) (2001) 40–42.
[47] S. Kumar, S. Pityana, Laser-based additive manufacturing of metals, In Adv. Mater. Res. 227 (2011) 92–95.
[48] B. Khoda, T. Benny, P.K. Rao, M.P. Sealy, C. Zhou, Applications of laser-based additive manufacturing, in: Laser-Based Additive Manufacturing of Metal Parts, CRC Press, 2017, pp. 239–284.
[49] M. Schmidt, M. Merklein, D. Bourell, D. Dimitrov, T. Hausotte, K. Wegener, L. Overmeyer, V. Frank, G.N. Levy, Laser based additive manufacturing in industry and academia, Cirp Ann. 66 (2) (2017) 561–583.
[50] Y. Wang, J. Choi, J. Mazumder, Laser-aided direct writing of nickel-based single-crystal super alloy (N5), Metall. Mater. Trans. 47 (12) (2016) 5685–5690.
[51] Z. Yan, et al., Review on thermal analysis in laser-based additive manufacturing, Optic Laser. Technol. 106 (2018) 427–441, https://doi.org/10.1016/j.optlastec.2018.04.034.
[52] R. Sreenivasan, A. Goel, D.L. Bourell, Sustainability issues in laser-based additive manufacturing, Phy. Proc. 5 (2010) 81–90, https://doi.org/10.1016/j.phpro.2010.08.124. PART 1.
[53] K.P. Davidson, S. Singamneni, Magnetic characterization of selective laser-melted Saf 2507 duplex stainless steel, J. Occup. Med. 69 (3) (2017) 569–574.
[54] J. Mazumder, L. Song, Advances in direct metal deposition, in: ASME International Mechanical Engineering Congress and Exposition, vol. 56185, 2013. V02AT02A012.
[55] H. Zhang, H. Zhu, T. Qi, Z. Hu, X. Zeng, Selective laser melting of high strength Al-Cu-Mg alloys: processing, microstructure and mechanical properties, Mater. Sci. Eng. 656 (2016) 47–54, https://doi.org/10.1016/j.msea.2015.12.101.

[56] M. Leary, Surface Roughness Optimisation for Selective Laser Melting (SLM): Accommodating Relevant and Irrelevant Surfaces, Elsevier Ltd, 2017, pp. 99–118, https://doi.org/10.1016/B978-0-08-100433-3.00004-X. Volume Issue.
[57] O.D. Neikov, Powders for Additive Manufacturing Processing, second ed., Elsevier Ltd., 2019.
[58] I. Gibson, D. Rosen, B. Stucker, Powder bed fusion processes, in: Additive Manufacturing Technologies, Springer, 2015, pp. 107–145.
[59] K. Davidson, S. Singamneni, Selective laser melting of duplex stainless steel powders: an investigation, Mater. Manuf. Process. 31 (12) (2016) 1543–1555.
[60] K.P. Davidson, S.B. Singamneni, Metallographic evaluation of duplex stainless steel powders processed by selective laser melting, Rapid Prototyp. J. 23 (6) (2017) 1146–1163, https://doi.org/10.1108/RPJ-04-2016-0053.
[61] J.H. Tan, W.L.E. Wong, K.W. Dalgarno, An overview of powder granulometry on feedstock and part performance in the selective laser melting process, Addit. Manuf. 18 (2017) 228–255.
[62] K.A. Mumtaz, P. Erasenthiran, N. Hopkinson, High density selective laser melting of Waspaloy®, J. Mater. Process. Technol. 195 (1–3) (2008) 77–87.
[63] E. Yasa, J.-P. Kruth, Application of laser re-melting on selective laser melting parts, Adv. Prod. Eng. Manag. 6 (4) (2011) 259–270.
[64] L.E. Loh, et al., Selective Laser Melting of aluminium alloy using a uniform beam profile: the paper analyzes the results of laser scanning in Selective Laser Melting using a uniform laser beam, Virtual Phys. Prototyp. 9 (1) (2014) 11–16.
[65] W.E. Frazier, Metal additive manufacturing: a review, J. Mater. Eng. Perform. 23 (6) (2014) 1917–1928.
[66] B. Vayre, F. Vignat, F. Villeneuve, Metallic additive manufacturing: state-of-the-art review and prospects, Mec. Ind. 13 (2) (2012) 89–96.
[67] A.R. Nassar, E.W. Reutzel, Beyond layer-by-layer additive manufacturing—Voxel-Wise directed energy deposition, in: Solid Freeform Fabr Symp Proc, vol. 26, 2015, pp. 273–283.
[68] B. Dutta, F.H.S. Froes, The additive manufacturing (AM) of titanium alloys, in: Titanium Powder Metallurgy, Elsevier, 2015, pp. 447–468.
[69] S.M. Yusuf, N. Gao, Influence of energy density on metallurgy and properties in metal additive manufacturing, Mater. Sci. Technol. 33 (11) (2017) 1269–1289.
[70] R. Anderson, J. Terrell, J. Schneider, S. Thompson, P. Gradl, Characteristics of Bi-metallic interfaces formed during direct energy deposition additive manufacturing processing, Metall. Mater. Trans. B 50 (4) (2019) 1921–1930.
[71] I. Gibson, D. Rosen, B. Stucker, Directed energy deposition processes, in: Additive Manufacturing Technologies, Springer, 2015, pp. 245–268.
[72] M. Javidani, J. Arreguin-Zavala, J. Danovitch, Y. Tian, M. Brochu, Additive manufacturing of AlSi10Mg alloy using direct energy deposition: microstructure and hardness characterization, J. Therm. Spray Technol. 26 (4) (2017) 587–597.
[73] J. Mazumder, L. Song, in: Advances in Direct Metal Deposition BT - Proceedings of the 36th International MATADOR Conference, 2010, pp. 447–450.
[74] Z. Wang, T.A. Palmer, A.M. Beese, Effect of processing parameters on microstructure and tensile properties of austenitic stainless steel 304L made by directed energy deposition additive manufacturing, Acta Mater. 110 (2016) 226–235.
[75] D. Hu, R. Kovacevic, Modelling and measuring the thermal behaviour of the molten pool in closed-loop controlled laser-based additive manufacturing, Proc. Inst. Mech. Eng. Part B J. Eng. Manuf. 217 (4) (2003) 441–452, https://doi.org/10.1243/095440503321628125.

[76] E.W. Reutzel, A.R. Nassar, A survey of sensing and control systems for machine and process monitoring of directedenergy, metal-based additive manufacturing, Rapid Prototyp. J. 21 (2) (2015) 159–167, https://doi.org/10.1108/RPJ-12-2014-0177.
[77] D.A. Kriczky, J. Irwin, E.W. Reutzel, P. Michaleris, A.R. Nassar, J. Craig, 3D spatial reconstruction of thermal characteristics in directed energy deposition through optical thermal imaging, J. Mater. Process. Technol. 221 (2015) 172–186.
[78] F. Feuerhahn, A. Schulz, T. Seefeld, F. Vollertsen, Microstructure and properties of selective laser melted high hardness tool steel, Phys. Proc. 41 (2013) 843–848, https://doi.org/10.1016/j.phpro.2013.03.157.
[79] F. Hashimoto, R.G. Chaudhari, S.N. Melkote, Characteristics and performance of surfaces created by various finishing methods (invited paper), Proc. CIRP 45 (2016) 1–6, https://doi.org/10.1016/j.procir.2016.02.052.
[80] C. Li, J.F. Liu, Y.B. Guo, Prediction of residual stress and Part Distortion in selective laser melting, Proc. CIRP 45 (2016) 171–174, https://doi.org/10.1016/j.procir.2016.02.058.
[81] C.E. Protasov, V.A. Safronov, D.V. Kotoban, A.V. Gusarov, Experimental study of residual stresses in metal parts obtained by selective laser melting, Phys. Proc. 83 (2016) 825–832, https://doi.org/10.1016/j.phpro.2016.08.085.
[82] R. Molaei, A. Fatemi, Fatigue design with additive manufactured metals: issues to consider and perspective for future research, Proc. Eng. 213 (2017) (2018) 5–16, https://doi.org/10.1016/j.proeng.2018.02.002.
[83] D. Greitemeier, C. Dalle Donne, F. Syassen, J. Eufinger, T. Melz, Effect of surface roughness on fatigue performance of additive manufactured Ti–6Al–4V, Mater. Sci. Technol. 32 (7) (2016) 629–634, https://doi.org/10.1179/1743284715Y.0000000053.
[84] J. Pegues, M. Roach, R. Scott Williamson, N. Shamsaei, Surface roughness effects on the fatigue strength of additively manufactured Ti-6Al-4V, Int. J. Fatig. 116 (2018) 543–552, https://doi.org/10.1016/j.ijfatigue.2018.07.013.
[85] N. Kalentics, et al., 3D laser shock peening – a new method for the 3D control of residual stresses in selective laser melting, Mater. Des. 130 (2017) 350–356, https://doi.org/10.1016/j.matdes.2017.05.083.
[86] R. Sun, et al., Microstructure, residual stress and tensile properties control of wire-arc additive manufactured 2319 aluminum alloy with laser shock peening, J. Alloys Compd. 747 (2018) 255–265, https://doi.org/10.1016/j.jallcom.2018.02.353.
[87] F. Zhihao, L. Libin, C. Longfei, G. Yingchun, Laser polishing of additive manufactured superalloy, Proc. CIRP 71 (2018) 150–154, https://doi.org/10.1016/j.procir.2018.05.088.
[88] C.P. Ma, Y.C. Guan, W. Zhou, Laser polishing of additive manufactured Ti alloys, Optic Laser. Eng. 93 (2017) 171–177, https://doi.org/10.1016/j.optlaseng.2017.02.005. January.
[89] M. Salmi, J. Huuki, I.F. Ituarte, The ultrasonic burnishing of cobalt-chrome and stainless steel surface made by additive manufacturing, Prog. Addit. Manuf. 2 (1–2) (2017) 31–41, https://doi.org/10.1007/s40964-017-0017-z.
[90] N.S.M. El-Tayeb, K.O. Low, P.V. Brevern, Influence of roller burnishing contact width and burnishing orientation on surface quality and tribological behaviour of Aluminium 6061, J. Mater. Process. Technol. 186 (2007) 272–278, https://doi.org/10.1016/j.jmatprotec.2006.12.044.
[91] B. AlMangour, J.M. Yang, Improving the surface quality and mechanical properties by shot-peening of 17-4 stainless steel fabricated by additive manufacturing, Mater. Des. 110 (2016) 914–924, https://doi.org/10.1016/j.matdes.2016.08.037.
[92] L. Hackel, J.R. Rankin, A. Rubenchik, W.E. King, M. Matthews, Laser peening: a tool for additive manufacturing post-processing, Addit. Manuf. 24 (2018) 67–75, https://doi.org/10.1016/j.addma.2018.09.013.

[93] S. Marimuthu, A. Triantaphyllou, M. Antar, D. Wimpenny, H. Morton, M. Beard, Laser polishing of selective laser melted components, Int. J. Mach. Tool Manufact. 95 (2015) 97–104, https://doi.org/10.1016/j.ijmachtools.2015.05.002.

[94] S. Dadbakhsh, L. Hao, C.Y. Kong, Surface finish improvement of LMD samples using laser polishing, Virt. Phys. Prototyp. 5 (4) (2010) 215–221, https://doi.org/10.1080/17452759.2010.528180.

[95] X. Wang, S. Li, Y. Fu, H. Gao, Finishing of additively manufactured metal parts by abrasive flow machining, in: Proc. 27th Annu. Int. Solid Free. Fabr. Symp, 2016, pp. 2470–2472.

[96] K.L. Tan, S.H. Yeo, Surface modification of additive manufactured components by ultrasonic cavitation abrasive finishing, Wear 378 (379) (2017) 90–95, https://doi.org/10.1016/j.wear.2017.02.030.

[97] R.K. Jain, V.K. Jain, Stochastic simulation of active grain density in abrasive flow machining, J. Mater. Process. Technol. 152 (2004) 17–22, https://doi.org/10.1016/j.jmatprotec.2003.11.024.

[98] P. Zhang, Z. Liu, Effect of sequential turning and burnishing on the surface integrity of Cr-Ni-based stainless steel formed by laser cladding process, Surf. Coating. Technol. 276 (2015) 327–335, https://doi.org/10.1016/j.surfcoat.2015.07.026.

[99] A. Maamoun, M. Elbestawi, S. Veldhuis, Influence of shot peening on AlSi10Mg parts fabricated by additive manufacturing, J. Manuf. Mater. Process 2 (3) (2018) 40, https://doi.org/10.3390/jmmp2030040.

[100] N.E. Uzan, S. Ramati, R. Shneck, N. Frage, O. Yeheskel, On the effect of shot-peening on fatigue resistance of AlSi10Mg specimens fabricated by additive manufacturing using selective laser melting (AM-SLM), Addit. Manuf. 21 (2010) (2018) 458–464, https://doi.org/10.1016/j.addma.2018.03.030.

# CHAPTER 15

# Nanofunctionalized 3D printing

**Maria P. Nikolova[1], K. Karthik[2], Murthy S. Chavali[3]**
[1]Department of Material Science and Technology, University of Ruse "A. Kanchev", Ruse, Bulgaria; [2]Department of Physics, Bharathidasan University, Tiruchirappalli, Tamil Nadu, India; [3]Shree Velagapudi Rama Krishna Memorial College (PG Studies; Autonomous), Guntur, Andhra Pradesh, India

## 15.1 Introduction

Additive manufacturing (AM) is a computer-guided fabrication of 3D structures with high shape complexity using multifunctional materials such as polymers, metals, and ceramics. Compared with some conventional formative techniques such as CNC machining or injection molding, the computer-aided design (CAD)−dependent fabrication is slower, but it enables layer-by-layer digital-controlled manufacturing of customized parts tailored to meet various demands on specific applications and functionalities, including biosystems that are unmatched by conventional processes. Moreover, it was calculated that the light-weighted AM design contributes to usage savings reaching up to 63% in $CO_2$ emissions and energy savings over the whole life cycle of the product [1]. The AM technology can combine different materials to fabricate 3D structures with a spatial resolution of the most commercially available AM processes of 20−50 μm [2] although there exists the necessity of nanoscale resolution for various engineering materials.

According to ASTM Standard Terminology for AM technology [3], the AM processes are classified into seven main groups tabulated in Table 15.1, together with the advantages and disadvantages of each of these seven groups. Among these seven types of different AM technologies, a significant disparity in productivity, resolution, and build rate exists. Despite the significant progress in the development of these easy-to-use technologies with decreased prices, many challenges are still needed to be faced such as thermomechanical properties enhancement, long-term stability, printing accuracy of CAD reproduction, surface quality, porosity, anisotropy, and so on. In the 21st century, the focus on AM technology is shifted from concept modeling and precise geometry reproduction toward improvement in functional properties of AM-produced objects. For most of the existing AM techniques, pure materials with unsuitable properties for user applications are usually employed. The mechanical and functional properties,

*Additive Manufacturing*
ISBN 978-0-12-818411-0
https://doi.org/10.1016/B978-0-12-818411-0.00006-9

Table 15.1 Classification of various type of AM processes together with their advantages and shortcomings.

| AM process | Characteristics | Techniques included | Materials | Advantages | Disadvantages |
| --- | --- | --- | --- | --- | --- |
| Material extrusion | AM process where the material is dispensed through a nozzle | **Without material melting**<br>- FFF (fused filament fabrication)<br>- Robocasting<br>- Direct writing<br>- 3D dispensing<br>- 3D bioplotting | Soft materials (monomer resins, polymer solutions, metals, ceramics, thermoplastics, organic–inorganic mixture, cements, gels, hydrogels, cells, etc.) | - Easy operation<br>- Mixing material<br>- Patterning at the micro- and mesoscale<br>- Control on thickness (in robocasting)<br>- Good printing resolution<br>- Heating is not applied<br>- Wide material selection<br>- Soft materials capability<br>- Incorporation of biomolecules | - Inferior mechanical properties<br>- Well-defined viscosity ink required with narrow viscosity window<br>- Nozzle clogging occurrence<br>- Low shape and strength maintenance<br>- Postprocessing is sometimes needed<br>- Low resolution (~100 μm)<br>- Low printing speed<br>- Generating smooth surfaces[a] |

| | | | | | |
|---|---|---|---|---|---|
| | | **With material melting**<br>- FDM (fused deposition modeling)<br>- 3D fiber deposition | Thermoplastics, composites | - Versatile pattern design<br>- Good strength<br>- Low price printers<br>- Simultaneous deposition of diverse materials | - Rough surface<br>- Anisotropy<br>- Only thermoplastics (including nanofilled)<br>- Nozzle clogging occurrence<br>- HHigh process temperature<br>- Limited material range<br>- "Stair-stepping" effect |
| Vat polymerization | AM process that uses liquid polymer in a vat that is selectively light-activated to polymerize. This technology groups different lithography-based methods | - SLA (stereolithography) | Photocurable resin (epoxy- or acrylate-based), ceramics, composites | - High resolution (up to 100 nm) and precision<br>- Complex structures and thin wall features reproduction<br>- Uncured resin is easily removed<br>- Incorporation of different NPs, proteins, cells, etc. in the resin | - Limited mechanical properties<br>- Limited materials;<br>- Posttreatment required<br>- "Stair-stepping" effect<br>- Cytotoxicity; |

*Continued*

**Table 15.1** Classification of various type of AM processes together with their advantages and shortcomings.—cont'd

| AM process | Characteristics | Techniques included | Materials | Advantages | Disadvantages |
|---|---|---|---|---|---|
| | | - DOP-SLA (dynamic optical projection SLA) | Polymers, ceramics, composites | - Fast fabrication speed;<br>- Fabrication of large areas;<br>- High mass production<br>- Good surface finishing; | - Limited materials (photosensitive resins) |
| | | - 2 PP (two-photon polymerization)<br>- MPP (multi-photon polymerization) | Polymers, composites | - Materials with high surface-to-volume ratio<br>- Highly interconnected porous materials or helices<br>- Up to 100 nm resolution | - Materials limitations (low viscosity resin)<br>- High-cost equipment |

| | | | | | |
|---|---|---|---|---|---|
| Powder bed fusion | AM process where external thermal energy source such as laser or electron beam selectively fused different areas of materials in the powder bed | - SLS (selective laser sintering)<br>- SLM (selective laser melting)<br>- DMLS (direct metal laser sintering) | Metals, ceramics, glasses, polymers, composites, hybrids | - The particle size determines the precision;<br>- Easy removal of the support powder;<br>- Miniature or porous implants fabrication with no need for supports;<br>- No post-processing<br>- Large range of materials | - a Large amount of material required at a high cost (for SLS and SLM)<br>- Slow<br>- Rough surface<br>- Poor reusability<br>- Component defects and powdery surface<br>- Size limitations<br>- High cost<br>- The resolution depends on laser beam diameter and powder particle size. |
| | | - EBM (electron beam melting) | High melting temperature materials | - Processing of high melting temperature materials;<br>- Fast working | - High capital cost<br>- Rough surface;<br>- Limited materials<br>- High power required; |

*Continued*

**Table 15.1** Classification of various type of AM processes together with their advantages and shortcomings.—cont'd

| AM process | Characteristics | Techniques included | Materials | Advantages | Disadvantages |
|---|---|---|---|---|---|
| Direct energy deposition | AM technology that uses focused thermal energy such as plasma arc or laser to fuse deposited metal materials by melting. | - Laser deposition<br>- Laser consolidation<br>- Direct metal deposition<br>- Plasma-arc melting | Metals, hybrid | High-quality parts with—controlled density<br>- Controlled grain structure<br>- Suitable for repair applications | - Limited materials (metals/hybrids in form of powder or wires)<br>- Surface quality depends on the processing speed |
| Material jetting | AM process in which droplets of photopolymer or thermoplastic materials are selectively deposited and cured by exposure to light | - Polyjet inkjet<br>- Thermojet | Powders (ceramic, polymers, composites) | - Different resin materials can be printed in one part<br>- High resolution (about 30 μm) because of polymerization (polyjet) of deposited drops<br>- High printing speed (thermojet) | - Limited materials (photopolymers, wax)<br>- Posttreatment (UV light curing/ annealing) for solidification<br>- Thermal stress and nonuniform droplet size (thermojet)<br>- Weaker strength than SLS and SLM produced parts |

| | | | | | |
|---|---|---|---|---|---|
| Binder jetting | AM process that uses powder materials selectively fused by liquid bonding agents | - Ink jetting<br>- 3D printing<br>- S-print<br>- M-print | Variety of inks (ceramics, metals, water-soluble polymers, composites) or powder hybrid materials connected with different binders | - Absence of high temperature<br>- Fast printing<br>- Design freedom<br>- The flexibility of materials selection<br>- Free of supports/ substrates<br>- Relatively low cost | - Limited strength<br>- Clogging of binder jet<br>- Rough surface<br>- High heterogeneity of morphology<br>- May include postprocessing treatment |
| Sheet lamination | AM process where sheets of material are bonded together to form a deposited object | - Ultrasonic consolidation<br>- Laminated object manufacture | Variety of materials (polymers, hybrids, ceramic, metals) | - High speed<br>- Low-cost production<br>- Facile material handling | - Limited materials<br>- High anisotropy<br>- May require postprocessing finishing<br>- Different strength depending on the adhesive used<br>- Low resolution |

[a]For orthopedic implants, rough surface is preferred.

such as toughness, load-bearing capacity, strength, conductivity, bactericidal or fungicidal properties, and so on, of conventional printing materials, such as eco-friendly but low-strength thermoplastic materials, low temperature resistant and low-strength photopolymers, inadequately microstructured 3D printed metal alloys, or brittle thermosets, could be substantially improved by adding certain nanoreinforcements. The integration of AM advantages with nanotechnology achievements enables the creation of entirely new materials with multiple functions that extend the application and reduce the limitation of AM technologies.

Current studies focus on developing new heterogeneous materials intended to address the disadvantages of the pure material and eliminate cost and time for posttreatment. The nanofunctional concept is available to many fields such as electronics, optics, energy conversion, aerospace, biosystems, and so on. When shifting from micro- to nanoscale, certain fundamental properties of materials, such as reactivity, thermal, electrochemical, magnetic, and optical, could be completely changed as opposed to bulk counterparts. The obtained material is called nanocomposite because of the nanodimension of a rigid and stronger reinforcement phase, which is dispersed in a usually weaker continuous phase termed as the matrix.

The elaboration of advanced materials with "smart" properties and enhanced performance as well as new techniques for their processing presents environmental and economic interest. Consequently, in this chapter, a holistic study on advanced hybrid inorganic or inorganic–organic materials with nanoadditives produced by AM processes is performed to reveal their competitive advantages, enhanced properties, and progress in this area together with the problems faced and future perspective. Usually, the incorporation of the nanoadditives into the host matrix is implemented via three ways: (1) premixing of the NPs into the host material followed by 3D processing; (2) introducing the nanomaterial by interrupting the printing process and manual or automatic addition (Fig. 15.1); and (3) attaching to the surface of already printed materials [4]. In the view of the obtained properties and nanoparticle (NP) incorporation, this chapter focuses on four main topics:

**a.** nanoenhancement of structural materials;
**b.** 3D-printed electronics, optics, and energy conversion devices;
**c.** 3D-printed nanocomposites used in medicine; and
**d.** additive-manufactured nanofunctionalized surfaces.

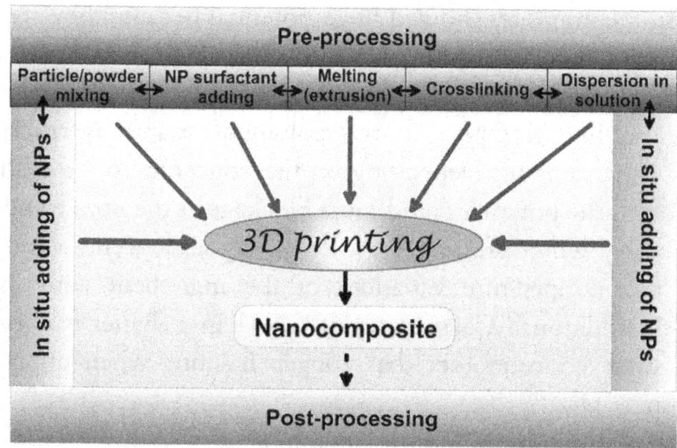


**Figure 15.1** Simplified schematic depicting the processes of premixing and in situ adding of NPs for nanocomposite fabrication. *NPs*, nanoparticles.

## 15.2 Nanoenhancement of structural materials

Recently, nanotechnology offers different options to make new printing materials for constructing complex 3D structures used in aerospace, automotive, microelectronics, buildings, medicine, and so on. Compared with products fabricated by formative or subtractive technologies, parts produced by AM usually displayed inferior performance concerning mechanical properties. This weakness could be due to the lack of suitable materials, their porosity, layer structure anisotropy, etc. Through the insertion of nanomaterials such as carbon nanotubes (CNTs) or nanofibers, ceramics, or metal nanoparticles, the mechanical and wear properties of the final parts are improved. The properties of the nanofunctionalized system are dependent on the NP size, shape, distribution, filler fraction, matrix nature, and its interaction with NPs. Some limitations when NPs are introduced could be faced such as aggregation in liquid media, an increase in viscosity for SLA resins and inkjet inks, nanopowder aggregation, nozzle clogging, and so on. For instance, the main requirements for a filament feedstock are high strength, strain, flexibility, and viscosity at elevated temperatures. By mixing feedstock filament containing nanocomponent powder with an organic binder system, FDM processing could be performed. CNTs are among the stiffest and strongest materials with extraordinary electrical and thermal properties [5]. For that reason, due to the low mechanical properties of pure thermoplastic materials, Shofner et al. [6] tried to reinforce ABS plastics with nanofibers of single-walled carbon nanotubes (SWCNTs). Using similar FDM processing conditions, 10 wt% nanofiber-load material increased its modulus, stiffness, and tensile strength with about 60%, 68%, and 40%, respectively, compared with plain ABS. Fiber-reinforced thermoplastics composites showed huge potential because they are light but rigid, chemically inert, nonflammable and could retain their properties under high-temperature conditions (up to 400°C) [7]. Nevertheless, the homogeneous dispersion of CNTs in the polymeric matrix is crucial toward properties' enhancement. Depending on the concentration, the addition to carbon fillers to the polymer could cause blockage at the nozzle and clogging (flux instability) while printing. Problems may also occur because of the NPs' size and temperature variations of the ingredient with the matrix material [8]. Additionally, aggregates of CNTs in polymer nanocomposites could be stress concentrators that trigger fracture when applying even moderate stress [9].

Because of its unique properties, graphene is positioned as indispensable in applications such as optoelectronics, electronics, and energy storage but also a structural material because graphene is about a hundred times stronger than steel [7]. However, in polymer composites, the sheets of the graphene phase tend to separate because of hindered interaction with polymer chains. In contrast, the oxygenated functional groups in graphene oxide (GO) were found to assist graphene dispersion and its adhesion to polymer matrix [10] achieving stable and homogeneous mixtures. For instance, nanocomposites of GO with concentrations 0.02, 0.04, and 0.06 wt% in ABS polymer matrix mixed by initial sonication in acetone-based liquid state and extrusion were FFF-printed in the form of tensile samples in upright, facedown, and edge-up mode. The addition of GO gradually made the material more ductile, and it progressively compensated the initial strength loss because of acetone processing. At concentration 0.06% GO, the nanocomposites had higher strength-to-failure (29% for upright and 14% for facedown), toughness (55% upright and 20% for facedown), and fracture strength (10% upright and 3.5% for facedown) at the expense of a decrease in stiffness (15% for upright and 6% for facedown), which made the material suitable for applications with static and dynamic loading [11].

To be used for structural application, the properties of the SLA-printed objects have to be tailored. When using a liquid resin as a precursor for the printing of nanocomposite scaffolds, the main goal is to achieve a homogeneous mixture with appropriate viscosity for high-quality printing. Despite applying different mixing methods, when controlling the resin viscosity, a homogeneity could be achieved by preventing particles' agglomeration in a stable dispersion. Once a stable and homogeneous NPs—resin mixture with appropriate viscosity for producing flat surfaces is obtained, a successful functional print could be achieved. Manapat et al.

found out that an SLA-printed nanocomposite containing 1% GO added in commercially available SLA resin demonstrated a tremendous increase in tensile strength at 673.6% after annealing at 100°C for 12 h. The authors explained this effect together with the enhanced thermal stability with the polymer—nanofiller cross-linking using acid-catalyzed esterification and removal of intercalated water [12]. Focusing on the enhancement of mechanical properties, Sandoval et al. elaborated nanocomposite material of photosensitive epoxy-based resin with CNTs filler suitable for SLA printing. The MWCNTs with a mean diameter of 30 nm and length 5—20 μm in small concentration (from 0.025% to 0.1%) homogeneously dispersed in the epoxy-based resin were found to increase hardness, fracture strength, and ultimate tensile stress when compared with unfilled SLA resins because of assisted adhesion between the layers due to the random distribution and orientation of MWCNTs. This indicates a reduction in mechanical property anisotropy. However, the brittle type of fracture was observed for the nanoembedded composite because of the constraints imposed by the nanotubes in the polymer upon deformation [13]. Except for carbon-containing nanofillers, inorganic NPs had also demonstrated an improvement in the mechanical and thermal properties of 3D-printed nanocomposite materials. For instance, the addition of 0.25 wt% anatase nano-$TiO_2$ (30 nm) particles to photosensitive SLA resin increased the tensile strength, tensile modulus, hardness, and flexural strength of the final part by 89%, 18%, 5%, and 6%, respectively, compared with nonloaded SLA-produced objects [14]. Moreover, the plasticity and thermal stability of the modified resin were also improved. Negative thermal expansion in three directions over a range of 170°C manifested micro-SLA-printed PEGDA nanocomposite lattices when the content of copper (Cu) NPs with an average size of 50—80 nm was increased. The authors explained the observed effect with the structural interaction of two materials with different thermal expansion coefficients [15]. It is worth noticing that depending on their nature, size, and concentration, nanofillers may bring SLA printing issues such as scattering the laser that worsens the quality and accuracy of the printed part. Usually, when using SLA technology, the ideal wavelength for solidification of the resin with a certain NP has to be experimentally determined. The presence of NPs could also affect the curing width and depth because of UV light absorption. The penetration depth determining the printing details depends on the size, loading concentration of NPs, and refractive index of both liquid resin and NPs [16]. Also, the presence of NPs could cause porosity (void defects) due to the difference in the nature of the materials used. However, when loading NPs into a resin, the content should be high enough to obtain the intended result.

Adding oxide NPs to organic inks could not only modulate their rheological properties but also bring functionality to composite materials. As the surface nature of the nanooxides is hydrophilic, their embedding in the hydrophobic polymeric matrix should be accompanied by different interfacing methods such as sonication, ultrasonication, shear, or mechanical mixing, stirring, centrifugation, and so on as well as introducing linkers or polymer coatings. Additionally, solvents with a polarity index higher than 4.0 were found to suspend the oxide NPs, thus hindering the NPs' clustering [17]. Using direct writing for the fabrication of 3D scaffolds, Cai et al. produced a series of organic vaseline inks doped with $TiO_2$ nanoparticles (21 nm) with different concentrations (2.5, 5, 7.5, and 10 wt%). The researchers demonstrated that inks with $TiO_2$ concentration higher than 5 wt% exhibited a tendency toward increased elastic properties, lower deformation and rod bending, and superior performance to retain shape after stretching and repeated bending on a flexible substrate [18]. According to Volyanskii et al. [19], the isolation of NPs into inert materials could control not only aggregation, oxidation, and corrosion but also NP release that could influence the chemical and phase composition obtained. For instance, the addition of $BaTiO_3$ NPs (60 nm) coated with PAA in high concentration (>45%) in inks for direct writing increased the modulus by two orders of magnitude when divalent salt was used because of fluid-to-gel transition [20]. The addition of nanosilica in PDMS ink was found not only to improve the mechanical strength without risk of collapsing of the ink but also to enable the formation of desired topographical structures during 3D printing which provided superhydrophobic properties of the printed porous membrane for water—oil separation [21]. It should be also noticed that when high NP-loaded material is printed, nozzle clogging because of aggregates' formation and decreased flowability could occur. According to Kosmala et al. to avoid nozzle clogging, the size of the NPs should be less than 100 times smaller than the jetting nozzle diameter [22].

Similar to microparticles, NPs increase tensile modulus and strength but without reducing the impact resistance [23] and increasing the object weight. As mentioned earlier, the nanofillers could change the physical and thermodynamic characteristics of the matrix material. Within a polymer matrix, the nanofillers develop a huge number of interfacial contacts causing superior properties than the bulk polymer phase. When mechanically deformed, the comparable stiffer nanofillers bear a significant amount of the overall load as opposed to the relatively soft polymer. Another explanation could be the stiffening effect on the polymer chains caused by the nanofillers. The interaction volume with the adjacent polymer chains could be additionally tailored by covalent or noncovalent functionalization. However, if nanoclays are functionalized by organic substances (organoclays), such NPs are thermally unstable and could degrade at about 170°C due to the exchange of metal cations in clay with organic ammonium salts. Nevertheless, by reinforcing engineering polymers with nanoclays, a significant improvement in a wide range of physical–mechanical properties was demonstrated. For instance, adding 5 wt% clay NPs in polyamide 6 (PA6) increased the values of melt of fusion and heat of crystallization during SLS processing because of polymer–clay interaction. Simultaneously, the crystallization temperature and mobility of polymer molecules were reduced, whereas the viscosity of the composite increased with clay content [24]. In contrast to neat PA11 polymer, SLS-produced PA11–clay nanocomposite parts and PA11-carbon nanofiber composites demonstrated better thermal properties and flammability [25]. The addition of 4 wt% carbon black NPs in thermoplastic PA12 made the nanocomposite not only electrically conductive ($1 \times 10^{-4}$ S/cm) but enhanced the absorption behavior and facilitated laser sintering at room temperature without affecting crystallinity [26]. PA11/nanosilica (15 nm) composites prepared by SLS exhibited a nonlinear variation of properties as a function of filler volume fraction (2%–10%). By adjusting the printing parameters, the object density reached over 90% for different silica loading. As silica NPs volume rose, the tensile modulus, compressive modulus, and strain at yield increased, whereas the strain at break decreased because the parts became stiffer but more brittle. At a critical composition, these characteristics reversed in trend [27]. Similar to SLA and inkjet techniques, to resolve the issue of NP agglomeration, organic coatings or linkers that keep particles away from each other in the printing media are added. For example, polystyrene-coated $Al_2O_3$ NPs process by SLS printers demonstrated a 300% increase in tensile strength compared with SLS-produced samples by untreated particles [28]. Moreover, multilayered glass–fiber nanocomposites demonstrated self-healing effect driving and segregating the NPs into crack because of the polymer-induced "depletion effect" between the surface and NPs. A crucial prerequisite for achieving this effect was the homogeneous dispersion of the NPs [19]. More recent studies on polymer rheology discover that polymer laser sintering of nanocomposites was hindered by difficulties in repeatability and process control because the melt viscosity of the polymer was increased by adding an even low amount of NPs. Bai et al. found that CNTs—PA12 nanocomposites with CNTs content of 0.1 and 0.2 wt% had increased viscosity resulting from incomplete melting of the powder particles during SLS. It was proposed that CNTs and polymer interaction hindered the chain motions, which resulted in higher storage modulus, loss modulus, and decreased thermal expansion coefficient [29].

The combination of high-performance polymers such as PEEK, PARA, and PAEK and special nanoingredients such as CNTs as composite materials for 3D printing has the potential to obtain high-performance polymer nanocomposites with exceptional strength-to-weight, fatigue, impact resistance, dimensional stability, flame retarding, semiconducting, corrosion-resistant, and biocompatible properties with application in aerospace, automotive, electronics, and medical industry. For example, in more demanding automotive applications, high-performance polymer nanocomposites are a suitable replacement to metals for external body parts, fuel, and coating system components, giving safety, fuel efficiency, enhanced durability, and noise reduction [30]. Such nanocomposites are expected to substantially improve the functionality and applicability of polymer nanocomposites soon.

Despite improving polymer properties, NP infiltration in powder injection-molded or SLS-produced metal objects could improve dimensional accuracy by reducing the creep deformation under self-weight and sintering shrinkage. Crane et al. used a dispersion of metal (iron) NPs (7—10 nm) to decrease the dimensional changes during sintering of SLS-produced 410 stainless steel parts because NPs were able to densify the structure by enlarging the bonds between bigger particles while the surrounding skeleton of micron-scaled particles (63—90 μm) sintered very little. By adding 2.7 wt% nanoiron in four application cycles of sintering in the magnetic field, the creep deflections were reduced up to 95% while the sintering shrinkage decreased down to 60%. Moreover, the nanobinder was able to heel the cracks formed during sintering shrinkage [31]. Except for steel parts, ceramic NPs—reinforced aluminum matrix composites also exhibited superior properties. For example, SLM-produced nanocomposite parts containing homogeneously dispersed TiC nanoscale particles with average size 77—93 nm embedded in AlSi10Mg matrix experienced high microhardness (181.2 HV0.2), low coefficient of friction (0.36), and decreased wear rate ($2.94 \times 10^5$ $mm^3/N/m$) [32].

## 15.3 3D-printed electronics, optics, and energy conversion devices

3D printing is increasingly useful in prototyping novel wearable electronics and microelectronics components such as 3D antennas, acoustic imaging, mission-specific satellite components, LEDs, batteries, stretchable, and other (strain, thermal, electrochemical, piezoresistive, capacitive, etc.) sensors. Nanocomposite systems demonstrating large specific surface area and porosity could be even a part of catalysis systems or hydrogen storage devices [33]. SLS-produced alternating ferromagnetic Ni NPs (30–50 wt%)–polycarbonate (PC) and nonmagnetic Cu NPs (10–30 wt%)–PC layers formed functionally graded 3D parts indicating hysteresis phenomena which could be useful in MEMS or (nano)-NEMS applications [34].

Printed electronics technologies are promising in realizing energy-, time-, and cost-efficient devices on flexible substrates. 3D-printed conductive structures derived from metal-based materials exhibit the highest electrical conductivity. Nanoinks from metal NP suspensions offer ease of dispersion in different solutions (although unstable), fast drying, and high conductivity close to that of bulk metal. They are suitable materials for the fabrication of connectors, electrodes, springs, or conductors with small sizes. Conductive paths made by inkjet technologies using conductive inks for the incorporation of electrical components in microfluidic devices demonstrated a resolution from 500 μm to several microns [19]. Together with noble metals, promising NPs for conductive nanoinks are carbon-based nanomaterials such as graphene, CNTs, and carbon black. For example, highly conductive inks containing PLA/CNTs nanocomposites demonstrated sensitivity as 3D-printed structures of repeated patterns of filaments with potential applications as smart sensors in the textile or electronics field [35]. Hybrid conductive and stretchable nanocomposites of MWCNTs with self-assembled silver NPs and micrometer-sized silver flakes were also suitable for the production of effective electrical networks after hot rolling printing [36]. However, when printing droplets on hydrophobic surfaces, they tend to merge because of the affinity of nanoink fluid increases, which affect the quality, width, and resolution of printed lines. Vafaei et al. demonstrated that applying microstructuring and coating on the substrate was an alternative method of increasing the spreading of nanofluid silver ink and decreasing the solid surface tension that lessened the width and enhanced the cross-sectional area of the printed lines [37]. In other research, aqueous silver nanoink–printed on PET substrate and annealed at 180°C showed resistivity of 3.7 μΩ or twice lower than that of bulk silver. The PAA-coated silver NPs with mean size less than 40 nm were dissolved in deionized water and directly inkjet-printed on the flexible substrate. Shen at al. demonstrated the potential of the proposed low-cost and short-processing method by synthesizing highly conductive LED device circuits sinter at a temperature below 100°C [38]. As a result of their thermodynamic size effect, NPs with controlled size and shape demonstrate melting temperature depression. The low annealing temperature used makes the process compatible with most of the plastic substrates. However, the temperature for annealing frequently creates cracks and voids in the conductive circuits. In 2016, Yamada et al. proposed a new printing-based method that used a dense suspension of 40–60 wt% nanometal ink containing weakly alkylamide-encapsulating nanosilver. When printed on the photoactivated polymer surface, the occurring amine–carboxylate (chemisorption) conversion triggered the spontaneous formation of a self-fused solid silver layer under ambient conditions with strong adherence (over 5 MPa) to the substrate. The highest resolution obtained was equal to 800 nm line width, while the conductivity of the layers was measured to be $1 \times 10^5$ S/cm [39].

ZnO is a semiconductor that could be used as a channel conducting material for transparent thin-film transistors (TTFTs) that are alternative to organic semiconductors for OLED or conventional indium-tin-oxide (ITO) and Si-based materials. Ceramic inks containing pure and manganese-doped ZnO NPs (with size less than 200 nm) in ethylene glycol were found to be stable against sedimentation up to a month. By commercial inkjet printer and inkjet writer, after annealing at 200°C, the printed patterns showed uniform deposition and a high transmittance (92%) in the visible region [40]. Using piezoelectric inkjet nozzle, Lui et al. synthesized ZnO thin film deposited on a PVP/ITO/glass from ZnO NP (30–50 nm size) containing ink with dispersant and butylamine. After annealing, the transmittance in the visible spectrum was about 60%–70%, whereas the leakage current was less than $10^{-8}$ A/cm$^2$ under a bias of 100 V. The $I_D$–$V_D$ characteristics of ZnO-NPs-based top contact-TFT showed typical electrical behavior, current saturation, and pinch-off. However, the saturation field mobility was lower than the published. Such type of printable nanomaterials enables the production of low-cost flexible displays, OLED, and other transparent printed electronic applications [41]. Similarly, barium titanate ($BaTiO_3$), which is an important ferroelectric material with application especially as a ceramic capacitor, could be included in inks for inkjet printing. Using NPs with a size of 3–7 nm in sols that contain acetic acid and glycerin, inks were deposited on a silicon substrate with a precision of 75 μm. The films with 200 and 300 nm thickness were crack-free with good adhesion to the substrate [42]. Multilayered printed device consists of a transparent electrode of silver NP ring, subsequent PEDOT/PSS layer, followed by PolyTPD charge transport layer and semiconductor emission layer of CdSe/ZnS quantum dots, covered by eutectic GaIn liquid metal formed quantum dot LED that achieved color emission and a maximum brightness of 250 cd/$m^2$ at 5 V [43]. Moreover, aerosol inkjet-printed SWCNTs inks proved to be useful for the production of high-performance thin-film transistors with high yield [44]. It follows that inkjet printing offers precision down to 10 μm, precise control over printing position and production of complex multilayered structures with different thickness. Nevertheless, the increased viscosity or surface tension of the ink could result in difficulties in small-size drop formation and, therefore, a reduction in the resolution of inkjet printing. Other potential detrimental effects could be sedimentation, adhesion to nozzle or crystallization, NP conglomeration, etc.

In terms of accessibility and low cost, the creation of highly conductive polymer filament for direct use in FDM or FFF printers would be ideal for the production of interconnectors or electronic components without the requirement for postprocessing [45]. However, difficulties in the processability of metal NPs–polymer composites usually affect the quality and resolution of 3D objects. These NPs are susceptible to corrosion and aggregation usually because of van der Waals forces and/or magnetic attraction. The high weight percentage of the filler is needed to obtain workable conductivity at the expense of increased viscosity, density, and changed rheological properties of the polymer. Highly conductive polymer composites containing CNTs or graphene fillers with relatively low amounts allow one to build a mechanically stable object with good electrical and thermal conductivity. The minimum volume fracture of the nanofiller called percolation threshold determines the electrical conductivity or NP content that converts the insulating polymer into a conductive composite. The percolation threshold is strongly influenced by the size and shape distribution of nanofillers and their interactions [46]. To control these factors, the printing conditions such as speed, temperatures of bed and nozzle, and residence time should be carefully selected to obtain printability and functionality. Gnanasekaran et al. used semicrystalline PBT as a polymer matrix for the development of FDM-printed conductive nanocomposite containing uniformly distributed CNTs and graphene platelets. For fabrication of well-conductive filament, at least 0.49 wt% CNTs and 5.2 wt% graphene were required—a content above the percolating volume fraction. However, the resistivity changed as a function of the angle of measurement because of the limited connectivity at the interface of the printed lines and 3D printed layers. CNTs were predominately located within the crystalline regions of the polymer matrix indicating a tendency of PBT to crystallize on CNT walls. As a consequence, the thermal stability of the PBT/CNTs composite was higher than the rougher and more brittle PBT/graphene composite [47]. A detrimental effect upon prolonged printing of abrasive materials is wearing out the brass nozzle causing degradation in functional and esthetic properties of the printed material. Therefore, for printing abrasive material harder materials for manufacturing, the nozzles should be used together with optimizing the printing conditions.

Carbon nanomaterials are also potential candidates for energy storage and conversion application in favor of renewable energy sources due to their high absorption coefficients, scalability, and ability to be printed on flexible substrates. Piezoelectric inkjet printing was used for the production of an organic solar cell of P3HT/fullerene (1:0.8 w:w ratio) nanocomposite printed on glass/ITO substrate. The organic photovoltaic (OPV) displayed a power conversion efficiency of 3.07% [48]. Moreover, graphene and CNTs could ideally replace ITO as a transparent electrode in OPVs when roll-to-roll processing and continuous printing techniques are utilized [49].

Miniaturized 3D-printed sensors successfully developed in laboratories can provide reliable measurements working in harsh environmental conditions more effectively. For instance, highly conductive filaments from nanocarbon materials with PLA extruded through FDM printers with a volume resistivity of 1 Ω cm provided an excellent opportunity for the construction of 3D-printed capacitive and circuitry touch sensors [8]. Using polymorph PCL, Leigh et al. synthesized a composite with 15 wt% carbon black for the piezoresistive sensor. When bent, the sensor changed its resistivity with 4%, while the conductivity of the filament printed was 11.1 S/m or in the range of semiconductors [50]. As the conventional thin film, electromechanical sensors do not show high sensitivity; using optimization by 3D geometry, improvement in parasitic disturbances from the substrate material during measuring could be expected. When using carbon black–PP nanocomposite, the thermal stability of the heteromaterial, as opposed to carbon black–PLA (~60°C), increased to ~130°C.

The resistivity of the FDM-printed nanocomposite material with over 30 wt% carbon black was below $10^{-2}$ Ω, whereas the conductive nanofilled thermoplastic material with >25 wt% carbon black showed low variability in electrical resistance under different stress (thermal, electrical, UV) conditions [51]. However, the major drawback of the conductive PP-based nanocomposites is their low adhesion strength with other thermoplastics that are used in FDM printing. Using vat photopolymerization, Bodkhe et al. printed smart 3D nanocomposite structures with high piezoelectric and dielectric properties. The nanocomposite consisting of PVDF (piezoelectric material) and $BaTiO_3$ NPs was used for fabrication of 3D contact microsensor printed into fibers with about 56 μm size. The sensor generated up to 4 V upon finger tap has potential application in aerospace, biomedicine, or robotics [52].

Because of their increased metallic character, MWCNTs are usually used in sensor applications. For example, Christ et al. produced highly elastic and piezoresistive FDM-printed nanocomposite from TPU and MWCNTs (up to 5 wt%) that performed with no decay under applied strain as large as 100% and could be successfully used for complex-designed strain sensors [53]. However, Vatani et al. reported production of a conductive sensor for large strain detection from SWCNTs (1.5 nm diameter and 1–5 μm length)-loaded polymer nanocomposite within a matrix of two photocurable polymers (CTFA and acrylate ester) that was printed using direct writing on PU (polyurethane) substrate. The wires sustained 90% elongation, while the resistivity raised proportionally with the strain [54]. Solvent-cast 3D-printed nanocomposites of PLA/CNTs demonstrated transparency varying from 0% to 75% together with electrical conductivity up to 5000 S\m by modifying the interfilamentous spacing and printing pattern. The specific electromagnetic interference shielding effectiveness of the printed nanocomposite scaffolds demonstrated significant improvement (around 70 dB/g $cm^3$) compared with solid hot-pressed PLA/CNTs (about 37 dB /g $cm^3$). These properties could be useful for lightweight electromagnetic interference shields in airplanes, portable and flexible electronic devices, or even antistatic coatings [55].

To impart magnetic or conductivity properties to 3D-printed parts, nanofunctionalization with magnetite ($Fe_3O_4$) NPs that are known to be biocompatible and chemically stable is commonly applied. In this respect, by direct loading with magnetite NPs in a photopolymer matrix containing SR349 monomer and Lucirin-TPO photoinitiator, ferromagnetically responsive nanocomposites were SLA fabricated. The diameter of the oxide

NPs varied from 50 up to 100 nm. The sensing performance of the cantilever-based beams tested in terms of static deflection versus applied magnetic field showed good functional properties that made this material promising for actuator and sensor applications [56]. Other studies used micro-SLA to develop a flow sensor device by introducing magnetite NPs (25 wt%) with an average diameter of 50 nm into photopolymer resin containing acrylic oligomers. The printed internal impeller assembled in flow sensor device showed a directly proportional increase in rotation between 1 and 2.5 PSIG from 16.5 to 28 rpm, respectively, which proved its functionality for magnetometry in microfluidics or micropneumatics application [57]. All these functional sensors are some examples of the applicability of 3D printing technologies in promising applications of electronic devices.

In the field of optics, getting optical nanostructures applicable in quantum communication require uniform photon transport for a certain wavelength in a thin layer. The ability to fabricate submicron periodic lattice structures makes AM suitable for the fabrication of photonic crystals consisting of well-ordered structures with spacing equal to the length scale of light (400–800 nm) [58]. However, the deposition of transparent dielectric structures with a thickness of several nanometers is still a challenging task for the AM technologies. Obtaining inks with high refractive index and transparency at low cost includes the application of several oxides such as $TiO_2$, ZnO, $ZrO_2$, $Fe_3O_4$, and some mixed oxides, whereas silicon and aluminum oxides in polymer matrix demonstrate a low refractive index (RI). The photon flow is usually controlled when layers with high RI wrapped a low RI core. For that reason, alternating layers produced by 3D printing should be applied [19]. 2PP technology using femtosecond laser offers fabrication of periodic structures with features down to 100 nm in two or three dimensions. This resolution enabled the synthesis of woodpile-type photonic crystals within hybrid matrices that could be used for light filters, resonators, switches, or optical amplifiers [59]. Wrinkler et al. were able to synthesis 3D metallic resonant nanoarchitectures with nanometer resolution via electron-stimulated reactions composed of highly compact pure gold that reveal strong plasmonic activity [60].

Over the past few years, 3D printing technologies have been used for the fabrication of electromagnetic structures integrated as antenna parts. An approach to produce passive, low-weight, wearable, and flexible antenna was proposed by Matyas et al. They used PET substrate with a thickness of 150 μm and inkjet-printed silver NPs both incorporated into the plastic casing for protection, to construct microstrip antenna. The silver NPs with a size of 20–200 nm were obtained by solvothermal synthesis method and dispersed in deionized water at 25 wt% concentration. After drying at 120°C for 20 min, the printed antenna with a weight of 0.208 g had a homogeneous and electrically conductive surface Ag layer and was found to operate in two frequency bands of 2.3 GHz (−19.3 dB) and 2.02 GHz (−16.02 dB) [61]. Printing on concave and convex hemispherical glass substrates of silver NPs allowed for the fabrication of electrically small antenna offering nearly an order of magnitude improvement over monopole design [62]. AM antennas and RF circuit components demonstrate the potential of delivering multifunctional RF materials to support simple and low-cost 3D printing technology requiring minimum labor time.

Another application where 3D printing technologies implementing nanocomposite materials is starting to impact is energy storage including electric vehicles or portable electronic or renewable energy storage. Additive-manufactured microbatteries have the potential to enlarge energy density in limited space. By using GO-based composite inks, all component 3D-printed lithium-ion microbatteries with solid-state gel polymer electrolytes were produced. The 3D-printed device exhibited stable cycling performance with specific capacities and high electrode mass loading of 18 mg/cm$^2$ normalized to the overall area of the battery [63]. Similarly, ultrahigh rate performance was obtained for carbon-coated 3D-printed cathode with $LiMn_{0.21}Fe_{0.79}PO_4$ nanocrystals. LIBs with 3D-printed cathodes showed high conductivity and impressive electrochemical performance: a capacity of 108.45 mAh/g at 100°C and reversible capacity 150.21 mAh/g at 10°C after 1000 cycles [64]. Sun et al. were even able to print a microbattery of $LiFePO_4$ (LFP) and $Li_3Ti_4O_{12}$ (LTO) as anode and cathode, respectively, using a coprinting method. The device achieved high areal energy and power density (9.7 J/cm$^2$ at a power density of 2.7 mW/cm$^2$) [65]. Wei et al. printed a flexible LIB using water-based ink, comprised of $TiO_2$ (anatase) NPs and modified graphene sheets with *p-* or *n-type* anionic groups. LIBs displayed stable cycle performance at 240 mAh/g after 100 cycles [66]. Except for batteries, aerogel 3D-printed periodic and porous graphene-containing microlattices exhibited excellent electrochemical properties, in particular, exceptional capacitive retention (ca 90% from 0.5 to 10 A/g) and power density (over 4 kW/kg), which enabled the fabrication of high-performance and fully integrable energy storage devices [67]. It follows that besides lightening the overall weight and volume of the cell, fully printed energy storage devices with increased volumetric energy density could be developed using flexible and conductive polymer electrolytes. However, the potential of AM technologies for batteries and supercapacitors is still away from being fully realized and widespread.

The use of NPs could overcome the resolution limitations faced by AM technologies. For instance, Vyatskikh et al. proposed a reproducible method for the production of complex 3D metal geometries with resolution varying from 25 to 100 nm without sacrifice in mechanical strength. They used hybrid organic–inorganic material to develop UV-curable metal-based photoresist, and using two-photon lithography and pyrolysis, periodic Ni octet nanolattices with 2 μm unit cell size and 30 nm layers were fabricated. The beam diameters were equal to 300–400 nm with 20 nm mean grain size and 10%–30% porosity. The nanoscaled metals proposed huge potential applications in complex sub-mm devices such as MEMS, microbattery electrodes, tools for invasive medical procedures, or microrobots [68].

## 15.4 3D-printed nanocomposites used in medicine

Utilizing material extrusion, vat photopolymerization or cell printing techniques, AM could be used for the fabrication of organs, blood vessels, ligaments, soft tissue, and bone prosthesis. For example, end-user parts complying with tomographic patient data could be produced directly in prosthetic dentistry, orthodontics, and osteoplastic [69]. Moreover, 3D bioplotting is one of the most promising approaches for the printing of fully functioning multicellular organs that could thrive in the human organism [70]. Various biomedical materials such as metals, alloys, bioactive glasses, plastics, and ceramics have been used for implant 3D manufacturing [71]. Bioceramics used as nanofillers or artificial bone matrix could be divided into three main groups: *bioinert* such as alumina ($Al_2O_3$) and zirconia ($ZrO_2$); *bioactive* such as bioglasses and sintered HA; and *bioresorbable* like α- and β-TCP, TTCP, etc. [72]. Adding nanosubstances (carbon nanotubes, Ca–P, CHA, TCP, and bioactive glasses) in biodegradable or biocompatible naturally derived or synthetic polymers, the biocompatibility of the nanocomposite system could be enhanced by increasing the roughness, hydrophilicity, or mechanical strength of the scaffold. Table 15.2 summarizes the effect of the different NPs embedded in ceramic, polymer, or hydrogel scaffold on the biological performance of the nanocomposites. The crucial parameters for 3D-printed hard implant materials are

**Table 15.2** Summary of material composition, type of scaffold, 3D printing technology used for production, and biological properties of different nanocomposites.

| Type | Technology | Nanofiller | Size | Matrix material | Scaffold | Biological effect | References |
|---|---|---|---|---|---|---|---|
| Ceramic | SLS | Graphene | 0.7–1.2 nm in thickness and 0.8–3 μm in diameter | 58S bioactive glass (58 mol% $SiO_2$, 33 mol% CaO, 9 mol% $P_2O_5$) | 3D network of interconnected pores from powder 58S (30–60 nm)-0.5 wt% graphene; pore size 0.8 μm | On 58S-0.5 wt%, graphene human osteoblast stem cells (MG-63) indicated round shape with mineralized nodules and elongated filopodia suggesting good cell biocompatibility | [73] |
| | SLS | HA and β-TCP | HA—60–100 nm TCP—100–300 nm | HA and β-TCP | Scaffolds with different compositions (TCP/HA: 0/100, 10/90, 30/70, 50/50, 70/30, and 100/0); porosity 61%; pore size ranged in 0.8–1.2 mm | In vitro studies with MG-63 indicated that TCP/HA (30/70) and TCP/HA (50/50) revealed higher biological efficiency because of the optimal balance between stability and dissolution rate of the scaffold | [74] |

*Continued*

**Table 15.2** Summary of material composition, type of scaffold, 3D printing technology used for production, and biological properties of different nanocomposites.—cont'd

| Type | Technology | Nanofiller | Size | Matrix material | Scaffold | Biological effect | References |
|---|---|---|---|---|---|---|---|
| | SLS | Nano 58S bioactive glass | 60 nm | Forsterite | Scaffolds with interconnected pores with different 58S content | MG-63 cell adhesion enhanced with increase in the amount of nano-58S bioactive glass | [75] |
| | Microdroplet jetting | β-TCP and ibuprofen | 400 mesh and 1.0 −0.6 nm | Silica ($SiO_2$) | Highly interconnected porous scaffold with macropores (350−450 μm) and mesopores (3.65 nm) | Long-term inhibiting effect on *Escherichia coli* growth; | [76] |
| Biodegradable polymers | SLS | CS (calcium silicate, $CaSiO_3$) | 0.2 −20 μm | PVA | No scaffolds fabricated because of PVA particle fused when CS reached 20 wt% | Good bioactivity and cytocompatibility | [77] |
| | SLS | Ca−P and CHAp (carbonated HA) | 10−30 nm | PHBV and PLLA | Porous scaffolds of Ca−P (15 wt%)/PHBV (62.6% porosity) and CHA (10 wt%)/PLLA (66.8% porosity) with strut size and pore size 0.5 | - Ca−P incorporation facilitates the proliferation of human osteoblast-like cell line (SaOS-2) and ALP expression | [78] |

| | | | | | | | |
|---|---|---|---|---|---|---|---|
| | | | | | and 0.8 mm, respectively | - No statistical difference in cell proliferation was found for CHA/PLLA and PLLA indicating concentration dependence | |
| | SLS | Ca–P | 10–30 nm | PHBV | A tetragonal porous scaffold of biomolecule-loaded microspheres that protect their biological activity and temporally released them | - Initial burst released of the protein was observed because of diffusion from the surface<br>- After that, nearly constant release for 28 days was observed<br>- The matrix slightly degraded | [79] |
| | Pressure-assisted microsyringing | HA | 20–90 nm and 20–80 μm | PCL | Scaffolds with interconnected macropores; 72%–73% porosity; pore size 500 μm | - Nanodoped scaffolds were more hydrophilic than micro-HA/PCL scaffolds | [80] |

*Continued*

**Table 15.2** Summary of material composition, type of scaffold, 3D printing technology used for production, and biological properties of different nanocomposites.—cont'd

| Type | Technology | Nanofiller | Size | Matrix material | Scaffold | Biological effect | References |
|---|---|---|---|---|---|---|---|
| | | | | | | - The attachment and MG-63 cell proliferation on nanoHA/PCL was better than on micro-HA/PCL scaffolds | |
| | Pressure-assisted microsyringing | CNTs | — | PCL | Scaffolds with hexagonal square, octagonal cell grids, and 3D bone-like morphology and changing ration of CNTs to PCL (from 5.9 to 83 relative wt% CNTs) | - The majority of CNTs/PCL nanocomposites showed good MG-63 compatibility with values higher than 75% as opposed to the bare PCL<br>- Osteoblast viability depended on the intrinsic rigidity of the substrate | [81] |

| | | | | | | | |
|---|---|---|---|---|---|---|---|
| | FDM | TCP | – | PCL | Wound healing PCL-20% TCP scaffold with honeycomb-like pattern loaded with gentamicin sulfate (GS—5, 15, and 25 wt%); 85% porosity | - 15 wt% GS-loaded scaffold was able to eliminate both G+ and G– bacteria (*Pseudomonas aeruginosa* and *Staphylococcus aureus*) within 2 h and demonstrated good biocompatibility with dermal fibroblast cells<br>- The wounds in mice model healed rapidly and were earlier reepithelized with no signs of overall infection by day 7<br>- 15 wt% GS-loaded scaffold demonstrated antiadhesive properties | [82] |

*Continued*

**Table 15.2** Summary of material composition, type of scaffold, 3D printing technology used for production, and biological properties of different nanocomposites.—cont'd

| Type | Technology | Nanofiller | Size | Matrix material | Scaffold | Biological effect | References |
|---|---|---|---|---|---|---|---|
| | FDM | TCP | — | PP | Different cylindrical porous scaffolds with 20.5 vol% TCP, the pore size of 160 μm, and porosity varying from 36%, 48% and 52% | - Cells of modified human osteoblast cell line (HOB) attached and anchored on the composite substrate<br>- The samples were nontoxic for the osteoblast cells | [83] |
| | Aerosol-based 3D printing | $TiO_2$ (80% anatase/20% rutile) | 32 nm | PLGA | Composite cubic scaffolds with cubic pores (100 μm) and 32% porosity | - The scaffolds promoted osteoblast infiltration<br>- The cell was well spread and attached on the surface of the scaffolds<br>- Osteoblast infiltration in the pores was 4.2 times greater than adhesion on the surface | [84] |

| | | | | | | | |
|---|---|---|---|---|---|---|---|
| Hydrogel | Robocasting | Nanobioactive glass ($70SiO_2 \cdot 25CaO \cdot 5P_2O_5$) | Few hundred nanometers | Chitosane | Predesigned scaffold with 0.1 wt% nanobioactive glass and micropores (10 μm in size). As a result, the micro—macro dual pores structure was formed | - In vitro apatite-forming ability in SBF<br>- Preosteoblastic cells (MC3T3-E1) adhered and spread well-forming intimate contact with micropore surface | [85] |

**a.** high permeability and good resorption characteristics;
**b.** appropriate surface roughness;
**c.** mechanical characteristics close to that of the natural tissue; and
**d.** osteoconductive scaffold (providing a path for intergrowth and biological flows) for cell attachment, proliferation, and differentiation.

Synthetic 3D-printed scaffolds provide a template and suitable biomimetic environment for cell adhesion and proliferation since printed fibers and pores could be scaled from tens of micrometers to hundreds of nanometers which dimensions are close to the living cells scale. Teixeira et al. observed that fibers with diameters smaller than the size of the cells stimulated cell attachment growth and proliferation [86]. Besides mechanical strength, pores' volume, size, and interconnectivity, as well as material chemistry, are critical for the scaffold performance. In contrast to conventional techniques, employing AM technologies precise control over the shape, size location, and interconnectivity of pores in a scaffold could be exercised.

Nano-HA is one of the most widely used inorganic components in biocomposites of polymers and nanoceramics because of its similarity to natural bone mineral, biocompatibility, and bioactivity [87]. Combined with appropriate open porosity, the bioscaffold could enhance bone regeneration and allow the ingrowth of blood vessels. The most critical issue for porous ceramic scaffolds remains their mechanical properties. However, ceramic interconnected porous scaffolds fabricated directly using direct inkjet printing and sintering of nanopowder (550 nm) of TCP with silica ($SiO_2$—0.5 wt%) and ZnO (0.25 wt%) additives demonstrated enhanced mechanical and biological properties. Compared with pure TCP scaffold, the doped samples revealed decreased porosity and a 250% increase in compressive strength. MTT analysis suggested that the doped samples triggered increased cellular attachment and proliferation [88].

Similarly, Lin et al. demonstrated that CNTs (0.2%) used as nanofillers in β-TCP scaffold produced by SLS reached an 85.7% increase in strength compared with nondoped TCP scaffolds. However, in 0.1% and 0.3% CNT concentration, the material strength of the composite was reduced [89]. Nevertheless, these results, low strength along with brittleness, make the ceramic porous scaffolds difficult to process, manipulate, or even handle. To combine osteoconductivity, strength and stiffness of HA, and biodegradability and flexibility of biopolymers, nanocomposite scaffolds have been developed for tissue engineering of 3D scaffolds. For example, a mixture of bioresorbable and easily available PCL, montmorillonite (MMT), and HA nanofiller was successfully produced as new feedstock filament for the FDM process [90]. Except for enhanced cell attachment and proliferation, nanofillers are found to be able to control the biodegradability and bioactivity because of their high surface-to-volume ratio as well as other functional properties such as material strength. The addition of bioactive ceramic is also considered beneficial because it reduces the hydrophobicity of the polymer matrix and is therefore favorable for cell attachment and enhanced scaffold degradation [91]. Moreover, the presence of nano-HA up to 20 wt% in a porous scaffold PLGA fabricated via SLS was found to improve the compressive stress and modulus [92]. The mechanical properties and biological performance of 3D-printed scaffolds for hard tissue regeneration may be changed by controlling the wall thickness, pore size, and interconnectivity, together with the material selection. The main issue for polymer/nanoceramic composites remains the agglomeration of the ceramic NPs in the polymer matrix. As previously discussed, regarding SLS technology coating, the nanopowder particles with polymer binders could improve their distribution into the polymer.

Hydrogels are commonly synthesized in 3D bioprinting process by hydrophilic polymer chains in a water-based environment. These natural ductile biomaterials have poor mechanical performance, lack of bioactive properties, and show variations depending on the type of tissue they originated. For improvement of their bioactivity, hydrogel blends and embedded NPs were further developed. NPs loaded in a hydrogel matrix could also generate multiplexed spatial gradients, while the particles' properties could be adjusted to control the payload release kinetic. Similar to articular cartilage, many soft tissues could be regarded as NP-reinforced hydrogel composites. Sachlos et al. used a 3D wax printer to synthesize collagen-based polymer containing CHA nanocrystals with an approximate size of $180 \times 80 \times 20$ nm. They produced 3D biocomposite scaffolds that incorporated microchannels made to accommodate blood vessels and nutrient channels [93]. Because of its inherited photosensitivity, PEGDA is commonly used as synthetic hydrogel material for SLA printing. To enhance the biocompatibility of this intrinsic hydrophobic polymer, Zhou et al. synthesized a 3D scaffold of PEGDA bioink containing HA NPs with an average width of 25 nm and cell adhesive termed Arginine—glycine—aspartic acid serene. The optimal geometric pattern of the SLA-synthesized nanocomposites for human mesenchymal stem cell (HMSC) attachment was found to be a scaffold with small square pores. The authors combined this optimal scaffold geometry with low-intensity pulsed ultrasound (LIPUS) and achieved improved HMSCs function evaluated by increased proliferation, calcium deposition, total protein content, and alkaline phosphatase activity [94]. Apart from the advantages of hydrogels that most closely simulate the native tissue environment concerning water content, texture, morphology, and biodegradability, for some highly loaded hard tissue applications, they have inadequate mechanical performance. Usually, nanoembedded artificial or natural hydrogels such as chitosan, fibrin, collagen, agar, gelatin, alginate, and so on are used for wound healing or regeneration of articulating cartilages or tendons. For example, the 3D-printable nanocellulose—alginate porous hydrogel was found to be a suitable platform for the development of drug-releasing material for wound healing. Moreover, the 3D-printed hydrogels cross-linked with $CaCl_2$ showed the potential to be functionalized with small bioactive molecules such as biotin, growth factors, antibodies, antibiotics, or peptides that are required as specific adhesion-mediated signals. Another essential property of nanocellulose-containing hydrogel was its ability to regulate the moisture level of the wound [95]. Other research group fabricated programmable material of clay, nanofibrillated cellulose, enzyme/glucose, monomer, photoinitiator, and water. The 3D-printed composite hydrogel architecture was encoded with localized anisotropic swelling behavior determined by the cellulose nanofiber alignment along prescribed 4D printing paths. The flower structure with a filament size of 100 μm performed different swelling ratio of active and rigid materials on immersion in water [96]. The potential of nanocomposites containing noble metal NPs and hydrogels in tissue regeneration and sensing makes them promising biocompatible multifunctional materials. Using chondrocyte cell-seeded hydrogel matrix along with silver NPs—embedded intertwined conducting silicone, Mannoor at al. enhanced in vitro cartilage tissue culturing while giving auditory sensing for radio frequency (up to 5 GHz) reception of 3D-printed bionic ear [97].

The regeneration of damaged neurogenic tissue remains a challenging task because most of the implanted scaffolds demonstrate a lack of electrical properties. One option is to print an electroactive scaffold from liquid ink. A group of researchers used extrusion-based 3D printing to produce graphene composite consisting mainly of graphene and minority PLGA. The flexible and mechanically robust material retained its electrical conductivity (higher than 800 S/m), and in the absence of neurogenic stimuli, it supported HMSC adhesion, proliferation, viability, and neurogenic differentiation. These features highly mimicked axons and presynaptic terminals [98].

When looking for a solution to reduce drug toxicity or induce cancer therapy, magnetic NP—containing nanospheres or nanocapsules in the high-frequency magnetic field could be used for directing toward the target position. Accumulating in the intended location, magnetic NPs can be used for initiation of drug desorption, hyperthermia therapy in cancer cells under alternating magnetic field [99] or even visualizing the place by magnetic resonance imaging. Santis et al. synthesized magnetic scaffolds to examine their cartilage and bone regeneration activity. By combining FDM- and SLA-printing techniques, hybrid coaxial and bilayer magnetic scaffolds embedded with 200 nm magnetite particles were produced. For bone application, highly loaded polymer scaffolds were obtained by FDM processing. The scaffolds demonstrated a detailed structure and a good quality interface between PEGDA and PCL. PEGDA/magnetite cartilage scaffold and PCL/magnetite bone scaffolds indicated good adhesion and spreading of HMSCs without applying an external magnetic field that was thought to guide bone regeneration [100]. However, the uncontrolled release of metal or oxide NPs from scaffolds is still an unresolved issue because they may not be taken by the surrounding cell, and if translocated to other organs, they can cause undesired toxicity or other adverse effects. In other research, distortion-free 3D scaffolds fabricated of acrylate-base ink and nonmagnetic alumina platelets with adsorbed iron oxide NPs built heterogeneous composites with unparalleled microstructural features by magnetically assisted 3D printing. It was found out that the magnetic alignment of 15 wt% of platelets in the tensile loading direction increased the strength, elastic modulus, and swelling strain by 49%, 52%, and 30%, respectively, compared with perpendicularly aligned platelets. The ability to change the nanocomposite properties made magnetically assisted 3D printing a promising approach for printing shape-changing parts such as autonomously triggering flexible joint or selective pick-and-place systems in soft robotics [101].

## 15.5 Additive-manufactured nanofunctionalized surfaces

To make active materials beneficial for a component performance, surface control methods including AM could be applied. Nanoparticles such as TiC, $Al_2O_3$, $TiB_2$, nanospinel, and so on could be used as reinforcement materials during laser beam melting or direct metal laser sintering to increase the mechanical properties, heat, and wear resistance. Moreover, the laser sintering of NPs provides facile fabrication of functional layers on heat-sensitive polymer layers important for flexible electronics [102]. Nano-to-nano techniques such as laser sintering, laser-assisted electrophoretic deposition, direct nanoparticle ink writing, and electrohydrodynamic printing are used for direct deposition of nanoparticles. The difference in the position of NPs in blended nanocomposites and printed scaffolds with nanofunctionalized surfaces is illustrated in Fig. 15.2.

As previously discussed, noble metallic NPs have tunable electrical and optical properties due to their surface plasmonic resonance. The direct metal patterning of a variety of small-volume manufacturing paradigms by laser-induced local melting of metal nanoparticle ink proposes low-temperature, low-cost, fast, and simple fabrication methods. Commonly metal NP inks such as Ag, Au, or Cu are printed on a flexible substrate, while the binders and dispersive agents included in the ink have to be eliminated to recover the conductivity. In contrast to the thermal sintering process, the laser-based patterning provides uniformity and resolution down to a couple of microns in a fast-single step at room temperature and ambient pressure. Using nanoparticle Ag ink, a metal electrode with low-resistivity patterning (2.1 μΩ cm) was fabricated for flexible electronic devices with electronic and mechanical functionalities [103]. Not to waste the remaining nanoparticles, Chung et al. integrated a drop-on-demand jetting system where the moving substrate displaced the nanoparticle Au ink in front of and around the continuous Gaussian laser spot to conduct deposition and sintering simultaneously [104].

To make superhydrophobic surfaces that are known to possess functionalities such as corrosion protection, water collection, drag reduction, self-cleaning, antiicing, and so on, low-surface-energy and nanoscale surfaces are commonly fabricated. To make superhydrophobic FDM-printed porous PLA membrane for oil—water separation, Xing et al. used subsequent chemical etching with acetone followed by coating of polystyrene (PS) nanospheres with dopamine added Tris-buffer solution. The measured

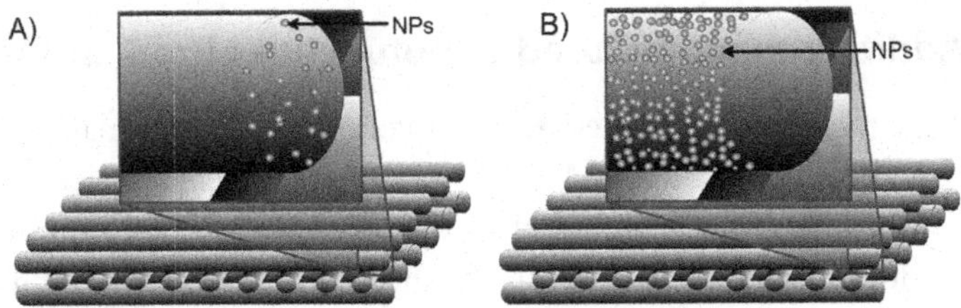


**Figure15.2** Schematic of NP location in (A) blended nanocomposite; (B) nanofunctionalized surface of 3D-printed scaffold.

water contact angle was 151.7 degrees, and water adhesion force was 21.8 μN. When using a pore size of 250 μm, this two-dimensional membrane part showed maximum oil—water separation efficiency of 99.4% at a flux of 60 kL/m$^2$/h [105]. Similarly, using FDM-printed PLA scaffold, hydrophobic nanosilica ($SiO_2$) was dip-coated to form micro—nanoscale surface using MEK dipping. The flat PLA surfaces after dip coating demonstrated a 114.1 degree contact angle, whereas using an additional user-defined grid pattern, the contact angle increased up to over 150 degrees. Line-patterned surfaces showed anisotropic superhydrophobic behavior, whereas the grid-patterned surfaces were found to have isotropic superhydrophobic behavior. However, a small reduction in the mechanical properties was observed because of the effect of MEK on PLA, while silica coating did not affect them [106].

Talking about metal parts, fused coating—based AM offers potential advantages such as obtaining of microvoid-free surfaces with low residual stresses and distortions of metal materials such as Pb—Sn, aluminum alloys, copper alloys, magnesium alloys, and steel. Such 3D-manufactured coating should have good sealing ability to prevent gas leak [107] or the ability to form complex geometries such as honeycomb or lattice structures. Although not realized so far, various suitable NPs could be initially added in molten metal to pass through the fused coating head. This will affect not only the crystallization process but also the properties of the metal layers. It should also be noticed that the small particle size significantly reduces the melting temperature of NPs allowing the high-temperature process to increase their size. Consequently, NPs such as CNTs, graphene, GO, or some metal oxides will be suitable for consideration.

The limited surface quality and smoothness of 3D-printed materials usually require additional procedures for sealing the surface or enhancement of properties without deforming the parts. Autoclave sterilization process of fine scaffolds produced with ABS modeling material usually undergoes deformation because of the high temperature used. To reduce the bacterial growth on 3D-printed ABS-produced medical devices, Stratasys tested commercially available water-based antimicrobial coating containing 10 wt% silver sulfadiazine. They found out that pure ABS and silver-loaded ABS devices with uniform coatings showed a nearly 100% reduction in the growth of *Escherichia coli* and *Staphylococcus aureus* after 48 h incubation, whereas the addition of a small amount of silver into the thermoplastic polymer provided only limited sterility [108]. Moreover, 3D-printed scaffolds can be used in drug or growth factor delivery. For example, bioceramic powder—printed scaffold with adsorbed vancomycin, ofloxacin, and tetracycline showed a linear correlation with drug concentration in immersion solution. The differences in the amount of the adsorbed drug were correlated with the specific surface area of the scaffolds. With additional polymer impregnation of the drug-loaded matrix, the release kinetics could be delayed, and sustained released could be achieved [109]. This postprinting modification enables not to compromise the viability of the drugs or growth factors during printing. In other interesting research, Gupta and his group printed stimuli-responsive core—shell capsules from aqueous hydrogel core for loading with biomolecules embedded in PLGA shell. Loading the shell with plasmonic gold nanorods enabled selective rupturing of the capsule upon laser irradiation that rapidly heated the nanorods and melted the polymer capsule which allowed precise control over time, space, and selectivity. The laser-triggered rupture was successfully tested for the programmable release of horseradish peroxidase into a hydrogel ambient [110].

In bone tissue engineering to achieve mechanical and biochemical properties close to that of the bone, the printed scaffold should have not only suitable architecture but also osteoconductive properties. In this respect, 3D-printed polymer—ceramic composites are good candidates, but the limited exposure of ceramic could diminish their osteoconductive properties. For that reason, coatings with osteogenic particles such as HA are used to make the osteoconductive layer. Numerous studies showed successful coating of PEEK with HA by cold spray, spin coating technique, aerosol deposition, RF magnetron sputtering, and so on [111]. However, PEEK is a nonbiodegradable polymer. As for biodegradable polymers, the coprecipitation of CHA—gelatin composite on rapid prototyped PCL/TCP scaffolds proved to uniformly distribute bone marrow stromal cells (BMSCs) and accumulate extracellular matrix (ECM) in the interior of the modified scaffold. The proliferation rate of BMSCs on the coated scaffolds was about 2.3 times higher as opposed to the bare PCL/TCP scaffold [112]. Other authors proposed surface functionalization of SLS-produced PCL scaffold with the first layer of gelatin and second of the glycoprotein fibronectin that is known to enhance the initial cell-adhesion contact. The coating was found to increase the surface hydrophilicity without changing the mechanical or morphological properties of the scaffold. In vitro tests with human osteoprogenitor cells showed that the double protein coating gave rise to cell differentiation and metabolic activity compared with the untreated PCL scaffolds [113]. Similarly, FDM-printed PLA material, which is typically a hydrophobic polymer, was surface-modified by direct immersion in mussel-inspired PDA. The bioinspired coating promoted the adhesion, proliferation, angiogenetic, and osteogenetic differentiation of human adipose-derived stem cells related to pure PLA scaffold, which made the modification suitable for regulation of stem cell behavior [114]. Last but not least, FDM-printed scaffolds of PCL/PLGA and PCL/PLGA with β-TCP nanofiller (~100 nm size) were biologically functionalized by surface ornamenting with cell-laid ECM composed of inorganic mineral phases and collagenous and noncollagenous proteins. Compared with bare scaffolds, the ECM-ornamented scaffold induced greater bone formation in rat models in vivo [115].

By surface modification of unique high-volume-to-area 3D-printed structures, the sensitivity and selectivity of electrode materials could be vastly enhanced. For example, plating with nanothick gold film of 3D-printed metal electrodes was proved to be an efficient tool in the detection of Pb and Cd in aqueous solution [116], electrochemical detection of nitroaromatic compounds [117], phenolic compounds [118], and even DNA biosensing with a detection range of 1–1000 nM [119].

## 15.6 Conclusions

New nanocomposite materials represent the next major innovation in the field of AM, and they are believed to provide radically new properties for the products and broaden their functionality. The ability to tune material properties through incorporating nanoingredients indicates a revolution step in the functionalization of AM materials. On one hand, using nanoparticle inks, the size of the printed features could be substantially minimized. On the other hand, the printing of reinforced nanocomposites may enhance the mechanical, electrical, optical, etc. properties and develop functionalized sensors, energy storage, or electronic devices. Using 3D printing techniques, it is also possible to print graded nanocomposite materials by changing the NPs and nanostructures produced during synthesis. The graded structures could be single nanocomposite printed but with graded geometry or varying the concentration or type of constituents upon printing across the geometry. Nevertheless, the fabrication procedures with nanocomposite materials are still longer, more complex, whereas the obtained microstructure is difficult to be controlled. Moreover, to improve the competitiveness of AM techniques, the elimination of postprocessing steps is essential.

3D technologies can produce highly porous scaffolds with controlled architecture and nanofiller content, which could strengthen the nanocomposite and facilitate cell proliferation and differentiation because of their high surface-to-volume ratio and other unique properties. The CAD/CAM (computer-aided manufacturing) modeling allows precise fabrication of anatomically and biochemically designed implants compatible with personalized medicine. As a new multidisciplinary approach in 3D tissue engineering, nanofunctionalization seems to be promising in terms of using biological nanosubstances that could reduce implant rejection and offer treatment of diseases, tissue repairing, and elimination reoperations. Despite all achievements in the biomaterial area, there are still gaps in the knowledge related to surface chemistry, morphology, permeability, and growth factor release that accelerate tissue formation. The development of bioprinting of materials should focus more on controllable biodegradation that has to meet the requirements of proportionality of different regeneration stages in vivo together with metabolization and/or excretion. In this respect, 4D-printed smart nanocomposites with specific responsiveness could be the key for shape shifting or controlled degradation upon a time- or stimuli-related changes.

However, there are still issues in controlling the particle size distribution which hinders the volume production at a great extent. Moreover, the versatility in the development of new nonfunctionalized material for AM may necessitate the elaboration of new 3D printing nanofabrication techniques, approaches, sets of technical standards for characterization, and in-line process monitoring to provide feedback for material processing. As most of the printing methods are time-consuming, fabrication of parts in large volumes is difficult, which hinders industry adoption. Furthermore, an additional issue to overcome is printing resolution without complicating the printing technologies, extending time, or scarifying product geometry.

Numerous issues concerning nanofilled composite production via 3D printing have been identified together with possible solutions that have been assessed. It is not a matter of indifference that AM processes using NPs as ingredients can produce a lot of waste from supports, rafts, and failed prints. The issue of recycling of these materials is becoming more urgent because recycling could not only reduce their cost but also decrease the environmental and human health threat. To minimize the toxicity in each production stage, printing step, final nanocomposite product, usage, and disposal should be assessed, and solutions to reduce or eliminate the potential hazard or risk should be given.

## Abbreviations

| | |
|---|---|
| **(G+)** | gram-positive bacteria |
| **(G−)** | gram-negative bacteria |
| **2 PP** | two-photon polymerization |
| **3D** | three-dimensional |
| **ABS** | acrylonitrile−butadiene−styrene |
| **ALP** | alkaline phosphatase |
| **AM** | additive manufacturing |
| **ASTM** | American Society for Testing and Materials |
| **BMSC** | bone marrow stromal cells |
| **Ca−P** | calcium phosphate |
| **CAD** | computer-aided design |
| **CAM** | computer-aided manufacturing |
| **CHA** | carbonated hydroxyapatite |
| **CNC** | computer numerical control |
| **CNT** | carbon nanotube |
| **CS** | calcium silicate |
| **CTFA** | cyclic trimethylpropane formal acrylate |
| **DMLS** | direct metal laser sintering |
| **DOP-SLA** | dynamic optical projection stereolithography |
| **EBM** | electron beam melting |
| **ECM** | extracellular matrix |
| **FDM** | fused deposition modeling |
| **FFF** | fused filament fabrication |
| **GO** | graphene oxide |
| **HA** | hydroxyapatite |
| **HMSC** | human mesenchymal stem cell |
| **HOB** | human osteoblast cell line |
| **ITO** | indium−tin oxide |
| **LED** | light-emitting diode |
| **LIB** | lithium-ion battery |
| **LIPUS** | low-intensity pulsed ultrasound |
| **MC3T3-E1** | preosteoblastic cell |
| **MEK** | methyl ethyl ketone |
| **MEMS** | microelectromechanical systems |
| **MG-63** | human osteoblast stem cells |
| **mm** | millimeter |
| **MMT** | montmorillonite |
| **MPP** | multiphoton polymerization |
| **MTT** | 3-(4,5-dimethylthiazol-2-yl)-2,5-diphenyl tetrazolium bromide |
| **MWCNT** | multiwalled carbon nanotube |
| **NEMS** | nanoelectromechanical systems |
| **NP** | nanoparticle |
| **OLED** | organic light-emitting diode |
| **OPV** | organic photovoltaic |
| **P3HT** | poly(3-hexylthiophene) |

| | |
|---|---|
| **PA** | polyamide (nylon) |
| **PAA** | polyacrylic acid |
| **PAEK** | polyarylether ketone |
| **PARA** | polyacrylamide |
| **PBT** | polybutylene terephthalate |
| **PC** | polycarbonate |
| **PCL** | poly(ε-caprolactone) |
| **PDA** | polydopamine |
| **PDMS** | polydimethylsiloxane |
| **PEDOT:PSS** | poly(3,4-ethylene dioxythiophene) polystyrene sulfonate |
| **PEEK** | polyetheretherketone |
| **PEG** | polyethylene glycol |
| **PEGDA** | polyethylene glycol diacrylate |
| **PET** | polyethylene terephthalate |
| **PHBV** | poly(hydroxybutyrate-co-hydroxyvalerate |
| **PLA** | polylactide acid |
| **PLGA** | poly(lactic-co-glycolic acid) |
| **PLLA** | poly-L-lactide acid |
| **polyTPD** | poly(N,N′-bis-4-butylphenyl-N,N′-bisphenyl) benzidine |
| **PP** | polypropylene |
| **PS** | polystyrene |
| **PU** | polyurethane |
| **PVA** | polyvinyl acetate |
| **PVDF** | polyvinylidene fluoride |
| **PVP** | polyvinylpyrrolidone |
| **SaOS-2** | human osteoblast-like cell line |
| **SLA** | stereolithography |
| **SLM** | selective laser melting |
| **SLS** | selective laser sintering |
| **SR349** | startomer 349 |
| **SWCNT** | single-walled carbon nanotube |
| **TCP** | tricalcium phosphate |
| **TPU** | thermoplastic polyurethane |
| **TTCP** | tertracalcium phosphate |
| **TTFTs** | transparent thin film transistors |

## References


[1] E.P. Koumoulos, E. Gkartzou, C.A. Charitidis, Additive (nano)manufacturing perspectives: the use of nanofillers and tailored materials, Manuf. Rev. 4 (12) (2017) 1–9, https://doi.org/10.1051/mfreview/2017012.

[2] M. Vaezi, H. Seitz, S. Yang, A review on 3D micro-additive manufacturing technologies, Int. J. Adv. Manuf. Technol. 67 (2013) 1721–1754, https://doi.org/10.1007/s00170-013-4962-5.

[3] ASTM F2792-12a Standard Terminology for Additive Manufacturing Technology, ASTM International, West Conshohocken, PA, 2012. https://www.astm.org/Standards/F2792.htm.

[4] R.D. Farahani, M. Dubé, Printing polymer nanocomposites and composites in three dimensions, Adv. Eng. Mater. 20 (2) (2018) 1700539, https://doi.org/10.1002/adem.201700539.

[5] J. N. Coleman, U. Khan, Y. K. Gun'ko, Mechanical reinforcement of polymers using carbon nanotubes, Adv. Mater. 18 (6) (2006) 689–706, https://doi.org/10.1002/adma.200501851.

[6] M.L. Shofner, K. Lozano, F.J. Rodríguez-Macías, E.V. Barrera, Nanofiber-reinforced polymers prepared by fused deposition modelling, J. Appl. Polym. Sci. 89 (11) (2003) 3081–3090, https://doi.org/10.1002/app.12496.

[7] A.C. De Leon, Q. Chen, N.B. Palaganas, J.O. Palaganas, J. Manapat, R.C. Advincula, High-performance polymer nanocomposites for additive manufacturing applications, React. Funct. Polym. 103 (2016) 141–155, https://doi.org/10.1016/j.reactfunctpolym.2016.04.010.

[8] S.F.A. Acquah, B.E. Leonhardt, M.S. Nowotarski, J.M. Magi, K.A. Chambliss, T.E.S. Venzel, S.D. Delekar, L.A. AlHariri, Carbon nanotubes and graphene as additives in 3D printing, in: M. Berber (Ed.), Carbon Nanotubes Current Progress of Their Polymer Composites, InTech, 2016, https://doi.org/10.5772/63419.

[9] M. Wong, M. Paramsothy, X.J. Xu, Y. Ren, S. Li, K. Liao, Physical interactions at carbon nanotube-polymer interface, Polymer 44 (25) (2003) 7757–7764, https://doi.org/10.1016/j.polymer.2003.10.011.

[10] X.J. Wei, D. Li, W. Jiang, Z.M. Gu, X. Wang, Z. Zhang, Z. Sun, 3D printable graphene composite, Sci. Rep. 5 (2015) 11181, https://doi.org/10.1038/srep11181.

[11] B.E. Yamamoto, A.Z. Trimble, B. Minei, M.N.G. Nejhad, Development of multifunctional nanocomposites with 3-D printing additive manufacturing and low graphene loading, J. Thermoplas. Compos. Mater. 32 (3) (2019) 383–408, https://doi.org/10.1177/0892705718759390.

[12] J.Z. Manapat, J.D. Mangadlao, B.D. Tiu, G.C. Tritchler, R.C. Advincula, High-strength stereolithographic 3D printed nanocomposites: graphene oxide metastability, ACS Appl. Mater. Interface. 9 (11) (2017) 10085–10093, https://doi.org/10.1021/acsami.6b16174.

[13] J.H. Sandoval, K.F. Soto, L.E. Murr, R.B. Wicker, Nano-tailoring photo crosslinkable epoxy resins with multi-walled carbon nanotubes for stereolithography layered manufacturing, J. Mater. Sci. 42 (1) (2007) 156–165, https://doi.org/10.1007/s10853-006-1035-2.

[14] D. Yugang, Z. Yuan, T. Yiping, L. Dichen, Nano $TiO_2$ modified photosensitive resin for RP, Rapid Prototyp. J. 17 (4) (2011) 247–252, https://doi.org/10.1108/13552541111138360.

[15] Q. Wang, J.A. Jackson, Q. Ge, J.B. Hopkins, C.M. Spadaccini, N.X. Fang, Lightweight mechanical metamaterials with tunable negative thermal expansion, Phys. Rev. Lett. 117 (17–21) (2016) 175901, https://doi.org/10.1103/PhysRevLett.117.175901.

[16] J.Z. Manapat, Q. Chen, P. Ye, R.C. Advincula, 3D printing of polymer nanocomposites via stereolithography, Macromol. Mater. Eng. 302 (9) (2017) 1600553, https://doi.org/10.1002/mame.201600553.

[17] S. Lewis, L., Influence of nanocomposite materials for next generation nano lithography piccirillo, in: B. Reddy (Ed.), Advances in Diverse Industrial Applications of Nanocomposites, InTech Publisher, Rijeka, Croatia, 2011.

[18] K. Cai, J. Sun, Q. Li, R. Wang, B. Li, J. Zhou, Direct-writing construction of layered meshes from nanoparticles-vaseline composite inks: rheological properties and structures, Appl. Phys. A 102 (2) (2011) 501–507, https://doi.org/10.1007/s00339-010-5955-y.

[19] I. Volyanskii, I.V. Shishkovsky, Laser-assisted 3D printing of functionally graded structures from polymer covered nanocomposites: a self-review, in: I. Shishkovsky (Ed.), New Trends in 3D Printing, InTech, 2016, https://doi.org/10.5772/63565.

[20] Q. Li, J.A. Lewis, Nanoparticle inks for directed assembly of three-dimensional periodic structures, Adv. Mater. 15 (19) (2003) 1639–1643, https://doi.org/10.1002/adma.200305413.

[21] J. Lv, Z. Gong, Z. He, J. Yang, Y. Chen, C. Tang, Y. Liu, M. Fan, W.-M. Lauc, 3D printing of a mechanically durable superhydrophobic porous membrane for oil-water separation, J. Mater. Chem. 5 (2017) 12435–12444, https://doi.org/10.1039/C7TA02202F.

[22] A. Kosmala, R. Wright, Q. Zhang, A. Kirby, Synthesis of silver nanoparticles and fabrication of aqueous Ag inks for inkjet printing, Mater. Chem. Phys. 129 (2011) 1075–1080, https://doi.org/10.1016/j.matchemphys.2011.05.064.

[23] R.A. Vaia, R. Krishnamoorti (Eds.), Polymer Nanocomposites, vol. 804, ACS Series, 2001, p. 1.

[24] J. Kim, T.S. Creasy, Selective laser sintering characteristics of nylon 6/clay-reinforced nanocomposite, Polym. Test. 23 (6) (2004) 629–636, https://doi.org/10.1016/j.polymertesting.2004.01.014.

[25] J.H. Koo, S. Lao, W. Ho, K. Nguyen, J. Cheng, L. Pilato, G. Wissler, M. Ervin, Polyamide nanocomposites for selective laser sintering, in: Proc. 2006 SFF Symposium, Austin, TX, Aug 14-16, 2006.

[26] S. Ram Athreya, K. Kalaitzidoua, S. Das, Processing and characterization of carbon black-filled electrically conductive Nylon-12 nanocomposite produced by selective laser sintering, Mater. Sci. Eng. 527 (10–11) (2010) 2637–2642, https://doi.org/10.1016/j.msea.2009.12.028.

[27] H. Chung, S. Das, Functionally graded Nylon-11/silica nanocomposites produced by selective laser sintering, Mater. Sci. Eng. 487 (1–2) (2008) 51–257, https://doi.org/10.1016/j.msea.2007.10.082.

[28] H. Zheng, J. Zhang, S. Lu, G. Wang, Z. Xu, Effect of core-shell composite particles on the sintering behaviour and properties of the nano-$Al_2O_3$/polystyrene composite prepared by SLS, Mater. Lett. 60 (9) (2006) 1219–1223, https://doi.org/10.1016/j.matlet.2005.11.003.

[29] J. Bai, R.D. Goodridge, R.J.M. Hague, M. Song, M. Okamoto, Influence of carbon nanotubes on the rheology and dynamic mechanical properties of polyamide-12 for laser sintering, Polym. Test. 36 (2014) 95–100, https://doi.org/10.1016/j.polymertesting.2014.03.012.

[30] Nanowerk.com Reports on Polymer Nanocomposites Drive Opportunities in the Automotive Sector, 2016. http://www.nanowerk.com/spotlight/spotid=23934.php.

[31] N.B. Crane, J. Wilkes, E. Sachs, S.M. Allen, Improving accuracy of powder sintering-based SFF processes by metal deposition from nanoparticle dispersion, Rapid Prototyp. J. 12 (5) (2006) 266–274, https://doi.org/10.1108/13552540610707.

[32] D. Gu, H. Wang, F. Chang, D. Dai, P. Yuan, Y.-C. Hagedorn, W. Meiners, Selective laser melting additive manufacturing of TiC/AlSi10Mg bulk-form nanocomposites with tailored microstructures and properties, Phys. Proc. 56 (2014) 108–116, https://doi.org/10.1016/j.phpro.2014.08.153.

[33] I.V. Shishkovsky, A.V. Bulanova, Y.G. Morozov, Porous polycarbonate membranes with Ni and Cu nano catalytic additives fabricated by selective laser sintering, J. Mater. Sci. Eng. B 2 (12) (2012) 634–639.

[34] I.V. Shishkovsky, Y.G. Morozov, Electrical and magnetic properties of multilayer polymer structures with nano-inclusions as prepared by selective laser sintering, J. Nanosci. Nanotechnol. 13 (2) (2013) 1440–1443.

[35] K. Chizari, M.A. Daoud, A.R. Ravindran, D. Therriault, 3D printing of highly conductive nanocomposites for the functional optimization of liquid sensors, Small 12 (44) (2016) 6076–6082, https://doi.org/10.1002/smll.201601695.
[36] K.-Y. Chun, Y. Oh, J. Rho, J.-H. Ahn, Y.-J. Kim, H.R. Choi, S. Baik, Highly conductive, printable and stretchable composite films of carbon nanotubes and silver, Nat. Nanotechnol. 5 (2010) 853–857, https://doi.org/10.1038/nnano.2010.232.
[37] S. Vafaei, C. Tuck, I. Ashcroft, R. Wildman, Surface microstructuring to modify wettability for 3D printing of nano-filled inks, Chem. Eng. Res. Des. 109 (2016) 414–420, https://doi.org/10.1016/j.cherd.2016.02.004.
[38] W. Shen, X. Zhang, Q. Huang, Q. Xu, W. Song, Preparation of solid silver nanoparticles for inkjet printed flexible electronics with high conductivity, Nanoscale 6 (3) (2014) 1622–1628, https://doi.org/10.1039/c3nr05479a.
[39] T. Yamada, K. Fukuhara, K. Matsuoka, H. Minemawari, J. Tsutsumi, N. Fukuda, K. Aoshima, S. Arai, Y. Makita, H. Kubo, T. Enomoto, T. Togashi, M. Kurihara, T. Hasegawa, Nanoparticle chemisorption printing technique for conductive silver patterning with submicron resolution, Nat. Commun. 7 (2016) 11402, https://doi.org/10.1038/ncomms11402.
[40] S. Sharma, S.S. Pande, P., Swaminathan Top-down synthesis of zinc oxide based inks for inkjet printing, RSC Adv. 7 (2017) 39411, https://doi.org/10.1039/c7ra07150g.
[41] C.-T. Liu, W.-H. Lee, T.-L. Shih, Synthesis of ZnO nanoparticles to fabricate a mask-free thin-film transistor by inkjet printing, J. Nanotechnol. 710908 (2012) 1–8, https://doi.org/10.1155/2012/710908.
[42] J. Vukmirovic, Đ. Tripković, Đ. Tripković, B. Bajac, B. Bajac, V.V. Srdić, Comparison of barium titanate thin films prepared by inkjet printing and spin coating, Process. Appli. Ceramic. 9 (3) (2015) 151–156, https://doi.org/10.2298/PAC1503151V.
[43] Y.L. Kong, I.A. Tamargo, H. Kim, B.N. Johnson, M.K. Gupta, T.W. Koh, H.A. Chin, D.A. Steingart, B.P. Rand, M.C. McAlpine, 3D printed quantum dot light-emitting diode, Nano Lett. 14 (2014) 7017–7023, https://doi.org/10.1021/nl5033292.
[44] J. Zhao, Y. Gao, J. Lin, Z. Chen, Z. Cui, Printed thin-film transistors with functionalized single-walled carbon nanotube inks, J. Mater. Chem. 22 (2012) 2051–2056, https://doi.org/10.1039/C1JM14773K.
[45] P.F. Flowers, C. Reyes, S. Ye, M.J. Kim, B.J. Wiley, 3D printing electronic components and circuits with conductive thermoplastic filament, Addit. Manufact. 18 (2017) 156–163, https://doi.org/10.1016/j.addma.2017.10.002.
[46] K. Gnanasekaran, G. de With, H. Friedrich, On packing, connectivity, and conductivity in mesoscale networks of polydisperse multiwalled carbon nanotubes, J. Phys. Chem. C 118 (51) (2014) 29796–29803, https://doi.org/10.1021/jp5081669.
[47] K. Gnanasekaran, T. Heijmans, S. van Bennekom, H. Woldhuis, S. Wijnia, G. de With, H. Friedrich, 3D printing of CNT- and graphene-based conductive polymer nanocomposites by fused deposition modelling, Appli. Mater. Today 9 (2017) 21–28, https://doi.org/10.1016/j.apmt.2017.04.003.
[48] M. Neophytou, W. Cambarau, F. Hermerschmidt, C. Waldauf, C. Christodoulou, R. Pacios, S.A. Choulis, Inkjet-printed polymer-fullerene blends for organic electronic applications, Microelectron. Eng. 95 (2012) 102–106, https://doi.org/10.1016/j.mee.2012.02.005.
[49] S. Lawes, A. Riese, Q. Sun, N.C. Cheng, X.L. Sun, Printing nanostructured carbon for energy storage and conversion applications, Carbon 92 (2015) 150–176, https://doi.org/10.1016/j.carbon.2015.04.008.

[50] S.J. Leigh, R.J. Bradley, C.P. Purssell, D.R. Billson, D.A. Hutchins, A simple, low-cost conductive composite material for 3D printing of electronic sensors, PloS One 7 (11) (2012) e49365, https://doi.org/10.1371/journal.pone.0049365.

[51] S.W. Kwok, K.H.H. Goh, Z.D. Tan, S.T.M. Tan, W.W. Tjiu, J.Y. Soh, Z.J.G. Ng, Y.Z. Chan, H.K. Hui, K.E.J. Goh, Electrically conductive filament for 3D-printed circuits and sensors, Appli. Mater. Today 9 (2017) 167–175, https://doi.org/10.1016/j.apmt.2017.07.001.

[52] S. Bodkhe, G. Turcot, F.P. Gosselin, D. Therriault, One-step solvent evaporation-assisted 3D printing of piezoelectric PVDF nanocomposite structures, ACS Appl. Mater. Interface. 9 (24) (2017) 20833–20842, https://doi.org/10.1021/acsami.7b04095.

[53] J.F. Christ, N. Aliheidari, A. Ameli, P. Pötschke, 3D printed highly elastic strain sensors of multiwalled carbon nanotube/thermoplastic polyurethane nanocomposites, Mater. Des. 131 (2017) 394–401, https://doi.org/10.1016/j.matdes.2017.06.011.

[54] M. Vatani, Y. Lu, K.-S. Lee, H.-C. Kim, J.-W. Choi, Direct-write stretchable sensors using single-walled carbon nanotube/polymer matrix, J. Electron. Packag. 135 (1) (2013) 011009, https://doi.org/10.1115/1.4023293.

[55] K. Chizari, M. Arjmand, Z. Liu, U. Sundararaj, D. Therriault, Three-dimensional printing of highly conductive polymer nanocomposites for EMI shielding applications, Mater. Today commun. 11 (2017) 112–118, https://doi.org/10.1016/j.mtcomm.2017.02.006.

[56] C. Credi, A. Fiorese, M. Tironi, R. Bernasconi, L. Magagnin, M. Levi, S. Turri, 3D printing of cantilever-type microstructures by stereolithography of ferromagnetic photopolymers, ACS Appl. Mater. Interface. 8 (39) (2016) 26332–26342, https://doi.org/10.1021/acsami.6b08880.

[57] S.J. Leigh, C.P. Purssell, J. Bowen, D.A. Hutchins, J.A. Covington, D.R. Billson, A miniature flow sensor fabricated by micro-stereolithography employing a magnetite/acrylic nanocomposite resin, Sensor Actuator Phys. 168 (1) (2011) 66–71, https://doi.org/10.1016/j.sna.2011.03.058.

[58] J.-H. Lee, C.Y. Koh, J.P. Singer, S.-J. Jeon, M. Maldovan, O. Stein, E.L. Thomas, 25th anniversary article: ordered polymer structures for the engineering of photons and phonons, Adv. Mater. 26 (2014) 532–569, https://doi.org/10.1002/adma.201303456.

[59] R. Houbertz, P. Declerck, S. Passinger, A. Ovsianikov, J. Serbin, B.N. Chichkov, Investigations on the generation of photonic crystals using two-photon polymerization (2PP) of inorganic-organic hybrid polymers with ultra-short laser pulses, Phys. Status Solidi 204 (2007) 3662–3675, https://doi.org/10.1002/pssa.200776416.

[60] R. Winkler, F.P. Schmidt, U. Haselmann, J.D. Fowlkes, B.B. Lewis, G. Kothleitner, P.D. Rack, H. Plank, Direct-write 3D nanoprinting of plasmonic structures, ACS Appl. Mater. Interface. 9 (2017) 8233–8240, https://doi.org/10.1021/acsami.6b13062.

[61] J. Matyas, P. Slobodian, L. Munster, R. Olejnik, P. Urbanek, Microstrip antenna from silver nanoparticles printed on a flexible polymer substrate, Materials today Proceedings 4 (2017) 5030–5038, https://doi.org/10.1016/j.matpr.2017.04.110.

[62] J.J. Adams, E.B. Duoss, T.F. Malkowski, M.J. Motala, B.Y. Ahn, R.G. Nuzzo, J.T. Bernhard, J.A. Lewis, Conformal printing of electrically small antennas on three dimensional surfaces, Adv. Mater. 23 (2011) 1335–1340, https://doi.org/10.1002/adma.201003734.

[63] K. Fu, Y. Wang, C. Yan, Y. Yao, Y. Chen, J. Dai, S. Lacey, Y. Wang, J. Wan, T. Li, et al., Graphene oxide-based electrode inks for 3D-printed lithium-ion batteries, Adv. Mater. 28 (2016) 2587–2594, https://doi.org/10.1002/adma.201505391.

[64] J. Hu, Y. Jiang, S. Cui, Y. Duan, T. Liu, H. Guo, L. Lin, Y. Lin, J. Zheng, K. Amine, et al., 3D-Printed cathodes of LiMn1 xFexPO4Nanocrystals achieve both ultrahigh rate and high capacity for advanced lithium-ion battery, Adv. Energy Mater 6 (2016) 6030–6037, https://doi.org/10.1002/aenm.201600856.

[65] K. Sun, T.S. Wei, B.Y. Ahn, J.Y. Seo, S.J. Dillon, J.A. Lewis, 3D printing of interdigitated Li-ion microbattery architectures, Adv. Mater. 25 (2013) 4539–4543, https://doi.org/10.1002/adma.201301036.

[66] D. Wei, P. Andrew, H. Yang, Y. Jiang, F. Li, C. Shan, W. Ruan, D. Han, L. Niu, C. Bower, T. Ryhänen, M. Rouvala, G.A.J. Amaratunga, A. Ivaska, Flexible solid state lithium batteries based on graphene inks, J. Mater. Chem. 21 (26) (2011) 9762–9767, https://doi.org/10.1039/C1JM10826C.

[67] C. Zhu, T.Y. Liu, F. Qian, T.Y.J. Han, E.B. Duoss, J.D. Kuntz, C.M. Spadaccini, M.A. Worsley, Y. Li, Supercapacitors based on three-dimensional hierarchical graphene aerogels with periodic macropores, Nano Lett. 16 (2016) 3448–3456, https://doi.org/10.1021/acs.nanolett.5b04965.

[68] A. Vyatskikh, S. Delalande, A. Kudo, X. Zhang, C.M. Portela, J.R. Greer, Additive manufacturing of 3D nano-architected metals, Nat. Commun. 9 (2018) 593, https://doi.org/10.1038/s41467-018-03071-9.

[69] S. Ligon, R. Liska, J. Stampfl, M. Gurr, R. Mulhaupt, Polymers for 3D printing and customized additive manufacturing, Chem. Rev. 117 (2017) 10212–10290.

[70] M.C. O'Brien, B. Benjamin, F. Scott, L.G. Zhang, Three-dimensional printing of nanomaterial scaffolds for complex tissue regeneration, Tissue Eng. B Rev. 21 (1) (2015) 103–114, https://doi.org/10.1089/ten.teb.2014.0168.

[71] R.B. Osman, M.V. Swain, A critical review of dental implant materials with an emphasis on titanium versus zirconia, Materials 8 (3) (2015) 932–958, https://doi.org/10.3390/ma8030932.

[72] M. Vaezi, S. Yang, Freeform Fabrication of Nanobiomaterials Using 3D Printing, Book: Rapid Prototyping of Biomaterials, 2014, pp. 16–74, https://doi.org/10.1533/9780857097217.16.

[73] C. Gao, T. Liu, C. Shuai, S. Peng, Enhancement mechanisms of graphene in nano-58S bioactive glass scaffold: mechanical and biological performance, Sci. Rep. 4 (2014) 4712, https://doi.org/10.1038/srep04712.

[74] C. Shuai, P. Li, J. Liu, S. Peng, Optimization of TCP/HAP ratio for better properties of calcium phosphate scaffold via selective laser sintering, Mater. Char. 77 (2013) 23–31, https://doi.org/10.1016/j.matchar.2012.12.009.

[75] J. Deng, P. Li, C. Gao, P. Feng, C. Shuai, S. Peng, Bioactivity improvement of forsterite-based scaffolds with nano-58 S bioactive glass, Mater. Manuf. Process. 29 (2014) 877–884, https://doi.org/10.1080/10426914.2014.921712.

[76] F. Chen, Z. Song, L. Gao, H. Hong, C. Liu, Hierarchically macroporous/mesoporous poc composite scaffolds with ibu-loaded hollow SiO2 microspheres for repairing infected bone defects, J. Mater. Chem. B 4 (2016) 4198–4205, https://doi.org/10.1039/C6TB00435K.

[77] C.-J. Shuai, Z.-Z. Mao, Z.-K. Han, S.-P. Peng, Preparation of complex porous scaffolds via selective laser sintering of poly (vinyl alcohol)/calcium silicate, J. Bioact. Compat. Polym. Biomed. Appl. 29 (2014) 110–120, https://doi.org/10.1177/0883911514522570.

[78] B. Duan, M. Wang, W.Y. Zhou, W.L. Cheung, Z.Y. Li, W.W. Lu, Three-dimensional nanocomposite scaffolds fabricated via selective laser sintering for bone tissue engineering, Acta Biomater. 6 (2010) 4495–4505, https://doi.org/10.1016/j.actbio.2010.06.024.

[79] B. Duan, M. Wang, Encapsulation and release of biomolecules from Ca-P/PHBV nanocomposite microspheres and three-dimensional scaffolds fabricated by selective

laser sintering, Polym. Degrad. Stabil. 95 (2010) 1655–1664, https://doi.org/10.1016/j.polymdegradstab.2010.05.022.

[80] S.J. Heo, S.E. Kim, J. Wei, Y.T. Hyun, H.S. Yun, D.H. Kim, J.W. Shin, Fabrication and characterization of novel nano and micro-HA/PCL composite scaffolds using a modified rapid prototyping process, J. Biomed. Mater. Res. 89A (2009) 108–116, https://doi.org/10.1002/jbm.a.31726.

[81] M. Mattioli-Belmonte, G. Vozzi, Y. Whulanza, M. Seggiani, V. Fantauzzi, G. Orsini, A. Ahluwalia, Tuning polycaprolactone-carbon nanotube composites for bone tissue engineering scaffolds, Mater. Sci. Eng. C-Mater. Biol. Applicat. 32 (2012) 152–159, https://doi.org/10.1016/j.msec.2011.10.010.

[82] E.Y. Teo, S.-Y. Ong, M.S.K. Chong, Z. Zhang, J. Lu, S. Moochhala, B. Ho, S.-H. Teoh, Polycaprolactone-based fused deposition modeled mesh for delivery of antibacterial agents to infected wounds, Biomaterials 32 (2010) 279–287, https://doi.org/10.1016/j.biomaterials.2010.08.089.

[83] S.J. Kalita, S. Bose, H.L. Hosick, A. Bandyopadhyay, Development of controlled porosity polymer-ceramic composite scaffolds via fused deposition modelling, Mater. Sci. Eng. C- Biomimetic Supramol. Syst. 23 (2003) 611–620, https://doi.org/10.1016/S0928-4931(03)00052-3.

[84] H. Liu, T. Webster, From nano to micro: nanostructured titania/PLGA orthopedic tissue engineering scaffolds assembled by three-dimensional printing, in: AIChE Annual Meeting, San Francisco, CA, 2006, 56002/1–8, https://www.ncbi.nlm.nih.gov/pmc/articles/PMC4932007/.

[85] B. Dorj, J.H. Park, H.W. Kim, Robocasting chitosan/nano bioactive glass dual-pore structured scaffolds for bone engineering, Mater. Lett. 73 (2012) 119–122, https://doi.org/10.1016/j.matlet.2011.12.107.

[86] A.I. Teixeira, G.A. Abrams, P.J. Bertics, C.J. Murphy, P.F. Nealey, Epithelial contact guidance on well-defined micro- and nanostructured substrates, J. Cell Sci. 116 (2003) 1881–1892, https://doi.org/10.1242/jcs.00383.

[87] H.J. Zhou, J. Lee, Nanoscale hydroxyapatite particles for bone tissue engineering, Acta Biomater. 7 (7) (2011) 2769–2781, https://doi.org/10.1016/j.actbio.2011.03.019.

[88] G.A. Fielding, A. Bandyopadhyay, S. Bose, Effects of silica and zinc oxide doping on mechanical and biological properties of 3D printed tricalcium phosphate tissue engineering scaffolds, Dent. Mater. 28 (2) (2012) 113–122, https://doi.org/10.1016/j.dental.2011.09.010.

[89] L. Lin, Y. Shen, J. Zhang, M. Fang, Microstructure and mechanical properties analysis of β -tricalcium phosphate/carbon nanotubes scaffold based on rapid prototyping, J. Shanghai University 13 (2009) 349–351, https://doi.org/10.1016/j.apsusc.2011.03.004.

[90] R.H.A. Haq, M.S.B. Wahab, N.I. Jaimi, Fabrication Process of Polymer Nano-Composite Filament for Fused Deposition Modeling, Applied Mechanics and Materials, 2013, pp. 465–466, https://doi.org/10.4028/www.scientific.net/AMM.465-466.8, 8-12.

[91] J. Chlopek, A. Morawska-Chochol, C. Paluszkiewicz, J. Jaworska, J. Kasperczyk, P. Dobrzyński, FTIR and NMR study of poly(lactide-co-glycolide) and hydroxyapatite implant degradation under in vivo conditions, Polym. Degrad. Stabil. 94 (9) (2009) 1479–1485, https://doi.org/10.1016/j.polymdegradstab.2009.05.010.

[92] C. Shuai, B. Yang, S. Peng, Z. Li, Development of composite porous scaffolds based on poly(lactide-co-glycolide)/nano-hydroxyapatite via selective laser sintering, Int. J. Adv. Manuf. Technol. 6 (1–4) (2013) 51–57, https://doi.org/10.1007/s00170-013-5001-2.

[93] E. Sachlos, D. Gotora, J.T. Czernuszka, Composite nano-sized carbonate-substituted hydroxyapatite crystals and shaped by rapid prototyping to contain internal micro-channels, Tissue Eng. 12 (9) (2006) 2479–2487, https://doi.org/10.1089/ten.2006.12.2479.
[94] X. Zhou, N.J. Castro, W. Zhu, H. Cui, M. Aliabouzar, K. Sarkar, L.G. Zhang, Improved human bone marrow mesenchymal stem cell osteogenesis in 3D bioprinted tissue scaffolds with low-intensity pulsed ultrasound stimulation, Sci. Rep. 6 (2016) 32876, https://doi.org/10.1038/srep32876.
[95] J. Leppiniemi, P. Lahtinen, A. Paajanen, R. Mahlberg, S. Metsä-Kortelainen, T. Pinomaa, H. Pajari, I. Vikholm-Lundin, P. Pursula, V.P. Hytönen, 3D-Printable bioactivated nanocellulose–alginate hydrogels, ACS Appl. Mater. Interface. 9 (26) (2017) 21959–21970, https://doi.org/10.1021/acsami.7b02756.
[96] A.S. Gladman, E.A. Matsumoto, R.G. Nuzzo, L. Mahadevan, J.A. Lewis, Biomimetic 4D printing, Nat. Mater. 15 (2016) 413–418, https://doi.org/10.1038/nmat4544.
[97] M.S. Mannoor, Z. Jiang, T. James, Y.L. Kong, K.A. Malatesta, W.O. Soboyejo, N. Verma, D.H. Gracias, M.C. McAlpine, 3D printed bionic ears, Nano Lett. 13 (6) (2013) 2634e9, https://doi.org/10.1021/nl4007744.
[98] A.E. Jakus, E.B. Secor, A.L. Rutz, S.W. Jordan, M.C. Hersam, R.N. Shah, Three-dimensional printing of high-content graphene scaffolds for electronic and biomedical applications, ACS Nano 9 (4) (2015) 4636–4648, https://doi.org/10.1021/acsnano.5b01179.
[99] J. Zhang, S. Zhao, M. Zhu, Y. Zhu, Y. Zhang, Z. Liu, C. Zhang, 3D-printed magnetic $Fe_3O_4$/MBG/PCL composite scaffolds with multifunctionality of bone regeneration, local anticancer drug delivery and hyperthermia, J. Mater. Chem. B 2 (43) (2014) 7583–7595, https://doi.org/10.1039/C4TB01063A.
[100] R. De Santis, U. D'Amora, T. Russo, A. Ronca, A. Gloria, L. Ambrosio, 3D fibre deposition and stereolithography techniques for the design of multifunctional nanocomposite magnetic scaffolds, J. Mater. Sci. Mater. Med. 26 (10) (2015) 250, https://doi.org/10.1007/s10856-015-5582-4.
[101] D. Kokkinis, M. Schaffner, A.R. Studart, Multimaterial magnetically assisted 3D printing of composite materials, Nat. Commun. 6 (2015) 8643, https://doi.org/10.1038/ncomms9643.
[102] S. Hong, Selective Laser Sintering of Nanoparticles, Sintering of Functional Materials, Igor Shishkovsky, IntechOpen, 2017, https://doi.org/10.5772/intechopen.68872.
[103] J. Yeo, S. Hong, D. Lee, N. Hotz, M.-T. Lee, C.P. Grigoropoulos, S.H. Ko, Next generation non-vacuum, maskless, low-temperature nanoparticle ink laser digital direct metal patterning for a large area flexible electronics, PloS One 7 (8) (2012) e42315, https://doi.org/10.1371/journal.pone.0042315.
[104] J. Chung, N.R. Bieri, S. Ko, C.P. Grigoropoulos, D. Poulikakos, In-tandem deposition and sintering of printed gold nanoparticle inks induced by continuous Gaussian laser irradiation, Appl. Phys. A 79 (4) (2004) 1259–1261, https://doi.org/10.1007/s00339-004-2731-x.
[105] R. Xing, R. Huang, W. Qi, R. Su, Z. He, Three-dimensionally printed bioinspired superhydrophobic PLA membrane for oil water separation, AIChE J. 64 (10) (2018) 3700–3708, https://doi.org/10.1002/aic.16347.
[106] K.-M. Lee, H. Park, J. Kim, D.-M. Chun, Fabrication of a superhydrophobic surface using a fused deposition modelling (FDM) 3D printer with polylactic acid (PLA) filament and dip coating with silica nanoparticles, Appl. Surf. Sci. 467–468 (15) (2019) 979–991, https://doi.org/10.1016/j.apsusc.2018.10.205.
[107] X. Fang, J. Du, Z. Wei, X. Wang, P. He, H. Bai, B. Wang, J. Chen, R. Geng, B. Lu, Study on metal deposit in the fused-coating based additive manufacturing, Proc. CIRP 55 (2016) 115–121, https://doi.org/10.1016/j.procir.2016.08.034.

[108] J. Zhu, Water-based antimicrobial coating for 3D printed medical devices, published on Stratasys website at http://www.stratasys.com/resources/white-papers/antimicrobial-coating.
[109] U. Gbureck, E. Vorndran, F.A. Müller, J.E. Barralet, Low-temperature direct 3D printed bioceramics and biocomposites as drug release matrices, J. Contr. Release 122 (2) (2007) 173–180, https://doi.org/10.1016/j.jconrel.2007.06.022.
[110] M.K. Gupta, F. Meng, B.N. Johnson, Y.L. Kong, L. Tian, Y.-W. Yeh, N. Masters, S. Singamaneni, M.C. McAlpine, 3D printed programmable release capsules, Nano Lett. 15 (2015) 5321–5329, https://doi.org/10.1021/acs.nanolett.5b01688.
[111] R. Ma, T. Tang, Current strategies to improve the bioactivity of PEEK, Int. J. Mol. Sci. 15 (4) (2014) 5426–5445, https://doi.org/10.3390/ijms15045426.
[112] M.T. Arafat, C.X.F. Lam, A.K. Ekaputra, S.Y. Wong, X. Li, I. Gibson, Biomimetic composite coating on rapid prototyped scaffolds for bone tissue engineering, Acta Biomater. 7 (2) (2011) 809–820, https://doi.org/10.1016/j.actbio.2010.09.010.
[113] S. Van Bael, T. Desmet, Y.C. Chai, G. Pyka, P. Dubruel, J.-P. Kruth, J. Schrooten, In vitro cell-biological performance and structural characterization of selective laser sintered and plasma surface functionalized polycaprolactone scaffolds for bone regeneration, Mater. Sci. Eng. C 33 (6) (2013) 3404–3412, https://doi.org/10.1016/j.msec.2013.04.024.
[114] C.-T. Kao, C.-C. Lin, Y.-W. Chen, C.-H. Yeh, H.-Y. Fang, M.-Y. Shie, Poly(-dopamine) coating of 3D printed poly(lactic acid) scaffolds for bone tissue engineering, Mater. Sci. Eng. C 56 (2015) 165–173, https://doi.org/10.1016/j.msec.2015.06.028.
[115] F. Pati, T.-H. Song, G. Rijal, J. Jang, S.W. Kim, D.-W. Cho, Ornamenting 3D printed scaffolds with a cell-laid extracellular matrix for bone tissue regeneration, Biomaterials 37 (2015) 230–241, https://doi.org/10.1016/j.biomaterials.2014.10.012.
[116] K.Y. Lee, A. Ambrosi, M. Pumera, 3D-printed metal electrodes for heavy metals detection by anodic stripping voltammetry, Electroanalysis 29 (11) (2017), https://doi.org/10.1002/elan.201700388.
[117] C. Tan, M.Z.M. Nasir, A. Ambrosi, M. Pumera, 3D printed electrodes for detection of nitroaromatic explosives and nerve agents, Anal. Chem. 89 (17) (2017) 8995–9001, https://doi.org/10.1021/acs.analchem.7b01614.
[118] T.S. Cheng, M.Z.M. Nasir, A. Ambrosi, M. Pumera, 3D-printed metal electrodes for electrochemical detection of phenols, Appl. Mater. Today 9 (2017) 212–219, https://doi.org/10.1016/j.apmt.2017.07.005.
[119] A.H. Loo, C.K. Chua, M. Pumera, DNA biosensing with 3D printing technology, Analyst 142 (2017) 279–283, https://doi.org/10.1039/C6AN02038K.

# CHAPTER 16

# Additive manufacturing for the automotive industry

Joel C. Vasco[1,2]
[1]School of Technology and Management, Polytechnic of Leiria, Leiria, Portugal; [2]Institute for Polymers and Composites, University of Minho, Guimarães, Portugal

## 16.1 Introduction

The automotive industry is one of the most competitive industries worldwide. New market and design trends emerge continuously, requiring new manufacturing approaches to comply with the automotive market demands [5,11]. Additive manufacturing (AM) provides an important competitive edge to this industry, acting as a disruptive approach by shortening product's design and development, delivering flexibility on production, and producing optimized automotive components and customized vehicle products upon request [39,56,85].

The use of AM on the automotive production started with the use of soft assembly tools produced by fused deposition modeling (FDM), selective laser sintering (SLS), or 3D printing. These tools can be produced on site, according to vehicle's assembly requirements (tolerances, positioning, or spacing between body panels) and operator/assembly process needs (ergonomics and assembly time reduction) [3,27].

AM processes also appeared on the automotive supply chain, providing significant benefits on the production of specialized tools to produce vehicle components. Examples of optimized production tools for plastic injection were presented in previous studies. In these cases, the use of selective laser melting (SLM) for molding tools provided significant improvements on part quality and production costs [41,48,55]. Another competitive edge of AM use can be found on conventional body panels produced by stamping process where time-to-tooling can play a significant role on production costs by reducing the supply chain [50].

The freeform capacity offered by AM enables the design and direct production of optimized automotive components, focused on vehicle's performance [69]. Automotive competitions, such as Formula 1, WTC, or NASCAR started as its main field of application due to higher-level

*Additive Manufacturing*
ISBN 978-0-12-818411-0
https://doi.org/10.1016/B978-0-12-818411-0.00010-0

demands [32]. Nowadays, the current costs on the use of AM technologies still may be an obstacle for wide application, making it feasible only on small production lots, such as high-end models. However, market demands are pushing hard toward customized parts, and therefore, AM feasibility increases [66,87].

Another related AM technical advantage is the possibility of producing lightweight components, supported by generative design algorithms. Lattice structures were already used to achieve lightweight components, replacing solid sections where tensions are less demanding [51,88]. Generative design provides topological optimized geometries, enabling material saving and production costs reduction [89,90].

The time-to-market on AM parts decreased significantly, enabling the concept of mass customization. This trend of mass customization is pushing AM equipment manufacturers to make their equipment competitive enough to dispute a market share with conventional manufacturing processes [5,15,31].

The strong trend on energy consumption decrease poses new challenges regarding vehicle's design, performance, and regulations compliance. AM is one of the key-enabling technologies for lightweight components for new vehicles considering the actual paradigm shift from the common combustion engine to alternative motion systems [30,87].

## 16.2 Complementary methods and techniques

The increasing application of AM processes throughout several industrial domains could only be achieved with complementary methods and techniques to support this manufacturing approach. The automotive industry is no exception to this, and it benefits from several processes and products optimizations, concerning focused design approaches, building process control, and related postprocessing operations.

### 16.2.1 Design for additive manufacturing

The design for manufacturing method gathers the best practices for general-purpose manufacturing processes. However, the paradigm shift operated on manufacturing provided by AM requires additional knowledge despite the freeform building capacity [68,80].

The best practices for design for additive manufacturing (DfAM) are focused on resource reduction such as raw material and energy as well as

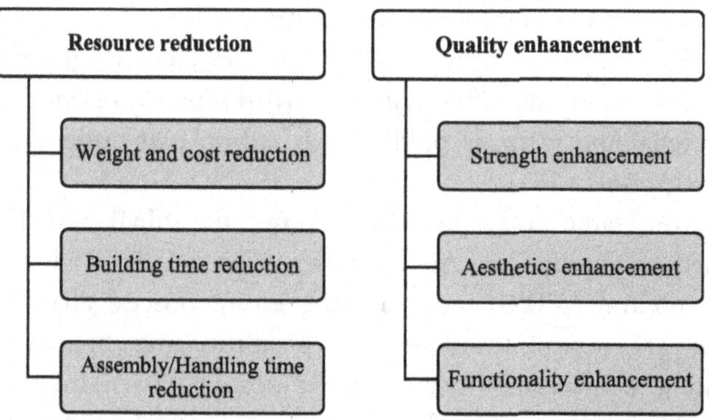


**Figure 16.1** Focus of best practices for DfAM. *DfAM*, design for additive manufacturing.

quality enhancement of AM components supported by preliminary numerical studies and process monitoring [80]. The main focus of DfAM principles is summarized and depicted in Fig. 16.1.

Application examples of DfAM were extensively used on extreme situations where mechanical performance competes with lightweight design and harsh operation conditions. The Bloodhound SSC land vehicle that was developed to beat the world speed record presents several of these challenges. The use of computational fluid dynamics (CFD) enabled the optimization of the vehicle's body, by achieving an engineering solution to the problem of excessive supersonic lift at the rear of the vehicle [59]. Among many other aspects, the air brake hinges of the vehicle as well as the driver's steering wheel were optimized considering the constraints imposed [35,72].

The continuous evolution on AM materials as well as the consequent development of newer and optimized AM processes will surely lead to the ongoing improvement of the DfAM principles, to meet more and more demanding market requirements.

#### *16.2.1.1 Simulation of additive manufacturing building processes*

The simulation of the building process enables to anticipate problems on specific geometries that cannot be easily anticipated. The numerical simulation can be applied to AM building processes since it is a layer-based process and material interaction with energy sources and/or chemical agents can be mathematically modeled [23,43,53]. Some of the currently relevant market solutions are detailed briefly:

- The DIGIMAT-Additive Manufacturing software is a product from MSC Software, recently acquired by the Hexagon AB group (Sweden). This software bundle offers enhanced workflow for FDM and SLS processes, enabling warpage prediction based on part's orientation on the building process [82].
- From the same software house comes the SIMUFACT Additive, focused on powder-bed AM processes, such as SLS, SLM, and electron beam melting (EBM), covering the entire process chain, from part design and optimization to the postprocessing operations usually required. Concerning metal powder-bed processes, the major issues on building parts may include distortion, residual stresses, and overall part quality [36].
- The software package Magics 3D Print Suite from Materialise NV (The Netherlands) offers a complete set of software tools, starting on part design and build preparation. The Magics e-stage pack enables an important optimization procedure for metal AM processes which is support's geometry optimization, by reducing preparation time. Concerning the building and postprocessing operations, this suite contains an entire set of platforms to control the process workflow [54].
- The 3DXPERT is also a software bundle for metal AM processes, from 3D Systems (USA). This set of solutions is focused on part design optimization and lead time reduction, enabling successful builds for metal AM parts and overall costs reduction [2].

The prior mentioned solutions for AM building simulation deliver a streamlined AM process, which is a critical factor to fully include AM on the production chain, ensuring product's quality and reliability.

#### *16.2.1.2 Generative design*

The need for lightweight components led to the development of methodologies that could take advantage of the freeform building capacity of the AM processes [24,44].

The weight reduction achieved through generative design becomes advantageous from the point of view of both the manufacturing process and the application itself. Therefore, targets are defined to meet specific application requirements such as volume reduction, allowable displacement, deformation energy, or other. Furthermore, constraints such as fitting, bearing, and load regions should be preserved to ensure final results feasibility [44,45].

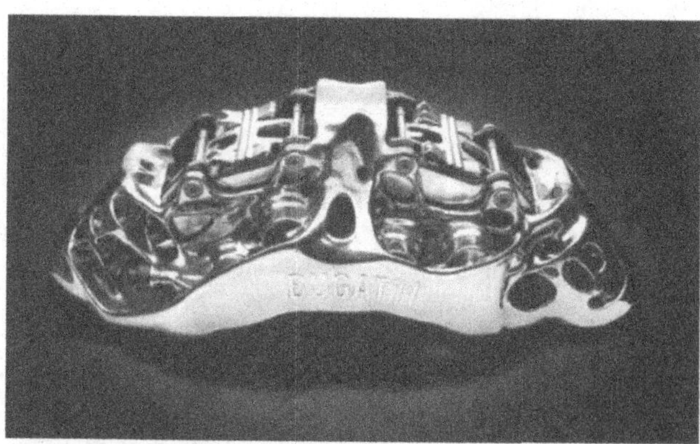

**Figure 16.2** AM Brake caliper built on TiAl6V4 for the Bugatti Chiron. *(Reproduced with kind permission of BUGATTI AUTOMOBILES S.A.S.)*

Besides automotive competition, another major field for lightweight design is high-end and customized vehicles or supercars that can also benefit from this design approach to comply with high performance levels and/or esthetic purposes [42]. One of these examples is the air intake manifold developed by the high-performance division of Ford Performance. The air intake manifold is built on aluminum alloy powder, and AM provides the possibility for lattice structures and other geometric features that enhance air homogeneous distribution within the manifold [1].

Another example is the brake caliper manufactured in TiAl6V4 for the Bugatti Chiron shown in Fig. 16.2, due to its improved strength to weight ratio. The geometry of this brake caliper was obtained by generative design, requiring that an AM process would be mandatory [86].

### 16.2.2 Reverse engineering

The need for a 3D model that an AM process can materialize makes reverse engineering as an excellent complementary technique for AM [4,62]. Whether it is used on the scanning of a physical model for reengineering or as a noncontact metrology method of an AM product, reverse engineering capabilities have evolved in the past decade to overcome dimensional and geometrical limitations [92]. While AM may become limited due to material features (grain size or melting temperature) or process capacities (minimum feature size or processing temperature), reverse engineering is only limited by optical and computation capacities [4].

In the automotive industry, reverse engineering plays an important role on continuous improvement on production lines or on product's design,

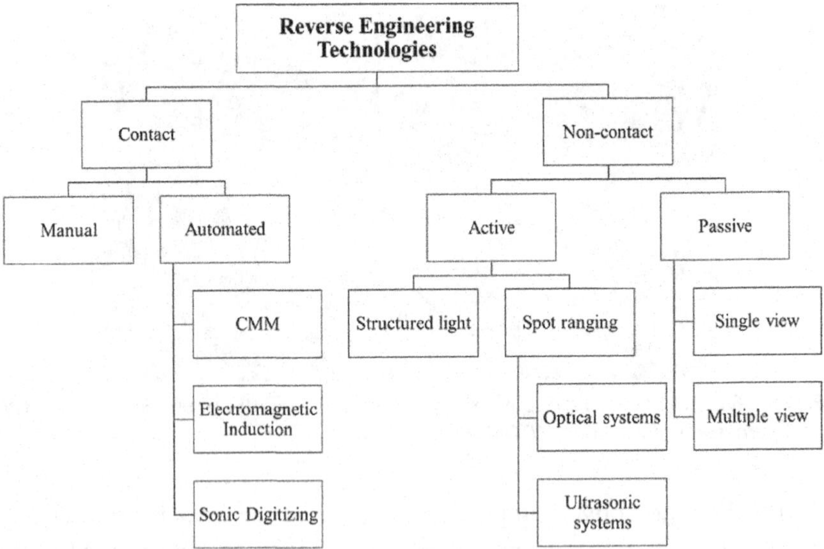


**Figure 16.3** Classification of reverse engineering technologies. *(Reproduced with kind permission from B. Bidanda, Z. Geng, Emerging trends in reverse engineering. Proc. 2nd Int. Conf. Prog. Addit. Manuf. (Pro-AM 2016), 2016, pp. 2–7.)*

providing a better edge on a very competitive market. Contact methods through the use of a probe have been used throughout the past decade for its high accuracy and analysis quality. Noncontact methods use light or laser beams to scan the object's surface and X-ray radiation to provide a more comprehensive and detailed tomographical analysis without the need of destroying the component. The classification of reverse engineering technologies is depicted in Fig. 16.3, providing an insight on contact and noncontact technologies [7].

## 16.2.3 Inline building process control for metal additive manufacturing systems

The building process control provides a higher added value on metal AM processes, due to material costs and processing costs involved [14]. The powder-based metal AM processes, such as SLM, EBM, or Laser Engineered Net Shaping (LENS), are prone to distortions caused by thermal asymmetries on the part that, ultimately, may cause the loss of an entire build [26,40,46]. A proper design approach for an AM part may avoid building problems, although the building process itself requires inline monitoring to prevent defects. Power fluctuations or optical distortions on the energy beam or abnormal layer roughness caused by a poor recoating procedure can be monitored to assess if the build can be corrected by a quality closed-loop control or, ultimately, if the build should be interrupted to reduce material and energy waste [26,37].

Current state-of-the-art monitoring equipment for inline melt pool measurement and control include high-speed complementary metal oxide semiconductor (CMOS) cameras, high-speed charge-coupled device (CCD) cameras, pyrometers, and coherent imaging are used to provide feedback the energy source, adjusting laser power according to the building requirements [26].

### 16.2.4 Postprocessing for additive manufacturing components

The final properties of an AM component can be significantly enhanced by postprocessing procedures. Although these may be time-consuming or even costly, it can be easily integrated into the supply chain since most procedures are also applicable to conventional parts [49]. Considering the AM processes classification proposed by Chua et al. (2003) in terms of raw material supply form [13], the postprocessing procedures are summarized in Table 16.1.

## 16.3 Overview of additive manufacturing applications on automotive production tools

The shortening of product's design and development as well as the increase on production flexibility requires that manufacturing units are flexible and agile to enable production equipment and/or layout reconfigurations. AM provides agility to the shop floor, enabling the development of soft tools to support and speed up the production process as well as the improvement of hard tools to enhance mass-replication processes.

### 16.3.1 Soft tools for assembly lines

The definition of soft tools refers to customized tools developed and built by AM processes to provide tooling solutions for assembly line operators, bringing flexibility to the shop floor. Aluminum jigs and fixtures have been replaced by AM tools, reducing handling weight and tool customization [58,84]. The production costs and lead time were seriously reduced by the adoption of entry-level AM equipment to overcome the dependence from third-party companies that used to be included in this supply chain, affecting the assembly line's workflow.

Table 16.1 Postprocessing procedures.

| | Raw material supply form | Process interaction with raw material | Postprocessing procedures applicable |
|---|---|---|---|
| Liquid-based | Photosensitive resin, wax | Drop-on-demand, selective photocure or entire layer photocure | UV postcure or thermal postcure for nano-charged resins, support removal, polishing |
| Solid-based | Polymer filament, metal or ceramic feedstock filament, metal sheets | Fused filament deposition or material sheet binding | Support removal, debinding for feedstock materials, polishing |
| Powder-based | Polymer, ceramic, or metal powder | Binding agent printed over polymer, ceramic or metal powder, selective sintering of polymer or ceramic powder, selective fusion of metal powder, directed energy deposition for metal fusion | Debinding, CIP/HIP for ceramic and metal materials, support removal for metal parts, shot peening for surface enhancement, heat treatment, and CNC machining for physical properties enhancement |

The shift to AM tools became possible due to the increased mechanical performance and geometrical stability of AM materials, supported by the precision building capacity of the AM processes used. Thus, automotive manufacturers can take advantage of AM to increase component's functionality, reducing weight and increasing the product's potential across the supply chain [73]. Markforged (United States) was one of the early adopters of the introduction of fiber within the filament for the FDM process, enabling in-house production of fixtures for coordinate-measuring machine (CMM) inspection, replacing the machined aluminum parts provided by external suppliers (Fig. 16.4).

## 16.4 Hard tools for mass-scale replication of automotive components

The mass-scale production capacity required for the automotive industry is typically provided by conventional processes such as hot stamping, die casting, and injection molding. Despite the customization trend, the mass-scale production still provides a systematic approach for process optimization, which can be added by AM due to lead time reduction.

**Figure 16.4** Fixture for CMM inspection produced by FDM incorporating fiber reinforcement produced on site. *CMM*, coordinate-measuring machine; *FDM*, fused deposition modeling. *(Reproduced with kind permission of Markforged Inc.)*

The optimization of the production process benefits from the shortening or even total replacement of the current supply chain provided by AM technologies. Thus, that means that automotive manufacturers using AM would have to control the entire value chain to optimize their workflow. However, the shortening of the value chain requires in-house expertise in the relevant skills-related design such as material selection, monitoring of the built part, and postprocessing [81].

Concerning metal parts, in particular, body panels, studies have been conducted to evaluate the impact of AM on production. It was noticed that AM could significantly change the supply chain of hot stamping inserts, providing additional time for reengineering iterations, thus minimizing or even eliminating rework operations [50].

For automotive plastic parts, the mass-scale production is supported by injection molding which is widely known for its precision and repeatability [61]. The injection molding process can benefit from several optimizations provided by AM, in terms of cycle time reduction, lead time to tooling, and increase of the overall quality of the injected parts [12,38,83].

## 16.5 Emerging additive manufacturing applications on automotive components

The application of AM processes to produce automotive components was limited by material properties, such as mechanical, thermal, and chemical behaviors under operation or surface finishing on esthetic components [71].

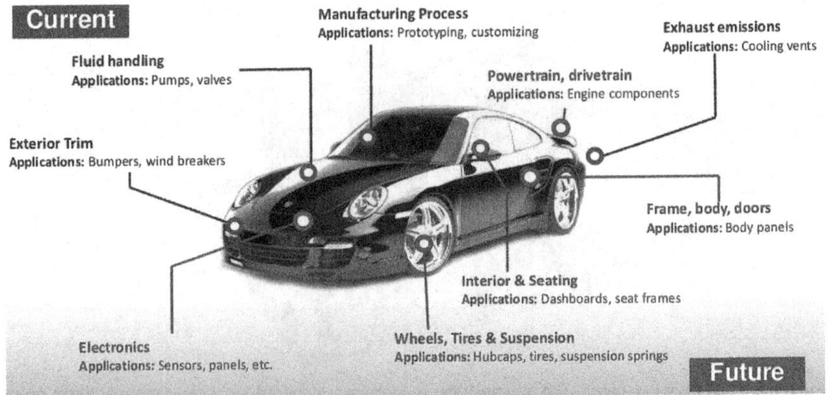


**Figure 16.5** Current and future applications of AM on automotive. *AM*, additive manufacturing. *(Reproduced with kind permission from S. Radisavljevic, Additive manufacturing the automotive industry, Slideshare (2014). https://www.slideshare.net/StefanRadisavljevic/additive-manufacturing-43047855 (Accessed December 28, 2019).)*

Fiber reinforcements have been added to materials to improve their mechanical properties [93]. Current developments on filament materials reinforced with carbon fiber brought a new competitive edge to the FDM process. Different approaches have been made by AM equipment manufacturers concerning the length of the carbon fiber deposited within the filament [8]. Fig. 16.5 depicts some of the current and prospective applications of AM on automotive components [70].

## 16.5.1 Engine compartment, powertrain, and exhaust components

The engine compartment presents several challenges to material selection due to the harsh environment in which the components must operate. Service temperatures above 90°C for long periods of time and lubricants or fuel that may provide chemical agents to react the component's material are prone to reduce the component's lifetime if an unsuitable material is selected [52,71].

Stratasys recently released a high-performance thermoplastic material, a PEKK-based material for FDM. This range of available materials provides chemical resistance and ultralow outgassing, enabling fuel and lubricant contact. Furthermore, these materials offer high-performance levels concerning service temperature and wear [75].

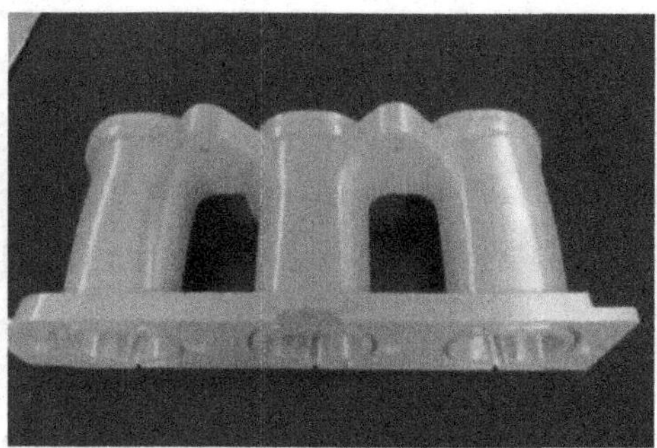

**Figure 16.6** Air duct produced in ULTEM 9085. *(Reproduced with kind permission of A. Pearson, Lamborghini Accelerates and Perfects Automotive Engineering with Stratasys 3D Printed Prototypes and Track-Ready Parts, Strat Blog, 2020. https://www.stratasys.com/explore/blog/2015/lamborghini-3d-printing (Accessed April 29, 2019).)*

Another successful material used on the FDM technology is the ULTEM material. Developed originally by GE, the material is now provided by SABIC, and it is in use by Stratasys since 2008. Currently, ULTEM 9085 possesses a certificate for aircraft cabin components, which offers an insight on the demands that this material can sustain [77]. Lamborghini uses the FDM technology for several applications such as high-performance esthetic parts, profiles, and air ducts, taking advantage of the freeform capacity provided by AM. Fig. 16.6 shows an air duct produced in ULTEM 9085 [63].

Concerning powertrain components, these are often combined with fiber reinforced composites to enhance their mechanical properties. An excellent application example was performed at the Oak Ridge National Laboratories (United States) with the Big Area Additive Manufacturing (BAAM) process. Based on the FDM operation principle, the filament used to build the Shelby Cobra chassis/powertrain was ABS incorporating chopped carbon fiber, processed on a high build-rate single-screw extruder. By adding carbon fiber, warpage of the products is seriously reduced, and mechanical properties are enhanced. Although carbon fiber-reinforced ABS plastic cannot yet be considered for direct use for the commercial market, since it does not comply with the engineering requirements of a vehicle frame, there is a significant interest on new materials for BAAM to speed up design to integration for new vehicles [16]. The final frame for the Shelby Cobra is shown in Fig. 16.7.

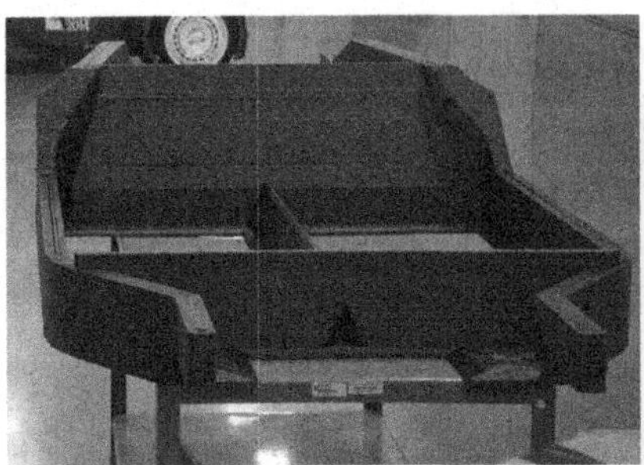

**Figure 16.7** Finished frame after the building process (left image) and void filling with an epoxy-based resin (right image). *(Reproduced with kind permission of Scott Curran, Oak Ridge National Laboratory.)*

A different and somehow disruptive approach to automotive powertrain design was presented at the 2019 edition of Formnext by Divergent 3D (United States) at the SLM Solutions Group AG booth (Germany). The front-quarter section of the Blade chassis, the supercar jointly developed by Divergent 3D and SLM Solutions, was depicted, showing the several AM components included [25].

Divergent 3D developed and patented the Divergent Manufacturing Platform, focused on the transformation of economics and adverse environmental impact of designing and manufacturing complex structures such as cars. SLM Solutions is involved on the development of this platform due to its experience on the production/distribution of SLM systems and expertise on the delivery of strong and lightweight AM parts. The two companies are working together to further develop this platform and to accelerate its scaling-up for cost-effective and high-volume production of vehicles, enabling to reduce manufacturing costs, time-to-market, and carbon emissions [22]. Divergent 3D and SLM Solutions are extending their partnership, by establishing the Divergent's intent of acquiring five next-generation SLM Solutions machines to meet the production demand expected from other automotive manufacturers [17, 18].

Exhaust components are a very demanding application due to the high temperatures developed on the combustion process. Although it may not be fully feasible for mass production, it is certainly a solution for high-performance levels or even customization. The Bugatti's Chiron provides the field for another AM application example. The exhaust ends were built in a titanium alloy given its excellent thermomechanical behavior combined with its low specific weight (Fig. 16.8). The freeform capacity provided by AM enabled optimized exhaust aerodynamics and increased down force, improving vehicle's handling at high speed [57].

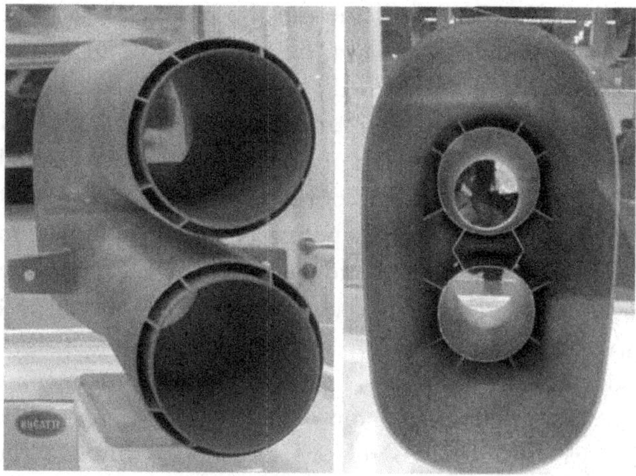

**Figure 16.8** AM titanium exhaust for the Bugatti's Chiron. *AM*, additive manufacturing. *(Reproduced with kind permission of BUGATTI AUTOMOBILES S.A.S.)*

## 16.5.2 Exterior body panels and lighting

The production of automotive body panels is also a reality. Front grill and bumpers can be produced by large volume AM systems, such as stereolithography (Fig. 16.9). Although this is a resin-based process, the surface finish and a high level of detail provide a smooth surface, suitable for vinyl coating or painting, enabling all possible colors and decoration for product-on-demand [29].

**Figure 16.9** Front bumper of an automobile at [29]. *(Reproduced with kind permission of Mesago Formnext, Experience the Magic of Additive manufacturing at the Formnext 2018, Twitter, 2018. https://twitter.com/formnext_expo/status/1062309304454262785 (Accessed December 22, 2019).)*

The SLS and the FDM processes can also produce body panels despite its lower surface finishing. However, the products from these technologies are fully able to be postprocessed, and they can provide an excellent alternative for this application domain. The SLS process, despite its process accuracy and support-free building capacity, is somewhat limited in terms of direct product dimensions and porosity on the current state-of-the-art equipment. However, it enables postprocessing to enhance mechanical properties, increasing the application level [78,91]. Product bonding to obtain bigger components such as body panels is also possible [28].

The BAAM process mentioned on the previous topic is also applicable on the production of body panels. Process inaccuracies can be minimized or even solved with machining, sanding, filling, and polishing procedures. In this case, it was necessary to remove loose fibers through the sanding process, followed by preliminary surface preparation and coating procedures to fill the ridges created by the filament deposition process. Fig. 16.10 shows the components of the Shelby Cobra built at the Oak Ridge National Laboratories, after finishing and just before painting.

Concerning functional components, the BMW Group has been using AM technologies at its Additive Manufacturing Center at Munich. Two AM components are used on the i8 Roadster model, the first is a fixture for the soft-top attachment, much lighter than the replaced original aluminum fixture, with a similar level of performance. The second component is a window guide rail that allows the window to operate smoothly, and it took just 5 days to be developed and was integrated into series production shortly after using the HP's Multijet Fusion process [9]).

**Figure 16.10** Body panels for the Shelby Cobra manufactured by BAAM. *(Reproduced with kind permission of Scott Curran, Oak Ridge National Laboratory.)*

**Figure 16.11** Automotive taillight produced by the PolyJet technology. *(Reproduced with kind permission of Stratays.)*

Audi has been testing some AM processes on its Pre-Series Center at Ingolstadt. The main use of AM technologies has its focus on lighting due to the complexity of these components as well as the required tooling cost to produce them. Currently, Audi is using these advantages to make the design verification process faster and more accurate, enabling a significant reduction on lead time. The PolyJet technology is being extensively used by Audi for automotive components such as lighting housings, multicolored taillights, and other components that need to be assembled [21]. These components can now be produced on a single component, with multicolored and multimaterial capabilities like the taillight shown in Fig. 16.11, produced by Stratasys J-series equipment.

The BMW Group stepped up the customization level on its Mini model, by proposing the MINI Yours Customised program, enabling customers to select, design, and order the parts available in the product range of the customization program at the online shop. Customizable parts such as side scuttle depicted on Fig. 16.12 are combined with direction indicators and supplied in pairs so that the customer can order it and assemble it with ease. The AM equipment used in this process has been aimed by the BMW Group through strategic partnerships with AM players such as Hewlett–Packard Inc., Carbon Inc., and EOS GmbH [10].

### 16.5.3 Cockpit applications

Esthetic concerns are particularly important for cockpit components, since this is the environment in which the driver and his passengers travel. Therefore, surface finish of AM parts within the car's cockpit should be properly enhanced before assembly to comply with esthetical requirements. Considering build rate and size limitations, the FDM process is one of the least affected AM processes among the nonmetal processing techniques.

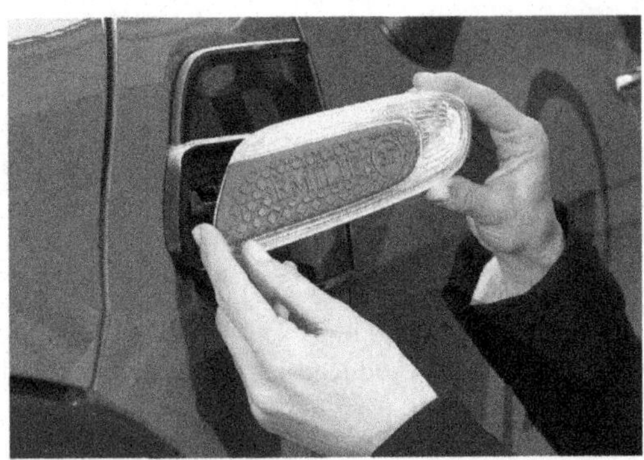

**Figure 16.12** Side scuttle available on the MINI Yours Customised program. *(Reproduced with kind permission of BMW Group, MINI Yours Customised: From the Original to the Personalised Unique Special, Press Kit, 2017. https://www.press.bmwgroup.com/global/article/detail/T0276990EN/mini-yours-customised:-from-the-original-to-the-personalised-unique-special (Accessed January 2, 2020).)*

However, the fused filament deposition process is typically prone to a poor surface finishing if some preventive actions are not implemented. Suitable part orientation, self-supporting features, image texturing, and abrasion/surface coating are among the best design practices and techniques required to help minimize the amount of time required for post-processing [74].

On the other hand, Stratasys offers the PolyJet technology, fully capable of reproducing textures and colors with high surface finishing, which can be used for both design validation or customized interiors. The cockpit interior possesses a possible wide range of materials, colors, and textures which creates additional complexity that can be overcome with the PolyJet technology. A traditional approach to mock up the central console, similar to the one depicted on Fig. 16.13, would require each part to be produced separately. The combination of wood surfaces and leather or fabric requires complicated and time-consuming procedures, or, in other words, highly specialized expertise to achieve the final form [76].

The interior trim of the Mini model is also one of the customizable parts available on MINI Yours Customised program. The trim is integrated on the passenger side dashboard where it replaces the original trim, which was factory-mounted (Fig. 16.14). This trim is designed to be exchanged at any time, if desired, with any other customized version [10].

Lamborghini S.p.A. (Italy) and Carbon 3D (United States) have been collaborating on the development of AM parts, exploring the properties

**Figure 16.13** Automotive console produced by a Stratasys J-series printer. *(Reproduced with kind permission of Stratays.)*

**Figure 16.14** Interior trim available on the MINI Yours Customised program. *(Reproduced with kind permission of BMW Group, MINI Yours Customised: From the Original to the Personalised Unique Special, Press Kit, 2017. https://www.press.bmwgroup.com/global/article/detail/T0276990EN/mini-yours-customised:-from-the-original-to-the-personalised-unique-special (Accessed January 2, 2020).)*

of the EPX82 resin, processed by the Digital Light Synthesis (DLS) technology. This epoxy resin enables the production of more durable parts that can withstand high pressure and high temperature requirements, and it has already been used on the new textured fuel cover cap and on a clip component for an air duct for its Urus model [19]. Carbon's printing technology is also being used by Lamborghini on interior parts, in particular, to produce the central and lateral dashboard air vents of its Sián FKP 37 hybrid sports car [20].

## 16.6 Economic impact of additive manufacturing applications on the automotive industry

The initial impact of the AM introduction onto the automotive industry has been noticed since the early days of rapid prototyping on product development. AM has been in use since the late 1980s. At that time, AM application was limited to the production of prototypes, and the goal was to materialize parts on an affordable and fast way so that it could provide feedback for the product development process. AM provides the means to speed up product development and to enhance the quality of future products, reducing tooling costs already on the product design phase [33,67].

The AM introduction of AM onto the automotive manufacturing process was noticed directly on tooling due to the significant changes that AM provided to the tooling supply chain. Whether it is soft tooling to support the assembly lines or hard tooling to improve the manufacturing process capacities, AM proves to be an inescapable process, delivering flexibility on the shop floor and producing optimized automotive components [6].

Concerning future business models for the automotive industry, mass customization plays the lead role. The possibility of purchasing a customized vehicle upon request puts the costumer in close collaboration with the manufacturer, thus shortening or simplifying the supply chain. AM enables customers and manufacturers to codesign products to fit their mutual requirements. Therefore, the range of products can grow almost infinitely; however, the manufacturing cost is not significantly increased, unless the customer is able to pay an additional price for a product with additional features. AM is a powerful tool to achieve a high level of customization for an almost endless market potential [85].

The ownership model for vehicles may be decaying, depending on the country and its national regulations concerning mobility limitations within cities. The use of a private vehicle is, therefore, seriously restrained to avoimissions and traffic jamming in cities with low urbanization density development, which might make private vehicle ownership an obsolete model [65]. Thus, customization could be led in other directions and integrate new business models for medium- or long-term lease. The customer may desire to customize the vehicle at his taste as added value despite the fact that it is a leased vehicle [60]. Options like the one provided by the MINI Yours Customised program enable the customer to order specific parts to be produced by AM technologies with all the inherent design freedom that can be modular enough to be easily assembled or disassembled.

On the automotive manufacturer's point of view, AM may provide a different approach to stocks management and to the supply chain. Nowadays, automotive manufacturers are obliged to keep a replacement parts stock for each vehicle model during a legal period of 10 years. For that purpose, they need to rely on third-party companies to produce those replacement parts and support a reliable spare parts supply chain at a high service level. The assurance of the supply chain helps to minimize downtime, but on the industrial focus, a supply service failure may impact significantly on production. The capacity of reproducing a part without specific and expensive tools is provided by AM, enabling that the stock of physical parts is converted to a digital warehouse, where all the 3D model files and manufacturing methods and strategies are stored to be used when required [47,67,85].

The feasibility of AM on industrial use will be increased with the entry of third-party raw material suppliers. This allows AM users to become independent of proprietary materials and forces AM equipment manufacturers to make their equipment compatible with third-party materials. Therefore, the cost of raw materials for parts production is potentially reduced, fostering competition on the market [47].

Another AM economic impact related to manufacturers, and also with a significant environmental impact, is decentralized manufacturing. Logistics can be simplified through digitalization, enabling location-independent manufacturing. This way, the physical flow of material and products can be reduced significantly and leads to a substantial reduction of emissions [33,79,85].

The benefits provided by AM concerning, in particular, to the capacity of materializing a digital product is, on the other hand, one of most critical issues regarding the respect for property rights of product design, and it may become one of the most severe economic consequences of AM [67]. In the digital age, 3D models of products are transmitted for industrial or end-user purposes, by the contractual confidence established between supplier and customer. The need to protect these files relates also with the fact that it may contain not only the product's geometry but also material information, building parameters, and strategies, among other possible content, required to produce a high-quality product [64]. A proposition to promote protection of manufacturing files for AM is to include features that can be used for security and identification of genuine products [34].

## 16.7 Conclusions

The automotive industry has been one of the main drivers for AM processes development, both in terms of recent technological developments or related to new materials with enhanced properties for more demanding applications. Automotive competition has been the test field for most of these new developments due its wide range of demands and applications.

AM is a key-enabling technology for the digital age, bridging the gap between consumers and manufacturers. Future flexible manufacturing systems will provide automotive end users the tools to produce almost anything for vehicle's performance enhancement or a simple customization of a body panel. Collaborative product design will enhance mass customization, enabling customers and manufacturers to share a similar satisfaction level.

From the industrial point of view, the trade-off between freeform building capacity and operation costs of an AM process plays an important role for decision-makers concerning the adoption of AM. The reduction of the overall processing time is still the main challenge; however, huge steps are being taken to overcome this, whether by increasing building volumes, including multiple printing heads or energy sources, integrating building units with postprocessing units, among others.

Although that AM still lacks maturity, especially concerning material behavior during processing, this manufacturing approach offers significant advantages for many manufacturers, including lighter-weight parts, reduced inventories, and improved profitability.

Concerning the mass customization concept, AM is definitely a game-changer. The massification of industrial AM provides the necessary production flexibility and freeform design, to answer to market requirements. Examples of these applications go from customized body panels to optimized ducting components.

The current digitalization trend is another important breakthrough that puts AM onto the spotlight. Virtual models can be discussed and validated by both manufacturers and end users, using virtual and augmented reality, exploring all aspects of the product before its materialization through an AM process. Direct digital manufacturing enables the follow-up of all product stages, ensuring that all stakeholders involved play their role to obtain a fully optimized product in the end.

The final note goes to AM materials which are, in fact, the main sustainability pillar for AM processes. AM has evolved a lot from the early days of rapid prototyping, where material properties could hardly be able to produce final parts. Nowadays, AM material properties are fully comparable with conventional material properties. The strong trend on energy consumption decrease poses new challenges regarding vehicle's design, performance, and regulations compliance. AM is also the key-enabling technology for this new generation vehicles with alternative motion systems, enabling the production of lightweight components, using generative design and introducing lattice structures to minimize weight, replacing metal parts by AM components incorporating fiber reinforcements or components using high operation temperature and harsh environment-compatible AM materials.

## References


[1] 3D Natives, Ford Produces the Largest Ever 3D Printed Metal Automotive Part, 2019. https://www.3dnatives.com/en/ford-3d-printed-metal-part-050220195/ (accessed April 29, 2019).

[2] 3D Systems, Metal Additive Manufacturing Software: A Critical Element for Successful and Profitable Metal 3D Printing, 2018.

[3] H. Ahuett-Garza, T. Kurfess, A brief discussion on the trends of habilitating technologies for Industry 4.0 and Smart manufacturing, Manuf. Lett. 15 (2018) 60–63, https://doi.org/10.1016/j.mfglet.2018.02.011.

[4] N. Anwer, L. Mathieu, From reverse engineering to shape engineering in mechanical design, CIRP Ann. Manuf. Technol. 65 (2016) 165–168, https://doi.org/10.1016/j.cirp.2016.04.052.

[5] M. Attaran, Additive manufacturing: the most promising technology to alter the supply chain and logistics, J. Serv. Sci. Manag. 10 (2017) 189–206, https://doi.org/10.4236/jssm.2017.103017.

[6] M. Baumers, P. Dickens, C. Tuck, R. Hague, The cost of additive manufacturing: machine productivity, economies of scale and technology-push, Technol. Forecast. Soc. Change (2015), https://doi.org/10.1016/j.techfore.2015.02.015.

[7] B. Bidanda, Z. Geng, Emerging trends in reverse engineering, in: Proc. 2nd Int. Conf. Prog. Addit. Manuf. (Pro-AM 2016), 2016, pp. 2–7.

[8] L.G. Blok, M.L. Longana, H. Yu, B.K.S. Woods, An investigation into 3D printing of fibre reinforced thermoplastic composites, Addit. Manuf. 22 (2018) 176–186, https://doi.org/10.1016/j.addma.2018.04.039.

[9] BMW Group, A Million Printed Components in Just Ten Years: BMW Group Makes Increasing Use of 3D Printing, Press Release, 2018. https://www.press.bmwgroup.com/global/article/detail/T0286895EN/a (accessed December 31, 2019).

[10] BMW Group, MINI Yours Customised: From the Original to the Personalised Unique Special, Press Kit, 2017. https://www.press.bmwgroup.com/global/article/detail/T0276990EN/mini-yours-customised:-from-the-original-to-the-personalised-unique-special (accessed January 2, 2020).

[11] D. Breitschwerdt, A. Cornet, L. Michor, N. Müller, L. Salmon, Performance and Disruption – A Perspective on the Automotive Supplier Landscape and Major Technology Trends. Munich, 2016.

[12] J. Carreira, J. Vasco, H. Almeida, AM tooling for the mouldmaking industry, in: H.A. Almeida, J.C. Vasco (Eds.), Prog. Digit. Phys. Manuf., Springer International Publishing, Cham, 2020, pp. 162–170.

[13] C.K. Chua, K.F. Leong, 3D Printing and Additive Manufacturing: Principles and Applications, fourth ed., World Scientific Publishing Company, 2014.
[14] S. Coeck, M. Bisht, J. Plas, F. Verbist, Prediction of lack of fusion porosity in selective laser melting based on melt pool monitoring data, Addit. Manuf. 25 (2019) 347–356, https://doi.org/10.1016/j.addma.2018.11.015.
[15] C. Costa, J. Aguzzi, Temporal shape changes and future trends in European automotive design, Machines 3 (2015) 256–267, https://doi.org/10.3390/machines3030256.
[16] S. Curran, P. Chambon, R. Lind, L. Love, R. Wagner, S. Whitted, et al., Big Area additive manufacturing and hardware-in-the-loop for rapid vehicle powertrain prototyping: a case study on the development of a 3-D-printed Shelby Cobra, SAE Tech. Pap. (2016-April), https://doi.org/10.4271/2016-01-0328.
[17] S. Davies, Divergent to Purchase Five Pre-production SLM Solutions Metal Additive Manufacturing Machines as Partnership Expands, TCT Mag, 2019. https://www.tctmagazine.com/3d-printing-news/divergent-to-purchase-five-pre-production-slm-solutions-meta/ (accessed January 3, 2020).
[18] Divergent 3D, Joint Development Partnership between Divergent 3D and SLM Solutions, LinkedIn (2019). https://www.linkedin.com/company/divergent3d/ (accessed January 3, 2020).
[19] S. Davies, Lamborghini Unveils End-Use Components Produced with Carbon 3D Printing Technology, TCT Mag, 2019. https://www.tctmagazine.com/3d-printing-news/lamborghini-end-use-components-carbon-3d-printing/ (accessed January 3, 2020).
[20] S. Davies, Carbon 3D Printing Technology Used to Produce Lamborghini Dashboard Air Vents, TCT Mag, 2019. https://www.tctmagazine.com/3d-printing-news/carbon-3d-printing-technology-used-to-produce-lamborghini-da/ (accessed January 3, 2020).
[21] S. Davies, Audi Pre-series Center Adopts Stratasys J750 for Full-Colour Taillight Housing Prototypes, TCT Mag, 2018. https://www.tctmagazine.com/3d-printing-news/audi-pre-series-center-j750-full-colour-taillight-prototypes/ (accessed December 31, 2019).
[22] S. Davies, Divergent 3D and SLM Solutions Partner to Accelerate Production of Road Vehicles, TCT Mag, 2017. https://www.tctmagazine.com/3d-printing-news/divergent-3d-slm-solutions-accelerate-production-vehicles/ (accessed January 3, 2020).
[23] T. Debroy, H.L. Wei, J.S. Zuback, T. Mukherjee, J.W. Elmer, M. JO, et al., Progress in materials science additive manufacturing of metallic components – process, Struct. Prop. 92 (2018) 112–224, https://doi.org/10.1016/j.pmatsci.2017.10.001.
[24] V. Dhokia, W.P. Essink, J.M. Flynn, CIRP Annals - manufacturing Technology A generative multi-agent design methodology for additively manufactured parts inspired by termite nest building, CIRP Ann. Manuf. Technol. 66 (2017) 153–156, https://doi.org/10.1016/j.cirp.2017.04.039.
[25] Divergent 3D, Presence of Divergent 3D on the SLM Solutions Booth at Formnext2019, LinkedIn, 2019. https://www.linkedin.com/company/divergent3d/ (accessed January 3, 2020).
[26] S.K. Everton, M. Hirsch, P. Stravroulakis, R.K. Leach, A.T. Clare, Review of in-situ process monitoring and in-situ metrology for metal additive manufacturing, Mat. Des. 95 (2016) 431–445, https://doi.org/10.1016/j.matdes.2016.01.099.
[27] M. Fera, R. Macchiaroli, F. Fruggiero, A. Lambiase, A new perspective for production process analysis using additive manufacturing—complexity vs production volume, Int. J. Adv. Manuf. Technol. 95 (2018) 673–685, https://doi.org/10.1007/s00170-017-1221-1.

[28] FormLabs, Guide to Selective Laser Sintering (SLS) 3D Printing, Eng Guid, 2020. https://formlabs.com/eu/blog/what-is-selective-laser-sintering/ (accessed January 2, 2020).
[29] Formnext, Experience the Magic of Additive manufacturing at the Formnext 2018, Twitter, 2018. https://twitter.com/formnext_expo/status/1062309304454262785 (accessed December 22, 2019).
[30] H.E. Friedrich, E. Beeh, C.S. Roider, Solutions for next generation automotive lightweight concepts based on material selection and functional integration BT, in: D. Orlov, V. Joshi, K.N. Solanki, N.R. Neelameggham (Eds.), Magnesium Technology 2018, Springer International Publishing, Cham, 2018, pp. 343–348.
[31] H. Gaub, Customization of mass-produced parts by combining injection molding and additive manufacturing with Industry 4.0 technologies, Reinforc. Plast 60 (2016) 401–404, https://doi.org/10.1016/j.repl.2015.09.004.
[32] I. Gibson, D. Rosen, B. Stucker, Direct digital manufacturing BT, in: I. Gibson, D. Rosen, B. Stucker (Eds.), Additive Manufacturing Technologies: 3D Printing, Rapid Prototyping, and Direct Digital Manufacturing, Springer New York, New York, NY, 2015, pp. 375–397, https://doi.org/10.1007/978-1-4939-2113-3_16.
[33] C.A. Giffi, B. Gangula, P. Illinda, 3D Opportunity in the Automotive Industry: Additive Manufacturing Hits the Road, 2014.
[34] N. Gupta, F. Chen, N.G. Tsoutsos, M. Maniatakos, ObfusCADe: obfuscating additive manufacturing CAD models against counterfeiting, in: Proc Des Autom Conf, 2017, https://doi.org/10.1145/3061639.3079847. Part 12828.
[35] C. Hannon, B. Evans, D. Johns, Design for additive manufacture of the bloodhound SSC steering wheel, in: 13th Rapid Des. Prototyp. Manuf. Conf. Proc, 2012.
[36] A.B. Hexagon, Simufact Additive - A Simulation Tool for Distortion Prediction in Powder Bed Additive Manufacturing, 2019. https://www.mscsoftware.com/product/simufact-additive.
[37] P.A. Hooper, Melt pool temperature and cooling rates in laser powder bed fusion, Addit. Manuf. 22 (2018) 548–559, https://doi.org/10.1016/j.addma.2018.05.032.
[38] R. Huang, M.E. Riddle, D. Graziano, S. Das, S. Nimbalkar, J. Cresko, et al., Environmental and economic implications of distributed additive manufacturing: the case of injection mold tooling, J. Ind. Ecol. 21 (2017) S130–S143, https://doi.org/10.1111/jiec.12641.
[39] S.H. Huang, P. Liu, A. Mokasdar, Additive manufacturing and its societal impact: a literature review, Int. J. Adv. Manuf. Technol. 67 (2013) 1191–1203, https://doi.org/10.1007/s00170-012-4558-5.
[40] O. Illies, G. Li, J.-P. Jürgens, V. Ploshikhin, D. Herzog, C. Emmelmann, Numerical modelling and experimental validation of thermal history titanium alloys in laser beam melting, Proc. CIRP 74 (2018) 92–96, https://doi.org/10.1016/j.procir.2018.08.046.
[41] S.A. Jahan, T. Wu, Y. Zhang, J. Zhang, A. Tovar, H. Elmounayri, Thermo-mechanical design optimization of conformal cooling channels using design of experiments approach, Proc. Manuf. 10 (2017) 898–911, https://doi.org/10.1016/j.promfg.2017.07.078.
[42] D. Jankovics, A. Barari, Customization of automotive structural components using additive manufacturing and topology optimization, IFAC-Pap. OnLine 52 (2019) 212–217, https://doi.org/10.1016/j.ifacol.2019.10.066.
[43] S. Jayanath, A. Achuthan, A computationally efficient finite element framework to simulate additive manufacturing processes, J. Manuf. Sci. Eng. 140 (2018), https://doi.org/10.1115/1.4039092.

[44] S. Junk, B. Klerch, U. Hochberg, Structural in lightweight design for additive manufacturing, Proc. CIRP 84 (2019) 277–282, https://doi.org/10.1016/j.procir.2019.04.277.

[45] D. Kang, S. Park, Y. Son, S. Yeon, S. Hoon, I. Kim, Multi-lattice inner structures for high-strength and light-weight in metal selective laser melting process, Mater. Des. 175 (2019) 107786, https://doi.org/10.1016/j.matdes.2019.107786.

[46] J.A. Kanko, A.P. Sibley, J.M. Fraser, Journal of Materials Processing Technology in situ morphology-based defect detection of selective laser melting through inline coherent imaging, J. Mater. Process. Technol. 231 (2016) 488–500, https://doi.org/10.1016/j.jmatprotec.2015.12.024.

[47] H.S. Khajavi, J. Holmström, J. Partanen, Additive manufacturing in the spare parts supply chain: hub configuration and technology maturity, Rapid Prototyp. J. 24 (2018) 1178–1192, https://doi.org/10.1108/RPJ-03-2017-0052.

[48] S. Kitayama, H. Miyakawa, M. Takano, S. Aiba, Multi-objective optimization of injection molding process parameters for short cycle time and warpage reduction using conformal cooling channel, Int. J. Adv. Manuf. Technol. 88 (2017) 1735–1744, https://doi.org/10.1007/s00170-016-8904-x.

[49] N.N. Kumbhar, A.V. Mulay, Post processing methods used to improve surface finish of products which are manufactured by additive manufacturing technologies: a review, J. Inst. Eng. Ser. C 99 (2018) 481–487, https://doi.org/10.1007/s40032-016-0340-z.

[50] R. Leal, F.M. Barreiros, L. Alves, F. Romeiro, J.C. Vasco, M. Santos, et al., Additive manufacturing tooling for the automotive industry, Int. J. Adv. Manuf. Technol. 92 (2017), https://doi.org/10.1007/s00170-017-0239-8.

[51] M. Leary, M. Mazur, J. Elambasseril, M. McMillan, T. Chirent, Y. Sun, et al., Selective laser melting (SLM) of AlSi12Mg lattice structures, Mater. Des. 98 (2016) 344–357, https://doi.org/10.1016/j.matdes.2016.02.127.

[52] J.Y. Lee, J. An, C.K. Chua, Fundamentals and applications of 3D printing for novel materials, Appl. Mater. Today 7 (2017) 120–133, https://doi.org/10.1016/j.apmt.2017.02.004.

[53] M. Livesu, D. Cabiddu, M. Attene, slice2mesh, A meshing tool for the simulation of additive manufacturing processes, Comput. Graph. 80 (2019) 73–84, https://doi.org/10.1016/j.cag.2019.03.004.

[54] N.V. Materialise, The Complete Software Suite for Professional 3D Printing, 2019. https://www.materialise.com/en/software/magics-3d-print-suite (accessed December 12, 2019).

[55] M. Mazur, P. Brincat, M. Leary, M. Brandt, Numerical and experimental evaluation of a conformally cooled H13 steel injection mould manufactured with selective laser melting, Int. J. Adv. Manuf. Technol. 93 (2017) 881–900, https://doi.org/10.1007/s00170-017-0426-7.

[56] S. Mellor, L. Hao, D. Zhang, Additive manufacturing: a framework for implementation, Int. J. Prod. Econ. 149 (2014) 194–201, https://doi.org/10.1016/j.ijpe.2013.07.008.

[57] A.M. Metal, Record breaking 300mph Bugatti Chiron with AM exhaust components, AM 5 (2019) 28.

[58] S. Mollett, Innovation Takes a Front Seat at TS Tech, 2018.

[59] B.E.T. Morton, L.S.O. Hassan, K.M.J.W.J.M. Chapman, R.A.I. Niven, Design optimisation using computational fluid dynamics applied to a land – based supersonic vehicle, The BLOODHOUND SSC, Struct. Multidis. Opt. (2013) 301–316, https://doi.org/10.1007/s00158-012-0826-0.

[60] T. Muhammad, K.E. Bang, Integration of value adding services related to financing and ownership: a business model perspective, Proc. Des. Soc. Int. Conf. Eng. Des. 1 (2019) 2279–2286, https://doi.org/10.1017/dsi.2019.234.

[61] O. Ogorodnyk, K. Martinsen, Monitoring and control for thermoplastics injection molding A review, Proc. CIRP 67 (2018) 380–385, https://doi.org/10.1016/j.procir.2017.12.229.
[62] M. Paulic, T. Irgolic, J. Balic, F. Cus, A. Cupar, T. Brajlih, et al., Reverse engineering of parts with optical scanning and additive manufacturing, Proc. Eng 69 (2014) 795–803, https://doi.org/10.1016/j.proeng.2014.03.056.
[63] A. Pearson, Lamborghini Accelerates and Perfects Automotive Engineering with Stratasys 3D Printed Prototypes and Track-Ready Parts, Strat Blog, 2020. https://www.stratasys.com/explore/blog/2015/lamborghini-3d-printing. (Accessed 29 April 2019).
[64] E. Pei, M. Ressin, R.I. Campbell, B. Eynard, J. Xiao, Investigating the impact of additive manufacturing data exchange standards for re-distributed manufacturing, Prog. Addit. Manuf. 4 (2019) 331–344, https://doi.org/10.1007/s40964-019-00085-7.
[65] G. Perboli, F. Ferrero, S. Musso, A. Vesco, Business models and tariff simulation in car-sharing services, Transport. Res. Part A Policy Pract. 115 (2018) 32–48, https://doi.org/10.1016/j.tra.2017.09.011.
[66] S. Peters, G. Lanza, J. Ni, J. Xiaoning, Y. Pei-Yun, M. Colledani, Automotive manufacturing technologies – an international viewpoint, Manuf. Rev. 1 (2014) 10, https://doi.org/10.1051/mfreview/2014010.
[67] F.T. Piller, C. Weller, R. Kleer, Business models with additive manufacturing—opportunities and challenges from the perspective of economics and management, in: C. Brecher (Ed.), Adv. Prod. Technol. - Lect. Notes Prod. Eng, Springer Cham Heidelberg, 2015, pp. 39–47.
[68] P. Pradel, Z. Zhu, R. Bibb, J. Moultrie, Investigation of design for additive manufacturing in professional design practice, J. Eng. Des. 29 (2018) 165–200, https://doi.org/10.1080/09544828.2018.1454589.
[69] K.S. Prakash, T. Nancharaih, V.V.S. Rao, Additive manufacturing techniques in manufacturing -an overview, Mater. Today Proc. 5 (2018) 3873–3882, https://doi.org/10.1016/j.matpr.2017.11.642.
[70] S. Radisavljevic, Additive manufacturing - the automotive industry, Slideshare (2014). https://www.slideshare.net/StefanRadisavljevic/additive-manufacturing-43047855 (accessed December 28, 2019).
[71] S. Singh, S. Ramakrishna, R. Singh, Material issues in additive manufacturing: a review, J. Manuf. Proc. 25 (2017) 185–200, https://doi.org/10.1016/j.jmapro.2016.11.006.
[72] C.J. Smith, M. Gilbert, I. Todd, F. Derguti, Application of layout optimization to the design of additively manufactured metallic components, Struct. Multidiscip. Optim. (2016) 1297–1313, https://doi.org/10.1007/s00158-016-1426-1.
[73] O.O. Spencer, O.T. Yusuf, T.C. Tofade, Additive manufacturing technology development: a trajectory towards industrial revolution, Am. J. Mech. Ind. Eng. 3 (2018) 80, https://doi.org/10.11648/j.ajmie.20180305.12.
[74] Stratasys Ltd, Designing for optimal FDM surface finish, Appl. Br (2019). https://www.stratasys.com/-/media/files/application-brief/AB_FDM_SurfaceFinish_0819a (accessed December 29, 2019).
[75] Stratasys Ltd, Stratasys Adds PEKK-Based, High-Performance Thermoplastic: Antero 800NA, for FDM Process, Invest Relations, 2018. https://investors.stratasys.com/news-releases/news-release-details/stratasys-adds-pekk-based-high-performance-thermoplastic-antero (accessed December 29, 2019).
[76] Stratasys Ltd, The Road Ahead | Rapid Prototyping in the Automotive Industry, Eden Prairie, 2018.
[77] Stratasys Ltd, Stratasys to Add SABIC's ULTEM 9085 High Performance Thermoplastic for Direct Digital Manufacturing & Rapid Prototyping, Invest Relations, 2008. https://investors.stratasys.com/news-releases/news-release-details/stratasys-add-sabics-ultem-9085-high-performance-thermoplastic (accessed January 2, 2020).

[78] D. Strobbe, P Van Puyvelde, J. Kruth, B Van Hooreweder, Laser sintering of PA12/PA4,6 polymer composites, in: Solid Free. Fabr. 2018 Proc. 29th Annu. Int. Solid Free. Fabr. Symp, 2018, pp. 1550–1559.
[79] F. Thiesse, M. Wirth, H.G. Kemper, M. Moisa, D. Morar, H. Lasi, et al., Economic implications of additive manufacturing and the contribution of MIS, Bus. Inf. Syst. Eng. 57 (2015) 139–148, https://doi.org/10.1007/s12599-015-0374-4.
[80] M.K. Thompson, G. Moroni, T. Vaneker, G. Fadel, R.I. Campbell, I. Gibson, et al., Design for additive manufacturing: trends, opportunities, considerations, and constraints, CIRP Ann. Manuf. Technol. 65 (2016) 737–760, https://doi.org/10.1016/j.cirp.2016.05.004.
[81] S.A.M. Tofail, E.P. Koumoulos, A. Bandyopadhyay, S. Bose, L. O'Donoghue, C. Charitidis, Additive manufacturing: scientific and technological challenges, market uptake and opportunities, Mater. Today 21 (2018), https://doi.org/10.1016/j.mattod.2017.07.001.
[82] M. Toth, Digimat Holistic Simulation Solution for Additive Manufacturing, 2018.
[83] J. Vasco, F. Barreiros, A. Nabais, N. Reis, Additive manufacturing applied to injection moulding: technical and economic impact, Rapid Prototyp. J. 25 (2019) 1241–1249, https://doi.org/10.1108/RPJ-07-2018-0179.
[84] C de Vries, Volkswagen Autoeuropa: Maximizing Production Efficiency with 3D Printed Tools and Fixtures, Ultimaker, 2017. https://ultimaker.com/learn/volkswagen-autoeuropa-maximizing-production-efficiency-with-3d-printed (accessed March 12, 2019).
[85] C. Weller, R. Kleer, F.T. Piller, Economic implications of 3D printing: market structure models in light of additive manufacturing revisited, Int. J. Prod. Econ. 164 (2015) 43–56, https://doi.org/10.1016/j.ijpe.2015.02.020.
[86] T.M. Wischeropp, H. Hoch, F. Beckmann, C. Emmelmann, Opportunities for braking technology due to additive manufacturing through the example of a Bugatti brake caliper, in: R. Mayer (Ed.), Proceedings. Springer Vieweg, Springer Berlin Heidelberg, Berlin, Heidelb., Berlin, Heidelberg, 2019, pp. 181–193.
[87] T. Wohlers, Wohlers Report 2019, Part 5 - Final Part Production. Fort Collins, Colorado, 2019.
[88] C. Yan, L. Hao, A. Hussein, P. Young, D. Raymont, Advanced lightweight 316L stainless steel cellular lattice structures fabricated via selective laser melting, Mater. Des. 55 (2014) 533–541, https://doi.org/10.1016/j.matdes.2013.10.027.
[89] S. Yang, Y.F. Zhao, Additive manufacturing-enabled design theory and methodology: a critical review, Int. J. Adv. Manuf. Technol. 80 (2015), https://doi.org/10.1007/s00170-015-6994-5.
[90] T. Zegard, G.H. Paulino, Bridging topology optimization and additive manufacturing, Struct. Multidiscip. Optim. 53 (2016) 175–192, https://doi.org/10.1007/s00158-015-1274-4.
[91] Z. Zeng, X. Deng, J. Cui, H. Jiang, S. Yan, B. Peng, Improvement on selective laser sintering and post-processing of polystyrene, Polymers 11 (2019), https://doi.org/10.3390/polym11060956.
[92] Z. Zhu, N. Anwer, L. Mathieu, Deviation modeling and shape transformation in design for additive manufacturing, Proc. CIRP 60 (2017) 211–216, https://doi.org/10.1016/j.procir.2017.01.023.
[93] D. Zindani, K. Kumar, An insight into additive manufacturing of fiber reinforced polymer composite, Int. J. Light Mater. Manuf. 2 (2019) 267–278, https://doi.org/10.1016/j.ijlmm.2019.08.004.

CHAPTER 17

# Additive manufacturing of large parts

**G.A. Turichin, O.G. Klimova-Korsmik, K.D. Babkin, S. Yu. Ivanov**
Saint-Petersburg State Marine Technical University, Institute of Laser and Welding Technologies, Saint-Petersburg, Russian Federation

## 17.1 Direct laser deposition process simulation

Direct laser deposition (DLD) is the most advanced technology of additive manufacturing for production of large-scale parts for aerospace industry, as mentioned in Ref. [1], ship fittings, propellers and water jet motion systems, large-scale brackets and machinery parts, high-pressure vessels, and so on. DLD process is one of the DED (directed energy deposition)-type technologies and analog of LMD (laser metal deposition) and DMD (direct metal deposition). The essence of this technology is the formation of the product from metal powder, fed by a gas—powder jet, coaxial or lateral to the focused laser beam, and directed to the processing area, with controlled heating and melting of the powder particles and substrate. High productivity of DLD process, which currently reaches up to 2 kg/h or even more [2], require use of high laser power and melting speed, as well as appropriate level of powder flow. Long melt pools with high speed of motion, which forms in this case, can lead to hydrodynamic instability appearance [3]. This fact together with the large number of parameters, which are necessary for the determination of the treatment mode (and required to provide a smooth and stable formation of melt pools), makes the selection of the technological mode parameters very difficult. Also, for the generation of the cladding head, trajectory with consideration of the product thermal distortion requires the utilization of a thermomechanical model. In such a model, a melt pool shape and temperature distribution around it are the main input parameters. Therefore, the physical adequate and fast mathematical model of melt pool formation in DLD process is necessary for design of technological modes of DLD of large parts. A number of articles devoted to the process simulation of DLD, in general, based on the usage of different numerical schemes, do not allow the analysis of the influence of different physical effects (for example, Marangoni

*Additive Manufacturing*
ISBN 978-0-12-818411-0
https://doi.org/10.1016/B978-0-12-818411-0.00001-X

convection), or the mutual influence of different physical processes (such as hydrodynamics and heat transfer), on the shape and size of melt pool. Physical adequate process model for DLD must be based on connected solution of several tasks, among them, task about powder jet transfer and impingement with substrate, heat transfer in liquid and solid phases, and hydrodynamics of melt pool. For the main purpose—description of the heat source for thermomechanical simulation, it is enough to restrict this model to a steady-state case. The process model in this case consists of model of powder flux transfer to product surface, powder and surface heating by laser beam, melt hydrodynamic in melt pool, and heat transfer in melt pool and surrounded material.

### 17.1.1 The model of powder particles transfer and heating in gas—powder jet with consideration of a jet impingement to product surface

The main approaches for steady-state theoretical description of material transfer by gas—powder jet during DLD are described in Ref. [4] with consideration of gas dynamics of gas—powder jet impingement on the substrate.

The problem of heating and melting of powder during transportation allows the association of a value of energy flux from powder jet to melt pool surface with characteristics of laser radiation, dependent on the technological process parameters. The task of moving the powder allows to determine the distribution of mass density flow in DLD that defines the profile of the formed surface. The solution to this problem is also necessary to determine the heating of the powder particles, because of their trajectories passing through areas with different value of laser radiation intensity. Finally, the purpose of the flow of vapor jets, falling on the surface of the substrate, defines the field of gas velocity, affecting the transfer of powder. For the solution of this problem, the following simplifications, based on the peculiarities of real technological process, can be used:

**(a)** The mass ratio of the gas flow out from the nozzle and powder moved by gas jet allows to use the model of "dust-laden gas jet" and ignore the influence of the powder on the velocity distribution of the gas jet.

**(b)** Reynolds number for the jet during cladding is within $10^2-10^3$, and Mach number is within 0.01—0.3. This allows to ignore the influence of the gaseous viscosity and to use the approach of potential flow.

**(c)** The particles of powder are isotropically heated (this corresponds to the case of chaotic rotation of the sprayed particles in the jet), and the influence of the interaction between the particles is ignored.

It is possible to assume that at the end of the nozzle, the process of mixing of the powder with a jet is completed, and the velocity of the particles is equal to the local gas velocity. Using cylindrical coordinates, the volumetric flow rate is introduced as

$$Q = V_0 \pi \frac{d^2}{4}, \tag{17.1}$$

where "*d*" is the nozzle diameter. Using a potential gas flow with constant gas density, and solving the Laplace equation for potential flow, with zero condition for the velocity component perpendicular to the product surface, the distribution of velocity of the flow is obtained as

$$v_z = -\frac{4Q}{\pi H d^2} z; \; v_r = \frac{2Q}{\pi H d^2} r, \tag{17.2}$$

where "*H*" is the distance from the nozzle to the substrate. The resulting flow pattern is presented in Fig. 17.1.

For simplification, it is possible to assume that the particles of powder are spheres; therefore, the equations of motion for a particle with allowance approximation "dust-laden gas jet" can be written as

$$m\ddot{\vec{r}} = m\vec{g} - \vec{F}_s, \tag{17.3}$$

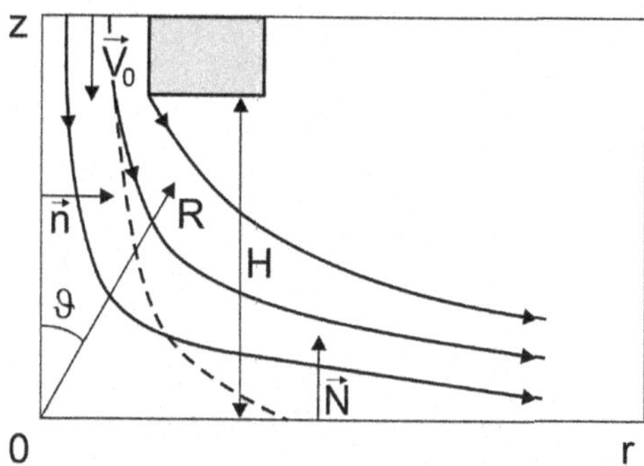


**Figure 17.1** The scheme of transportation of powder during the cladding. The *solid line* indicates the line of gas flow, and *dashed line* indicates the path of the particle of powder.

where $\vec{F}_s$ is the force acting on the particle from the gas flow. According to Ref. [4], the force $\vec{F}_s$ can be written as

$$\vec{F}_s \sim \frac{1}{2} k_f \rho_g \left(\dot{\vec{r}} - \vec{v}\right)^2 \frac{\dot{\vec{r}} - \vec{v}}{\left|\dot{\vec{r}} - \vec{v}\right|} \cdot \pi a^2 \; k_f \approx \frac{1}{2}, \tag{17.4}$$

where "$a$" is the radius of the particle of powder and $k_f$ is a coefficient of shape; accordingly [5], for sphere it is equal to 1/2. The projection on the axis of the coordinate system converts this equation into a system of two nonlinear differential equations of second order. The solution of this system gives us the expression for the radius of a jet of powder on the surface of the substrate:

$$r = r_0 ch\left(\frac{H}{V_0}\sqrt{\frac{\alpha}{2m}\frac{V_0}{H}}\right) \approx r_0\left(1 + \frac{H}{V_0}\frac{\alpha}{4m}\right). \tag{17.5}$$

To solve the problem of the heating of the particles, it is possible to make a number of assumptions regarding the behavior of particles in the gas flow under the action of laser radiation. One can assume that, when transported, the particle flow spins; then, this rotation ensures the uniform heating of the particles. To find the temperature field in the particle, it is necessary to solve a symmetric thermal problem in spherical coordinates. Using Laplace transformation for solving this problem and after simple transformations, it is possible to obtain a temperature field in the powder particle $T(r, t)$ as

$$T(r,t) = T_0 + \frac{3q\cdot\chi^{3/2}\cdot\pi^{1/2}}{\lambda\cdot R\cdot r}\int_0^t \frac{\tau}{(t-\tau)^{1/2}}\cdot\exp\left(\frac{r^2}{4\chi(t-\tau)}\right)\cdot erf\left(\frac{r}{2\sqrt{\chi(t-\tau)}}\right)d\tau. \tag{17.6}$$

Requirements of formation of poreless melt layer on target surface determine the required depth of melting for the particles. We assume that the powder forms a dense packing with a factor of packaging equal to 0.74. Melting point of the powder particles $T_p$ is used. From Eq. (17.6), we obtain the equation for the necessary depth of melting $\delta$ for the particles:

$$\delta = R - \frac{3q\cdot\chi^{3/2}\cdot\pi^{1/2}}{\lambda\cdot R\left(T_p - T_0\right)}\int\limits_0^t \frac{\tau}{(t-\tau)^{1/2}}\cdot\exp\left(\frac{r^2}{4\chi(t-\tau)}\right)\cdot erf\left(\frac{r}{2\sqrt{\chi(t-\tau)}}\right)d\tau. \tag{17.7}$$

We find the melting depth of the particles of a powder from the condition that the amount of the produced melt is able to completely fill discontinuities between particles. Simple geometrical considerations lead in this case $\delta = 0.074$a. This condition together with expressions for the time-of-flight particles from the nozzle to the substrate, depth of melting of the powder particles, and the known formula for the density of heat flow at heating of a particle in a laser radiation field allows to define the processing parameters of the laser cladding of metal powders. They will provide the conditions required to obtain dense coatings with consideration of the condition of conservation of solid core of particles, which are necessary for the DLD production of fine grain structures in the deposited material. Obtained with the help of the developed model, values for the broadening of the jet of powder due to the action of aerodynamic forces, as the jets impacts on the surface, it is necessary to agree with the experimental data on the measurement of the width of the clad paths.

### 17.1.2 Melt flow model

In the typical conditions of DLD, it is possible to restrict the analysis to that case, where the melt pool length "L" is much larger than its width "b" and depth "H." In this case, the one-dimensional boundary layer approximation (i.e., when the "longitudinal" velocity $V_x$, directed along the beam axis, is much larger than the "transverse" velocities $V_y$, $V_z$) can be used. Therefore, Navier–Stokes equation can be written as:

$$V_x\frac{\partial V_x}{\partial x} = -\frac{1}{\rho}\frac{\partial p}{\partial x} + \nu\frac{\partial^2 V_x}{\partial z^2} \tag{17.8}$$

The boundary conditions are as follows: at the "bottom" of the melt pool is given as $v_x|_{z=H} = 0$, on the "top" surface: $-\eta\frac{\partial v_x}{\partial z}\Big|_{z=H} = \frac{\partial \sigma}{\partial x}$, requirement of the stress tensor continuity, where $\sigma$ is the surface tension coefficient. Assuming that the temperature along the melt pool surface changes much less than the average surface temperature, defining $T_t$ as the maximum surface temperature, $T_m$ and $T_v$ melting and evaporation temperatures correspondingly, and taking a linear law for the temperature drop up to the "tail" of the melt pool, we can rewrite the boundary condition at the top surface as

$$-\eta \frac{\partial V_x}{\partial z}\bigg|_{z=H} = \frac{\sigma}{L}\frac{T_t - T_m}{T_v - T_m} = \frac{\sigma^*}{L} \tag{17.9}$$

To satisfy the boundary conditions and the condition of mass flux conservation along the "$x$" axis, it is possible to accept the hypothesis of a "parabolic" distribution of melt speed with respect to depth:

$$V_x(z) = V_x(\alpha + \beta z + \gamma z^2).$$

The boundary conditions give us in this case the following expressions for the coefficients:

$$\alpha = 0$$

$$\gamma = -\frac{3}{4H}\left(\frac{2}{H} - \frac{\sigma}{L\eta v_x}\right)$$

$$\beta = \frac{2}{H}\left(1 + \frac{H}{4}\left(\frac{2}{H} - \frac{\sigma}{L\eta v_x}\right)\right) = \frac{2}{H}\left(\frac{3}{2} - \frac{\sigma}{4\eta v_x}\frac{H}{L}\right) = \frac{3}{H} - \frac{\sigma}{2\eta v_x L} \tag{17.10}$$

Accordingly, Eq. (17.1) can be written as

$$V_x\frac{\partial V_x}{\partial x} = -\frac{1}{\rho}\frac{\partial \rho}{\partial x} - 3\nu\frac{V_x}{H^2} + \frac{3\sigma^*}{2\rho LH} \tag{17.11}$$

On the other hand, we can use the continuity equation to link the melt velocity $v_x$ with the melt pool surface position. In our case, it is necessary to take into account the mass flow coming to the surface of the melt with gas—powder jet. On the surface of the melt, the incident mass flow density, which can be calculated as it was described in the previous section, can be denoted as *j(x)*. Therefore, it is possible to write the flow continuity equation as

$$\frac{\partial V}{\partial x}(V_x H) = \frac{j(x)}{\rho} \tag{17.12}$$

Let us determine the pressure in the melt pool "$p$," taking into account, that $p = \frac{\sigma}{R}$, where $R$ is the surface curvature radius (Fig. 17.2).

When $H < b$, we have $R \approx b + \frac{H^2}{2b}$, and it is possible to write the pressure as $p \approx \frac{\sigma}{b} - \frac{\sigma H^2}{2b^3}$.

Therefore, the Navier–Stokes equation term, connected with changes of "transverse" curvature radius of surface, can be written as $\frac{\sigma H^2}{2b^3} \frac{H \partial H}{\partial x}$.

Pressure produced by changes of the "longitudinal" surface curvature radius can be written as $\sigma \frac{\partial^2 H}{\partial x^2}$.

Then, we can write

$$V_x \frac{\partial V_x}{\partial x} = \frac{\sigma}{\rho b^3} \frac{H \partial H}{\partial x} - 3v \frac{V_x}{H^2} + \frac{3\sigma^*}{2\rho L H} \tag{17.13}$$

After integration of the continuity equation, we can get

$$H(x) = \frac{1}{\rho V_x} \int_0^x j(x) dx \tag{17.14}$$

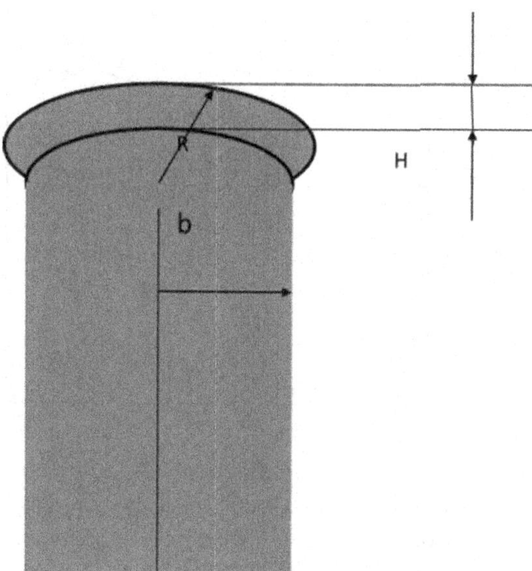


**Figure 17.2** Scheme of the cross section for the growing wall.

The initial and boundary conditions for this problem can be represented as

$$H=0 \text{ when } x=0, \left.\frac{\partial H}{\partial x}\right|_{x=0}=0, \left.\frac{\partial H}{\partial x}\right|_{y=L}=0.$$

Substitution of Eqs. (7.14) to (7.13) and numerical solution of this differential equation give us a profile for the top surface of the melt pool with consideration of Marangoni effect. Examples of such shapes for different values of powder jet distributions and laser clad head velocity are shown in Fig. 17.3. It is evident that the surface shape, as well as melt flow speed, depends on the melt pool length "$L$" and depth "$H$," which are given by the solution of the heat transfer problem.

### 17.1.3 Heat transfer model

In steady-state case of heat transfer during DLD process in the zone of laser action on metal, the heat transfer equation can be represented as

$$v_x \frac{\partial T}{\partial x} = \chi\left(\frac{\partial^2 T}{\partial x^2} + \frac{\partial^2 T}{\partial z^2}\right)$$

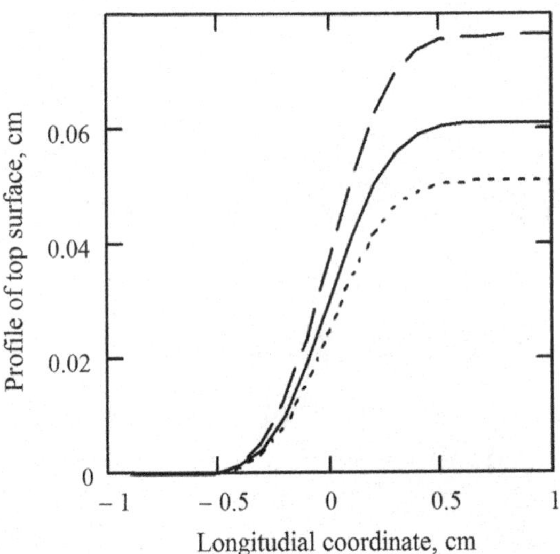


**Figure 17.3** Shape of melt pool top surface for Inconel 625 deposition. Motion velocity 2 cm/s (solid), 2.4 cm/s (dot), and 1.6 cm/s (long dot), laser beam radius on the surface 2 mm, beam power 1770 W, powder jet diameter 3 mm, powder mass rate 3 kg/h, absorption coefficient 0.7.

Boundary conditions for this case can be written as

$$-\lambda \frac{\partial T}{\partial z}\bigg|_{z=0} = q(x), \ T|_{z=\infty} \to 0,$$

where $\lambda$ and $\chi$ correspond heat transfer and temperature transfer coefficients.

For its solution, we can, as given in Ref. [6], introduce a new unknown function $\Theta$ such as

$$T(x, z) = \Theta(x, z) exp\left(\frac{vx}{2\chi}\right).$$

In this case, for $\Theta$, we obtain the Helmholz equation:

$$\frac{v^2}{4\chi^2}\Theta = \frac{\partial^2 \Theta}{\partial x^2} + \frac{\partial^2 \Theta}{\partial z^2},$$

with boundary conditions of second kind: $-\lambda \frac{\partial \Theta}{\partial z}\Big|_{z=0} = q(x) exp\left(-\frac{vx}{2\chi}\right)$, $\Theta|_{z\to\infty} \to 0$.

To solve this Helmholz type equation, it is possible to use a Fourier integral transformation. After several calculations, one can obtain

$$T(x, z) = \frac{exp\left(-\frac{vx}{2\chi}\right)}{\lambda} \int_{-\infty}^{\infty} q(x') exp\left(-\frac{vx'}{2\chi}\right) K_0\left(\frac{v}{2\chi}\sqrt{z^2 + (x - x')^2}\right) dx' + T_h, \tag{17.15}$$

where $T_h$ is a value of initial temperature after previous cladding path. This expression gives the temperature field in the active zone in DLD process but neglecting the effect of the latent heat of melting. An example of calculation with this formula for distribution of superficial temperature along the melt pool surface is shown in Fig. 17.4.

Calculations show that relatively small (from 1.7 to 2.1 kW) increments in the laser beam power lead to overheating of surface to evaporation. This produces the formation of a cavity on the surface and a bed melt pool. With the determination of $T = T_m$, we can transform the Eq. (17.8) into an equation for the distribution of the melting depth with respect to the melt pool length. Results of the numerical solution of this equation are shown in Fig. 17.5.

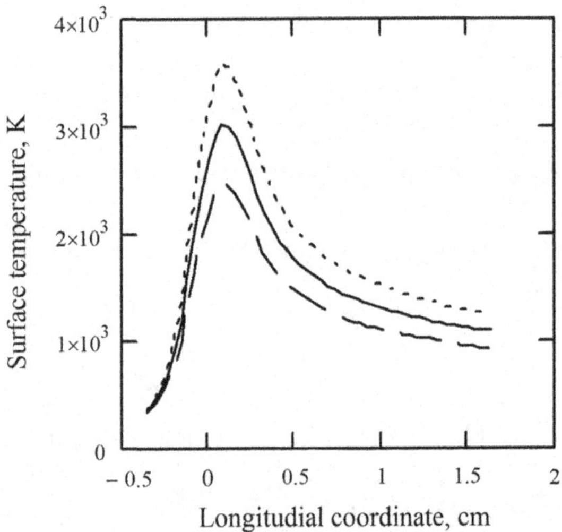


**Figure 17.4** Temperature distribution along melt pool length for Inconel 625 deposition. Motion velocity 2 cm/s, laser beam radius on the surface 2 mm, beam power 1770 W (solid), 2130 W (dot), and 1420 W (long dot).

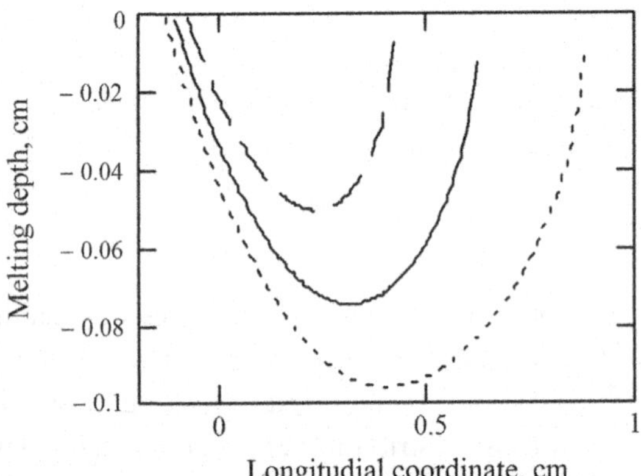


**Figure 17.5** Melt pool shape for Inconel 625 deposition. Motion velocity 2 cm/s, laser beam radius on the surface 2 mm, beam power 1770 W (solid), 2130 W (dot), and 1420 W (long dot).

This example shows that relatively small (from 1.7 to 1.4 kW) reductions of the laser beam power lead to a case where the melting depth became lower than the melt pool thickness, which was in these experiments equal to 0.6 mm. This leads to the appearance of faulty fusion between layers.

Taking into account that on the melt pool surface, $z = 0$, it is possible to get an equation for the determination of the melt pool length "$L$":

$$T_m = \frac{exp\left(-\frac{vL}{2\chi}\right)}{\lambda} \int_{-\infty}^{\infty} q(x')exp\left(-\frac{vx'}{2\chi}\right) K_0\left(\frac{v(L-x')}{2\chi}\right) dx' + T_h. \quad (17.16)$$

Numerical solution of this equation allows to find the length of the melt pool in dependence of material and processing parameters. Example of such solution is shown in Fig. 17.6.

Now this value, together with melting depth "$H$," can be used in the hydrodynamic model as input parameters. After that, improved value of

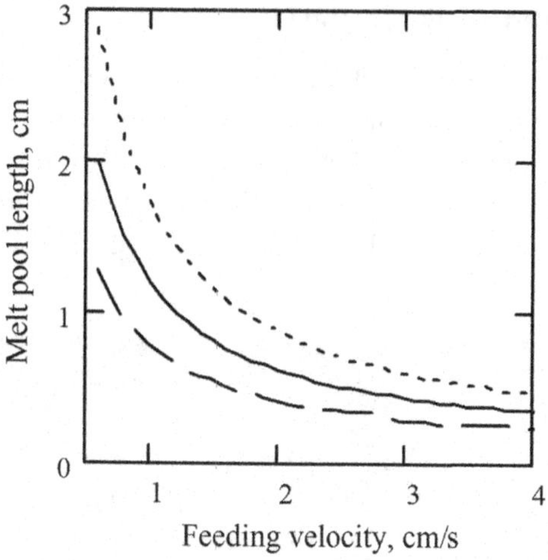


**Figure 17.6** Melt pool length for Inconel 625 deposition. Motion velocity 2 cm/s, laser beam radius on the surface 2 mm, beam power 1770 W (solid), 2130 W (dot), and 1420 W (long dot).

melt velocity "$v$" can be used in heat transfer task. Mathcad 15 was used for numeric solutions of these model equations. Calculations show that after three to four iterations, all values became stable, so that such numeric procedure allows to get self-consistent solution of heat transfer and hydrodynamic tasks for size and shape of melt pool in DLD process.

### 17.1.4 Model implementation

The developed model has been used for the determination of DLD processing parameters at the Institute of Laser and Welding Technologies SPbSMTU (ILWT) during preliminary tests of new large industrial installation for fabrication of fan body ring with diameter 2100 mm and same scale body of turbine support for prospective aircraft engine (Fig. 17.7). Also the model has been used as a heat source model for simulation of stress distribution for the parts. Comparison of simulation results with experimental data has shown mismatch less than 20%, which is quite good for fast semianalytic process model.

## 17.2 Residual stresses and distortion in direct laser deposition of large parts

DLD process is based on multiple local heating events in the manufactured parts, up to the melting temperature and then cooling them down. The temperature distribution is highly nonuniform, both in spatial coordinates and in time. Areas close to the deposited layer are heated up to several thousand degrees Celsius and then cooled down. The heat is conducted

**Figure 17.7** Production of fan body.

into the bulk of the buildup part. The transient and residual stresses developed are based on the volumetric changes of the part during its production. With the heat propagating through the body and the temperature equalization, the metal redistribution continues. Kinetics of the stress field is a process of stress development lasting the whole period of buildup. The stress kinetics investigation is quite a complicated task because of the numerous variables affecting the process. The main factors affecting the stress–strain fields are the shape and the size of the buildup, the process parameters, the thermomechanical and the thermophysical properties of the deposited material, and the interlayer dwell time. The study of the origin of the stress field is essential for solving a number of problems in additive manufacturing, including the assessment of cold and hot cracking, prediction of the fatigue resistance, and the stress corrosion cracking.

Although there are several techniques that can measure residual stresses, many of them are destructive techniques that can be used on academic samples, but not on industrial parts. Some of the simpler methods to apply, such as center hole drilling, only measure the residual stress at the surface, giving no information about the residual stresses in the bulk of the buildup. The most accurate methods, such as neutron diffraction, are very expensive research techniques that are not applied commercially. A solution to this problem is a numerical simulation. It helps to understand and analyze the effects of the different parameters on the transient stress and strain fields in the three-dimensional geometry. Making use of the similarity between the additive manufacturing and the fusion welding processes, more precisely, the cause of the resulted stresses and distortion, well-developed approaches of computational welding mechanics are used for the simulation of the DLD process.

### 17.2.1 Residual stresses in direct laser deposition of large structures

Most of the reported studies [7–11] have been focused on the measurement of residual stresses in buildups deposited on a flexible substrate practically without dwell time between deposited passes/layers. During fabrication of large-size parts, there is enough time for cooling of the buildup down to 70–100°C before the deposition of each layer. It leads to high-temperature gradients in a narrow area including several layers near to the top of the deposited pass/layer. Usually, a rigid substrate is used to prevent additional displacement of the buildup due to the bending of the substrate. Thus, to use small academic samples to study stress distribution in large-sized parts, the rigid substrate and pronounced dwell time must be used.

According to this, a four-pass 50-layer wall of Ti–6Al–4V was deposited on the edge of a 12-mm-thick plate. The mechanical properties of the Ti–6Al–4V at room temperature are the following: yield stress of 1100 MPa and an elastic modulus of 111 GPa. The layer height resulted in 0.56 mm leading to a total build height of 28 mm and a length of 70 mm. After deposition of each pass, the buildup was cooled down by natural convection and heat conduction to the temperature close to 60–80°C. Simulation and neutron diffraction study revealed that the longitudinal tensile stress is concentrated on almost the entire length of the several last layers near the top of the buildup (Fig. 17.8A). During the deposition of a pass, the narrow area, including the deposited metal and heat-affected zone (HAZ), tends to increase its volume due to the thermal expansion. Since the surrounding cold metal has higher rigidity, preventing the expansion of the heating area, transient compressive longitudinal stress and plastic strain appear in this area. During the redistribution of the thermal strain during cooling, the plastically deformed zone cannot lengthen freely. As a result, the compressive stress is replaced by tensile stress, and compressive transient plastic strain is partially compensated by the tensile plastic strain. Near the edges of the buildup, there is a region where all the three components of the stress field are tensile, and normal stress significantly exceeds the yield stress. Moreover, normal plastic strain in this area is also tensile and reaches up to 2.5%–3%. In the case of large dwell time as during the fabrication of large-scale parts, only a narrow area of buildup undergoes high-temperature heating followed by rapid cooling. This leads to the formation of a banded microstructure with low ductility. The combination of low ductility and the presence of defects, such as pores and lacks of fusion, significantly increases the probability of fracture at the beginning and the end of buildup.

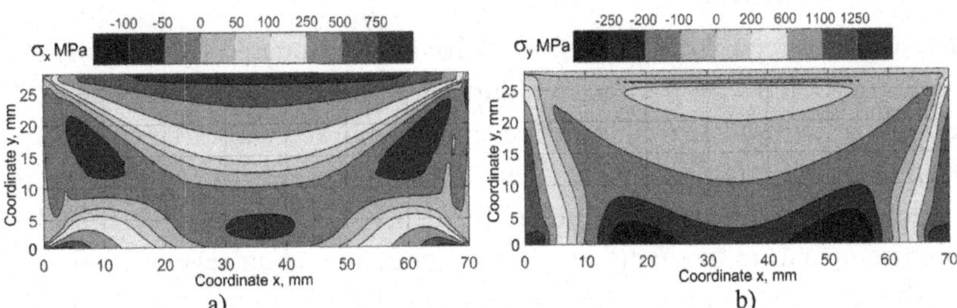


**Figure 17.8** Longitudinal $\sigma_x$ (A) and normal $\sigma_y$ (B) residual stress in the built-up wall. (Rigid substrate is not shown).

The nature and pattern of residual stresses in real parts are similar to the previously described sample. The highest tensile stress corresponds to the normal stress, i.e., stress in the direction of build-up. In the case of the axisymmetric parts, it is attended just near the substrate around the perimeter of the buildup. During the fabrication of arbitrary-shaped parts, normal stress is peaked along the ribs. For example, it can be seen in the DLD of the box-shaped shell (Fig. 17.9C). The part of IN625 alloy of size 500 × 100 × 300 mm (length × width × height) was deposited on a rigid substrate. Each layer has a cross section of 2 mm in width and 0.8 mm in height. Tensile longitudinal stresses oriented in the direction of movement of the processing head in plane of the deposited layer are peaked near the top of the built-up part (Fig. 17.8A and B). However, thickness stress is highly dependent on the thickness of the part wall. In the case of moderate thick parts, it has a negligibly small value.

### 17.2.2 Distortion of large structures during direct laser deposition

The thermal expansion and contraction stresses in the deposited layer (which are initially greater than the yield strength) cause localized yielding to occur around the layer. Permanent and significant deformations may result from this localized yielding, known as distortion. In thinner section parts, or in alloys with a low yield strength or large thermal expansion coefficient, the residual stresses are sufficient to significantly distort the whole buildup. Buckling distortion occurs when compressive residual

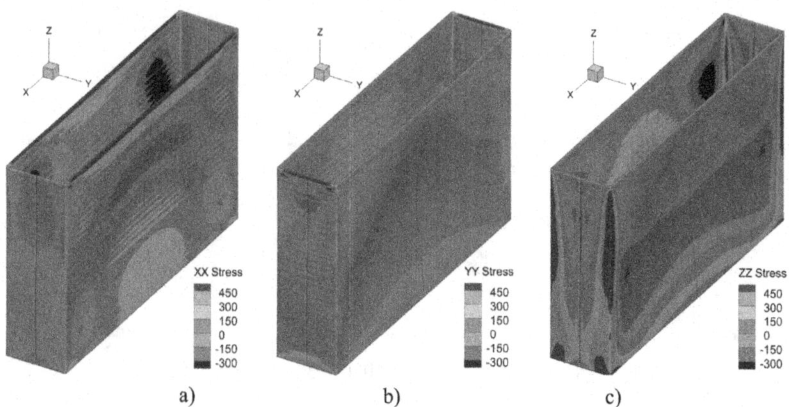


**Figure 17.9** Residual stress field in built-up box-shaped shell. (Rigid substrate is not shown).

stresses generated by deposited layers exceed the critical buckling strength of the buildup. Buckling distortion is one of the hardest distortion modes to control. Furthermore, once buckling distortion occurs, the magnitude of the distortion is very large.

Contrary to the residual stresses, the pattern of the distortion field is highly dependent on the shape and size of the fabricated part. As an example of axisymmetric part, the buildup of an aircraft engine part of Ti–6Al–4V alloy consisting of a 7.5-mm wall thick cylinder of inner radius 977 mm and the height 250 mm with two flanges 10.5 and 13.5 mm thick was studied. To reduce the magnitude of normal stresses near the substrate, a flexible substrate was used. An 8-mm-thick titanium substrate was fixed on a 20-mm-thick steel plate, which in turn was attached directly to the rotary table. Simulation of the buildup using the original computer-aided design (CAD) model showed that the cylinder wall is significantly distorted (Fig. 17.10). The peak radial displacement $U_r = -10.8$ mm is achieved at a distance of 150 mm from the bent substrate. It is seen that essential change of wall curvature occurs over a length of approximately 40 and 50 mm near the substrate and the top of cylinder, respectively. Between these areas, the wall is almost straight (radial displacement varies

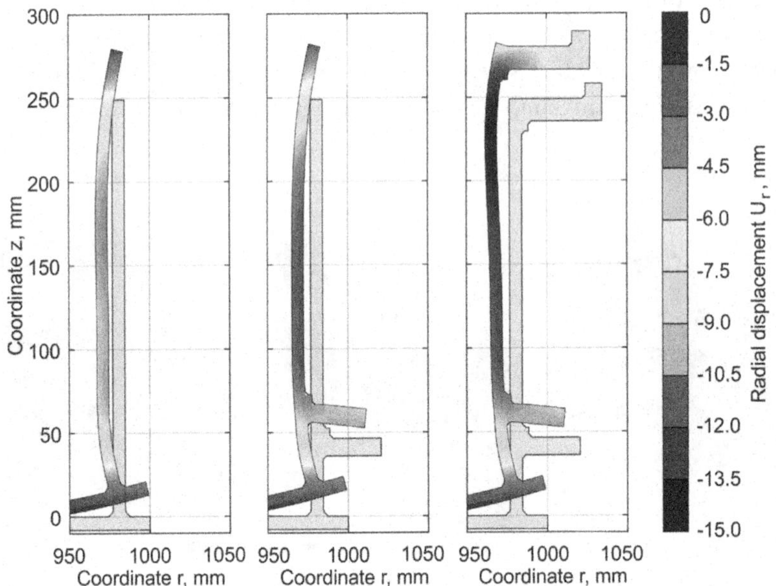


**Figure 17.10** Simulated buildup shape and radial displacement field during the fabrication of the part using original CAD model. *CAD*, computer-aided design.

within ±1.5 mm). Deposition of the bottom flange results in its significant displacement along *z*-axis. The angle between the center cross section of the flange and the horizontal axis is minus 5.8 degrees corresponded to the vertical displacement of the flange end by 2.9 mm. Such an effect is not observed during processing the top flange. Significant shrinkage forces generated during deposition of the bulky top flange lead to the bending of the upper part of the cylinder wall. Thus, the value of a peak radial displacement increases to minus 15.6 mm, and its location is shifted to the top of the buildup by 45 mm. The final shape and dimensions of the part manufactured using the original CAD model have an unacceptable distortion.

In the case of previously described box-shaped shell, the sidewalls have a different stiffness that depends on the size and mechanical properties. The low stiffness of the lengthy wall causes its buckling (Fig. 17.11B). The clear visible bulges of different sign and magnitude arise. The shorter in length wall has a different pattern of the bulges of a rather lower magnitude due to higher stiffness (Fig. 17.11A).

### 17.2.3 Distortion control and compensation in direct laser deposition

There are several methods for controlling distortion in additive manufacturing. Some methods may reduce one distortion mode but may increase another. Buckling can be eliminated by ensuring that the compressive longitudinal residual stress is lower than the critical buckling stress of the buildup. This can be achieved by one of the following methods: increasing the critical buckling stress of the part and reducing the residual stress. Increasing the critical buckling stress in a part involves

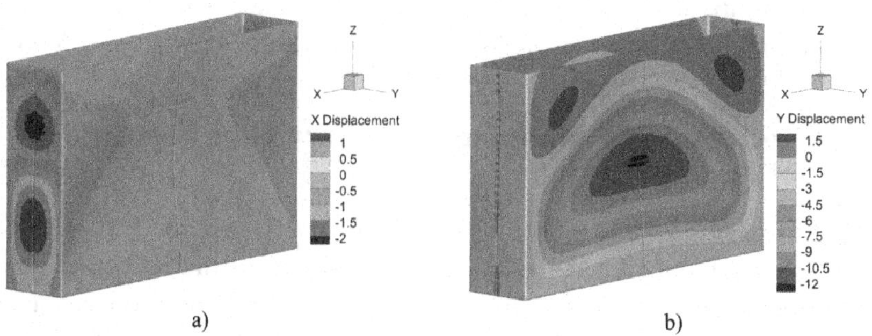


**Figure 17.11** Residual displacement field in built-up box-shaped shell. (Rigid substrate is not shown).

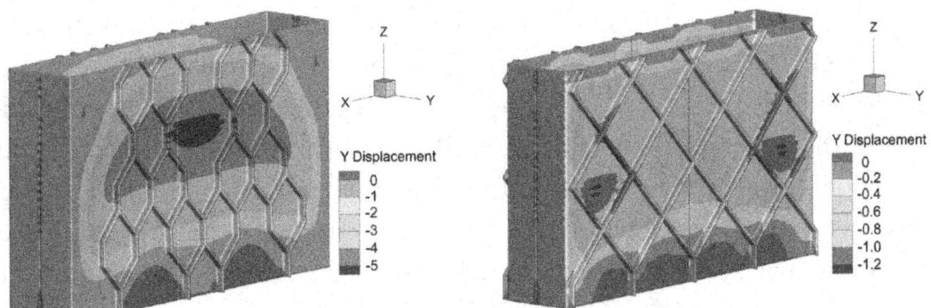


**Figure 17.12** Residual displacement $U_y$ field in the stiffened built-up box-shaped shell. (Rigid substrate is not shown).

modifying the design of the part by either increasing its thickness or adding stiffeners. The bulging of the sidewalls of the box-shaped shell can be eliminated using stiffeners (Fig. 17.12). The effect of the stiffeners depends on its shape and pattern.

Another approach is to compensate distortion using an initial predistorted CAD model. The final shape of the part manufactured using the original model is required to do this. The distortion compensation consists of inverting the deformations predicted by FE simulation followed by modifying the original model. Let us illustrate the effectiveness of this approach using described above buildup of axisymmetric aircraft part. The correction of the CAD model was made according to the simulated residual distortion field shown in Fig. 17.10. The predistorted shape (Fig. 17.13A) of the cylinder wall takes into account both shrinkage and deformation associated with buildup of the flanges. Values of peak radial displacement during manufacturing of the part were the following: cylinder $U_r = -11.5$ mm; bottom flange $U_r = -12.4$ mm; top flange $U_r = -15.4$ mm. Comparison of the simulated shape of the part with experimentally obtained has showed a satisfactory agreement (Fig. 17.13B). The average deviation is $\pm 1.5$ mm, and the peak deviation observed near the substrate is 3.5 mm.

## 17.3 Technological equipment in direct laser deposition of large parts

DLD process is based on laser deposition of metal powder. Part is made through combining single beads, placed according to the tool trajectory. Commonly DLD machine consists of a manipulator (based on robot or linear axes), a processing head, laser source, powder feeder, process

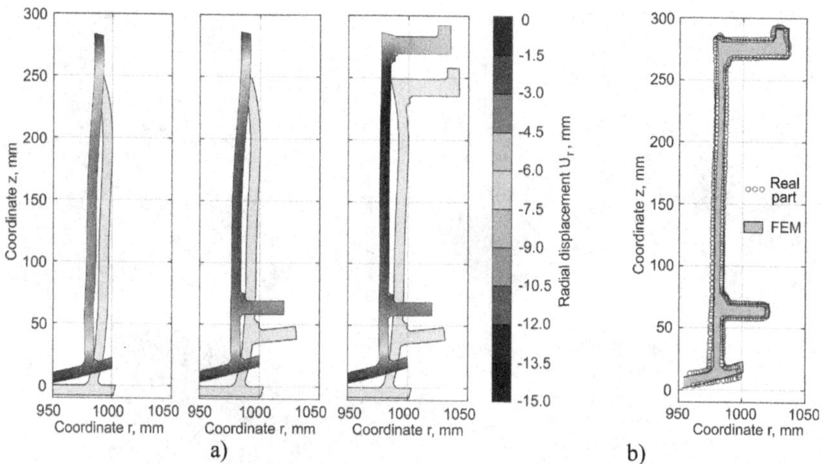


**Figure 17.13** Simulated buildup shape and radial displacement field during the fabrication of the part using predistorted CAD model (A) and comparison of measured and simulated shape of manufactured part (B). *CAD*, computer-aided design.

chamber, and additional periphery equipment. Manipulator is used to move the processing head relative to the workpiece (Fig. 17.14).

Manipulator workspace defines the maximum size of the workpiece. Nowadays, CNC machines are used for parts less than 1000 mm diameter and 600 mm height. For bigger parts, robot-based approach is more common. The number of interpolated axis is usually from 5 up to 8. Processing heads are used for deposition of metal powder and consist of two main parts—laser head, which focuses emission on treated surface, and a powder nozzle, which brings powder into melt pool and provides local gas shielding. Laser is used to provide energy for melting of metal and has a maximum power from 250 up to 5 kW. Deposition productivity is limited mainly by the maximum laser power, so deposition of large parts needs power of about 2 kW or more. Powder feeder is also a very important part of the machine, which provides a reliable feeding of metal powder into the nozzle. The last important part of DLD machines is a process chamber. It has several functions—first to protect personnel from laser radiation and metal powder and second to provide inert atmosphere for preventing part oxidation. Manufacturing of large parts means that appropriate chamber size is needed, but on the other hand, large chamber needs more argon for purging.

DLD machine is designed according to features of technological process. Focusing on large parts brings additional process complexity and equipment

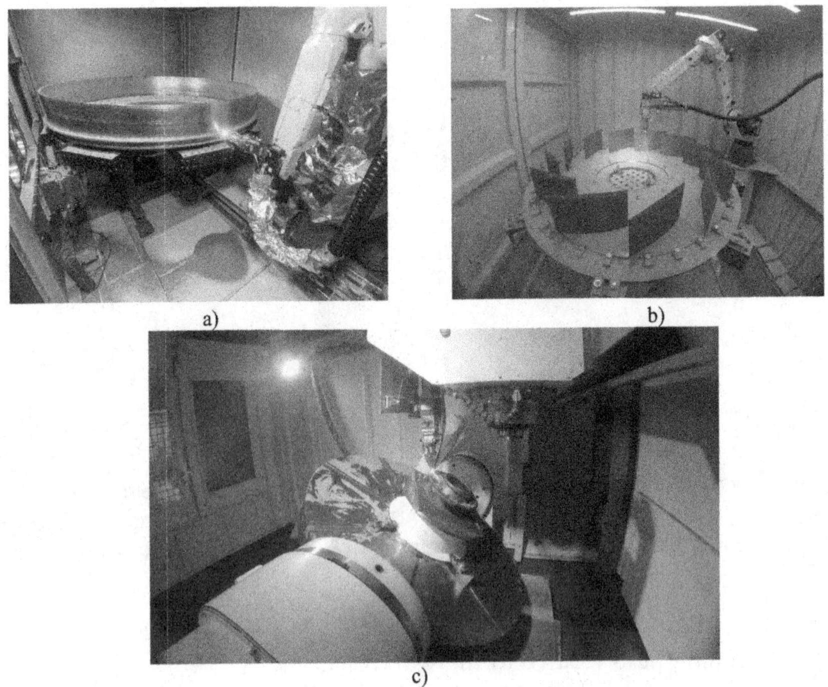


Figure 17.14 (A), (B), (C) Interior decoration of the machine.

requirements. Deposition of large parts needs productivity of 0.2–1.5 kg/h which can be achieved with laser power of 800–4000 W. Even with such productivity, it takes up to 200–400 h on nonstop deposition for 100+ kg part. DLD machine must be capable of maintenance-free work with no breakdowns for a long time.

### 17.3.1 Equipment reliability

There are several parts, which can break down during long processes, and all of them should be designed in such way that deposition process will not be aborted. First of all—the manipulator must be dustproof for metal powder, and it is also necessary to take into account that during deposition, a very fine metal dust is produced through condensing of evaporated metal. This condensate has a particle size less than 2.5 μm and can clog moving parts with unappropriated protection. Another issue is that this process uses a lot of powder which is dispersed inside process chamber. It brings additional heating of manipulator, influencing its accuracy and reliability. Passive cooling may not be enough if take into account that commonly used for inserting Argon gas has 30% less heat capacity than air.

Reliability of processing head is vital for process stability. There are several problems in DLD process; some of them are discussed as follows:

**(a)** Head overheating due to high average laser power, increased ambient temperature, and heating through thermal radiation of deposited part: Overheating brings thermal distortion in lenses and cover slides, which can affect spot size and power distribution. If heating is not so critical—and monitoring system does not shut down whole machine, it can change the melt pool geometry and bring lack-of-fusion defects, powder sticking on side surfaces of part, and changing of powder catchment ratio.

**(b)** Improper shielding gas flow or nozzle design can provide insufficient protection for cover slide, so during process, some pollution on the protection glass occurs. If polluting level is high—it can be noticed by head monitoring system, so operator is able to change cover slide and restart the process. This is not critical for part deposition but should be taken into account and reworked after process completion. Problems that are more serious occur if protection glass pollution is not critical and is not detected by head monitoring system. This can bring glass overheating and appearing of thermal lensing effect, which changes spot size and consequently melt pool geometry. Problems are same to head overheating: lack-of-fusion, powder sticking, and worse process stability due to changed powder catchment ratio. Improper shielding gas flow also can bring sticking of molten metal droplets to the nozzle. Mechanics of their formation is following: If nozzle tip is soiled, some powder particles can stick to the surface. This is frequently acquiring on nozzle tip, near central channel, where zone of turbulent shielding gas flow is formed. These particles are heated through reflected laser radiation and thermal radiation from melt pool, and after they are melted, they "gather" surrounding particles, reflected from workpiece. After a while, a melted droplet is formed, which can fall down and spoil workpiece. Two main factors influence this process—central channel design and nozzle soiling.

**(c)** Last, but most important—nozzle degradation during process. Powder feeding nozzles are made in different designs—with coaxial slit or with discrete injectors, but anyway, pure copper is the most commonly material used for producing nozzles. It has outstanding thermal conductivity, so heat can be transferred to a water cooling system. On the other hand, copper has very poor wear resistance, so focusing ability may deteriorate over time. If this happens during process, it can influence process passive stability and even stop deposition process. Even if this is noticed by personnel and adjusted by feed rate increasing, these manual adjustments influence process repeatability, which is very bad in terms of multiple parts productions.

Previously mentioned problems are most vital, but they do not make an exhaustive list. It is also necessary that laser power should be stable over time, powder feeder will not wear down too fast and so on. But these factors are simpler to prevent—just by using adequate equipment, produced by reliable companies. Factors "a", "b," and "c" are "process-specific" and should be taken into account.

### 17.3.2 Process passive stability

Ability to build 3D parts in DLD process is based on equality of bead height and manipulator Z step. In real world, there are a lot of fluctuations that affect bead geometry—instability of process parameters, such as speed, laser power, spot size, powder flow rate, substrate temperature, cooling rates, and so on. These parameters can vary through time—for example, when starting at cool substrate the melt pool will be small and after some passed accumulated heat get melt pool wider, so single bead width is rising. That is why the utilization of a precalculated set of processing parameters, optimized for powder catchment efficiency, which is common in laser cladding process, will lead to failure, when bead height will become less then manipulator Z step. Common approach is to use process passive stability to maintain bead height exactly equal to Z step [12–14]. This is done by feeding excess powder through the nozzle, so working distance between nozzle and substrate is stabilized at distance, less then focal distance of powder nozzle. This extra powder makes process stable, so if working distance for some reason appears to be slightly bigger, then powder catchment ratio will rise and bead height will rise. After several passes, working distance will stabilize again.

Maintaining passive stability is key for deposition of large parts; it can provide fully automatic deposition process without any influence of machine operator. It also brings best surface finish and defect-free parts. The so-called stability margin depends on the amount of excess powder and must be adjusted according to overall process stability. On the other hand, excess powder means that overall powder efficiency is not maximized, and part cost rises. On some parts, powder losses can be counted in hundreds of kilograms. Key to efficient deposition process is to search and minimize process instabilities and consequently amount of excess powder, holding process stability margin at comfort level. This can only be done by improvement both equipment and technology. Several common ways are described in the following:

**(a)** Constant powder feeding rate is vital for stable process. Most of industrial grade powder feeders provide only volumetric feeding, which is very inaccurate and may vary not only with different powders, but also within same powder during constant feeding. For example, standard rotary disk feeder can vary mass flow rate in a range of ±10% depending on the amount of powder, filled in bulb. This means that about 10% of extra powder should be fed to prevent process instability. Solution to this problem is transition to mass feeding devices.

**(b)** Development of nozzle efficiency and durability: Powder feeding nozzle design is key to overall efficiency. Proper nozzle should provide 80%+ catchment ratio at chosen bead width. But high catchment ratio can be only realized if nozzle design is durable and nozzle does not degrade during deposition process. Any reduction of efficiency brings process instability and should be prevented by increasing of excess powder amount. Frequent nozzle measurement can provide necessary information about catchment ratio, so feed rate can be adjusted, or nozzle replaced.

**(c)** Stable bead width is also very important for process stability. If width is decreased, then powder catchment radio is also dropped down, which can lead to process instability. For example, changing of bead width in 0.2 mm (for 2.5 mm bead) changes amount of deposited material in 5%. This means that if there is any possibility that bead width can rise up in 0.2 mm (f.e. due to heat accumulation), then powder feed rate should have addition 5% of excess material. Technically, this problem can be solved via melt pool monitoring system that will adjust process parameters to maintain constant bead width.

As shown before process stability is key aspect of DLD process. Equipment designed to provide stable, controlled process will also demonstrate perfect powder catchment efficiency, minimal number of defects, and predictable part quality.

### 17.3.3 Process economics

Talking about passive stability another issue of DLD process should be described. Production cost of parts must be reduced as much as possible.

Cost calculation depends the on country and company standards, but some common principles can be described:

**(a)** Powder utilization is ratio of finished product weight to weight of spent powder. Cost of metal powder makes up a significant part of total part cost, so amount of powder, used for part deposition, should be reduced as much as possible. Powder utilization depends on catchment efficiency of nozzle, amount of excess powder providing passive stability, and amount of idle movements. As first two parameters were described earlier, idle movements should be reviewed. The motion of the processing head consists of trajectory segments, during which laser is emitted and deposition process is provided. These segments are connected via links—idle passes when the laser is turned off. Also due to finite acceleration of real manipulator, segments have so-called lead-ins and lead-outs to provide turning laser on and off in movement. In real process, from 10% up to 15% of the powder can be wasted on idle movements. Some of this material can be saved by using powder switch, situated at technological head. Powder switch can redirect powder to hopper when emission is turned off. In conclusion, powder utilization of about 80%–85% can be achieved through proper equipment and process adjusting

**(b)** Another important issue is machine productivity, as machine time and labor also make up a large part in part cost. Deposition of 100+ kilogram parts means that reasonable productivity should be at least 0.5 g/h. Increasing of productivity can decrease part cost in tens of percent, but it commonly connected with increasing of laser power, which leads to higher heat impact on machine. Reasonable approach is designing machine according to target products, so chamber size and laser power can be matched. In terms of equipment, usage of laser head with programmable spot size can increase productivity. Usually, laser spot size is fixed and can be adjusted only manually, so whole part can be deposited only with preselected bead width. Programmable spot size can provide adjusting of bead width according to part geometry, for example, thick parts can be deposited with large beads and high productivity, and on thin walls, spot size reduces to provide thin walls and good surface finish.

**(c)** Talking about process economics, it also should be considered how many attempts are needed to produce valid parts. DLD is a welding-based process, so residual stresses and distortion are common and should be considered. Two approaches are commonly used—one is "simulation-based" as described previously, and another is based on consistent approach to the required geometry through the number of part depositions. After each deposition, part is 3D scanned and trajectory rebuilds to minimize distortion. This approach can provide accuracy up to ±0.2 mm on 500 mm size parts. This brings new requirement to the equipment. DLD machine must provide outstanding repeatability of deposition process. Any slight change in the processing parameters, any breakdown, or maintenance during process can produce significant changes into the part geometry. It is necessary to understand that any part of equipment, flaw of technology, or improper personnel training will affect repeatability and reduce achievable level of precision. For example, if powder nozzle is degrading during deposition process, that powder catchment ratio will go down. If unnoticed, it will lead to loss of passive stability, so the deposition will stop. After that machine operator will make an attempt to restart NC program with rebuilding of part of trajectory. Part will be saved, but pause and changing of heat distribution will lead to changing of part geometry. And final geometry will differ from "no-problem" deposition, and consistent approach will fail.

As shown before, reliability of equipment, process stability, and overall powder utilization ratio are key properties of high-performance DLD machine. DLD of large parts is a difficult process, which makes highest demands to equipment, technology, and personnel. On the other hand, ability to produce complex large-scale flawless parts in matter of days will bring additive manufacturing to a whole new level.

## 17.4 Structure and properties of products obtained by direct laser deposition technology

The technology of DLD has shown high productivity and significant economic effect on high-alloy steels and on nickel and titanium-based alloys. However, due to extreme cooling rates, high internal stress levels, and high demands on the products, detailed elaboration of material requirements, selection of process parameters, and subsequent processing are required to avoid manufacturing low-quality products.

### 17.4.1 Requirements for powders

Structure formation in metal alloys at DLD process depends on the quality of the powder, properly selected processing parameters, volume heat investment, cooling rate, and subsequent thermal cycling, which is observed in obtaining large-size products. Powder materials used in DLD technology have the following requirements:

**(a)** The content of impurities, especially light elements, is minimally and strictly regulated, for the manufacture of large products; there are more requirements for powders than, for example, for materials for laser powder cladding.

**(b)** A powder with a high degree of sphericity and not inclined to agglomerate is required to ensure good powder fluidity so that the powder delivery process is not affected by the interaction between the powder and the machine parts.

**(c)** Foreign inclusions in the powder are prohibited, as this will inevitably lead to macrodefects in the product and unsatisfactory product characteristics.

**(d)** The number of satellites on the powder surface should be minimized, as satellites usually contain more light elements, as their increased content, as mentioned, will lead to poor characteristics.

Macrostructures obtained using poor quality powder are given in Fig. 17.15.

## 17.4.2 Macrostructure formation during direct laser deposition process

The selection of DLD process parameters includes many experiments to find the "process window." The appearance of macrodefects is usually

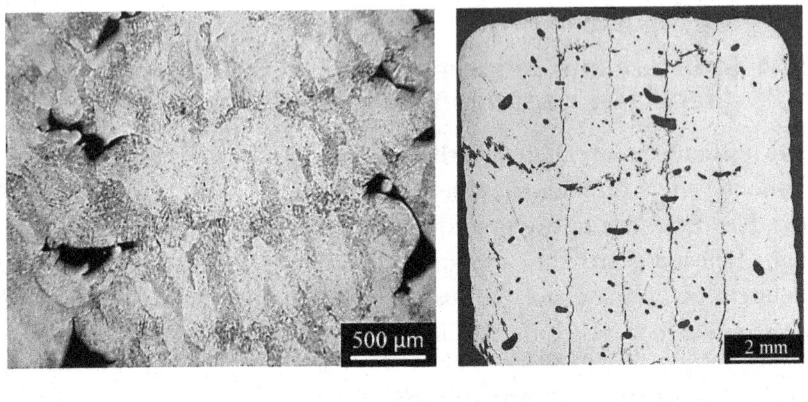


**Figure 17.15** Sample produced using low-quality powder (A); inappropriate process parameters, titanium alloy (B).

associated with improperly selected DLD parameters. The main macrodefects that can be detected are lack of fusion and pores, and less frequently cracks. Pores in the DLD process are almost always single defects, and their size is typically in the range of 100–300 μm. The appearance of pores can be a consequence of both technological and metallurgical defects (for example, humidity in the protective gas, and overheating of the melt). As mainly in the obtained products, pores have a single character of appearance, and special methods of struggle against them are not required. The maximum permissible pore size is 400 μm.

Lack of fusion usually occurs systematically, i.e., in multipass deposition, it appears in each layer above each other at the fusion boundary between adjacent tracks of the same layer. This macrodefect is primarily due to incorrectly selected process parameters but can also occur due to the presence of oxide films on the powder. Lack of fusion can also cause cracks. The presence of an accumulation of lack of fusion defects in obtained products leads to a drop in strength characteristics and almost zero plasticity [15–17]. Hot isostatic pressing can be used for removal of this defect from products.

Cracks may appear in the DLD products. The use of a low-quality powder containing foreign matter in the total mass of the powder or a high oxide content in the powder can lead to cracks (Fig. 17.15A). Nevertheless, primarily, it is due to the chemistry of the alloy and its tendency to appear hot cracks. In this case, it is necessary to select process parameters ensuring low welding stresses and minimal deformations associated with both shrinkage in the process of crystallization and deformations of the structure itself in the process of deposition [18]. Cracks are an unacceptable defect in DLD products, and their absence is strictly regulated.

As a defect of the macrostructure, we should also note a high degree of anisotropy of crystals in the direction of X and Z, the result of which is a drop in strength and plastic characteristics in the transverse direction relative to growth. For alloys with a tendency to anisotropy, for example, most steels and nickel-based alloys, it is impossible to determine a window of processing parameters to obtain isotropic structure during growth. This type of defect for DLD products is solved by selection of a complete heat treatment (HT).

Among surface defects, the surface roughness of the resulting wall plays an important role. In case of high roughness and presence of high level of

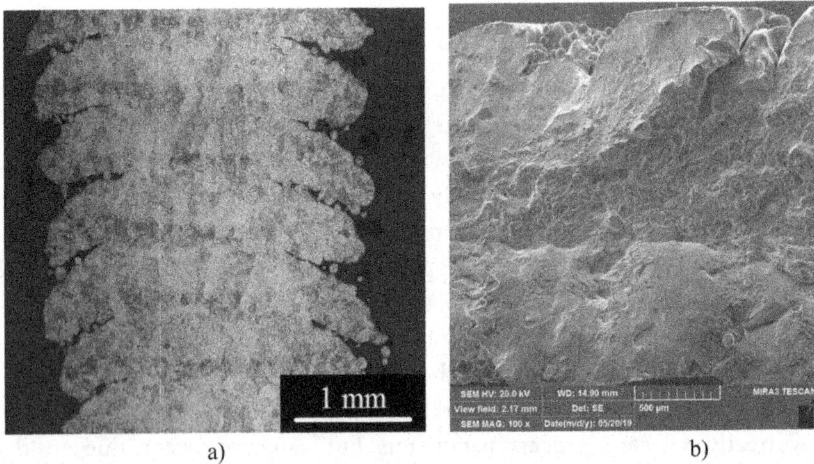


**Figure 17.16** Ti-64 alloy: produced wall with high roughness (A), brittle fracture of this wall (B).

welding tensions, cracking in the process of cooling of grown product can occur (Fig. 17.16). To avoid cracking, it is necessary to reduce the cooling rate and reduce the roughness, for example, by reducing the step layer.

### 17.4.3 Microstructure and properties

Evaluation of the influence of DLD parameters on the microstructure should be carried out in a complex, as in real conditions of produced items change several parameters at once, trying to maintain a satisfactory fragment formation and temperature regime of deposition. For alloys that have martensitic transformation, the microstructure of the alloy is in nonequilibrium state, such as steel and titanium alloys, which leads to high strength characteristics, but low plasticity. These alloys require HT for higher ductility. For most welded and limited-welded nickel-based alloys, the microstructure in the DLD state, in contrast, is a homogeneous solid solution with minimal hardness, which provides high ductility but low ultimate strength.

It should be noted that the selection of processing parameters for production of large-size items is primarily focused on the satisfactory formation of the product and the high productivity of the process. Attempts to receive equilibrium structure in large-sized product after DLD process led to productivity decrease.

Thus, in 95% of cases, it is necessary to carry out complex thermal processing aimed at removal of welding tensions, removal of anisotropy, obtaining of microstructure in equilibrium state, and strengthening of microstructure due to separation of dispersed phases. Since the HT presented in the handbooks is primarily focused on castings and rolled products, it is necessary to adjust the temperature of each operation, the holding time, and the cooling rate for products in the DLD state. For some alloys, HT modes have been developed (Table 17.1).

Some materials (316L, Inconel 625, Ti−5Al−2V) after the DLD process have the same mechanical characteristics as the materials in rolled or heat-treated casting. From the point of view of using such materials in DLD, they are the most successful choice, as they do not require additional technological operations on finished products. These alloys have good weldability and can be used without additional HT, but such materials are rather an exception to the rule.

### 17.4.4 Heat treatment of direct laser deposition products

Most steels require a complete HT after the DLD process. Due to the need for complete maintenance, it makes no sense to do a stress-relief annealing. The HT includes a homogenization to remove anisotropy in the structure, subsequent hardening and tempering. For example, 09CrNi2MoCu steel in its original state after deposition has anisotropic and rough structure and a plasticity difference of more than twice in the longitudinal and transverse directions.

After homogenization, the characteristics are leveled in two directions (Fig. 17.17), but now it is necessary to carry out hardening with cooling into oil and the subsequent high tempering; as a result, we get a bainite structure combining high strength and satisfactory plasticity.

Most of the nickel alloy parts have been developed considering the need for subsequent HT: hardening to a solid solution and aging. The example of Inconel 718 shows that a material with good ductility but a very low ultimate strength has been obtained in DLD state (Fig. 17.18).

For such alloys, processing on a solid solution at temperatures in the range of 1080−950°C is used to obtain a homogeneous solid solution and the subsequent double aging at temperatures of 720 and 620°C with periods of 8−10 h. As a result, of aging in Inconel 718 alloy, precipitation of disperse phases and intermetallic $Ni_3Ti$ and $Ni_3Nb$ is occurred.

**Table 17.1** Heat treatment (HT) modes for selected alloys.

| Materials | DLD/HT | Heat treatment | Yield strength, MPa | Ultimate strength, MPa | Relative elongation % | Impact tough., J | Hardness, HV |
|---|---|---|---|---|---|---|---|
| 316L | Rolled metal | | 205–310 | 515–580 | 40–55 | 160 | 165 |
| | X; Z | | 448 | 609 | 46 | 162 | 180 |
| 06Cr15Ni4 –CuMo | Cast | | 620 | 790 | 19 | 40 | 290 |
| | Z | | 858 | 1093 | 7.4 | 12.5 | 353 |
| | X | | 831 | 1089 | 8.5 | 14.8 | 353 |
| | HT | **1.** T = 1200°C/t = 6 h cooling in furnace<br>**2.** T = 1060°C/t = 3 h air cooling<br>**3.** T = 620°C/2 h × 2 air cooling | 766 | 832 | 14.5 | 28.8 | 260 |
| 08MnCuNiV | Cast | | 380 | 480 | 20 | 63 | 159–192 |
| | Z | | 447 | 524 | 7.7 | 21.3 | 210–230 |
| | X | | 463 | 550 | 18.6 | 44.2 | 210–230 |
| | HT | **1.** T = 1100°C/t = 6 h cooling in furnace<br>**2.** T = 940°C/t = 2 h air cooling<br>**3.** T = 940°C/30 min water cooling<br>**4.** T = 640°C/5 h air cooling | 417 | 503 | 23.5 | 56.7 | 244 |
| 09CrNi2MoCu [19–21] | Hot rolled metal | | 588–686 | 637 | 18 | 78 | 187–241 |
| | Z | | 563 | 626 | 9.7 | 29.8 | 212 |
| | X | | 515 | 686 | 21 | 57.5 | 212 |
| | HT | **1.** T = 1100°C/t = 6 h cooling in furnace<br>**2.** T = 920°C/t = 2 h oil cooling<br>**3.** T = 650°C/5 h air cooling | 585 | 660 | 23.2 | 130 | 220 |

| | | | | | | | |
|---|---|---|---|---|---|---|---|
| CrNi73MoNbTiAl | E | | 705 | 1150 | 16 | 39 | |
| | Z | **1.** T = 1160°C/t = 4 h air cooling<br>**2.** T = 1120°C/t = 8 h air cooling<br>**3.** T = 1000°C/ t = 4 h, air cooling<br>**4.** T = 775°C/ t = 16 h, air cooling<br>**5.** T = 700°C/ h = 16 h, air cooling | 779 | 886.4 | 38.6 | — | |
| | X | | 522.4 | 770.5 | 43.2 | — | |
| | HT | | 875.2 | 1256.9 | 16,4 | — | |
| Inconel 718 | ASTM | | 920–940 | 1240 | >12 | — | |
| | X; Z | **1.** T = 1050°C/1 h, air cooling<br>**2.** T = 720°C/8 h, cooling rate 50°C/h to 620°C<br>**3.** T = 620°C/8 h, cooling rate 50°C/h) | 339 | 452 | 19 | 63 | |
| | HT | | 1094 | 1230 | 16.3 | 31.5 | |
| In 625 [14] | ASTM | | 345 | 760 | 30 | — | |
| | X; Z | | 479 | 855 | 28 | — | |
| Ti–5Al–2V | OST 5P.907 1–88 | | 588.6 | 637.6 | 8.0 | — | — |
| | X; Z | | 745 | 823 | 12.2 | 72.5 | — |

*Continued*

**Table 17.1** Heat treatment (HT) modes for selected alloys.—cont'd

| Materials | DLD/HT | Heat treatment | Yield strength, MPa | Ultimate strength, MPa | Relative elongation % | Impact tough., J | Hardness, HV |
|---|---|---|---|---|---|---|---|
| Ti-64 [22,23] | ASTM B348 | | 810 | 950–970 | 10 | 40 | 300–360 |
| | X; Z | Annealing T = 900°C/ t = 2 h, cooling in furnace | 982 | 1047 | 6 | 27.7 | 420 |
| | HT | | 899 | 970 | 14 | – | 350 |
| Ti–6Al–1V –1Mo–2Zr [24] | GOST hot rolled metal 22178-76 | | – | 930–980 | 6–12 | – | – |
| | DLD | | 882 | 968 | 6.6 | | 420 |
| | DLD + HIP | | 885 | 954 | 9.7 | | 345 |

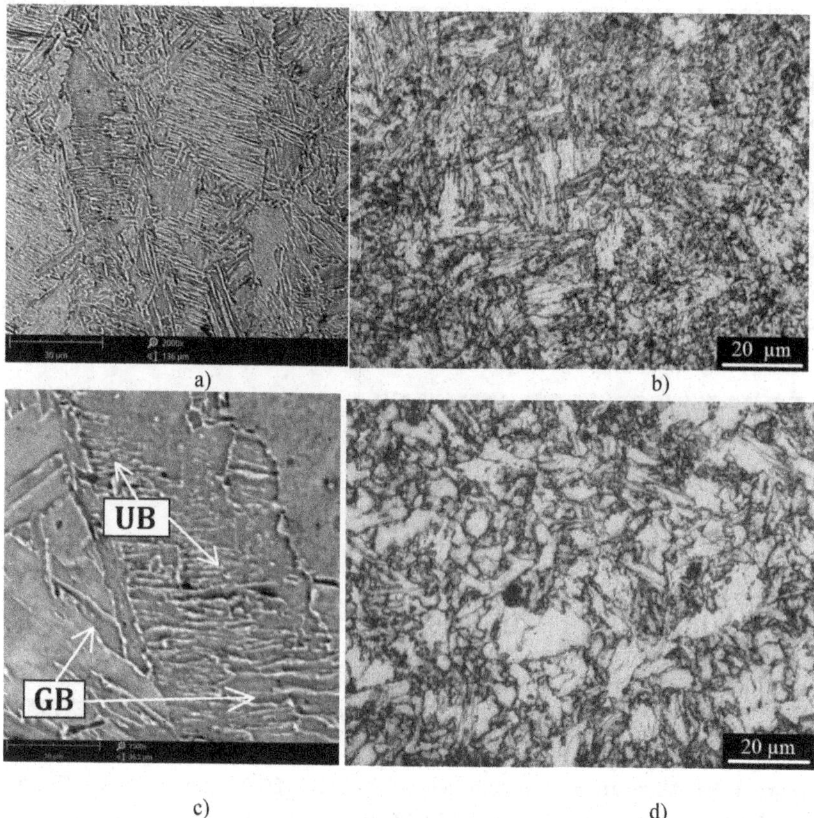


**Figure 17.17** Microstructure of 09CrNi2MoCu, (A) optical microscope (DLD—bainite); (B) (HT—bainite); (C), (D) electron microscope upper bainite (UB), granular bainite (GB).

Most $\alpha+\beta$ titanium alloys, which are most used in industry, have a martensitic structure at high cooling rates. In contrast to steels, martensitic structures in titanium have little ductility, but in equilibrium, the ductility will be almost twice as high at satisfactory yield strength and ultimate strength.

Titanium alloys do not require a homogenization after surfactant, as they do not have the harmful effects of dendritic liquation, as is typical for steels and nickel alloys. To obtain the equilibrium state for most $\alpha+\beta$ titanium alloys after DLD, it will not be necessary to carry out an aging quenching process, but it will be sufficient to carry out an annealing process

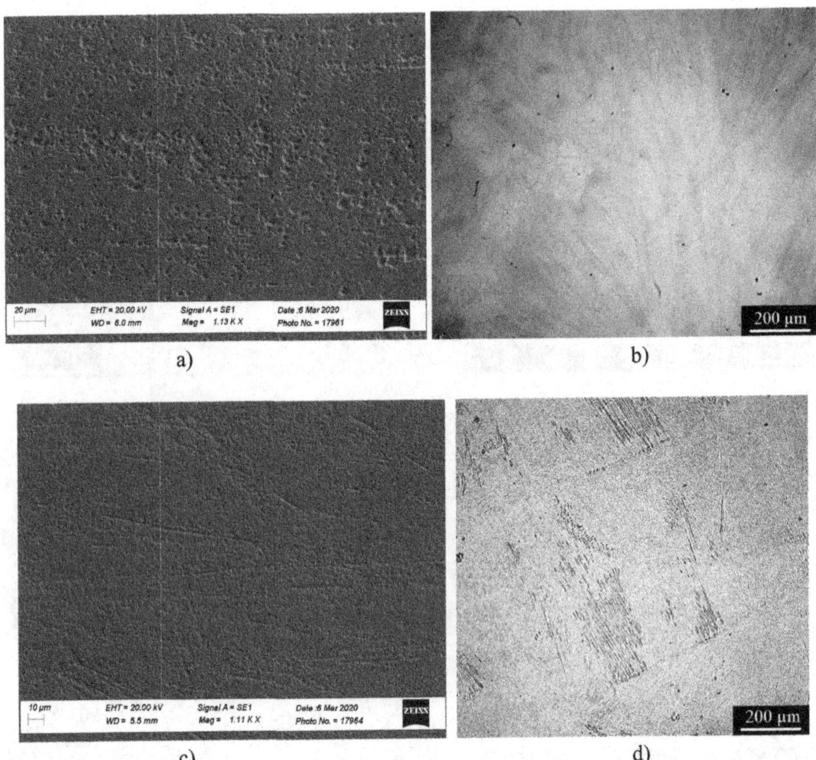


**Figure 17.18** Microstructure of Inconel 718: (A), (B)—DLD state, (C), (D)—HT state. *DLD*, direct laser deposition; *HT*, heat treatment.

at 700–950°C and withstands of 1–10 h. For Ti-64 alloy (grade 5 analogue) to obtain the equilibrium structure and, consequently, high mechanical characteristics, annealing at 900°C for 2 h with subsequent cooling in the air was performed (Fig. 17.19).

If after DLD the microstructure Ti-64 is martensitic, after annealing the equilibrium two-phase structure with the content of β phase up to 9 wt% was obtained. Some pseudoalpha alloys will require a long annealing period at 700–800 degrees to stabilize the microstructure, providing a larger equiaxial grain to increase ductility.

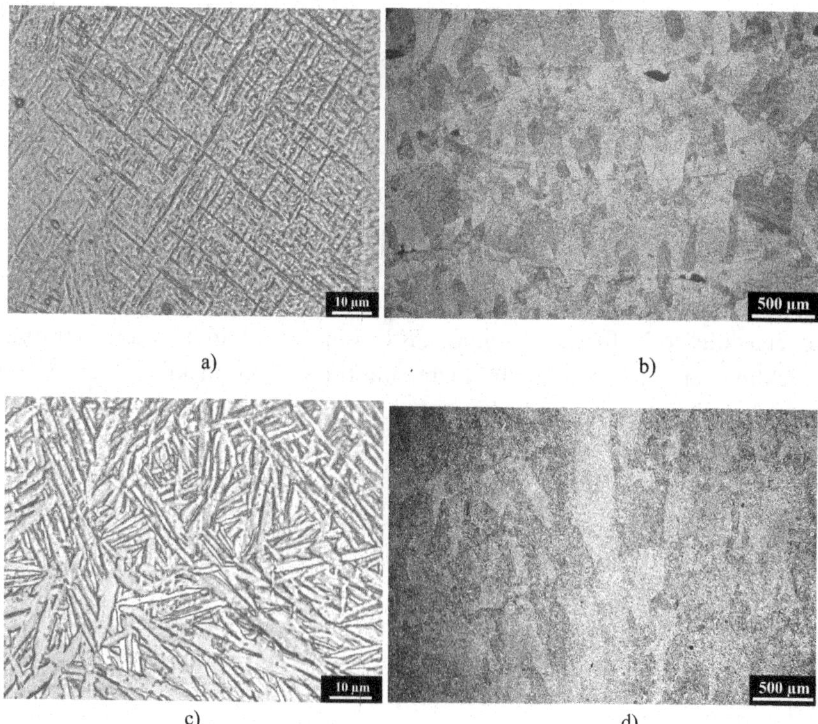


**Figure 17.19** Microstructure of Ti-64 (grade 5): (A), (B) DLD state, (C), (D) HT state. *DLD*, direct laser deposition; *HT*, heat treatment.

## 17.5 Conclusions

DLD is a complex process with many processing parameters involved, which can dramatically affect the result. Choice of the mode of fabrication is advantageously carried out with the help of mathematical modeling; experimental selection modes can be extremely time-consuming.

Semianalytical process model, which allows to calculate melt pool depth, length, and surface profile, based on connected solutions of heat transfer and hydrodynamic tasks, has been developed. The model input data include material physical properties, processing head velocity, laser beam power, and radius on the deposited surface, powder rate, and radius of powder jet. Very short calculation time less than 1 second makes this model convenient for technological use.

Simulation results show that for high productive technological modes, the process parameters window is very narrow; any modification of the parameters in a range of 20% can lead to dramatic changes of clad pad formation and the appearance of defects.

The developed model can be used both for the calculation of the active zone parameters and for the heat source modeling in the thermomechanical simulation.

There are two main critical challenges that impede the advancement of additive technologies for manufacturing of large parts. These are the residual stresses and the part distortion. In thinner section parts, the residual stresses are sufficient to significantly distort the whole buildup. There are several methods for controlling distortion in additive manufacturing. Buckling can be eliminated by increasing the critical buckling stress of the part or reducing the residual stress. Another approach is to compensate distortions using a predistorted initial CAD model.

DLD of large parts is highly dependent on the equipment used. Unreliable, inaccurate, and wearing machine makes the process unrepeatable. On other hand, proper equipment can provide perfect quality with outstanding accuracy and stability.

The formation of a satisfactory macro- and microstructure in the DLD process requires the use of high-quality powder materials and a large experimental study on the search for processing parameters.

Products made using DLD process mainly require HT. HT parameters may differ greatly from those developed for products obtained by classical metallurgical technologies. On heat-treated products, it is possible to achieve mechanical characteristics exceeding the values of hot-rolled materials and heat-treated castings.

## References


[1] G. Turichin, E. Zemlyakov, K. Babkin, Analysis of distortion during laser metal deposition of large parts, Phys. Proc, conference series 1 (2018).

[2] D. Boisselier, S. Simon, T. Engel, Improvement of the laser direct metal deposition process in 5-axis configuration, Phy. Proc. 56 (2014) 239–249.

[3] G. Turichin, E. Zemlyakov, O. Klimova, K. Babkin, Hydrodynamic instability in high-speed direct laser deposition for additive manufacturing, Phy. Proc. 83 (2016) 674–683.

[4] G. Turichin, E. Valdaytseva, E. Pozdeeva, E. Zemlyakov, Influence of aerodynamic force on powder transfer to flat substrate in laser cladding, beam technology & laser application, in: P.G. Turichin (Ed.), Proceedings of the Six International Scientific and Technical Conference, 2009, pp. 42–47. Saint-Petersburg.

[5] V.G. Levich, Physicochemical Hydrodynamics, Prentice-Hall, Englewood Cliffs, N.J., 1962.

[6] V. Lopota, Y. Sukhov, G. Turichin, Computer simulation of laser beam welding for technological applications, Izvestiya Akademii Nauk. Ser. Fizicheskaya 61 (Issue 8) (1997) 1613–1618.
[7] W. Woo, D.-K. Kim, E.J. Kingston, V. Luzin, F. Salvemini, M.R. Hill, Effect of interlayers and scanning strategies on through-thickness residual stress distributions in additive manufactured ferritic-austenitic steel structure, Mater. Sci. Eng. 744 (2019) 618–629.
[8] Z. Wang, E. Denlinger, P. Michaleris, A.D. Stoica, D. Ma, A.M. Beese, Residual Stress Mapping in Inconel 625 Fabricated through Additive Manufacturing: Method for Neutron Diffraction Measurements to Validate Thermomechanical Model Predictions, Mater Des, 2017, pp. 113169–113177.
[9] L. Sochalski-Kolbus, E.A. Payzant, P.A. Cornwell, T.R. Watkins, S.S. Babu, R.R. Dehoff, M. Lorenz, O. Ovchinnikova, C. Duty, Comparison of residual stresses in Inconel 718 simple parts made by electron beam melting and direct laser metal sintering, Metall. Mater. Trans. 46 (3) (2015) 1419–1432.
[10] V. Luzin, N. Hoye, Stress in thin wall structures made by layer additive manufacturing, Mater. Res. Proc. 2 (2016) 497–502.
[11] B.A. Szost, S. Terzi, F. Martina, D. Boisselier, A. Prytuliak, T. Pirling, M. Hofmann, D.J. Jarvis, A comparative study of additive manufacturing techniques: residual stress and microstructural analysis of CLAD and WAAM printed Ti-6Al-4V components, Mater. Des. 89 (2016) 559–567.
[12] J.C. Haley, B. Zheng, U.S. Bertoli, A.D. Dupuy, J.M. Schoenung, E.J. Lavernia, Working distance passive stability in laser directed energy deposition additive manufacturing, Mater. Des. 161 (2019) 86–94.
[13] Y.H. Xiong, J.E. Smugeresky, J.M. Schoenung, The influence of working distance onlaser deposited WC-Co, J. Mater. Process. Technol. 209 (2009) 4935–4941, https://doi.org/10.1016/j.jmatprotec.2009.01.016.
[14] G.A. Turichin, O.G. Klimova, E.V. Zemlyakov, K.D. Babkin, D.Y. Kolodyazhnyy, F.A. Shamray, A.Y. Travyanov, P.V. Petrovskiy, Technological aspects of high speed direct laser deposition based on heterophase powder metallurgy, Phy. Proc. 78 (2015) 397–406.
[15] R. Mendagaliyev, R.S. Korsmik, O.G. Klimova-Korsmik, S.A. Shalnova, Effect of powder fraction 09CrNi2MoCu on the structure and properties of the obtained samples using the direct laser deposition, Solid State Phenom. 299 (2019) 571–576.
[16] R. Mendagaliyev, G.A. Turichin, O.G. Klimova-Korsmik, O.G. Zotov, A.D. Eremeev, Microstructure and mechanical properties of laser metal deposited cold-resistant steel for arctic application, Proc. Manufact. 36 (2019) 249–255.
[17] S.A. Shalnova, O.G. Klimova-Korsmik, M.O. Sklyar, Influence of the roughness on the mechanical properties of ti-6al-4v products prepared by direct laser deposition technology, Solid State Phenom. 284 (2018) 312–318.
[18] N. John, C. DuPont John, D. Lippold Samuel, Kiser, Welding Metallurgy and Weldability of Nickel-Base Alloys, John Wiley & Sons, Inc, 2009.
[19] R. Mendagaliyev, S.Y. Ivanov, S.G. Petrova, Effect of process parameters on microstructure and mechanical properties of direct laser deposited cold-resistant steel 09CrNi2MoCu for arctic application, Key Eng. Mater. 822 (2019) 410–417.
[20] Y.A. Bistrova, E.A. Shirokina, R. Mendagaliev, M.O. Gushchina, A. Unt, Research of mechanical properties of cold resistant steel 09CrNi2MoCu after direct laser deposition, Key Eng. Mater. 822 (2019) 418–424.
[21] G.G. Zadykyan, R.S. Korsmik, R.V. Mendagaliev, G.A. Turichin, formation of bead shape, structure and mechanical properties of cold resistant high-strength steel produced by direct laser deposition method, Solid State Phenom. 299 (2019) 345–350.

[22] O.G. Klimova-Korsmik, G.A. Turichin, S.A. Shalnova, M.O. Gushchina, V.V. Cheverikin, Structure and properties of Ti-6Al-4V titanium alloy products obtained by direct laser deposition and subsequent heat treatment, J. Phys. Conf. 1109 (2018) conference 1.
[23] M.O. Sklyar, O.G. Klimova-Korsmik, G.A. Turichin, S.A. Shalnova, Influence of technological parameters of direct laser deposition process on the structure and properties of deposited products from alloy Ti-6Al-4V, Solid State Phenom. 284 (2018) 306.
[24] M.O. Sklyar, O.G. Klimova-Korsmik, V.V. Cheverikin, Formation structure and properties of parts from titanium alloys produced by direct laser deposition, Solid State Phenom. 265 (2017) 535–541.

# CHAPTER 18

# 3D printing of pharmaceutical products

**Iria Seoane-Viaño[1,4], Francisco J. Otero-Espinar[1,4], Álvaro Goyanes[2,3]**

[1]Departamento de Farmacología, Farmacia y Tecnología Farmacéutica, Facultade de Farmacia, Universidade de Santiago de Compostela, Santiago de Compostela, Spain; [2]FabRx Ltd., Ashford, Kent, United Kingdom; [3]Departamento de Farmacología, Farmacia y Tecnología Farmacéutica, I+D Farma Group (GI-1645), Universidade de Santiago de Compostela, Santiago de Compostela, Spain; [4]Paraquasil Group. Health Research Institute of Santiago de Compostela (IDIS), Santiago de Compostela, Spain

## 18.1 Introduction

Current pharmaceutical manufacturing is based on the mass production of medicines with a limited number of dose strengths of each drug. This traditional approach has been successful providing good-quality drug products to the patients and reducing the overall per unit production cost. The most common medicines are solid dosage forms for oral drug delivery, being tablets the most widely used. These formulations consist of a mixture of one or more drugs (active pharmaceutical ingredient) and excipients (inactive ingredients), showing different shapes (cylindrical, oblong, etc.) and designed to fulfill specific drug delivery requirements [1].

Although this conventional approach offers advantages for both patients and manufacturers, it also presents some drawbacks, such as limitations in the range of doses and dose combinations in marketed products. With the advent of digital technologies, significant opportunities have emerged to improve the development and manufacture of pharmaceutical products. In this sense, three-dimensional printing (3DP) technology has represented a breakthrough in the pharmaceutical sector since it constitutes an effective strategy to overcome some challenges of current pharmaceutical manufacturing. 3DP is an additive manufacturing method that enables the production of bespoke objects in a layer-by-layer fashion. The physical object to be printed can be designed by a computer-aided design (CAD) software and the 3D model exported to be printed layer-by-layer, creating an individualized object with the desired shape and size. These unique manufacturing process could make 3DP a revolutionary technology to prepare formulations and medical devices that could be tailored to meet the

*Additive Manufacturing*
ISBN 978-0-12-818411-0
https://doi.org/10.1016/B978-0-12-818411-0.00022-7

individual needs of each patient [2]. As an example, printlets (3D-printed tablets) can be prepared with a defined drug dose, size, and shape combining one or more drugs into a single formulation, improving the efficacy of the treatment and reducing side effects [3]. It is forecasted that these innovative printlets will be prepared on-demand, close to the patient and at the dispensing point [4].

3D printing of medicines becomes a hot topic in the pharmaceutical industry since the US Food and Drug Administration (FDA) approved Spritam, the first 3D-printed tablet, in 2015 [5]. The formulation created by Aprecia Pharmaceuticals for the treatment of epilepsy is an oral medicine that rapidly disintegrates in the mouth with a sip of liquid to facilitate swallowing. Spritam includes high dose of drug so it is especially useful for patients who have swallowing difficulties, such as children, the elderly, and people with neurological disorders. However, this medicine is manufactured in industrial facilities, and it is not an example of personalized dose medicines.

The big pharmaceutical companies and regulatory authorities like the Food and Drug Administration (FDA) in the United States or the European Medicines Agency (EMA) in the European Union are in the process of evaluating the 3DP technology and adopting it for the manufacture of medicines. The aim of this chapter is to provide an overview on the current state-of-the-art in 3DP technology within pharmaceutical industry and the potential benefits and challenges that could arise from its implementation in medical practice, highlighting the major role that 3DP will play in the era of personalized medicine.

## 18.2 Pharmaceutical 3D printing technologies

3D printing, otherwise known as additive manufacturing, is a generic term that encompasses a range of different printing technologies, which have in common the construction of objects in a layer-by-layer manner. Independently of used technique, the whole process starts by designing the object to be printed with a CAD software package. Afterward, the 3D design is divided into a series of layers and exported to the 3D printer. As a result, a bespoke object of virtually any shape and size can be produced. Since the entire manufacturing process is a computer-controlled procedure, any desired change in the final object can be achieved by a modification of the CAD file. Based on the American Society for Testing and Materials (ASTM) scheme, the different 3DP technologies can be classified into seven main categories [6]. Since three of the technologies are not very suitable for drug products, in this chapter, the pharmaceutical application of only four of them is described: binder jetting, material extrusion, powder bed fusion, and vat polymerization.

### 18.2.1 Binder jetting

Binder jetting was first invented and patented in 1989 by Sachs et al. [7] at the Massachusetts Institute of Technology (MIT). In this type of printing, a liquid binding solution is selectively deposited with a printer nozzle over a thin layer of powder (powder bed). The powder particles wetted by the binder solution adhere together, causing layer solidification. Once the layer is printed, a new layer of powder is spread over the bed, often employing a blade or a roller, and the process is repeated sequentially to produce successive layers of selected regions of bonded powder until the printing process is completed. Finally, the 3D-printed object is extracted from the powder bed, and excess unbound powder is removed [8].

The applications of binder jetting in healthcare were licensed to Therics Inc., trademarked as TheriForm in 1994 [9]. The initial pharmaceutical development of TheriForm focused primarily on the deposition of drug-loaded liquids onto powder bed made of pharmaceutical excipients to form immediate, extended, or multirelease tablets. The main disadvantage of this approach is the limited options to obtain drug-loaded inks that could perform consistent jetting. An alternative approach involves the deposition of liquid binder onto a powder beds incorporating not only excipients but also the drug. This technique is the basis of ZipDose technology, which was used to manufacture the first FDA-approved 3D-printed tablet (Spritam by Aprecia Pharmaceuticals) in 2015 (Fig. 18.1) [10]. Spritam tablets are specially designed to quickly dissolve in the mouth, in less than 11 s, needing only small amount of saliva. This technology allows the production of high-dose medications, up to 1000 mg, which is a very high dose for an orodispersible tablet. ZipDose technology was scaled up using multiple nozzles that deposit the binding solution onto a powder bed transported on a conveyor belt. Tablets are gradually built layer-by-layer until they are completed; then, the tablets are removed from the powder bed, dried, and packed [11]. ZipDose technology represents an alternative to manufacture dosage forms that cannot be manufactured by conventional processes.

Other examples of fast dissolving tablets developed using binder jetting include drug delivery devices with a predefined inner structure of unbound

**Figure 18.1** Spritam by Aprecia Pharmaceuticals (first FDA-approved 3D-printed tablet). *(Reproduced with kind permission of Aprecia Pharmaceuticals, LLC. from A. Pharmaceutials, Manufactured Using 3D Printing. 2015. Available from: http://www.spritam.com/-/hcp/zipdose-technology/manufactured-using-3d-printing.)*

powder surrounded by an external region of bound powder [12]. Binder jetting has also been used to create zero-order drug release formulations composed of an immediate release core and a shell that control the drug release [13]. Other types of delayed release tablets were also developed with excellent content uniformity and also demonstrating that 3DP is capable of accurately construct dosage forms containing a few micrograms of drug [14].

The main commercial application of this technique in the pharma field is its ability to formulate highly porous, fast-dissolving tablets with high drug loadings, as in Spritam. However, multiple tablet structures could be produced, enabling the creation of 3DP objects with highly complex geometries. Binder jetting also has some drawbacks, for instance, the high porosity of the tablet could affect the mechanical strength and friability of the formulation. Another limitation is that it is not possible to print hollow objects as the powder is filling all the spaces. Furthermore, additional excess powder removal and drying steps are required to obtain tablets without lose particles and to evaporate any residual solvent [15].

## 18.2.2 Material extrusion

### *18.2.2.1 Fused deposition modeling*

Fused deposition modeling (FDM), also known as fused filament fabrication (FFF), is possibly one of the most common and affordable printing technologies. In this 3DP technique, a polymer filament is heated and extruded through a small heated tip. The soften polymer is deposited onto a building plate to harden, creating one layer of the object to be fabricated, and subsequently, the build plate moves down and the next layer is deposited and so on until finally creating the 3D-printed object [16].

Goyanes et al. [16] were one of the first researchers to demonstrate the feasibility of FDM to fabricate drug-loaded tablets. Polyvinyl alcohol (PVA) filaments were loaded with fluorescein (used as a model drug) by placing them into an ethanolic solution of fluorescein. It was possible to print tablets with different drug release profiles by modifying printing parameters like the infill percentage of the formulations. PVA filaments were also loaded by diffusion using different model drugs including prednisolone and aminosalicylates [17,18]. During the printing process of PVA filaments loaded with two aminosalicylate isomers used in the treatment of colonic conditions (5-ASA and 4-ASA), 4-ASA (thermally labile) was significant degraded, while 5-ASA (nonthermally labile) did not suffer degradation [18]. This shows that FDM may not be suitable for the manufacture of drugs with degradation temperatures lower than the printing temperature. This finding leads to the search of new excipients that can print at lower temperatures than PVA to avoid degradation of the drug [19–21].

On the other hand, drug-loaded filaments are commonly manufactured by hot melt extrusion (HME), which involves the mixing excipients and/or drugs together and the application of heat and pressure to force the mixture to pass through a die [22]. This highly versatile technology is widely used within the pharmaceutical sector, and the FDA has already approved several hot melt extruded products [23]. The combination of HME and FDM was used to fabricate printlets with different shapes (cube, pyramid, cylinder, sphere, and torus) and sizes. Drug release was dependent on the surface area-to-volume ratio of the formulations, so changing the shape and/or the size of printlets will result in a different drug release profile [24]. The shape of the formulations can change drug release from the printlets but also modify medicine acceptability, as shown in the first patient acceptability study in humans, conducted with 3D-printed tablets manufactured by FDM 3DP [25]. The results showed that patients have preference for torus shape printlets, and other factors, such as color and size, could affect the willingness of people to shallow the printlets. This fact gains importance in certain age groups such as pediatric patients or the elderly, where a more attractive dosage form could improve patients' acceptability of medications and adherence to the treatment.

HME and FDM were combined with fluid-bed coating to create a new modified release dosage form loaded with budesonide, also used in the treatment of colonic conditions [26]. The 3D-printed cores were coated with an enteric polymer to provide them with enteric properties. The resultant capsule-shaped tablets (caplets) were compared with commercial budesonide products, demonstrating the potential of FDM combined with established pharmaceutical processes to manufacture oral dosage forms. To avoid the need for a separate coating process, different alternatives were processed like printing a shell using a dual 3D printer [27,28] or the use of 3D printed capsular devices made with different excipients [29–31]. The manufacturing process was also improved by developing a single filament that incorporates different grades of drug loading, enteric polymer, and infill, obtaining printlets (3D-printed tablets) with different delayed release patterns, making it possible to target different regions of the gastrointestinal tract [3]. The effect of the microstructure of the extruded filament and the 3D-printed tablets on drug dissolution rate was also investigated [32], concluding that the porosity of the tablets did not affect the drug release, which could ultimately be controlled by diffusion/erosion mechanisms.

Another asset of FDM is the possibility of incorporating different drugs in the same 3D-printed tablet (polypill). This could be achieved through the construction of tablets containing distinct regions, where multiple combinations of drugs and polymers are possible. The same drug could be incorporated in different polymers, or different drugs could be incorporated in the same polymer, so that the drug release profiles could be modulated [33,34]. Furthermore, it is possible to develop 3D-printed tablets loaded with polymeric nanocapsules by soaking the printed devices in a nano-particle liquid suspension. This strategy offers a useful approach to convert nanocapsules liquid suspensions into solid drug dosage forms with tailored dose and drug release profiles [35]. Apart from tablets, another FDM application that is worth mentioning is the manufacture of personalized wound dressings through the combination of 3D scanning and 3D printing [36]. In this work, metal ions were incorporated into PCL filaments to print dressings against scanned templates of a target wound, creating anatomically adaptable dressings that can be tailored to the particular needs of each patient.

In the pharmaceutical field, FDM is one of the most feasible 3DP technologies due to the low cost of the printers and the high range of useable materials. The 3D-printed devices show high mechanical strength, which makes them stable throughout different forms of processing. Moreover, several materials could be integrated into the same printing layer, making FDM a highly efficient process. Nevertheless, FDM also has some limitations, such as the heat needed for the extrusion and printing. This could be partially avoided by the selection of polymers with lower melting temperatures and, if possible, polymers with melting temperatures lower than the melting temperature of the drugs [37,38]. Another limitation is the need for the preparation of drug-loaded filaments as previous step for 3D printing.

#### *18.2.2.2 Direct powder extrusion*

Direct powder extrusion involves the extrusion of material in the form of powder through the nozzle of the printer, which uses a single screw extruder. Unlike FDM, direct powder extrusion is a single-step technology that does not require the preparation of filaments using HME. Since the HME step is avoided, this technique may be especially useful for preparing formulations as amorphous dispersions using small amounts of drugs and excipients. This could be advantageous for preclinical and clinical studies, where the quantity of drugs is often limited. By using a direct powder extruder 3D printer, sustained release printlets containing a poorly soluble drug were prepared, and the effect of different hydroxypropylcellulose (HPC) grades on the final printlets was also evaluated (Fig. 18.2) [39].

#### *18.2.2.3 Gel/paste extrusion*

Gel extrusion, also known as semisolid extrusion (SSE), is a manufacturing technique that employs a syringe-like system to deposit semisolids, such as gel or pastes, onto a build plate to create a solid object. As in FDM, the extruded material hardens following cooling or by solvent evaporation, which allows the material to support the weight of subsequent layers [40,41].

**Figure 18.2** Printlets produced by direct powder extrusion technique containing four different grades of hydroxypropylcellulose (HPC). *(Reproduced with kind permission of Elsevier from A. Goyanes, et al., Direct powder extrusion 3D printing: fabrication of drug products using a novel single-step process, Int. J. Pharm. (567) (2019) 118471.)*

Semisolids commonly used in pharmaceuticals can be formulated by mixing the polymers, drugs, and solvents in the correct ratio to produce formulations of an adequate viscosity for printing. The viscosity of the material to be extruded and the position of the extrusion head with respect to the bedplate can significantly impact the printing process. Moreover, to facilitate the removal of the object from the printing bed, the surface of the bed could be heated to minimize the adherence of the base layers thereto [40,41].

SSE technique has been used to produce a number of formulations, such as bilayer tablets [42] and polypills capable of delivering multiple drugs in an immediate or sustained release manner through different release mechanisms, such as osmosis and diffusion [43,44]. Another application of SSE technology is found in the fabrication of fast-dissolving tablets. As an example, the manufacture of high drug loading paracetamol oral tablets with an immediate release profile that achieve disintegration in less than 60 s and almost complete release of paracetamol within 5 min could be mentioned [45]. One interesting approach in this field of orodispersible printlets is the use of cyclodextrins as excipients in the fabrication of printlets loaded with poorly soluble drugs. This is the case of the rapid release formulations of the hydrophobic drug carbamazepine prepared with hydroxypropil-β-cyclodextrin (HPβCD), which acts as a soluble filler and forms drug–HPβCD complexes [46]. This technology was used to fabricate chewable formulations in a hospital setting for patients with a rare metabolic disease [4]. This was the first clinical study using 3D printing to prepare personalized dose medicines. The printlets had a good acceptance among patients and were able to maintain the optimum range of isoleucine levels in the blood of the patients.

Furthermore, this technology has also been applied to prepare lipid-based formulations, including suppositories [47,48] (Fig. 18.3). Self-microemulsifying drug delivery systems (SMEDDSs) are liquid formulations that form oil-in-water (O/W) emulsions in the gastrointestinal tract allowing the solubilization of lipophilic drugs in the small lipid droplets [49]. SSE techniques have made possible the transformation of these liquid formulations into solid self-microemulsifying drug delivery systems (S-SMEDDSs) with different geometrical shapes (cylindrical, prism, cube, and torus) [50]. These printlets with different geometries were evaluated under simulated gastric conditions, and the obtained results indicated that geometry affects the dispersion time, obtaining the shortest dispersion time for torus shapes.

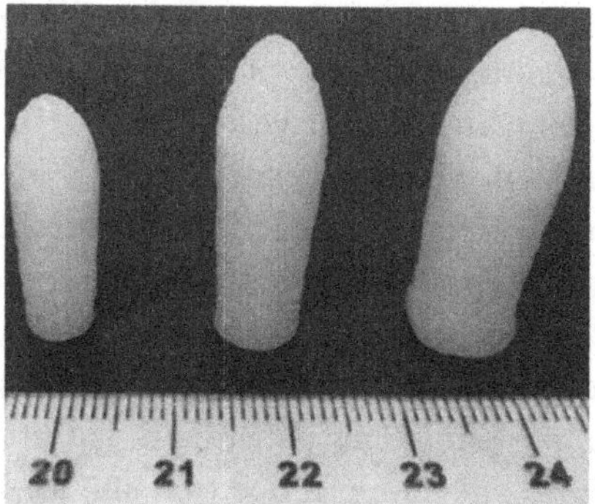


**Figure 18.3** Lipid-based suppositories with different sizes produced by semisolid extrusion (gel extrusion) technique. *(Reproduced with kind permission of Elsevier from I. Seoane-Viaño, et al., 3D printed tacrolimus suppositories for the treatment of ulcerative colitis, Asian J. Pharm. Sci. (2020).)*

SSE stands out for its simplicity since drugs and excipients can be mixed directly forming the gel/paste base for printing. Moreover, the heat applied to obtain a suitable viscosity of the feedstock (if any needed) is much lower than in other 3DP techniques, being thus possible to utilize thermolabile drugs. Due to the necessity to accommodate the viscous material, the nozzle heads need wider orifices, which affects the resolution of the printer. However, this low resolution allows to achieve higher printing speed rates than other 3DP technologies [40,41].

## 18.2.3 Power bed fusion

Power bed fusion is a selective thermal process that involves the fusion of powder particles by the application of a heat source, such as a laser. This technology includes selective laser sintering (SLS), which employs a laser to build up a 3D-printed object from a powder bed. The laser binds the powder particles together forming a solid structure with a specific pattern. Once the first layer is completed, a new layer of powder is deposited on top of the previous one and the process is repeated sequentially, building the object in a layer-wise manner. Finally, the object is recovered from underneath the powder bed [51].

SLS technique has been used in a variety of medical applications including the production of patient-specific anatomical models [52], implantable devices [53], and in the field of tissue engineering [54]. Commonly used materials for SLS are polymeric powdered forms of plastics, metal alloys, and ceramics that require high temperatures to be sintered, which may cause drug or excipient degradation. Because of these harsh printing conditions, the entry of SLS technology in the pharmaceutical field has been hampered for years [55].

In 2017, oral printlets were first manufactured using an SLS 3D printer [51]. The printlets were prepared using pharmaceutical grade polymers, and no degradation of the drug was observed. The 3D printer used in the study contained a diode laser that emits a lower intensity laser compared with the more potent infrared $CO_2$ lasers, which have been tested before to prepare drug delivery systems [53,56]. This initial work led to the development of printlets with orally disintegrating properties and accelerated drug release [57], as well as printlets with gyroid lattice structures having customizable drug release characteristics [58]. SLS technique was also used to produce small 3D-printed pellets (miniprintlets) containing two different drugs with customized drug release patterns [59]. These studies demonstrate that by simply changing the 3D design, the drug release profiles of different polymers can be modified, which allows to tailor the dosage form to the individual needs of each patient.

SLS technology offers some advantages over other printing techniques. For instance, SLS is a solvent-free process and offers faster production as compared with binder jetting, which may require around 48 h post-manufacture to allow the solvent to evaporate [60]. Moreover, SLS does not require the prior production of drug-loaded filaments, as in the case of FDM, only a suitable powder mixture is required [19]. In addition, SLS produces objects of higher resolution due to the precision of the laser compared with those produced by other methods, such as FDM or semi-solid extrusion.

### 18.2.4 Vat polymerization

Vat polymerization process selectively cures a vat of liquid photopolymer, transforming it into a solid through the application of a light source. In this technology, it included the technology called stereolithography (SLA), which uses a laser to induce the solidification of a liquid resin by photopolymerization. In this process, the photoinitiator and the photopolimerizable resin are placed in a build tank exposed to a high energy light source focused to a particular depth. The photoinitiator undergoes a reaction producing initiating species, such as free radicals, which attack the monomer units of the resin adding more monomers/oligomers, and consequently, cross-linking takes place. When the first layer of the object is solidified, the build plate moves up to a defined distance according to the desired thickness of each layer, and more liquid resin is redistributed on the top of the previous layer. The process is repeated until the object is built layer-by-layer, and finally, the 3D-printed object is washed to remove any excess liquid resin [61,62].

The possibilities for using SLA in the biomedical area are numerous [63]. For instance, SLA can be applied to tissue engineering for fabricating very precise molds using 3D models obtained by micro—computed tomography (CT), as well as scaffolds [64,65] and 3D-printed biomimetic hydrogels that simulate tissues with detoxification properties [66] due to its high accuracy for manufacturing complex structures. Moreover, SLA printing can be used to produce medical devices, such as personalized antiacne masks loaded with drugs such as salicylic acid [67].

In the field of oral drug delivery, the use of SLA is more recent [62]. This printing technique has been used to produce oral-modified release dosage forms loaded with paracetamol and 4-aminosalicylic acid (4-ASA) [68]. The drug 4-ASA is known to be thermally labile; however, no drug degradation occurred during the printing process, which makes SLA an interesting alternative to print objects loaded with thermolabile drugs. SLA was also used to fabricate ibuprofen-loaded hydrogels with different water content using riboflavin as a nontoxic photoinitiator. Hydrogels with higher water content showed faster drug release rates [69]. Other study investigated the effect of geometric parameters on the drug release kinetics of SLA printlets, having the greatest influence on them the constant surface area/volume ratio (SA/V). This implies that the specific drug release profile can be maintained by changing the SA/V ratio (Fig. 18.4) [70]. One limitation of SLA is the production of polypills due to the difficulty of printing spatially separated layers, although some printlets incorporating up to six drugs were prepared [71].

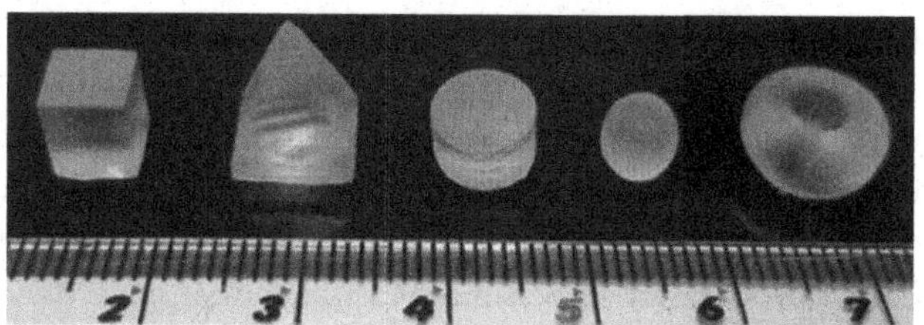


**Figure 18.4** Printlets with different geometries produced by SLA technique. *SLA*, stereolithography. *(Reproduced with kind permission of Springer Nature from P.R. Martinez, et al., Influence of geometry on the drug release profiles of stereolithographic (SLA) 3D-printed tablets, AAPS Pharm. Sci. Tech. 19(8) (2018) 3355–3361.)*

Among the advantages of this technology, the possibility of dissolving or dispersing the drugs directly in the resin to incorporate them into the printing object could be mentioned. Moreover, SLA is capable of producing objects with complex geometries and allows to achieve higher accuracy and resolution than other printing technologies [72]. The photopolymerization reaction is considered to be a green technology since it is solvent-free and requires low electrical input and low temperature of operation. Nevertheless, the use of SLA for biomedical and pharmaceutical purposes has some limitations, mainly due to the appreciable toxicity that photopolymerizable chemicals tend to have, along with the limited number of biocompatible resins that are commercially available for SLA [61]. Another drawback is that some drugs may chemically react with the monomers during the printing process preventing any drug release from the printlets [73].

## 18.3 3D printing: a new era of personalized medicine

The numerous applications of 3DP to pharmaceuticals are evident. The advent of personalized medicine involves a shift away from the traditional one-size-fits-all approach, to one in which the management of a patient's health is based on the individual patient's specific characteristics. But to make this possible, it is essential to understand how a person's unique genetic profile makes them susceptible to a certain disease [74]. Since the first draft of the human sequence has been announced in 2000 [75], there has been a dramatic drop in DNA sequencing costs, mainly due to the new sequencing technology and the development of high speed computing needed for analysis, which has allowed this technology to be considered as part of routine healthcare. The integration of genetic information and other clinical data results in the discovery of new pathways involved in diseases, and in extension, helps to understand the mode of action of the pharmaceutical products [76].

Based on this approach, different subtypes of patients within a given condition can be identified, and treatment can be tailored to match an individual's underlying cause. However, this tailoring of medical treatment to the patient's unique genetic makeup has not yet been fully achieved, and one of the major issues that has contributed to this delay is manufacturing technology. Traditional pharmaceutical manufacturing processes are unsuitable for the production of personalized medicines, since they are based on mass production of fixed-dose medicines, which limits the commercial availability of doses or dose combinations [11]. Therefore, it is evident that before this new era of personalized medicine can truly begin, the pharmaceutical industry has to adapt and embrace new innovative technologies for tailored therapy production. 3DP technologies have the potential to reshape the way that medicines are designed and manufactured. This technology is forecast to play a major role in the production of highly flexible and customized dosage forms on-demand, overcoming the limitations of conventional manufacturing processes.

### 18.3.1 The unlimited possibilities of 3D printing in healthcare

The therapeutic doses of the drugs that are used in clinical practice are selected in early-phase clinical trials based on the doses that exerted a therapeutic effect in most patients. This dosing approach focuses on the average patient and, to certain extent, does not normally take into account the variability between patients based on their genetic profiles, disease state, and other factors such as age, weight, and gender [77]. Hence, it is expected that the response to drug therapy and the susceptibility to side effects may differ between each patient. This is of particular importance for drugs with narrow therapeutic index where the interindividual variability of the response increases the likelihood of serious adverse effects or otherwise inadequate therapeutic levels [78]. This understanding laid the foundations for the development of personalized medicine, which aims to tailor the treatment according to each patient's individual characteristics, needs, and preferences. The ideal dosing method for personalization should be accurate, simple, cheap, and suitable for the maximum number of patients, starting from young children to the elderly [79].

3DP technologies have the potential to lead the way for this era of personalized medicine, revolutionizing the way medicines are made. Instead of using conventional large batches processes, 3DP offers the opportunity to produce customized 3D-printed tablets (printlets) with a tailored dose, shape, size, and release characteristics, far from the traditional "one-size-fits-all" approach. In a digital pharmacy era, the specific prescription for a particular patient could be sent directly to the compounding pharmacy or the hospital pharmacy setting, and small batches of individualized printlets could be produced with the appropriate dosage, drugs combinations, and formulation type specially designed to suit the patient [80,81]. Furthermore, the high automation of the printing process and the lower number of production steps reduce the possibility of human errors, for example, weighting errors, and increasing the safety of the formulations prepared in pharmacies and hospitals.

Quality control analysis in solid oral dosage forms is commonly performed by high-performance liquid chromatography (HPLC) and UV spectroscopy, which are destructive methods that require sample preparation and are not suitable for small batches produced by 3DP. In this regard, analytical alternatives to test content uniformity and dose verification in 3D-printed dosage forms have been explored. For instance, near-infrared (NIR) and Raman spectroscopy have shown to be capable of performing quality control measures of medicines in a nondestructive manner [82,83]. In particular, NIR spectroscopy was used as nondestructive method for quality control tests in printlets, obtaining a sensitivity comparable with that of chromatographic and UV spectroscopy methods [84]. Other interesting technique is terahertz pulsing imaging (TPI), which allows the acquisition of single depth-resolved scans in a few milliseconds. Terahertz radiation penetrates through polymeric materials, which makes it an attractive tool for the nondestructive analysis of 3D-printed products [85]. Moreover, inkjet printing has been used to produce dosage forms in the pattern of a quick response (QR) codes. This edible printed pattern contains the drug itself and information relevant for the patient or healthcare professionals readable by a smartphone [86]. The tracking of printlets through the supply chain is also possible by printing quick response (QR) codes and data matrices onto the surface of the printlet. Anticounterfeiting strategies could be also implemented through the deposition in the printlet of inks detectable by Raman spectroscopy [87].

Overall, the digitalization of the printing process together with the new analytical approaches and tracking methods can make a significant contribution to the safety of 3D printed medicines. However, the combined efforts of industry, pharmacies, and regulatory agencies are necessary to set a path to full 3DP implementation in clinical practice.

#### *18.3.1.1 Manufacturing of personalized printlets on-demand*

The demand for personalized medicines is expected to increase in the coming years, especially for pediatric and geriatric patients. These patients differ in many aspects from the "standard patient," in which pharmaceutical industry focuses during the drug development process. Dose requirements in children and older patients can be markedly different compared with the average patient, mainly due to physiological changes, differences in physical characteristics (e.g., age, body weight, and surface area), and pharmacokinetics (e.g., changes in metabolic functions and in renal clearance) [88].

Drug dosing in children is commonly extrapolated linearly from adult doses with adjustments based on body weight, length, and age [89]. They receive medicines initially designed for adults despite the important differences in pharmacokinetics and pharmacodynamics. This practice is called "unlicensed" or "off label use" and may result in serious toxicity and inadequate clinical responses [90]. In the case of the elderly, it is common to find patients aged 65 years or more who are prescribed more than one medicine and also have comorbidities that further complicate the situation [91]. The high rate of prescribed drugs increases the likelihood of drugs interaction and thus the risk of drug-related hospitalizations. According to a systematic review, up to 10% of hospitalizations in adults are drug-related [92]. Older adults reported a higher prevalence, comprising around 30% of all hospitalizations [93].

Another problem arises in the case of medicines with few available strengths or in a single-type formulation. It is common among patients and carers to achieve the target dose by means of splitting the tablets, which results in variations in the drug content of each part [77]. Moreover, when there are no suitable dosage forms available for specific patients, they must be extemporaneously compounded by pharmacists or clinical staff, for example, by crushing the tablets or using the contents of the capsules. This leads to a number of risks, such as compounding errors and inaccurate dosing [94].

Most of these challenges could be easily overcome by the implementation of 3DP in clinical practice. Pharmacists and clinical staff could design a personalized printlet containing an exact dosage of drug or drug combination for each patient [80]. A great example of the benefits derived from the use of this technology is found in the first clinical study using 3D-printed medicines in pediatric patients [4]. In this study, a pharmaceutical 3D printer was integrated into the Pharmacy Department of the Clinic University Hospital of Santiago de Compostela (Spain), to produce chewable printlets containing personalized dosages of isoleucine intended to treat children with maple syrup urine disease (MSUD). The dosage forms were printed in a variety of colors and flavors that were well accepted among all patients (Fig. 18.5). The results have shown that a tighter control over target blood levels is obtained compared with the extemporaneous formulations prepared by clinical staff, which were the standard therapy. Other examples of the potential applicability of 3DP could be found in the manufacture of liquid capsules [28] or immediate release tablets [21] containing individualized doses of theophylline (a narrow therapeutic index drug), as well as controlled release tablets loaded with the corticosteroids prednisolone [17] and budesonide [26].

**Figure 18.5** Chewable printlets of isoleucine with different colors and savors. *(Reproduced with kind permission of Elsevier from A. Goyanes, et al., Automated therapy preparation of isoleucine formulations using 3D printing for the treatment of MSUD: first single-centre, prospective, crossover study in patients, Int. J. Pharm. 567 (2019) 118497.)*

### *18.3.1.2 Patient-friendly formulations*

An appropriate pharmaceutical design of a dosage form is a key aspect to improve acceptability and patient outcomes. This is of especial importance in pediatric and geriatric populations, as they differ in many aspects from the other age subsets of population and, therefore, require particular considerations regarding the physical characteristics of formulations intended for them [95,96]. For instance, the ability to swallow the dosage form intact (e.g., capsules or tablets) largely determines the acceptability of medicines in older patients, whereas formulation factors such as taste, smell, and viscosity are important features for pediatric medicines. Dysphagia (difficulty in swallowing) can occur in both populations, being associated with an age-related decline in swallowing function in the elderly, and with developmental, behavioral, and psychological disorders in children [95].

Oral liquid formulations are regarded as appropriated dosage forms for children and the geriatric patients due to advantages such as dose flexibility and ease of swallowing. Nevertheless, palatability is a critical factor for the acceptance of these formulations. Many drugs and excipients are known to have an aversive taste, and other attributes such as viscosity, particle size, and smell also affect the acceptability of the product [97]. In this regard, orally disintegrating tablets (ODTs) and chewable tablets represent an interesting approach especially if they could be administered without water (Fig. 18.6).

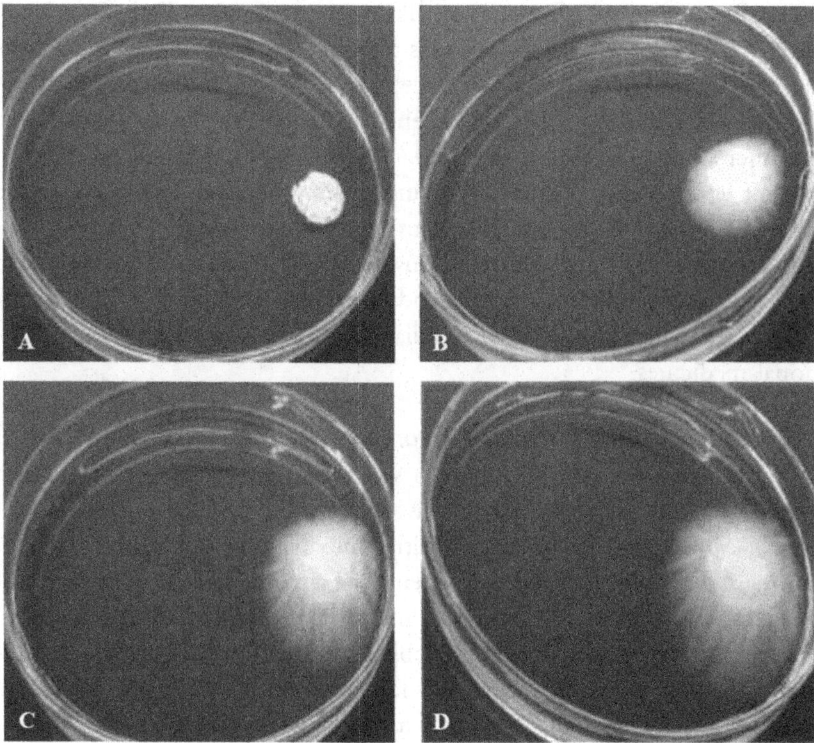


**Figure 18.6** Disintegration of an orodispersible printlet in distilled water: (A) after 5 s; (B) 120 s; (C) 360 s; and (D) 540 s. *(Reproduced with kind permission of Elsevier from J. Conceição, et al., Hydroxypropyl-β-cyclodextrin-based Fast Dissolving Carbamazepine Printlets Prepared by Semisolid Extrusion 3D Printing, Carbohydrate Polymers, 2019.)*

For instance, 3D-printed orodispersible films (ODFs) for children were successfully prepared in a hospital pharmacy setting [98] as an alternative to the established method of producing patient-tailored warfarin doses. The 3D-printed ODFs displayed improved drug content along with other advantages, such as the administration directly into the patient's mouth without the need for water. On the contrary, the unit dose sachets prepared by the conventional method need to be dissolved in a liquid prior to administration. Moreover, 3D-printed ODTs could be especially useful for patients who are purposely nonadherent, such as patients with psychiatric disorders, since it is more difficult to hide the formulation in the mouth or to split it out [99].

Apart from palatability and size [100], other characteristics of oral drug products, such as the shape and color of the dosage form, could strongly affect the end user willingness to take their medication [96]. One study found that patients consider polypills to be an adequate way to reduce the number of medicines. Moreover, acceptability of 3D printed medicines was determined based on whether patients relied on 3DP technology [101]. Another study investigated the influence of shape of different printlets on patient acceptability regarding picking and swallowing [25]. Among the different printed shapes (sphere, torus, disc, tilted, capsule, pentagon, diamond, heart, triangle, and cube), torus shape was found to be an acceptable novel shape together with capsule and disc shapes, which were found to be acceptable mainly due to their familiar geometry associated with conventional medicines.

#### 18.3.1.3 3D-printed polypills for complex dosage regimes

Printing technologies could make a significant contribution to treatment adherence and safety of patients with complex dosage regimes. Polypharmacy (the concurrent use of multiple medicines) is a common problem among elderly and hospitalized patients, which are affected by high tablet burdens with different dosing regimens. 3DP offers the possibility of loading different drugs in a single tablet, called polypill [102]. These polypills offer an innovative approach for personalized medicine, designing multiple drug containing devices with specific pharmacokinetics characteristics that would not otherwise be created using conventional manufacturing methods. Different 3DP technologies have already been used to manufacture polypills with varying designs and release profiles [44,71,103]. In the same way, safety issues arising from the use of narrow therapeutic index (TI) drugs could also be overcome using 3DP. Narrow TI drugs show a small range between their effective and toxic dose [78], which could lead to an increased risk of adverse effects. By the use of 3DP, printlets with an exact dose of the drug could be created, thus reducing the risk of medication errors arising from the incorrect manipulation of conventional fixed-strength tablets to create the customized formulation [27]. Furthermore, conventional fabrication techniques are ineffective in controlling the release of two or more drugs independently in different areas of the GI tract. However, using 3DP techniques, the manufacture of polypills containing multiple drugs accommodated in multiple polymers becomes possible (Fig. 18.7). These formulations could be tailored to the specific characteristics of each patient, adjusting the formulation to the individual's gastric transit time, pH, and even microflora depending on the disease state [104,105].

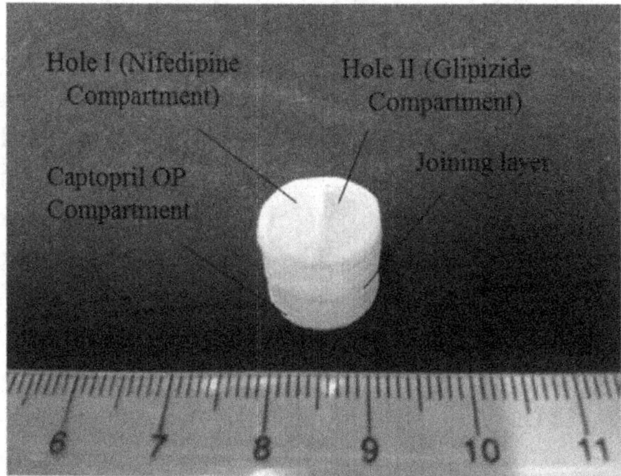


**Figure 18.7** 3D design of polypills with different compartments in which each compartment contains a different drug. *(Reproduced with kind permission of Elsevier from Khaled, S.A., et al., 3D printing of tablets containing multiple drugs with defined release profiles, Int. J. Pharm. 494(2) (2015) 643–650.)*

### *18.3.1.4 4D printing*

4D printing (4DP) is a relatively new concept originated by integrating a new dimension to 3DP technology, namely time. 4DP involves the use of smart materials with the ability to change their configuration over time in response to an external stimulus such as pH, light, heat, moisture, and magnetic or electric forces [106]. In the field of bioengineering, the production of dynamic 3D-printed structures by 4DP has been applied in tissue engineering, for example, in constructing patient specific scaffolds [107]. Moreover, medical device's such as stents can be printed, deformed into its temporary shape, inserted into the body, and then deployed back into its permanent shape through the use of a thermal stimulus. These 4D-printed stents better match the particular structure, improving the prognosis of the patients [108]. Despite the high potential of 4DP, this technique has not yet been fully exploited in the pharmaceutical field. One study explored the viability of 4DP approach to fabricate retentive intravesical delivery systems [109]. The devices have water-induced shape memory, which allows them to be retained in the bladder with no need for being removed due to its erosion/dissolution over time. In addition, the release rate of the drug could be modified by changing the molecular weight of the polymer. Future applications of 4DP within pharmaceuticals could help to manufacture medicines that were previously challenging to produce, opening new perspectives in personalized drug manufacturing.

### 18.3.2 The potential role of 3D printing in drug development

Early-phase drug development comprises several stages, from drug discovery and preclinical studies in animals to first-in-human (FIH) clinical trials. At the end of this timeline, the financial burden for the pharmaceutical industry created by the cost of bringing a new compound into commercialization is really high, while the clinical approval success rate is very low [110]. For this reason, the development of innovative technologies that could optimize the drug development process becomes essential, allowing the identification of new drug candidates as early as possible at a minimal cost [111]. When a compound with therapeutic potential is discovered, a series of steps must be followed until the final formulation is obtained [112]. The selection of an appropriate formulation and its optimization is one of the main steps in the development process. It is at this stage where 3DP could play a major role by enabling the production of small batches of formulations with unique characteristics at low cost. These requirements are not often met by traditional manufacturing methods based on mass production of, for example, oral dosage forms [113]. The implementation of 3DP as an alternative manufacturing tool could be the solution to overcome the current challenges in manufacturing formulations for the early stages of drug development.

#### *18.3.2.1 3D printing in preclinical studies*

The drug discovery process begins with the identification of an unmet medical need, that is, a medical condition whose treatment is not satisfactorily addressed with currently available treatments or these are nonexistent. Broadly, the drug development process can be segregated into preclinical and clinical development stages. In preclinical development, a molecular target is selected and validated, with a high-throughput screening of compound libraries performed to identify potential drug candidates. The candidates are subsequently optimized to exhibit adequate potency and selectivity toward the molecular target in vitro before testing its efficacy in vivo [114]. Preclinical studies are designed to assess the efficacy and safety of the new compounds in relevant animal models to select suitable drug candidates to be tested in humans [115,116]. Apart from safety pharmacology and toxicology studies in animals, the preclinical development program includes other activities, such as formulation development and quality control measures. If the candidate successfully completes the preclinical phase, its clinical development begins after requesting permission from the drug regulatory agencies [117].

Oral dosage forms, such as tablets and capsules, are commonly used in preclinical research to administer drugs to animals. However, the manufacture of tablets is a lengthy process, and the tablets are often produced in a fixed dose. Capsules also have some disadvantages, such as the need for staff to manually fill the capsules with the exact dose of drug, which is a time-consuming task [113]. On the other hand, liquid formulations permit a higher degree of dose flexibility, but their use is limited since the solubility and stability of the drug could be compromised [116]. Technological advances have led to a sharp drop in the price of 3D printers over the past years, which made 3DP an affordable technology that could be easily integrated into a laboratory setting. With a small and compact 3D printer, researchers could produce printlets with different sizes and geometries adapted to meet the animal requirements. In preclinical studies, small devices such as capsules (size 9 and 9h), minitablets, and pellets are often used to administer drugs to rodents [118]. With a 3D printer, these devices could be manufactured on demand in a rapid manner and containing the exact dosage for the animal model, from rodents to larger animals, such as primates [119]. Moreover, 3D-printed devices could be filled with contrast agents or radiotracers to investigate their pass through the gastrointestinal tract of rats using imaging techniques [118,120]. The size could be rapidly modified to improve the gastric emptying of the devices [121]. A study in beagle dogs investigated the delayed release of drug-form 3D-printed capsules by modulating the wall thickness [122].

#### *18.3.2.2 3D printing in first-in-human clinical trials*

Clinical drug development consists of four temporal phases (I—IV). First-in-human (FIH) trials are performed as part of phase I and represent the first opportunity to investigate the drug in humans [123]. The investigated substance is sequentially administered to a small group of healthy volunteers with an appropriate interval of observation between dosing of individual subjects [124]. The study begins with the administration of the drug at low doses based on preclinical toxicological data [125], which are then escalated incrementally following different schemes (commonly single ascending dose studies and multiple ascending dose studies) [124]. The primary goal of FIH trials is the identification of an appropriate dose and dosing interval for testing efficacy in phase II trials, based on the safety and tolerability of the drug [126].

Considering the schemes commonly followed by FIH trials, it is evident that multiple dose strengths are required, especially in multiple ascending dose studies. Traditional manufacturing processes are lengthy and multistep processes that require expensive equipment and large workspaces [127]. The implementation of 3D printing in the early-phase drug development could allow the production of small batches of printlets with any desired drug dose [113]. Dosage forms with varying sizes and many geometrical outlines could be produced to increase the acceptance and make the administration easier [128,129].

In the case of poorly soluble drugs, HME coupled with FDM could be used to create a solid dispersion of the poor soluble drug within a polymer matrix, thereby increasing drug bioavailability [130,131]. In blinded trials, printlets could be designed to meet the requirements of the study [132] by masking the presence of the drug using, for example, a DuoCaplet design [34], which is a two-compartment device where the drug core is embedded within a larger tablet (Fig. 18.8). Moreover, the possibility of manufacturing printlets on demand immediately before administration avoids the need for long-term stability studies. Stability studies are necessary for long storage and usually delay the beginning of the clinical trial, being only necessary a short-term evaluation [133].

It is likely that the integration of 3DP in the field of early-phase drug development is easily attainable under current regulatory pathways. However, both its use in later phases of drug development, such as phase II and III trials, and its implementation in clinical practice are not currently possible. Current 3DP platforms are not suitable for scale-up and do not meet the standards required by good manufacturing practice (GMP) guidelines [80]. Regulatory initiatives are needed to move this innovative technology from the research level to its application as a manufacturing tool to support the pharmaceutical industry.

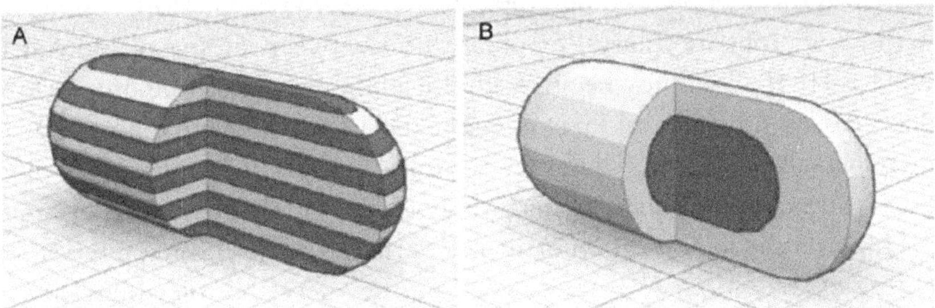


**Figure 18.8** (A) Multilayer device and (B) two-compartment device (DuoCaplet). *(Reproduced with kind permission of ACS Publications from A. Goyanes, et al., 3D printing of medicines: engineering novel oral devices with unique design and drug release characteristics, Mol. Pharm. 12(11) 2015 4077–4084.)*

### 18.3.3 Concluding remarks

The healthcare industry is evolving toward the development of a more personalized medicine. The traditional approach of "one-size-fits-all" is becoming a thing of the past, and new innovative platforms are required to produce tailored medicines. Conventional manufacturing processes based on mass manufacture are unsuitable to produce personalized medicines, creating a need for the pharmaceutical industry to adapt and embrace new technologies for tailored therapy production. In this sense, 3DP could reshape the way that medicines are manufactured due to its capability to produce bespoke medicines tailored to the individual needs of each patient. The unprecedented flexibility of 3DP allows the production of printlets with different shapes, sizes, and geometries on-demand directly at the point of care. In early-phase drug development, this technology could be exploited to produce small batches of formulations adapted to the requirements of the study. However, several regulatory and quality control requirements must be overcome before the full implementation of 3DP into practice. Meanwhile, 3DP technologies are still evolving, and the advent of 4DP will allow the development of "smart medicines" that provide new strategies to improve targeted therapies. In near future, the opportunities of 3DP as a manufacturing platform can be fully exploited to produce customized pharmaceutical products both in the pharmaceutical industry and in clinical practice.

## References

[1] R.M. Pacheco, Tratado de Tecnología Farmacéutica, in: R.M. Pacheco (Ed.), Formas de Dosificación, vol. 3, 2017. Síntesis. 458.

[2] J. Norman, et al., A new chapter in pharmaceutical manufacturing: 3D-printed drug products, Adv. Drug Deliv. Rev. 108 (2017) 39–50.

[3] A. Goyanes, et al., Development of modified release 3D printed tablets (printlets) with pharmaceutical excipients using additive manufacturing, Int. J. Pharm. 527 (1–2) (2017) 21–30.

[4] A. Goyanes, et al., Automated therapy preparation of isoleucine formulations using 3D printing for the treatment of MSUD: first single-centre, prospective, crossover study in patients, Int. J. Pharm. 567 (2019) 118497.

[5] A. Pharmaceutials, Manufactured Using 3D Printing, 2015. Available from: http://www.spritam.com/-/hcp/zipdose-technology/manufactured-using-3d-printing.

[6] C.M. Madla, et al., 3D printing technologies, implementation and regulation: an overview, in: A.W. Basit, S. Gaisford (Eds.), 3D Printing of Pharmaceuticals, Springer International Publishing, 2018, pp. 21–40.

[7] E.M. Sachs, et al., Three-dimensional Printing Techniques, 1993. US5204055A.

[8] S.J. Trenfield, et al., Binder jet printing in pharmaceutical manufacturing, in: A.W. Basit, S. Gaisford (Eds.), 3D Printing of Pharmaceuticals, Springer International Publishing, Cham, 2018, pp. 41–54.
[9] J. Goole, K. Amighi, 3D printing in pharmaceutics: a new tool for designing customized drug delivery systems, Int. J. Pharm. 499 (1–2) (2016) 376–394.
[10] W.-K. Hsiao, et al., 3D printing of oral drugs: a new reality or hype? Expet Opin. Drug Deliv. 15 (1) (2018) 1–4.
[11] S. Gaisford, 3D printed pharmaceutical products, in: D. Kalaskar (Ed.), 3D Printing in Medicine, Woodhead Publishing, 2017, pp. 155–165.
[12] D.G. Yu, et al., A novel fast disintegrating tablet fabricated by three-dimensional printing, Drug Dev. Ind. Pharm. 35 (12) (2009) 1530–1536.
[13] C.C. Wang, et al., Development of near zero-order release dosage forms using three-dimensional printing (3-DP (TM)) technology, Drug Dev. Ind. Pharm. 32 (3) (2006) 367–376.
[14] W. Katstra, et al., Oral dosage forms fabricated by three dimensional Printing™, J. Contr. Release 66 (1) (2000) 1–9.
[15] D.G. Yu, et al., Three-dimensional printing in pharmaceutics: promises and problems, J. Pharmaceut. Sci. 97 (9) (2008) 3666–3690.
[16] A. Goyanes, et al., Fused-filament 3D printing (3DP) for fabrication of tablets, Int. J. Pharm. 476 (1–2) (2014) 88–92.
[17] J. Skowyra, K. Pietrzak, M.A. Alhnan, Fabrication of extended-release patient-tailored prednisolone tablets via fused deposition modelling (FDM) 3D printing, Eur. J. Pharmaceut. Sci. 68 (2015) 11–17.
[18] A. Goyanes, et al., 3D printing of modified-release aminosalicylate (4-ASA and 5-ASA) tablets, Eur. J. Pharm. Biopharm. 89 (2015) 157–162.
[19] G. Kollamaram, et al., Low temperature fused deposition modeling (FDM) 3D printing of thermolabile drugs, Int. J. Pharm. 545 (1) (2018) 144–152.
[20] W. Kempin, et al., Immediate release 3D-printed tablets produced via fused deposition modeling of a thermo-sensitive drug, Pharmaceut. Res. 35 (6) (2018) 124.
[21] T.C. Okwuosa, et al., A lower temperature FDM 3D printing for the manufacture of patient-specific immediate release tablets, Pharmaceut. Res. 33 (11) (2016) 2704–2712.
[22] M. Wilson, et al., Hot-melt extrusion technology and pharmaceutical application, Ther. Deliv. 3 (6) (2012) 787–797.
[23] B. Lang, J.W. McGinity, I. Williams, O. Robert, Hot-melt extrusion – basic principles and pharmaceutical applications, Drug Dev. Ind. Pharm. 40 (9) (2014) 1133–1155.
[24] A. Goyanes, et al., Effect of geometry on drug release from 3D printed tablets, Int. J. Pharm. 494 (2) (2015) 657–663.
[25] A. Goyanes, et al., Patient acceptability of 3D printed medicines, Int. J. Pharm. 530 (1) (2017) 71–78.
[26] A. Goyanes, et al., Fabrication of controlled-release budesonide tablets via desktop (FDM) 3D printing, Int. J. Pharm. 496 (2) (2015) 414–420.
[27] T.C. Okwuosa, et al., Fabricating a shell-core delayed release tablet using dual FDM 3D printing for patient-centred therapy, Pharmaceut. Res. 34 (2) (2017) 427–437.
[28] T.C. Okwuosa, et al., On demand manufacturing of patient-specific liquid capsules via co-ordinated 3D printing and liquid dispensing, Eur. J. Pharmaceut. Sci. 118 (2018) 134–143.
[29] A. Melocchi, et al., Industrial Development of a 3D-Printed Nutraceutical Delivery Platform in the Form of a Multicompartment HPC Capsule, AAPS PharmSciTech, 2018.

[30] A. Melocchi, et al., Hot-melt extruded filaments based on pharmaceutical grade polymers for 3D printing by fused deposition modeling, Int. J. Pharm. 509 (1–2) (2016) 255–263.
[31] A. Melocchi, et al., 3D printing by fused deposition modeling (FDM) of a swellable/erodible capsular device for oral pulsatile release of drugs, J. Drug Deliv. Sci. Technol. 30 (B) (2015) 360–367.
[32] A. Goyanes, et al., Fused-filament 3D printing of drug products: microstructure analysis and drug release characteristics of PVA-based caplets, Int. J. Pharm. 514 (1) (2016) 290–295.
[33] X. Xu, et al., 3D printed polyvinyl alcohol tablets with multiple release profiles, Sci. Rep. 9 (1) (2019) 12487.
[34] A. Goyanes, et al., 3D printing of medicines: engineering novel oral devices with unique design and drug release characteristics, Mol. Pharm. 12 (11) (2015) 4077–4084.
[35] R.C.R. Beck, et al., 3D printed tablets loaded with polymeric nanocapsules: an innovative approach to produce customized drug delivery systems, Int. J. Pharm. 528 (1–2) (2017) 268–279.
[36] Z. Muwaffak, et al., Patient-specific 3D scanned and 3D printed antimicrobial polycaprolactone wound dressings, Int. J. Pharm. 527 (1) (2017) 161–170.
[37] A. Awad, S. Gaisford, A.W. Basit, Fused deposition modelling: advances in engineering and medicine, in: A.W. Basit, S. Gaisford (Eds.), 3D Printing of Pharmaceuticals, Springer International Publishing, Cham, 2018, pp. 107–132.
[38] E. Fuenmayor, et al., Material considerations for fused-filament fabrication of solid dosage forms, Pharmaceutics 10 (2) (2018) 44.
[39] A. Goyanes, et al., Direct powder extrusion 3D printing: fabrication of drug products using a novel single-step process, Int. J. Pharm. 567 (2019) 118471.
[40] J. Firth, A.W. Basit, S. Gaisford, The role of semi-solid extrusion printing in clinical practice, in: A.W. Basit, S. Gaisford (Eds.), 3D Printing of Pharmaceuticals, Springer International Publishing, 2018, pp. 133–151.
[41] I. Seoane-Viaño, et al., Semi-solid extrusion 3D printing in drug delivery and biomedicine: personalised solutions for healthcare challenges, J. Control. Release (2021), https://doi.org/10.1016/j.jconrel.2021.02.027. In press.
[42] S.A. Khaled, et al., Desktop 3D printing of controlled release pharmaceutical bilayer tablets, Int. J. Pharm. 461 (1–2) (2014) 105–111.
[43] S.A. Khaled, et al., 3D printing of tablets containing multiple drugs with defined release profiles, Int. J. Pharm. 494 (2) (2015) 643–650.
[44] S.A. Khaled, et al., 3D printing of five-in-one dose combination polypill with defined immediate and sustained release profiles, J. Contr. Release 217 (2015) 308–314.
[45] S.A. Khaled, et al., 3D extrusion printing of high drug loading immediate release paracetamol tablets, Int. J. Pharm. 538 (1–2) (2018) 223–230.
[46] J. Conceição, et al., Hydroxypropyl-β-cyclodextrin-based Fast Dissolving Carbamazepine Printlets Prepared by Semisolid Extrusion 3D Printing, Carbohydrate Polymers, 2019.
[47] I. Seoane-Viaño, et al., 3D printed tacrolimus suppositories for the treatment of ulcerative colitis, Asian J. Pharm. Sci. 16 (1) (2021) 110–119.
[48] I. Seoane-Viaño, et al., 3D printed tacrolimus rectal formulations ameliorate colitis in an experimental animal model of inflammatory bowel disease, Biomedicines 8 (12) (2020) 563, https://doi.org/10.3390/biomedicines8120563.
[49] K. Vithani, et al., An overview of 3D printing technologies for soft materials and potential opportunities for lipid-based drug delivery systems, Pharm. Res. (N. Y.) 36 (1) (2018) 4.

[50] K. Vithani, et al., A proof of concept for 3D printing of solid lipid-based formulations of poorly water-soluble drugs to control formulation dispersion kinetics, Pharm. Res. (N. Y.) 36 (7) (2019) 102.

[51] F. Fina, et al., Selective laser sintering (SLS) 3D printing of medicines, Int. J. Pharm. 529 (1) (2017) 285–293.

[52] D. Silva, et al., Dimensional error in selective laser sintering and 3D-printing of models for craniomaxillary anatomy reconstruction, J. Cranio-Maxillof. Surg. 36 (2008) 443–449.

[53] K.F. Leong, et al., Building porous biopolymeric microstructures for controlled drug delivery devices using selective laser sintering, Int. J. Adv. Manuf. Technol. 31 (5) (2006) 483–489.

[54] Y. Du, et al., Selective laser sintering scaffold with hierarchical architecture and gradient composition for osteochondral repair in rabbits, Biomaterials 137 (2017) 37–48.

[55] M.A. Alhnan, et al., Emergence of 3D printed dosage forms: opportunities and challenges, Pharm. Res. (N. Y.) 33 (8) (2016) 1817–1832.

[56] K.F. Leong, et al., Fabrication of porous polymeric matrix drug delivery devices using the selective laser sintering technique, Proc. Inst. Mech. Eng. H 215 (2) (2001) 191–201.

[57] F. Fina, et al., Fabricating 3D printed orally disintegrating printlets using selective laser sintering, Int. J. Pharm. 541 (1–2) (2018) 101–107.

[58] F. Fina, et al., 3D printing of drug-loaded gyroid lattices using selective laser sintering, Int. J. Pharm. 547 (1–2) (2018) 44–52.

[59] A. Awad, et al., 3D printed pellets (miniprintlets): a novel, multi-drug, controlled release platform technology, Pharmaceutics 11 (4) (2019).

[60] C.W. Rowe, et al., Multimechanism oral dosage forms fabricated by three dimensional printing, J. Contr. Release 66 (1) (2000) 11–17.

[61] P. Robles Martinez, A.W. Basit, S. Gaisford, The history, developments and opportunities of stereolithography, in: A.W. Basit, S. Gaisford (Eds.), 3D Printing of Pharmaceuticals, Springer International Publishing, 2018, pp. 55–79.

[62] X. Xu, et al., Vat photopolymerization 3D printing for advanced drug delivery and medical device applications, J. Control. Release 329 (2021) 743–757, https://doi.org/10.1016/j.jconrel.2020.10.008.

[63] S. Derakhshanfar, et al., 3D bioprinting for biomedical devices and tissue engineering: a review of recent trends and advances, Bioact. Mat. 3 (2) (2018) 144–156.

[64] F.P. Melchels, J. Feijen, D.W. Grijpma, A poly(D,L-lactide) resin for the preparation of tissue engineering scaffolds by stereolithography, Biomaterials 30 (23–24) (2009) 3801–3809.

[65] L. Elomaa, et al., Preparation of poly(ε-caprolactone)-based tissue engineering scaffolds by stereolithography, Acta Biomater. 7 (11) (2011) 3850–3856.

[66] M. Gou, et al., Bio-inspired detoxification using 3D-printed hydrogel nanocomposites, Nat. Commun. 5 (2014) 3774.

[67] A. Goyanes, et al., 3D scanning and 3D printing as innovative technologies for fabricating personalized topical drug delivery systems, J. Contr. Release 234 (2016) 41–48.

[68] J. Wang, et al., Stereolithographic (SLA) 3D printing of oral modified-release dosage forms, Int. J. Pharm. 503 (1–2) (2016) 207–212.

[69] P.R. Martinez, et al., Fabrication of drug-loaded hydrogels with stereolithographic 3D printing, Int. J. Pharm. 532 (1) (2017) 313–317.

[70] P.R. Martinez, et al., Influence of geometry on the drug release profiles of stereolithographic (SLA) 3D-printed tablets, AAPS Pharm. Sci. Tech. 19 (8) (2018) 3355–3361.

[71] P. Robles-Martinez, et al., 3D printing of a multi-layered polypill containing six drugs using a novel stereolithographic method, Pharmaceutics 11 (6) (2019).
[72] F.P. Melchels, J. Feijen, D.W. Grijpma, A review on stereolithography and its applications in biomedical engineering, Biomaterials 31 (24) (2010) 6121–6130.
[73] X. Xu, et al., Stereolithography (SLA) 3D Printing of an Antihypertensive Polyprintlet: Case Study of an Unexpected Photopolymer-Drug Reaction, Additive Manufacturing, 2020, p. 101071.
[74] H.K. Brittain, R. Scott, E. Thomas, The rise of the genome and personalised medicine, Clin. Med. 17 (6) (2017) 545–551.
[75] E.S. Lander, et al., Initial sequencing and analysis of the human genome, Nature 409 (6822) (2001) 860–921.
[76] A.A. Seyhan, C. Carini, Are innovation and new technologies in precision medicine paving a new era in patients centric care? J. Transl. Med. 17 (1) (2019) 114.
[77] M. Alomari, et al., Personalised dosing: printing a dose of one's own medicine, Int. J. Pharm. 494 (2) (2015) 568–577.
[78] H.S. Blix, et al., Drugs with narrow therapeutic index as indicators in the risk management of hospitalised patients, Pharm. Pract. 8 (1) (2010) 50–55.
[79] K. Wening, J. Breitkreutz, Oral drug delivery in personalized medicine: unmet needs and novel approaches, Int. J. Pharm. 404 (1–2) (2011) 1–9.
[80] S.J. Trenfield, et al., 3D printing pharmaceuticals: drug development to frontline care, Trends Pharmacol. Sci. 39 (5) (2018) 440–451.
[81] M.R.P. Araujo, et al., The digital pharmacies era: how 3D printing technology using fused deposition modeling can become a reality, Pharmaceutics 11 (3) (2019).
[82] M. Edinger, et al., Visualization and non-destructive quantification of inkjet-printed pharmaceuticals on different substrates using Raman spectroscopy and Raman chemical imaging, Pharmaceut. Res. 34 (5) (2017) 1023–1036.
[83] A. Cournoyer, et al., Quality control of multi-component, intact pharmaceutical tablets with three different near-infrared apparatuses, Pharmaceut. Dev. Technol. 13 (5) (2008) 333–343.
[84] S.J. Trenfield, et al., 3D printed drug products: non-destructive dose verification using a rapid point-and-shoot approach, Int. J. Pharm. 549 (1) (2018) 283–292.
[85] D. Markl, et al., Analysis of 3D prints by X-ray computed microtomography and terahertz pulsed imaging, Pharmaceut. Res. 34 (5) (2017) 1037–1052.
[86] M. Edinger, et al., QR encoded smart oral dosage forms by inkjet printing, Int. J. Pharm. 536 (1) (2018) 138–145.
[87] S.J. Trenfield, et al., Track-and-trace: novel anti-counterfeit measures for 3D printed personalized drug products using smart material inks, Int. J. Pharm. 567 (2019) 118443.
[88] J. Breitkreutz, J. Boos, Paediatric and geriatric drug delivery, Expet Opin. Drug Deliv. 4 (1) (2007) 37–45.
[89] B. Al-Metwali, H. Mulla, Personalised dosing of medicines for children, J. Pharm. Pharmacol. 69 (5) (2017) 514–524.
[90] J.C. Visser, et al., Personalized medicine in pediatrics: the clinical potential of orodispersible films, AAPS Pharm. Sci. Tech. 18 (2) (2017) 267–272.
[91] I. Cascorbi, Drug interactions–principles, examples and clinical consequences, Deutsch. Arztebl. Int. 109 (33–34) (2012) 546–556.
[92] A. Al Hamid, et al., A systematic review of hospitalization resulting from medicine-related problems in adult patients, Br. J. Clin. Pharmacol. 78 (2) (2014) 202–217.
[93] T.J. Oscanoa, F. Lizaraso, A. Carvajal, Hospital admissions due to adverse drug reactions in the elderly. A meta-analysis, Eur. J. Clin. Pharmacol. 73 (6) (2017) 759–770.

[94] T.E. Kairuz, et al., Quality, safety and efficacy in the 'off-label' use of medicines, Curr. Drug Saf. 2 (1) (2007) 89–95.
[95] F. Liu, et al., Patient-centred pharmaceutical design to improve acceptability of medicines: similarities and differences in paediatric and geriatric populations, Drugs 74 (16) (2014) 1871–1889.
[96] P. Januskaite, et al., I spy with my little eye: a paediatric visual preferences survey of 3D printed tablets, Pharmaceutics 11 (12) (2020) 1100, https://doi.org/10.3390/pharmaceutics12111100.
[97] J. Walsh, et al., Patient acceptability, safety and access: a balancing act for selecting age-appropriate oral dosage forms for paediatric and geriatric populations, Int. J. Pharm. 536 (2) (2018) 547–562.
[98] H. Oblom, et al., Towards printed pediatric medicines in hospital pharmacies: comparison of 2D and 3D-printed orodispersible warfarin films with conventional oral powders in unit dose sachets, Pharmaceutics 11 (7) (2019).
[99] W. Jamróz, et al., 3D printed orodispersible films with Aripiprazole, Int. J. Pharm. 533 (2) (2017) 413–420.
[100] F. Liu, et al., Acceptability of oral solid medicines in older adults with and without dysphagia: a nested pilot validation questionnaire based observational study, Int. J. Pharm. 512 (2) (2016) 374–381.
[101] M.M. Fasto, et al., Perceptions, preferences and acceptability of patient designed 3D printed medicine by polypharmacy patients: a pilot study, Int. J. Clin. Pharm. 41 (5) (2019) 1290–1298.
[102] B.J. Park, et al., Pharmaceutical applications of 3D printing technology: current understanding and future perspectives, J. Pharm. Invest. 49 (6) (2019) 575–585.
[103] A.P. Haring, et al., Programming of multicomponent temporal release profiles in 3D printed polypills via core–shell, multilayer, and gradient concentration profiles, Adv. Healthcare Mat. 7 (16) (2018) 1800213.
[104] N.B. Charbe, et al., Application of three-dimensional printing for colon targeted drug delivery systems, Int. J. Pharm. Investig. 7 (2) (2017) 47–59.
[105] V. Linares, M. Casas, I. Caraballo, Printfills: 3D printed systems combining fused deposition modeling and injection volume filling. Application to colon-specific drug delivery, Eur. J. Pharm. Biopharm. 134 (2019) 138–143.
[106] S.J. Trenfield, et al., Shaping the future: recent advances of 3D printing in drug delivery and healthcare, Expet Opin. Drug Deliv. 16 (10) (2019) 1081–1094.
[107] S. Miao, et al., Four-dimensional printing hierarchy scaffolds with highly biocompatible smart polymers for tissue engineering applications, Tissue Eng. C Methods 22 (10) (2016) 952–963.
[108] M. Zarek, et al., 4D printing of shape memory-based personalized endoluminal medical devices, Macromol. Rapid Commun. 38 (2) (2017).
[109] A. Melocchi, et al., Retentive device for intravesical drug delivery based on water-induced shape memory response of poly(vinyl alcohol): design concept and 4D printing feasibility, Int. J. Pharm. 559 (2019) 299–311.
[110] C.H. Wong, K.W. Siah, A.W. Lo, Estimation of clinical trial success rates and related parameters, Biostatistics 20 (2) (2019) 273–286.
[111] A. Awad, et al., Reshaping drug development using 3D printing, Drug Discov. Today 23 (8) (2018) 1547–1555.
[112] N. Shah, et al., Structured development approach for amorphous systems, in: R.O. Williams III, A.B. Watts, D.A. Miller (Eds.), Formulating Poorly Water Soluble Drugs, Springer New York, New York, NY, 2012, pp. 267–310.
[113] S. Trenfield, et al., The Shape of Things to Come: Emerging Applications of 3D Printing in Healthcare, 2018, pp. 1–19.

[114] S. Sinha, D. Vohora, Chapter 2 - drug discovery and development: an overview, in: D. Vohora, G. Singh (Eds.), Pharmaceutical Medicine and Translational Clinical Research, Academic Press, Boston, 2018, pp. 19–32.
[115] I. Seoane-Viaño, et al., Evaluation of the therapeutic activity of Melatonin and Resveratrol in inflammatory bowel disease: a longitudinal PET/CT study in an animal model, Int. J. Pharm. (2019) 118713.
[116] Y. Gao, C. Gesenberg, W. Zheng, Chapter 17 - oral formulations for preclinical studies: principle, design, and development considerations, in: Y. Qiu, et al. (Eds.), Developing Solid Oral Dosage Forms, second ed., Academic Press, Boston, 2017, pp. 455–495.
[117] G. Singh, Chapter 4 - preclinical drug development, in: D. Vohora, G. Singh (Eds.), Pharmaceutical Medicine and Translational Clinical Research, Academic Press, Boston, 2018, pp. 47–63.
[118] N. Gómez-Lado, et al., Gastrointestinal tracking and gastric emptying of coated capsules in rats with or without sedation using CT imaging, Pharmaceutics 12 (1) (2020) 81, https://doi.org/10.3390/pharmaceutics12010081.
[119] K. Pietrzak, A. Isreb, M.A. Alhnan, A flexible-dose dispenser for immediate and extended release 3D printed tablets, Eur. J. Pharm. Biopharm. 96 (2015) 380–387.
[120] I. Seoane-Viano, et al., Longitudinal PET/CT evaluation of TNBS-induced inflammatory bowel disease rat model, Int. J. Pharm. 549 (1–2) (2018) 335–342.
[121] A. Goyanes, et al., PET/CT imaging of 3D printed devices in the gastrointestinal tract of rodents, Int. J. Pharm. 536 (1) (2018) 158–164.
[122] D. Smith, et al., 3D printed capsules for quantitative regional absorption studies in the GI tract, Int. J. Pharm. 550 (1) (2018) 418–428.
[123] C. Buoen, O.J. Bjerrum, M.S. Thomsen, How first-time-in-human studies are being performed: a survey of phase I dose-escalation trials in healthy volunteers published between 1995 and 2004, J. Clin. Pharmacol. 45 (10) (2005) 1123–1136.
[124] U. Derhaschnig, B. Jilma, Phase-I studies and first-in-human trials, in: M. Müller (Ed.), Clinical Pharmacology: Current Topics and Case Studies, Springer Vienna, Vienna, 2010, pp. 89–99.
[125] FDA, Estimating the Maximum Safe Starting Dose in Initial Clinical Trials for Therapeutics in Adult Healthy Volunteers, 2005. Available from: https://www.fda.gov/regulatory-information/search-fda-guidance-documents/estimating-maximum-safe-starting-dose-initial-clinical-trials-therapeutics-adult-healthy-volunteers.
[126] J. Shen, et al., Design and conduct considerations for first-in-human trials, Clin. Transl. Sci. 12 (1) (2019) 6–19.
[127] K.J. Bittorf, T. Sanghvi, J.P. Katstra, Design of solid dosage formulations, in: D.J. am Ende (Ed.), Chemical Engineering in the Pharmaceutical Industry, 2010.
[128] Y. Yang, et al., 3D printed tablets with internal scaffold structure using ethyl cellulose to achieve sustained ibuprofen release, Eur. J. Pharmaceut. Sci. 115 (2018) 11–18.
[129] A. Isreb, et al., 3D printed oral theophylline doses with innovative 'radiator-like' design: impact of polyethylene oxide (PEO) molecular weight, Int. J. Pharm. 564 (2019) 98–105.
[130] J. Zhang, et al., Coupling 3D printing with hot-melt extrusion to produce controlled-release tablets, Int. J. Pharm. 519 (1–2) (2017) 186–197.
[131] N.G. Solanki, et al., Formulation of 3D printed tablet for rapid drug release by fused deposition modeling: screening polymers for drug release, drug-polymer miscibility and printability, J. Pharmaceut. Sci. 107 (1) (2018) 390–401.
[132] S.J. Page, A.C. Persch, Recruitment, retention, and blinding in clinical trials, Am. J. Occup. Ther. 67 (2) (2013) 154–161.
[133] EMA, ICH Q1A (R2) Stability Testing of New Drug Sustances and Drug Products, 2003. Available from: https://www.ema.europa.eu/en/ich-q1a-r2-stability-testing-new-drug-substances-drug-products.

CHAPTER 19

# 3D bioprinting: a step forward in creating engineered human tissues and organs

**O. Alheib[1,2], L.P. da Silva[1,2], Yun Hee Youn[1,2,3], Il Keun Kwon[3], R.L. Reis[1,2,3], V.M. Correlo[1,2]**

[1]3B's Research Group, I3Bs – Research Institute on Biomaterials, Biodegradables and Biomimetics, University of Minho, Headquarters of the European Institute of Excellence on Tissue Engineering and Regenerative Medicine, AvePark, Parque de Ciência e Tecnologia, Zona Industrial da Gandra, Barco, Portugal; [2]ICVS/3B's–PT Government Associate Laboratory, Braga, Portugal; [3]Department of Dental Materials, School of Dentistry, Kyung Hee University, Dongdaemun-gu, Seoul, Republic of Korea

## 19.1 Introduction

According to Langer and Vacanti, "tissue engineering applies the principles of biology and engineering to the development of functional substitutes for damaged tissue" [1]. Thus, tissue engineering involves the combination of cells, biomaterial, and growth factors toward developing a tissue-like construct. Usually, this construct is either maintained for therapeutic studies outside the body or transplanted in vivo for regenerative purposes.

Scaffold design is a crucial determinant for building a novel tissue. It works as a platform where cells are able to adhere, proliferate, and differentiate, as naturally occurring in the native tissue. Among other scaffold properties, such as mechanical features, physicochemical properties, and bioactivity, the topography and internal design of the scaffold have a critical impact on the cellular spreading and maturation toward the preferred phenotype. Multiple methods have been used to fabricate scaffolds with reasonable macrostructure and feasible network of interconnectivity. Those methods including, but not limited to, solvent casting, salt leaching, and gas foaming [2,3]. Each method contributes to fabricate scaffolds with variable properties. However, none is able to design a porous structure with precision [4].

3D bioprinting is a promising technology, so it is expected to overcome the aforementioned limitations and improve the output of tissue engineering field in terms of complexity and reproducibility. It holds within a notable efficacy for a rapid, precise, and personalized fabrication of tissue engineering constructs by simply adding several layers of one or different inks or bioinks. A 3D modeling software is used to design a variety of

*Additive Manufacturing*
ISBN 978-0-12-818411-0
https://doi.org/10.1016/B978-0-12-818411-0.00016-1

complex geometries. The 3D models or "CAD files" can be digitally designed through specific software, or more precisely, obtained by computerized tomography (CT) scans of the desired existing scaffold. Next, the designs are exported to the bioprinter in a CAD (computer-aided design) file format to build the corresponding 3D structure [5]. Different 3D bioprinting techniques have appeared in the market, each one presenting specific features, drawbacks, and advantages. Thus, each method provides specific conditions of pressure, temperature, and depositing speed to ensure the optimum atmosphere for cell survival, and configure the best printability and cellular processing. In addition to the design, the type of bioink extruded through the bioprinter plays a major role in protecting the cells from harsh conditions, enhancing their spreading and proliferation [6–8].

## 19.2 Bioprinting technologies

Up till now, several methods of bioprinting were adopted for extruding inks and/or bioinks, including laser, inkjet, and extrusion-based bioprinting [6]. Each approach has different features and was studied widely (Table 19.1).

### 19.2.1 Laser bioprinting

Laser bioprinting is a nozzle-free method used to extrude bioink with milder stress on the cells, higher speed, and enhanced resolution. It consists of a precise focused laser beam pulsed onto a layer-shaped bioink coated with energy-sensitive metal. The process of bioprinting consists of using a laser beam to irradiate a transparent glass coated with a thin film of an absorbing material (such as gold or aluminum), and that has been later on covered with a bioink layer. The laser beam heats up the absorbing film forming a shock wave that releases a bioink droplet.

Extreme details of the final construct are one of the most attractive properties. Thus, this technique offers a microscale resolution (10–100 μm) within a short time. Compared with other techniques, laser bioprinting controls the geometry of the extruded construct on a cell-scale level, allowing the design of highly organized 3D structures with elevated percentage of pores, higher speed, and better spatial positioning (Fig. 19.1). Moreover, it involves the sequential deposition of layers consisting of two or more materials that bond together to build a considerable thick 3D construct [12,27].

**Table 19.1** Type of 3D bioprinting technologies and their main features.

| | Laser bioprinting | Inkjet bioprinting | Extrusion-based bioprinting |
|---|---|---|---|
| Main parts | 1. Laser source<br>2. Release layer (bioink layer) | 1. Nozzle<br>2. Piezoelectric crystal/ electric coil | 1. Nozzle<br>2. Piston/screw |
| Extruding power | Laser beam | Thermal, electrical energy | Pneumatic/screw force |
| Bioink type | Na-alg, hydrogel, acrylated PEG, nano HAp | Collagen, ceramics, PEGDA, Na-alg | PCL, hydrogel, HPC, PLA, PEG, Na-alg |
| Bioprinting speed | 200–2000 mm/s | 300 μm/s | 100 μm/s–30 mm/s |
| Droplet diameter | 40–180 μm | 10–60 μm | >500 μm |
| Cell density | $1 \times 10^8$ cell/mL, high | $2 \times 10^6$ cell/mL | $1 \times 10^6$ cell/mL |
| Pulse rate | 5 kHz | 10 kHz | 1–10 Hz |
| Bioprinting resolution | 10–100 μm Single-cell resolution | 100–400 μm | 10–250 μm |
| Cell viability | 90%–98% | 94%–98% | >90% |
| Ink viscosity | 120 mPas | 4–20 mPas | 3000 mPas |
| Purchase cost | High | Low | Moderate (cost-effective) |
| Other features | - Up to seven living cells/ droplet<br>- Scaffold porosity 90%<br>- In vivo bioprinting in defective site | - Dot density: 360 dpi<br>- Nozzle size: 20–30 μm | - Scaffold porosity: 80%,<br>- Nozzle size: 100–500 μm diameter,<br>- Flow rate 1.5 mL/h |
| References | [9–14] | [15–22] | [23–26] |

*Hap*, hydroxyapatite; *HPC*, hydroxypropyl cellulose; *Na-ALG*, sodium alginate; *PCL*, polycaprolactone; *PEG*, polyethylene glycol; *PEGDA*, polyethylene glycol diacrylate; *PLA*, polylactic acid.

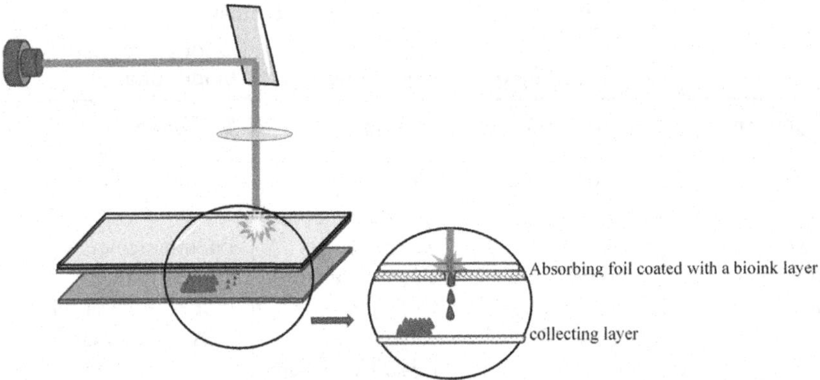


**Figure 19.1** Laser bioprinter is a nozzle-free technique, utilizing laser energy to direct droplets of bioink into a collective substrate.

Most remarkably, laser-equipped bioprinting is an ejector free technique; thus, cells can survive high pressure, and death rate is remarkably low. The rate of viable bioprinted cells is related to process parameters used that allow producing a nondeformed droplet. Thus, the association between the cell viability and the laser power is primarily due to the thermal energy generated from the laser beam, leading to irreversible deformation of the cell droplet formation and, consequently, compromising cellular survival [28]. More recently, it was optimized to be compatible with in vivo bioprinting during miniature surgical operation, showing high positioning and precision of the bioprinted living cells [29].

## 19.2.2 Inkjet bioprinting

Inkjet bioprinting technology, as in the traditional jet printing, works by delivering the bioink as small droplets into a predesigned digital construct. The bioprinter is mainly composed of a nozzle equipped with a thermal/electrical source of power to control the release of bioink droplets into a proper substrate or culture dish. The nozzle is fully controlled by a software to regulate the structure, dimensions, and positioning of the bioprinted construct [17] (Fig. 19.2).

Within the broad category, inkjet bioprinters can be classified into thermal or piezoelectric bioprinters. **Thermal inkjet** bioprinters depend on electrical current to pass through the coil to warm the nozzle (bioink deposit). Once heated, air bubbles create enough pressure to propel bioink droplets out of the tip of the nozzle. In contrast to thermal inkjet bioprinters, which depend on high temperature to eject the bioink out of the

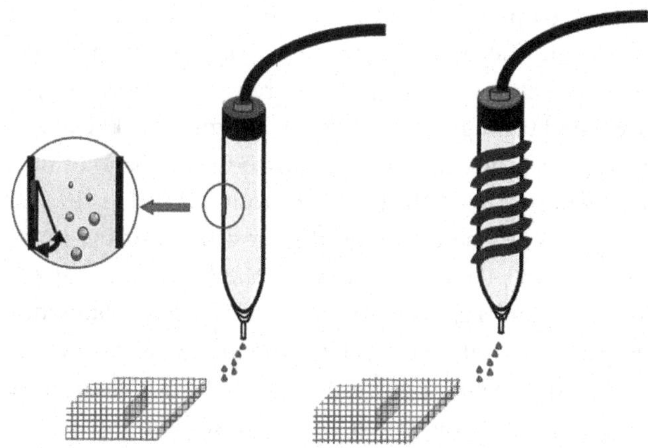

**Figure 19.2** Inkjet bioprinters depend on electric coil around the head which produces enough heat to eject the droplets (right) or depend on the piezoelectric crystal (left) to make droplets.

nozzle, **piezoelectric** bioprinters depend on electrosensitive layers of ceramic or crystal, which can morphologically move or bend in response to an electrical potential, to push the droplets out of the tapped head according to the force that was generated [19].

Different cell types/bioinks were successfully extruded through this type of technique to build heterogeneous tissue with more sophisticated forms, or complex shape. Hence, Xu et al. [22] utilized jet bioprinter to extrude fibroblast-laden alginate into tubular construct (3 mm diameter); importantly, the tubular shape was overhang, including sharp turns, "zigzag shape", $CaCl_2$ was used to cross-link the extruded construct. Cells were still viable ($\sim$80%) after 3 days in culture. As a result, this proof-of-concept study opens the door toward engineering, not only vascular-like tissue, but also complex structure and topography in the future. Nevertheless, inkjet bioprinting suffers many limitations, including nozzle obstruction and low cell viability [25]. However, to eliminate the risk of bioink clogging, cross-linkers can be used during the postbioprinting stage to ensure complete gelation out of the bioprinter head [16]. Other parameters, including nozzle diameter, or cell concentration, can also be adjusted to enhance cellular viability rate [30].

### 19.2.3 Extrusion-based bioprinting

Extrusion-based bioprinting is the most commonly used approach. This is possibly because extrusion-based bioprinters are cheaper, easier to use, and becoming more commercially available. Extrusion-based bioprinters have one or multiple printheads that comprise a syringe where a bioink is loaded. By means of mechanical force (piston/screw) or pneumatic pressure, the bioink is dispensed through a nozzle to a platform [31]. The closely attached drops of bioink keep flowing out the orifice in the shape of fibers to form multilayered thick tissues with tunable porosity [23] (Fig. 19.3). Layer-by-layer bioprinting is attained through the movement of the nozzle in the z axis. This bioprinting technology enables the bioprinting of complex tissues, as different bioinks can be bioprinted in the same/different layers, and a wide spectrum of bioinks with different properties (viscosities, mechanical properties, bioactivity, etc.) can be used to bioprint the construct. Nevertheless, the settings to extrude each bioink need optimization to ensure the accuracy of the desired construct and to get the highest level of stability, complexity, and uniformity of distributed cells.

## 19.3 Bioinks

Over the past two decades, the term bioink has evolved, alongside with bioprinting development. It refers to cells encapsulated into biomaterials,

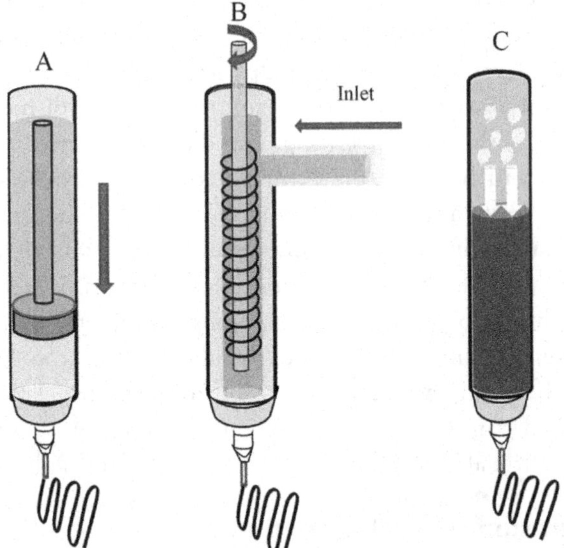


**Figure 19.3** Extrusion-based bioprinters use mechanical force generated by either piston (A), or screw (B), or pneumatic pressure (C) to extrude beads of biomaterial through the nozzle.

which can be extruded through bioprinting technology [32]. Ideally, an extrudable bioink should present flexibility in converting from a solution to a gel-like phase, out of the bioprinter head to avoid any clogging within the nozzle. The gel-like phase allows the maintenance of construct stability and shape fidelity. The maximum control of the transition between the two phases depends on biomaterial physical properties and the particular method by which the biomaterials cross-link after being extruded (ionic, temperature, UV).

Importantly, the biomaterial (ink), as a cell holder, should provide the ultimate conditions of biocompatibility, which outline the need for maintaining bioprinted cells' life [33]. Unlike conventional thermoplastic polymers or ceramics, the selected biomaterial should be able to encase cells, provide a natural hosting environment necessary for proliferation, communication, and maintain the cellular stress to the minimum during the whole extruding process [34,35]. Biomaterials (inks) used in bioprinting can be of synthetic origin (i.e., polyethylene glycol [PEG] or natural origin (agarose, alginate, chitosan, collagen, fibrin, gelatin, gellan gum, hyaluronic acid, and silk fibroin). In Table 19.2, a list of biomaterials (inks) commonly used in bioprinting is outlined.

### 19.3.1 Natural-based inks and bioinks

#### *19.3.1.1 Agarose*

Agarose is a D-galactose-based polysaccharide that can be obtained from marine algae. It is widely used as an ink for bioprinting purposes due to its thermal cross-linking traits, as it jellifies below the temperature of 35°C [66,67]. The gelation time, mesh strength, and elasticity are proportional to its molecular weight, and its mechanical properties can be tuned according to its polymeric concentration [68]. Although the possibility of tailoring an ink may represent a valuable feature, alteration of its rheological properties leads to recurrent optimization of printing parameters. To overcome this hurdle, agarose has been chemically functionalized. Accordingly, carboxyl group was introduced to the C-6 D-galactose; as such, the carboxylated agarose represented a mechanically tunable bioink with stable viscous properties and lower shear stress. Consequently, hMSCs-based bioink featured enhanced biocompatibility and high viability rate (95%) [37].

Agarose, as most of the polysaccharides, does not contain cell-adhesive sites. As a result, only around one-third of cells have been reported to be alive after the bioprinting process [36]. Different approaches have been explored to counteract this drawback. Gu et al. (2016) [38] blended agarose

**Table 19.2** Summary of various bioinks and their extruding methods and their main features.

| Source | Ink | Chemical nature | Ink properties | Bioprinting method | Method to improve mechanical properties | Method to improve cell adhesion | Cells | References |
|---|---|---|---|---|---|---|---|---|
| Natural | Agarose | Poly saccharide | Rapid gelation kinetics, low cell affinity | EBB | Carboxylation, combination with alginate, incorporation with nanosilicate | Combination with Matrigel | HCT116, hMSCs, hNSCs, fibroblasts | [36–39] |
| Natural | Alginate | Poly saccharide | Biocompatibility, limited biodegradation | EBB | N/R | Functionalization with RGD peptide, blending with nanofibrillated cellulose | hADSCs, CPCs, iPSCs | [40–42] |
| Natural | Chitosan | Poly saccharide | Hard to control viscosity, weak stability after extrusion | EBB | Methacrylation, combination with alginate | Conjugation to gelatin | CPCs, fibroblasts | [43–46] |
| Natural | Gellan gum | Poly saccharide | Gelation capacity at body temperature, tunable mechanical properties, lack of cellular adhesive traits | EBB | Blending with PEGDA | Functionalization with RGD peptide, functionalization with methacrylic anhydride | hMSCs, BMSCs | [47–49] |
| Natural | Hyaluronic acid | Amino-poly saccharides | Not cell adhesive, lack of immunogenicity | EBB | Methacrylation | Combination with collagen or gelatin | Chondrocytes, VICs, hAVICs | [50–52] |

| | | | | | | | | |
|---|---|---|---|---|---|---|---|---|
| Natural | Fibrin | Protein | Controlled degradation | IBB, EBB | Combination with alginate and genipin | N/R | HMVEC, hGBM | [30], [53,54] |
| Natural | Collagen | Protein | Mechanical properties are proportional to the collagen concentrations | DBB, EBB | Combination with alginate | N/R | SMCs, MFCs, chondrocytes | [55–57] |
| Natural | Gelatin | Protein | Weak mechanical properties, enriched with cellular adhesive motifs | EBB | Methacrylation | N/R | HUVEC, skeletal muscle cells | [58,59] |
| Natural | Silk fibroin | Protein | Biocompatibility, controlled degradability, low viscosity | DLP, EBB | Methacrylation, combination with alginate | N/R | Fibroblasts | [60,61] |
| Synthetic | PEG | Polyether | Poor mechanical strength, lack of cellular-adhesive traits | EBB | Methacrylation | Fibronectin and vitronectin coating | Endothelial cells, hMSCs | [62–65] |

*BMSCs*, bone marrow stem cells; *CPCs*, cartilage progenitor cells; *DBB*, droplet-based bioprinting; *DLP*, digital light processing; *EBB*, extrusion-based bioprinting; *hADSCs*, human adipose-derived stem cells; *hAVICs*, human aortic valvular interstitial cells; *HCT116*, human colonic epithelial cells; *hGBM*, human glioblastoma multiforme; *hMSCs*, human mesenchymal stem cells; *HMVECs*, human microvascular endothelial cells; *hNSCs*, human neural stem cells; *HUVEC*, human umbilical vein endothelial cells; *IBB*, inkjet-based bioprinting; *iPSCs*, induced pluripotent stem cells; *MFCs*, meniscus fibrochondrocytes; *N/R*, not required; *PEGDA*, poly(ethylene glycol) diacrylate; *RGD*, arginine–glycine–aspartic acid; *SMCs*, smooth muscle cells; *VICs*, valvular interstitial cells.

with alginate and carboxymethyl-chitosan for ionic cross-linking and cellular encapsulation purposes, respectively. Induced human-derived neural stem cells (hNSCs) were successfully encapsulated in this blend of inks and used to bioprint a stable 3D neural-like tissue. Construct biofunctionality was demonstrated by calcium imaging of active neurons expressing gamma-aminobutyric acid (GABA) receptors. More recently, a nanocomposite agarose ink was developed by incorporating laponite nanosilicate into agarose, to enhance its physical structure, printability, and cell attachment [39]. Remarkably, specific percentage of nanosilicate (2 wt%) impacted the agarose rheology, improved mechanical stability, as well as the viability and spreading of encapsulated fibroblasts.

#### 19.3.1.2 Alginate

Alginates are polysaccharides extracted from brown seaweeds that consist of multiple residues of mannuronic and guluronic acids. The repulsion between its negatively charged group results in a highly viscous solution, even at low concentrations. Moreover, alginate is biocompatible and presents tunable stiffness, making it a widely used ink in 3D bioprinting [40]. In fact, the impact of alginate stiffness on cellular behavior has been previously studied [41]. Gradual concentration (2%, 4%, and 6% w/v solution) of alginate inks were evaluated with cartilage progenitor cells (CPCs) and extruded with coaxial nozzle. Bioprinting of alginate alone, at the higher alginate concentration (6%), resulted in higher viscosity and sharp cell death (up to 69%). Hence, the stiffer alginate had a higher impact on cell damage during bioprinting due to the increased shear stress inside the nozzle.

Various strategies have been applied to overcome the lack of cell adhesion motifs in alginate. Accordingly, alginate was oxidized with sodium periodate and reacted with G4RGDSP-OH peptides to improve cell adhesiveness [40]. Inks combined with hADSCs were bioprinted, and the cells showed a higher living rate and were able to spread homogeneously in 3D lattice structures. Moreover, to enhance its biological cues, alginate was also blended with nanofibrillated cellulose (NFC) to provide the best microenvironment for pluripotent stem cells (iPSCs) toward cartilage regeneration. The bioprinted construct showed good biocompatibility and differentiation after 5 weeks of culture [42].

#### 19.3.1.3 Chitosan

Chitosan is a biodegradable, biocompatible, and nontoxic glucosamine polysaccharide. It is obtained from chitin by means of deacetylation with alkaline solutions, which leads to an increase of its water absorbance capacity and hydrophilic properties. Moreover, thanks to the cationic amino side group, chitosan is capable of blending with negatively charged polymers (e.g., alginate), thus providing an alternative method of cross-linking during bioprinting [44].

Chitosan has been suffering from chemical modifications in an attempt to get a more stable chitosan inks. Accordingly, chitosan was chemically modified with methacrylic anhydride to induce photo-cross-linking, posteriorly, conjugated with β-glycerol phosphate salt (β-GP) for thermal gelation [45]. The combined effect of dual cross-linking mechanism had the potential of cytocompatibility, shape fidelity, fine-detailed bioprinting, and stability. In addition, cellular studies on fibroblast demonstrated a high viability rate without cytotoxicity, as well as cellular proliferation.

Chitosan has also been combined with cell-adhesive polymers taking in consideration its weak biocues for printability purposes. Hence, to enhance its potential for bioprinting, chitosan was conjugated with gelatin to yield polyelectrolyte gelatin—chitosan [46]. Extrusion of fibroblasts loaded in this ink showed enhanced cellular viability compared with the chitosan alone.

#### *19.3.1.4 Gellan gum*

Gellan gum is a nonbranched polysaccharide produced by bacterial fermentation. Although it is already widely used in food industry and pharmaceutical purposes, only recently, it has become an attractive biomaterial in the field of tissue engineering owing to its tailored physical properties, moderate gelation temperature close to body temperature, and the ability to cross-link with low levels of mono or divalent cationic molecules ($Na^+$, $Ca^{2+}$). However, as a polysaccharide, it lacks the superficial anchoring cell-adhesive motifs that can be added though chemical tethering of short cell-adhesive sequences, such as RGD (arginine—glycine—aspartic acid) moieties [69]. Accordingly, Lozano et al. [48] reported the use of functionalized gellan gum (GG) with RGD to bioprint a 3D brain-like structure. For this purpose, GG was activated by carbodiimide derivatives and sulfo-NHS compounds to yield an intermediate compound, which is posteriorly linked to the peptide. The GG-RGD bioink showed tunable features enough to bioprint the primary cortical cells with a notable cell positioning, which developed neuronal networks in the bioprinted 3D structure. This study demonstrated that the modification with RGD overcame the cell adherence properties of GG. Moreover, no shear stress damage was detected.

In a different approach [49], GG was blended with UV-curable poly(ethylene glycol) diacrylate (PEGDA) to enhance its compressive modulus and stability. Murine bone marrow stem cells (BMSCs) were encapsulated into GG/PEGDA ink and bioprinted using an extrusion-based bioprinter. After 21 days of culture, cells were alive, spreading, and homogenously distributed throughout the bioprinted construct.

#### 19.3.1.5 Hyaluronic acid or hyaluronan

Hyaluronic acid (HA) is a nonsulfated glycosaminoglycan polysaccharide spread out through many body tissues including, but not limited to, connective and epithelial tissues. Thus, it is an important component of the cartilage, skin, or even the vitreous fluid of the eye. HA plays a significant role in tissue formation, as the cells are capable of binding to HA through cell receptors, such as CD44 [70], and mediate different cellular responses. Nevertheless, HA is still considered a weak cell-adhesive polymer [71]. To increase its cellular attachment property, HA has been chemically functionalized with cell-adhesive peptides (RGD), by using a CDI (1,1′-carbonyldiimidazole) coupling agent [72], or blended with collagen or fibronectin to obtain the tunable adhesive formula [73].

HA also presents viscoelastic properties that make it a great ink candidate to be explored for bioprinting purposes. However, using inks of sole HA does not present suitable mechanical properties and proper gelation behavior [74]. To overcome this limitation, blended compositions of HA have been used. HA was blended with gradual concentrations of methacrylated gelatin to set the required viscosity for maximum encapsulation [75]. Likewise, HA was also combined with alginate for nerve tissue engineering applications to support the stiffness of the extruded construct [76]. Moreover, a blend of methacrylated HA and methacrylated gelatin was used to bioprint a human aortic valvular interstitial cells for trileaflet valve engineering purposes [52]. While HA was methacrylated to be photo-cross-linkable, HA was also blended with gelatin to tailor its physical properties (rheology, viscosity). The final construct was well established, and the cells were spreading and properly integrated.

#### 19.3.1.6 Fibrin

Fibrin is a protein involved in the coagulation cascade during tissue injuries and wound sealing, by forming the fibrin network. It is a natural, biocompatible, and nontoxic biomaterial owing to its autologous merit. In the context of tissue engineering, it has been successfully used to fabricate tissue-like constructs with different progenitor cells (e.g., smooth muscle cells, skeletal muscle cells, chondrocytes) [30].

Fibrin has been widely used for bioprinting owed to its biological features and native cross-linking mechanisms. Accordingly, a modified thermal inkjet bioprinter was used to construct human microvasculature-like tissue using a fibrin-based bioink [53]. Briefly, an HMVEC-laden thrombin bioink was extruded with thermal jet bioprinter and collected on a fibrinogen substrate. Once the thrombin became in contact with fibrinogen, a grid of fibrin was formed, resembling a vasculature net. Although fibronectin presents cell-adhesive features that make it highly attractive for bioprinting, it lacks mechanical stability. As an attempt to overcome its weak mechanical properties and stability after bioprinting, fibrin was combined with alginate and genipin to bioprint a glioblastoma tumor model [54]. The bioink was composed of human glioblastoma multiforme (GBM) cells suspended in fibrin solution, which was incorporated with alginate and genipin for cross-linking purposes. After 12 days of culture, cells were highly viable (over 80%), able to form spheroids and produce neuronal markers. Hence, the bioprinted fibrin-based tumor model was suggested as a 3D model for personalized drug screening.

#### *19.3.1.7 Collagen*

Collagen is the main structural protein in the mammalian kingdom. So far, collagen is considered the main biomaterial in the connective tissues, usually obtained from the skin of domestic mammals, such as cows and pigs. However, due to issues related with possible transmitted disease, fish is considered a reliable alternative source for collagen [77]. Recently, the recombinant human collagen is getting more attention due to its reduced antigenicity [78]. Collagen has more than 20 types reflecting its distribution in the body. Among the different collagen types, collagen type I is considered the most prevalent protein in the body, including skin, tendons, and bones.

In the context of tissue engineering, collagen has been broadly used to encapsulate cells, providing cells with a suitable substrate for adhesion and proliferation [55,56]. In view of this, collagen was used as an ink to extrude smooth muscle cells. A multilayered 3D construct was established by depositing multiple layers successively, waiting between each two layers to allow thermal cross-linking (at 37°C) to occur. Cells were homogeneously distributed within the 3D construct and remained highly viable (over 90%) along 14 days of culture [25]. As collagen has slow jellifying properties (takes up to 30 min to cross-link), it was also blended with alginate to speed its ionic cross-linking rate. As such, chondrocytes were encapsulated in a blend of collagen and alginate, bioprinted for cartilage tissue engineering purposes, and the cells were highly viable along 14 days, well spreading, and differentiating [57].

#### 19.3.1.8 Gelatin

Gelatin is a natural polypeptide originated from collagen by an irreversible hydrolysis reaction. It is a cheap biomaterial that can be easily obtained from skin, bone, or connective tissues of cuttlefish. However, due to the increased risk of any transmitted disease, and batch-to-batch variation, recombinant gelatin is attracting more attentions for biomedical purposes [79]. Gelatin is widely used in tissue engineering applications due to its naturally embedded RGD motifs, as derived from collagen [80], and aqueous solubility, providing a proper environment for cell encapsulation and proliferation with a high viability rate [81]. On the contrary, based on its weak mechanical properties, gelatin has been covalently interacted with extra supportive fillers such as hydroxyapatite [82], or chemically modified with higher numbers of tyrosine-derived phenol residuals. Accordingly, gelatin was combined with desaminotyrosine (DAT) and desaminotyrosyl tyrosine (DATT) that increased the tensile properties and stiffness despite decreasing water affinity [82,83]. Moreover, gelatin was used as an ink after chemically modification with methacrylate groups, as it increased bioprintability and tuned its mechanical properties without compromising its biocompatibility [84,85]. In another study [59], gelatin methacryloyl (GelMA) and alginate were used to bioprint a skeletal muscle-like tissue, where the skeletal muscle cells (C2C12) were highly viable after extrusion, spreading, and differentiating after 12 days of culture.

#### 19.3.1.9 Silk fibroin

Silk fibroin (SF) is a natural fibrous protein, the main component of the silk and spider web. It is usually purified by removing the silk gum (sericin protein) from the raw silk in a process called degumming. SF-based biomaterials have been widely explored for biomedical applications owing to its biocompatibility, biodegradability, construct stability, and flexible transition between sol–gel state [8]. However, the extracted SF itself is hardly controllable in 3D bioprinting process due to batch–batch variation, as different molecular weights and concentration are detected between batches, which in turn impact the gelation time and cross-linking methods [61,86,87].

Different approaches have been used to better control SF gelation process and cross-linking. Therefore, SF was chemically modified with glycidyl methacrylate (GMA) to become photo-cross-linkable. Human chondrocytes were encapsulated into the silk fibroin methacrylate (SF-MA) ink [60]. After extrusion, the construct was mechanically stable; cells were homogenously distributed in the 3D construct and remained highly viable and proliferative over 14 days of culture. Thus, it could serve in engineering different tissues and organs with higher stability and complexity (heart, trachea, vessels). In the same context, SF was blended with alginate, mixed with NIH/3T3, bioprinted, and cross-linked in a multistep process [61]. In particular, alginate was rapidly cross-linked with calcium chloride to maintain the 3D structure; then, SF was jellified with enzyme horseradish peroxidase (HRP). Alginate played a role as a sacrificial agent, and it was removed by chelating agents such as sodium citrate or ethylenediamine-tetraacetate (EDTA). The successful cell-laden SF construct demonstrated living, spreading, and proliferating cells after 5 weeks in culture.

## 19.3.2 Synthetic-based inks and bioinks

### *19.3.2.1 Polyethylene glycol*

Polyethylene glycol (PEG) is a polymeric biomaterial consisting of multiple units of ethylene oxide. It is available in different molecular weights according to the chain length, which contributes to its versatile physical properties. PEG is a biocompatible material with a water-soluble trait, but due to its low viscosity, it is difficult to be applied for 3D bioprinting. Therefore, it is commonly conjugated with other polymers to enhance its mechanical strength. Accordingly, PEG was modified into methacrylated poly(ethylene) glycol (PEGDMA) to induce its photopolymerization and reacted with GRGDS peptides to enhance its cellular adhesive properties for bioprinting purposes. Thus, human mesenchymal stem cells (hMSCs) were successfully encapsulated in PEGDMA, conjugated with GRGDS peptides [64], and extruded through an inkjet bioprinter. Cells were viable, spreading, and differentiating to different cell lineages, including the osteogenic and chondrogenic lineages. Moreover, the final construct featured robust mechanical properties and uniformly cellular distribution. For more stability purposes, as an ink, PEG microgel was developed by click chemistry [65], PEG was reacted with PEG-dithiol, photopolymerized, and electrosprayed to produce microspheres, eventually mixed with hMSCs for bioprinting purposes. The extruded construct resulted in smooth, stable filaments for complex 3D structures; cells were 90% viable and proliferating after 10 days in culture.

## 19.4 Tissue and organ bioprinting

3D bioprinting has insightful promise toward large-scale organ production. The list of medical models that have already been successfully bioprinted in this field reveals the capacity that this technology holds for the medical field in the near future. Although the main goal is to produce a full functional and realistic tissue/organ, bioprinting progress is constrained by tissue/organ complexity. Multiple types of bioinks and high-throughput bioprinters have been used to bioprint tissues, such as the skin, bone, cartilage, and the skeletal muscle. Nevertheless, complex organs, such as the brain and kidney, are still to be explored.

### 19.4.1 Skin

Large burns and chronic skin injuries represent a hurdle for patients and the healthcare system. Current treatments rely on the use of autologous or allogenic skin grafts that pose constraints of availability and rejection. Thus, 3D-bioprinted skin analogs are a promising alternative to skin grafts.

The skin is a layered organ composed of the epidermis, the dermis, and the hypodermis. Keratinocytes (KCs) are spread on the epidermis, fibroblasts (FBs) are loaded in the dermis, and adipocytes are loaded in the hypodermis. The skin is also innervated, vascularized, and pigmented and presents appendages, such as hair follicles and sweat glands. In spite of the notable progress achieved within tissue engineering to develop skin-like tissue, further steps are required to design stable structure comprising all the previously described components.

3D bioprinting technique has been used to bioprint the complexity of skin tissue architect, by precisely deploying multiple distinct layers with versatile cell types. In view of this, a pneumatic pressure-based bioprinter was utilized to fabricate a 3D skin model [88]. Layer-by-layer stacking of collagen was supplemented to make a 3D structure, intervened with a superimposed layer of KCs over the FBs layers, to mimic the natural skin morphologically. However, histological studies revealed some drawbacks of the final constructs, including missing the proper stacking of its layers and low FB density in dermis. These results were attributed to many factors, including the unique culture conditions for heterogeneous types of cells and the simplified structure, which did not imply with the sophisticated "in vivo" model.

Laser bioprinter was used to fabricate skin-like tissue using a similar printing design [89]. 40 layers of KCs and FBs embedded in collagen were deployed alternatively to mimic the dermis and epidermis pattern of the real skin. Cells were viable and proliferating, and more importantly, the structure was stable over 10 days, not blending or fusing into each other. Thus, as a platform, this skin model can be used to integrate extra cell types, i.e., melanocytes, hair follicles, and design more complex tissue structures with a vascularized inner structure in the future.

An inkjet bioprinter integrated with a laser-based scanning system was used by Yoo's group to take a full scan of a burn or a wound and directly extrude the bioink in position (Fig. 19.4) [90]. Once the topography of the wound was scanned, a pressure-based inkjet bioprinter was used to introduce two diverse types of skin cells (KCs and FBs) directly into the wound. Accordingly, this group was able to successfully demonstrate a viable piece of skin, to serve as a template for a lab study or implantation.

Other studies have shown the ability of bioprinting skin analogues, with higher resemblance with the native skin. Besides KCs and FBs, pericytes (PCs) and endothelial cells (ECs) were bioprinted in the dermis layer as an attempt to construct a vascularized skin [91]. The bioprinted vascularized tissue was able to interconnect with microvascular networks from host microvessels after implantation into mice full-thickness wounds [91]. Melanocytes were printed in the epidermis layer to provide the skin analog with pigmentation [92]. Thus, multilayers of FBs-laden collagen were hierarchically deposited to mimic dermis layer, and KCs and melanocytes were deposited on the surface of the last construct to resemble a full skin layer with pigmentation. Histologically, two distinct skin layers were observed with differentiated KCs and notable melan-A expression. In addition, a blend of gelatin and alginate was used to bioprint epithelial-derived stem cells and epidermal growth factor toward achieving sweat glands [93]. The tissue was able to restore the sweat glands of a burned animal foot pad in vivo.

The bioprinted skin models proved to be analogs of the skin tissue, which were unreachable with traditional clinical approaches. However, in spite of the huge progress, up till now, a full functional skin analog has yet to be achieved, as innervation and all basic appendages are not yet completely integrated into the 3D-bioprinted structures.

### 19.4.2 Bone and cartilage

Bone disorders, including fractures, trauma, and necrosis, are becoming a global concern among physicians and orthopedic surgeons. Besides the

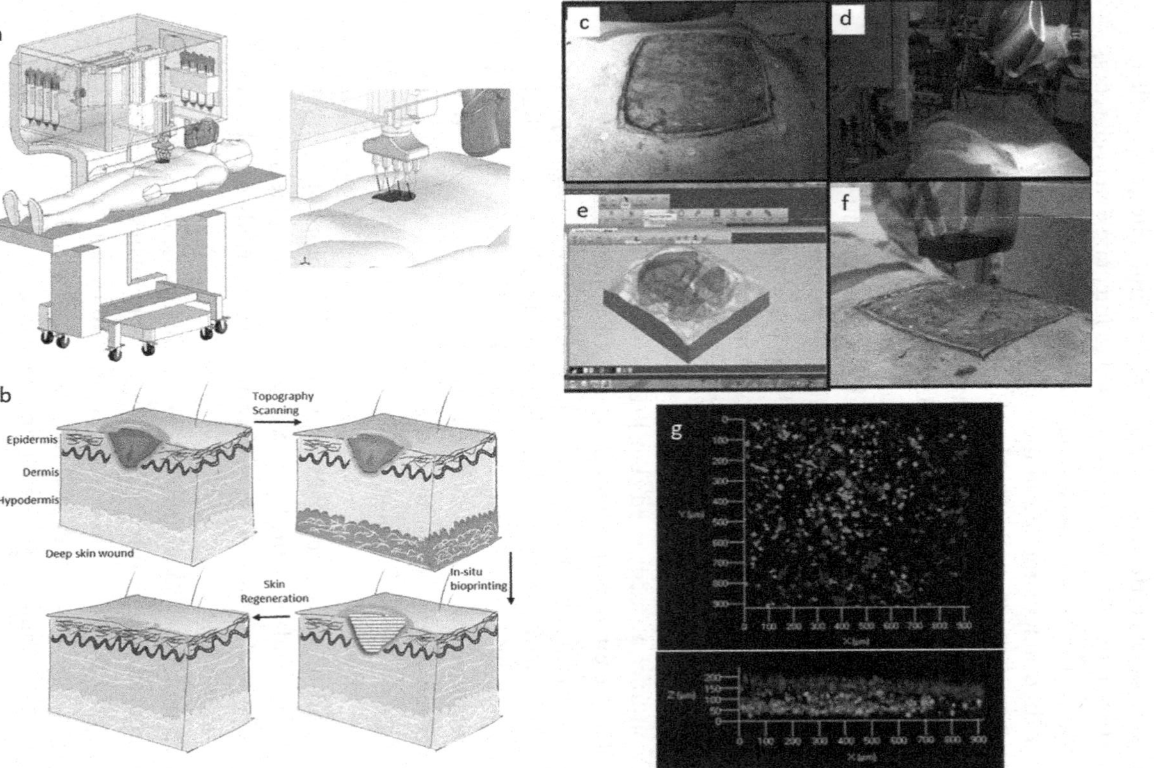


**Figure 19.4** In situ skin bioprinting principle. (A) Illustrative design of the skin bioprinter and the bioprinting process. (B) Skin bioprinting process: Laser-based scanning technology is used to specify the damaged area of the skin, and spatial information is transformed into STL files, which then give the proper information to the bioprinter head to extrude the bioink and fill the corresponding missing volume accordingly. (C-F) Example of bioprinting procedure: wound lesion is outlined with a marker, scanned with a Zscanner, designed and stored in an STL file format by CAD software, finally, uploaded into the bioprinter to be extruded (G) cytoskeletal staining of layering two cell types with high precision, fibroblasts (green [light gray in print version]), and keratinocytes (red [dark gray in print version]). *(Image adapted from Reference M. Albanna et al., "In situ bioprinting of autologous skin cells accelerates wound healing of extensive excisional full-thickness wounds," Sci. Rep., vol. 9, no. 1, pp. 1–15, 2019 and reproduced under creative commons attribution 4.0 international license (CC BY 4.0).)*

limitation of bone grafts and their following immune outcome, autohealing process within bone tissue relies on gender, age, and defect size. Thus, the regenerative potential falls behind its constituent capacity in severe cases of injuries, infection, and tumors.

Bone is a rigid heterogenous tissue with remarkable mechanical properties thanks to hydroxylapatite salts. In addition to its cortical, mineralized, and dense outer layer, it is highly porous core structure enriched with vascular network. Several biomaterials with versatile mechanical properties were recruited to regenerate a realistic bone tissue. Although mechanically stable, cells were not homogenously distributed or encapsulated within the construct, which resulted in a limited cellular connectivity, hence failing in cellular maturation.

Recently, 3D bioprinting has opened the door to create tissues with more bone mimicking traits. Bendtsen et al. [94] adopted 3D bioprinting technique to engineer such a layered tissue-like bone. An alginate—polyvinyl alcohol (PVA)—hydroxyapatite ink was tuned to get the maximum mechanical properties and osteoconductivity to resemble the bone tissue. Alginate was considered the main ink for cellular encapsulation, PVA was used as a thickener agent, and hydroxyapatite was incorporated to mimic the mineralized microenvironment of the bone tissue. MC3T3, osteoblast precursor, was successfully bioprinted with a high viability rate (~95%), and the final construct showed proper mechanical stability. Likewise, a blend of alpha-tricalcium phosphate (α-TCP) and collagen was used to bioprint a ceramic-like construct with proper mechanical properties and porosity, via a pneumatic pressure extruder [95]. Thereafter, preosteoblasts (MC3T3-E1)-laden collagen was bioprinted onto the aforementioned construct, and the cells were alive (more than 90%), spreading, and differentiating toward the osteogenic lineage, as demonstrated by the mineralization assay after 14 days.

3D bioprinting took a step forward on mimicking the bone tissue. Complex constructs with a porous structure and homogenous cellular distribution were all attained by implementing hydrogels, combined with mineralized and osteoconductive biomaterials. Highly viable constructs with reasonable mechanical properties were also achieved. Even though bone tissue bioprinting is still limited to a small-sized construct, efficiently vascularized tissue and mechanical stability are yet to be achieved.

Multiple cartilage injuries, including trauma, rupture, osteoarthritis, and sport injuries, are the most common defects among elderly, which are identified with irreversible degeneration or ossification. At present, medical interventions are limited to pain killers or to surgical intervention in the worst cases.

Articular cartilage is an aneural and vascular-free tissue that encompasses chondrocytes embedded in a matrix of polysaccharides and protein fibers. 3D bioprinting has allowed the development of the cartilage tissue, with the capability to regenerate, enhance, or maintain the function of injured cartilage. Simonsson group [42] successfully bioprinted a cartilage-like tissue using human-derived induced pluripotent stem cells (iPSCs), loaded in a blend of nanofibrillated cellulose NFC and alginate. The ink provided important cues for chondrogenesis; alginate is well known for its encapsulation properties and fast cross-linking, and NFC offers high mechanical properties. As an attempt to influence the differentiation of iPSCs toward the chondrogenesis lineage, chondrocytes were electronically irradiated at 25 KGy and mixed with growth factors (BMP2, TGFβ1, and TGFβ3) to provoke iPSCs differentiation. Overall, printed cells were 73% viable, evenly distributed, and the cartilage-like tissue was well established, since iPSCs were committed to chondrogenesis. Similarly, another group from Sweden [96] managed to bioprint a cartilage-like tissue for in vivo implantation, by using a composite of NFC/alginate-based ink loaded with both human nasal chondrocytes (hNCs) and human bone marrow–derived mesenchymal stem cells (hBMSCs). Once extruded through a pneumatic pressure extruder, the construct was implanted into a nude mouse. Histopathological tests after 2 months of implantation revealed that the combination of the two different types of cells and biomaterial contributed to a higher glycosaminoglycans deposition owing to hBMSCs and its role in promoting chondrogenesis.

3D bioprinting demonstrated advantages on building engineered tissues for chondrogenesis purposes. Specific blends of biomaterials were tuned to provide the essential mechanical properties and biocues for cellular proliferation and differentiation. Nevertheless, multiple challenges are still existing, namely, the need of a huge amount of cells [97] with high regeneration capacity to mimic the chondrogenic development during early embryonic stages. In addition, long-term construct stability with enhanced mechanical properties, construct biocompatibility, and integration into the host cartilage are also required.

### 19.4.3 Skeletal muscle

Skeletal muscles are contractible organs with highly organized units (myofibers). The myofiber consists of a long, cylindrical, multinucleated cell resulted from fusion of multiple myotubes. Each bundle of myofibers is composed of not only myofibers but also motor neurons, blood vessels, and more importantly, satellite cells. Since the capacity of their self-regeneration is limited to small-scale injury, tissue engineering represents a reliable technique in restoring muscular tissue in case of immune rejection or massive loss. Nevertheless, the regeneration of striated muscles is still challenging due to the need for aligned, twitching structure. According to literature, the key factors to get aligned myofibers in a 3D scaffold rely on different approaches, including the use of electrical stimulation, nanocomposites, or finely patterned scaffolds.

Kim et al. (2019) [98] bioprinted a skeletal muscle-like tissue by applying variable bioprinting conditions (nozzle speed, dispensing rate). The combination of gold nanowires (GNWs) with collagen had a positive impact on growth and differentiation of bioprinted c2c12. While the collagen-reinforced cellular encapsulation, GNWs had a remarkable effect on cellular alignment. Moreover, application of an electrical field (5 V, 1 H, along 14 days) onto the bioprinted structure promoted GNWs orientation, which in turn greatly influenced myotubes alignment. Moreover, Costantini's group [99] managed to bioprint a skeletal muscle tissue analog, using c2c12 encapsulated in fibrinogen-PEG/Alginate (FP/Alginate). While alginate was used to enhance quick ionic cross-linking and cellular encapsulation, FP, a photocurable ink, was used to maintain construct stability in the postprinting stage. To avoid nozzle clogging, a double inlet feeder connected to coaxial dispensing head was used to extrude (PF/Alginate) in an inner needle and the cross-linker ($CaCl_2$) in the outer needle. As a result, the engineered construct featured an organized structure with mature and aligned myotubes alongside the deployed biomaterial fibers (Fig. 19.5).

Therefore, 3D bioprinting proved to be a reliable application for engineering muscle-like tissue, with high rate of cellular viability and differentiation. Moreover, myotubes were patterned into aligned fibers based on the extruding design. Nonetheless, further studies are required to fabricate a fully functional tissue with enhanced vascular and innervated network, by incorporating endothelial cells in addition to neural cells. Moreover, in vivo studies are required to understand construct integrity, vascular anastomosis, and metabolic activity.

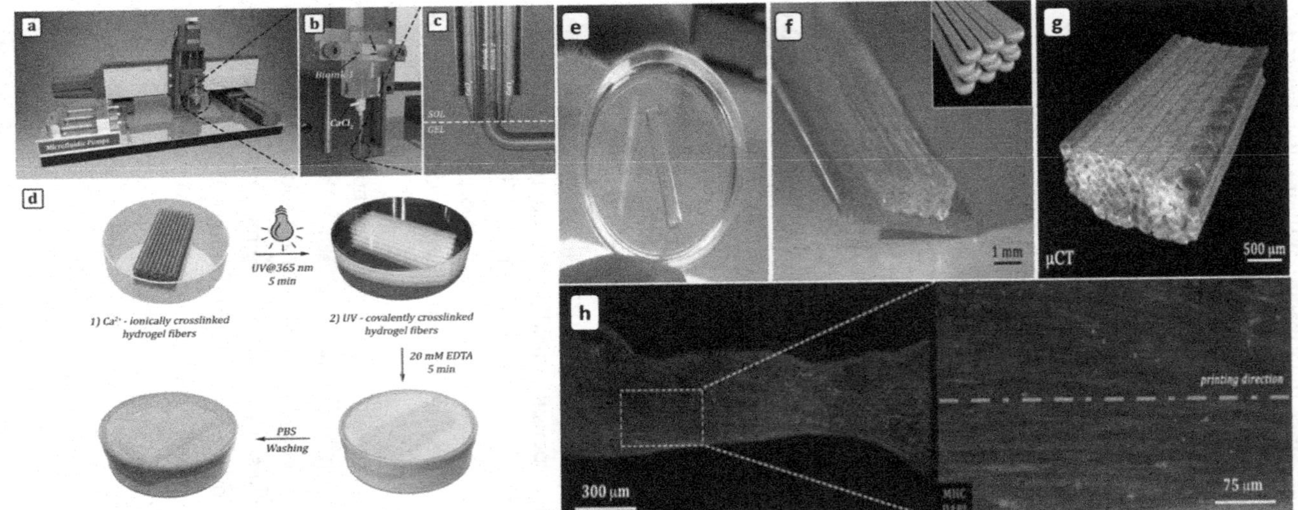


**Figure 19.5** 3D bioprinting schematic. (A), (B) Customized 3D bioprinter equipped with microfluidic dispensing head. Coaxial extruder tip showing inner needle (bioink), and outer needle (cross-linker). (D) UV cross-linking of bioprinted construct and the following alginate removal procedure. 3D-bioprinted structure of aligned PEG–fibrinogen fibers as shown by (F) optical micrographic, and (G) X-ray micro-CT images. (I) Myotube alignment was evidenced in 3D-bioprinted construct after 15 days of in vitro culture by immunofluorescence. MHC is stained in red [dark black in print version], and nuclei are stained with DAPI in blue [light black in print version]. *CT*, computerized tomography; *PEG*, polyethylene glycol. *(Reproduced from Reference M. Costantini et al., "Microfluidic-enhanced 3D bioprinting of aligned myoblast-laden hydrogels leads to functionally organized myofibers in vitro and in vivo," Biomaterials, vol. 131, pp. 98–110, 2017; Reproduced with permission Copyright 2017, Elsevier.)*

## 19.4.4 Other tissues, organs, and structures

**Vascularization** is one of the most-cited issues in tissue engineering applications, since vessels are a vital component of any living tissue and, hence, responsible for mass trafficking of food, metabolites, wastes, and oxygen. Up till now, many attempts have fallen behind building an engineered tissue exceeding few hundreds of micrometers in thickness, owing to the lack of active grid of vessels [100]. In terms of 3D bioprinting, multiple tailored approaches have been adopted to engineer vascularized tissues. Accordingly, a vascular model [101] was engineered using a coaxial bioprinter where the bioink was extruded through the outer part of the coaxial nozzle, building the shell part of the tubular construct, and simultaneously, $Ca^{2+}$ ions were pumped out through the inner part of the coaxial nozzle to cross-link the ink and give shape to a lumen structure. The bioink was composed of human umbilical vein endothelial cells (HUVECs) encapsulated in a blend of alginate and vascular tissue–derived extracellular matrix (VdECM) ink; as such, VdECM collagen content provided a proper microenvironment for HUVECs maturation and enhanced postprinting thermal cross-linking. After 7 days in culture, a thin layer of HUVECs was confluently lining the entire vessel construct. The tubular-endothelialized tissue featured a decreased adhesion rate of blood platelets based on blood perfusion test. Moreover, it showed a lower diffusion rate than the acellular vessel model, according to fluorescein isothiocyanate (FITC)–conjugated 70-kDa dextran perfusion study, which reflects its selective permeability. Therefore, this model could serve as a platform to study vascular-related diseases, for drug screening, or to build heterogenous organ-on-ship.

An emerging approach [102] was also used to produce a vascularized cardiac-like tissue construct using a coaxial microfluidic-assisted bioprinting head. HUVECs and induced pluripotent cell-derived cardiomyocytes (iPSC-CMs) were encapsulated in a blend of PF/alginate and extruded through Y-junction microchannels linked to an inner needle. The $CaCl_2$ was pumped through an outer needle to cross-link the extruded construct. Cellular studies confirmed that HUVECs were spreading, lining the inner structure, and expressing VEGF after 7 days in culture. Furthermore, the prevascularized structure was implanted in vivo, and after 15 days, a number of blood vessels were found within the engineered construct. Moreover, iPSC-CMs cells were aligned and expressing heart-specific genes (i.e., cardiac alpha myosin heavy-chain $\alpha$-MHC and troponin).

Intensive steps are still undergoing to fabricate more complex organs. An in vitro **pulmonary epithelial tissue barrier** [103] was bioprinted for studying the gas trafficking membrane in alveoli. Pressurized air-assisted bioprinter was used to extrude a thin layer of Matrigel, which interfaces two different kinds of well-distributed cells: the alveolar epithelial type II cells and endothelial cells. Viability assays revealed a high survival rate, demonstrated by more than 95% of alive cells after 3 days of culture. The barrier efficiency was confirmed by blue dextran molecule trafficking. This model was intended to mimic the microstructure of the natural tissue; furthermore, it could be used in toxicological and pharmacokinetic screening studies (Fig. 19.6).

In terms of prosthetic replacement, much effort has been spent to engineer a **heart valve** by Butcher's research team [104]. MicroCT images were used as a source for CAD blueprints of the heart valves. A blend of alginate/gelatin was used to encapsulate two different types of cells responsible for the root and the leaflet parts of the valve. As such, two separate nozzles were used to dispense aortic root sinus smooth muscle cells (SMCs) and aortic valve leaflet interstitial cells (VIC), to match the macrostructure of the valve. As a result, the team was able to build a viable aortic valve that mimics the geometry of a natural one with a high viability rate (over 80%) and proliferation after 7 days in culture. Hence, these heart valves can serve as a platform for any personalized valve replacement in the future.

In a recent study [105], Chen research team managed to mimic the **hepatic parenchyma** microenvironment. CAD files were digitally designed in a hexagonal architect ($\sim$900 μm) to match the hepatic unit (lobule) in few millimeter squares. Bioinks were composed of, not only, personalized human-induced pluripotent stem cells (hiPSCs), but also adipose stem cells and endothelial cells, encapsulated in gelatin methacrylate (GelMA). The hepatic model was bioprinted in two consecutive steps: hiPSCs were extruded firstly into hexagons, and after that, the other cell types were extruded in the borderline between hexagonal units to get a honeycomb-shaped construct. The hepatic bioprinted construct showed an organized microstructure, and the cells survived the extruding process along 20 days of culture. Moreover, the key biomarkers of liver function, such as α-fetoprotein, albumin, and hepatocyte nuclear factor 4α, were increased, suggesting cellular differentiation. Members of the cytochrome superfamily of enzymes responsible for drugs metabolism were also highly expressed. With the aforementioned model similar to the human liver, the door is open for pilot studies to assess drug metabolism and detoxification.

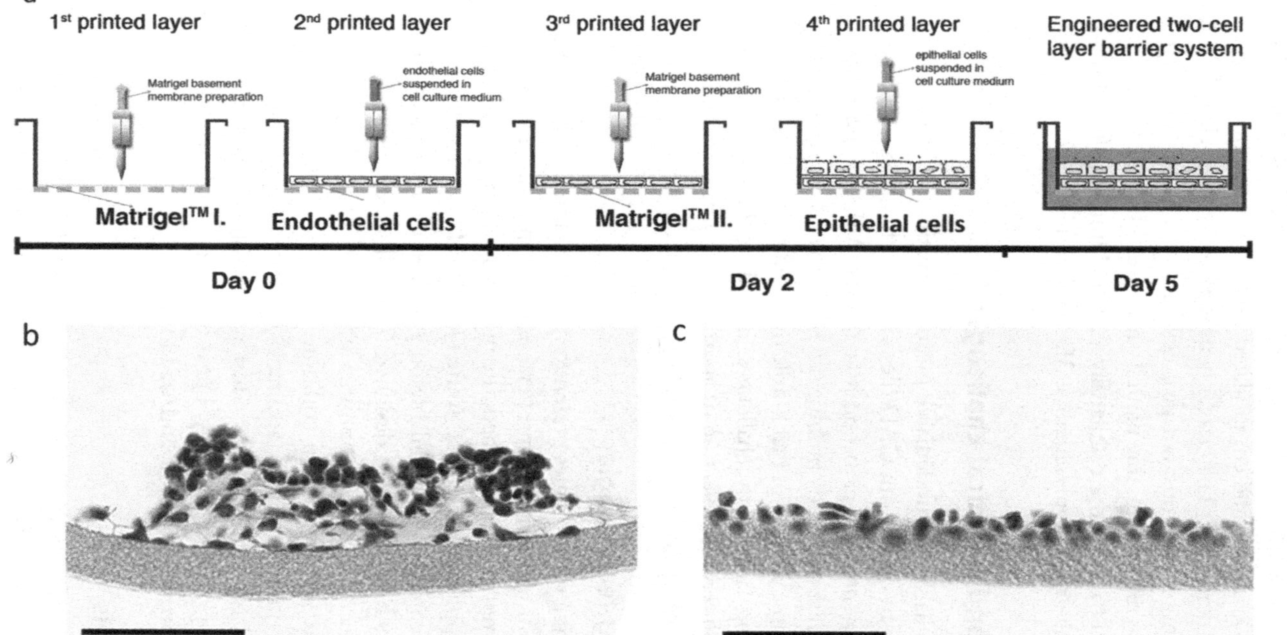


**Figure 19.6** Schematic illustration of pulmonary epithelial tissue barrier processing. (A) The chronological bioprinting of double cell-layer analog was prepared by depositing a thin layer of epithelial/endothelial cells on a Matrigel barrier system. Brightfield micrographs of (B) manually seeded and (C) bioprinted barrier analog as shown by Masson-Goldner trichrome staining. Cytoplasm is stained in red [light gray in print version], collagen fibers of Matrigel are shown in green (mild gray in printed version), and cell nuclei are shown in brown [dark black in print version]. Scale bars are 100 μm. *(Reproduced from Reference L. Horváth, Y. Umehara, C. Jud, F. Blank, A. Petri-Fink, and B. Rothen-Rutishauser, "Engineering an in vitro air-blood barrier by 3D bioprinting.," Sci. Rep., vol. 5, p. 7974, 2015; Reproduced under creative commons attribution 4.0 international license (CC BY-NC-ND 4.0).)*

Most remarkably, Lozano et al. [48] fabricated a **cerebral cortex-like tissue** by a manual (hand-held)-based extruding system. Neural stem cells were isolated from mouse brain and encapsulated into RGD-modified gellan gum and extruded accordingly. The results demonstrated viable, elongated axons and well-distributed neurons and glial cells into the modified gellan gum-based construct, resembling the natural structure. However, further studies are still needed to attain the cellular diversity and functionality of the cerebral cortex, to have more representative 3D models (Fig. 19.7).

## 19.5 Limitation and technical challenges

3D bioprinting is a reliable and undeniable mean of creating test models by stacking bioinks layer-by-layer from CAD files source. It has been attractive due to its immense potential and global positive impact. Although novel approaches and advances are constantly pushing this field forward, there are some shortcomings and technical obstacles that hurdle the rapid pace of bioprinting. In fact, 3D bioprinting faces many challenges, including the technological potential of 3D bioprinters, the selection of appropriate biomaterials and cell types, and the process of fabricating more realistic tissue analogues.

### 19.5.1 Bioprinting technologies

Current 3D bioprinting technologies present critical limitations, still having space for improvement. Although bioprinters' prices keep coming down as the technology becomes more established, the price of bioprinters is still too high for daily basis applications. In fact, some types of 3D bioprinters, i.e., laser bioprinters, are commercially available on a small scale. Moreover, the majority of nowadays bioprinters are designed for small-scale uses, and the production of larger objects or sophisticated inner structures would compromise the time of printing. In addition, the most recent 3D bioprinters claim they present spatial accuracy to fabricate any digitally modeled object in a micrometer scale. However, resolution is not only dependent on the bioprinter technology, but it is also highly dependent on the rheological properties of the ink and/or bioink.

### 19.5.2 Biomaterials

Undoubtedly, the selection of the biomaterial is one of the most important factors toward succeeding in 3D bioprinting. Although numerous kinds of biomaterials, i.e., polymers, hydrogels, and ceramics, are commercially

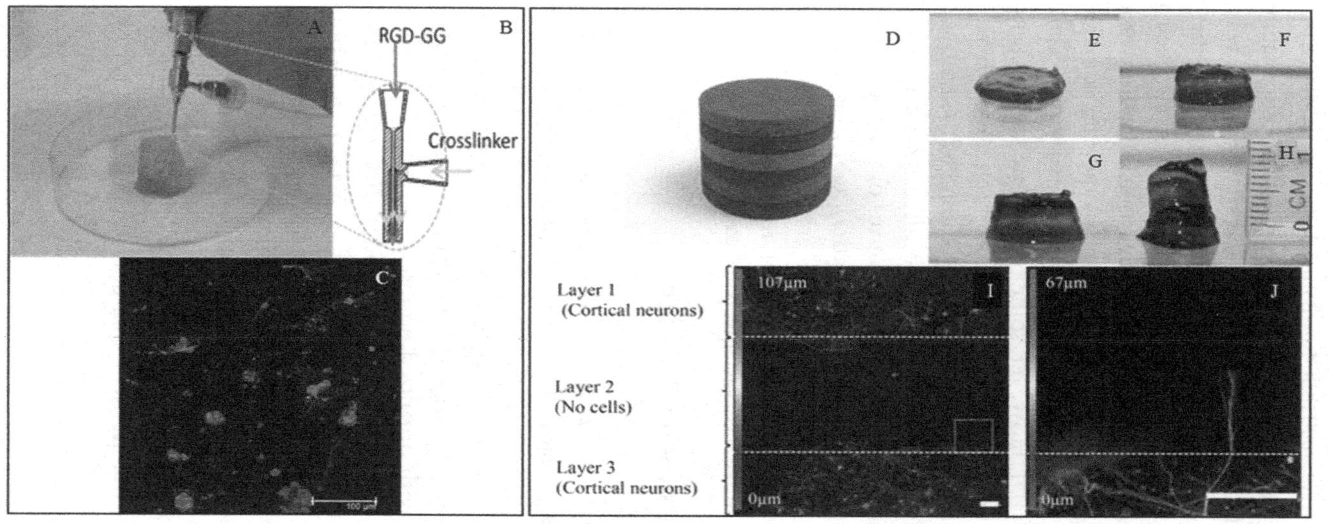


**Figure 19.7** Bioprinted layered brain–like tissue. (A) Hand-held printing method of cortical neurons and (B) closer view illustration of the extruding nozzle. (C) Cortical neuron encapsulated in RGD-GG (0.5% w/v) 7 days postprinting. Nuclei are stained with DAPI (blue [light gray in print version]), glial cells are stained with GFAP (green [dark black in print version]), and cortical neurons are stained with β-III tubulin (red [dark gray in print version]). (D) Model of the brain-like structure depicting in each color a different layer of the tissue. (E)–(H) Process of printing the brain-like structure model layer-by-layer. (I), (J) Localization of neuronal cells within the 3D structure 5 days postprinting as confirmed by confocal, neuronal cells remain in the same layer, while their axons start crossing to the acellular middle layer. Scale bars represent 100 μm. *GG*, gellan gum; *RGD*, arginine–glycine–aspartic acid. *(Reproduced with permission from Reference R. Lozano et al., "3D printing of layered brain-like structures using peptide modified gellan gum substrates," Biomaterials, vol. 67, pp. 264–273, 2015; Copyright 2015, Elsevier.)*

available, each bioprinting technique is limited to a specific list of biomaterials. Also, the range of available biomaterials is supposed to fulfill multiple terms, including, rheological/mechanical properties, processability, printability, biological features, biocompatibility, and safety. Today, we are limited to hydrogels as a favorable biomaterial for 3D bioprinting, due to their high water content and intrinsic properties similar to ECM. They are also easily processed during bioprinting due to their quick phase inversion from liquid to gel state by cross-linking. Furthermore, they are considered safe and nontoxic to the human being and can easily encapsulate cells at mild conditions [45,106–109]. However, some of them are polysaccharides without any cell adhering motifs limiting cellular response. Others are proteins and lack postprinting mechanical stability leading to the collapse of the bioprinted structure. More importantly, as most hydrogels are naturally derived materials, they show batch variations in terms of physicochemical, rheological, mechanical, and biological properties, which can impact printability and cell behavior. Hence, researchers are making headway on polymer modification of natural or synthetic polymers to get the desired properties and minimize the risk of variable batches [110].

### 19.5.3 Complex microstructural designs

3D bioprinting in its current state is good enough to comply with the geometrical shape, although only at a limited level of complexity. Current approaches are still having many hurdles to acquire topographic fine-tuning, high resolution, and cellular-scale bioprinting. In addition, engineering of fully vascularized, innervated, and heterogeneous constructs is still to be attained. A decorated microstructure exceeding few hundreds of micrometers in thickness with organized pore size is crucial, as it provides routes for cell–cell contact, oxygen perfusion, food trafficking, and vascular diffusion density. The assembly of a microfluidic system to a bioprinter brought a higher level of sophistication to bioprinting, as it allows the deposition of the bioink on a submilliliter scale, allowing the formation of complex structures like the vascular tree [111]. Moreover, the assembly of a coaxial nozzle to a bioprinter allowed the use of sacrificial material (inner part) and a bioink (outer part) to extrude microsized hollow tubes [100] or to provide a higher space for nutrient transport (Fig. 19.8) [112]. After printing, the sacrificial material can be easily leached out, and only the bioprinted bioink remains.

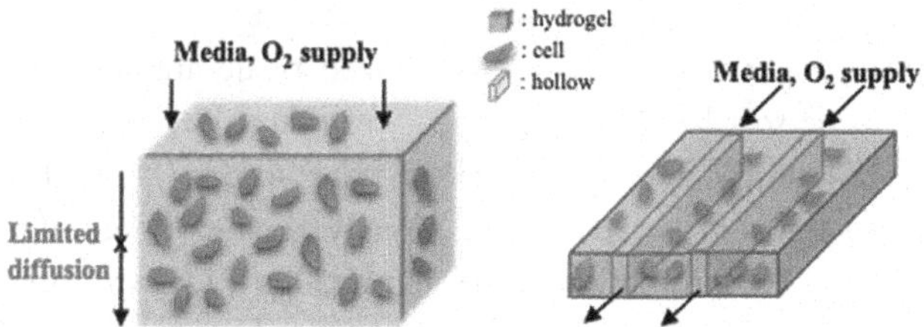


**Figure 19.8** Schematic illustration of nutrients and oxygen diffusion depending on construct design.

## 19.6 Conclusions and future directions

3D bioprinting is a multistep process that transforms digital designs into a 3D object by means of bioinks. Each element of the bioprinting process is subject to growth onward. Thus, the incredible promise of evolving this technology could one day save lives, heal wounds and injuries, and revolutionize healthcare and disease control in ways that the world has never seen. However, counter to common perception, 3D bioprinting at present will not be replacing traditional ways in the medical field at the push of a button. Instead, we shall take it as a supplement technology ready to contribute to gradual advances in healthcare by customizing the design work, enhancing the prototyping process to get the complete and comprehensive outcome that no one ever could make it.

3D bioprinting offers innumerous opportunities to evolve in the near future. The demand of highly dynamic and responsive constructs has prompted the evolution of 3D bioprinting to 4D bioprinting, by integrating the factor of time into the bioprinting process [113]. On one hand, the constructs can change their morphology in time once they are in contact with specific stimuli such as water, heat, and light. On the other hand, the embedded cells can proliferate, migrate, and differentiate along the time, forming more mature tissues with higher resemblances with the native tissue.

Moreover, the future bioprinted 3D models are envisaged to be miniaturized, as bioprinting has progressed in miniatures designing, by extruding parts of tissues or organs into "organ-on-a-chip" model. This, in turn, may inspire recruiting different miniaturized organs on one small chip, to have a comprehensive platform for drug screening, by analyzing pharmacodynamics and pharmacokinetics, focusing on minimizing clinical trials [114]. In the same context, the well-known L'Oreal cosmetic company has pioneered in the drug screening field by producing a lab-born skin to test their products before marketing [115].

The perspective is that 3D bioprinting will be further explored in a personalized manner, to develop personalized tissue analogs for implantation, as well as personalized 3D models to provide a deepen in vitro study of patient's diseases and screen potential effective drugs. Thus, the development of personalized tissue analogs has already been explored in the field of prosthetic limbs. Accordingly, scanner-embedded 3D bioprinters allowed the scanning of a defective limb and consequential printing of a customized prosthetic limb [116]. Although highly attractive, personalized bioprinting of transplantable tissues is still estimated to be costly. It requires the elaboration and high-tech processes from scanning, designing, patient's cellular isolation, and harvesting, as well as the cost of 3D bioprinting machine itself.

On average, thus, the idea is to one day be able to assess how a realistic tissue will react to specific drugs. While this is still in an entry-level, bioprinting is still holding revolutionary promises in the near future.

## References

[1] R. Langer, J.P. Vacanti, Tissue engineering, Science 80 (1993).

[2] A.G. Mikos, J.S. Temenoff, Formation of highly porous biodegradable scaffolds for tissue engineering, Electron. J. Biotechnol. 3 (August, 2000), pp. 0–0.

[3] R.T. Tran, E. Naseri, A. Kolasnikov, X. Bai, J. Yang, A new generation of sodium chloride porogen for tissue engineering, Biotechnol. Appl. Biochem. 58 (5) (2011) 335–344.

[4] Y.-J. Seol, T.-Y. Kang, D.-W. Cho, Solid freeform fabrication technology applied to tissue engineering with various biomaterials, Soft Matter 8 (6) (2012) 1730.

[5] W.-Y. Yeong, et al., Rapid prototyping in tissue engineering: challenges and potential, Trends Biotechnol. 22 (12) (2004) 643–652.

[6] A.B. Dababneh, I.T. Ozbolat, Bioprinting technology: a current state-of-the-art review, J. Manuf. Sci. Eng. 136 (6) (October, 2014) 61011–61016.

[7] M. Hospodiuk, M. Dey, D. Sosnoski, I.T. Ozbolat, The bioink: a comprehensive review on bioprintable materials, Biotechnol. Adv. 35 (2) (March, 2017) 217–239.

[8] Q. Wang, G. Han, S. Yan, Q. Zhang, 3D printing of silk fibroin for biomedical applications, Materials 12 (3) (2019).

[9] F. Guillemot, et al., High-throughput laser printing of cells and biomaterials for tissue engineering, In Acta Biomaterialia 6 (7) (2010) 2494–2500.

[10] B. Guillotin, et al., Laser assisted bioprinting of engineered tissue with high cell density and microscale organization, Biomaterials 31 (28) (2010) 7250–7256.

[11] A. Ovsianikov, et al., Laser printing of cells into 3D scaffolds, Biofabrication 2 (1) (2010) 014104.

[12] L. Koch, et al., Laser printing of skin cells and human stem cells, Tissue Eng. C Methods 16 (5) (2010) 847–854.

[13] B.R. Ringeisen, et al., Laser printing of pluripotent embryonal carcinoma cells, Tissue Eng. 10 (3–4) (2004) 483–491.

[14] V. Keriquel, et al., In vivo bioprinting for computer- and robotic-assisted medical intervention: preliminary study in mice, Biofabrication 2 (1) (March, 2010) 014101.

[15] E.A. Roth, T. Xu, M. Das, C. Gregory, J.J. Hickman, T. Boland, Inkjet printing for high-throughput cell patterning, Biomaterials 25 (17) (2004) 3707–3715.
[16] N. Makoto, et al., Ink jet three-dimensional digital fabrication for biological tissue manufacturing: analysis of alginate microgel beads produced by ink jet droplets for three dimensional tissue fabrication, J. Imag. Sci. Technol. 52 (1) (2008) 1–15.
[17] M.M. Mohebi, J.R.G. Evans, Combinatorial ink-jet printer for ceramics: calibration, J. Am. Ceram. Soc. 86 (10) (2003) 1654–1661.
[18] Y. Nishiyama, et al., Development of a three-dimensional bioprinter: construction of cell supporting structures using hydrogel and state-of-the-art inkjet technology, J. Biomech. Eng. 131 (3) (December, 2008) 35001–35006.
[19] R.E. Saunders, J.E. Gough, B. Derby, Delivery of human fibroblast cells by piezoelectric drop-on-demand inkjet printing, Biomaterials 29 (2) (2008) 193–203.
[20] T. Xu, W. Zhao, J.M. Zhu, M.Z. Albanna, J.J. Yoo, A. Atala, Complex heterogeneous tissue constructs containing multiple cell types prepared by inkjet printing technology, Biomaterials 34 (1) (2013) 130–139.
[21] X. Cui, G. Gao, T. Yonezawa, G. Dai, Human cartilage tissue fabrication using three-dimensional inkjet printing technology, J. Vis. Exp. 88 (2014) 1–5.
[22] C. Xu, W. Chai, Y. Huang, R.R. Markwald, Scaffold-free inkjet printing of three-dimensional zigzag cellular tubes, Biotechnol. Bioeng. 109 (12) (2012) 3152–3160.
[23] S. Khalil, W. Sun, Bioprinting endothelial cells with alginate for 3D tissue constructs, J. Biomech. Eng. 131 (11) (2009) 111002.
[24] I. Zein, D.W. Hutmacher, K.C. Tan, S.H. Teoh, Fused deposition modeling of novel scaffold architectures for tissue engineering applications, Biomaterials 23 (4) (2002) 1169–1185.
[25] S. Moon, et al., Layer by layer three-dimensional tissue epitaxy by cell-laden hydrogel droplets, Tissue Eng. C Methods 16 (1) (2010) 157–166.
[26] K. Iwami, T. Noda, K. Ishida, K. Morishima, M. Nakamura, N. Umeda, Bio rapid prototyping by extruding/aspirating/refilling thermoreversible hydrogel, Biofabrication 2 (1) (2010) 014108.
[27] J.A. Barron, P. Wu, H.D. Ladouceur, B.R. Ringeisen, "Biological laser printing: a novel technique for creating heterogeneous 3-dimensional cell patterns," *Biomed*, Microdevices 6 (2) (2004) 139–147.
[28] Y. Lin, Y. Huang, D.B. Chrisey, Metallic foil-assisted laser cell printing, J. Biomech. Eng. 133 (2) (2011) 25001.
[29] V. Keriquel, et al., In situ printing of mesenchymal stromal cells, by laser-assisted bioprinting, for in vivo bone regeneration applications, Sci. Rep. 1778 (2017).
[30] X. Cui, T. Boland, D.D. D'Lima, M.K. Lotz, Thermal inkjet printing in tissue engineering and regenerative medicine, Recent Pat. Drug Deliv. Formulation 6 (2) (2012) 149–155.
[31] J.H. Lee, et al., Fabrication and characterization of 3D scaffold using 3D plotting system, Chin. Sci. Bull. 55 (1) (2010) 94–98.
[32] J. Groll, et al., A definition of bioinks and their distinction from biomaterial inks, Biofabrication 11 (2019) 013001.
[33] J. Malda, et al., 25th anniversary article: engineering hydrogels for biofabrication, Adv. Mater. 25 (36) (2013) 5011–5028.
[34] B. V Slaughter, S.S. Khurshid, O.Z. Fisher, A. Khademhosseini, N.A. Peppas, Hydrogels in regenerative medicine, Adv. Mater. 21 (32–33) (Sep. 2009) 3307–3329.
[35] A.G. Tabriz, M.A. Hermida, N.R. Leslie, W. Shu, Three-dimensional bioprinting of complex cell laden alginate hydrogel structures, Biofabrication 7 (4) (2015) 045012.
[36] R. Fan, M. Piou, E. Darling, D. Cormier, J. Sun, J. Wan, Bio-printing cell-laden Matrigel-agarose constructs, J. Biomater. Appl. 31 (5) (2016) 684–692.

[37] A. Forget, et al., Mechanically tunable bioink for 3D bioprinting of human cells, Adv. Healthc. Mater. 6 (20) (2017) 1–7.
[38] Q. Gu, et al., Functional 3D neural mini-tissues from printed gel-based bioink and human neural stem cells, Adv. Healthc. Mater. 5 (12) (2016) 1429–1438.
[39] A. Nadernezhad, O.S. Caliskan, F. Topuz, F. Afghah, B. Erman, B. Koc, Nanocomposite bioinks based on agarose and 2D nanosilicates with tunable flow properties and bioactivity for 3D bioprinting, ACS Appl. Bio Mater. 2 (2) (2019) 796–806.
[40] J. Jia, et al., Engineering alginate as bioink for bioprinting, Acta Biomater. 10 (10) (2014) 4323–4331.
[41] Y. Yu, Y. Zhang, J. a Martin, I.T. Ozbolat, Evaluation of cell viability and functionality in vessel-like bioprintable cell-laden tubular channels, J. Biomech. Eng. 135 (9) (2013) 91011.
[42] D. Nguyen, et al., Cartilage tissue engineering by the 3D bioprinting of iPS cells in a nanocellulose/alginate bioink, Sci. Rep. 7 (1) (2017) 658.
[43] Y. Zhang, Y. Yu, I.T. Ozbolat, Direct bioprinting of vessel-like tubular microfluidic channels, J. Nanotechnol. Eng. Med. 4 (2013) 020902. May 2013.
[44] Q. Liu, Q. Li, S. Xu, Q. Zheng, X. Cao, Preparation and properties of 3D printed alginate-chitosan polyion complex hydrogels for tissue engineering, Polymers 10 (6) (2018).
[45] C. Tonda-Turo, et al., Photocurable chitosan as bioink for cellularized therapies towards personalized scaffold architecture, Bioprinting 18 (2020) e00082. December 2019.
[46] W.L. Ng, W.Y. Yeong, M.W. Naing, Polyelectrolyte gelatin-chitosan hydrogel optimized for 3D bioprinting in skin tissue engineering, Int. J. Bioprinting 2 (1) (2016) 53–62.
[47] A. Akkineni, T. Ahlfeld, A. Funk, A. Waske, A. Lode, M. Gelinsky, Highly concentrated alginate-gellan gum composites for 3D plotting of complex tissue engineering scaffolds, Polymers 8 (5) (2016) 170.
[48] R. Lozano, et al., 3D printing of layered brain-like structures using peptide modified gellan gum substrates, Biomaterials 67 (2015) 264–273.
[49] D. Wu, et al., 3D bioprinting of gellan gum and poly (ethylene glycol) diacrylate based hydrogels to produce human-scale constructs with high-fidelity, Mater. Des. 160 (2018) 486–495.
[50] J.Y. Park, et al., A comparative study on collagen type I and hyaluronic acid dependent cell behavior for osteochondral tissue bioprinting, Biofabrication 6 (3) (2014) 035004.
[51] K.S. Masters, D.N. Shah, L.A. Leinwand, K.S. Anseth, Crosslinked hyaluronan scaffolds as a biologically active carrier for valvular interstitial cells, Biomaterials 26 (15) (2005) 2517–2525.
[52] B. Duan, E. Kapetanovic, L.A. Hockaday, J.T. Butcher, 3D printed trileaflet valve conduits using biological hydrogels and human valve interstitial cells, Acta Biomaterialia 10 (5) (May-2014) 1836–1846.
[53] X. Cui, T. Boland, Human microvasculature fabrication using thermal inkjet printing technology, Biomaterials 30 (31) (2009) 6221–6227.
[54] C. Lee, E. Abelseth, L. de la Vega, S.M. Willerth, Bioprinting a novel glioblastoma tumor model using a fibrin-based bioink for drug screening, Mater. Today Chem. 12 (2019) 78–84.
[55] F. Xu, et al., Cell proliferation in bioprinted cell-laden collagen droplets, Bioeng. Proc. Northeast Conf. (2009) 1–2.
[56] S. Rhee, J.L. Puetzer, B.N. Mason, C.A. Reinhart-King, L.J. Bonassar, 3D bioprinting of spatially heterogeneous collagen constructs for cartilage tissue engineering, ACS Biomater. Sci. Eng. 2 (10) (2016) 1800–1805.

[57] X. Yang, Z. Lu, H. Wu, W. Li, L. Zheng, J. Zhao, Collagen-alginate as bioink for three-dimensional (3D) cell printing based cartilage tissue engineering, Mater. Sci. Eng. C 83 (2018) 195–201.
[58] S.A. Irvine, et al., Printing cell-laden gelatin constructs by free-form fabrication and enzymatic protein crosslinking, Biomed. Microdevices 17 (2015) 16.
[59] R. Seyedmahmoud, et al., Three-dimensional bioprinting of functional skeletal muscle tissue using gelatin methacryloyl-alginate bioinks, Micromachines 10 (10) (2019) 679.
[60] S.H. Kim, et al., Precisely printable and biocompatible silk fibroin bioink for digital light processing 3D printing, Nat. Commun. 9 (1) (2018) 1–14.
[61] A.M. Compaan, K. Christensen, Y. Huang, Inkjet bioprinting of 3D silk fibroin cellular constructs using sacrificial alginate, ACS Biomater. Sci. Eng. 3 (8) (2017) 1519–1526.
[62] J.S. Miller, et al., Rapid casting of patterned vascular networks for perfusable engineered three-dimensional tissues, Nat. Mater. 11 (9) (September, 2012) 768–774.
[63] K. Zhang, C.G. Simon, N.R. Washburn, J.M. Antonucci, S. Lin-Gibson, In situ formation of blends by photopolymerization of poly(ethylene glycol) dimethacrylate and polylactide, Biomacromolecules 6 (3) (2005) 1615–1622.
[64] G. Gao, T. Yonezawa, K. Hubbell, G. Dai, X. Cui, Inkjet-bioprinted acrylated peptides and PEG hydrogel with human mesenchymal stem cells promote robust bone and cartilage formation with minimal printhead clogging, Biotechnol. J. 10 (10) (2015) 1568–1577.
[65] S. Xin, D. Chimene, J.E. Garza, A.K. Gaharwar, D.L. Alge, Clickable PEG hydrogel microspheres as building blocks for 3D bioprinting, Biomater. Sci. 7 (3) (2019) 1179–1187.
[66] V. Normand, D.L. Lootens, E. Amici, K.P. Plucknett, P. Aymard, New insight into agarose gel mechanical properties, Biomacromolecules 1 (4) (2000) 730–738.
[67] P. Zarrintaj, et al., Agarose-based biomaterials for tissue engineering, Carbohydr. Polym. 187 (2018) 66–84.
[68] J. Wei, et al., 3D printing of an extremely tough hydrogel, RSC Adv. 5 (99) (2015) 81324–81329.
[69] C.J. Ferris, et al., Peptide modification of purified gellan gum, J. Mater. Chem. B 3 (6) (2015) 1106–1115.
[70] D. Jiang, J. Liang, P.W. Noble, Hyaluronan in tissue injury and repair, Annu. Rev. Cell Dev. Biol. 23 (1) (2007) 435–461.
[71] Z. Zhu, Y.-M. Wang, J. Yang, X.-S. Luo, Hyaluronic acid: a versatile biomaterial in tissue engineering, Plast. Aesthetic Res. 4 (2017) 219–227.
[72] F.Z. Cui, W.M. Tian, S.P. Hou, Q.Y. Xu, I.S. Lee, Hyaluronic acid hydrogel immobilized with RGD peptides for brain tissue engineering, J. Mater. Sci. Mater. Med. 17 (12) (2006) 1393–1401.
[73] A. Ramamurthi, I. Vesely, Smooth muscle cell adhesion on crosslinked hyaluronan gels, J. Biomed. Mater. Res. 60 (1) (2002) 195–205.
[74] J. Gopinathan, I. Noh, Recent trends in bioinks for 3D printing, Biomater. Res. 22 (1) (2018) 1–15.
[75] G. Camci-Unal, D. Cuttica, N. Annabi, D. Demarchi, A. Khademhosseini, Synthesis and characterization of hybrid hyaluronic acid-gelatin hydrogels, Biomacromolecules 14 (4) (2013) 1085–1092.
[76] M.-D. Wang, et al., Novel crosslinked alginate/hyaluronic acid hydrogels for nerve tissue engineering, Front. Mater. Sci. 7 (3) (September, 2013) 269–284.
[77] N. Muralidharan, R. Jeya Shakila, D. Sukumar, G. Jeyasekaran, Skin, bone and muscle collagen extraction from the trash fish, leather jacket (Odonus Niger) and their characterization, J. Food Sci. Technol. 50 (6) (2013) 1106–1113.

[78] C. Dong, Y. Lv, Application of collagen scaffold in tissue engineering: recent advances and new perspectives, Polymers 8 (2) (2016) 42.
[79] D. Olsen, et al., Recombinant collagen and gelatin for drug delivery, Adv. Drug Deliv. Rev. 55 (2003) 1547–1567.
[80] N. Davidenko, et al., Evaluation of cell binding to collagen and gelatin: a study of the effect of 2D and 3D architecture and surface chemistry, J. Mater. Sci. Mater. Med. 27 (10) (2016).
[81] X. Wang, et al., Gelatin-based hydrogels for organ 3D bioprinting, Polymers 9 (9) (2017) 401.
[82] A.T. Neffe, A. Loebus, A. Zaupa, C. Stoetzel, F.A. Müller, A. Lendlein, Gelatin functionalization with tyrosine derived moieties to increase the interaction with hydroxyapatite fillers, Acta Biomater. 7 (4) (April, 2011) 1693–1701.
[83] A.T. Neffe, A. Zaupa, B.F. Pierce, D. Hofmann, A. Lendlein, Knowledge-based tailoring of gelatin-based materials by functionalization with tyrosine-derived groups, Macromol. Rapid Commun. 31 (17) (2010) 1534–1539.
[84] J.W. Nichol, S.T. Koshy, H. Bae, C.M. Hwang, S. Yamanlar, A. Khademhosseini, Cell-laden microengineered gelatin methacrylate hydrogels, Biomaterials 31 (21) (2010) 5536–5544.
[85] Y. Jiang, et al., Preparation of cellulose nanocrystal/oxidized dextran/gelatin (CNC/OD/GEL) hydrogels and fabrication of a CNC/OD/GEL scaffold by 3D printing, J. Mater. Sci. 55 (6) (2019) 2618–2635.
[86] Z. Wang, H. Yang, W. Li, C. Li, Effect of silk degumming on the structure and properties of silk fibroin, J. Text. Inst. 110 (1) (2019) 134–140.
[87] Y. Qi, et al., A review of structure construction of silk fibroin biomaterials from single structures to multi-level structures, Int. J. Mol. Sci. 18 (3) (2017) 237.
[88] V. Lee, et al., Design and fabrication of human skin by three-dimensional bioprinting, Tissue Eng. C Methods 20 (6) (2014) 473–484.
[89] L. Koch, et al., Skin tissue generation by laser cell printing, Biotechnol. Bioeng. 109 (7) (2012) 1855–1863.
[90] M. Albanna, et al., In situ bioprinting of autologous skin cells accelerates wound healing of extensive excisional full-thickness wounds, Sci. Rep. 9 (1) (2019) 1–15.
[91] T. Baltazar, et al., Three dimensional bioprinting of a vascularized and perfusable skin graft using human keratinocytes, fibroblasts, pericytes, and endothelial cells, Tissue Eng. 26 (2019) 227–238.
[92] D. Min, W. Lee, I.H. Bae, T.R. Lee, P. Croce, S.S. Yoo, Bioprinting of biomimetic skin containing melanocytes, Exp. Dermatol. 27 (5) (2018) 453–459.
[93] S. Huang, B. Yao, J. Xie, X. Fu, 3D bioprinted extracellular matrix mimics facilitate directed differentiation of epithelial progenitors for sweat gland regeneration, Acta Biomater. 32 (2016) 170–177.
[94] S.T. Bendtsen, S.P. Quinnell, M. Wei, Development of a novel alginate-polyvinyl alcohol-hydroxyapatite hydrogel for 3D bioprinting bone tissue engineered scaffolds, J. Biomed. Mater. Res. 105 (5) (2017) 1457–1468.
[95] W.J. Kim, H.S. Yun, G.H. Kim, An innovative cell-laden α-TCP/collagen scaffold fabricated using a two-step printing process for potential application in regenerating hard tissues, Sci. Rep. 7 (1) (2017) 3181.
[96] T. Möller, et al., In vivo chondrogenesis in 3D bioprinted human cell-laden hydrogel constructs, Plast. Reconstr. Surg. - Glob. Open 5 (2) (2017) e1227.
[97] W.C. Puelacher, S.W. Kim, J.P. Vacanti, B. Schloo, D. Mooney, C.A. Vacanti, Tissue-engineered growth of cartilage: the effect of varying the concentration of chondrocytes seeded onto synthetic polymer matrices, Int. J. Oral Maxillofac. Surg. 23 (1994) 49–53.

[98] W. Kim, C.H. Jang, G. Kim, A myoblast-laden collagen bioink with fully aligned Au nanowires for muscle-tissue regeneration, Nano Lett. 19 (2019) p. acs.nanolett.9b03182.
[99] M. Costantini, et al., Microfluidic-enhanced 3D bioprinting of aligned myoblast-laden hydrogels leads to functionally organized myofibers in vitro and in vivo, Biomaterials 131 (2017) 98–110.
[100] M. Lovett, K. Lee, A. Edwards, D.L. Kaplan, Vascularization strategies for tissue engineering, Tissue Eng. B Rev. 15 (3) (2009) 353–370.
[101] G. Gao, J.Y. Park, B.S. Kim, J. Jang, D.W. Cho, Coaxial cell printing of freestanding, perfusable, and functional in vitro vascular models for recapitulation of native vascular endothelium pathophysiology, Adv. Healthc. Mater. 7 (23) (2018) e1801102.
[102] F. Maiullari, et al., A multi-cellular 3D bioprinting approach for vascularized heart tissue engineering based on HUVECs and iPSC-derived cardiomyocytes, Sci. Rep. 8 (2018) 13532.
[103] L. Horváth, Y. Umehara, C. Jud, F. Blank, A. Petri-Fink, B. Rothen-Rutishauser, Engineering an in vitro air-blood barrier by 3D bioprinting, Sci. Rep. 5 (2015) 7974.
[104] B. Duan, L.A. Hockaday, K.H. Kang, J.T. Butcher, 3D Bioprinting of heterogeneous aortic valve conduits with alginate/gelatin hydrogels, J. Biomed. Mater. Res. 101 (5) (2013) 1255–1264.
[105] X. Ma, et al., Deterministically patterned biomimetic human iPSC-derived hepatic model via rapid 3D bioprinting, Proc. Natl. Acad. Sci. Unit. States Am. 113 (8) (2016), 201524510.
[106] T.R. Hoare, D.S. Kohane, Hydrogels in drug delivery: progress and challenges, Polymer 49 (8) (2008) 1993–2007.
[107] Y.-H. Tsou, J. Khoneisser, P.-C. Huang, X. Xu, Hydrogel as a bioactive material to regulate stem cell fate, Bioact. Mater. 1 (1) (September, 2016) 39–55.
[108] P.B. Malafaya, G.A. Silva, R.L. Reis, Natural–origin polymers as carriers and scaffolds for biomolecules and cell delivery in tissue engineering applications, Adv. Drug Deliv. Rev. 59 (4–5) (May, 2007) 207–233.
[109] G. Camci-Unal, J.W. Nichol, H. Bae, H. Tekin, J. Bischoff, A. Khademhosseini, Hydrogel surfaces to promote attachment and spreading of endothelial progenitor cells, J. Tissue Eng. Regen. Med. 7 (5) (2013) 337–347.
[110] D.A. Lobo, P. Ginestra, Cell bioprinting: the 3D-bioplotter™ case, Materials 12 (23) (2019) 4005.
[111] J. Ma, Y. Wang, J. Liu, Bioprinting of 3D tissues/organs combined with microfluidics, RSC Adv. 8 (2018) 21712–21727.
[112] L. Shao, Q. Gao, C. Xie, J. Fu, M. Xiang, Y. He, Synchronous 3D bioprinting of large-scale cell-laden constructs with nutrient networks, Adv. Healthc. Mater. 1901142 (2019) 1–9.
[113] Y.-C. Li, Y.S. Zhang, A. Akpek, S.R. Shin, A. Khademhosseini, 4D bioprinting: the next-generation technology for biofabrication enabled by stimuli-responsive materials, Biofabrication 9 (1) (2016) 012001.
[114] A. Skardal, et al., Drug compound screening in single and integrated multi-organoid body-on-a-chip systems, Biofabrication 12 (2) (2020) 025017.
[115] S. Vijayavenkataraman, W.F. Lu, J.Y.H. Fuh, 3D bioprinting of skin: a state-of-the-art review on modelling, materials, and processes, Biofabrication 8 (3) (2016) 032001.
[116] D. Radenkovic, A. Solouk, A. Seifalian, Personalized development of human organs using 3D printing technology, Med. Hypotheses 87 (2016) 30–33.

CHAPTER 20

# Additive processing of biopolymers for medical applications

**Rajkumar Velu[1,4], Dhileep Kumar Jayashankar[2], Karupppasamy Subburaj[3]**
[1]Centre for Laser Aided Intelligent Manufacturing, University of Michigan, Ann Arbor, MI, United States; [2]Digital Manufacturing and Design Centre, Singapore University of Technology and Design, Singapore; [3]Engineering Product Development Pillar, Singapore University of Technology and Design, Singapore; [4]Department of Mechanical Engineering, Indian Institute of Technology Jammu, Jammu & Kashmir, India

## 20.1 Introduction

In recent years, there has been a demand for developing bioartificial organs/tissues, drug delivery systems, and medical devices, especially in the field of organ transplantation and in vitro toxicological drug screening [1]. The primary challenge toward clinical translation is difficult to scale up complex, biologically effective tissues and organs to the size relevant for humans [2]. The conventional fabrication processes such as solvent casting-particulate leaching, phase separation, gas foaming, emulsion freeze drying, and fiber meshes [3] have been used to generate engineered scaffolds of foam-like internal structure with a random architecture and limited control of scale. However, in tissue engineering, there is a requirement for structures that can bond cell growth and also support physiological environment such as geometrical, topographical, and physical features of the targeted applications [4]. These existing fabrication techniques have been designed to use highly porous structures with interconnected pores or microchannel that provides space for penetration and growth of cells [5]. Even though such processing techniques are fast, scalable, and economical, they have their limitations such as dimensional control on microarchitectures such as pore size, geometry, their interconnections, and distributions with the structures [6]. Researchers have implemented layer-by-layer assembly techniques using polymers, patterned by molding and embossing processes to overcome these limitations [7], and these fabrication methods achieved the formation of channels with accurate dimensions. On the other hand, the requirement for microfabrication's master molds and manual alignment

*Additive Manufacturing*
ISBN 978-0-12-818411-0
https://doi.org/10.1016/B978-0-12-818411-0.00019-7

of layers implies a slow and tedious process for achieving a multilayer three-dimensional construction [8].

To conquer these difficulties in developing three-dimensional structures for medical necessities, recently, the researchers propelled toward 3D printing as a rapid prototyping technique to fabricate controlled meso- and microlevel porous structures of any desired complexities [9]. The outcome of 3D fabrication techniques has numerous advantages with great potential, for example, the incorporation of vascular beds would allow for larger constructs than could be supported by nutrient diffusion alone, and the combination of clinical imaging data with CAD-based freeform techniques allows the fabrication of replacement tissues that are customized to the desired shape [10]. Eventually, the development enlarges to large-scale fabrication of identical functional medical units, which is used to medical devices, bone replacement structures, dental surgery, cell-based assays for drug delivery, or fundamental biological studies [11,12].

There are various methods for creating 3D structures, which differentiated on their modes of fabrication, using heat (fusion deposition modeling), light (selective laser sintering and stereolithography), and chemical adhesives (3DP) [13–16]. Each 3D printing technology has its unique processing systems and influences the selection of materials. Moreover, currently available commercial printers are compatible with a limited choice of materials, and finally, it significantly limits the variations in the physical and chemical properties of 3D printed objects. Therefore, to overcome these limitations, composite materials and multimaterial printers have been developed. However, to improve the significant properties of the 3D printed part, researchers are still focusing on the appropriate composition of materials for 3D printing techniques. There are different types of materials such as metals, ceramics, natural and synthetic polymers, wood, magnetic, and liquid are used in 3D printing techniques [17,18]. However, this chapter focuses on polymers, particularly biopolymers and their composites for most promising medical applications. Healthcare products, implants, customized prostheses, scaffolds, and bioengineered tissues have extensively used 3D printing techniques for fabrication. For example, in the beginning, researchers developed the 3D structures using stereolithography based on PLGA solution, whereas they achieved microstructures for cell attachment and proliferation [19].

Furthermore, the development reached to process and combine polymeric powder materials using a selective laser sintering process into desired shapes [20]. Lee and Barlow investigated with polymer-coated calcium phosphate powders to fabricate oral implants and demonstrated extensive bone tissue ingrowth in dog models [21]. In addition, Yeong and others have widely investigated selective laser sintering for various biopolymer applications [22]. Similarly, other 3D printing processing techniques such as FDM have also been investigated by researchers such as Zein and Hutmacher who investigated the PCL scaffolds exhibiting various honeycomb geometries with finely tuned pore and channel dimensions of 250–700 μm [23]. However, FDM capabilities are expanding with new developments such as multiphase jet solidification (MJS), a technique that allows simultaneous extrusion of multiple melted materials [24]. Berry et al. fabricated SLS models of two skulls and a normal femur using nylon, with the dimensions in good agreement with the CT data [25]. Though there are significant research efforts in the field of 3D printing technology using biopolymers, it is still challenging to fabricate products in the area of biomedical engineering, including human health. The technology is expected to proliferate for the human organ regenerations, implantable devices, and precise drug development.

In this chapter, the classification of biopolymers and their conventional processing methods for producing medical devices, dental implants, drug delivery systems, and bone implants are comprehensively reviewed. Subsequently, the need for additive manufacturing (AM) to process biopolymers, which overwhelm the limitation of conventional fabrication techniques for medical applications, is discussed. Then the 3D printing fabrication strategies and their challenges are introduced concerning medical applications. Eventually, the future directions of 3D printing techniques used for medical applications are identified.

## 20.2 Biopolymers and their biomedical applications

### 20.2.1 Classification of biopolymers

In the early days, the biomaterials focused on proving biological indifference and the development improved biological indifference to bioactivity. Consequently, the recent research and development efforts are envisioned to be a temporary structure stimulating an organism to reproduce its own tissues or even entire organs [26]. The existing biomaterials are classified into the following divisions: metal, ceramic, polymer, and composite biomaterials. These materials have their own physical and mechanical properties, which could be suitable for medical applications such as bone replacement and repair tasks, heart valves, dental implants, and so on. In

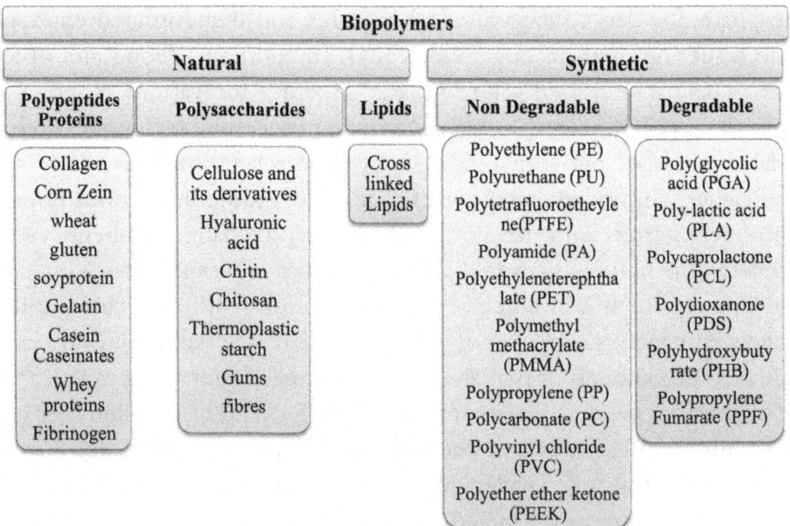


**Figure 20.1** Classification of biopolymers.

addition, the appropriate selection of biomaterial for an implant should consider all its functional characteristics [27].

However, recently, polymers ascertain as a promising alternative biomaterial for metal and ceramic. Biopolymers are able to achieve medical application tasks and activities. Initially, plastics were used as disposable medical products, but now they are used as self-dependent implants, which can sustain functions of the human body system [28]. The classification of biopolymers depends on their chemical composition, origin and synthesis method, processing method, economic importance, applications, etc. [29]. Significantly, the biopolymers are divided into two groups: naturally resourced biopolymers and synthesized biopolymers, as shown in Fig. 20.1 [30–33].

Biopolymers are polymers formed in nature during the growth cycles of all organisms; hence, they are also referred to as natural polymers. Their synthesis generally involves enzyme-catalyzed, chain-growth polymerization reactions of activated monomers, which are typically formed within cells by complex metabolic processes [34]. They are polypeptide proteins and polysaccharides. Collagen is the most widely used natural protein biopolymer in medical applications, a way ahead of fibrinogen and elastin, which are reasonably used. Most polysaccharides are produced from starch for photosynthetic tissues and various plant storage organs such as seed and swollen stems [35]. The general classification of polysaccharides is cellulose and its derivatives, hyaluronic acid, chitin, chitosan, thermoplastic starch, gums, fibers, etc. [36,37]. Natural biopolymers have their unique properties such as high water vapor permeability, excellent oxygen barrier, not electrostatically chargeable, and low thermal stability [38].

Synthesized biopolymers are nondegradable and degradable. Nondegradable biopolymers such as silicones, polyethylene (PE), polypropylene (PP), polyamide (PA), polyurethane (PU), polymethylmethacrylate (PMMA), polytetrafluoroethylene (PTFE), polycarbonates (PC), acrylic resins, and polyvinyl chloride (PVC) are resistant to bioactivity [39,40]. In contrast, the degradable biopolymers such as polylactic acid (PLA), polyglycolide (PGA) and its copolymers with lactides, polycaprolactone (PCL), polydioxanone (PDS), polyhydroxybutyrate (PHB), and polypropylene fumigate (PPF) are not resistant. These types of biopolymers are extensively used for various surgical procedures including, plastic, reconstructive, vascular, and trauma. However, the current focus of these biopolymers and their applications is in the fields of orthopedics, tissue engineering, drug delivery, and dental implants [41–44].

### 20.2.2 Biopolymer medical applications based on conventional fabrication techniques

Synthetic biodegradable polymers are investigated for medical applications such wound closure (sutures, staples) and orthopedic fixation devices (pins, rods, screws, tacks, and ligaments). These synthetic polymers hold a significant advantage compared with the natural polymers since these polymers can be tailored for a broader range of properties for appropriate biological functions [44]. The notable factors involved in synthesis are monomer selection, initiator selection, process conditions, and also the composition of additives [45]. The most chemical functional groups used for biodegradation to accomplish by synthesizing are esters, anhydrides, orthoesters, and amides. Consequently, polyesters composed of homopolymers or copolymers of glycolide and lactide are commercially available biodegradable devices [42].

### 20.2.3 Medical devices

Medical devices are varied from simple tongue depressors to sophisticated programmable pacemakers with microchip technology and laser surgical devices [46]. Besides, it includes various in vitro diagnostic products, such as lab equipment, reagents, and test kits, which include monoclonal antibody technology. Other electronic devices also play an essential role in medical devices such as diagnostic ultrasound products, X-ray machines, and medical lasers. Manufacturing of these medical devices includes all relevant aspects of the fabrication process including designing a manufacturing process, scaling up to reduce lead time, ongoing process improvements, sterilization, and packaging for shipment [47]. For example, polymers such as polyurethanes are used for cardiovascular applications such as catheters, vascular prostheses, and heart valves. The injection molding process is used to manufacture the polymer-based cardiovascular application medical devices. In this process, polymer resins/materials are heated, melted, and injected under high pressure into a mold and allowed to cool for product extractions [48]. The final product requires high fatigue strength for cardiovascular application materials because they are subjected to repeated loading and unloading; else, it would lead to catastrophic results [49]. Medical grade PVC is used for fabricating medical device applications such as tubing, solution bags, syringe stoppers, dropper bulbs, facemasks, films drip chambers, gaskets, valves, and even medical device cable insulation and jacketing.

In recent developments, thermoplastic elastomers (TPEs) are found to be a viable candidate for replacing PVC considering their chemical structure, toxicology, solubility migration, crazing, sterilization, mechanical properties, processing design, and price. TPEs are nothing but rubbers, which is a physical mix of polymers (usually plastic and rubber) [50]. Plastic component molding and injection molding methods are used for manufacturing the aforementioned medical devices [51]. Also, the polymers are commonly used to produce devices that come in contact with blood as mentioned in Table 20.1.

### 20.2.4 Tissue engineering and drug delivery system

In tissue engineering, the materials for the cells' support, called scaffolds and growth factors, are vital. They need to offer a similar function as an artificial extracellular matrix onto which cells attach, grow, and form new tissues [52]. These disclosures have led to tremendous advancement in the design of a new generation of multifunctional biomaterials able to mimic the molecular regulatory characteristics and the three-dimensional architecture of the native extracellular matrix [53]. The tissue engineering further developed to mimic the complex temporal and spatial microenvironment presented in vivo, an increased interaction of material engineering, drug delivery technology, and cell and molecular biology which ultimately leads to biomaterials that encode the necessary signals to guide and control developmental process in tissue- and organ-specific differentiation and morphogenesis [54].

**Table 20.1** Biopolymer conventional processing techniques for producing medical devices.

| Device | Polymers | Manufacturing techniques |
|---|---|---|
| Membranes for dialysis | Cellulose and derivatives, polyacrylonitrile, polyamides, polysulfones | **1.** Injection molding<br>**2.** Plastic component molding<br>**3.** Extrusion<br>**4.** Compression molding<br>**5.** Rotational molding<br>**6.** Thermoforming<br>**7.** Casting<br>**8.** Blow molding |
| Plasma expanders | Cross-linked collagen, dextran | |
| Heart valves | Polyacetal, polyamides, polycarbonates | |
| Catheters | Polyamides, polycarbonates, polypropylene, poly(tetrafluoroethylene), polyurethanes, poly(vinyl chloride) | |
| Syringe | Polycarbonates, polypropylene, | |
| Vascular prostheses | Poly(tetrafluoroethylene), polyurethanes, poly(ethylene terephthalate) | |
| Artificial hearts | Silicones, polyurethanes, poly(tetrafluoroethylene), | |

In the past decade, various fabrication techniques were developed to fabricate three-dimensional scaffolds including electrospinning, self-assembly, freeze-drying, solvent casting, fiber bonding, phase separation, gas-induced foaming, salt leaching, and others [3]. Electrospinning methods fabricate three-dimensional scaffolds using fibers with diameter ranges varying from nanoscale or microscale. The process is controlled by varying parameters such as viscosity, conductivity, surface tension, and operational conditions. Various synthetic polymers are used to fabricate 3D nanofibrous scaffolds using these methods: poly(lactic acid) [PLA], poly(glycolic acid) [PGA], poly(lactic-co-glycolic acid) [PLGA] and polycaprolactone [PCL], and natural polymers (e.g., collagen, chitosan, silk fibroin, and chitin) [55]. Similarly, these polymers have also undergone phase separation method that is induced thermally or by a nonsolvent and has been utilized to fabricate porous membranes [56]. Besides, this method is used for fabricating drug on microencapsulation technique without changing its physical and chemical properties [57]. Masayoshi et al. [58] established the effect of coacervation-inducing agents such as polyisobutylene (PIB), butyl rubber, or polyethylene in the microencapsulation of ascorbic acid by a phase separation method. Followed by self-assembly method, freeze-drying method was developed for scaffold fabrication with various polymers. Autissier et al. [59] prepared a porous polysaccharide-based scaffold using the freeze-drying method and demonstrated that the freeze-drying pressure regulates pore diameter (55−243 μm) and porosity (33%−68%) in the scaffolds. Adhesion and proliferation of human mesenchymal stem cells on both porous and nonporous polysaccharide-based scaffolds have also been investigated.

### 20.2.5 Dental applications

Polymers are primarily used in the dental application, by considering its ease of forming and final cure by simple techniques and equipment available in a dental laboratory or dentist habitation [60]. The polymers used in dentistry are as follows: vinyl acrylics are used for relining material; epoxy resins are used for die material; polyether, polysulfide, and silicone are used for impression material; polycarbonates are used for temporary crown material; polyacrylic acid is used for denture base material, and other various polymers involved are polystyrene, polyethylene, and polyvinyl acetate [61−64]. Generally, the thermoplastic polymers and resins are softened and molded under heat and pressure to form a required structure without any chemical change. Subsequently, the products made of thermoset polymers cannot be softened by reheating, they are all chemically reacted called polymerized for required form for dental applications, and they are insoluble in organic solvents [52,65]. The polymers prepared via polymerization reactions are two types, as shown in Fig. 20.2.

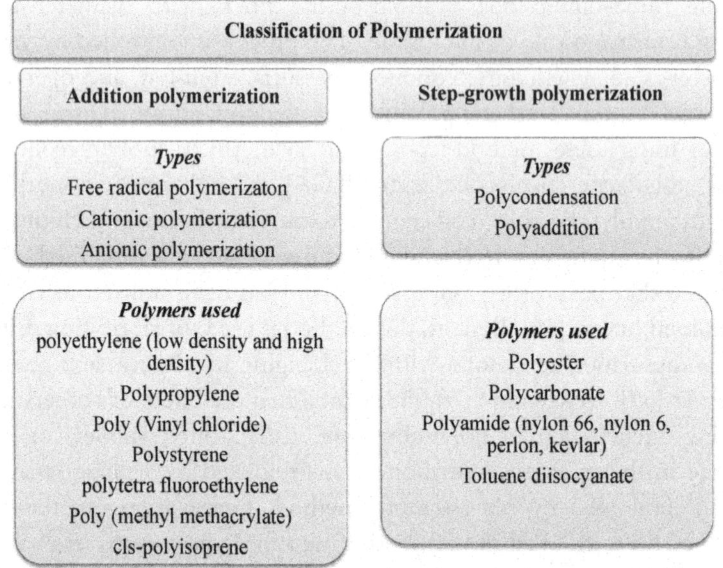


**Figure 20.2** Classification of polymerization.

## 20.2.6 Bone repair and replacement tasks

For the past 50 years, polymer bone cements, particularly PMMA cement, have been widely used as the anchoring/grouting agent in total joint replacements of the hip, knee, ankle, elbow, and shoulder [66]. To ascertain the quality of bone cement for prolonged life, one should adhere to right proportions, mixing ratio, and application technique to reduce the rate of loosening. The rate of loosening is the important factor that is associated with the life of replacement, the lower the rate of loosening, the higher the life of replacement. PMMA bone cements are the most common cements used for joint arthroplasty procedures. The main functions of the cement are to immobilize the implant, transfer body weight, and service loads from the prosthesis to the bone and increase the load-carrying capacity of the prosthesis—bone cement—bone system. Bone cement acts such as a grout, filling in space to create a tight space to hold the implant against the bone. Then the femoral tunnel was drilled through the bone void filler and native bone, allowing anatomic positioning. The autologous graft was then placed and fixed with a bioabsorbable interference screw or any other implants [67]. PMMA bone cement produced is based on two-component systems; they are powder and liquid, as shown in Table 20.2. These two components are mixed at an appropriate ratio of 2:1 to start a chemical reaction called polymerization [68].

John et al. [43] studied the use of synthetic materials and their properties for orthopedic applications; poly(lactides) and poly(glycolides) have been investigated; and the chemistry of polymers including synthesis and degradation, the tailoring of properties by controlling copolymer composition, processing, handling, and mechanisms of biodegradation have been depicted. Table 20.3 shows a list of generic degradable orthopedic devices and their material composition.

**Table 20.2** Classification of components used to produce PMMA bone cements.

| Powder components | Liquid components |
|---|---|
| • Copolymers beads based on PMMA<br>• Initiator: benzoyl peroxide<br>• Contrast agents: zirconium dioxide ($ZrO_2$) or barium sulfate ($BaSO_4$)<br>• Antibiotics: gentamicin, tobramycin | • A monomer, methylmethacrylate<br>• Accelerator (*N,N*-dimethyl para-toluidine)<br>• Stabilizers (or inhibitors)<br>• Chlorophyll or artificial pigment |

*PMMA*, polymethylmethacrylate.

**Table 20.3** Commercial biodegradable polymer composition for orthopedic devices [43].

| Polymer and its compositions | Orthopedics applications |
|---|---|
| • Poly(D,L-lactide-*co*-glycolide) SR-LPLA<br>• Poly(glycolide) (LPLA)<br>• Poly(D,L-lactide-co-L-lactide) (PDO) | Fracture fixation |
| • Poly(glycolide) (LPLA)<br>• Poly(glycolide-*co*-trimethylene carbonate)85/15 DLPLG<br>• Poly(L-lactide-*co*-glycolide) PGA–TMC | Interference screws |
| • Poly(D,L-lactide-*co*-glycolide) SR-LPLA<br>• Poly(glycolide) (LPLA)<br>• Poly(L-lactide-*co*-glycolide) PGA–TMC<br>• Poly(?-caprolactone)82/18 LPLG<br>• Poly(glycolide-*co*-trimethylene carbonate)85/15 DLPLG | Suture anchors |
| • Poly(?-caprolactone)82/18 LPLG | Craniomaxillofacial fixation |
| • Poly(DL-lactide-*co*-glycolide) SR-LPLA<br>• Poly(glycolide) (LPLA) | Meniscus repair |

## 20.3 Summary and need for 3D printing techniques

From the review mentioned before, the significant factors observed based on conventional fabrication techniques for medical applications are as follows [1]: it requires an enormous time to build the final product [2], it creates more wastage, and the subtractive process will compromise on precision [3], innovative designs are limited due to cost constraints and complications in fabrications, and [4] finally, the higher cost of manufacturing and shipping is required. 3D printing could be the best technology option for medical fields to overcome these challenges. 3D printing techniques are required to save time and cost. These two factors can be attained because 3D printing saves on energy by 40%–65% as it eliminates shipping and other logistics activities and enables users to produce objects with lesser material. Also, it allows inexpensive innovation design and high precision with layer-by-layer manufacturing.

## 20.4 Additive manufacturing/3D printing strategies and challenges

Rapid prototyping, AM, or solid freeform fabrication is a digital manufacturing technique for producing a part or prototype, in which cost no longer depends on the complexity of its geometry. As the fabrication process becomes uncomplicated, and the cost becomes reasonable, it has been steadily evolving in recent decades and has the potential to compete against traditional manufacturing techniques [87]. The advancements in the field are rapidly growing across the world, which may lead to a massive or radical restructuring of industrial systems in the near future [89]. Throughout the journey of AM, several technologies were developed primarily for engineering applications but have manifested the possibility for bioprinting and biofabrication. Notably, the additive processing of biopolymers or bioprinting is of more concern as it is predominantly focused on patient-specific treatment concepts. However, coined this field as an integrated AM and monopolized five unique challenges in bioprinting, they are to build with complicated shapes that match human anatomy; build with porous microarchitecture from biocompatible materials; build with living cells, genes, and proteins; build with multiple biomaterial and cells/genes/proteins together and separately on the same platform; and build at resolutions below 10 μm over structure greater than 1 cm in size. These unique challenges have drawn the engineers and scientists to work on producing artificial cells, tissues, organs, bone regeneration, and so on through AM for various medical applications [86,88]. One such leverage of AM is that it mainly focused on increasing the quality of human life or life extensions by repairing the damaged organs in the human body through interaction with the living system/cells. Generally, AM in medical applications follows a reverse engineering method in which the patient's damaged or affected organ/body parts are captured through various imaging techniques such as CT (computed tomography) scan, MRI (magnetic resonance imaging), ultrasound, and through SPECT (single photon emission computed tomography). These three-dimensional models are then analyzed thoroughly to find the most appropriate automated fabrication method (3D printing) to reproduce the damaged organ/bone through natural/synthetic biopolymers. Most often, several processing modes are combined to produce a part/organ with higher-dimensional accuracy, especially the one with optics that can produce the part at high resolution [69]. As plenty of fabrication processes and materials are available, it is a challenging task to choose the right one for a specific application. To get an overview, AM techniques and their classification for biopolymers are discussed in this section.

### 20.4.1 Classification of additive manufacturing techniques for biopolymers

The broad classification of the AM techniques is based on both natural biopolymers, i.e., collagen, gelatin, starch, cellulose, chitosan, etc., and synthetic biopolymers [90] such as PCL, PLA, PGA, PPF, PEEK, CPP, etc. However, it is hard to classify the AM techniques: it depends on the material as it involves combined techniques for some specific applications. So, the most common and widely used techniques are shown in Fig. 20.3 (see also Table 20.5).

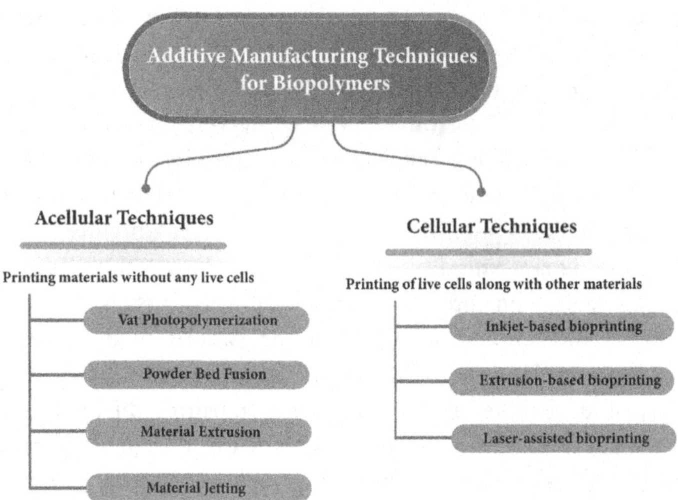


**Figure 20.3** Additive manufacturing techniques for biopolymers [70].

**Table 20.4** Pros and cons of various driving mechanisms in extrusion-based bioprinting.

| Various driving mechanism | Advantages | Disadvantages |
|---|---|---|
| Piston-driven | • Direct control over the flow of the printing material from the nozzle<br>• High cell viability | • Subjected to minor pressure drop at the nozzle |
| Pneumatic | • Works better for high viscous material<br>• High cell viability | • Delay in dispensing due to the compressed gas |
| Screw-driven | • High spatial control and valuable for the dispensing,high viscosity materials | • Generate large pressure drop at the nozzle |

**Table 20.5** AM Techniques and associated materials [70–85].

| AM techniques | Description | Examples | Biopolymers | Cells |
|---|---|---|---|---|
| Vat photopolymerization | Liquid photopolymer in a vat is selectively cured by light-activated polymerization | Stereolithography (SLA), microstereolithography (μ-SLA), Digital, light processing (DLP), two-photon polymerization (2PP) | PCL, PLA, PPF, PDLLA, PEO, PEG, gelatine, Collagen, alginate, PPF/HA, PEG/HA | No |
| Powder bed fusion | Thermal energy selectively fuses regions of a powder bed | Selective laser Sintering (SLS) | PCL, PLA, PLLA, PVA, PEEK, HA/PCL, (Ca-P)/PHBV, HA/PVA | No |
| Material extrusion | Material is selectively dispensed through a nozzle or orifice | Fused Deposition modeling | PCL, PEGT/PBT composite, PLL/TCP, PCL/HA, PCL/TCP | No |
| Material jetting | Droplets of build material are selectively deposited | Polyjet, binder jetting | Dextran, Full Cure 930 Tango Plus, gelatine, PLA/PCL | No |
| Inkjet-based bioprinting | Analogous to commercially available inkjet printers but with live cells as ink | Thermal and piezoelectric inkjet bioprinting | PLGA, DMSO, fibronectin, gelatine | Yes |
| Extrusion-based bioprinting | Dispense of material through micronozzles | Piston, pneumatic and screw-driven bioprinting | Gelatin, HA, dextran, alginate | Yes |
| Laser-assisted bioprinting | Uses nanosecond UV laser as energy source | LAB bioprinting for skin constructs | Hydroxyapatite, keratinocytes | Yes |

### 20.4.2 Acellular techniques

Acellular techniques are a variation of 3D-printing the materials without any live cells that are analogous to general-purpose 3D printing techniques or application-oriented techniques. Nevertheless, in recent advancements, a compelling development has been achieved by making custom-made 3D printers that could combine both living and nonliving materials, which are also known as hybrid printing techniques. However, those techniques are not standardized, so a general classification technique has been discussed here.

### 20.4.3 Vat photopolymerization

Vat photopolymerization (VP) is the process of 3D-printing the geometrical part layer-by-layer from a vat of liquid photopolymer resin through the use of heat source in terms of light, often ultraviolet or visible light. Photopolymer resins are also known as light-activated resins that crystallize and stiffen when exposed to UV light or otherwise known as curing/hardening. The mechanism behind the VP is based on monomers and oligomers in the resin, which gets polymerized when exposed to a heat source. The photoinitiators presented in the resin will be decomposed after exposure to heat and become reactive. This decomposed photoinitiators are solely responsible for polymerizing the functional groups of monomers and oligomers together. One of the significant drawbacks of this AM technique is the limited choice of materials, as the resin should be photopolymerizable [70,71]. This process is the basic strategy for many such printing techniques such as stereolithography (SLA), microstereolithography (μ-SLA—an extension of stereolithography to achieve high-resolution prints), digital light processing (DLP), and multiphoton polymerization (MPP). VP processes have been used to print a wide variety of materials, including natural and synthetic biopolymers and polymer—ceramic composites such as PCL, PLA, PPF, PDLLA, PEO, PEG, gelatine, collagen, alginate, PPF/HA, and PEG/HA. The detailed information about each technique is already well established and published in high impact and reputed journals [72]. Anthony et al. researched about different photopolymer resins and nanoscale reinforcements (cellulose nanocrystals [CNCs], hydroxyapatite [HA], silica, polyvinyl pyrrolidone—coated gold, sepiolite, and graphene) to print nanocomposites through VP technique. The investigation also involves preprocessing (especially their effects on agglomeration and viscosity of resin), effects of different postprocessing techniques, and disparate material properties (mechanical, electrical, magnetic, and biocompatibility).

The research outcome has demonstrated that there is still a lack of development in equipment and material, which is potentially seen as an area that needs to be developed in future endeavors [70]. Donald et al. researched on formulating new biopolymers that feature acid-cleavable linkages through the characterization of tunable 3D polymer networks using various cross-linkers. In his work, a VP AM process photocured an acid-labile cross-linker within a methacrylate-terminated poly(ethylene glycol) polymer network that yielded a biocompatible 3D structure with diverse architectures, which could be used for tissue engineering applications [71]. His research concluded with one major challenge of using the VP process for biomedical applications. In essence, this is finding the optimal combination of design, properties of polymer, its processing abilities for its specific application, and also numerous polymer properties that need to be studied for specific biological environments. Another study by Mondschein et al. [73] briefly reviewed using the VP in biomedical applications and revealed the polymer—structure—property relationship requirements for specific tissue engineering applications.

### 20.4.4 Powder bed fusion

Powder bed fusion (PBF) technology works based on utilizing the micron-level powder particles that are fused by the effect of heat source, i.e., either laser or electron beam, to build the complex array of geometrical parts. This technique enables the manufacturers to produce the parts with high precision and with remarkable design freedom achieved through available technologies and materials. PBF comprises of selective laser sintering (SLS), selective laser melting (SLM), and electron beam melting (EBM) technologies. Still, however, the only nonmetallic process with biocompatible and biodegradable technology in PBF is SLS, which is used to produce scaffolds for tissue engineering applications. SLS machine consists of a computer-controlled laser to sinter the powder particles, a mirror to project the laser beam, a construction zone where the model is built, a material vat to recover the nonprocessed powders, a powder reservoir, and mechanical roller to supply a new layer. The properties of the powder, such as particle size and powder morphology, and the operational parameters such as laser power, spot diameter, layer thickness, and scanning speed, have a significant influence over the printed scaffolds. The applicability of the SLS technique not just is limited to the biopolymers such as PCL, PLA, PLLA, PVA, and polyether ether ketone but also includes composite materials such as HA/PCL, Ca-P/PBHV, HA/polyamide, and so on [72].

### 20.4.5 Material extrusion

Material extrusion, otherwise known as fused deposition modeling (FDM) or free-form fabrication (FFF), is the most reliable and economical type of AM technique to print thermoplastic polymers. This technique uses a heated printhead in which the thermoplastic filaments get melted and extruded through the nozzle and build the part in a layer-by-layer pattern. A wide variety of biomaterials can be printed in FDM, and the most common biopolymers are PLA and PCL with a low melting point, which are well known for their biocompatible and biodegradable properties and hence, widely used for medical applications [74]. One main advantage of using FDM technology is that the machine does not need much knowledge and high skilled labor, so even a nonspecialist could operate and utilize its full functionality. By taking leverage from this most straightforward technology, it has found a wide area of applications using polymers and or polymer/ceramic composites such as PCL, PEGT/PBT composite, PLL/TCP, PCL/HA, and PCL/TCP to produce scaffolds for bone, cartilage, and osteochondral and vascular tissues [72].

### 20.4.6 Material Jetting

Material jetting (MJ), otherwise known as inkjet printing, uses a printhead with hundreds of tiny nozzles in a range of 18–25 μm to jet photopolymers or wax onto the build platform to build the parts in the presence of light (ultraviolet) or heat that ensures the physical objects are built one layer at a time. MJ process can fabricate a complex multimaterial object by combining different materials and print it together in a single part. With these capabilities, this method is mostly used to print precise anatomical models for surgical planning and preoperative simulations [74]. This technique often requires a dissolvable support material that is printed along with the main part, which would be removed during the postprocessing. Since this is a relatively new technique, only a handful of materials can be printed. Another related method is called binder jetting, which is a powder-based technology similar to PBF. However, instead of using a heat source, a binding agent is used to bind the powders together, which are deposited from the printhead. A slightly modified version of this technique has been used for the printing of live cells along with other materials.

### 20.4.7 Cellular techniques

Cellular techniques are also known as bioprinting, which is the process of printing the live cells—either cells alone or in combination with other materials to print the tissues, organs, and scaffolds using natural and synthetic biopolymers. The most common bioprinting techniques that combine both living cells and nonliving materials are inkjet-based bioprinting, extrusion-based bioprinting, and laser-assisted bioprinting.

### 20.4.8 Inkjet-based bioprinting

Inkjet-based bioprinting is the first technique used to print live cells. The architecture of the bioprinter is analogous to the commercially available 2D desktop inkjet printers. However, instead of using the inks in the cartridge, it was replaced by biological materials, i.e., live cells and a build platform with control axes replaced the paper. This technique comes with two types of material ejection system, the thermal and piezoelectric method to eject the material onto the bed to build parts such as scaffolds, tissues, and so on. Thermal inkjet bioprinters work based on electrically heating the printhead to produce pulses of pressure that force droplets from the nozzle. The major advantages of using thermal inkjet bioprinters include relatively high performance, economical, and broad availability. However, there are some downsides in using this method, as it exposes the cells and materials to mechanical stresses, and the presence of thermal energy has a significant impact on the part being built. Other disadvantages include low directional droplets, variation in droplet size, nozzle clogging, and unreliable cell encapsulation. On the other hand, piezoelectric bioprinters are most widely used than inkjet bioprinter due to the following advantages: in essence, it can generate and control a uniform droplet size and ejection directionality as well as to avoid exposure of cells to heat and pressure stressors. The piezoelectric crystal in the printhead creates an acoustic wave that breaks the liquid into droplets at regular intervals, which in turn ejects the uniform droplets of material. Both thermal and piezoelectric bioprinters have some limitations on material viscosity owing to the excessive force required to eject drops using solutions at higher viscosities. Another drawback of inkjet bioprinting is that the biological material has to be in a liquid form to enable droplet formation; as a result, the printed liquid must then form a solid and stable 3D structure with a structural organization and functionality [75].

### 20.4.9 Extrusion-based bioprinting

Extrusion-based bioprinting is also known as microextrusion printing works on the basis of temperature-controlled dispensing of biopolymers/hydrogels via micronozzle through a three-axis computer-controlled translational motion. This method uses three types of driving mechanisms, such as piston-driven, pneumatic, or screw-driven mechanism, to extrude the printing material. This print method offers a higher printing speed but comes with low resolution in the order of 180–200 um with more flexibility in the material that can be printed. The most common downside of using this technique is that the cell viability is much lower when compared with inkjet-based bioprinting with a range of 40%–86% [75,76]. There are several advantages and disadvantages associated with the driving mechanism, which are tabulated in Table 20.4.

### 20.4.10 Laser-assisted bioprinting

Laser-assisted bioprinting (LAB) uses the laser as the energy source to dispense materials onto the substrate. It consists of a pulsed laser source, a ribbon (usually coated with liquid biomaterial), and a receiving substrate. The liquid biomaterials reach the receiving substrate in the form of droplets through evaporation by the effect of laser irradiation of the ribbon. The cell culture medium in the receiving substrate preserves the adhesion and its growth once the cells are transferred to the ribbon. LAB uses nanosecond lasers with UV to print cells, proteins, hydrogels, and some ceramic materials. This method has higher resolution in the order from pico-to-microscale compared with other two bioprinting techniques [77], and it can be affected by many factors such as laser pulse, substrate wettability, the distance between substrate and ribbon, the thickness of a biological layer, surface tension, and viscosity of the biological layer. Another significant advantage of using this system over the other two is that it is nozzle free, so the major drawback is that nozzle clogging is avoided. The disadvantages of this technique are as follows: it is relatively high in cost, it requires additional technologies such as cell recognition scanning to speed up the printing time, and it is time-consuming [75].

Since 3D printing technology was not originally developed for biological applications, it poses many challenges in place when dealing with live cells, tissues, and organs. It has become quite challenging with the advancements in material technologies by providing the researchers with a wide range of materials that made them infer in-depth details and effects of each material before initiating the printing. Each material and printing technique combination has its advantages, disadvantages, and difficulties and to present all the AM challenges is quite complicated and inconceivable. Still, however, the most primitive AM/bioprinting challenges for biopolymers are presented. The primary challenge or one limitation is that even though the material cost for 3D printing is affordable, due to the prolonged processing time of producing, the model increases the cost of equipment that significantly reduces the number of consumers in the market. The secondary challenge involves not only finding the materials that are compliant with biological applications and the printing process but also finding its capability that could provide the part being built with the desired mechanical and functional properties.

In terms of design and fabrication challenges, there are still a lot of complications and adversity in replicating the exact models by 3D printing/bioprinting due to the resolution of the printer and its inaccuracies. For example, the pulmonary alveolar in lungs is made up of most arduous biological design, and replicating such intricate design needs a high accuracy meshing and slicing system. Even though the advanced 3D laser scanning systems are used to obtain the digital images of body contours, the conversion to a tremendous form of buildable and heterogeneous model representation is required to define the precise geometrical information. Therefore, upgrading to a higher imagery scanning system and improvement in *.STL file conversion with subsequent development in Bio-CAD modeling and machine architecture can enhance the resolution, accuracy, and the printing ability of the printers and the parts being printed.

## 20.5 Future and prospects

The future direction of the AM for biopolymers does not directly rely on the material, technology, or applications but commercialization. So far, the advancements in the research are at an early stage, and development in the past two to three decades had led us into a stringent path, which is just the beginning of the evolution of AM for biological applications. When comparing with other fields such as aerospace or engineering, there has been a significant development in a short period of time, and the 3D printing systems and machines are being commercialized and pose as a serious contemporary competitor for conventional manufacturing methods. These powerful rapid developments in the aforementioned field had happened only because of revolutionizing the industrial age, and its effects were resulted in bringing the products into the market by breaking the research boundary and promoting the commercialization by deploying suitable norms and regulations. And so, the first step in the development of AM in biopolymers should be by taking an auxiliary effort and become fecund to implement the norms and regulations in a short period of time by promoting more funding and by agile evaluation of the advancements.

Every year, the number of people who die or become handicapped or at the risk of losing their life has been growing persistently due to organ damage or organ failure. In the United States alone, more than 110,000 people need organs and are at the risk of losing their life due to damaged organs. The recent decline in organ donors had a severe impact on society and increased the number of people in need of organs over the decade. Once the bioprinting systems and machines are commercialized and well established, it will radically eradicate their needs and help in improving the people's life extension.

The first and foremost step in improving anything or any process can be achieved by improving the materials (or bioinks) being used for its specific purpose, which is a most captivating trend that the material scientists should be incorporated with new scientific methods and approaches to increase the height of mimicking any tissues or organs. Some of the requirements of new biopolymer materials should have and comply with the following properties:

1. Better printability (i.e., viscosity and rheological properties)—For example, materials with low viscosity are more enticing for bioprinting as with the low-pressure environment, the cells can grow enormously well.
2. Good biocompatibility—The material should enhance the biological functions and increase the survival of the printed cells/merged tissues.
3. Mechanical properties and structural integrity—The material should have the ability to withstand the construct once it is printed and be able to integrate with other material if printed together.
4. Biomimicry—The material should mimic the exact replication of the original tissue/organ with the comparable structural and dynamic material properties.

Additionally, improvements in the current field should be carried onto the next level by developing new techniques by making portable 3D printers that can be directly printed/integrated into the patient's body. Another approach involves growing the organs from patients' own cells and tissues, which has already been initiated in some places, but still lack of facilities pushes this technique down, and so it must be promoted. Lastly, a different combination of current AM techniques and disparate materials may link to the birth of a new arena to produce the body parts/segments, and by doing so, there are higher chances of obtaining a positive outcome. Thus, instead of focusing on the exploitation of one single technique, it would be most advantageous to combine the positive effects of different techniques into one operational procedure.

The most prominence application of AM of biopolymers may include in the development of (bio-based) smart polymers or smart materials, which is the new state-of-the-art in the current modern world with a vibrant technological era. Smart biomaterials have significant properties that could change from one state to another state by external stimuli such as a variation in temperature, pressure, light, electrical and or magnetic fields, and so on. This area of research might be a gateway to bring in new materials, and it could potentially lead to new scientific advancements called self-healing biopolymers.

## References

[1] P.K. Yarlagadda, M. Chandrasekharan, J.Y.M. Shyan, Recent advances and current developments in tissue scaffolding, Bio Med. Mater. Eng. 15 (3) (2005) 159–177.

[2] S.V. Murphy, A. Atala, 3D bioprinting of tissues and organs, Nat. Biotechnol. 32 (8) (2014) 773.

[3] E. Lavik, R. Langer, Tissue engineering: current state and perspectives, Appl. Microbiol. Biotechnol. 65 (1) (2004) 1–8.

[4] D.W. Hutmacher, Scaffold design and fabrication technologies for engineering tissues—state of the art and future perspectives, J. Biomater. Sci. Polym. Ed. 12 (1) (2001) 107–124.

[5] P.O. Bagnaninchi, Y. Yang, N. Zghoul, N. Maffulli, R.K. Wang, A.E. Haj, Chitosan microchannel scaffolds for tendon tissue engineering characterized using optical coherence tomography, Tiss. Eng. 13 (2) (2007) 323–331.

[6] S.A. Bencherif, T.M. Braschler, P. Renaud, Advances in the design of macroporous polymer scaffolds for potential applications in dentistry, J. Period. Imp. Sci. 43 (6) (2013) 251–261.

[7] M.B. Chan-Park, W.K. Neo, Ultraviolet embossing for patterning high aspect ratio polymeric microstructures, Microsyst. Technol. 9 (6–7) (2003) 501–506.

[8] S. Thian Chen Hai, Three Dimensional Microfabrication and Micromoulding, 2009 (Doctoral Dissertation).

[9] R. Velu, S. Singamneni, Selective laser sintering of polymer biocomposites based on polymethyl methacrylate, J. Mater. Res. 29 (17) (2014) 1883–1892.

[10] R. Velu, F. Raspall, S. Singamneni, 3D printing technologies and composite materials for structural applications, in: Green Composites for Automotive Applications, Woodhead Publishing, 2019, pp. 171–196.

[11] R. Velu, B.P. Kamarajan, M. Ananthasubramanian, T. Ngo, S. Singamneni, Post-process composition and biological responses of laser sintered PMMA and β-TCP composites, J. Mater. Res. 33 (14) (2018) 1987–1998.

[12] L. Hao, M.M. Savalani, Y. Zhang, K.E. Tanner, R.A. Harris, Selective laser sintering of hydroxyapatite reinforced polyethylene composites for bioactive implants and tissue scaffold development, Proc. IME H J. Eng. Med. 220 (4) (2006) 521–531.

[13] B. Huang, S.B. Singamneni, Curved layer adaptive slicing (CLAS) for fused deposition modelling, Rapid Prototyp. J. 21 (4) (2015) 354–367, https://doi.org/10.1108/RPJ-06-2013-0059.

[14] R. Velu, S. Singamneni, Evaluation of the influences of process parameters while selective laser sintering PMMA powders, Proc. IME C J. Mech. Eng. Sci. 229 (4) (2015) 603–613.

[15] P.J. Bártolo (Ed.), Stereolithography: Materials, Processes and Applications, Springer Science & Business Media, 2011.
[16] E.M. Hamad, S.E.R. Bilatto, N.Y. Adly, D.S. Correa, B. Wolfrum, M.J. Schöning, A. Yakushenko, Inkjet printing of UV-curable adhesive and dielectric inks for microfluidic devices, Lab Chip 16 (1) (2016) 70–74.
[17] S. Singamneni, R. Velu, M.P. Behera, S. Scott, P. Brorens, D. Harland, J. Gerrard, Selective laser sintering responses of keratin-based bio-polymer composites, Mater. Des. 183 (2019) 108087.
[18] R. Velu, A. Fernyhough, D.A. Smith, M. Guen, S. Singamneni, Selective Laser Sintering of Biocomposite Materials. Lasers in Engineering, Old City Publishing, 2016, p. 35.
[19] N.J. Castro, J. O'brien, L.G. Zhang, Integrating biologically inspired nanomaterials and table-top stereolithography for 3D printed biomimetic osteochondral scaffolds, Nanoscale 7 (33) (2015) 14010–14022.
[20] J.P. Kruth, P. Mercelis, J. Van Vaerenbergh, L. Froyen, M. Rombouts, Binding mechanisms in selective laser sintering and selective laser melting, Rapid Prototyp. J. 11 (1) (2005) 26–36, https://doi.org/10.1108/13552540510573365.
[21] J.W. Barlow, G. Lee, R.H. Crawford, J.J. Beaman, H.L. Marcus, R.J. Lagow, U.S. Patent No. 6,540,784, U.S. Patent and Trademark Office, Washington, DC, 2003.
[22] W.Y. Yeong, N. Sudarmadji, H.Y. Yu, C.K. Chua, K.F. Leong, S.S. Venkatraman, L.P. Tan, Porous polycaprolactone scaffold for cardiac tissue engineering fabricated by selective laser sintering, Acta Biomaterialia 6 (6) (2010) 2028–2034.
[23] D.W. Hutmacher, T. Schantz, I. Zein, K.W. Ng, S.H. Teoh, K.C. Tan, Mechanical properties and cell cultural response of polycaprolactone scaffolds designed and fabricated via fused deposition modeling, J. Biomed. Mater. Res. 55 (2) (2001) 203–216.
[24] C.B. Williams, F. Mistree, D.W. Rosen, A functional classification framework for the conceptual design of additive manufacturing technologies, J. Mech. Des. 133 (12) (2011).
[25] E. Berry, J.M. Brown, M. Connell, C.M. Craven, N.D. Efford, A. Radjenovic, M.A. Smith, Preliminary experience with medical applications of rapid prototyping by selective laser sintering, Med. Eng. Phys. 19 (1) (1997) 90–96.
[26] K.F. Farraro, K.E. Kim, S.L. Woo, J.R. Flowers, M.B. McCullough, Revolutionizing orthopaedic biomaterials: the potential of biodegradable and bioresorbable magnesium-based materials for functional tissue engineering, J. Biomech. 47 (9) (2014) 1979–1986.
[27] P. Parida, A. Behera, S.C. Mishra, Classification of Biomaterials Used in Medicine, 2012.
[28] L.G. Griffith, Polymeric biomaterials, Acta Materialia 48 (1) (2000) 263–277.
[29] C. Agrawal, R.B. Ray, Biodegradable polymeric scaffolds for musculoskeletal tissue engineering, J. Biomed. Mater. Res. 55 (2) (2001) 141–150.
[30] S.K. Bhasin, R. Mann, Introductory Polymer Science, Dhanpat Rai Publications, 2015.
[31] J.A. Brydson, Plastics materials, Butterworth-Heinemann, 1999.
[32] J.A. Hubbell, Synthetic biodegradable polymers for tissue engineering and drug delivery, Current Opinion in Solid State and Materials Science 3 (3) (1998) 246–251.
[33] B.L. Seal, T.C. Otero, A. Panitch, Polymeric biomaterials for tissue and organ regeneration, Materials Science and Engineering: R: Reports 34 (4) (2001) 147–230.
[34] A.K. Mohanty, M. Misra, L.T. Drzal (Eds.), Natural Fibers, Biopolymers, and Biocomposites, CRC press, 2005.
[35] H.N. Englyst, S.M. Kingman, Dietary fiber and resistant starch, in: Dietary Fiber, Springer US, 1990, pp. 49–65.
[36] H. Englyst, Classification and measurement of plant polysaccharides, Anim. Feed Sci. Technol. 23 (1) (1989) 27–42.

[37] H.N. Englyst, J.H. Cummings, Non-starch polysaccharides (dietary fiber) and resistant starch, in: New Developments in Dietary Fiber, Springer US, 1990, pp. 205–225.
[38] L. Averous, Biodegradable multiphase systems based on plasticized starch: a review, J. Macromol. Sci. Polym. Rev. 44 (3) (2004) 231–274.
[39] J.P. Santerre, K. Woodhouse, G. Laroche, R.S. Labow, Understanding the biodegradation of polyurethanes: from classical implants to tissue engineering materials, Biomaterials 26 (35) (2005) 7457–7470.
[40] J.M. Wozney, H.J. Seeherman, Protein-based tissue engineering in bone and cartilage repair, Curr. Opin. Biotechnol. 15 (5) (2004) 392–398.
[41] A. Sarasam, S.V. Madihally, Characterization of chitosan–polycaprolactone blends for tissue engineering applications, Biomaterials 26 (27) (2005) 5500–5508.
[42] P.A. Gunatillake, R. Adhikari, Biodegradable synthetic polymers for tissue engineering, Eur. Cell. Mater. 5 (1) (2003) 1–16.
[43] J.C. Middleton, A.J. Tipton, Synthetic biodegradable polymers as orthopedic devices, Biomaterials 21 (23) (2000) 2335–2346.
[44] L.S. Nair, C.T. Laurencin, Biodegradable polymers as biomaterials, Prog. Polym. Sci. 32 (8) (2007) 762–798.
[45] K.A. Athanasiou, C.M. Agrawal, F.A. Barber, S.S. Burkhart, Orthopaedic applications for PLA-PGA biodegradable polymers, Arthrosc. J. Arthrosc. Relat. Surg. 14 (7) (1998) 726–737.
[46] D.M. Zuckerman, P. Brown, S.E. Nissen, Medical device recalls and the FDA approval process, Arch. Intern. Med. 171 (11) (2011) 1006–1011.
[47] G.H. Llanos, P. Narayanan, M.B. Roller, A. Scopelianos, U.S. Patent No. 6,746,773, U.S. Patent and Trademark Office, Washington, DC, 2004.
[48] S.T. Mejlhede, S. Gyrn, U.S. Patent No. 7,431,876, U.S. Patent and Trademark Office, Washington, DC, 2008.
[49] N.M. Lamba, K.A. Woodhouse, S.L. Cooper, Polyurethanes in Biomedical Applications, CRC press, 1997.
[50] M.M. Tunney, S.P. Gorman, S. Patrick, Infection associated with medical devices, Rev. Med. Microbiol. 7 (4) (1996) 195–206.
[51] http://www.madehow.com/Volume-3/Syringe.html.
[52] M. Gajendiran, J. Choi, S.J. Kim, K. Kim, H. Shin, H.J. Koo, K. Kim, Conductive biomaterials for tissue engineering applications, J. Ind. Eng. Chem. (2017).
[53] S. Mohanty, L.B. Larsen, J. Trifol, P. Szabo, H.V.R. Burri, C. Canali, A. Wolff, Fabrication of scalable and structured tissue engineering scaffolds using water dissolvable sacrificial 3D printed moulds, Mater. Sci. Eng. C 55 (2015) 569–578.
[54] S. Mohanty, K. Sanger, A. Heiskanen, J. Trifol, P. Szabo, M. Dufva, A. Wolff, Fabrication of scalable tissue engineering scaffolds with dual-pore microarchitecture by combining 3D printing and particle leaching, Mater. Sci. Eng. C 61 (2016) 180–189.
[55] H. Yoshimoto, Y.M. Shin, H. Terai, J.P. Vacanti, A biodegradable nanofiber scaffold by electrospinning and its potential for bone tissue engineering, Biomaterials 24 (12) (2003) 2077–2082.
[56] R. Zhang, P.X. Ma, Poly (α-hydroxyl acids)/hydroxyapatite porous composites for bone-tissue engineering. I. Preparation and Morphology, J. Biomed. Mat. Res. (1999).
[57] S. Freitas, H.P. Merkle, B. Gander, Microencapsulation by solvent extraction/evaporation: reviewing the state of the art of microsphere preparation process technology, J. Contr. Release 102 (2) (2005) 313–332.
[58] M. Samejima, G. Hirata, Y. Koida, Studies on microcapsules. I. Role and effect of coacervation-inducing agents in the microencapsulation of ascorbic acid by a phase separation method, Chem. Pharm. Bull. 30 (8) (1982) 2894–2899.

[59] A. Autissier, C. Le Visage, C. Pouzet, F. Chaubet, D. Letourneur, Fabrication of porous polysaccharide-based scaffolds using a combined freeze-drying/cross-linking process, Acta Biomat. 6 (9) (2010) 3640–3648.
[60] B.D. Halpern, Dental polymers, Ann. N. Y. Acad. Sci. 146 (1) (1968) 106–112.
[61] S. Jarby, E. Andersen, Dental fillings of polycarbonate, J. Dent. Res. 41 (1) (1962), 214-214.
[62] J. Slais, Die Einheilung der porösen Polymethylmethakrylate. Zbl. allg. Path. path, Anatolia 98 (1958) 571.
[63] E.A. Martins, F.A. Peyton, R.H. Kingery, Properties of custom-made plastic teeth formed by different techniques, J. Prosthet. Dent 12 (6) (1962) 1059–1065.
[64] J.L. Ferracane, Developing a more complete understanding of stresses produced in dental composites during polymerization, Dent. Mater. 21 (1) (2005) 36–42.
[65] M. Atai, M. Ahmadi, S. Babanzadeh, D.C. Watts, Synthesis, characterization, shrinkage and curing kinetics of a new low-shrinkage urethane dimethacrylate monomer for dental applications, Dent. Mater. 23 (8) (2007) 1030–1041.
[66] J.G.F. Santos Jr., V.J.R.R. Pita, P.A. Melo, M. Nele, J.C. Pinto, Production of bone cement composites: effect of fillers, co-monomer and particles properties, Braz. J. Chem. Eng. 28 (2) (2011) 229–241 (X).
[67] Z.D. Vaughn, J. Schmidt, D.P. Lindsey, J.L. Dragoo, Biomechanical evaluation of a 1-stage revision anterior cruciate ligament reconstruction technique using a structural bone void filler for femoral fixation, Arthrosc. J. Arthrosc. Relat. Surg. 25 (9) (2009) 1011–1018.
[68] L. Lemos, M. Nele, P. Melo, J.C. Pinto, Modeling methyl methacrylate (MMA) polymerization for bone cement production, in: Macromolecular Symposia, vol. 243, WILEY-VCH Verlag, November 2006, pp. 13–23. No. 1.
[69] F.P.W. Melchels, M.A.N. Domingos, T.J. Klein, et al., Additive manufacturing of tissues and organs, Prog. Polym. Sci. 37 (2012) 1079–1104, https://doi.org/10.1016/j.progpolymsci.2011.11.007.
[70] A. 2Medellin, W. Du, G. Miao, et al., Vat photopolymerization 3d printing of nanocomposites: a literature review, J. Micro Nano-Manuf. 7 (2019), https://doi.org/10.1115/1.4044288.
[71] D.C. Aduba, E.D. Margaretta, A.E.C. Marnot, et al., Vat photopolymerization 3D printing of acid-cleavable PEG-methacrylate networks for biomaterial applications, Mater. Today Commun. 19 (2019) 204–211, https://doi.org/10.1016/j.mtcomm.2019.01.003.
[72] R.F. Pereira, P.J. Bártolo, Recent advances in additive biomanufacturing, Compr. Mater. Process 10 (2014) 265–284, https://doi.org/10.1016/B978-0-08-096532-1.01009-8.
[73] R.J. Mondschein, A. Kanitkar, C.B. Williams, et al., Polymer structure-property requirements for stereolithographic 3D printing of soft tissue engineering scaffolds, Biomaterials 140 (2017) 170–188, https://doi.org/10.1016/j.biomaterials.2017.06.005.
[74] K. Tappa, U. Jammalamadaka, Novel biomaterials used in medical 3D printing techniques, J. Funct. Biomater. 9 (2018), https://doi.org/10.3390/jfb9010017.
[75] S.V. Murphy, A. Atala, 3D bioprinting of tissues and organs, Nat. Biotechnol. 32 (2014) 773–785, https://doi.org/10.1038/nbt.2958.
[76] F. Pati, J. Jang, J.W. Lee, D.W. Cho, Extrusion bioprinting, Essent. 3D Biofab. Transl. (2015) 123–152, https://doi.org/10.1016/B978-0-12-800972-7.00007-4.
[77] J. Li, M. Chen, X. Fan, H. Zhou, Recent advances in bioprinting techniques: approaches, applications and future prospects, J. Transl. Med. 14 (2016) 1–15, https://doi.org/10.1186/s12967-016-1028-0.
[78] X. Li, R. Cui, L. Sun, et al., 3D-Printed biopolymers for tissue engineering application, Int. J. Polym. Sci. 1–13 (2014), https://doi.org/10.1155/2014/829145.

[79] S. Iwanaga, K. Arai, M. Nakamura, Inkjet Bioprinting, Elsevier Inc, 2015.
[80] S.C. Ligon, R. Liska, J. Stampfl, et al., Polymers for 3D printing and customized additive manufacturing, Chem. Rev. 117 (2017) 10212–10290, https://doi.org/10.1021/acs.chemrev.7b00074.
[81] P. Ahangar, M.E. Cooke, M.H. Weber, D.H. Rosenzweig, Current biomedical applications of 3D printing and additive manufacturing, Appl. Sci. 9 (2019), https://doi.org/10.3390/app9081713.
[82] L. Chaunier, S. Guessasma, S. Belhabib, et al., Material extrusion of plant biopolymers: opportunities & challenges for 3D printing, Addit. Manuf. 21 (2018) 220–233, https://doi.org/10.1016/j.addma.2018.03.016.
[83] S. Bose, D. Ke, H. Sahasrabudhe, A. Bandyopadhyay, Additive manufacturing of biomaterials, Prog. Mater. Sci. 93 (2018) 45–111, https://doi.org/10.1016/j.pmatsci.2017.08.003.
[84] A.A. Mäkitie, J. Korpela, L. Elomaa, et al., Novel additive manufactured scaffolds for tissue engineered trachea research, Acta Otolaryngol. 133 (2013) 412–417, https://doi.org/10.3109/00016489.2012.761725.
[85] Y.L. Cheng, F. Chen, Preparation and characterization of photocured poly (ε-caprolactone) diacrylate/poly (ethylene glycol) diacrylate/chitosan for photopolymerization-type 3D printing tissue engineering scaffold application, Mater. Sci. Eng. C 81 (2017) 66–73, https://doi.org/10.1016/j.msec.2017.07.025.
[86] R. Velu, et al., A comprehensive review on bio-nanomaterials for medical implants and Feasibility studies on fabrication of such implants by additive manufacturing technique, Materials 13 (1) (2020) 92, https://doi.org/10.3390/ma13010092.
[87] R. Velu, 3D printing technologies and composite materials for structural applications. Green Composites for Automotive Applications, in: Arlindo Silva (Ed.), in: Composites Science and Engineering, Woodhead Publishing, 2019, pp. 171–196.
[88] R. Whenish, Velu R., Design and performance of additively manufactured lightweight bionic hand, AIP Conf. Proc. 2317 (1) (2021) 020028, https://doi.org/10.1063/5.0036119.
[89] S. Anandkumar, Velu R., Single crystal metal deposition using laser additive manufacturing technology for repair of aero-engine components, Mater. Today (2021), https://doi.org/10.1016/j.matpr.2021.02.083.
[90] A. Selvam, Velu R., Preparation and evaluation of the tensile characteristics of carbon fiber rod reinforced 3D printed thermoplastic composites, J. Compos. Sci. 5 (1) (2021) 8, https://doi.org/10.3390/jcs5010008.

CHAPTER 21

# Additive manufacturing using space resources

**Athanasios Goulas[1], Daniel S. Engstrøm[1], Ross J. Friel[2]**
[1]Loughborough University, Wolfson School of Mechanical, Electrical & Manufacturing Engineering, Loughborough, United Kingdom; [2]ITE, Halmstad University, Halmstad, Sweden

## 21.1 Additive manufacturing: a tool to support future activities on another planet

Additive manufacturing [14,44] (AM) and its associated techniques could be combined together with the in situ resource utilization (ISRU) concept, for building a range of physical assets off-world [10,15,30,32,37,43,57]. This could happen by effectively using the abundant and readily available natural resources onsite [13,31].

Additionally, AM techniques have the distinct advantage of being able to function in an autonomous way, without the need for continuous supervision or direct input from a machine operator. This ability to operate autonomously is one of the main reasons why AM is considered as being ideal for remote and harsh environments, such as the surface of a planet like Mars or the Moon. Aside from a remote operation scenario, AM has great potential to serve as an enabling technology within a space manufacturing laboratory, for either low-volume production of replacement parts or one-off manufacture of bespoke components. The materials for supporting such a facility could be either the locally excavated regoliths directly [29], its beneficiated products such as extracted metals (i.e., titanium from ilmenite reduction [39]), or by-products of the beneficiation processes that would normally be useless for any other application.

## 21.2 Indigenous space material resources

A planetary surface typically consists of a thin and loose layer of granular material called "regolith." The term originates from the combination of the two Greek words ρέγος [translation: cover] and λίθος [translation: rock] and refers to the complex layer of unconsolidated rock debris that has been

*Additive Manufacturing*
ISBN 978-0-12-818411-0
https://doi.org/10.1016/B978-0-12-818411-0.00018-5

derived through the fragmentation of the bedrock due to meteoritic or micrometeoroid impacts [40], especially in the case of a nonatmospheric body such as the Moon.

Based on their igneous origin and formation history, regoliths are considered as ceramic materials and can be further characterized as multicomponent ceramics, comprising of a mixture of different silicate minerals, mostly embedded in a glass matrix [24].

Regoliths' inherent nature of being in granular form classifies them as a candidate feedstock for several AM techniques, as many of them use powders, either directly or in the form of a colloidal suspension. However, given their dielectric nature, not being electrically conductive, regoliths cannot pose as a compatible feedstock for processing techniques that specifically require electrically conductive materials, such as electron beam melting (EBM).

## 21.2.1 Lunar resources

By observing the Moon with the naked eye, it can be seen that its surface consists of a combination of bright and dark gray areas called "highlands" and "marias," respectively. The latter term originates from the Latin word *mare* [translation: sea] named by the ancient astronomers who mistook those dark plains for actual seas.

Later studies revealed that marias are the more recently formed areas of the Moon and were generated after large impacts that penetrated the lunar crust and excavated basins. That was followed by subsequent volcanic episodes that filled the basins with basaltic magma [47]. Since these regions were formed more recently, they are less affected from meteorite impacts in comparison with the early-formed highland regions [9]. Impact processes, estimated to have occurred over a period of a few billion years, have formed a subcentimeter grain size, powdered fraction of the lunar bedrock, called the lunar regolith [26]. Depending on the location, the overall thickness of the regolith layer may differ due to past impact flux on the lunar surface. The current consensus is that the regolith spans between 5 m deep in the Mare regions and 15 m in the much older highland regions [40].

The lunar regolith consists of a multicomponent material with high morphological variability (from spherical to extremely angular), including five basic material and particle types: mineral fragments, pristine crystalline rock fragments, breccias, and glasses of various kinds [9,52,55]; an example is shown in Fig. 21.1. This nature of variable particle size and shape is

**Figure 21.1** Selected coarse fines from 10,085 lunar soil sample, collected during the Apollo 11 mission, showcasing the morphological variability of lunar soil, from coarse fragmented bedrock to spherical glass agglutinates caused from micrometeoroid impacts. *(Reproduced with kind permission of NASA from Reference Wood, J.A., Dickey, J.S., Marvin, U.B., Powell, B.N., 1970. Lunar anorthosites and a geophysical model of the moon. In: Apollo 11 Science Conference. pp. 965–988).*

something that is not considered as standard among most materials used in AM (in traditional AM, particles' size and shape are often well controlled and purposefully varied for the application—it is not random). Such morphological variance is very likely to hinder the performance of the material during processing, such as reduced flowability, compaction, and so on.

Information on the chemistry, mineralogy, and petrology that we have [45] is dependent on the actual lunar mission landing sites and the specifics of the material in that area. Apollo 11 and 12 landed inside mare basalt regions, and collected samples were characterized by abundance in mare-derived basaltic rock clasts and mafic minerals such as olivine [$(Mg,Fe)_2SiO_4$], pyroxene [$CaSiO_3$ and $(Mg,Fe)SiO_3$] and the oxide mineral ilmenite [$FeTiO_3$] [28]. On the other hand, samples collected from the Apollo 16 mission that landed within the highland regions consisted of highland-derived anorthositic rocks, breccias, and anorthositic feldspars [48]. In addition to the mineral and lithic debris, the regolith consists of glass that exists either in isolated glassy particles caused by localized impact melting or as the matrix in the agglutinates[1] that is a combination of mineral and lithic fragments embedded in a glassy matrix [47].

[1] Agglutinates: localized melts caused from meteorite impacts.

Lunar mare basalts are depleted of volatile elements (potassium, sodium, rubidium, led, carbon, and hydrogen) and exhibit higher concentrations of titanium/chromium and iron/manganese when compared with terrestrial basalts [52]. For example, lunar plagioclase minerals at both the highland or Marian planes are depleted in sodium and consist mostly of calcium-rich anorthite or bytownite mineral species [3].

This is practically interpreted as a change in the mineral's physical properties since; according to *Bowen*, lower calcium levels in plagioclase minerals are associated with lower melting temperatures [6]. This is shown in the phase diagram in Fig. 21.2.

The aforementioned research findings mean that the regolith materials on the lunar surface are variable in their compositional texture and mineral-to-glass ratios over the entire lunar surface. Such compositional variance is naturally expected to carry over to the material/s thermophysical properties, such as variable or multiple melting temperatures, multiple glass transitions, and so on. Again, this is considered atypical for a material that is to be used for AM, since most materials have very well-defined thermophysical properties with small ranges of those properties. This highlights the fact that using a naturally occurring material vs. an "engineered" material in AM is a significant challenge; in most cases, a material is often specifically developed to be compatible with a specific AM process/technique.

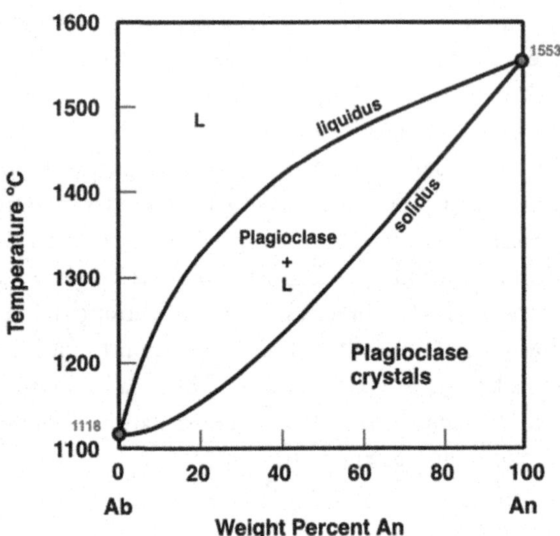


**Figure 21.2** The anorthite–albite plagioclase feldspar phase diagram.

## 21.2.2 Regolith simulants/analogues

### *21.2.2.1 Lunar simulants*

To date, the Moon is the only celestial body beyond Earth, which has been systematically sampled and analyzed by scientists back on Earth. Results from analyzing several meteorite fragments [28] and over 381 kg of actual rock and soil samples brought back from the Moon during the six NASA Apollo and nine USSR Luna missions have provided a wealth of information regarding the Moon's surface geological composition and mineral chemistry. This abundant information has aided scientists and engineers in developing a range of simulant materials, which are necessary and available in sufficient quantity to assist engineering studies for future Lunar or Martian missions.

Lunar regolith simulants are synthesized from terrestrial components for the purpose of simulating the main physical, morphological, and chemical properties of the lunar regolith [8]. When compared with the real material, they comprise of similar but not identical mineralogy and the same rock forming elements; however, they differ from the actual extraterrestrial matter, since terrestrial materials are typically formed under more oxidizing conditions with significant access to water and were altered by a wider range of weathering processes [9], which are not present on the Moon, affecting their final physical properties. For example, according to studies performed by Papike et al., actual lunar regolith samples exhibited an increased melting temperature of approximately 100–150 °C due to the absence of water in the materials' microcrystalline structure [45].

Many lunar simulants are available today, to be used for research and development. However, since most of them were developed with specific applications in mind, scientists must be very careful when selecting a simulant, so it is made sure that the material possesses the right properties for the lunar location desired, to provide with meaningful data as per each case. Lately several studies, such as the ones from Metzger et al., have become available, discussing the fidelity of the simulants, implicating factors such as their chemistry, mineralogy, particle size distribution, and so on [41,42]. Such data should be taken into consideration prior to selecting any simulant. The list given in Table 21.1 contains all known lunar regolith simulants that have been available to study to date (January 2020).

**Table 21.1** List of lunar regolith simulants.

| Simulant | Simulant complete name | Detail | Category |
|---|---|---|---|
| MLS-1 | MLS-2 Minnesota Lunar Simulant | Highlands | Geotechnical |
| NAO-1 | NAO-1 National Astronomical Observatories | Highlands | General |
| NU-LHT | NU-LHT-1M | Highlands | General |
| OB-1 | OB-1/CHENOBI Olivine Bytownite | Highlands | General |
| ALS | ALS Arizona Lunar Simulant | Mare | Geotechnical |
| ALRS-1 | ALRS-1 Australian Lunar Regolith Simulant | Mare | Geotechnical |
| BP-1 | BP-1 Black Point W172158 | Mare | Geotechnical |
| BP-1 | BP-1 Black Point W172154 | Mare | Geotechnical |
| CAS-1 | CAS-1 Chinese Academy of Sciences | Mare | General |
| CLRS-1 | CLRS-1/2 Chinese Lunar Regolith Simulant | Mare | Unknown |
| CSM-CL | CSM-CL Colorado School of Mines Colorado Lava | Mare | Geotechnical |
| CUG-1A | CUG-1A China University of Geosciences | Mare | Geotechnical |
| DNA-1 | DNA-1 De NoArtri | Mare | Geotechnical |
| FJS-1 | FJS-1 Fuji Japanese Simulant | Mare | General |
| FJS-2 | FJS-2 Fuji Japanese Simulant | Mare | General |
| FJS-3 | FJS-3 Fuji Japanese Simulant | Mare | General |
| GSC-1 | GSC-1 Goddard Space Center | Mare | General |
| JSC-1 | JSC-1/1A/1AF/1AC/2A Johnson Space Center | Mare | General |
| KLS-1 | KLS-1 Korea Lunar Simulant | Mare | Geotechnical |
| KOHLS-1 | KOHLS-1/KAUMLS Korean Lunar Simulants | Mare | Geotechnical |
| | Maryland–Sanders Lunar Simulant | Mare | Geotechnical |
| MLS-1 | MLS-1/1P Minnesota Lunar Simulant | Mare | General |
| MKS-1 | MKS-1 Lunar Simulant | Mare | Unknown |
| NEU-1 | NEU-1 Northeastern University Lunar Simulant | Mare | Unknown |
| | Oshima Simulant | Mare | General |
| TJ-1 | TJ-1/2 Tongji University | Mare | Geotechnical |
| CMU-1 | CMU-1 Carnegie Mellon University | Other | Spectral |
| CRC-1 | GRC-1/3 Glenn Research Center | Other | Geotechnical |
| | Kohyama Simulant | Other | General |
| BLHD20 | BHLD20 Lunar Dust Simulant | Dust | General |
| CLDS-i | CLDS-i Lunar Dust Simulant | Dust | General |

Class Exolith Lab; University of Central Florida.

#### 21.2.2.2 Martian simulants

Martian regolith simulants are also synthesized out of terrestrial components and are tailored to match the properties of the regolith found on the Martian surface.

The main objectives for several studies conducted in the past, toward characterizing the chemistry and mineralogy of the actual Martian regolith, have concluded that it has considerable variation (higher that the Moon's) over the entire planet's surface, in respect to those characteristics [12,49]. Therefore, a single regolith simulant is not capable of representing the entire Martian surface. With the exception of Perko et al., studying the mechanical performance of the JSC-MARS-1A Martian simulant, there are no other available studies expanding on relevant properties of this simulant [46]. Table 21.2 contains a list of all the known Martian simulants developed to date and have been available for studies.

When compared with the lunar simulant analogues, development of Martian regolith simulants has been investigated significantly less, and only a small number of research-grade simulants, as evident through the previous list, are available to the scientific community [1,2,50,51,56].

However, their properties have only been based on the remote sensing and in situ analyzed data from the Martian landers; therefore, they have not been compared with actual samples that have been returned to Earth. This suggests that the close matching bulk chemistry, the simulant's

**Table 21.2** List of Martian regolith simulants.

| Simulant | Simulant complete name | Detail | Category |
|---|---|---|---|
| ES-X | ES-X Mars Simulants | | Geotechnical |
| JMSS-1 | JMSS-1 Jining Mars Soil Simulant | | Geotechnical |
| JSC Mars-1/1A | Johnson Space Center | Dust | Spectral |
| KMS-1 | KMS-1 Korean Mars Simulant | | Geotechnical |
| MGS-1 | MGS-1 Mars Global | | General |
| MMS | MMS Mojave Mars Simulant | | Geotechnical |
| MMS-1 | MMS-1/2 The Martian Garden | | General |
| | MMS Rocknest Augmented MMS | | General |
| | Salten Skov 1 | Dust | Magnetic |
| | UF Acid-Alkaline-Salt Basalt Analog Soils | | General |
| | UC Mars1 | | Geotechnical |

Class Exolith Lab; University of Central Florida.

mineralogy is likely to be inaccurate. This translates to different, spectral and thermophysical properties that are of fundamental importance to the processing style.

#### *21.2.2.3 Asteroid simulants*

Over the past few years, a number of asteroid simulants have become available in the scientific community. That is to serve and provide support to fundamental studies for developing technology for resource prospecting and further beneficiation. Table 21.3 contains a list of the asteroid simulants that have been discussed in the literature and have been available for studies.

## 21.3 Additive manufacturing using indigenous space resources

Over the past two decades, a great number of researchers have studied several potential ways, including using conventional manufacturing or AM, for production activities, using regoliths as a building material.

AM techniques are able to use a wide range of different materials, such as metals, polymers, or ceramics. Such materials might come either in liquid

**Table 21.3** List of asteroid regolith simulants.

| Simulant | Simulant full name | Body | Category |
|---|---|---|---|
| HCCL-1 | HCCL-1 Hydrated Carbonaceous Chondrite Lithologies | Asteroid | General |
| CI | UCF/DSI CI Carbonaceous Chondrite Simulant | Asteroid | General |
| CR | UCF/DSI CR Carbonaceous Chondrite Simulant | Asteroid | General |
| C2 | UCF/DSI C2 Carbonaceous Chondrite Simulant | Asteroid | General |
| CM | UCF/DSI CM Carbonaceous Chondrite Simulant | Asteroid | General |
| | Carbonaceous Chondrite Based Simulant of Phobos | Phobos | General |
| PCA-1 | PCA-1 Phobos Captured Asteroid | Phobos | General |
| PGI-1 | PGI-1 Phobos Giant Impact | Phobos | General |
| UTPS-TB | UTPS-TB | Phobos | General |
| UTPS-IB | UTPS-IB | Phobos | General |
| MPACS | MPACS Mechanical Porous Ambient Comet Simulant | Comet | Geotechnical |

Class Exolith Lab; University of Central Florida.

(i.e., photocurable resin, slurry/paste, ink) or in solid form (i.e., powder, filaments, sheets). The choice of the feedstock type is process specific, as it has to be compatible with the operating principle of each technique.

It should be noted, however, that in most cases for AM, several process-related additives are required (i.e., plasticizers, photoinitiators, viscosity modifiers, etc.), to allow for successful operation of the technique, as they govern the properties and behavior of the material during processing. However, this is a less desirable scenario under the ISRU concept; as in principle, all raw materials should be sourced locally, and it hinders the economical/technical feasibility that the combination of AM and ISRU is expected to offer. The following subsections present examples from envisaged methods, where AM techniques are proposed for processing regolith matter to make various engineering and mission assets. The sections are broken down to examples where additional process additives are either used (indirect) or not (direct).

### 21.3.1 Indirect usage of indigenous resources for additive manufacturing

Khoshnevis et al. investigated the probable use of an extrusion-based additive layer manufacturing technology named "contour crafting" intended for constructing full-scale habitats in the extraterrestrial environment [33]. The project's intentions were originally to provide low-income housing, or military dormitories using cementitious materials, but were further supported by NASA due to their applicability for lunar construction [32,34,35]. The contour crafting printer consists of a three-axis gantry system fitted with an extruder that utilizes slurries as feedstock. The process showed very successful results with a variety of materials, such as plastics, composites, concrete, and ceramics [7]. To date, this is the only AM approach for extraterrestrial construction classified by NASA as at Technology Readiness Level (TRL) 6. The investigators managed to fabricate a variety of thin/thick wall and dome structures using a sulfur-based concrete using the JSC-1A lunar mare regolith simulant as the aggregate. The resultant additively manufactured structures exhibited an average compressive strength of 3.65 MPa [34] which is slightly less than the performance of a conventionally manufactured regolith simulant brick; via sintering routes. The success of this extrusion-based process relies on the use of slurry feedstocks. This would require liquid binders to be transported off-planet to the construction site, since they cannot be found onsite, thus dramatically increasing mission costs and complexity.

Cesaretti et al., by utilizing a full-scale manufacturing method named D-Shape, proposed a 3D printing mechanism for the fabrication of one-off medium-sized (less than 2 m approximately height) complete building structures, or building blocks, by extruding a regolith-based concrete-type material using their own developed lunar regolith simulant DNA-1. This approach was further investigated by the European Space Agency (ESA) as a potential technology for building infrastructure on the Moon using the local regolith [10]. The researchers investigated the ability of this type of extruded material to "survive" under replicated lunar environment–simulated conditions. As with the contour crafting, this approach requires liquid (water) for the reticulation reactions to take place [11,38].

Jakus et al. investigated the AM of robust and elastic lunar/martian regolith simulant architectures using an extrusion-based technique [27]. The investigators synthesized two variants of novel regolith simulant-loaded inks using the JSC-1A lunar mare regolith simulant and JSC-MARS-1A Martian regolith simulant, in a 70–30 volume ratio (70: regolith—30: carrier). A variety of lattice-like three-dimensional structures were successfully printed using a commercially available material extrusion printer (bioplotter) and were further assessed for their mechanical performance via compression testing (see Fig. 21.3). The resultant test samples exhibited rubber-like and quasi-static cyclic material properties with elasticity moduli ranging from 1.8 up to 13.2 MPa. The samples also marked a 250% extension ability until failure. The samples internal porosity was ranging

**Figure 21.3** Examples of additively manufactured parts with indirect use of space resources (A). *(Reproduced with kind permission of Springer Nature from Reference Jakus, A.E., Koube, K.D., Geisendorfer, N.R., Shah, R.N., 2017. Robust and elastic lunar and martian structures from 3D-printed regolith inks. Sci. Rep. 7, 44931).*

from 20% to 40%. Their study concluded that the inks had identical rheological and mechanical properties, regardless of the loading material being used, suggesting that ink formulation would not rely on the specific properties of either lunar or Martian regolith and highlights the transferability of the approach. It should be noted that this approach requires complex chemicals and materials sourced from Earth. However, they highlighted the possibility of those ink constituents to be synthesized in a limited-resources environment (i.e., space station, lunar base, Martian base) using recycled biological waste products, such as compost and human urine. Those extruded structures could then be sintered, to be densified and achieve further properties [54].

## 21.3.2 Direct usage of indigenous resources for additive manufacturing

Balla et al. presented a directed energy deposition method in a freeform environment using Laser Engineering Net Shaping (LENS) and JSC-1A lunar mare regolith simulant. The fabricated parts were characterized to evaluate the influence of laser processing on the microstructure, constituent phases, and chemistry of the material (see Fig. 21.4A) [4]. The researchers used a laser-based system equipped with a 50W $CO_2$ laser, where a combination of process parameters delivering a laser energy density of 2.12 J/mm$^2$ resulting in successful fabrication of three-dimensional structures without any apparent macroscopic defects. This study also revealed

**Figure 21.4** Examples of additively manufactures parts using with direct use of space resources; (A), (B) laser additively manufactured part from lunar mare regolith simulant JSC-1A (this work) (C). *(A) Reproduced with kind permission of Emerald Publishing Ltd. from Reference Balla, V.K., Roberson, L.B., O'Connor, G.W., Trigwell, S., Bose, S., Bandyopadhyay, A., 2012. First demonstration on direct laser fabrication of lunar regolith parts. Rapid Prototyp. J. 18, 451–457; (C) Reproduced with kind permission of ASCE from Reference Fateri, M., Ph, D., Meurisse, A., Ph, D., Sperl, M., Ph, D., Urbina, D., Madakashira, H.K., Govindaraj, S., Gancet, J., Ph, D., Imhof, B., Ph, D., Hoheneder, W., Waclavicek, R., Preisinger, C., Ph, D., Podreka, E., Ph, D., 2019. Solar sintering for lunar additive manufacturing. Am. Soc. Civ. Eng. 32, 1–10).*

that laser processing caused microstructural changes in the simulant's crystallinity by turning the material into nanocrystalline and/or amorphous. Despite the promising experimental results, there were no data on the laser-processed material's internal microstructure and no data on their mechanical performance, to assess the potential of the resultant structures to be used in any engineering application. Additionally, this LENS approach requires blowing of the powdered material via compressed air/nitrogen making it problematic for implementation within the lunar environment (there is no air or nitrogen, and the carrier gas would be expanding into a very high vacuum environment instead of an approximately 1 atm environment) or the Martian surface (again at significantly less pressure than 1 atm).

Based on lunar regolith' s propensity to absorb microwave radiation [53], a modified printing head equipped with a microwave heating module operating at 2.45 GHz (the same frequency found in household microwave equipment) was discussed by researchers *Barmatz and Steinfeld*. This could directly fuse the regolith for building infrastructure and other ISRU assets [5]. This was shown to be, theoretically, a potential approach to bonding the regolith material; however, it would require multiple stages of processing.

Fateri et al. were the first to investigate the feasibility of using a selective laser melting (SLM) powder bed fusion (PBF) AM process to create functional parts using lunar and Martian regolith simulant [16,19]. In this preliminary study, the researchers successfully managed to manufacture and present three-dimensional objects, such as cubical, cylindrical, and thin wall structures, using a bespoke SLM machine, equipped with a 100W Yb:YAG laser, operating at a 1.07 μm central emission wavelength while using the JSC-1A lunar mare regolith simulant (see Fig. 21.4C). Successful builds were made using a 45W laser output and two different scanning speeds of 190 mm/s and 400 mm/s for the solid hatch and contour, respectively.

Following up on their first published study, Fateri and Gebhard focused on developing process parameters of the SLM technique for the lunar mare regolith simulant JSC-1A [17]. In this study, the investigators used the same equipment as in their previous published work [18] and presented three-dimensional structures such as cubes, gears, and fasteners.

Later studies by Fateri et al. investigated the advent of layerwise solar sintering using JSC-1A lunar regolith simulant, under both ambient and vacuum conditions, as a more energy-efficient approach. The researchers performed manufacturing experiments using a bespoke solar 3D printer, fitted with xenon lamps to simulate solar radiation. Fabricated test samples were evaluated for their compression and bending properties. Samples produced under ambient conditions yielded a compressive strength of 2.49 MPa and Young's modulus of 0.21 GPa. However, samples produced under vacuum conditions appeared to be too foamy, porous, and brittle and did not allow for any sufficient mechanical data [20].

## 21.4 The potential of laser-based powder bed fusion additive manufacturing

PBF is one of the main process categories in the family of AM, which is using materials in powder form as feedstock; such as regolith matter [25], to fabricate bespoke and highly complex three-dimensional objects by exposing the powder to a controlled heat source that otherwise would be impractical to build with conventional production methods (see Fig. 21.5).

PBF is considered as the most versatile AM technology, since it allows processing of virtually any material [36]. However, it must be mentioned that the process is not compatible to many alloys, since it is altering the alloy blend and its properties and it is also incompatible with thermoset polymers.

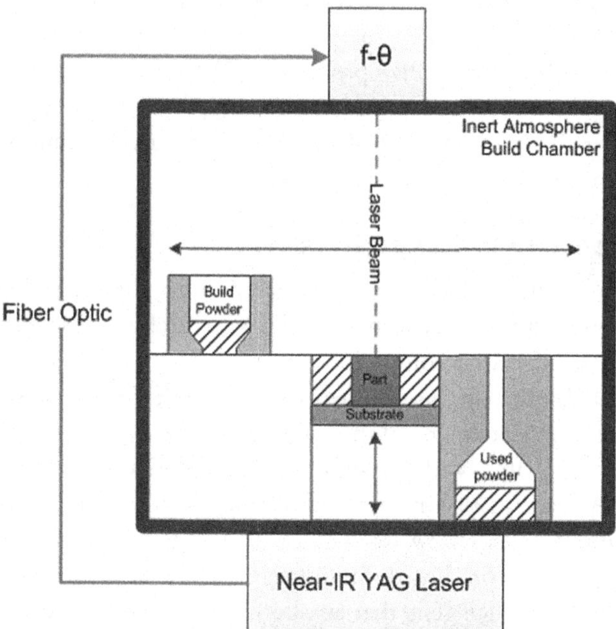


**Figure 21.5** Schematic diagram of the PBF AM process. *PBF*, powder bed fusion; *AM*, additive manufacturing. *(Reproduced with kind permission of Emerald Publishing Ltd Goulas, A., Friel, R.J., 2016. 3D printing with moondust. Rapid Prototyp. J. 22(6), 864–870).*

The building process consists of depositing a fine layer of powder and scanning the sliced x–y section of the geometry by using a thermal source (i.e., laser, electron beam, etc.) to locally fuse together the granular material, forming a fully incorporated layer. The platform is then lowered in the z axis to allow a new layer of material to be deposited and the scanning process to repeat until the desired geometry is fully formed.

A PBF approach, such as that discussed in this chapter, could utilize thermal power to fuse the indigenous material that exists on the Moon or Mars. The thermal power needed for this purpose would not necessarily have to be provided by a laser source but could be provided from an alternative means, such as harnessing the thermal power from the sun through a system of lens (i.e., Fresnel optics) and mirrors [21,22].

Additionally, there would be no need for transporting any binders, liquids, or gasses to the lunar surface that would be necessary for other AM techniques. Moreover, a lunar-based PBF process would not require resources that are also extremely valuable to humans for survival such as water and from the material aspect; a greater range of particle sizes could be utilized.

However, regoliths are not typical materials used for AM, and it is necessary to study their key material properties to assess their potential to be used for laser-based PBF processing. The following subsections will briefly cover a series of experimental investigations, using the lunar mare (JSC-1A) and Martian (JSC-MARS-1A) simulants, outlining those properties mentioned earlier.

## 21.4.1 Spectral/optical performance

The spectral/optical performance of materials for PBF is of high significance, as such knowledge can provide with useful information toward choosing a suitable laser source.

Spectral absorbance measurements have revealed a high absorption for both lunar and Martian regolith simulants in the visible to near-infrared band of 0.4–1.1 μm, with maximum values of 92% @ $\lambda = 1.06$ μm for the JCS-1A lunar mare simulant and 60% @ $\lambda = 1.06$ μm for its Martian counterpart JSC-MARS-1A (see Fig. 21.6). This is compatible with current laser systems used on board most laser additive manufacturing (LAM) equipment but also suggests that an alternative system, based on solar power, is very likely to be a good candidate energy source, since both materials exhibited a high absorption performance, close to 90% in the visible band of 0.45–0.65 μm.

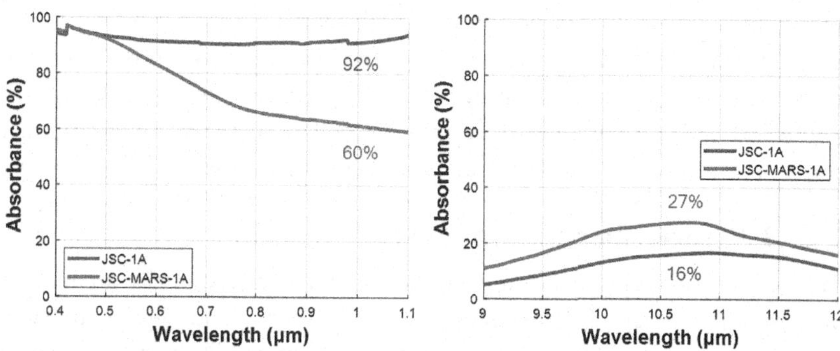


**Figure 21.6** Absorbance of regolith simulants in the (left) visible to near-infrared and (right) far-infrared spectra.

## 21.4.2 Thermal stability and melting behavior

After spectral performance analysis, the thermal behavior of a material is the second major factor that needs to be taken into account for PBF and is of utmost importance due to the fundamental requirement of the material to be stable under a range of thermal loads induced during laser irradiation [23].

Thermal analysis results have shown a complex melting behavior for both simulants, characterized by a plethora of thermal events associated to phenomena such as glass transition, crystallization, and finally melting (see Fig. 21.7). That was attributed to their complex multicomponent and mineral-based chemistry and mineralogy of igneous origin. Additionally, the Martian simulant has shown very poor thermal stability when subjected to thermal loads with an approximate mass loss of 30% when heated up to 1000°C, suggesting the material as highly incompatible for direct laser

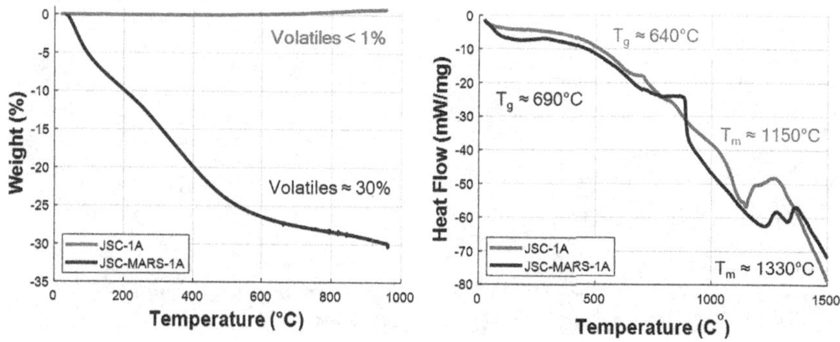


**Figure 21.7** (left) TGA curves showing the weight loss during heating of the JSC-1A mare lunar and (right) JSC-MARS-1A Martian regolith simulants.

processing. This is a considerable material behavior aspect that should be taken into account when assessing the suitability of extraterrestrial materials to be used for thermal-based processing; the feedstock should be thermally stable and will not decompose or easily evaporate, or alternatively thermal processing is less likely to be suitable for such a feedstock.

### 21.4.3 Particle shape/morphology

Regoliths, both actual and their simulants, are characterized by an angular and irregular external morphology (see Fig. 21.8), and significant sub-granular porosity (see Fig. 21.9). Such external morphology characteristics are expected to have a negative effect toward the materials' flowability and packing performance within the powder bed. Together with the evident, subparticle porosity is very likely to limit densification activity.

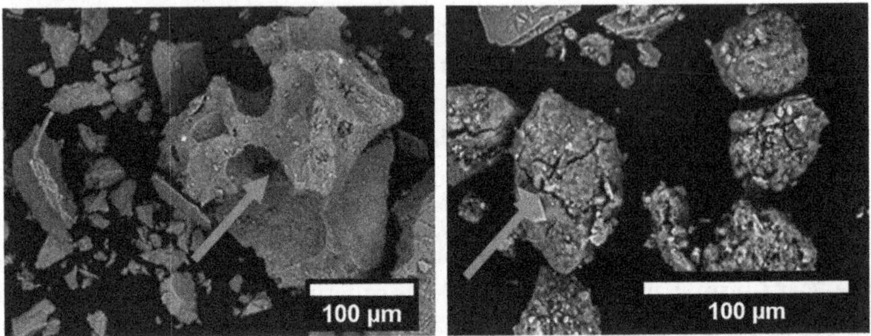


**Figure 21.8** (left) Electron micrographs showing (left) the external particle morphology of the JSC-1A lunar regolith and (right) JSC-MARS-1A Martian regolith simulant.

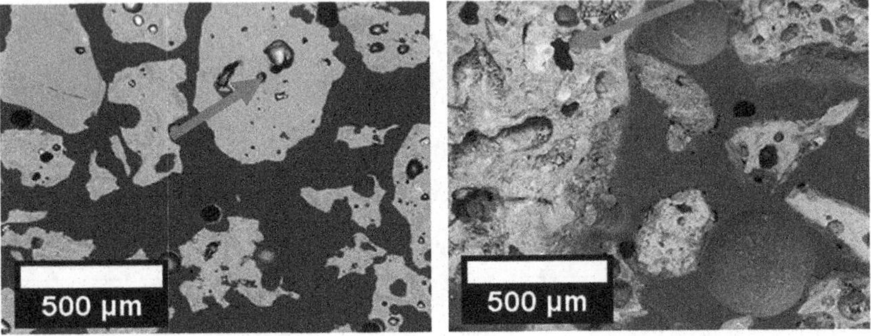


**Figure 21.9** Electron micrographs showing the subgranular porosity of (left) JSC-1A lunar mare and (right) JSC-MARS-1A Martian regolith simulant materials.

### 21.4.4 Laser processing lunar regolith simulants

A series of manufacturing experiments, using a continuous-wave ytterbium-doped fiber laser (YLR-50, IPG Photonics, Oxford, MA, USA) operating on a central emission wavelength of $\lambda = 1.06\ \mu m$, have shown that the laser energy input, as a result of the combination of process parameters, has a direct effect on the quality of single tracks and the microstructure of the multilayer test geometries. Too fast scanning speeds will result in insufficient fusion between particles due to less laser energy being delivered onto the powder bed, where relatively slower scanning speeds will result in excess input and is likely to cause process-related irregularities such as powder particle vaporization. A laser power output of 50 W, together with a scanning speed of 320 mm/s, corresponding to an energy density input of $0.92\ J/mm^2$, provided with single tracks of stable width and good morphological characteristics (see Fig. 21.10).

The hatch spacing (the distance between the individual laser tracks forming a monolayers) was observed to be a critical factor that influenced the surface quality; when not optimized, it hindered the optimum deposition of powder layers, which, in turn, negatively affected the resulting porosity. Hatch spacing settings from 170 to 250 μm resulted in successful and repeatable three-dimensional geometries with no macroscopic build failures and with a final measured porosity of 44%–49% and density of $1.76–2.3\ g\ cm^{-3}$ as measured by gas expansion pycnometry (see Fig. 21.11).

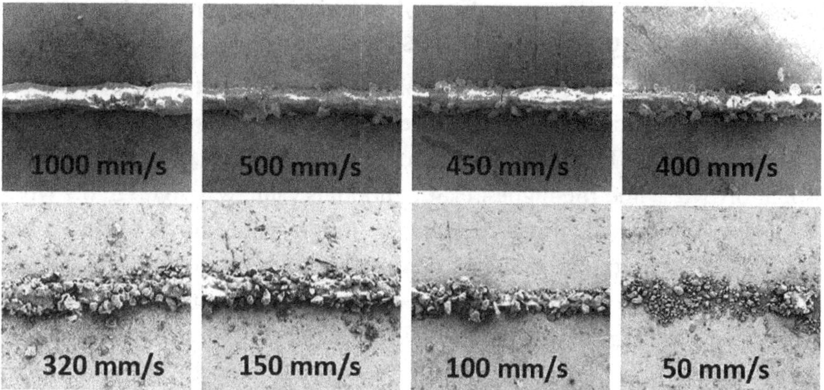


**Figure 21.10** - Optical micrographs showing the morphology of single tracks of JSC-1A lunar mare regolith simulant obtained over 50–1000 mm/s laser scanning speeds.

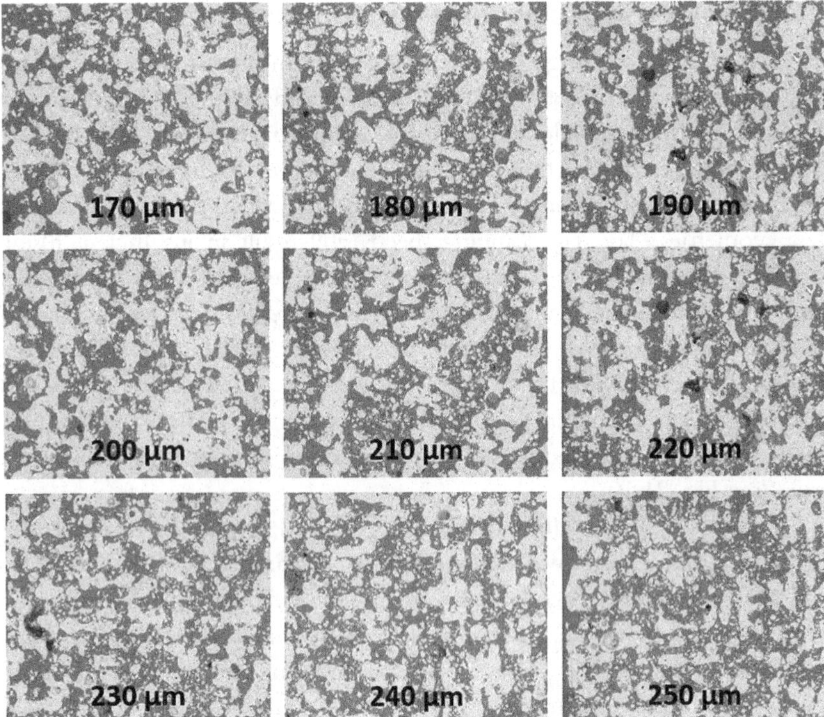


**Figure 21.11** Electron micrographs showing the inherently porous microstructure of the laser additively manufactured test samples using JSC-1A, processed with $\lambda = 1.06\ \mu m$ and 50 W laser power, 320 mm/s scanning speed, and 170–250 μm hatch spacing.

### 21.4.5 Mechanical properties of additively manufactured lunar regolith structures

Laser energy input had a direct and linear effect on the mechanical performance of the additively manufactured structures, a result of the microstructure formed. The maximum compressive strength value measured was $4.2 \pm 0.1$ MPa, and the highest elastic modulus value was $287.3 \pm 6.7$ MPa; both were obtained using an energy input value of $0.9\ J/mm^2$ (see Fig. 21.12). The strength value is comparable with that of a common masonry brick and so should be adequate for constructing buildings on the Moon, especially given the absence of storms and the low gravity. It was also shown that the laser energy input had no significant impact on the material's hardness, with an average hardness calculated value of $657 \pm 14\ HV_{0.05/15}$.

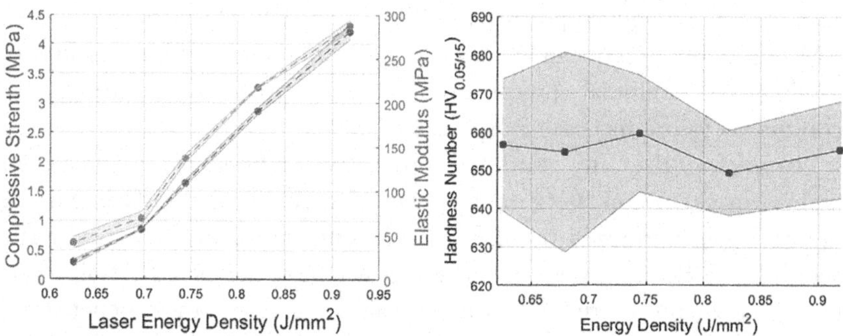


**Figure 21.12** (left) Influence of energy density on the compressive strength and elastic modulus of additively manufactured parts from JSC-1A lunar regolith simulant. The shaded areas represent the standard error (right) influence of energy density on the hardness of additively manufactured parts from lunar regolith simulant. The shaded areas represent the standard error (n = 5).

Further analysis of the elemental/chemical makeup of the resultant structures revealed a small reduction in the concentration of magnesium, phosphorus, and potassium as a result of processing. That was identified as the result of vaporization of the finer particles, in combination of the low dissociation points of their original mineral phases.

Additionally, calorimetric data acquired via DSC revealed a percentile reduction of the fused material's crystallinity, as a function of laser energy density used during laser additive processing. Test samples that were processed with the maximum laser energy input of 0.92 $J/mm^2$ were calculated to a reduced crystalline content down to an approximately 14%. That is comparable with the feedstock powder samples that are known to comprise of 50% crystalline content, due to their igneous origin. That identified vitrification reactions favored by the very high cooling rates occurring after thermal processing.

## 21.5 Conclusions

The concept of ISRU has been put forward since humans first set foot on the surface of the Moon. The goal of ISRU is to exploit the readily available natural space resources (e.g., lunar regolith) instead of sending raw materials from Earth. The long-term vision is for equipment to be sent off-planet and allow the construction of complete outposts and the manufacture of any necessary operational physical assets, to help sustain space exploration. AM technologies have been envisioned as a highly promising tool for serving and supporting the ISRU concept. This is mainly due to AM being able to work in an autonomous manner and with no additional tooling and minimal postprocessing while potentially utilizing the commonly found materials in space.

It is clear that any candidate processing technologies intended for making space mission assets, on the lunar, Martian, or other surface, must be designed and optimized to take into account the inherent physical characteristics of the indigenous materials that will be either directly available on the space manufacturing/construction lab's locale, or their beneficiated products. Unlike a terrestrial manufacturing scenario, where raw materials are "engineered" to work optimally with the fabrication process, the processing equipment should be developed accordingly to match the response of those feedstocks available in situ. A look into the physical properties of the regolith samples, which have been returned to Earth for analysis, has revealed a rather challenging processing nature. Better knowledge of those extraterrestrial materials will improve our ability to design and optimize potential solutions. Therefore, it is of paramount importance for space manufacturing scientists and engineers, to gain access to real samples retrieved from the Moon or the Martian surface. Such data will aid into developing technology and pushing up its status and readiness.

Envisaged approaches for processing space resources have been looking into using process additives (such as binders) for the successful manufacture of engineering components and structures. This may not be strictly in line with the ISRU concept; however, the use of such materials could open up a path for developing space-worthy AM process additives. It has been demonstrated by research that a laser-based AM approach can function without the use of any additives—which fully satisfies the ISRU concept—and can successfully deliver functional engineering assets by directly using regolith matter as feedstock.

## References

[1] C.C. Allen, K.M. Jager, R.V. Morris, D.J. Lindstrom, M.M. Lindstrom, J.P. Lockwood, Martian Soil Stimulant Available for Scientific, Educational Study, Eos, Trans. Am. Geophys. Union, 1998.

[2] M.-C.C. Allen, R.V. Morris, K.M. Jager, D.C. Golden, D.J. Lindstrom, M.M. Lindstrom, J.P. Lockwood, L. Martin, C.C. Allen, R.V. Morris, M.M. Lindstrom, J.P. Lockwood, M.J. Karen, D.C. Golden, M.M. Lindstrom, J.P. Lockwood, Martian regolith simulant JSC MARS-1, in: Lunar and Planetary Science Conference XXIX, 1998, p. 1690.

[3] V. Badescu, Moon: Prospective Energy and Material Resources, Springer, Romania, 2012.

[4] V.K. Balla, L.B. Roberson, G.W. O'Connor, S. Trigwell, S. Bose, A. Bandyopadhyay, First demonstration on direct laser fabrication of lunar regolith parts, Rapid Prototyp. J. 18 (2012) 451–457.
[5] M. Barmatz, D. Steinfeld, M. Anderson, D. Winterhalter, 3D microwave print head approach for processing lunar and mars regolith, in: 45th Lunar Planet. Sci. Conf, 2014, pp. 3–4.
[6] N.L. Bowen, The melting phenomena of the plagioclase feldspars, Am. J. Sci. s4–35 (1913) 577–599.
[7] S. Bukkapatnam, B. Khoshnevis, H. Kwon, J. Saito, Experimental investigation of contour crafting using ceramics materials, Rapid Prototyp. J. 7 (2001) 32–42.
[8] P. Carpenter, L. Sibille, S. Wilson, B. a E. Systems, S. Wilson, M. Space, Development of standardized lunar regolith simulant materials, in: Lunar Planet. Sci. Conf. XXXVII, 2006, pp. 2–3.
[9] W.D. Carrier, G.R. Olhoeft, W. Mendell, Physical properties of the lunar surface, Lunar Source (1991) 475–594.
[10] F. Ceccanti, E. Dini, X. De Kestelier, V. Colla, L. Pambaguian, 3D printing technology for a moon outpost exploiting lunar soil, 61st Int. Astronaut. Congr. Prague, CZ, IAC-10-D3 3 (2010) 1–9.
[11] G. Cesaretti, E. Dini, X. De Kestelier, V. Colla, L. Pambaguian, Building components for an outpost on the Lunar soil by means of a novel 3D printing technology, Acta Astronaut. 93 (2014) 430–450.
[12] V. Chevrier, P.E. Mathe, Mineralogy and evolution of the surface of Mars: a review, Planet. Space Sci. 55 (2007) 289–314.
[13] I.A. Crawford, Lunar resources: a review, Prog. Phys. Geogr. 39 (2015) 137–167.
[14] J. Edmunson, C. McLemore, In Situ manufacturing is a necessary part of any planetary architecture, in: Concepts and Approaches for Mars Exploration, 2012, pp. 2–3.
[15] E.J. Faierson, K.V. Logan, B.K. Stewart, M.P. Hunt, Demonstration of concept for fabrication of lunar physical assets utilizing lunar regolith simulant and a geothermite reaction, Acta Astronaut. 67 (2010) 38–45.
[16] M. Fateri, Experimental investigation of selective laser melting of lunar regolith for in-situ applications, ASME 2013 (2013) 1–6.
[17] M. Fateri, A. Gebhardt, Process parameters development of selective Laser Melting of lunar regolith for on-site manufacturing applications, Int. J. Appl. Ceram. Technol. 12 (2015) 46–52.
[18] M. Fateri, A. Gebhardt, M. Khosravi, Experimental investigation of selective laser melting of lunar regolith for in-situ applications, in: Proceedings of the 2013 ASME International Mechanical Engineering Congress and Exposition, 2013, pp. 1–6.
[19] M. Fateri, K. Maziar, On-site Additive Manufacturing by Selective Laser Melting of Composite Objects. Concepts Approaches Mars Explor, 2012.
[20] M. Fateri, D. Ph, A. Meurisse, D. Ph, M. Sperl, D. Ph, D. Urbina, H.K. Madakashira, S. Govindaraj, J. Gancet, D. Ph, B. Imhof, D. Ph, W. Hoheneder, R. Waclavicek, C. Preisinger, D. Ph, E. Podreka, D. Ph, Solar sintering for lunar additive manufacturing, Am. Soc. Civ. Eng. 32 (2019) 1–10.
[21] H.R. Fischer, In-situ resource utilization—feasibility of the use of lunar soil to create structures on the moon via sintering based additive manufacturing technology, Aeronaut. Aerosp. Open Access J. 2 (2018) 243–248.
[22] A. Ghosh, J.J. Favier, M.C. Harper, Solar sintering on lunar regolith simulant (JSC-1) for 3D printing, in: Proceedings of the International Astronautical Congress, IAC, 2016.
[23] A. Goulas, Investigating the Additive Manufacture of Extra-terrestrial Material Simulants, Loughborough University, 2018.

[24] A. Goulas, J.G.P. Binner, D.S. Engstrøm, R.A. Harris, R.J. Friel, Mechanical behaviour of additively manufactured lunar regolith simulant components, Proc. Inst. Mech. Eng. Part L J. Mater. Des. Appl. 233 (8) (2018) 1629–1644, https://doi.org/10.1177/1464420718777932.
[25] A. Goulas, R.J. Friel, 3D printing with moondust, Rapid Prototyp. J. 22 (6) (2016), 864–870.
[26] F. Hörz, R. Grieve, G. Heiken, P. Spudis, A. Binder, Lunar surface processes, Lunar Sourceb. A User's Guid. to Moon (1991) 61–120.
[27] A.E. Jakus, K.D. Koube, N.R. Geisendorfer, R.N. Shah, Robust and elastic lunar and martian structures from 3D-printed regolith inks, Sci. Rep. 7 (2017) 44931.
[28] R. Jaumann, H. Hiesinger, M. Anand, I.A. Crawford, R. Wagner, F. Sohl, B.L. Jolliff, F. Scholten, M. Knapmeyer, H. Hoffmann, H. Hussmann, M. Grott, S. Hempel, U. Köhler, K. Krohn, N. Schmitz, J. Carpenter, M. Wieczorek, T. Spohn, M.S. Robinson, J. Oberst, Geology, geochemistry, and geophysics of the Moon: status of current understanding, Planet. Space Sci. 74 (2012) 15–41.
[29] G.H. Just, K. Smith, K.H. Joy, M.J. Roy, Parametric review of existing regolith excavation techniques for lunar in Situ Resource Utilisation (ISRU) and recommendations for future excavation experiments, Planet. Space Sci. (2019) 104746.
[30] B. Kading, J. Straub, Utilizing in-situ resources and 3D printing structures for a manned Mars mission, Acta Astronaut. 107 (2015) 317–326.
[31] M. Kallerud, B. Nguyen, T. Paladin, A. Wilson, In-Situ Resource Utilization: Investigation of Melted Lunar Regolith Simulant JSC-1A, 2009, pp. 1–10.
[32] B. Khoshnevis, M. Bodiford, Lunar contour crafting—a novel technique for ISRU-based habitat development, in: 43rd AIAA Aerospace, 2005, pp. 1–12.
[33] B. Khoshnevis, D. Hwang, K.-T. Yao, Z. Yah, Mega-scale fabrication by contour crafting, Int. J. Ind. Syst. Eng. 1 (2006) 301–320.
[34] B. Khoshnevis, J. Zhang, Extraterrestrial construction using contour crafting, Solid Free. Fabr. (2012) 250–259.
[35] B. Khoshnevis, J. Zhang, M. Fateri, Z. Xiao, L. Angeles, Ceramics 3D printing by selective inhibition sintering, in: Solid Freeform Fabrication Symposium. Austin, TX, 2014, pp. 163–169.
[36] J.P.P. Kruth, X. Wang, T. Laoui, L. Froyen, Lasers and materials in selective laser sintering, Assemb. Autom. 23 (2003) 357–371.
[37] N. Labeaga-Martínez, M. Sanjurjo-Rivo, J. Díaz-Álvarez, J. Martínez-Frías, Additive manufacturing for a Moon village, Proc. Manuf. 13 (2017) 794–801.
[38] N. Leach, E. Dini, F. Partners, 3D printing in space, Architect. Des. 108–113 (2014).
[39] B.A. Lomax, M. Conti, N. Khan, N.S. Bennett, A.Y. Ganin, M.D. Symes, Proving the viability of an electrochemical process for the simultaneous extraction of oxygen and production of metal alloys from lunar regolith, Planet. Space Sci. (2019) 104748.
[40] D.S.D.S. Mckay, G. Heiken, A. Basu, G. Blanford, S. Simon, R. Reedy, B.M. French, J. Papike, The lunar regolith, Lunar Source B. A User's Guid. to Moon (1991) 285–356.
[41] P. Metzger, D. Britt, S. Covey, J.S. Lewis, Figure of merit for asteroid regolith simulants, in: European Planetary Science Congress, 2017.
[42] P.T. Metzger, D.T. Britt, S. Covey, C. Schultz, K.M. Cannon, K.D. Grossman, J.G. Mantovani, R.P. Mueller, Measuring the fidelity of asteroid regolith and cobble simulants, Icarus 321 (2019) 632–646.
[43] R.P. Mueller, S. Howe, D. Kochmann, H. Ali, C. Andersen, H. Burgoyne, W. Chambers, R. Clinton, X. De Kestellier, K. Ebelt, S. Gerner, D. Hofmann, K. Hogstrom, E. Ilves, A. Jerves, Automated additive construction (AAC) for Earth and space using in-situ resources, in: Proceedings of the Fifteenth Biennial ASCE Aerospace Division International Conference on Engineering, Science, Construction, and Operations in Challenging Environments (Earth & Space 2016), American Society of Civil Engineers, Reston, Virginia, USA, 2016.

[44] M.Z. Naser, Extraterrestrial construction materials, Prog. Mater. Sci. 105 (2019) 100577.
[45] J. Papike, L. Taylor, S. Simon, Lunar minerals, Lunar Source. (1991) 121–181.
[46] H.A. Perko, J.D. Nelson, J.R. Green, Mars soil mechanical properties and suitability of mars soil simulants, J. Aero. Eng. 19 (2006) 169–176.
[47] B.J. Pletka, Processing of lunar basalt materials, in: J.S. Lewis, M.S. Matthews, M.L. Guerrieri (Eds.), Resources of Near Earth Space, The University of Arizona Press, Tucson, 1993, pp. 325–350.
[48] E. Robens, A. Bischoff, A. Schreiber, A. Dεąbrowski, K.K. Unger, Investigation of surface properties of lunar regolith: Part I, Appl. Surf. Sci. 253 (2007) 5709–5714.
[49] S.W. Ruff, P.R. Christensen, Basaltic andesite, altered basalt, and a TES-based search for smectite clay minerals on Mars, Geophys. Res. Lett. 34 (2007) 1–6.
[50] A.N. Scott, C. Oze, Y. Tang, A. O'Loughlin, Development of a Martian regolith simulant for in-situ resource utilization testing, Acta Astronaut. 131 (2017) 45–49.
[51] K. Seiferlin, P. Ehrenfreund, J. Garry, K. Gunderson, E. Hütter, G. Kargl, A. Maturilli, J.P. Merrison, Simulating Martian regolith in the laboratory, Planet. Space Sci. 56 (2008) 2009–2025.
[52] G.J. Taylor, P.H. Warren, G. Ryder, J. Delano, C. Pieters, G. Lofgren, Lunar rocks. Lunar sourceb, A User. Guid. Moon (1991) 183–284.
[53] L.A. Taylor, T.T. Meek, Microwave processing of lunar soil, in: Science and Technology Series, 2004.
[54] S.L. Taylor, A.E. Jakus, K.D. Koube, A.J. Ibeh, N.R. Geisendorfer, R.N. Shah, D.C. Dunand, Sintering of micro-trusses created by extrusion-3D-printing of lunar regolith inks, Acta Astronaut. 143 (2018) 1–8.
[55] J.A. Wood, J.S. Dickey, U.B. Marvin, B.N. Powell, Lunar anorthosites and a geophysical model of the moon, in: Apollo 11 Science Conference, 1970, pp. 965–988.
[56] X. Zeng, X. Li, S. Wang, S. Li, N. Spring, H. Tang, Y. Li, J. Feng, JMSS-1: a new Martian soil simulant, Earth Planets Space 67 (2015) 72.
[57] H. Zhang, S. LeBlanc, Processing parameters on selective laser sintering or melting for oxide ceramics, in: Three-Dimensional Printing and Additive Manufacturing of High-Performance Metals and Alloys, 2018, pp. 89–124.

CHAPTER 22

# Modeling and simulation of additive manufacturing processes with metallic powders—potentials and limitations demonstrated on application examples

**Loucas Papadakis**
Department of Mechanical Engineering, Frederick University, Nicosia, Cyprus

## 22.1 Introduction

### 22.1.1 Background

Additive layer manufacturing processes are utilized for fabricating, repairing, and coating of three-dimensional components by deposing metallic powder or wire through melting and resolidification with the aid of layer-wise paths directly from CAD (computer-aided design) model [1–4]. The generation of complex solid geometries as well as thin-walled geometries is possible due to the narrow heat input provided by the laser beam during the deposition process [5,6]. However, this, in combination with relatively high laser transverse speed, creates high-temperature gradients across the processed layers. As a result, shrinkage of the built-up geometry is observed with especially noticeable deformations when removing parts from substrates or when removing lattice support structures [7,8]. Selective laser melting (SLM) and laser metal deposition (LMD) offer apparent advantages in modern manufacturing industries. These freeform manufacturing techniques provide system flexibility since they offer a just-in-time manufacture directly from CAD with the ability to produce components of high density [9]. Additionally, SLM and LMD offer the flexibility to process different metal powders (e.g., aluminum alloys, nickel-based steel alloys, and titanium-based alloys), thus enabling application in a wide range of industrial sectors, e.g., aerospace, aero-engine manufacturing, automotive, medicine, etc. [10–13].

*Additive Manufacturing*
ISBN 978-0-12-818411-0
https://doi.org/10.1016/B978-0-12-818411-0.00012-4

Despite the apparent advantages, SLM and LMD also exhibit process deficits. Due to the so-called temperature gradient mechanism (TGM), thermal stresses are induced during processing. Thereupon, depending on the plastification behavior of the material during heating and cooling to ambient temperature, stresses remain (i.e., residual stresses) in the component after process completion [8,14]. The final state formation of the residual stresses is closely related to postprocessing cutting or surface treatment operations, and thereupon, to the final component shape [15]. Finally, extensive stress formation during processing can lead to cracks, which affect the final quality of the built-up product [16].

The rising potentials of the SLM and LMD processes motivate industry and research to use simulation methods to support the process and product development by calculating structural characteristics in advance so as to streamline the design for experimentation by reducing time consuming trial-and-error. Several studies already exist in the literature in which the process feasibility and material behavior during additive manufacturing (AM) with metallic alloys are investigated. Hereby, the selection and control of suitable process parameters is described to reach process feasibility and manufactured part densities of over 99% compared with theoretical densities [17]. The most significant process parameters affecting the quality during melting of powder particles or wire during AM processing are the powder or wire alloy itself and the laser beam type/quality together with the laser power, the laser feed speed, and the powder/wire layer thickness. Based on the aforementioned parameter selection and the scanning sequence strategy, process effects such as the melt pool formation (width, depth and length) as well as the overall part shape accuracy are influenced. It is observed that a combination of the heat input and consequently the temperature gradient mechanism, which determines the melt pool characteristics together with the part geometry and posttreatment operations, i.e., cutting, are crucial for the final product shape accuracy [7,18].

### 22.1.2 Motivation

State-of-the-art SLM machines and LMD processing systems are already available in the market capable of providing solid freeform solutions without the use of tooling, direct from the CAD data and even with high strength properties from titanium and nickel-based steel powders. Nevertheless, despite the relatively high amount of information existing in the literature related to the process parameter selection depending on laser type, powders, production rate, and geometrical features, further research is necessary to collect more knowledge and experimental data on the process feasibility and final quality of SLM and LMD components in terms of their shape accuracy and strength. In these development efforts, modeling techniques can be very supportive; however, they have proved to be rather time-consuming from both the modeling point of view, since high level of expertise is required, and the computation time perspective due to high level of geometrical and time discretization. In this chapter, the potentials of modeling and simulation of metallic layer advanced manufacturing technologies for the production of complex components are highlighted using application examples from aeroengine component fabrication [19]. The transferability to broader industrial sectors is granted since the integration of suitable, time efficient, and reliable simulation methods in the process and product design phase is inevitable [7]. To this end, modeling strategies have to be evaluated with respect to their benefits and their reliability in replicating the involved nonlinear physical effects and providing useful results.

Moreover, the potentials and limitations of modeling and simulation for AM processes are demonstrated with emphasis on the heat effects and the prediction of the shape accuracy of industrial relevant applications. The findings of the presented cases act to encourage process and product developers to fabricate high-quality components by reducing time-intensive trial-and-error during process design with the aid of appropriate modeling and simulation methods. According to this, the benefits and restrictions of the presented modeling methods are exemplified on industrial aeroengine components. First, an analysis of the melt pool and a prediction of its geometrical characteristics are introduced based on findings from literature. Thereupon, a transient thermal model of powder deposition during SLM is implemented for a single scan vector during laser processing. A layer-wise heat input is deduced from these findings in such a way so that the bond, i.e., penetration, of the current processed layer into the underlying layer is provided similarly to real process. Thereafter, considering the process heat input, a transient thermomechanical simulation is introduced to calculate the stresses and the component's shape during the built-up process. Finally, posttreatment operation, i.e., the component removal from the base plate and from possible supporting structures, is performed to concur with the manufacture of the real demonstrator components.

Table 22.1 indicates the focus of this chapter considering particular AM process effects and the involved modeling challenges for four application examples: SLM melt pool formation (Fig. 22.1), SLM twin cantilever with support structures (Fig. 22.5), SLM turbine blade (Fig. 22.6), and LMD rotational symmetric thin-walled aeroengine combustor casing (Fig. 22.9). The manufacturing effects discussed in this contribution are as follows:

- Laser beam powder melting: The melt pool formation, i.e., the melt pool width and height, is an important aspect for the process feasibility and product quality. The density of the fabricated component and the associated crack formation are highly deep ending on the geometry of melt pool. In this study, the addressed analytical models for predicting the melt pool formation consider certain process parameters as the laser power, laser radius and absorptivity, the scanning speed, the powder thermal material properties, and change of phase from powder to liquid, i.e., latent heat.
- Powder deposition: Further related process parameters due to powder deposition as layer thickness and hatch distance are not included in the presented analytical models. Instead, these phenomena are replicated

**Table 22.1** Focus of the application examples in terms of the considered manufacturing process effects and the simulation challenge.

| | manufacturing process effect: | | | | | simulation challenges | | | | | |
|---|---|---|---|---|---|---|---|---|---|---|---|
| application example | laser beam powder melting | powder deposition | binding with underlying material | substrate or support structures removal | influence of scan strategy on shape accuracy | melt pool prediction with analytical models | melt pool prediction with FE models | FEA with geometrical reduction strategies | cut-off operation and spring-back effect | reverse engineering modelling | evaluation of model with experiment |
| SLM melt pool formation (Fig. 1.1) | ● | ◕ | ◕ | | | ● | ● | ● | | | ● |
| SLM Twin Cantilever with support structures (Fig. 1.5) | ◔ | ◔ | ◒ | ● | ● | | | ● | ● | ● | ● |
| SLM Turbine blade (Fig. 1.6) | ◔ | ◔ | ◒ | ● | | | | ● | ● | | ● |
| LMD rotational symmetric thin-walled aero-engine combustor casing (Fig. 1.9) | ◔ | ◔ | ◒ | ● | | | | ● | ● | | ● |

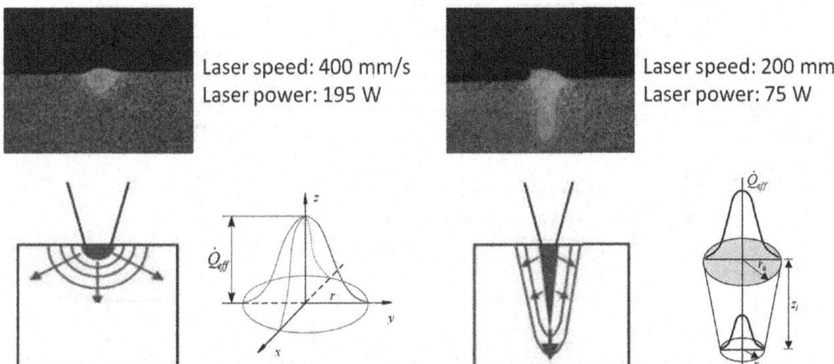


**Figure 22.1** Melt pool macrographs of single laser scans on Ti—6Al—4V titanium alloy powder with 30 μm layer thickness using an EOS M270 Direct Metal Laser Sintering (DMLS) with Yb-fiber laser and nominal maximum power 200 W [20]. On the left, a conduction mode is shown with lower depth-to-width ratio of the melt pool typically modeled by a 2D Gauss equivalent heat source. On the right, a keyhole mode is shown with a higher depth-to-width ratio typically replicated by a 3D Gauss equivalent heat source. *(Reproduced with kind permission from Reference Gong H, Gu H, Zeng K, D. JJS, Pal D, Stucker B, Christiansen D, Beuth J, Lewandowski JJ. Melt pool characterization for selective laser melting of Ti-6Al-4V pre-alloyed powder. In: Proceedings of the 25th Annual International Solid Freeform Fabrication Symposium: 2014 Aug 4-6; Austin, Texas, USA.)*

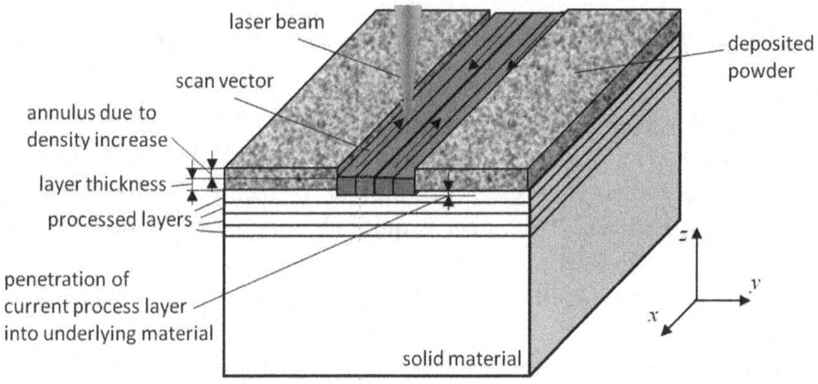


**Figure 22.2** Schematic illustration of powder deposition and processing during SLM. *SLM*, selective laser melting.

in the numerical approach by means by a finite element (FE) transient thermal analysis with temperature- and phase (solid, powder, liquid)-dependent material properties. In both the analytical and numerical solutions, the effect of annulus in the melt pool area, i.e., reduction of powder layer thickness, due to density increase during phase change form powder to liquid, proves to be very challenging to model (Fig. 22.2).

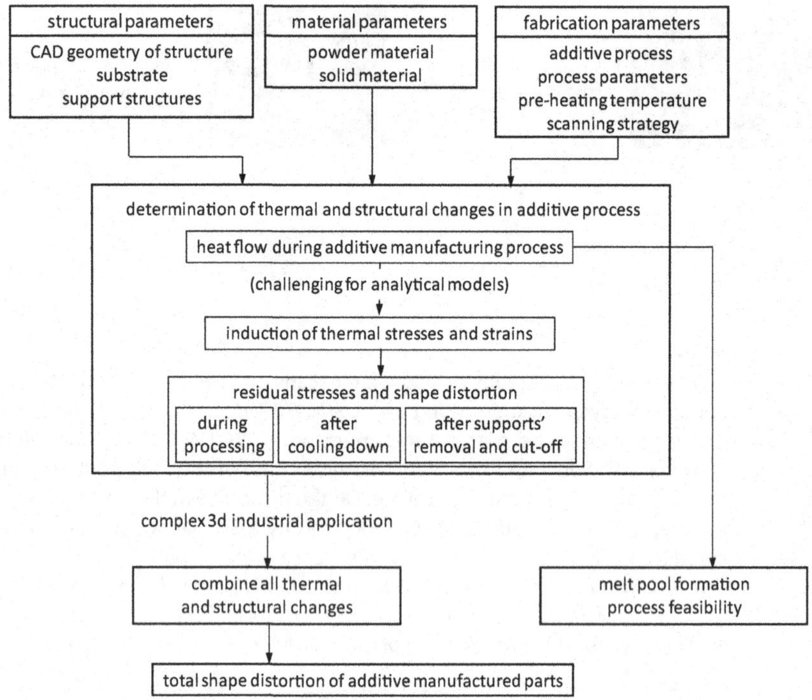


**Figure 22.3** Major parameters and sequences in additive manufacturing processes.

- Binding with underlying material: A further effect, which has a key impact on the melt pool propagation and consequently on the density and crack formation, is the penetration of the melt pool into underlying solidified material. This effect is quite challenging to be described by analytical models. For this reason, the FE models are facilitated to replicate this effect together with the hatch distance between laser scan vectors (Fig. 22.2).
- Cut-off from substrate: The main source of shape distortions of additively manufactured components is the cutting from substrates and the removal of support structures. For this reason, this manufacturing operation or process step has been included in this study to validate its significance in the manufacturing process chain and on the final component's geometry.
- Influence of scan strategy on shape accuracy: A further manufacturing process factor, which may have a significant influence on the shape accuracy of additively manufactured components, is the scan strategy.

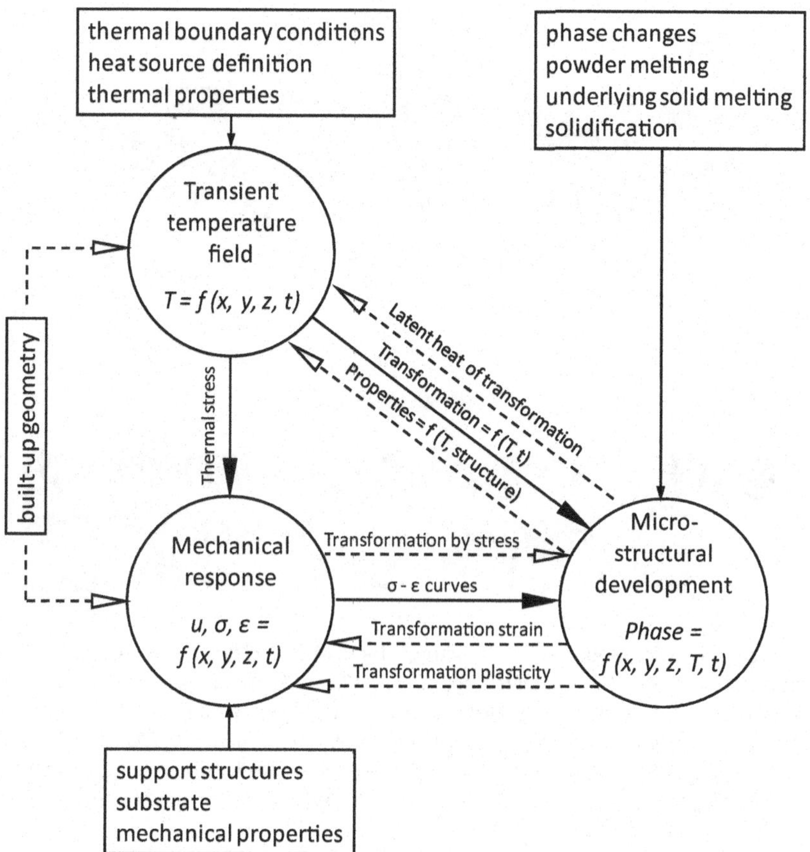


**Figure 22.4** Modeling and simulation steps including key influencing parameters and interactions.

The effect of the scan strategy is demonstrated in the following chapters by means of a practical experimental study on a twin cantilever beam with support structures.

The aforementioned AM process effects and parameters are mostly related to SLM and LMD of metallic powders and are illustrated comprehensively in Fig. 22.2. Due to the increasing role of modeling and simulation to accompany virtually the manufacturing process chain and the associated creation and operation of products, the author has identified the following simulation challenges for replicating the aforementioned effects:

- Melt pool prediction with analytical models: One of the continuous efforts of researchers is to develop analytical models for describing the

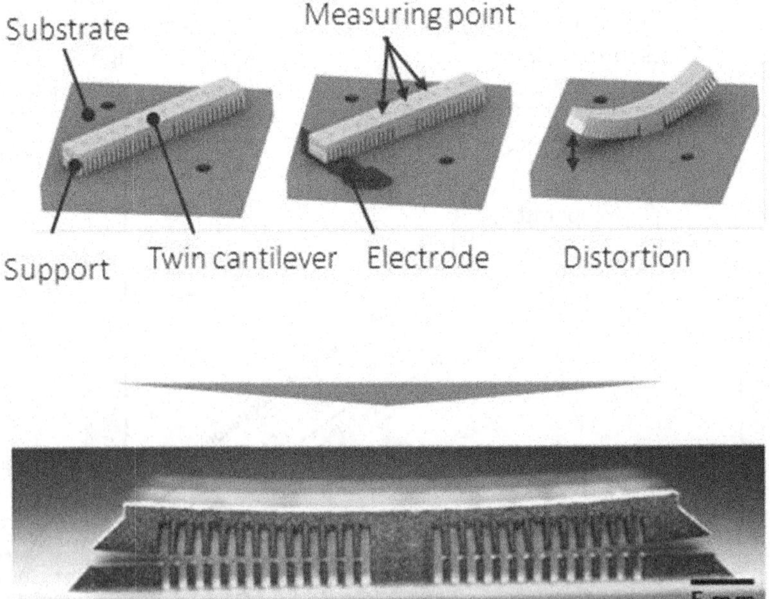


**Figure 22.5** Twin cantilever manufactured by SLM including cut-off operation from support structures. *SLM*, selective laser melting. *(Reproduced with kind permission of ILT from Reference MERLIN Project. Development of Aero Engine Component Manufacture Using Laser Additive Manufacturing. 7th Framework Programme FP7 2007–2013, Grant Agreement 266271 and of Elsevier from Reference Papadakis L, Loizou A, Risse J, Schrage J. Numerical computation of component shape distortion manufactured by selective laser melting. Procedia CIRP 2014;18;90–95.)*

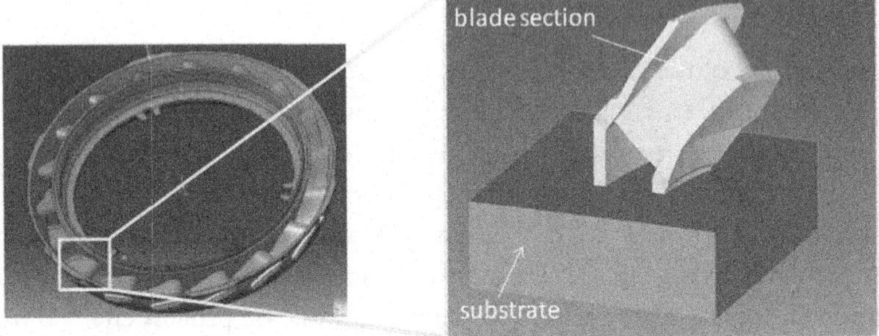


**Figure 22.6** Turbine blade section of SAFRAN Helicopter Engines manufactured by SLM. *SLM*, selective laser melting. *(Reproduced with kind permission of SAFRAN from Reference MERLIN Project. Development of Aero Engine Component Manufacture Using Laser Additive Manufacturing. 7th Framework Programme FP7 2007–2013, Grant Agreement 266271.)*

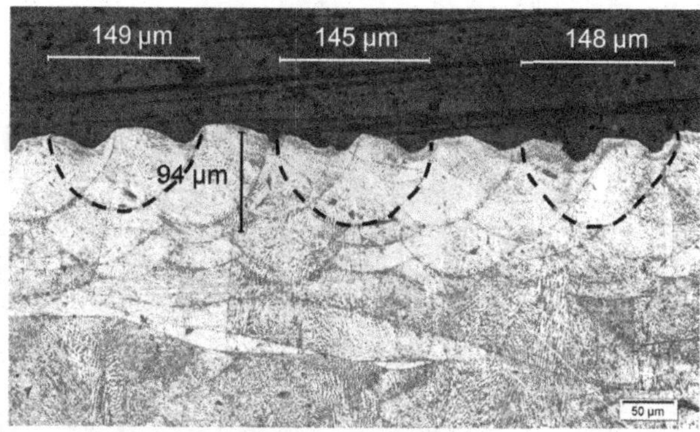


**Figure 22.7** Macrograph cross section of successive melt pools in top layer after SLM of IN718. *SLM*, selective laser melting. *(Reproduced with kind permission of ILT from Reference MERLIN Project. Development of Aero Engine Component Manufacture Using Laser Additive Manufacturing. 7th Framework Programme FP7 2007–2013, Grant Agreement 266271.)*

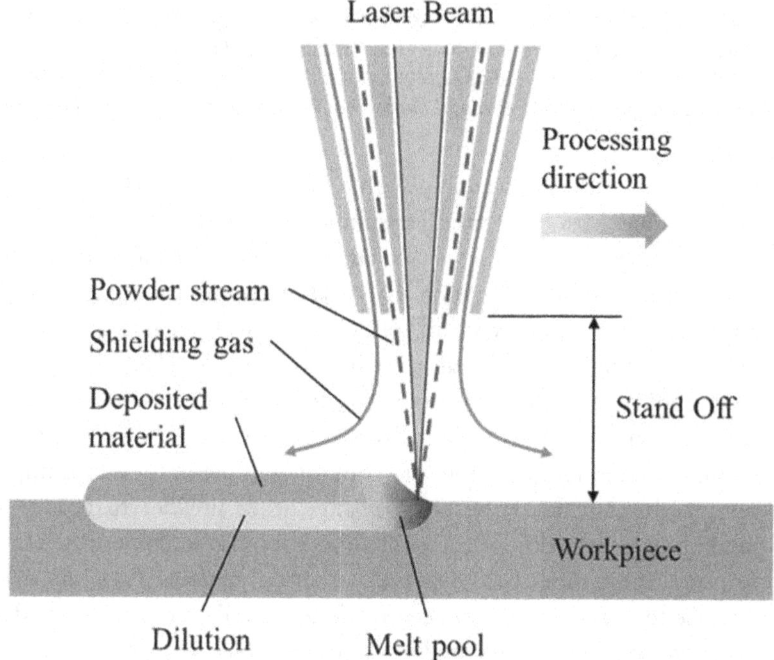


**Figure 22.8** Schematic illustration of the laser metal deposition process with powder (LMD/p). *(Reproduced with kind permission of Springer Nature from Reference Papadakis L, Hauser C. Experimental and computational appraisal of the shape accuracy of a thin-walled virole aero-engine casing manufactured by means of laser metal deposition. Prod. Eng. 2017;11(4–5):389–399.)*

**Figure 22.9** Thin-walled aero-engine combustor casing with sectioned CAD file with maximum diameter of 300 mm. *CAD*, computer-aided design. *(Reproduced with kind permission of Springer Nature from Reference Papadakis L, Hauser C. Experimental and computational appraisal of the shape accuracy of a thin-walled virole aero-engine casing manufactured by means of laser metal deposition. Prod. Eng. 2017;11(4–5):389–399.)*

thermal effects during scanning with a heat source in AM processes. In this manner, cost and time-consuming trial-and-error for selecting appropriate process parameters and attain process feasibility can be overcome to a certain degree. Such analytical models aim additionally to be integrated into process control solutions due to their fast, almost real-time, computation capabilities. Their main purpose is to compute the melt pool formation and secondarily to draw conclusions about the additively manufactured material density. Although they can provide useful results and information, they cannot consider or they oversimplify certain phenomena as the phase change from powder to liquid, the binding or penetration of the melted powder into underlying solidified material, the temperature-dependent thermal material data, the hatch distance to the adjacent scan vector, and the annulus created due to density increase of melt compared with powder. Nevertheless, it is worthy to present such analytical models and discuss their capabilities and boundaries.

- Melt pool prediction with FE models: To perform a transient thermal finite element analysis (FEA) of SLM and LMD additive processes to replicate the melt induced by high power concentration laser beams, mathematically defined heat sources are to be utilized. Depending on the degree of melt pool penetration, i.e., the depth-to-width ratio of the melt pool, different heat source models are used. For a low depth-to-width ratio, i.e., the melt pool is not as deep and the heat propagation is dominated by conduction effect, a circular 2D Gauss distribution heat source is suitable. In case of high depth-to-width ratio melt pools, their formation is dominated by the so-called "keyhole" mode, where the metal vapor plasma and the multiple reflections of the laser beam in the melt become relevant. In this case, a conical 3D Gauss distribution heat source is more suitable so that the melt pool depth can be modeled as in real process (Fig. 22.1). Such mathematical heat source models operate as "black box" since they comprise a number of complicated physical effects as the laser beam absorptivity, the laser beam focus or radius, the laser beam type and quality, and the interaction of the laser beam with the metallic powder and underlying solid material and its thermal properties. For this reason, experimental macrographs of the real process are necessary to validate this heat source models. Nevertheless, such thermal transient models, even though strongly dependent on experimental data, can be used in a later stage to compute thermal stresses, residual stresses, and shape distortions of complex components within a transient thermomechanical simulation.

- <u>FEA with geometrical reduction strategies</u>: Local melt pool dimensions are very small compared with overall component size. Since the component reaches its final dimensions additively with successive millions of tiny scan vectors, the computation time required to replicate this built-up process can be enormously long. For this reason, the effects of singe laser scans can be transferred as whole scan vectors, whole layers, or even volumes of the whole structure to be built, similar to a creation of a mosaic [23]. In this manner, the heat input can be discretized more roughly in the continuum, so that more useful computation durations can be realized. Such geometrical reduction strategies of the heat input will be presented in the following paragraphs.
- <u>Cut-off operation and spring-back effect</u>: Upon definition of the heat source model in the 3D FE component geometry which replicates the laser scanning during AM, a coupled or noncoupled thermomechanical analysis follows in which the computed transient temperature field including the phase transformation from powder to melt and, finally, to solid during cooling is applied as a thermal load on the

mesh. The thermomechanical model includes the elastic thermal strain which together with the von Mises' plasticity criterion for isotropic strain hardening is used for determining residual stresses and distortion in a nonlinear FEA after cooling to ambient temperature. During support structure removal and cut-off from substrate, the structure experiences a new equilibrium due to the existence and the formation of the elastic part of the residual stresses playing an important role in the development of final component's form.

- Reverse engineering modeling: It is essential to include reverse engineering methods within the virtual process chain to provide an optimized component final shape. Upon the computed final shape deviating from the ideal CAD component geometry, a reverse-engineered geometry, or so-called "negative" or "prebent" geometry, is deduced. The AM process is performed based on this predistorted geometry to achieve the ideal CAD shape after the AM process and cut-off operations.
- Evaluation of model with experiment: The four aforementioned application examples are modeled by using various methods addressing different manufacturing effects with certain degree of simplification. To evaluate the rendered simulation results and conclude on their utility, experimental measurements are performed regarding the melt pool formation and components' final shape accuracy.

The sequence of the aforementioned manufacturing procedures and the major computational steps that accompany it are summarized in Figs. 22.3 and 22.4.

## 22.2 Selective laser melting and laser metal deposition process cases

### 22.2.1 Selective laser melting process on twin cantilever and turbine blade

The SLM process considered in the frame of this book chapter concerns the fabrication of a twin cantilever (Fig. 22.5) and a turbine blade geometry of SAFRAN Helicopter Engines (Fig. 22.6). These geometries were created by SLM with a fiber laser in IN718 powder by using the process parameters listed in Table 22.2. The process was performed at Fraunhofer ILT in Aachen on an SLM280 HL machine by SLM Solutions plc in the frame of

**Table 22.2** Process parameters of SLM in IN718 powder.
SLM, *selective laser melting.*

| Process parameters | Value | Units |
|---|---|---|
| Laser type | Ytterbium fiber laser | – |
| Beam quality factor $M^2$ | $\approx 1$ | – |
| Laser power | 300 | W |
| Scan speed | 1600 | mm/s |
| Laser beam diameter | 90 (gauss) | μm |
| Layer thickness | 30 | μm |
| Hatch distance | 80 | μm |
| Hatch angle increment | 90 | degrees |

Reproduced with kind permission of ILT from Reference MERLIN Project. Development of Aero Engine Component Manufacture Using Laser Additive Manufacturing. 7th Framework Programme FP7 2007–2013, Grant Agreement 266271.

the MERLIN project [19]. The SLM process principle is presented in Fig. 22.2, whereas Fig. 22.7 illustrates a macrograph cross section of the formation of the successive melt pools after processing.

### 22.2.2 Laser metal deposition process on thin-walled aeroengine combustor casing

In LMD processes, a weld track is formed using metal powder or wire as a filler material which is fed, through a nozzle, to a melt pool created by a focused high power-density laser beam. Inert gas and powder are transported through the nozzle into the small area around the laser beam focus. By moving both the nozzle and laser simultaneously, a new material layer is deposited with high precision depending on the process parameters. In this manner, multilayering techniques allow for 3D structures to be formed. The principle of LMD is shown in Fig. 22.8.

The LMD thin-walled aeroengine combustor casing is an axis-symmetric cylindrical ring with a maximum diameter of 300 mm as presented in Fig. 22.9. The wall thickness measures 0.8 mm throughout, and the part has a height of 88.0 mm. The material powder used is an Inconel alloy IN718 and the final LMD component reached an overall density greater than 99% in relation to the wrought alloy. The main objective of this prototype fabrication was to obtain high-quality dimensional accuracy compared with ideal CAD geometry.

**Table 22.3** Process parameters of LMD/p with IN718 powder for the virole casing manufacture.

| **Process parameters** | **Value** | **Units** |
|---|---|---|
| Laser power | 925 | W |
| Beam quality factor $M^2$ | 1.1 | – |
| Scan speed | 1200 | mm/min |
| Laser beam focus diameter | 0.85 | μm |
| Argon shielding gas (1.0 bar) | 3.0 | L/min |
| Increment per revolution in building direction (layer thickness) | 0.18 | mm |
| Powder feed rate | 2.5 | grams/min |
| Argon gas for powder feed (1.5 bar) | 3.4 | L/min |
| Powder-gas focus diameter | 0.5 | mm |

Reproduced with kind permission of Springer Nature from Reference Papadakis L, Hauser C. Experimental and computational appraisal of the shape accuracy of a thin-walled virole aero-engine casing manufactured by means of laser metal deposition. Prod. Eng. 2017;11(4–5):389–399.

The manufacturing work was performed by TWI, United Kingdom, using a Trumpf DMD505 laser deposition system. This system comprised of a Trumpf 1.8 kW HQ $CO_2$ laser, a five-axis single cantilever Cartesian gantry system and a two-axis CNC positioning table (rotation and tilt capability), was developed in the frame of the MERLIN project [19,22]. Table 22.3 summarizes the process parameters of the LMD process with IN718 powder.

## 22.3 Heat input modeling in case of selective laser melting

In this section, the different modeling approaches for the heat input specifically for SLM are presented. Similar modeling methods and investigations can be transferred to LMD process by adapting the heat input models accordingly. For this purpose, mathematical heat sources are facilitated with different degree of complexity. Firstly, analytical models are introduced for replicating single scan vectors. Secondly, 3D thermal transient numerical analyses of single or successive scan vectors with the use of equivalent heat source models with different center of attention and degree of abstraction are discussed. Finally, the geometrical reduction modeling of the heat source in 3D macroscopic thermal FE models is presented with the aim to compute the residual stresses and shape distortion of industrial relevant components introduced in the previous section. The complexity and the undertaken simplification of each modeling strategy are presented in detail and discussed in the following paragraphs.

## 22.3.1 Analytical thermal models of melting phenomena of selective laser melting

Research work on mathematical models and analytical approaches describing melting phenomena in a solid continuum has been conducted in the past decades [24,25]. More specifically for the case of SLM, a three-dimensional analytical model has been adopted, which enables the calculation of the temperature distribution in powder for a Gaussian laser heat source [17]. Hereby, the determined melt pool shape was evaluated with experimental results on steel 316L and aluminum AlSi10Mg alloy powders [26,27]. Additionally, an association of the analytically calculated melt pool with the density of the printed specimens was performed concluding with a relationship of the process parameters on the process feasibility. Comparable research was conducted by further authors on different powders [28,29] based on the analytical model of a semiinfinite solid with a moving Gaussian heat source [30]. In this analytical modeling approach, the melt pool computation takes place assuming a single scan on a powder bed without considering the phase change and the associated latent heat. Further local effects as the powder sinking due to the density increase from powder to melt, the interaction with the underlying solid material or the adjacent scan vectors, i.e., the hatch space, are not included in this model. To this end, an alternative analytical solution is suggested to include the phase change from powder to melt as well as the interaction of the formed melt with the surrounding powders and the binding with underlying solid material. The interaction between the different phases (powder, melt, solid) was modeled with the use of the Laplace–Young boundary conditions on the interfaces. The sinking of the melt from the initial powder layer level was approximated by assuming the shape of circular arcs [31]. Hereby, the laser spot is, however, approximated as a line source yet again only considering a single scan. Analytical modeling of the melt pool formation can be enhanced by considering, for instance, the flow field in a moving spherical interface [32,33].

## 22.3.2 Numerical thermal transient models of melting phenomena in selective laser melting

The limitations in considering various effects of the melting process of SLM with the aid of analytical approaches encourage research on transient numerical models with increasing complexity. Here, it is worth mentioning an approach combining a finite difference method with a combined level set volume of fluid method [34]. The introduced model involves various physical effects as dynamic laser power absorption, buoyancy effect, Marangoni effect, capillary effect, evaporation, recoil pressure, and temperature-dependent material properties. It is validated for different process parameters using cubic samples of stainless steel 316L and nickel-based superalloy IN738LC. In a similar track, the melt pool dynamics were modeled by means of a FE simulation for an SLS/SLM process [35]. The proposed numerical model considered the interaction between laser beam and powder material and phase transformations, while submodels were developed to describe the capillary phenomena in the powder bed during processing. In addition to that, further research was conducted on the heat transfer and phase transition during an SLM process with a moving volumetric heat source using the finite difference method [36]. A model incorporating a phase function to differentiate the powder phase, melting liquid phase, dense solid phase, and vaporized gas phase was suggested that also includes the volume shrinkage induced by the density change during the melting process.

Furthermore the FE method was used to model an unsteady temperature field in the superalloy IN718 concentrating on the effect of varying scan length of the melt pool size [37]. A comprehensive discussion of FE models to span the scope of AM processes with a particular focus on the porosity and surface roughness prediction by considering a resolved powder model provided useful results on the melt pool formation and process feasibility during SLM [38]. In addition to that, a transient thermal prediction of the melt pool of SLM process with IN718 powders by means of the FEA proved a good correlation with the real process considering both the interface of melt into the underlying solid and the adjacent laser track shown in Fig. 22.10 [19]. Hereby, a 3D Gauss equivalent heat source was implemented as shown in Fig. 22.1 to replicate the SLM process with the parameters listed in Table 22.1.

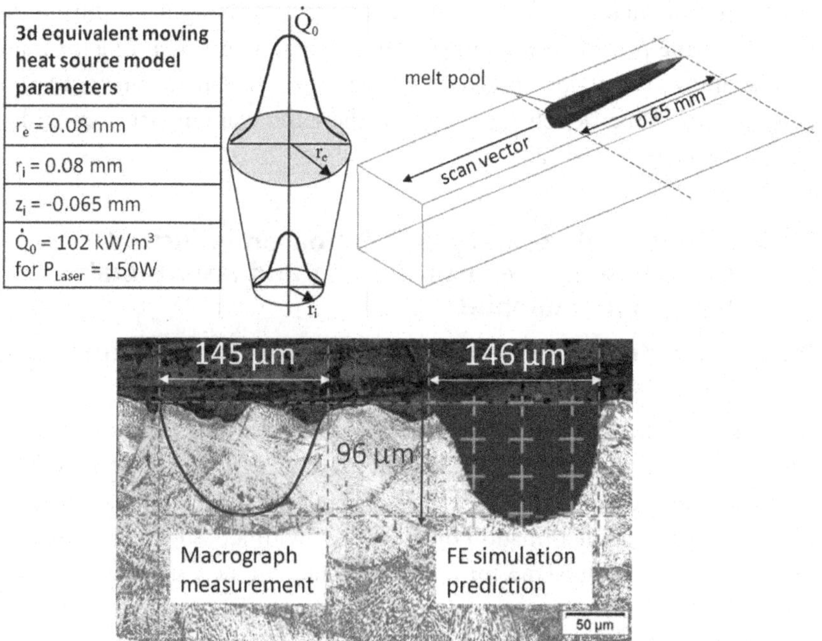


**Figure 22.10** Transient melt pool formation with a moving 3D Gauss equivalent heat source representing the laser beam (top) compared to experimental macrographs in cross section (bottom).

### 22.3.3 Geometrical reduction of 3D thermal finite element model

Beside the thermal model of the moving heat source replicating the laser beam and the transient development of the melt pool during scanning as shown in Fig. 22.10, following geometrical reduced models were proposed: a temperature profile on a whole length of scan vectors representing the total heat input per unit length, and a temperature profile in a whole layer or multiple layers, so-called global model, representing the total heat input per volume [23]. This model reduction was striven due to the extended computing capacity and duration when modeling every single scan vector for components of industrial relevance. Contrary to more detailed transient heat source models which consider the exact scanning strategy [18], the aim of the global model was to substitute the scanning vectors of each layer with volumes with an equivalent heat quantity as input. Therefore, based on this reduction approach, it was possible to calculate the residual stresses and deformations of complete parts. A global approach for the modeling of 3D components contained all transient physical effects taking place during the manufacturing process, e.g., comparable temperature gradients, melting and solidification, convection and cooling effects, material phase transformation, and material properties change, whereas the exact scanning strategy within each layer is neglected [23].

## 22.4 Geometrical accuracy calculation on industrial relevant selective laser melting and laser metal deposition examples

### 22.4.1 Transferability on larger geometries of industrial relevance

The results of the thermal temperature field presented in the previous section allow for the transferability of the heat input reduction model layer-by-layer or volume-by-volume (multiple layers). This allows the determination of the residual stresses and deformation of larger parts with industrial relevance after processing and eventual postprocessing operations, i.e., support structures removal. Based on the results rendered on a simplified geometry for the volume-by-volume heat input approach, the physical effects were transferred onto the FE mesh of a twin cantilever beam, whose geometry was built up with the aid of supports as shown in Fig. 22.5 [21]. The supports were cut off after the component had been fabricated.

The equivalent heat input for the thermal simulation of the twin-cantilever beam was defined in each layer in the form of a temperature function of time. In the FE model, each layer was defined with a thickness of 90 μm, i.e., threefold of the real process, to reduce simulation time and data storage. Between each process layer, the real process sequences were implemented, i.e., approximately 10 s for the powder deposition prior to the processing of the overlying layer. Additionally, the heat losses due to convection to environment and due to conduction to surrounding powder material were taken into account on the model with symmetry conditions shown in Fig. 22.11. The convection losses value was set to 25 $W/m^2K$ similarly to convectional welding processes as defined in past modeling works of the welding heat effects [38]. The coefficient for the losses to surrounding powder was estimated at 220 $W/m^2K$ based on test computations compared with experience values of the final temperature after the processing of each deposited powder layer [21].

Overlying, not-yet-deposited elements in the FE mesh were assigned with negligible thermal and mechanical properties, i.e., were deactivated.

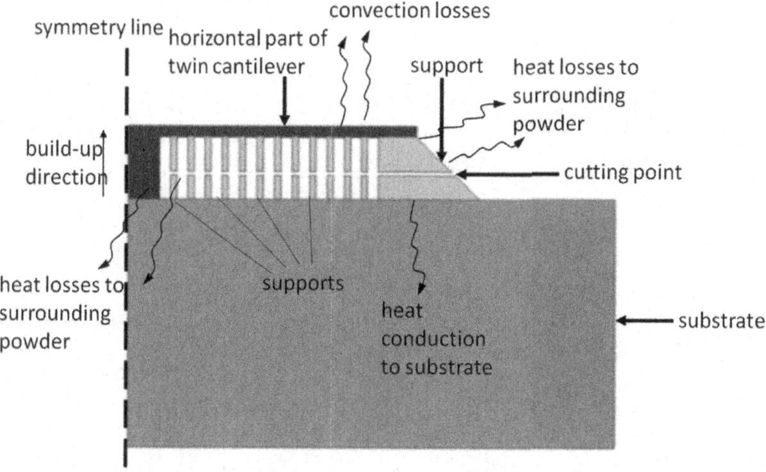


**Figure 22.11** Heat losses of the twin cantilever beam model with symmetry conditions. *(Reproduced with kind permission of Elsevier from Reference Papadakis L, Loizou A, Risse J, Schrage J. Numerical computation of component shape distortion manufactured by selective laser melting. Procedia CIRP 2014;18;90–95.)*

They were assigned with the powder material properties, i.e., activated, as soon as the layer they formed was to be processed. During processing, i.e., heating, and once reaching the melting point, the powder material was assigned with the material properties of IN718 with the aid of a phase transformation model during the thermal analysis [39,40]. The model layers were processed one after the other until the whole component was built. After the layer buildup, the phase transformation, and cooling, the model attained its final thermal and structural equilibrium. The results of the temperature field during processing of each model layer are exemplified in Fig. 22.12 in case of the combustion casing.

The temperature-dependent material properties of the solid IN718 implemented for the thermal and later for thermomechanical simulation were extracted from the literature [39], while the corresponding thermal values of the powder were deduced based on existing studies on powders [41].

## 22.4.2 Shape distortion computation of twin cantilever

The transient temperature field including the change of phase from powder into solid was used as a load for the thermomechanical FEA. After the buildup process and cooling to ambient temperature, the FE model reached its equilibrium. Due to the nature of the SLM process in conjunction with

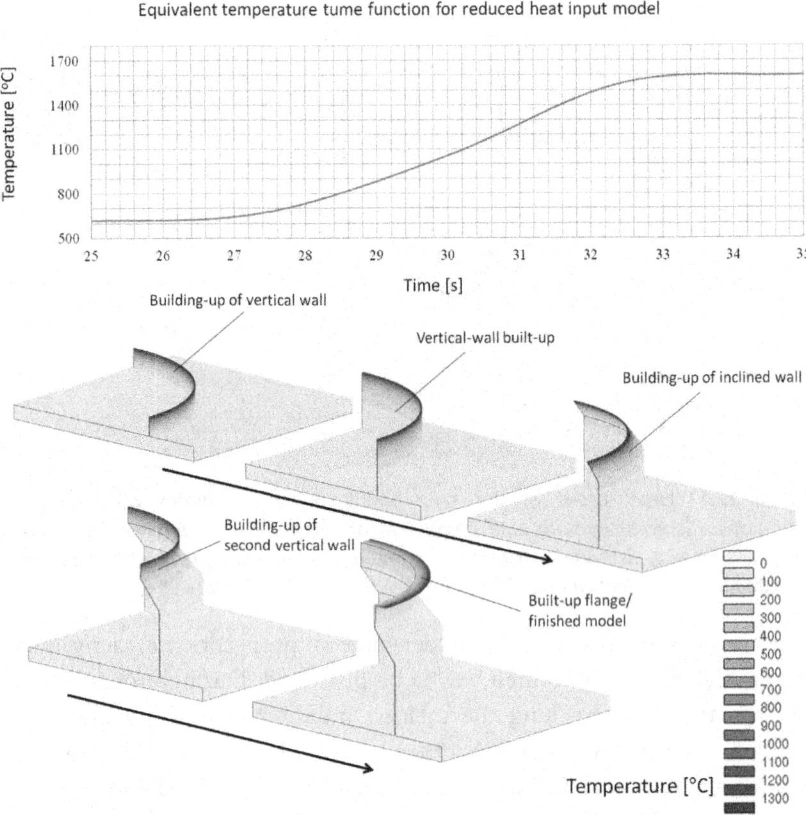


**Figure 22.12** Temperature field results of successive layers during the combustion casing model building process considering defined temperature time function as an equivalent heat input in each model layer. *(Reproduced with kind permission of Springer Nature from Reference Papadakis L, Hauser C. Experimental and computational appraisal of the shape accuracy of a thin-walled virole aero-engine casing manufactured by means of laser metal deposition. Prod. Eng. 2017;11(4–5):389–399.)*

the model geometry, the cantilever was manufactured with support structures, which were cut off with the aid of an electrode (wire-EDM) after fabricating and cooling to ambient temperature. It was significant to include this process step in the simulation chain, to replicate the spring back, i.e., geometrical changes and final stress state equilibrium in the structure.

To simulate the final distortion of the component after the material removal, the final state of equilibrium, i.e., spring back, was calculated. After the main process simulation, the elements at the point of cutting were removed, while the stress state and final temperature in the component at

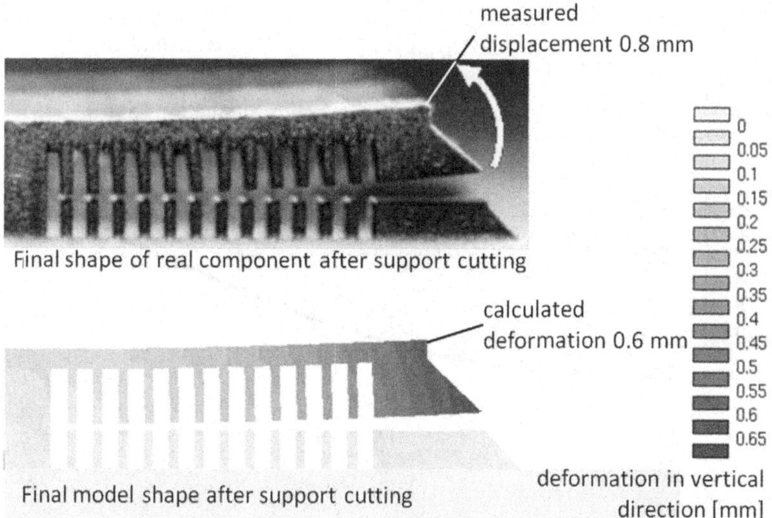


**Figure 22.13** **Calculated final shape distortion before and after support cut compared with real component measurement.** *(Reproduced with kind permission of Elsevier from Reference Papadakis L, Loizou A, Risse J, Schrage J. Numerical computation of component shape distortion manufactured by selective laser melting. Procedia CIRP 2014;18;90–95.)*

the end of the previous simulation were maintained on ambient temperature. At a later step, the simulation was restarted, and the component was forced to its final shape providing a relaxation of residual stresses due to the removal of 3D solid elements in the FE model in the area of the cut-off supports. The final model deformation after the supports removal is illustrated in Fig. 22.13. The higher distortions appear on the upper edges of the twin cantilever due to residual stresses release.

Not only are the computer capacity requirements of importance but also the reliability of the proposed method, which can be essentially applied in the design phase of the product development. The model was validated with experimental data. The vertical displacement was measured on a path at the top of the twin cantilever with a step of 2.5 mm, from the point of symmetry to the right side of the component as illustrated in Fig. 22.14.

As observed in Fig. 22.14, there is a 26% maximum result discrepancy on the far-right side of the component. This is acceptable, since it related to the modeling reduction approach, and the fact that three process layers were "fabricated" simultaneously in the simulation.

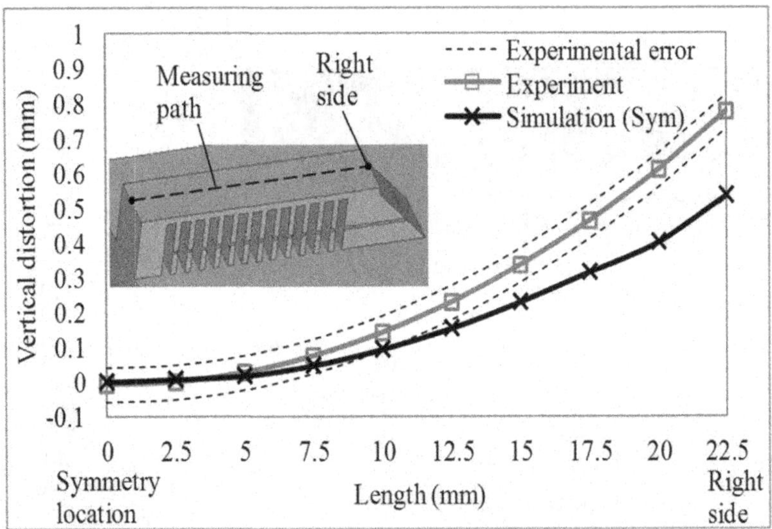


**Figure 22.14** Comparison of numerical and experimental results of twin cantilever after supports cut. *(Reproduced with kind permission of Elsevier from Reference Papadakis L, Loizou A, Risse J, Schrage J. Numerical computation of component shape distortion manufactured by selective laser melting. Procedia CIRP 2014;18;90–95.)*

### 22.4.3 Sensitivity study on shape distortion of twin cantilever

A parameter study on the twin cantilever was performed by running simulations with varying thicknesses of its horizontal part. The simulations included thicknesses of 1.0 mm (reference) 1.5 mm, 2.0 mm, and 3.0 mm. This added a +0.5 mm, +1.0 mm, and +2.0 mm, respectively, to the reference model of the 1.0 mm thickness. An identical symmetrical model shown in Fig. 22.11 was facilitated for the simulations and sensitivity analysis of the twin cantilever. The expected distortion behavior for increasing component thickness was verified with the measurements of the distortion on the SLM-manufactured components with the corresponding thicknesses by Fraunhofer ILT [19]. The provided measurements indicate a distortion decrease for higher component thicknesses.

The improved accuracy of simulation results of models with increasing component thickness compared with measurements along the measuring path on the twin cantilever upper side is exhibited in Fig. 22.15. Particularly for a sufficient increase of the component thickness to 2.0 and 3.0 mm, an adequate correlation with only marginal deviations was achieved with the

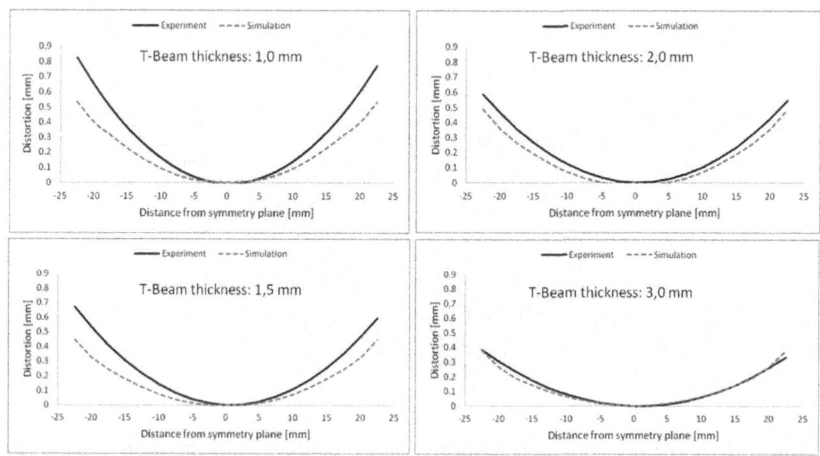


**Figure 22.15** Improved simulation result accuracy compared with experimental measurements on real fabricated components with increasing twin cantilever thickness.

use of the introduced model. This proves the reliability of the developed modeling approach especially for components with sufficient thicknesses, i.e., stiffness.

## 22.4.4 Shape distortion computation of turbine blade

A further industrial relevant component, a 3D turbine blade of a nozzle guide vane, was modeled and simulated with the aim to compute its final shape distortions. The component's geometry is presented in Fig. 22.6 manufactured by means of SLM with IN718 powder. The volume-by-volume modeling approach was used identical to the twin cantilever as described in Section 22.4.1. The real component was concurrently manufactured at the laboratories of Fraunhofer ILT [19]. Based on the transient temperature field results replicating the powder deposition and the heat input during SLM processing, a thermomechanical FEA was performed. The results of the 3D turbine blade final shape compared with the target CAD geometry are presented in Fig. 22.16.

The 3D turbine blade geometry was captured by means of optical 3D scans. The results of the fabricated component's final shape are presented in Fig. 22.17. The aim of the 3D turbine blade demonstrator manufacture was to evaluate the modeled shape distortions. The comparison of simulated and CAD data in used metrology software shows a good agreement with

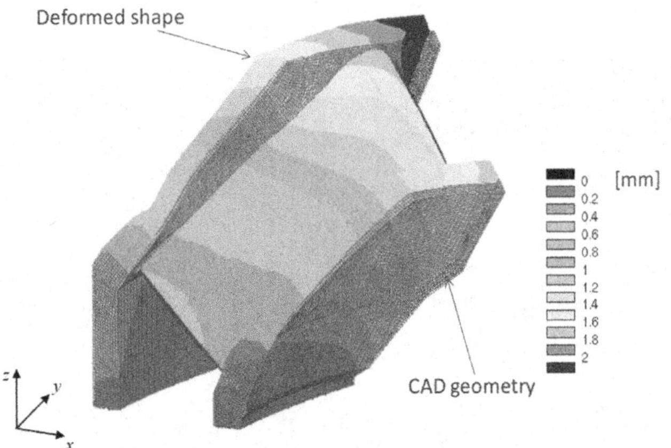


**Figure 22.16** Computed shape distortion of the 3D turbine blade model compared with the target CAD geometry. *CAD*, computer-aided design.

the FE simulation software. A maximum shape deviation of approximately −2 mm in upper part of demonstrator is observed. The simulation results of the shape distortion show a tendency of the direction indicated by the arrows. On the other hand, the fabricated geometry shows shape deviations of up to approximately 0.8 mm, while no general tendency of main distortion direction is noticeable.

In overall, the simulated and measured shape distortions in the region of the air foil indicate comparable tendencies. Part of the higher discrepancies in other regions of the demonstrator component, as for example at the edges of the stiffening webs, can be explained through the difficulty to fit the manufactured geometry with the CAD data at the base of the components, respectively, to create an identical reference area. Another reason is the fact that the material deposition during modeling takes place on the already-deformed component. In contrast to that, in the real SLM processing the powder layer is each time deposited on the target CAD region, thus acting as a correction to possible distortions during SLM processing. This could be the reason why the shape distortion in the real SLM part reveals lower distortion than in the simulation. This effect will be discussed in more detail in the following paragraph.

### 22.4.5 Shape distortion computation of rotational symmetric thin-walled aeroengine combustor casing

The aforementioned model reduction approach was also adopted for the case of the rotational symmetric thin-walled aeroengine combustor casing.

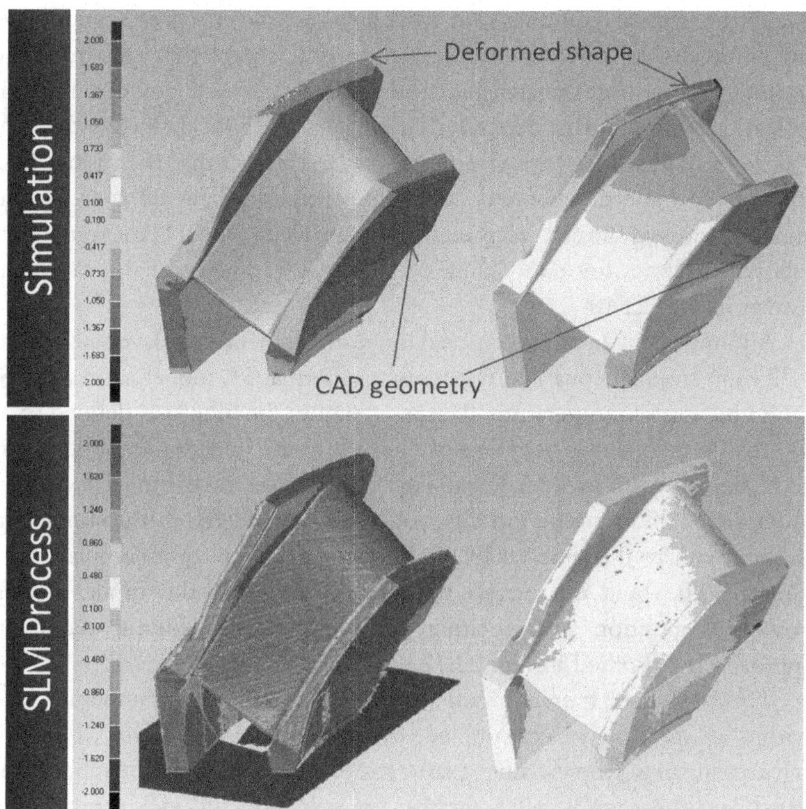


**Figure 22.17** Comparison of the simulation results (top) with optical 3D scans of the SLM manufactured turbine blade (bottom). *SLM*, selective laser melting. *(Reproduced with kind permission of ILT from Reference MERLIN Project. Development of Aero Engine Component Manufacture Using Laser Additive Manufacturing. 7th Framework Programme FP7 2007–2013, Grant Agreement 266271.)*

Due to the extended computing capacity and duration in case of modeling the transient circular motion of LMD process for the whole structural geometry, a model reduction was unavoidable. Contrary to more detailed transient heat source models which consider the exact scanning path, the aim of the reduced models is to substitute the deposition path of each layer with an equivalent heat quantity as thermal load. Therefore, based on this reduction approach, it was possible to calculate the residual stresses and deformations of the complete component with reasonable computations time.

LMD process differs compared with SLM since it is performed with slower scan speed. Additionally, the cooling process during LMD is caused due to the exposal to ambient air, while in SLM, the processed components are embedded in powder until the process is completed, and thus, the cooling process is dominated by conduction to adjacent powder. For this reason, a convection heat transfer coefficient of 25 $W/m^2K$ was defined, similar to models for welding processes [42]. The fact that the casing has a rotational symmetric geometry with the continuous helicoid laser motion creating no significant asymmetry, as proved by 3D measurements following, allows for a modeling of a quarter geometry of the casing, as shown in Fig. 22.18.

An equivalent heat input was defined on four process layers (equal to 0.72 mm) simultaneously in the transient thermal FE model as an average time function of temperature. The temperature function was defined so as to provide sufficient penetration of the melt in the underlying model layers according to Section 22.3.3 and Fig. 22.12. After each model layer was processed, a 30 s time period for cooling was considered in the simulation to approximate the time taken for the laser to complete a complete revolution. The built-up process of different stages of the model during powder deposition and melting in a volume-by-volume reduction approach is illustrated in Fig. 22.12 [22].

This completed transient temperature field was applied as a thermal load within a subsequent noncoupled thermomechanical analysis. For this calculation, an isotropic elastic–plastic material definition with temperature-dependent mechanical values for the Young's modulus, yield stresses, and strain hardening was implemented. Additionally, the change of phase from

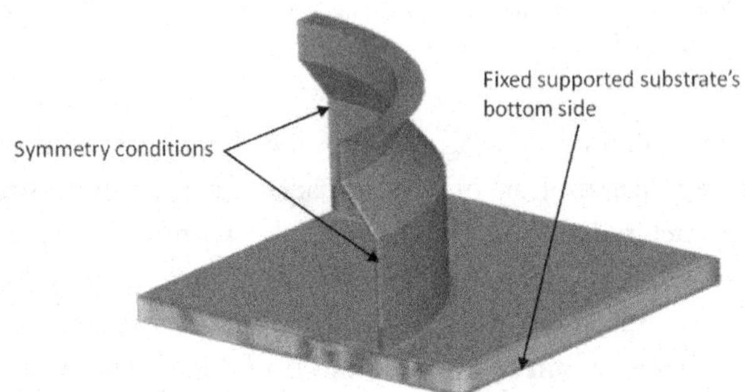


**Figure 22.18** Quarter model of the thin-walled aeroengine combustor casing. *(Reproduced with kind permission of Springer Nature from Reference Papadakis L, Hauser C. Experimental and computational appraisal of the shape accuracy of a thin-walled virole aero-engine casing manufactured by means of laser metal deposition. Prod. Eng. 2017;11(4–5):389–399.)*

powder into solidified material was considered by applying a phase change Leblond model in the FE solver SYSWELD similar to the previous demonstrator components [39–41].

Initially the shape distortion of the casing model attached on the base plate (i.e., prior to cut-off) was computed. Fig. 22.19 indicates the distortion results in the $x$-(radial) direction compared with the 3D scanned geometry, performed with the aid of ATOS measuring equipment, and the ideal CAD geometry. As evidenced by this figure, the FE model provides good results in the lower region of the casing with deformations in the negative $x$-direction, i.e., reduction of the component radius, reaching 1.2 mm. The 3D scan provides similar results with axial displacement in this area of approximately 1.1 mm. The results continue to demonstrate a quite good agreement for the area of the inclined wall with minor radial displacements for both the FE model and the scanned 3D geometry. On the contrary, in the upper vertical wall of the FE model, an increase of the casing radius of around 0.5 mm is observed, whereas the measurements

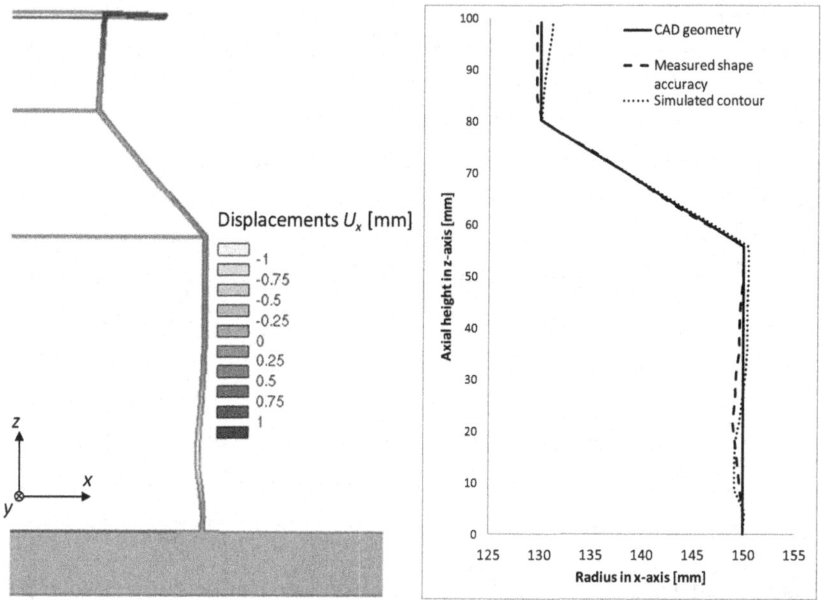


**Figure 22.19** Measured and simulated distortion results of the thin-walled aeroengine combustor casing in x-(radial) direction prior to cut-off from substrate compared with target CAD geometry. *CAD*, computer-aided design. *(Reproduced with kind permission of Springer Nature from Reference Papadakis L, Hauser C. Experimental and computational appraisal of the shape accuracy of a thin-walled virole aero-engine casing manufactured by means of laser metal deposition. Prod. Eng. 2017;11(4–5):389–399.)*

indicate about 0.4 mm decrease of the radius. The FE model fails to provide an adequate distortion trend in this segment of the casing. The reason for these discrepancies lies most probably in the motion of the overlying mesh representing the material to be deposited on the already-deformed created model. This was probably the cause why distortions with a wrong tendency were initiated, as the model was growing higher. On the contrary, in the real LMD process, initiated shape inaccuracies due to thermal shrinkage were compensated during the buildup process since the nozzle deposits new powder in the exact CAD target position. Fig. 22.20 visualizes the difference between the model buildup procedure and the buildup procedure in real AM processes.

As far as the distortions in $z$-(axial/build-up) directions prior to cut-off are concerned, the overall shrinkage of the FE model proves to be in good agreement with the measurements. A total shrinkage of approximately 1.4 mm is simulated by the numerical model, while the 3D scans capture an average total shrinkage of approximately 1.7 mm.

The finished combustion casing was thereafter removed from substrate at a height of 12 mm. This substrate cut-off operation step was also replicated in the simulation chain in a last "spring-back" calculation step

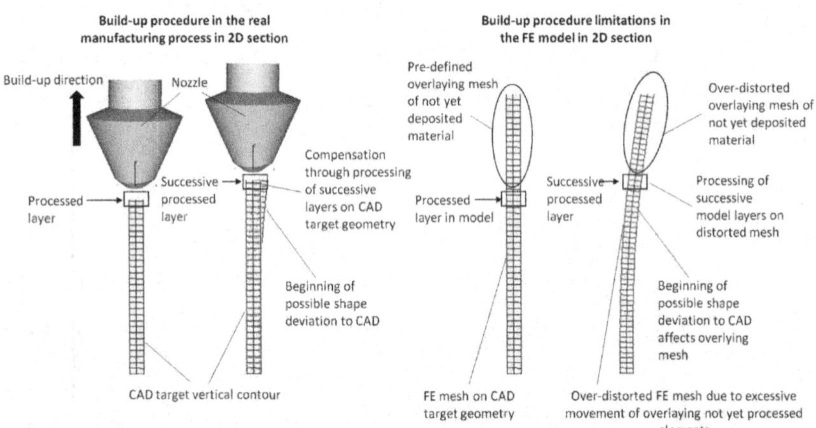


**Figure 22.20** Apparent source of shape accuracy discrepancies in the buildup procedure between simulation and experiment during the additive manufacturing processes in general. *(Reproduced with kind permission of Springer Nature from Reference Papadakis L, Hauser C. Experimental and computational appraisal of the shape accuracy of a thin-walled virole aero-engine casing manufactured by means of laser metal deposition. Prod. Eng. 2017;11(4–5):389–399.)*

(equilibrium after substrate removal) to compute the final component shape. The average von Mises residual stress distribution before and after cut-off in the casing is shown in Fig. 22.21. A significant variation of the stress distribution is noticeable due to the removal operation, especially in the cut-off area.

The final casing shape distortion results after cut-off from substrate in the $x$-(radial) direction compared with 3D scans are shown in Fig. 22.22. The FE model demonstrates adequate agreement with optical 3D scans in the lower region after the cut-off operation with deformations in the negative radial direction, i.e., reduction of the component radius, reaching 1.2 mm. Concurrently, measurements prove a comparable behavior with a radius reduction exceeding 1.0 mm. The results persist to display a very good agreement in the area of the inclined wall with minor radial changes. Contrariwise, in the upper vertical wall, the numerical model did not succeed to reproduce the radial distortions. As far as the axial distortions after cut-off are concerned, the overall shrinkage of the FE model is in a very good agreement with the 3D scans. An overall length decrease of about 1.2 mm was computed, while the 3D scans proved a total average length increase of approximately 1.4 mm.

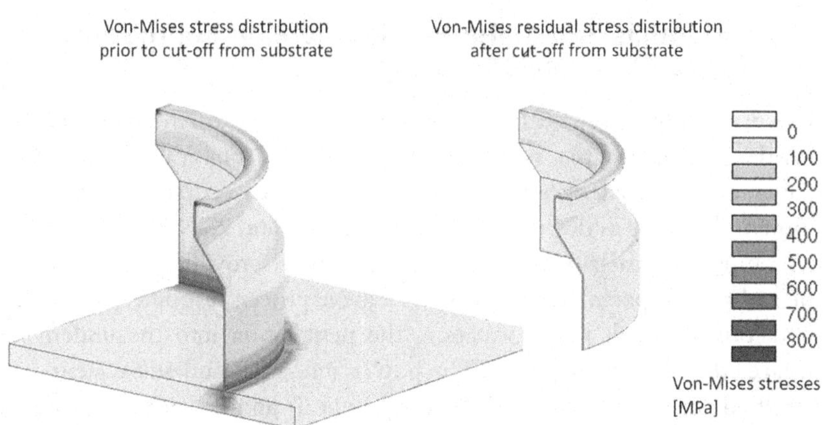


**Figure 22.21** Stress distribution after thermomechanical analysis before removal operation (left) and final stress relaxation after removal from base plate (right). *(Reproduced with kind permission of Springer Nature from Reference Papadakis L, Hauser C. Experimental and computational appraisal of the shape accuracy of a thin-walled virole aero-engine casing manufactured by means of laser metal deposition. Prod. Eng. 2017;11(4–5):389–399.)*

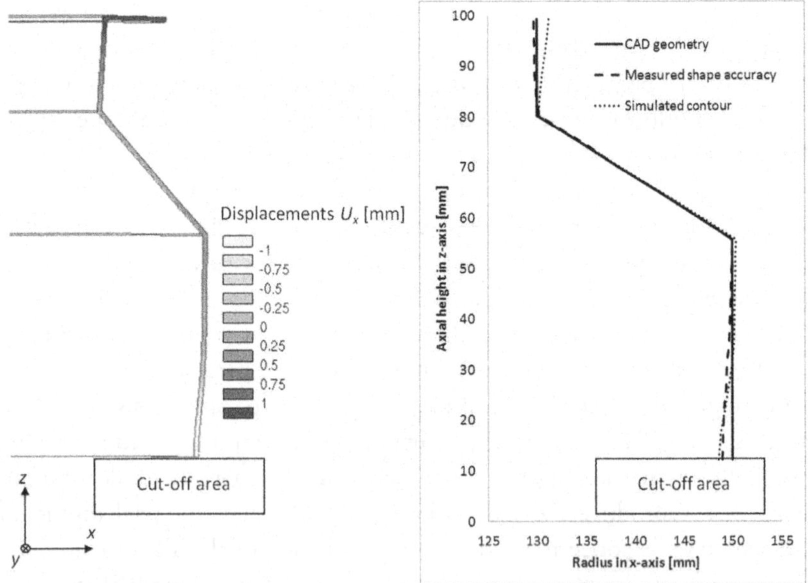


**Figure 22.22** Measured and simulated distortion results of the thin-walled aeroengine combustor casing in x-(radial) direction after cut-off operation. *(Reproduced with kind permission of Springer Nature from Reference Papadakis L, Hauser C. Experimental and computational appraisal of the shape accuracy of a thin-walled virole aero-engine casing manufactured by means of laser metal deposition. Prod. Eng. 2017;11(4–5):389–399.)*

## 22.5 Potentials and limitation of modeling approaches for additive manufacturing

The previously presented modeling methods related to AM with metallic powders manage to replicate the multiple physical effects and manufacturing challenges to a certain degree. Microscopic analytical and numerical models as described in Sections 22.3.1 and 22.3.2 concentrate in describing local melt pool phenomena and, thus, providing a feedback to the real process parameter selection to achieve process feasibility in terms of predicting the melt pool formation, the penetration into the underlying surface, the change of phase from power into melt and subsequently to solidified material, and the fluid dynamical and evaporation effects. Nevertheless, porosity and crack development prediction and high density attainability of fabricated parts can be only modeled to a limited number of powders with intensive support from experimental investigations, i.e., macrographs. Hereby, there is an increasing need to optimize processing times and simultaneously attain high densities [43]. Additionally, energy efficiency–related aspects have gained increasing interest in modeling and simulation [44,45].

Whereas complex modeling approaches concentrate on supporting process design for AM tasks at a microscopic melt pool level, their degree of complexity cannot be maintained for considering further macroscopic process issues such as the scan strategy. The scan strategy may influence local effects as the residual stress development along a processed layer at a first glance; however, it effects simultaneously the development of distortions with the majority of the investigations taking place on smaller samples [18,46]. Fig. 22.23 shows the influence of the scan strategy variation on the twin cantilever shape distortion after cut-off operation from the supports.

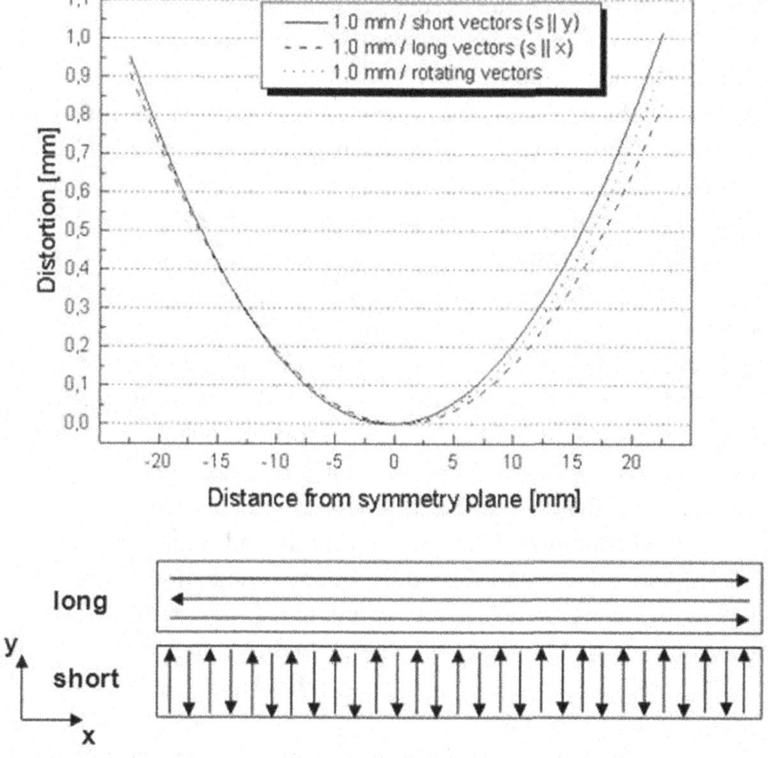


**Figure 22.23** Influence of the scan strategy on the final distortion development of the twin cantilever. *(Reproduced with kind permission of ILT from Reference MERLIN Project. Development of Aero Engine Component Manufacture Using Laser Additive Manufacturing. 7th Framework Programme FP7 2007–2013, Grant Agreement 266271.)*

Twin cantilever geometries as presented in Fig. 22.5 with a thickness of 1.0 mm were fabricated with three different scan vector orientations within each layer: along the long side of the twin cantilever parallel to x-axis; vertical to side of the twin cantilever parallel to y-axis; and rotating/alternating between the two directions for successive layers. The experimental measurements on the manufactured twin cantilevers with the aforementioned scan strategies showed that the scan vector orientation has an influence on the final distortion of parts after cut-off to a certain extend. It is observed that the longitudinal vector orientation leads to the lowest distortion development probably due to the increased rigidity development, i.e., lower temperature gradients in the transverse direction in y-axis, along the longer side of the twin cantilever. To include such local effects in the simulation of shape accuracy of complete components with industrial relevance will be very time-consuming although it proves to be of certain significance in terms of the accuracy improvement especially in case of removal of large supports areas as the twin cantilever example revealing a distortion variation of up to 17%.

In Section 22.4.1, a modeling reduction approach was discussed, which enables the shape accuracy computation of industrial relevant examples. The computed deformed shape of the twin cantilever, the turbine blade, and the combustion casing were proved to provide reasonable tendencies comparable with the experimental measurements. The discrepancies among the simulated distortions and the measured components' shape are justified on the one hand by the course mesh discretization in the built-up direction on the basis of a volume-by-volume reduction and the neglect of the scan strategy and the connected thermal gradients and stress development within each layer, and on the other hand by the overdistorted FE mesh due to excessive movement of overlaying not-yet-processed elements as described in Fig. 22.20. After identifying this source of discrepancy in the shape accuracy prediction between the simulation and measurements of the demonstration components related to the mesh definition, the introduced reduction modeling approach can be improved with respect to its mesh accuracy during the buildup of the model. In this context, an update of the deposited mesh in the thermomechanical simulation on the target CAD position, similar to real process, is suggested. Such a mesh recalculation can be realized with the aid of commands available in FE solvers, which allow for the modification of the node coordinates after each thermomechanical simulation step as illustrated in Fig. 22.24 [39]. The calculated displacements $\Delta u$ of merely the overlaying nodes, i.e., nodes above the processed layer, of

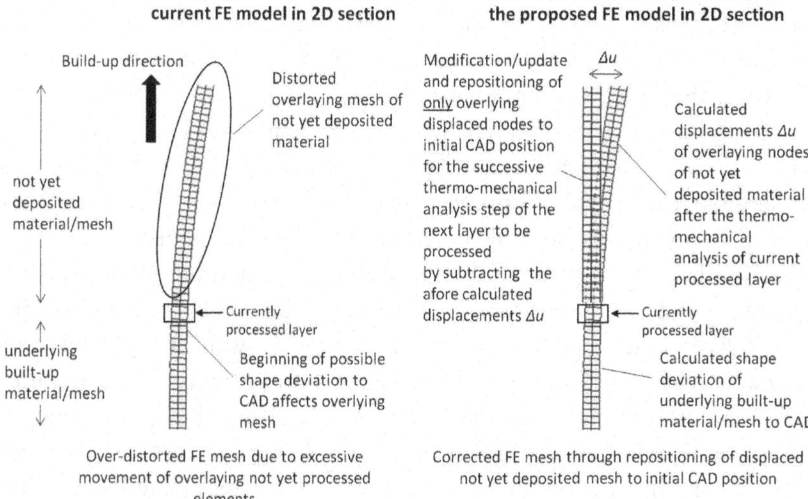


**Figure 22.24** Buildup procedure improvement potential through coordinate modification of the not-yet-deposited material to the initial CAD position after each simulation step. *CAD*, computer-aided design. *(Reproduced with kind permission of Springer Nature from Reference Papadakis L, Hauser C. Experimental and computational appraisal of the shape accuracy of a thin-walled virole aero-engine casing manufactured by means of laser metal deposition. Prod. Eng. 2017;11(4–5):389–399.)*

the not-yet-deposited material after each thermomechanical analysis of currently processed layer can be modified, i.e., repositioned, to the initial CAD position with the aid of specific subroutine. This operation will prepare the mesh for the successive thermomechanical analysis step of the next layer to be processed by subtracting the afore-calculated displacements $\Delta u$ similar to the corrective effect of the real AM process, in which the position of the last printed layer does not affect the position of the overlying following layer. This can prevent the incorrect motion of the overlying predefined mesh layers, which have not yet been processed, and, therefore, reduce undesirable accumulated shape distortions because of distorted positioning of the overlaying mesh during the model buildup procedure.

A further potential for the modeling and simulation is to assist the compensation of a part's shape distortion after AM. To compensate shape distortions and counteract the final shape, a "predistorted" geometry model can be used based on the FE-simulated shape distortions or optical 3D scans. Hereby, the calculated or measured distortions can be used to create an invert CAD mesh geometry. Based on the calculated deformation shape, a predistorted geometry was produced. The thermomechanical simulation on the basis of reduced heat input over a number of layers was performed on the predistorted geometry model with improvements on the final computed component shape accuracy compared with the 3D measurements on the studied component [47].

Finally, the last identified challenge of modeling and simulation in AM is to conduct a parameter sensitivity study of the studied geometries to test the sensitivity of the final computed shape compared with the experimental results. This will assist to evaluate the stability and reliability of the simulation method for computing the shape distortions and the residual stresses in AM processes. Specifically, modeling certain parameters as the geometrical dimensions and the mesh density as well as the equivalent thermal load definition over multiple layers in the FE model can be varied to identify possible model limits and provide an outcome of the computational time reduction benefit versus the result accuracy deficits compared with measurements. To this end, a database can be created where researchers and industrial partners can share their investigated geometrical models, used material data, FE mesh, subroutines, and rendered outcomes in specific forms. In this manner, further interested researchers may use existing available data, information, and knowledge to perform benchmark calculations with different software packages or further modeling approaches. This can help to evolve the modeling efficiency in different aspects of AM, from the process models for computing part's density, and avoid porosity with selection of appropriate machine parameters to the simulation of shape distortions and their compensation by means of reverse engineering strategies.

## Acknowledgment

The investigations and results presented in this chapter were accomplished in the framework of the MERLIN Project which has received funding from the European Commission's 7th Framework Programme FP7 2007–13 under the grant agreement 266271. Project website: www.merlin-project.eu.

## References

[1] Y. Huang, M.C. Leu, J. Mazumder, A. Donmez, Additive manufacturing: current state, future potential, gaps and needs, and recommendations, J. Manuf. Sci. Eng. 137 (1) (2015).

[2] S.A.M. Tofail, E.P. Koumoulos, A. Bandyopadhyay, S. Bose, L. O'Donoghue, C. Charitidis, Additive manufacturing: scientific and technological challenges, market uptake and opportunities, Mater. Today 21 (1) (2018) 22–37.

[3] K. Zhang, X.F. Shang, W.J. Liu, Laser metal deposition shaping system for direct fabrication of parts, Appl. Mech. Mater. 66–68 (2011) 2202–2207.
[4] S. Khademzadeh, N. Parvin, P.F. Bariani, Production of NiTi alloy by direct metal deposition of mechanically alloyed powder mixtures, Int. J. Precis. Eng. Manuf. 16 (11) (2015) 2333–2338.
[5] A. Zhang, B. Qi, B. Shi, D. Li, Effect of curvature radius on the residual stress of thin-walled parts in laser direct forming, Int. J. Adv. Manuf. Technol. 79 (1) (2015) 81–88.
[6] M. Dallago, F. Zanini, S. Carmignato, D. Pasini, M. Benedetti, Effect of the geometrical defectiveness on the mechanical properties of SLM biomedical Ti6Al4V lattices, Proc. Struct. Integ. 13 (2018) 161–167.
[7] M.F. Zaeh, G. Branner, Investigations on residual stresses and deformations in selective laser melting, Prod. Eng. 4 (1) (2010) 35–45.
[8] B. Vrancken, Study of Residual Stresses in Selective Laser Melting [dissertation], KU Leuven, 2016.
[9] A.M. Khorasani, I. Gibsond, U.S. Awana, A. Ghaderi, The effect of SLM process parameters on density, hardness, tensile strength and surface quality of Ti-6Al-4V, Addit. Manufact. 25 (2019) 176–186.
[10] M. Kamal, G. Rizza, Design for metal additive manufacturing for aerospace applications, Addit. Manufact. Aerospace Indus. (2019) 67–86.
[11] K.K. Dama, S.K. Malyala, V.S. Babu, R.N. Rao, I.J. Shaik, Development of automotive FlexBody chassis structure in conceptual design phase using additive manufacturing, Mater. Today 4 (9) (2017) 9919–9923.
[12] P. Han, Additive design and manufacturing of jet engine parts, Engineering 3 (2017) 648–652.
[13] M. Javaid, A. Haleem, Additive manufacturing applications in medical cases: a literature based review, Alexandria J. Med. 54 (2018) 411–422.
[14] P. Mercelis, J.P. Kruth, Residual stresses in selective laser sintering and selective laser melting, Rapid Prototyp. J. 12 (5) (2006) 254–265.
[15] A. Yaghi, S. Ayvar-Soberanis, S. Moturu, R. Bilkhu, S. Afazov, Design against distortion for additive manufacturing, Addit. Manufact. 27 (2019) 224–235.
[16] I. Koutiri, E. Pessard, P. Peyre, O. Amlou, T. De Terris, Influence of SLM process parameters on the surface finish, porosity rate and fatigue behavior of as-built Inconel 625 parts, J. Mater. Process. Technol. 255 (2018) 536–546.
[17] M. Letenneur, A. Kreitcberg, V. Brailovski, Optimization of laser powder bed fusion processing using a combination of melt pool modeling and design of experiment approaches: density control, J. Manuf. Mater. Process 3 (21) (2019).
[18] L. Mugwagwa, D. Dimitrov, S. Matope, I. Yadroitsev, Evaluation of the impact of scanning strategies on residual stresses in selective laser melting, Int. J. Adv. Manuf. Technol. 102 (2019) 2441–2450.
[19] MERLIN Project. Development of Aero Engine Component Manufacture Using Laser Additive Manufacturing. 7th Framework Programme FP7 2007–2013, Grant Agreement 266271.
[20] Gong H, Gu H, Zeng K, D. JJS, Pal D, Stucker B, Christiansen D, Beuth J, Lewandowski JJ. Melt pool characterization for selective laser melting of Ti-6Al-4V prealloyed powder. In: Proceedings of the 25th Annual International Solid Freeform Fabrication Symposium: 2014 Aug 4-6; Austin, Texas, USA.
[21] L. Papadakis, A. Loizou, J. Risse, J. Schrage, Numerical computation of component shape distortion manufactured by selective laser melting, Proc. CIRP 18 (2014) 90–95.
[22] L. Papadakis, C. Hauser, Experimental and computational appraisal of the shape accuracy of a thin-walled virole aero-engine casing manufactured by means of laser metal deposition, Prod. Eng. 11 (4–5) (2017) 389–399.

[23] L. Papadakis, A. Loizou, J. Risse, S. Bremen, J. Schrage, A computational reduction model for appraising structural effects in selective laser melting manufacturing: a methodical model reduction proposed for time-efficient finite element analysis of larger components in Selective Laser Melting, J. Virt. Phys. Prototyp. 9 (1) (2014) 17−25.
[24] S. Peterson, Propagation of a boundary of fusion, Proc. Glasgow Math. Assoc. 1 (1952) 42−47.
[25] H. Hu, S.A. Argyropoulos, Mathematical modelling of solidification and melting: a review, Model. Simulat. Mater. Sci. Eng. 4 (1996) 371−396.
[26] A. Foroozmehr, M. Badrossamay, E. Foroozmehr, Finite element simulation of selective laser melting process considering optical penetration depth of laser in powder bed, Mater. Des. 89 (2016) 255−263.
[27] Y. Li, D. Gu, Parametric analysis of thermal behavior during selective laser melting additive manufacturing of aluminum alloy powder, Mater. Des. 63 (2014) 856−867.
[28] M. Letenneur, V. Brailovski, A. Kreitcberg, V. Paserin, I. Bailon-Poujol, Laser powder bed fusion of water-atomized iron-based powders: process optimization, J. Manuf. Mater. Process 3 (21) (2017).
[29] A. Kreitcberg, V. Brailovski, S. Prokoshkin, New biocompatible near-beta Ti-Zr-Nb alloy processed by laser powder bed fusion: process optimization, J. Mater. Process. Technol. 252 (2018) 821−829.
[30] D. Schuöcker, Handbook of the Eurolaser Academy, Springer Science & Business Media, Vienna, Austria, 1998.
[31] Y. Ioannou, C. Doumanidis, M.M. Fyrillas, K. Polychronopoulos, Analytical model for geometrical characteristics control of laser sintered surfaces, Int. J. Nanomanufact. 6 (2010) 300−311.
[32] F. Font, T.G. Myers, S.L. Mitchell, A mathematical model for nanoparticle melting with density change, Microfluid. Nanofluidics 18 (2014) 233−243.
[33] M.M. Fyrillas, A.J. Szeri, Dissolution or growth of soluble, spherical, oscillating bubbles, J. Fluid Mech. 277 (1994) 381−407.
[34] T. Heeling, M. Cloots, K. Wegener, Melt pool simulation for the evaluation of process parameters in selective laser melting, Addit. Manufact. 14 (2017) 116−125.
[35] T. Polivnikova, Study and Modelling of the Melt Pool Dynamics during Selective Laser Sintering and Melting [dissertation], École Polytechnique Fédérale de Lausanne, 2015.
[36] Y. Li, K. Zhou, S.B. Tor, C.K. Chua, K.F. Leong, Heat transfer and phase transition in the selective laser melting process, Int. J. Heat Mass Tran. 108 (2017) 2408−2424.
[37] Cheng B, Chou K. Melt pool evolution study in selective laser melting. In: Proceedings of the 26th International Solid Freeform Fabrication Symposium: 2015 Aug 10-12; Austin, Texas, USA.
[38] M. Megahed, H.-W. Mindt, N. N'Dri, H. Duan, O. Desmaison, Metal additive-manufacturing process and residual stress modeling, Integrat. Mater. Manufact. Innov. 5 (4) (2016).
[39] E.S.I. Sysweld, User's Reference Manual, ESI Group, France, 2006.
[40] J.B. Leblond, A new kinetic model for anisothermal metallurgical transformations in steels including effect of austenite grain size, Acta Metall. 2 (1) (1984) 137−146.
[41] O. Biceroglu, A.S. Mujumdar, A.R.P. van Heiningen, W.J.M. Douglas, Thermal conductivity of sintered metal powders at room temperature, Lett. Heat Mass Tran. 3 (1979) 183−192.
[42] D. Radaj, Heat Effects of Welding, Springer, Berlin, 1992.
[43] P. Laakso, T. Riipinen, A. Laukkanen, T. Andersson, Jokinen, A. Revuelta, K. Ruusuvuori, Optimization and simulation of SLM process for high density H13 tool steel parts, Phys. Proc. 83 (2016) 26−35.

[44] W. Liu, H. Wei, C. Huang, F. Yuan, Y. Zhang, Energy efficiency evaluation of metal laser direct deposition based on process characteristics and empirical modeling, Int. J. Adv. Manuf. Technol. 102 (2019) 901–913.
[45] L. Papadakis, D. Chantzis, K. Salonitis, On the energy efficiency of pre-heating methods in SLM/SLS processes, Int. J. Adv. Manuf. Technol. 95 (1–4) (2018) 1325–1338.
[46] W. Xing, D. Ouyang, N. Li, L. Liu, Estimation of residual stress in selective laser melting of a Zr-based amorphous alloy, Materials 11 (8) (2018) 1480.
[47] S. Afazov, W.A.D. Denmark, B.L. Toralles, A. Holloway, A. Yaghi, Distortion prediction and compensation in selective laser melting, Addit. Manufact. 17 (2017) 15–22.

# Index

*Note*: 'Page numbers followed by "f" indicate figures and "t" indicate tables.'


## A

Abrasive blasting, 105–106
Abrasive flow machining (AFM), 104, 105f
Abrasive machining, 445–446, 446f–447f
Absorptivity, 431–432, 432f
ABS powder. *See* Acrylonitrile butadiene styrene (ABS) powder
Acellular techniques, 648
Acid-catalyzed esterification, 466–468
Acrylonitrile butadiene styrene (ABS), 33–34, 175
Additive manufacturing (AM)
  advantages, 159–160
  American Society for Testing and Materials (ASTM), 5
  applications, 1, 160, 383
  automotive industry. *See* Automotive industry
  biopolymers. *See* Biopolymers
  binder jetting, 10–11, 10f
  biomedical applications, 370
  ceramic components, 22–23, 23f
  classification, 5, 6t, 160, 160f, 373, 374f
  Computer-Aided Design (CAD) software, 373
  computer numerical control (CNC), 1
  damaged components, 18
  3D CAD data, 395
  dimensional accuracy, 160–161, 161f
  direct laser deposition (DLD). *See* Direct laser deposition (DLD)
  directed energy deposition (DED), 13, 14f
  4D printing (4DP) technology, 160. *See also* 4D Printing
  economic analysis, 383
  electron beam melting (EBM) process. *See* Electron beam melting (EBM) process
  engineered human tissues and organs. *See* Engineered human tissues and organs
  evolution of, 370
  extrusion process. *See* Extrusion process
  Fe–Cr–Ni composition designs, 20, 21f
  functionally graded material (FGM)
    design of, 15–17
    procedures, 15, 16f
    properties, 13–15
    residual stress, 15–17
    thermal conductivity, 15–17
    thermal spray, 13–15
  fused deposition modeling (FDM). *See* Fused deposition modeling (FDM)
  geometrical accuracy calculation, 702–713
  high-entropy alloys (HEAs), 21
  industrial applications, 370
  in-situ monitoring, 22–23
  ISO/ASTM 52900 standard, 372–373
  ISO/TC 261, 397–398, 398t
    ASTM F42, 399, 400f, 401, 401t
    implementation, 405
    Joint Steering Group (JAG), 400
    liaisons of, 401t, 402–403
    standards, 400–401, 402t, 403–404
    structure, 399, 399f
    technical working groups (WGs), 398
    University of Las Palmas de Gran Canaria (ULPGC), 405, 405f
  laser aided metal additive manufacturing. *See* Laser aided metal additive manufacturing
  laser-based additive manufacturing. *See* Laser-based additive manufacturing

Additive manufacturing (AM) (*Continued*)
laser-directed energy deposition (LDED). *See* Laser-directed energy deposition (LDED)
laser polishing. *See* Laser polishing
layer-by-layer fabrication process, 2
limitations, 26–28, 159, 383–384
metal matrix composites (MMCs). *See* Metal matrix composites (MMCs)
materials, 371–372
extrusion, 9–10, 9f
jetting, 8, 8f
metals, 384–385, 386t
nanofunctionalized 3D printing. *See* Nanofunctionalized 3D printing
non-contact scanning equipment, 18–19
optical metallography, 20, 22f
pharmaceutical 3D printing technologies. *See* Pharmaceutical 3D printing technologies
polymers, 121, 386–387, 387t. *See also* Polymers
Porous and Modifications for Engineering Surfaces (POMES), 370–371
postprocessing operations, 372–373
powder bed fusion (PBF), 11, 12f
pre-alloyed powder, 20
pre-mixed powder, 20
pre-repair heat treatment, 18
principle of, 372
process, 2, 3f
application, 5
classification, 457–464, 458t–463t
components, 1
computer-aided design (CAD) model, 2–3, 2f
G-code generation, 4
parameters, 4
post-processing, 5
printed objects removal, 4–5
standard tessellation language (STL), 4
toolpath, 4
prototyping technique, 372
remanufacturing process, 17–18, 17f
repair experiments, 17, 19
requirements and certifications, 396
reverse engineering digitizers, 18–19
selective laser melting (SLM). *See* Selective laser melting (SLM)
sheet lamination, 12
3D sintering machines, 1
space resources, 661. *See also* Space resources
standards, 396
stereolithography (SLA), 369–370
subcommittees (SCs), 395
subtractive machining techniques, 1, 19–20, 23–25
accuracy, 25
anisotropy, 23–24
cost, 25
design freedom, 24–25
hardness and wear resistance, 23–24
hybrid manufacturing, 25–26
materials, 23–24
speed, 24
*vs.* subtractive manufacturing process, 160–161, 373–375
surface quality, 370–371
surface roughness
AlSi10Mg alloy, 380
Inconel 625 samples, 380–381
layer-by-layer process, 378
layer thickness, 378–379, 379f
measurement methodology, 375–377
metal and polymeric parts, 379, 380t
polyether ether ketone (PEEK), 379–380
postfinishing, 381–383, 382f
quality of, 375, 377–378
staircase effect, 378–379, 378f
technical committees, 395
vat photopolymerization (VP), 7–8, 7f. *See also* Vat photopolymerization (VP) methods
wireless sensors, 22–23
worn geometry, 19
Additive manufacturing file format (AMF), 4

Additive processing, 644–645
Aerospace industry, direct laser deposition (DLD), 531–532
AFM. *See* Abrasive flow machining (AFM)
Alpha-tricalcium phosphate (α-TCP) matrix, 142
AM. *See* Additive manufacturing (AM)
American Society for Testing and Materials (ASTM), 5, 570–571
Amorphous thermoplastics, 43
Analytical thermal models, 699
Aprecia Pharmaceuticals, 571, 572f
Asteroid simulants, 668, 668t
ASTM. *See* American Society for Testing and Materials (ASTM)
ASTM F42, 399, 400f, 401, 401t
α-TCP. *See* Alpha-tricalcium phosphate (α-TCP) matrix
Autoclave sterilization process, 491–492
Automotive industry
  advantages, 506
  applications, 506
  automotive supply chain, 505
  design for additive manufacturing (DfAM), 507f
    computational fluid dynamics, 507
    generative design, 508, 509f
    resource reduction, 506–507
    simulation, 507–508
  economic impact, 522–523
  energy consumption, 506
  freeform capacity, 505–506
  fused deposition modeling (FDM), 505
  inline building process control, 510–511
  mass customization, 506
  postprocessing, 511, 512t
  production tools
    applications, 514, 514f
    automotive components, 513
    BAAM process, 518, 518f
    customizable parts, 519, 520f
    Digital Light Synthesis (DLS) technology, 520–521
    disruptive approach, 516
    Divergent Manufacturing Platform, 516
    engine compartment, 514–516
    esthetic components, 513
    exhaust components, 516, 517f
    exterior body panels, 517–519, 517f
    functional components, 518
    hard tools, mass-scale replication, 512–513
    lighting, 517–519, 517f
    mass-replication process, 511
    PolyJet technology, 519–520, 519f
    powertrain components, 515, 516f
    soft tools, assembly lines, 511–512, 513f
    Stratasys J-series printer, 520, 521f
    ULTEM material, 515, 515f
  reverse engineering, 509–510, 510f
  selective laser melting (SLM), 505
  selective laser sintering (SLS), 505

## B

Barrel finishing, 100–102, 101f
Beam deposition process, 411
Big area additive manufacturing (BAAM) system, 189, 191f, 194–196, 196f, 518, 518f
Binder-jet 3D printing process, 384
Bioinks
  features, 605, 606t–607t
  gel-like phase, 604–605
  natural-based inks. *See* Natural-based inks
  synthetic-based inks, 613
Biopolymers
  biomedical applications
    bone repair, 643
    classification, 637–639, 638f
    dental applications, 642, 642f
    drug delivery system, 640–642
    layer-by-layer manufacturing, 644
    medical applications, 637–638
    medical devices, 639–640, 641t
    nondegradable biopolymers, 639
    orthopedic devices, 643, 644t
    polysaccharides, 638–639

Biopolymers (*Continued*)
polymethylmethacrylate (PMMA)
bone cements, 643, 643t
replacement tasks, 643
self-dependent implants, 638
synthesized biopolymers, 639
synthetic biodegradable polymers, 639
tissue engineering, 640–642
fabrication methods, 635–636
layer-by-layer assembly techniques, 635–636
organ transplantation, 635–636
polymer-coated calcium phosphate powders, 636–637
3D printing strategies, 636
acellular techniques, 648
additive processing, 644–645
application, 655
cellular techniques, 650–651
classification, 646, 646f
extrusion-based bioprinting, 651–652
extrusion-based mechanisms, 646, 646t
fabrication processes, 644–645
inkjet-based bioprinting, 651
laser-assisted bioprinting (LAB), 652–653
material extrusion, 650
material jetting (MJ), 650
materials, 646, 647t
porous microarchitecture, 644–645
powder bed fusion (PBF) technology, 649
requirements, 654
solid freeform fabrication, 644–645
vat photopolymerization (VP), 648–649
selective laser sintering (SLS) process, 636–637
stereolithography (SLA), 636
three-dimensional structures, 636
Bone repair, 643
Bone tissue engineering, 492–493
Bowden systems, 249–252, 250f
Burnishing, 446–447, 447f

## C

Cable-driven system, 192
CAD model. *See* Computer-aided design (CAD) model
Calcium phosphates, 142–143, 142f
Calorimetric data, 679
Carbon black-reinforced polyurethane, 317–318
Carbon fiber–reinforced thermoplastics, 207–209, 208f
Carbon nanomaterials, 474
Cellular techniques, 650–651
Chemical flow polishing process, 104–105, 105f
CLIP. *See* Continuous liquid interface production (CLIP)
CNC. *See* Computer numerical control (CNC)
CoCrMo alloy, 138, 140–141, 146–147, 147f
Coefficient of thermal expansion (CTE), 200–201
Compressive residual stress (CRS), 443–444
Computational fluid dynamics, 507
Computer-aided design (CAD) model, 373, 569–570
process, 2–3, 2f
Computer numerical control (CNC), 1, 23
accuracy, 25
binder jetting, 24
cost, 25
design freedom, 24–25
materials, 23–24
multi-axis laser deposition process, 25–26
powder-fed feature, 25–26
wire and arc additive manufacturing (WAAM), 26
Conduction-mode melting, 80–81
Continuous liquid interface production (CLIP), 7–8, 166, 166f, 174
Copolymerization, 43
CRS. *See* Compressive residual stress (CRS)

CTE. *See* Coefficient of thermal expansion (CTE)
Cut-off from substrate, 690
Cut-off operation, 695–696

## D

Daylight polymer printing (DPP), 7–8
DED. *See* Directed energy deposition (DED)
Delamination, 291
Dental applications, 642, 642f
Design for additive manufacturing (DfAM), 507f
  computational fluid dynamics, 507
  generative design, 508, 509f
  resource reduction, 506–507
  simulation, 507–508
Desktop-scale gantry systems, 191–192
DfAM. *See* Design for additive manufacturing (DfAM)
Diamond HotEnd, 256, 257f
Dies manufacturing, 177, 178f
Differential scanning calorimetry (DSC), 44–45
Differential thermal analysis (DTA), 417–418, 418f
Digital light processing (DLP), 7–8, 164, 165f, 168–170
  4D printing, 314–315
  vat photopolymerization (VP), 4D printing, 327–329, 328f–329f
Digital Light Synthesis (DLS) technology, 520–521
Digital micromirror device (DMD), 314–315
Directed energy deposition (DED), 13, 14f, 121–122, 122f, 388
  direct laser deposition (DLD), 531–532
  laser aided metal additive manufacturing, 437–439, 438f
Direct ink writing printing
  hydrogel, 320, 321f
  liquid crystal elastomer (LCE), 320–322, 322f
  magnetoactive material, 322–324, 323f
  shape memory polymer, 318–320, 319f
Direct laser deposition (DLD)
  aerospace industry, 531–532
  compensation, 547–548
  directed energy deposition (DED), 531–532
  distortion, 542–543, 545–547, 546f–547f
    control, 547–548, 548f–549f
  physical effects, 531–532
  powder particles transfer and gas heating
    characteristics, 532–533
    during cladding, 533, 533f
    dust-laden gas jet, 533–534
    gas dynamics, 532
    gas velocity, 533
    heat transfer model, 538–542, 540f–541f
    Inconel 625 deposition, 538, 538f
    Laplace equation, 533
    Laplace transformation, 534
    laser radiation, 534
    Marangoni effect, 538
    mass ratio, 532
    melt flow model, 535–538, 537f
    melting point, 534–535
    model implementation, 542, 542f
    Reynolds number, 532
    volumetric flow rate, 533
  residual stress, 543, 544f–545f
  stress kinetics, 542–543
  structure and properties
    economic effect, 555
    heat treatment (HT) modes, 559–564, 560t–562t, 563f–565f
    macrostructure formation, 556–558, 558f
    microstructure, 558
    powders requirements, 555–556, 556f
    processing parameters, 558
  technological equipment, 550f
    deposition productivity, 549
    equipment reliability, 550–552
    features, 549–550
    manipulator, 548–549
    process economics, 553–555
    process passive stability, 552–553

Direct laser deposition (DLD) (*Continued*)
 thermomechanical model, 531–532
Direct laser writing (DLW), 330–331, 330f
Direct metal laser sintering, 385
Direct powder extrusion, 575, 575f
Direct-write (DW) technology, 187
Divergent Manufacturing Platform, 516
DLD. *See* Direct laser deposition (DLD)
DLP. *See* Digital light processing (DLP)
DLS technology. *See* Digital Light Synthesis (DLS) technology
DMD. *See* Digital micromirror device (DMD)
DPP. *See* Daylight polymer printing (DPP)
Drug delivery system, 640–642
DSC. *See* Differential scanning calorimetry (DSC)
DTA. *See* Differential thermal analysis (DTA)
Dual-material deposition, 189, 190f
DW technology. *See* Direct-write (DW) technology

## E

EBAM. *See* Electron beam additive manufacturing (EBAM)
EBM. *See* Electron beam melting (EBM)
EBSD. *See* Electron backscattered diffraction (EBSD)
E3D Hemera Direct Kit, 254, 255f
E3D V6 printing head, 253–254, 254f
Electrochemical polishing, 103
Electron backscattered diffraction (EBSD), 416–417
Electron beam additive manufacturing (EBAM), 122
Electron beam melting (EBM), 77, 662
 chemical composition, 291–292
 delamination, 291
 internal defects
  macroscopic density gradient, 289–290, 290f
  powder defects, 289, 289f
  residual impurities, 291
 materials and applications, 279–280
 numerical simulation, 277, 293–295, 295f
 physical mechanism, 277
  melting phase, 280, 280f
  neck formation, 281, 281f
  wetting characteristics, 281–282
 process control and process parameters, 282–285, 286t
 process description, 278–279, 278f
 process monitoring, 292–293
 surface roughness
  balling effect, 288–289, 288f
  dimensional accuracy, 285–287
  multibeam strategy, 285–287, 287f
  porosity, 288
  process parameters, 287–288
Electron beam powder-bed fusion (EB-PBF). *See* Electron beam melting (EBM) process
Electrospinning methods, 641–642
Energy storage devices, 477–478
Engineered human tissues and organs
 bioinks
  features, 605, 606t–607t
  gel-like phase, 604–605
  natural-based inks. *See* Natural-based inks
  synthetic-based inks, 613
 biomaterials, 624–626
 3D bioprinting, 624
  bone and cartilage, 615–618
  cerebral cortex-like tissue, 624, 625f
  extrusion-based, 603–604, 604f
  features, 600, 601t
  heart valve, 622
  hepatic parenchyma, 622
  inkjet, 602–603, 603f
  laser, 600–602, 602f
  limitations, 624–626
  pulmonary epithelial tissue barrier, 622, 623f
  scaffold design, 599
  skeletal muscle, 618–619, 620f
  skin, 614–615, 616f
  types, 614
  vascularization, 621

complex microstructural designs, 626, 627f
computerized tomography (CT), 599–600
European Medicines Agency (EMA), 570
European Space Agency (ESA), 670
Extrusion-based bioprinting, 651–652
Extrusion-based mechanisms, 646, 646t
Extrusion process, 177
applications, 210–212, 211f
big area additive manufacturing (BAAM) system, 189, 191f
build platform, 191–192
cable-driven system, 192
desktop-scale gantry systems, 191–192
dual-material deposition, 189, 190f
energy consumption, 192–193
feedstock material loading system. *See* Feedstock material loading system
fused deposition modeling (FDM), 183–185
G-code, 185–186
large format additive manufacturing (LFAM) systems, 183–185, 189
advantages, 193
hybrid systems, 193–194
large-scale additive manufacturing (LSAM) system, 189
materials development
anisotropic properties, 207–209, 208f
bead functionality, 200
bio-derived matrices, 197
coefficient of thermal expansion (CTE), 200–201
components, 197, 198f
composition properties, 201
fiber-reinforced systems, 199, 207
part distortion and cracking, 209–210
pelletized feedstock, 196–197
physical properties, 201
polymer degradation, 207
porosity, 205–210, 206f
printability criteria, 198–201, 199f
properties, 196–197
rheological properties, 203–205, 204f
thermophysical properties, 201–202, 202f
thermoset materials, 197–198
viscoelastic characteristics, 200
z-pinning approach, 209
polymeric materials, 183–185, 184f
print enclosure, 191–192
public–private partnership organizations, 185
resin–catalyst mixture, 189–191
smart manufacturing
big area additive manufacturing (BAAM) system, 194–196, 196f
glass transition temperature, 194
in-line monitoring systems, 194–196
lights-out operation, 194
thermoplastic composites, 183–185
thermoset-based composites, 183–185
three-axis gantry-based motion platform, 191–192
three-dimensional (3D) computer-aided design (CAD), 185–186

## F

FDM. *See* Fused deposition modeling (FDM)
FDM Maxum, 217–218, 220f
Fe–Cr–Ni composition designs, 20, 21f
Feedstock material loading system
direct-write (DW) technology, 187
filament-fed fabrication (FFF), 186
initiator-to-resin ratio, 187
pelletized feedstock, 186–187
polylactic acid (PLA) filament feedstock, 186, 186f
reactive extrusion, 187
thermoset materials, 187, 188f
Fiber-reinforced systems, 199, 207
Filament-fed fabrication (FFF), 183–186
Filament feeding mechanism, 249, 251f
Finite element analysis (FEA), geometrical reduction strategies, 695

First-in-human (FIH) clinical trials, 589–590, 590f
Flow continuity equation, 536
Full-scale manufacturing method, 670
Functionally graded biomaterials, 146–148, 147f–150f
Functionally graded materials (FGMs)
  laser-based additive manufacturing, 409–410
  laser-directed energy deposition (LDED), 127
    characteristics, 137
    composition, 136
    industrial applications, 135
    316L/Inconel 625, 136, 136f
    Ti6Al4V tribological properties, 137
Fused deposition modeling (FDM), 505, 572–575
  advantages, 271–272
  disadvantages, 272–273
  dumbbell specimen orientations, 231, 231f
  extrusion head structure
    Bowden systems, 249–252, 250f
    Diamond HotEnd, 256, 257f
    E3D Hemera Direct Kit, 254, 255f
    E3D V6 printing head, 253–254, 254f
    filament feeding mechanism, 249, 251f
    integrated dual-feed extruder, 256
    melting thermoplastic polymer, 248, 249f
    model materials, 254, 255f
    printing head, 249, 250f
    RepRap project, 253–254
    thermoplastic softens, 252, 252f
  FDM Maxum, 217–218, 220f
  Fiberlogy, 246–248, 247f
  finishing process, 265–270, 265f–271f
  fused filament fabrication (FFF), 218–219
  layer plastic deposition (LPD), 217–218
  material extrusion-based 3D printing, 313–315
  melted extrusion modeling (MEM), 217–218
  open systems, 256–260, 258f–263f
  plasticized polymer, 231
  plastic jet printing (PJP), 217–218
  polyvinyl alcohol (PVA), 246, 246f
  3D printing industry services, 228–229
  RepRap project, 222–228, 223f, 225f–227f
  Stratasys device, head construction, 260–265, 264f
  Stratasys FDM 3D printers, 219–221, 220f–221f, 230, 230f
    aqueous solutions, 239
    medical model, skull bones, 237–239, 238f
    model materials, 232–248, 232t–237t, 241t–244t
    polyetherketoneketone (PEEK), 237–239, 238f
    structures, 245, 245f
    Technical Data Sheets (TDS), 239
  Stratasys F123 Series printers, 219–221, 221t–222t
  thermoplastic model material, 229–230, 229f
  thermoplastic polymer, 217, 218f
Fused filament fabrication (FFF), 218–219
Fusion-based process, 27–28

## G

Gas dynamics, 532
Gel/paste extrusion, 575–577, 577f
Graphene, 466

## H

Hatch spacing, 677
Heat input modeling, 698–702
Heat shrinkage polymer, 315–316
Heat transfer model, 538–542, 540f–541f
Heat treatment (HT) modes, 559–564, 560t–562t, 563f–565f
High-entropy alloys (HEAs), 21
High-speed sintering (HSS), 33, 37
Hot melt extrusion (HME), 573
HPC grades. *See* Hydroxypropylcellulose (HPC) grades

HT modes. *See* Heat treatment (HT) modes
Hydrogels
 direct ink writing printing, 320, 321f
 nanofunctionalized 3D printing, in medicine, 487–488
 shape-programmable materials, 4D printing, 310–311
 stimuli-responsive, 308–310, 309f
Hydroxypropylcellulose (HPC) grades, 575
Hydrophobic polymeric matrix, 468
Hydroxyapatite (HA), 142

## I

IM. *See* Injection molding (IM)
Indigenous space material resources, 661–673
Indium-tin-oxide (ITO), 472–473
Inert gas injection system, 438–439
Injection molding (IM)
 process, 40–44
 tooling method, vat photopolymerization (VP) methods, 175–177, 176f
Inkjet-based bioprinting, 314, 468–470, 602–603, 603f, 651
In situ resource utilization (ISRU) concept, 661
Integrated dual-feed extruder, 256
International Organization for Standardization (ISO), 395. *See also* ISO/TC 261
Intuition-based design method, 333
Irregular external morphology, 676
ISO/ASTM 52900 standard, 372–373
ISO/TC 261, 397–398, 398t
 ASTM F42, 399, 400f, 401, 401t
 implementation, 405
 Joint Steering Group (JAG), 400
 liaisons of, 401t, 402–403
 standards, 400–401, 402t, 403–404
 structure, 399, 399f

## L

Laplace equation, 533
Laplace transformation, 534
Large format additive manufacturing (LFAM) systems, 183–185, 189
 advantages, 193
 components, 197, 198f
 hybrid systems, 193–194
Large-scale additive manufacturing (LSAM) system, 189
Laser additive manufacturing (LAM), 674
Laser aided metal additive manufacturing
 absorptivity, 431–432, 432f
 advantages, 427–428
 applications, 427
 characteristics, 429
 classification, 429, 430f
 laser irradiation, 431
 mechanical properties, 428–429
 metal-based process
  advantages, 434, 434f
  classification, 433–434, 433f
  components, 434
  cost-effective method, 434
  directed energy deposition (DED) process, 437–439, 438f
  parameters, 434
  powder bed fusion (PBF) process, 435–436
  selective laser melting (SLM) process, 435–437, 436f
 motivation, 427–428
 Nd:YAG laser *vs.* fiber laser *vs.* $CO_2$ laser, 429, 430t
 parameters, 430–433
 postprocessing techniques
  abrasive machining, 445–446, 446f–447f
  burnishing, 446–447, 447f
  classification, 441–449
  components, 440–441
  fatigue crack growth tests, 440–441

Laser aided metal additive manufacturing (*Continued*)
fatigue test, 440–441
frictional parameters, 440–441
heat treatment, 449, 450f
laser polishing, 444–445, 445f
laser shock peening (LSP), 443–444, 443f
limitations and effects, 441, 441t
operations, 441, 442t
shot peening, 448–449, 448f–449f
surface property, 440–441
tensile residual stresses (TRSs), 440–441
processing techniques, 427–428
respirators, 427
solid freeform fabrication (SFF), 427–428
solid-state lasers, 429
Laser-assisted bioprinting (LAB), 652–653
Laser-based additive manufacturing
Al matrix composites, 415
beam deposition process, 411
dissolution and reaction process, 414–415
functionally graded materials (FGMs), 409–410
hierarchical, 416
high-energy laser beam, 409–410
interfacial reactions, 414
microstructural characterization, 410–411, 416–418, 418f
modified metallic matrix, 416
powder bed fusion process, 409–411
properties and applications, 419–420
"pure" ex situ metal matrix composites, 410–411, 410f
ball milling, 414
chemical method, 413–414
"satellited" powder, 413–414
TiB 2/Al alloys system, 413
in situ synthesis
beam deposition, 411–412
Fe matrix composites, 412–413
metallic powders, 412
reactants, 411–412
steel matrix composites, 412–413
tungsten carbides, 414, 415f
Laser beam powder melting, 688
Laser bioprinting, 600–602, 602f
Laser cladding/alloying, 132–135, 134f, 143–145, 144f, 437–438
Laser-directed energy deposition (LDED), 123f
biomedical applications
ceramic biomaterials, 141–143, 142f–143f
dental implants, 137–138
functionally graded biomaterials, 146–148, 147f–150f
life expectancy, 137
metallic biomaterials, 138–141, 139f–141f
orthopedic implants, 137–138
properties, 137–138
surface treatments, 143–145, 144f–146f
coaxial metal wire, 125–126
components, 127
deposition rates, 127
feedstock material, 123–124
functionally graded materials (FGMs), 127
high-power laser source, 124
industrial applications
$Al_2O_3$ sample, grown microstructure, 131, 131f
anisotropy, 130
of ceramics, 130–132, 131f
components, 128, 129f
cost-effective method, 135
functionally graded materials (FGM). *See* Functionally graded materials (FGM)
laser cladding/alloying, 132–135, 134f
mechanical properties, 130
metallic components, 128
metal matrix composites (MMCs), 132

nickel-based superalloys, 128–130, 130f, 133
stellite alloys, 132–133
surface treatments, 132, 133f
Ti6Al4V, 128, 129f
turbine blades, 135, 135f
zirconia–alumina ($ZrO_2$–$Al_2O_3$) ceramics, 132
laser head setup types, 124–125, 126f
mechanical properties, 126–127
parameters, 124–125, 125t
technologies, 124, 124t
Laser Engineering Net Shaping (LENS), 671–672
Laser metal deposition (LMD), 696–698
geometrical accuracy calculation, 702–713
modeling approaches for, 714–718, 715f
motivation, 686–696
rotational symmetric thin-walled aeroengine combustor casing, 688–691, 694f, 708–713, 710f–713f
thin-walled aeroengine combustor casing, 697–698
transferability, 702–703, 703f
turbine blade, shape distortion computation, 707–708, 708f
twin cantilever
shape distortion computation, 703–705, 704f–706f
shape distortion sensitivity study, 706–707, 707f
Laser polishing
continuous-wave (CW) mode, 344
laser aided metal additive manufacturing, 444–445, 445f
laser energy, 343–344
layer-by-layer manufacturing technology, 343
LMD TC11. *See* LMD TC11 specimens
nickel-based superalloys, 344–345
selective laser sintered (SLS), 344
SLM IN718 components, 363, 363f
SLM inconel 718 superalloy, 364
electrochemical corrosion parameters, 361–362, 362t
mechanical properties, 361–363, 362f
numerical simulation and microstructure, 359–361, 360f
pulsed laser, 355–356
surface morphology, 356–359, 357t, 358f
SLM TC4 alloy. *See* SLM TC4 alloy
SLM Ti components, 355, 356f
Ti alloys, 344–345
Laser processing lunar regolith simulants, 677, 678f
Laser shock peening (LSP), 443–444, 443f
Laser sintering (LS)
acrylonitrile butadiene styrene (ABS) powder, 33–34
advantages, 36
anisotropy, 58
applications, 33
automation technology, 33
component characteristics, 58–59
development and implementation, 33–34
differential scanning calorimetry (DSC), 44–45
DIN EN ISO/ASTM 52900, 33
emissivity measurements, 49–51
energy density, 55–56
energy input, 48–60
isothermal process, 45, 51–52, 52f
materials, 39–40, 41t–42t
mechanical component properties, 55–56
melting/crystallization behavior, 44–45
melting temperature, 45
nanofunctionalized 3D printing, 489–490
optical properties, 49
orange peel effect, 46
postmelting processes, 54, 55f
postpolymerization, 45–46
powder parameter, 47–48
powder rheological properties, 47
process phases, 34, 35f

Laser sintering (LS) (*Continued*)
roller/recoating system, 34–36
semicrystalline plastics, 36
semicrystalline polymers, 51
shrinkage behavior, 52–53, 53f, 60
stress–strain curves, 56–57, 58f
structure, 54–56, 57f
system-specific threshold energy density, 59–60
temperature gradients, 52–53, 53f
thermal conductivity, 51
time-and temperature-dependent process, 53–54
viscosity, 45–47
wavelength-dependent emissivity, 49–51, 50f
Young's modulus, 57–58
Layer-by-layer assembly techniques, 635–636
Layer plastic deposition (LPD), 217–218
LDED. *See* Laser-directed energy deposition (LDED)
LFAM systems. *See* Large format additive manufacturing (LFAM) systems
Light-activated polymerization, 161
Liquid crystal display (LCD), 7–8
Liquid crystal elastomers (LCEs)
direct ink writing printing, 320–322, 322f
mesogens, 310
nematic-to-isotropic phase transition, 308–310, 309f
shape actuation, 308–310
stimuli-responsive hydrogels, 308–310, 309f
LMD TC11 specimens, 363
Gaussian heat source, 347
mechanical properties, 348–350
microstructure and properties, 348, 349f
numerical simulation, 347–348, 347f
optical microscope (OM), 348
scanning electron microscope (SEM), 348
surface morphology
chemical compositions, 345, 345t
laser polishing effects, 345, 346f
temperature distribution, 347–348
Load-bearing orthopedic implants, 139–140, 140f
LPD. *See* Layer plastic deposition (LPD)
LS. *See* Laser sintering (LS)
LSAM system. *See* Large-scale additive manufacturing (LSAM) system
LSP. *See* Laser shock peening (LSP)
Lunar regolith structures, mechanical properties of, 678–679, 679f
Lunar simulants, 665, 666t


## M


Magnetic abrasive finishing, 106, 108f
Magnetic-responsive microtransporter devices, 331
Magnetoactive soft material (MSM), 306
shape-programmable materials, 4D printing, 311–312, 311f
Martian simulants, 667–668, 667t
Mask sintering, 38
Mass finishing technologies, 110, 111t–112t
Material extrusion-based 3D printing
carbon black-reinforced polyurethane, 317–318
color-shifting blooming flower, 316–317, 317f
fused deposition modeling (FDM), 313–315
heat shrinkage polymer, 315–316
multimaterial printing, 316–317
multiwalled carbon nanotubes (MWCNTs), 317–318
polylactic acid (PLA) filaments, 315–316, 316f
sunlight-activated recovery, 317–318, 318f
thermoplastic bilayer composites, 316–317
Material jetting (MJ), 650
Melted extrusion modeling (MEM), 217–218
Melt flow model, 535–538, 537f
Melting behavior, 675–676
Melting thermoplastic polymer, 248, 249f
Melt pool prediction, 688–691, 689f

analytical models, 691–694
finite element (FE) models, 694–695
MEM. *See* Melted extrusion modeling (MEM)
Metal additive manufacturing systems, 510–511
Metal matrix composites (MMCs), 132
Al matrix composites, 415
beam deposition process, 411
dissolution and reaction process, 414–415
functionally graded materials (FGMs), 409–410
hierarchical, 416
high-energy laser beam, 409–410
interfacial reactions, 414
microstructural characterization, 416–418, 418f
microstructures, 410–411
modified metallic matrix, 416
powder bed fusion process, 409–411
properties and applications, 419–420
"pure" ex situ metal matrix composites, 410–411, 410f
ball milling, 414
chemical method, 413–414
"satellited" powder, 413–414
TiB 2/Al alloys system, 413
in situ synthesis
beam deposition, 411–412
Fe matrix composites, 412–413
metallic powders, 412
reactants, 411–412
steel matrix composites, 412–413
tungsten carbides, 414, 415f
MMCs. *See* Metal matrix composites (MMCs)
Multi-axis laser deposition process, 25–26
Multicomponent material, 662–663
Multi Jet Fusion process, 37
Multiwalled carbon nanotubes (MWCNTs), 317–318

## N

Nanofibrillated cellulose (NFC), 320
Nanofunctionalized 3D printing
advantages, 491
annealing temperature, 471–472
autoclave sterilization process, 491–492
bone marrow stromal cells (BMSCs), 492–493
bone tissue engineering, 492–493
carbon nanomaterials, 474
drug concentration, 491–492
energy storage devices, 477–478
human osteoprogenitor cells, 492–493
indium-tin-oxide (ITO), 472–473
laser sintering (LS), 489–490
limitations, 478
mechanical and functional properties, 457–464
in medicine, 479t–485t
biomedical materials, 478–486
hydrogels, 487–488
inorganic components, 486
nanocapsules, 489
neurogenic tissue regeneration, 488
osteoconductivity, 486–487
surface-to-volume ratio, 486–487
metal-based materials, 471–472
nanocomposite materials, 477–478
nanocomposites and printed scaffolds, 489–490, 490f
nano-to-nano techniques, 489–490
optical nanostructures, 476
particle size distribution, 494
percolation threshold, 473–474
photocurable polymers, 475
photopolymer matrix, 475–476
piezoresistive sensor, 474–475
refractive index (RI), 476
resistivity, 474–475
short-processing method, 471–472
Si-based materials, 472–473
"smart" properties, 464
solvothermal synthesis method, 476–477
structural materials, nanoenhancement
acetone processing, 466
acid-catalyzed esterification, 466–468
depletion effect, 468–470
formative/subtractive technologies, 465–466

Nanofunctionalized 3D printing (*Continued*)
graphene, 466
high-performance polymers, 470
hydrophobic polymeric matrix, 468
inkjet techniques, 468–470
microparticles, 468–470
nanooxides, 468
negative thermal expansion, 466–468
polyamide 6 (PA6), 468–470
polymer nanocomposites, 465–466
polymer properties, 470
properties, 465–466
single-walled carbon nanotubes (SWCNTs), 465–466
stiffening effect, 468–470
structural application, 466–468
superhydrophobic surfaces, 490–491
surface modification, 493
transparent thin-film transistors (TTFTs), 472–473
vat photopolymerization (VP), 474–475
Natural-based inks
agarose, 605–608
alginate, 608
chitosan, 608–609
collagen, 611–612
fibrin, 610–611
gelatin, 612
gellan gum, 609–610
hyaluronan, 610
hyaluronic acid (HA), 610
silk fibroin (SF), 612–613
Navier–Stokes equation, 535
Neurogenic tissue regeneration, 488
NFC. *See* Nanofibrillated cellulose (NFC)
Nickel-based superalloys, 128–130, 130f, 133, 344–345
Noncontact temperature measurement systems, 61–63
Numerical thermal transient models, 700

## O

Orally disintegrating tablets (ODTs), 584–585, 585f
Organ transplantation, 635–636
Osteoconductive bioceramics, 141
Osteoconductivity, 486–487

## P

Pandemic coronavirus (COVID-19), 427
PBF. *See* Powder bed fusion (PBF)
Pharmaceutical 3D printing technologies
American Society for Testing and Materials (ASTM) scheme, 570–571
Aprecia Pharmaceuticals, 571, 572f
binder jetting, 571–572
computer-aided design (CAD) software, 569–570
European Medicines Agency (EMA), 570
material extrusion
direct powder extrusion, 575, 575f
fused deposition modeling (FDM), 572–575
gel/paste extrusion, 575–577, 577f
power bed fusion, 577–578
semisolid extrusion (SSE), 575–577, 577f
vat polymerization, 578–580, 579f
personalized medicine
applications, 580
drug development, 588–590
first-in-human (FIH) clinical trials, 589–590, 590f
in healthcare, 581–587
patient-friendly formulations, 584–586, 585f
polypills, 586, 587f
preclinical studies, 588–589
4D printing (4DP), 587
printlets on-demand, 582–583, 584f
Photocuring, 318–320
Photopolymers

materials, 162–163, 163f–164f, 175
matrix, 475–476
types, 163, 164f
Photosensitive polymer resin, 7–8
Piezoelectric bioprinters, 602–603
Piezoresistive sensor, 474–475
PJP. *See* Plastic jet printing (PJP)
PLA filament feedstock. *See* Polylactic acid (PLA) filament feedstock
Plastic deformation, 306
Plastic jet printing (PJP), 217–218
Polyamide 6 (PA6), 468–470
Polyether ether ketone (PEEK), 379–380
Polyethylene glycol (PEG), 613
Polyjet printing
bilayer composite, direct printing, 326–327, 326f
elastomer layer, 326–327
shape memory polymer composites, 324–326, 325f
shape-programmable material, 326–327
PolyJet technology, 519–520, 519f
Polylactic acid (PLA) filament feedstock, 186, 186f
material extrusion-based 3D printing, 315–316, 316f
Polymer-coated calcium phosphate powders, 636–637
Polymer material jetting process, 387
Polymer nanocomposites, 465–466
Polymer polycaprolactone (PCL), 327
Polymer–polymer interactions, 310–311
Polymers, 386–387, 387t. *See also* Biopolymers; Extrusion process
amorphous thermoplastics, 43
copolymerization, 43
DTM/3D Systems, 36–37
electrical conductivity, 43–44
injection molding (IM), 40–44
isothermal process control, 43
laser sintering (LS). *See* Laser sintering (LS)
mask sintering, 38
mechanical properties, 39–40
Multi Jet Fusion process, 37
PA12 powder systems, 43
postprocessing, 38–39
process monitoring
laser scanner system, 60–61
noncontact temperature measurement systems, 61–63
pyrometer, 61
temperature distributions, 61–63, 62f
surface treatment, 38–39
thermoplastic behavior, 39
Polymethylmethacrylate (PMMA) bone cements, 643, 643t
Polysaccharides, 638–639
Polyvinyl alcohol (PVA), 246, 246f, 573
Porous and Modifications for Engineering Surfaces (POMES), 370–371
Powder bed fusion (PBF), 11, 12f, 77, 649, 672–679
laser-based additive manufacturing, 409–411
laser aided metal additive manufacturing, 435–436
pharmaceutical 3D printing technologies, 577–578
polymers. *See* Polymers
selective laser melting (SLM), 77, 79t
Powder deposition, 688–689
Pre-alloyed powder, 20
Pre-mixed powder, 20
4D Printing
applications, 312–313
concept, 303, 304f
definition, 303
development of, 305, 331, 331f
digital light processing (DLP), 314–315
digital micromirror device (DMD), 314–315
direct ink writing printing
hydrogel, 320, 321f
liquid crystal elastomer (LCE), 320–322, 322f
magnetoactive material, 322–324, 323f
shape memory polymer, 318–320, 319f
eigenstrain, 306

4D Printing (*Continued*)
inelastic strain/eigenstrain distribution, 303–305
inkjet 3D printing, 314
intuition-based design method, 333
lay-by-layer manner, 312–313
magnetoactive materials, 306
material extrusion-based 3D printing, 313–314, 313f. *See also* Material extrusion-based 3D printing
multiple finite-element modeling, 333
plastic deformation, 306
polyjet printing
bilayer composite, direct printing, 326–327, 326f
elastomer layer, 326–327
shape memory polymer composites, 324–326, 325f
shape-programmable material, 326–327
printed active composites (PACs), 303
self-assembly, 314
shape-programmable materials, 305
hydrogels, 310–311
liquid crystal elastomers (LCEs), 308–310, 309f
magnetoactive soft material (MSM), 311–312, 311f
shape memory polymers (SMPs), 307–308, 307f
viscoelastic/viscoplastic behaviour, 307–308
shape-shifting structure, 303–305, 304f
smart materials, 332
solid free-form fabrication, 313–314
stereolithography (SLA), 314–315
stimulus-responsive polymers, 305, 306f, 312–313
thermoplastic polymers, 313–314
time dimension, 305
two-stage curing approach, 314–315
vat photopolymerization (VP), 314–315
digital light processing (DLP), 327–329, 328f–329f
direct laser writing (DLW), 330–331, 330f
3D Printing strategies. *See also* individual entries
biopolymers
acellular techniques, 648
additive processing, 644–645
application, 655
cellular techniques, 650–651
classification, 646, 646f
extrusion-based bioprinting, 651–652
extrusion-based mechanisms, 646, 646t
fabrication processes, 644–645
inkjet-based bioprinting, 651
laser-assisted bioprinting (LAB), 652–653
material extrusion, 650
material jetting (MJ), 650
materials, 646, 647t
porous microarchitecture, 644–645
powder bed fusion (PBF) technology, 649
requirements, 654
solid freeform fabrication, 644–645
vat photopolymerization (VP), 648–649
synthesized, 639
4D Printing (4DP) technology, 160
Projection microstereolithography (PμSL), 327
PVA. *See* Polyvinyl alcohol (PVA)


## R


Refractive index (RI), 476
Regenerative bone treatments, ceramic material, 141
Remanufacturing process, 17–18, 17f
RepRap project, 222–228, 223f, 225f–227f, 253–254, 369–370
Residual stresses, 543, 544f–545f
selective laser melting (SLM), 82–83, 82f
Resorbable bioceramics, 141
Reverse engineering
digitizers, 18–19
tools, 3
Reynolds number, 532

RI. *See* Refractive index (RI)
Robocasting, 313–314
Rotational symmetric thin-walled aero-engine combustor casing, 708–713, 710f–713f

## S

S520 bioactive glass implants, 142, 143f, 146f
Selective laser melting (SLM), 126–127, 672
  abrasive blasting, 105–106
  abrasive flow machining (AFM), 104, 105f
  acceleration and deceleration phases, 94–96
  additive manufacturing, modeling approaches for, 714–718, 715f
  advantages, 79–80
  automotive industry, 505
  barrel finishing, 100–102, 101f
  binding with underlying material, 690
  chemical flow polishing process, 104–105, 105f
  conduction-mode melting, 80–81
  contour scanning, 93–94, 96f
  cut-off from substrate, 690
  cut-off operation, 695–696
  3D thermal finite element model, 701–702
  electrochemical polishing, 103
  fiber laser, 81
  geometrical accuracy calculation, 702–713
  geometrical reduction strategies, finite element analysis (FEA) with, 695
  heat input modeling, 698–702
  laser aided metal additive manufacturing, 435–437, 436f
  laser beam powder melting, 688
  laser metal deposition process cases, 696–698
  laser-polished region, 108–110, 109f
  laser re-melting, 93
  magnetic abrasive finishing, 106, 108f
  mass finishing technologies, 110, 111t–112t
  mechanical/metallurgical characteristics, 108–110
  mechanical surface treatments, 100, 101f
  melting phenomena
    analytical thermal models of, 699
    numerical thermal transient models, 700
  melt pool prediction, 688–691, 689f
    analytical models, 691–694
    finite element (FE) models, 694–695
  metallic materials, 96–97, 98f
  model evaluation, 696
  motivation, 686–696
  non-melted powder grains, 103, 104f
  optical microscopy, 89–92, 92f
  parameters, 77
    density, 80–81, 81f
  post-processes, 100–102, 102f
  powder-bed fusion, 77, 79t
  powder deposition, 688–689
  process parameters, 77, 78f, 89–92, 91f
  residual stresses, 82–83, 82f
  reverse engineering modeling, 696
  rotational symmetric thin-walled aero-engine combustor casing, 708–713, 710f–713f
  roughness, 92–93, 93f, 96f
  scanning electron microscopy, 89–92, 92f
  scan strategy, 82–83, 82f
  shape accuracy, scan strategy on, 690–691
  spring-back effect, 695–696
  static chemical polishing process, 104–105, 105f
  surface enhancement treatments, 89
  surface integrity
    balling effect, 88, 90f
    design parameters, 86–87, 87f
    dimensional accuracy, 83–84, 84f
    edge-effect, 87–88, 88f
    re-melting and erosion, 87–88, 89f
    stair effect, 84–85, 85f
    temperature gradient mechanism, 85–86, 86f

Selective laser melting (SLM) (*Continued*)
up-skin and down-skin surfaces, 84–85, 84f
volume fraction, 87
surface quality, 83, 96–97
surface topography, 97, 99f
surface treatments, 105–106, 107f
3D topographies, 97, 99f
topology optimization, 79–80, 80f
transferability, 702–703, 703f
turbine blade, 688–691, 692f, 696–697
geometries, 100–102
shape distortion computation, 707–708, 708f
twin cantilever, 696–697
shape distortion computation, 703–705, 704f–706f
shape distortion sensitivity study, 706–707, 707f
support structures, 688–691, 692f
up-skin and down-skin scanning, 93–94, 95f
Selective laser sintering (SLS), 77, 344, 577
automotive industry, 505
Self-dependent implants, 638
Self-healing shape memory polymer (SH-SMP), 327
Self-microemulsifying drug delivery systems (SMEDDSs), 576
Semicrystalline plastics, 36
Semicrystalline polymers, 51
Semisolid extrusion (SSE), 575–577, 577f
SFF. *See* Solid freeform fabrication (SFF)
Shape accuracy, scan strategy on, 690–691
Shaped metal deposition, 122
Shape memory polymers (SMPs)
direct ink writing printing, 318–320, 319f
polyjet printing, 324–326, 325f
shape-programmable materials, 4D printing, 307–308, 307f
Shape-programmable materials, 4D printing, 305
hydrogels, 310–311
liquid crystal elastomers (LCEs), 308–310, 309f
magnetoactive soft material (MSM), 311–312, 311f
shape memory polymers (SMPs), 307–308, 307f
viscoelastic/viscoplastic behaviour, 307–308
Shot peening, 448–449, 448f–449f
SH-SMP. *See* Self-healing shape memory polymer (SH-SMP)
Single-walled carbon nanotubes (SWCNTs), 465–466
SLA. *See* Stereolithography (SLA)
SLM. *See* Selective laser melting (SLM)
SLM IN718 components, 363, 363f
SLM inconel 718 superalloy, 364
electrochemical corrosion parameters, 361–362, 362t
mechanical properties, 361–363, 362f
numerical simulation and microstructure, 359–361, 360f
pulsed laser, 355–356
surface morphology, 356–359, 357t, 358f
SLM TC4 alloy, 364
mechanical properties, 353–354, 355f
numerical simulation, 352–353, 352f
surface morphology, 350, 351f
transmission electron microscope (TEM), 353, 354f
SLM Ti components, 355, 356f
SLS. *See* Selective laser sintering (SLS)
Smart manufacturing
big area additive manufacturing (BAAM) system, 194–196, 196f
glass transition temperature, 194
in-line monitoring systems, 194–196
lights-out operation, 194
SMEDDSs. *See* Self-microemulsifying drug delivery systems (SMEDDSs)
Solid freeform fabrication (SFF), 644–645
laser aided metal additive manufacturing, 427–428

Solid-state lasers, laser aided metal additive manufacturing, 429
Solvothermal synthesis method, 476–477
Space resources, 661
  indigenous resources, 661–673
    direct usage of, 671–673, 671f
    indirect usage of, 669–671
  laser-based powder bed fusion additive manufacturing, 673–679, 673f
    laser processing lunar regolith simulants, 677, 678f
    lunar regolith structures, mechanical properties of, 678–679, 679f
    melting behavior, 675–676
    particle shape/morphology, 676, 676f
    spectral/optical performance, 674, 675f
    thermal stability, 675–676, 675f
  lunar resources, 662–664, 663f–664f
  regolith simulants/analogues
    asteroid simulants, 668, 668t
    lunar simulants, 665, 666t
    martian simulants, 667–668, 667t
Spectral absorbance measurements, 674
Spectral/optical performance, 674, 675f
Spring-back effect, 695–696
SSE. *See* Semisolid extrusion (SSE)
Standard tessellation language (STL), 4
Stellite alloys, 132–133
Stereolithography (SLA), 5, 7–8, 24, 371, 578–579
  additive manufacturing (AM), 369–370
  4D printing, 314–315
  three-dimensional (3D) computer-aided design (CAD), 185–186
  vat photopolymerization (VP) methods, 163, 165f
STL. *See* Standard tessellation language (STL)
Stratasys FDM 3D printers, 219–221, 220f–221f, 230, 230f
  aqueous solutions, 239
  medical model, skull bones, 237–239, 238f
  model materials, 232–248, 232t–237t, 241t–244t
  polyetherketoneketone (PEEK), 237–239, 238f
  structures, 245, 245f
  Technical Data Sheets (TDS), 239
Stratasys F123 Series printers, 219–221, 221t–222t
Stratasys J-series printer, 520, 521f
Stress kinetics, 542–543
Subtractive machining techniques, 1, 19–20, 23–25, 160–161
  accuracy, 25
  *vs.* additive manufacturing (AM), 373–375
  anisotropy, 23–24
  cost, 25
  design freedom, 24–25
  hardness and wear resistance, 23–24
  hybrid manufacturing, 25–26
  materials, 23–24
  speed, 24
  tooling method, vat photopolymerization (VP) methods, 174–175
Sunlight-activated recovery, 317–318, 318f
SWCNTs. *See* Single-walled carbon nanotubes (SWCNTs)
Synthesized biopolymers, 639
Synthetic-based inks, 613
Synthetic biodegradable polymers, 639

## T

Technology Readiness Level (TRL), 669
Temperature gradient mechanism (TGM), 686
Tensile residual stresses (TRSs), 440–441
Thermal conductivity, 51
3D Thermal finite element model, 701–702
Thermal inkjet bioprinters, 602–603
Thermal stability, 675–676, 675f
Thermomechanical model, 531–532
Thermoplastic bilayer composites, 316–317
Thermoplastic elastomers (TPEs), 640
Thermoplastic polymers, 217, 218f
  4D printing, 313–314

Thermoset-based composites, 183–185
Three-dimensional (3D) computer-aided design (CAD), 185–186
Ti alloys, 344–345
Ti6Al4V alloy, 128
  aeroengine structural material, 135
  bioactive glass coatings, 145
  biocompatibility, 146
  CoCrMo over, 147, 147f
  functionally graded, 148, 148f
  implants, 138
  locations of, 128, 129f
  mechanical properties, 139
  metallic biomaterials, 138
  microstructures, 128
  tribological properties, 137
  wear resistance, 146
Tissue engineering, 640–642
Tooling method, vat photopolymerization (VP) methods, 174f
  acrylonitrile butadiene styrene (ABS), 175
  dies manufacturing, 177, 178f
  extrusion process, 177
  injection molding (IM), 175–177, 176f
  photopolymer materials, 175
  subtractive manufacturing methods, 174–175
TPP. *See* Two-photon polymerization (TPP)
Transparent thin-film transistors (TTFTs), 472–473
TRSs. *See* Tensile residual stresses (TRSs)
TTFTs. *See* Transparent thin-film transistors (TTFTs)
Turbine blade, 688–691, 692f, 696–697
  shape distortion computation, 707–708, 708f
Twin cantilever, 696–697
  shape distortion
    computation, 703–705, 704f–706f
    sensitivity study, 706–707, 707f
  structures, 688–691, 692f
Two-photon polymerization (TPP), 330–331

## V

Vacuum-assisted resin transfer molding (VARTM), 210
VARTM. *See* Vacuum-assisted resin transfer molding (VARTM)
Vat photopolymerization (VP) methods, 7–8, 7f, 160–161
  bottom-up and top-down orientations, 161–162, 162f
  continuous liquid interface production (CLIP), 166, 166f
  digital light processing (DLP), 164, 165f
  direct fabrication
    applications, 172–173
    components, 173
    continuous liquid interface production (CLIP), 174
    cost estimation, 168
    digital light processing (DLP), 168–170
    geometric features, 168
    hydrophobicity, 169
    microfeatures measurement, 172
    microgeometry, 173, 173f
    polymer microcomponents, 172
    printing samples parameters, 168–170, 169f, 171f
    size and geometries, 172, 172f
  light-activated polymerization, 161
  nanofunctionalized 3D printing, 474–475
  pharmaceutical 3D printing technologies, 578–580, 579f
  photoinduced cleavage reaction, 163
  photopolymers
    materials, 162–163, 163f–164f
    types, 163, 164f
  polymeric components, 161
  postprocessing process chain, 166–168, 167f
    printed structures, 167–168, 167f

3D printing strategies, 648–649
4D printing, 314–315
digital light processing (DLP), 327–329, 328f–329f
direct laser writing (DLW), 330–331, 330f
stereolithography (SLA), 163, 165f
tooling method, 174f
acrylonitrile butadiene styrene (ABS), 175
dies manufacturing, 177, 178f
extrusion process, 177
injection molding (IM), 175–177, 176f
photopolymer materials, 175
subtractive manufacturing methods, 174–175

## W

Wire arc additive manufacturing (WAAM), 26, 122

## X

X-Ray diffraction (XRD), 416–417

## Y

Young's modulus, 57–58, 138–139, 139f, 147

## Z

ZipDose technology, 571
Zirconia–alumina ($ZrO_2$–$Al_2O_3$) ceramics, 132
Z-pinning approach, 209

Printed in the United States
by Baker & Taylor Publisher Services